Paleolimnology IV

Developments in Hydrobiology 37

Series editor
H.J. Dumont

Paleolimnology IV

Proceedings of the Fourth International Symposium on Paleolimnology, held at Ossiach, Carinthia, Austria

Edited by
Heinz Löffler

With the assistance of:
Brigitta Ammann
Patrick De Deckker
Elizabeth Haworth
Frank Oldfield
Jonathan Richardson
Jürgen Schneider
Michael Sturm

Reprinted from Hydrobiologia, vol. 143

1987 **DR W. JUNK PUBLISHERS**
a member of the KLUWER ACADEMIC PUBLISHERS GROUP
DORDRECHT / BOSTON / LANCASTER

Distributors

for the United States and Canada: Kluwer Academic Publishers, P.O. Box 358, Accord Station, Hingham, MA 02018-0358, USA
for the UK and Ireland: Kluwer Academic Publishers, MTP Press Limited, Falcon House, Queen Square, Lancaster LA1 1RN, UK
for all other countries: Kluwer Academic Publishers Group, Distribution Center, P.O. Box 322, 3300 AH Dordrecht, The Netherlands

Library of Congress Cataloging in Publication Data

ISBN 90-6193-624-1 (this volume)
ISBN 90-6193-751-5 (series)

PRINTED IN THE NETHERLANDS

Preface

The Fourth International Symposium on Paleolimnology was held in 1985 from 2–7 September at Lake Ossiach and close to the Austrian village of Ossiach (mentioned in historical records for the first time in 1028). After the opening speeches a performance by a Carinthian folk-dance group gave the participants a taste of the rural atmosphere of Austria's southernmost federal land.

About 125 paleolimnologists presented and discussed papers during the meetings and many participated in the pre-congress excursion devoted to the Salzkammergut, and in the post-congress excursion to Neusiedlersee. There was a half-day trip during the symposium to Längsee which is well known from the pioneer work carried out by Professor D. G. Frey more than 30 years ago.

In addition to methodological papers most of the special fields of paleolimnology were covered. Special attention was given a.o. to geochemistry, pigments and meromictic lakes. The geographical range of the sites investigated was quite wide. In this connection it is regrettable that none of the expected Soviet colleagues was able to attent the symposium. Under the presidency of Professor D. G. Frey the present members of the international organizing committee (Prof. S. R. Brown, Prof. E. S. Deevey, Prof. F. Oldfield and Dr. M. Ralska-Jasiewiczowa) contributed considerably to the success of the symposium. Likewise our thanks goes to those who helped behind the scenes (Mr. M. Bobek, Mr. M. Fusko, Mrs. A. Klenner, Ms. B. Stipanits and Ms. K. Wolter).

Held under the patronage of SIL the symposium was sponsored by the Federal Ministry for Sciences and Research, the Government of Carinthia, the Carinthian Institute of Lake Research, the Mayor of Ossiach and by Creditanstalt-Bankverein. This support is gratefully acknowledged.

As our Finnish colleagues have rightly pointed out (HYDROBIOL. 133, preface) it is always the enthusiasm of the participants which gives such a meeting the final touch. Now we are looking forward having the next meeting in England in 1989.

Ossiacher See: View from its eastern end.

Contents

PART II. REGIONAL PALEOLIMNOLOGY

PART III. LAKE CHANGES RELATED TO THE INFLUENCE OF MAN

Hydrobiologia 143: 1–8, (1986)

A biographical sketch of David G. Frey

David S. Baas[1] & Mary M. Boucherle[2]
[1]*Department of Biology, Indiana University, Bloomington, IN 47405, USA*
[2]*Department of Biology, Queen's University, Kingston, Ontario K7L 3N6, Canada*

We have chosen this time to honor Dr. Frey because this is the year of his 70th birthday and the year he will retire from teaching. As a teacher, he has introduced countless undergraduates to limnology and paleolimnology, as well as having supervised many graduate students in paleolimnological research. We will return to this aspect of his career a little later.

David G. Frey was born on 10 October 1915, in Hartford, Wisconsin, USA. He studied at the University of Wisconsin at Madison, receiving his Bachelor's degree in 1936, his Master's degree in 1938, and his Ph.D. in 1940. Under his major professor, Chancey Juday, Frey studied the limnology of the three lower lakes at Madison, Wisconsin, as affected by the effluent from the Madison waste treatment plant.

After completing his graduate studies at Wisconsin, Frey worked for the United States Fish and Wildlife Service. For two years he worked on the salmon of the Columbia River system in the northwestern United States. He then spent three years working on the oysters of the Chesapeake Bay.

In 1945, as an Ensign in the U.S. Naval Reserve, Frey was assigned to the Naval Medical Research Unit, based on Guam. During this time he investigated the freshwater habitats on several groups of Pacific Islands including the Mariana, Yap, Palau, and Admiralty Islands. These investigations primarily looked at aquatic invertebrates as vectors of human disease. Near the end of his Naval duties, Frey was chosen to be part of a team that surveyed the war damage to the Phillippine fisheries. This was to be the first of several experiences he would have studying the freshwater biology of the Far East and Asia.

In 1947, Frey accepted a faculty position at the University of North Carolina. During his first summer there, he organized and managed a survey of the Carolina Bay Lakes of the Coastal Plain. This survey was under the auspices of the North Carolina Wildlife Resources Commission and was the means by which Frey began investigating lake sediments.

After just one year, the Bay Lakes survey was abruptly cancelled. Fortunately for us, Frey's curiosity had been aroused as to the origin and development of this unique series of lakes. So, but with a few hundred dollars of support, he hired a field assistant named H. T. Odum, and the two young men set out to core the sediments of some of the Bay Lakes. In his own words, Frey was 'plunging bravely into palynology without any prior expierence'. The resultant study showed, essentially for the first time, that the northern species *Pinus banksiana* and *Picea* had been displaced southward by the glaciers.

In 1950, Frey joined the faculty at Indiana Univesity, where he is still a member today. The palynological studies of the Bay Lakes were gradually phased out due to geographic constraints and the demands of his new position. As Director of the Indiana Lake and Stream Survey, Frey designed and coordinated studies that assisted the State of Indiana in their management of Indiana fisheries. During this time, Frey helped to translate Franz Ruttner's book 'Fundamentals of Limnology' and in so doing developed a great admiration for Ruttner as both a person and a limnologist.

In 1953, Frey was awarded a Guggenheim Fellowship and a Fullbright Senior Fellowship to Austria. These fellowships were to have made it possible for the Frey family to spend the academic year in Austria, allowing Frey to work with Ruttner at Lunz. Upon arriving at Vienna, it became apparent to Frey that his plans would change. Lunz was in the Russian zone of occupation and travel there was not permitted. On his way to Vienna, Frey attended the S.I.L. Congress in Cambridge. Here he met many people whose lasting friendships affected the course of his scientific development. This was the first of many international limnological meetings.

As an alternative to working in Lunz, Frey worked out of the University of Vienna. He studied the meromictic lakes of southern Austria, including Längsee, with the help of Findenegg, and also Schleinsee, in southern Germany, with the help of Kriegsmann. As a result of the peace treaty negotiations, the Frey family was finally allowed to travel to Lunz, after having lived in Austria for six months. In Lunz, they spend the month of April with the Ruttner family.

Upon looking at the pollen in the Längsee core, Frey also noticed many small bits of chitin that neither he, nor anyone he showed them to, could identify. Finally, at the British Museum, J. P. Harding suggested that these might be the remains of Cladocera, and indeed they were. Identifying these remains in the sediments led to Frey's dominant, lifelong research interests: paleolimnology and the Cladocera, especially the family Chydoridae, whose remains are the most abundant animal microfossils in the sediments. Frey was, in essence, the first person to show the utility of cladoceran remains in the sediments as a paleolimnological tool.

Cladoceran remains are found in the sediments as the disarticulated components of the animals, and because of this, their identification requires close examination of many fine details not necessary for the identification of whole animals. After having familiarized himself with the fine details of the European Cladocera, Frey began examining the Cladocera of North America. In so doing, he recognized that the North American species were not the same as their European counterparts, even though they bore the same names. This heightened his interest in the evolution and phylogeny of the Chydoridae. Frey realized that much time and careful attention would be required to straighten out the taxonomic confusion existing in the literature. Of his over 150 publications, 56 concern the Cladocera. Many of these describe new species and emphasize the non-cosmopolitan nature of the family *Chydoridae*.

Although the primary focus of Frey's research has been the phylogeny and evolution of the chydorids, his contributions to paleolimnology have been numerous and significant. Apart from his work on the Bay Lakes and his classic papers on Schleinsee and Längsee, he has published several excellent reviews of paleolimnology, both general reviews, and reviews focusing on animal microfossils and cladoceran remains in the sediments. These papers, too, are classic reference works in the literature.

In addition to his own research, Frey has contributed to paleolimnology by assisting graduate students. He has supervised the paleolimnological work of 14 of his own students and served on the thesis committees of many others. Many of these students are working in major North American universities, actively publishing on the paleolimnology of North America and Europe. In this way, too, Dr. Frey's influence on the science is far reaching. As we are here at the Fourth International Symposium on Paleolimnology, we should remember that Frey was a primary organiser of the First and Second International Symposia, in Hungary and Poland, respectively. It is fitting to tell you of some of the other honors David Frey has received, as we honor him today. As mentioned earlier, he received the Guggenheim and Fulbright Fellowship to Austria. In 1980, at the 21st Congres of the International Association of Limnology in Kyoto, Japan, Dr. Frey was awarded the Einar Nauman/August Thienemann Medal. This award is only given to one person each year, and in the 47 years since its inception, David Frey was but the fourth scientist from the United States to receive it. The citation reads *'De limnologia optime merito'* for 'outstanding research in paleolimnology and on the systematics and evolution of cladoceran zooplankton'.

In 1982, Frey was elected a Foreign Member of the Royal Danish Academy of Sciences and Letters, and in 1983, he was awarded the Distinguished Service Citation by the Ecological Society of America. Frey has also recently been appointed a Fullbright Professor at the University College of Dublin, Ireland, where he will spend this coming fall semester.

In addition to the above, Frey has served as a

visiting professor at the biological stations of the University of Virginia, University of Minnesota, and the University of Michigan. He has been Distinguished Visiting Professor at the Quaternary Research Center and the Department of Zoology at the University of Washington. There he taught a course in paleolimnology. Frey has also taught a course in paleolimnology at the University of Lund, Sweden, and has taught courses in tropical limnology in Guatemala, El Salvador, and Venezuala. He has worked as a consultant to the Ford Foundation in the Phillippines, to the U.S. Environmental Protection Agency, and was for four years a member of the Environmental Protection Agency Science Advisory Board. Since 1983, he has been a member of the science advisory board of the International Joint Commission for the Great Lakes. Dr. Frey is also currently serving his second term as senior vice-president of S.I.L.

David Frey was the founding editor of the journal Limnology and Oceanography, and has, throughout his career, served on the editorial staffs of Internationale Revue der gesamten Hydrobiologie, Freshwater Biology, Journal of Quaternary Research, and Monitore Zoologico Italiano.

His participation in study penals, editorships, invited talks, colloquia, conferences, and meetings is far too long to mention, but what we see here is a man devoted to his science, and through that devotion he has made many important contributions to various aspects of the science of freshwater biology. Suffice it to say he is a remarkable man, both as a scientist, and as a person.

Publications

1938

Frey, D. G. & Hubert Pedracine. Growth of the buffalo in Wisconsin lakes and streams. Trans. Wis. Acad. Sci. Arts. Lett. 31: 513–525.

1939

Frey, D. G., Hubert Pedracine & Lawrence Vike. Results of a summer creel census of lakes Waubesa and Kegonsa, Wisconsin. J. Wildl. Mgmt. 3: 243–254.

1940

Frey, D. G. Growth and ecology of the carp, *Cyprinus carpio* Linnaeus, in four lakes of the Madison region, Wisconsin. Summaries of Doctoral Dissertations Univ. Wis., Univ. Wisc. Press 5: 83–85.

1941

Frey, D. G. & Lawrence Vike. A creel census on lakes Waubesa and Kegonsa, Wisconsin in 1939. Trans. Wis. Acad. Sci. Arts. Lett. 33: 339–362.

1942

Frey, D. G. Studies on Wisconsin carp. 1. Influence of age, size, and sex on time of annulus formation by 1936 year class. Copeia 1942 (4): 214–223.

1946

Frey, D. G. Oyster bars of the Potomac River. U.S. Fish Wildl. Serv., Spec. Sci. Rept. 32: 1–93.

1947

Frey, D. G. The pond fisheries of the Philippines. J. Mar. Res. 6: 247–258.

1948

Frey, D. G. North Carolina's bay lakes. Wildlife in North Carolina, May, 1948, p. 10–17.

Frey, D. G. Lakes of the Carolina bays as biotic environments. Jour. Elisha Mitchell Sci. Soc. 64: 179 (Abstract).

Frey, D. G. A biological survey of Lake Waccamaw. Wildlife in North Carolina, July, 1948, p. 4–6, 23.

Frey, D. G. & Edward Hueske. A fish resources survey of North Carolina's bay lakes. N.C. Wildlife Resources Commission, p. 3–20.

1949

Frey, D. G. Morphometry and hydrography of some natural lakes of the North Carolina Coastal Plain: The bay lake as a morphometric type. J. Elisha Mitchell Sci. Soc. 65: 1–37.

Deevey, E. S., Murray Buell, H. C. Darlington, D. G. Frey, J. E. Potzger & E. S. Stone. Peat samples and the radiocarbon chronology. Bull. Ecol. Soc. Am. 30: 54–55, (Abstract).

Frey, D. G. A continuous pollen record from the sediments of Singletary Lake, North Carolina, extending well back into the Pleistocene. Bull. Ecol. Soc. Am. 30: 55, (Abstract).

1950

Frey, D. G. Carolina bays in relation to the North Carolina Coastal Plain. J. Elisha Mitchell Sci. Soc. 66: 44–52.

1951

Frey, D. G. The fishes of North Carolina's bay lakes and their intraspecific variation. J. Elisha Mitchell Sci. Soc. 67: 1–44.

Frey, D. G. Pollen succession in the sediments of Singletary Lake, North Carolina. Ecology 32: 518–533.

Bailey, J. R. & D. G. Frey. Darters of the genus *Hololepis* from some natural lakes of North Carolina. J. Elisha Mitchell Sci. Soc. 67: 191–204.

Frey, D. G. The use of sea cucumbers in poisoning fishes. Copeia, 1951. (2): 175–176.

1952

Frey, D. G. Late-glacial and post-glacial pollen diagrams from the North Carolina Coastal Plain. Bull. Ecol. Soc. Am. 33: 62, (Abstract).

Frey, D. G. Pollen analysis of the Horry Clay and a seaside peat deposit near Myrtle Beach, S.C. Am. J. Sci. 250: 212–225.

Frey, D.G. Fish need oxygen. Outdoor Indiana, Oct., 1952, p. 6–7, 23.

1953
Frey, D. G. The teaching of limnology in the United States. Sci. Monthly 76: 290–296.
Frey, D. G. Regional aspects of the late-glacial and post-glacial pollen succession of southeastern North Carolina. Ecol. Monogr. 23: 289–313.
Ruttner, F. Fundamentals of Limnology, 2nd ed. Translated from the German by D. G. Frey and F. E. J. Fry. Univ. Toronto Press, xi, 242 pp.

1954
Frey, D. G. Evidence for the recent enlargement of the 'bay' lakes of North Carolina. Ecology 35: 78–88.

1955
Frey, D. G. Stages in ontogeny of the Carolina Bays, Proc. Int. Assoc. Limnol. 12: 660–668.
Frey, D. G. Längsee: a history of meromixis. Mem. Ist. Ital. Idrobiol. Suppl. 8: 141–164.
Frey, D. G. A differential flotation technique for recovering microfossils from inorganic sediments. New Phytologist 54: 257–258.
Frey, D. G. The Winona Lake Biological Station. AIBS Bull. 5: 20–22.
Frey, D. G. Distributional ecology of the cisco (*Coregonus artedii*) in Indiana. Invest. Indiana Lakes and Streams 4: 177–228.
Frey, D. G. A time revision of the Pleistocene pollen chronology of southwestern North Carolina. Ecology 36: 762–763.

1956
Frey, D. G. Die Entwicklungsgeschichte des Längsees in Kärnten. Carinthia II, 66: 5–12.

1957
Frey, D. G. Book review: E. A. Birge, by G. C. Sellery. Limnol. Oceanogr. 2: 43.
Frey, D. G. Book review: Aquatic insects of California, by R. L. Usinger (ed.). Limnol. Oceanogr. 2: 44–46.
Frey, D. G. Book review: A Manual of aquatic plants, by N. C. Fassett. Limnol. Oceangr. 2: 297.

1958
Frey, D. G. The late-glacial cladoceran fauna of a small lake. Arch. Hydrobiol. 54: 209–275.
Frey, D. G. & J. B. Stahl. Measurements of primary production on Southampton Island in the Canadian Arctic. Limnol. Oceanogr. 3: 215–221.
Frey, D. G. Book review: Ergebnisse der deutschen Limnologischen Venezuela-Expedition 1952, Bd. I, F. Gessner and V. Vareschi (eds.). Limnol. Oceanogr. 3: 242–243.
Frey, D. G. Book review: A key to the British species of freshwater Cladocera, by D. J. Scourfield and J. P. Harding. Limnol. Oceanogr. 3: 482.
Frey, D. G. Book review: The systematics of North American *Daphnia*, by J. L. Brooks. Limnol. Oceanogr. 3: 482–483.

1959
Frey, D. G. The Two Creeks Interval in Indiana pollen diagrams. Invest. Ind. Lakes and Streams. 5: 131–139.
Frey, D. G. The taxonomic and phylogenetic significance of the head pores of the Chydoridae (Cladocera). Int. Rev. ges. Hydrobiol. 44: 27–50.
Frey, D. G. Occurrence of Cladocera remains in lake sediments. Science 130: 1417 (Abstract).
Frey, D. G. Book review: A treatise on limnology. Vol. 1. Geography, physics and chemistry, by G. E. Hutchinson. Quart. Rev. Biol. 34: 70.
Frey, D. G. Book review: Fresh-water biology, W. T. Edmondson (ed.). Limnol. Oceanogr. 5: 435–436.
Frey, D. G. Book review: Die Rolle der Mikroorganismen in Stoffkreislauf der Seen, by S. I. Kuznetsov. Limnol. Oceanogr. 5: 346–347.
Frey, D. G. 1.1. Hydrological appraisal; 2.11. Physical survey; 2.12. Chemical survey; 2.13. Biological survey; 2.21. Feeding; 3.11. Control of water level and flow, erosion and silting; 3.12. Lake and stream improvement practices; 3.13. Pollution control. Lectures presented at the Third International Inland Fisheries Training Center, Bogor, Indonesia, 31 October–10 December 1955. Vol. I. F.A.O., Rome, Publ. A 552.
Frey, D. G. The use of freshwater plant and animal microfossils in determining the paleoecology of lakes. Program Annual Meeting Amer. Assoc. Petroleum Geologists and Society Economic Paleontologists and Mineralogists, Atlantic City, p. 83–84, (Abstract).
Frey, D. G. Cladocera from the Eemian Interglacial of Denmark. Bull. geol. Soc. Am. 71: 1868, (Abstract).
Frey, D. G. The ecological significance of cladoceran remains in lake sediments. Ecology 41: 684–699.
Frey, D. G. On the occurrence of cladoceran remains in lake sediments. Proc. Nat. Acad. Sci. 46: 917–920.
Frey, D. G. Freshwater ecosystem. McGraw-Hill 'Encyclopedia of Science and Technology' 5: 521–534.

1961
Frey, D. G. Book review: The biology of polluted waters, by H. B. N. Hynes. Limnol. Oceanogr. 6: 98.
Frey, D. G. Differentiation of *Alonella acutirostris* (Birge, 1897) and *Alonella rostrata* (Koch, 1841) (Cladocera, Chydoridae). Trans. Am. Microsc. Soc. 79: 129–140.
Frey, D. G. Developmental history of Schleinsee. Proc. Int. Assoc. Limnol. 14: 271–278.

1962
Frey, D. G. Cladocera from the Eemian Interglacial of Denmark. J. Paleont. 36: 1133–1154.
Frey, D. G. Paleolimnology of freshwater lakes. Japanese Jour. Limnol. 23: 145–146, (Abstract).
Frey, D. G. Supplement to: The taxonomic and phylogenetic significance of the head pores of the Chydoridae (Cladocera). Int. Rev. Hydrobiol. 47: 603–609.

1963
Goulden, C. E. & D. G. Frey. The occurrence and significance of lateral head pores in the genus *Bosmina* (Cladocera). Int. Rev. ges. Hydrobiol. 48: 513–522.

Frey, D. G., editor. Limnology in North America. Univ. Wisconsin Press xviii, 734 pp.
Frey, D. G. Wisconsin: The Birge-Juday era, in 'Limnology in North America', Univ. Wisconsin Press, pp. 3–54.
Frey, D. G. & F. E. J. Fry, translators. Fundamentals of Limnology, 3rd edition, by F. Ruttner. Univ. Toronto Press. xvi, 295 pp.

1964
Frey, D. G. Remains of animals in Quaternary lake and bog sediments and their interpretation. Ergebnisse der Limnologie (suppl. Arch. Hydrobiol.) 2: 1–116.

1965
Frey, D. G. Differentiation of *Alona costata* Sars from two related species (Cladocera, Chydoridae). Crustaceana 8: 159–173.
Frey, D. G. Limnology in the Soviet Union. Limnol. Oceanogr. suppl. 10: i–xxix
Frey, D. G. A new genus of Chydoridae (Cladocera). Int. Rev. ges. Hydrobiol. 50: 153–168.
Wright, H. E. Jr. & D. G. Frey, editors. The Quaternary of the United States. Princeton University Press. x, 922 pp.
Wright, H. E. Jr. & D. G. Frey, editors. International studies on the Quaternary. Geol. Soc. Am., Spec. Paper 84. vi, 565 pp.
Frey, D. G. Other invertebrates: an essay in biogeography. The Quaternary of the United States, Princeton Univ. Press. pp. 613–631.
Frey, D. G. Gynandromorphism in the chydorid Cladocera. Limnol. Oceanogr., Alfred C. Redfield 75th Anniversary Volume, suppl. 10: R103–R114.

1966
Frey, D. G. Limnology, pp. 297–320 in: *Natural Features of Indiana*. In: A. A. Lindsey (ed.), Indiana Academy of Sciences, xxix, 579 pp.
Frey, D. G. (Book review) *Limnologie der Donau*, edited by R. Liepolt. Limnol. Oceanogr. 11: 648–649.
Frey, D. G. Cladocera in space and time. Mar. Biol. Assoc. India, Proc. Intl. Symposium on Crustacea, Part I: 1–9.
Frey, D. G. Phylogenetic relationships in the family Chydoridae (Cladocera). Mar. Biol. Assoc. India, Proc. Intl. Symposium on Crustacea, Part I: 29–37.

1967
Frey, D. G. Mi a paleolimnológia? Hidrölogiai Közlöny 6: 308–312.
Frey, D. G. Biological characteristics of meromictic lakes, In D. F. Jackson (ed.), Some aspects of meromixis. pp. 63–95. Dept. Civil. Eng., Syracuse University, 243 pp. (mimeograph).
Frey, D. G. Weekend in Cairo. The Mindanao Varsitariam, 61(1–2): 6–8, 18–19.
Frey, D. G. Reservoir research – objectives and practices with an example from the Soviet Union, pp. 26–36 in: Reservoir Fishery Resources Symposium, American Fisheries Society. Univ. Georgia, Athens. viii, 569 pp.

1968
Frey, D. G. Paleolimnology. Science 159: 1262–1264.
Shan, R. Kuo-cheng & D. G. Frey. Induced interbreeding between two stocks of a chydorid cladoceran. BioScience 18: 203–205.
Frey, D. G. (Book review) Hydrobiological Studies I, edited by J. Hrbáček. Limnol. Oceanogr. 13: 395–397.

1969
Frey, D. G. (Book review) Diatomen in Seesedimenten, by A. Juse, Quart. Rev. Biol. 44: 100–101.
Frey, D. G. (Book review) Algenreste in Seesedimenten. Zur Entwicklungsgeschichte der Seen und umliegenden Landschraften, by N. W. Korde. Quart. Rev. Biol. 44: 100.
Frey, D. G. (Book review) Pflanzensoziologie und Palynologie, edited by R. Tüxen. Quart. Rev. Biol. 44: 84.
Frey, D. G. (Book review) The systematics and evolution of the Moinidae, by C. E. Goulden. Syst. Zool. 18: 234–235.
Frey, D. G. (Book review) A treatise on Limnology. Volume II: Introduction to lake biology and limnoplankton, by G. E. Hutchinson. Limnol. Oceanogr. 14: 472–474.
Frey, D. G. Further notes on *Alona bicolor* Frey, 1965 (Cladocera, Chydoridae), with a description of the male. Trans. Am. Microsc. Soc. 88: 380–386.
Frey, D. G. A limnological reconnaissance of Lake Lanao. Verh. int. Verein. Limnol. 17: 1090–1102. (Reprinted in Mindanao Journal Vol. I (1): 81–101, 1974. Published by University Research Center, Mindanao State University, Marawi City, R.P., 1969).
Frey, D. G., editor. Symposium on paleolimnology. Mitt. Int. Verein. Limnol. 12: 1–448.
Frey, D. G. The rationale of paleolimnology. Mitt. Int. Verein. Limnol. 177: 18.
Frey, D. G. Evidence for Eutrophication from remains of organisms in sediments, pp. 594–613 in: Eutrophication: causes, consequences, correctives. G. A. Rohlich, chairman. Natl. Acad. Sciences, Washington. vii, 661 pp.

1970
Frey, D. G. (Book review) In R. B. Morrison & H. E. Wright, Jr. (eds.) Means of correlation of Quaternary successions. J. Geology Education 18: 48–50.
Flössner, Dietrich & D. G. Frey. A morphological comparison of European and North American populations of *Alona rustica* Scott and *Alona costata* Sars (Cladocera, Chydoridae). Limnologica (Berlin) 7: 325–327.

1971
Frey, D. G., editor. Symposium on ecology of Cladocera. Trans. Am. Microsc. Soc. 90: 100–121.
Frey, D. G. Zoography and evolution, pp. 119–121. In D. G. Frey (ed.) Symposium on ecology of Cladocera. Trans. Am. Microsc. Soc. 90: 100–121, (Abstract).
Frey, D. G. Worldwide distribution and ecology of *Eurycercus* and *Saycia* (Cladocera). Limnology and Oceanography (G. E. Hutchinson Celebratory Issue) 16: 254–308.
Frey, D. G. Concluding Remarks, p. 295–300. In J. Cairns, Jr. (ed.). The structure and function of fresh-water microbial communities. Res. Div. Monogr. 3, Virginia Polytech. Inst. vii + 301 p.
Smith, R. W. & D. G. Frey. Acid mine pollution effects on lake biology. EPA Water Pollution Control Research Series 18050 EEC 12/71, ix + 132 p.

1972
Frey, D. G. Asian waterways, pp. 179–179. In Yuh-jao Wu &

Wilfredo A. Clemente II (eds.). Environment and/or development in Asia: hope and frustration, Record of the Conference of Asian Environments, June 14–17, 1971, Ann. Arbor. Published by the Asian Environmental Society, ix, 266 p .

1973

Frey, D. G. Comparative morphology and biology of three species of *Eurycercus* (Cladocera, Chydoridae) with a description of *Eurycercus macrocanthus* sp. nov. Int. Rev. ges. Hydrobiol. 58: 221–267.

Frey, D. G. Corrections to: Comparative morphology and biology of three species of *Eurycercus* (Cladocera, Chydoridae) with a description of *Eurycercus macrocanthus* sp. nov. Int. Rev. ges. Hydrobiol. 58: 593.

1974

Frey, D. G. Reassignment of *Alona fitzpatricki* and *A. leei* Chien, 1970 (Cladocera, Chydoridae). Trans. Am. Microsci. Soc. 93: 162–170.

Frey, D. G. Paleolimnology, pp. 95–123. In W. Rodhe (ed.), Jubilee Symposium: 50 years of Limnological Research. Mitt. Int. Verein. Limnol. 20, 402 pp.

Frey, D. G. (Book review) Man-made lakes: Their problems and environmental effects, by W. G. Ackermann, G. R. White & E. B. Worthington. BioScience 24: 730.

Frey, D. G. The distribution of *Eurycercus glacialis* (Cladocera) in Western Europe. p. 58 in: Abstracts of XIX Congress International Association Limnology, Winnipeg, Canada (Abstract).

Frey, D. G. Limnologia. (Article from Ind. Acad. Sci., Nat. Features of Indiana, 1966: 207–320, translated into Hungarian by Olga Sebestyén). Magyar nyelvre fordította, Sebestyén Olga, Tihany, 35 p.

1975

Frey, D. G. Subgeneric differentiation within *Eurycercus* (Cladocera, Chydoridae) and a new species from northern Sweden. Hydrobiologia 46: 263–300.

Frey, D. G. Interpretation of Quaternary paleoecology from Cladocera and midges, and prognosis regarding usability of other organisms. Program and Abstracts, Quaternary Non-Marine Paleoecology Conference, Univ. Waterloo, Canada, 12–13 May 1975. (Abstract).

1976

Frey, D. G. Redescription of *Chydorus pigroides* Lilljeborg, 1900, and allocation to *Pleuroxus* (Cladocera, Chydoridae). Crustaceana 30: 89–97.

Frey, D. G. The distribution and ecology of *Eurycercus glacialis* (Cladocera, Chydoridae) in Western Europe. Verh. Int. Verein. Limnol. 19: 2922–2934.

Frey, D. G. (Book review) Textbook of limnology, by G. A. Cole. BioScience 26: 346.

Frey, D. G. Interpretation of Quaternary paleoecology from Cladocera and midges, and prognosis regarding usability of other organisms. Can. J. Zool. 54: 2208–2226.

Frey, D. G. Dynamics of the Lake Monroe system, p. 23–28. In H. E. McReynolds (ed.). Fish and Wildlife in Indiana, 1776–1976 and reports of fish and wildlife studies, Proc. Joint. Meeting Am. Fish. Soc. and Wildlife Soc., Bradford Woods, Ind. Mills Publ., Inc., Hoopeston, Ill. i–vi, 1–87.

1977

Chang, W. Y. B. & D. G. Frey. Monroe Reservoir, Indiana. Part II: Nutrient relations. I.U. Water Resources Center, Technical Rept. 87: 71–117.

Frey, D. G. Biological integrity of water – an historical approach, p. 127–140. In: R. K. Ballantine & L. J. Guarraia (eds.). The Integrity of Water. U.S. Environmental Protection Agency, Office of Water and Hazardous Materials. vii, 230 p. U.S. Govt. Printing Office.

1978

Frey, D. G. A new species of *Eurycercus* (Cladocera, Chydoridae) from the southern United States. Tulane Stud. Zool. Bot. 20: 1–25.

Frey, D. G. Lilljeborg's 'Offentliga Föreläsingar' of 1880. Crustaceana 34: 100–105.

Frey, D. G. Foreword, pp. 1–3 in: volume of papers deriving from the Second International Symposium on Paleolimnology held in Poland, 9/14–20/76. Polskie Archiwum Hydrobiologii 25: 1–498.

Frey, D. G. J. Hrbáček, V. Kořínek & D. G. Frey. Cladocera, pp. 189–195. In Joachim Illies (ed.), *Limnofauna Europaea*, 2nd edition, Gustav Fischer: Stuttgart.

1979

Frey, D. G. (Book review) The kindly fruits of the earth. Recollections of an embryo ecologist, by G. E. Hutchinson. Science 203: 1368–1369.

Fortunatov, M. D., D. G. Frey & N. N. Smirnov. Philaret Dimitrievich Mordukhai-Boltovskoi: Life and Scientific activities (7/7/10–8/20/78). Int. Rev. ges. Hydrobiol. 64: 827–836.

Frey, D. G. Cladocera Analysis, Chapt. 10.10, pp. 227–257 in: Paleohydrological changes in the temperate zone in the last 15000 years. Subproject B. Lake and Mire Environments. Project Guide, Vol. II, Bjorn E. Berglund (ed.). Dept. Quaternary Geol., Lund, Sweden.

Frey, D. G. Paleolimnology of Lake Valencia, Venezuela. International Project on Paleolimnology and Late Cenozoic Climate. Newsletter 2: 37–39. Contrib. on the Paleolimnology of Lake Biwa and the Japanese Pleistocene, No. 263.

1980

Frey, D. G. On the plurality of *Chydorus sphaericus* (O. F. Müller) (Cladocera, Chydoridae) and designation of a neotype from Sjaelsø, Denmark, Hydrobiologia 69: 83–123.

Frey, D. G. The heritage of our lakes. The Key Reporter (Phi Beta Kappa) 45: 2–4, 8.

Frey, D. G. The non-swimming chydorid Cladocera of wet forests, with descriptions of a new genus and two new species. Int. Rev. ges. Hydrobiol. 65: 613–641.

Landers, D. H. & D. G. Frey. The dieback role of *Myriophyllum spicatum* in Monroe Reservoir, Indiana. Purdue Univ. Water Resources Res. Center. Tech. Rept. No. 134, v, 105 p.

1981

Fryer, G. & D. G. Frey. Two-egged ephippia in the chydorid Cladocera. Freshwat. Biol. 11: 391–394.

Bradbury, J. P., B. Leyden, M. Salgado-Labouriau, W. M. Lewis, Jr., C. Schubert, M. W. Binford, D. G. Frey, D. W. Whitehead & F. H. Weibezahn. Late Quaternary environmental history of Lake Valencia, Venezuela. Science 214: 1299–1305.

Frey, D. G. The elusive male of *Chydorus faviformis* (Crustaceae, Cladocera). Am. Zool. 21: 935 (Abstract).

1982

Frey, D. G. Contrasting strategies of gamogenesis in northern and southern populations of Cladocera. Ecology 63: 223–241.

Relocation of *Chydorus barroisi* and related species (Cladocera, Chydoridae) to a new genus and description of two new species. Hydrobiologia 86: 231–269.

Frey, D. G. Questions concerning cosmopolitanism in Cladocera. Arch. Hydrobiol. 93: 484–502.

Frey, D. G. The reticulated species of *Chydorus* (Cladocera, Chydoridae): two new species with suggestions of convergence. Hydrobiologia 93: 255–279.

Frey, D. G. Cladocera Sueciae and Modern Nomenclature, pp. xv–xxi. In Wilhelm Rodhe & David G. Frey (eds.), 'Prologue to the Fascimile Reissue of Wilhelm Lilljeborg's Cladocera Sueciae', sponsored by the Royal Academy of Sciences of Uppsala, Almqvist and Wiksell, Stockholm.

Frey, D. G. The honeycombed species of *Chydorus* (Cladocera, Chydoridae): comparison of *bicornutus* and *bicollaris* n. sp. with some preliminary comments on *faviformis*. Can. J. Zool. 60: 1892–1916.

Forró, L. & D. G. Frey. Cladocera taxa described by E. Daday together with a catalogue of pertinent material in the Hungarian Natural History Museum. Misc. zool. hung. 1: 91–132.

Frey, D. G. Honeycombing of the carapace in the chydorid Cladocera: the elusive male of *Chydorus faviformis*. J. Crust. Biol. 2: 469–476.

Frey, D. G. & G.O. Sars and the Norwegian Cladocera: a continuing frustration. Hydrobiologia 96: 267–293.

Frey, D. G. & Cladocera. In S. H. Hurlbert & A. Villalobos-Figueroa (eds.), Aquatic Biota of Mexico, Central America and the West Indies, pp. 177–186. Aquatic Biota – SDSU Foundation, San Diego State Univ., xv, 529 p.

Frey, D. G. (Book review) Facsimile Reissue of Lilljeborg's Cladocera Sueciae. Bull. Ecol. Soc. Am. 63: 357.

1983

Frey, D. G. (Book review) Cladocera Sueciae, by Wilhelm Lilljeborg (1901). Facsimile reissue in 1982 of the original edition, in three volumes. . . with a prologue by editors Wilhelm Rodhe and David G. Frey. Almqvist & Wiksell, Stockholm. Arch. Hydrobiol. 96: 525–526.

Frey, D. G. (Book review) Cladocera Sueciae, by Wilhelm Lilljeborg, 1901. Facsimile reissue of the original edition. . . with a prologue by editors W. Rodhe & D. G. Frey. Almqvist & Wiksell, 1982. Int. Rev. ges. Hydrobiol. 68: 898.

Michael, R. G. & D. G. Frey. Assumed Amphi-Atlantic distribution of *Oxyurella tenuicaudis* (Cladocera, Chydoridae) denied by a new species from North America. Hydrobiologia 106: 3–35.

Frey, D. G. (Book review) Cladocera Sueciae, by Wilhelm Lilljeborg, 1901. Facsimile reissue in 1982 of the original edition. . . with a prologue by editors Wilhelm Rodhe and David G. Frey. Almqvist & Wiksell, Stockholm. Crustaceana 45: 224.

Frey, D. G. (Book review) Limnology, by C. R. Goldman & A. J. Horne. McGraw-Hill Book Co., New York, 1983, 464 pp. Limnol. Oceanogr. 29: 447–448.

1984

Frey, D. G. (Book review) The Ecological Century. A Personal Appraisal, by E. Barton Worthington. Clarendon (Oxford University Press), New York, 1983. xvi, 206 pp., Science 224: 734–735.

1985

Frey, D. G. The non-cosmopolitanism of chydorid Cladocera: implications for biogeography and evolution. In K. L. Heck & R. H. Gore (eds.), Crustaecean Issues, Vol. 4, Crustacean Biogeography, A. A. Balkema, Rotterdam. (In press).

Frey, D. G. The taxonomy and biogeography of the Cladocera. Proceedings of the First International Symposium on the Cladocera, Budapest, Hungary. Hydrobiologia. (In press).

Frey, D. G. & B. J. Hann. Growth in Cladocera. In: Crustacean Issues, Vol. 2, Crustacean Growth, ed. by A. M. Wenner, A. A. Balkema, Rotterdam. (In press).

1986

Frey, D. G. Comparison of *Chydorus faviformis* from North America with honeycombed taxa from other continents (Cladocera, Chydoridae). Phil. Trans. R. Soc. Lond. (In press).

Theses of graduate students of D. G. Frey

1958

Eberly, William R. The metalimnetic oxygen maximum in Myers Lake. Ph.D. thesis, Indiana Univ. 101 p.

Stahl, John B. The developmental history of the chironomid and *Chaoborus* faunas of Myers Lake, and the relationship of these faunas to microlaminae in the sediments. Ph.D. thesis, Indiana Univ. 142 p.

1962

Goulden, Clyde E. The history of the cladoceran fauna of Esthwaite Water (England) and its limnological significance. Ph.D. thesis, Indiana Univ. 94 p.

Megard, Robert O. The biostratigraphic history of Dead Man Lake, Chuska Mountains, New Mexico. Ph.D. thesis, Indiana Univ. 47 p.

Mueller, Wayne P. The distribution of cladoceran remains in surficial sediments from three northern Indiana lakes. Ph.D. thesis, Indiana Univ. 122 p.

DeCosta, John J. Latitudinal distribution of chydorid Cladocera in the Mississippi Valley, based on their remains in surficial lake sediments. Ph.D. thesis, Indiana Univ. 66 p.

Hobbie, John E. Limnological cycles and primary productivity of two lakes in the Alaskan Arctic. Ph.D. thesis, Indiana Univ. 124 p.

1965

Kalff, Jacob. Primary production rates and the effect of some environmental factors on algal photosynthesis in small arctic tundra ponds. Ph.D. thesis, Indiana Univ. 122 p.

Chandler, Clay M. Environmental factors affecting the local distribution and abundance of four species of stream-dwelling triclads. Ph.D. thesis, Indiana Univ. 104 p.

Clifford, Hugh F. The ecology of invertebrates in an intermittent stream. Ph.D. thesis, Indiana Univ. 84 p.

1966

Hosseinie, Farammarz. The ecology and reproductive cytology of *Daphnia middendorffiana* Fischer (Cladocera) from the Arctic. Ph.D. thesis, Indiana Univ. 78 p.

Smith, Judith A. A study of the winter factors associated with the occurrence of four species of *Daphnia* in one lake. M.A. thesis, Indiana Univ. 80 p.

1967

Harmsworth, Rodney V. The developmental history of Blelham Tarn (England) as shown by animal microfossils, with special reference to the Cladocera. Ph.D. thesis, Univ. London. 74 plus unnumbered pages.

Shan, Robert Kuo-cheng. Observation and experimentation on the life cycle of a chydorid cladoceran, *Pleuroxus denticulatus* Birge, with special reference to photoperiod and light intensity. Ph.D. thesis, Indiana Univ. 89 p.

1968

Whiteside, Melbourne, C. Danish chydorid Cladocera: species occurrence and lake typology. Ph.D. thesis, Indiana Univ. 129 p.

1971

Smith, Ronald W. Ecosystem changes associated with recovery from acid pollution in coal stripmine lakes in southern Indiana. Ph.D. thesis, Indiana Univ. 157 p.

Smith, Judith Morris. A study of the effect of environmental variables, particularly photoperiod, on four coexisting *Daphnia* species. Ph.D. thesis, Indiana Univ. 123 p.

Stimpfl, Kenneth J. The effects of four environmental variables on the induction of gamogenesis in *Daphnia pulex* Leydig. Ph.D. thesis, Indiana Univ. 74 p.

Williams, Alice Simms. Late-glacial – postglacial vegetational history of the Pretty Lake Region, northeastern Indiana, USA. Ph.D. thesis, Indiana Univ. ca. 150 p.

1972

Lewis, William M. Jr. The thermal regime, chemistry, and phytoplankton ecology of Lake Lanao, Phillippines. Ph.D. thesis, Indiana Univ. 263 p.

1973

Kubersky, Edward S. Worldwide distribution and ecology of *Alonopsis* (Cladocera, Chydoridae). Ph.D. thesis, Indiana Univ. 183 p.

Hobbs, Horton H. III. The population dynamics of cave crayfishes and their commensal ostracods from southern Indiana. Ph.D. thesis, Indiana Univ. 243 p.

1974

Quade, Henry W. The abundance and distribution of littoral Cladocera as related to sediments and plants. Ph.D. thesis, 165 p.

1975

Iovino, Anthony J. Extant chironomid larval populations and the representativeness and nature of their remains in lake sediments. Ph.D. thesis, Indiana Univ. 54 p.

1978

Brakke, David F. The effect of excluding fishes on the size structure and dynamics of chydorid Cladocera, with particular reference to *Camptocercus*. Ph.D. thesis, Indiana Univ. 108 p.

Santiago, Alejandro E. A preliminary study on the primary production of the lower basin, Lake Monroe, Indiana. M.A. thesis, Indiana Univ. viii, 107 p.

1979

Landers, Dixon H. The chemical and biological effects of natural dieback of *Myriophyllum spicatum* L., Monroe Reservoir, In. Ph.D. thesis, Indiana Univ. 109 p.

Binford, Michael W. Holocene paleolimnology of Lake Valencia, Venezuela: Evidence from animal microfossils and some chemical, physical, and geological features. Ph.D. thesis, 129 p.

1980

Hann, Brenda J. *Population differentiation in the Eurycercus (Bullatifrons)* species complex (Chydoridae, Cladocera) in eastern North America. Ph.D. thesis.

1981

Hartzell, Martin E. Microbial changes in and invertebrate utilization of leaf detritus in two contrasting streams of southern Indiana. M.A. thesis.

1982

Boucherle, Mary M. An ecological history of Elk Lake, Clearwater Co., Minnesota, based on Cladocera remains. Ph.D. thesis, 126 p.

1985

Vila, Peter B. The occurrence of Cladocera in some streams of central Indiana. M.A. thesis, Indiana Univ. 64 p.

Cotten, Clara A. Cladoceran assemblages related to lake conditions in eastern Finland. Ph.D. thesis, Indiana Univ. 97 p.

Accepted 18 March 1986.

Hydrobiologia 143: 9–11, (1986)

A biographical sketch of Seward R. (Ted) Brown

John P. Smol
Department of Biology, Queen's University, Kingston, Ontario, K7L 3N6 Canada

On July 7, 1985 over 300 people gathered at the Lake Opinicon Field Station of Queen's University to dedicate a new building. The opening of this modern laboratory facility was in honour of the retirement of Prof. Ted Brown – a fitting tribute to a man who had dedicated much of his life to research and teaching.

Prof. Brown's affiliation with Queen's University has been a long one. After having been born and raised in Nova Scotia, and following five years of decorated service in the Canadian Armed Forces, Ted Brown came to the Queen's Biology Department for his undergraduate training. He received his Bachelor of Arts in 1951 and, shortly thereafter, he completed his Master's research at Lake Opinicon on littoral zone production (Brown, 1952). He then returned to his secondary school, the prestigious Pictou Academy in Nova Scotia, as a high school teacher. Nonetheless, his interest in limnology had been kindled, and during his summers he would return to Kingston to work as a research assistant with the new limnology professor at Queen's – Dr. Jack Vallentyne.

In the early 1950's, Dr. Vallentyne was starting work on sediment cores from the small, meromictic Little Round Lake. His main interests were amino acids in lake sediments, but he was also finding the degradation products of what were believed to be photosynthetic pigments. It was these 'biochemical fossils' that would be the focus of Ted Brown's scientific career. In 1954 he was accepted to work towards a doctorate with Prof. G. E. Hutchinson at Yale University on the sedimentary chlorophylls of Little Round Lake.

It was during the Yale years (1954–1959) that Ted Brown did his original work on the algal and bacterial chlorophylls and their derivatives in Little Round Lake. The identification of the chlorophyll derivatives was the first challenge, and in many cases this was accomplished by comparison with experimentally produced derivatives. In order to quantify these results, extinction coefficients had to be determined from ^{14}C labelled chlorophylls and a series of derivatives. Previously, only six extinction coefficients were known. This clearly required an integrated approach. In realistic terms this meant that he received degrees in both the chemical and biological sciences (Brown, 1962). The Yale years were not spent working entirely on sedimentary pigments. Ted found time to collaborate on studies of mammalian skin pigments (Foster & Brown, 1957), as well as on the construction of a new piston corer that was specially designed for the collection of surficial lake sediments (Brown, 1956). His skills as a teacher continued to be used as a lecturer in zoology in the newly developed Master of Arts and Teaching Program.

Ted Brown's days at Yale were coming to an end. Back in Canada, Jack Vallentyne was leaving Queen's and in 1959 Ted was offered his position. In addition, he was given the responsibility of Director at the Queen's University Biology Station at Lake Opinicon. He would continue to be Director until 1972, a period of marked expansion for both Queen's and the Station. Despite these commitments, Ted continued his university teaching and research.

Chlorophylls and their diagenesis (e.g. Millson *et al.,* 1965; Brown, 1968a; Daley, 1973; Daley *et al.*, 1973a, b; Daley & Brown, 1973a, b) continued to be the main research focus in the Brown lab, and the pioneering work that he and his students did soon

Fig. 1. Prof. S. R. Brown assembling the Brown Corner at Little Round Lake, Ontario.

led to paleolimnological applications (e.g. Brown, 1969; Gray, 1974; Brown *et al.*, 1977; Daley *et al.*, 1977). In the early 1960's the specificity of many carotenoids was being established, and the study of their diagenesis was also underway (Beaulieu, 1969; Leavitt, 1985). Soon both bacterial (Brown, 1968) and algal (Brown & Colman, 1973) carotenoids were added to the suite of indicators that paleolimnologists could use to trace the history of organisms that did not leave morphological fossils. The full application of these holistic studies is only now being realized (McIntosh, 1983; Brown *et al.*, 1984; and many papers in preparation).

Biochemical paleolimnology was always a central theme in the Brown lab. Nonetheless, Ted realized that fossil pigments were only one aspect of our understanding of lake dynamics. He encouraged students to explore diverse areas. For example, it was under Ted's tutelage that Tippett (1963) completed his now classic study on the annual nature of varved sediments. Always realizing that one could never divorce a lake's present from its past, he supervised a wide array of projects encompassing both chemical (McNeely, 1973) and biological (e.g. Lewis, 1965; Daley, 1968; Daley *et al.*, 1973c; Munro, 1978; Jones, 1981; Rosemarin, 1982; Craig, 1983) facets of lakes. Moreover, he continued to foster the integrated use of many lines of evidence in interpreting lake histories (Smol, 1982; Smol *et al.*, 1983, 1984; Boucherle *et al.*, 1986).

Prof. Brown is retiring from teaching this year, but his research and writing will continue. Just as it was appropriate to dedicate a new laboratory to him, it is equally appropriate that this Symposium honours his pioneering and influential work in paleolimnology.

References

Beaulieu, C. E., Jr., 1969. Quantitative changes in carotenoid composition and content of a blue-green alga during its growth in culture. M.Sci. Thesis, Queen's Univ., Kingston, 99 pp.

Boucherle, M. M., J. P. Smol, T. C. Oliver, S. R. Brown & R. N. McNeely, 1986. Limnologic consequences of the decline in hemlock 4800 years ago in three Southern Ontario lakes. Hydrobiologia 143: 217–225.

Brown, S. R., 1952. Some aspects of lake soils in relation to the productivity of a lake. M. Arts Thesis, Queen's Univ., Kingston, 61 pp.

Brown, S. R., 1956. A piston sampler for surface sediments of lake deposits. Ecology 37: 610–613.

Brown, S. R., 1962. Diagenesis of chlorophylls in lacustrine sediments. Ph.D. Thesis, Yale Univ., New Haven, 52 pp.

Brown, S. R., 1968a. Absorption coefficients of chlorophyll derivatives. J. Fish. Res. Bd. Can. 25: 523–540.

Brown, S. R., 1968b. Bacterial carotenoids from freshwater sediments. Limnol. Oceanogr. 13: 233–241.

Brown, S. R., 1969. Paleolimnological evidence from fossil pigments. Mitt. Int. Ver. Limnol. 17: 95–103.

Brown, S. R. & B. Colman, 1963. Oscillaxanthin in lake sediments. Limnol. Oceanogr. 8: 352–353.

Brown, S. R., R. J. Daley & R. N. McNeely, 1977. Composition and stratigraphy of the fossil phorbin derivatives of Little Round Lake, Ontario. Limnol. Oceanogr. 22: 336–348.

Brown, S. R., H. J. McIntosh & J. P. Smol, 1984. Recent paleolimnology of a meromictic lake: Fossil pigments of photosynthetic bacteria. Verh. Int. Ver. Limnol. 22: 1357–1360.

Craig, S. R., 1983. Distribution and physiological ecology of picoplankton from Little Round Lake, Ontario. Ph.D. Thesis, Queen's Univ., Kingston, 144 pp.

Daley, R. J., 1968. Changes in the photosynthetic capacity of a blue-green alga during growth in a standardized culture system. M.Sci. Thesis, Queen's Univ., Kingston, 90 pp.

Daley, R. J., 1971. Methodology and characterization of lacustrine chlorophyll diagenesis. Ph.D. Thesis, Queen's Univ., Kingston, 144 pp.

Daley, R. J., 1973. Experimental characterization of lacustrine chlorophyll diagenesis. 2. Bacterial, viral and herbivore grazing effects. Arch. Hydrobiol. 72: 409–439.

Daley, R. J. & S. R. Brown, 1973a. Experimental characterization of lacustrine chlorophyll diagenesis. 1. Physiological and environmental effects. Arch. Hydrobiol. 72: 277–304.

Daley, R. J. & S. R. Brown, 1973b. Chlorophyll, nitrogen, and photosynthetic patterns during growth and senescence of two blue-green algae. J. Phycology 9: 395–401.

Daley, R. J., C. B. Gray & S. R. Brown, 1973a. A quantitative, semi-routine method for determining algal and sedimentary chlorophyll derivatives. J. Fish. Res. Bd. Can. 30: 345–356.

Daley, R. J., C. B. Gray & S. R. Brown, 1973b. Reversed-phase thin-layer chromatography of chlorophyll derivatives. J. Chromatography 76: 175–183.

Daley, R. J., G. P. Morris & S. R. Brown, 1973c. Phagotrophic ingestion of a blue-green alga by *Ochromonas*. J. Protozoology 20: 58–61.

Daley, R. J., S. R. Brown & R. N. McNeely, 1977. Chromatographic and SCDP measurements of fossil phorbins and the postglacial history of Little Round Lake, Ontario. Limnol. Oceanogr. 22: 349–360.

Foster, M. & S. R. Brown, 1957. The production of DOPA by normal pigmented mammalian skin. J. Biol. Chem. 225: 247–252.

Gray, C. B. J., 1974. Distributions of chlorophyll derivatives in sediments of the Great Lakes. M.Sci. Thesis, Queen's Univ., Kingston, 94 pp.

Jones, K. E., 1981. Potential heterotrophy in five species of the blue-green alga, *Oscillatoria*. M.Sci. Thesis, Queen's Univ., Kingston, 170 pp.

Leavitt, P. R., 1985. Effects of herbivore grazing and bacterial attack on the carotenoids of blue-green algae. M.Sci. Thesis, Queen's Univ., Kingston, 136 pp.

Lewis, C. A., 1965. The largemouth bass fishery, Lake Opinicon, Ontario. M.Sci. Thesis, Queen's Univ., Kingston, 69 pp.

McIntosh, H. J., 1983. A paleolimnological investigation of the bacterial carotenoids of Sunfish Lake. M.Sci. Thesis, Queen's Univ., Kingston, 185 pp.

McNeely, R. N., 1973. Limnological investigations of a small meromictic lake, Little Round Lake, Ontario. Ph.D. Thesis, Queen's Univ., Kingston, 185 pp.

Millson, M. F., D. S. Montgomery & S. R. Brown, 1966. An investigation of the vanadyl porphyrin complexes of the Athabascan oil sands. Geochim. Cosmochim. Acta 30: 207–221.

Munro, K. A., 1978. Phytoplankton dynamics in meromictic Little Round Lake, southeastern Ontario, with emphasis on the growth of *Cryptomonas rostratiformis* in nature and in culture. M.Sci. Thesis, Queen's Univ., Kingston, 154 pp.

Rosemarin, A. S., 1982. An analysis of the role played by growth and nutrient dynamics regarding the competitive success of *Cladophora glomerata* (L.) Kutz. in Lake Ontario with reference to a potential competitor *Stigeoclonium tenue* (Agargh) Kutz. Ph.D. Thesis, Queen's Univ., Kingston, 277 pp.

Smol, J. P., 1982. Postglacial changes in algal assemblages from three Canadian lakes. Ph.D. Thesis, Queen's Univ., Kingston, 151 pp.

Smol, J. P., S. R. Brown & R. N. McNeely, 1983. Cultural disturbances and trophic history of a small meromictic lake from central Canada. Hydrobiologia 103: 125–130.

Smol, J. P., S. R. Brown & H. J. McIntosh, 1984. A hypothetical relationship between differential algal sedimentation and diatom succession. Verh. Internat. Ver. Limnol. 22: 1361–1365.

Tippett, R., 1963. An investigation into the nature of layering of deep water sediments in two eastern Ontario lakes. M.Sci. Thesis, Kingston, 87 pp.

Hydrobiologia 143: 13–20, (1986)

The PIRLA project: Paleoecological investigations of recent lake acidification

Donald F. Charles & Donald R. Whitehead
Department of Biology, Indiana University, Bloomington, Indiana 47405, USA

Keywords: acid deposition, eastern North America, lake acidification, paleoecology, paleolimnology, sediment chemistry

Abstract

The PIRLA project is a broadly interdisciplinary paleolimnological investigation of five to fifteen comparable watershed/lake systems from each of four low-alkalinity regions in North America that are currently receiving acid deposition. The areas are the Adirondack Mountains (N.Y.), northern New England, northern Great Lakes states, and northern Florida. The primary objective of the study is to provide a detailed reconstruction of the recent acidification histories of a representative suite of lakes from each of the regions. The study will increase our understanding of the timing, rates, and magnitude of acidification (and other chemical changes), and the regional and inter-regional patterns of lake acidification.

Introduction

In the last decade a great deal of scientific attention has been focused on the effects of acid precipitation on aquatic and terrestrial ecosystems in acid-sensitive geographic regions. Most of the research has been concentrated in Scandinavia and in northeastern North America (Drabløs & Tollan, 1980; Altshuller & Linthurst, 1984; National Research Council, 1981, 1986). This research has increased our understanding of the temporal and spatial patterns of acid precipitation and ecosystem responses to acid deposition. However, many questions remain unanswered and there are significant gaps in our knowledge. Many low-alkalinity areas have been little-studied, temporal trends are often difficult to document, and in some cases seemingly similar ecosystems have responded differently to equivalent acid inputs. Accordingly, further research is needed so that we can develop a more complete understanding of the response of ecosystems to acid-loading. It is particularly important to understand the timing, rates, characteristics, and magnitude of changes within each acid-sensitive region and to be able to make comparisons among regions.

Careful stratigraphic studies of the recent sediments in lakes can provide valuable information concerning the response of watershed/lake systems to atmospheric inputs of strong acids and other components of wet and dry deposition (e.g., Norton, 1984). When these analyses are carried out on watershed/lake systems in which there have been no recent disturbances (e.g., logging, fire, development), the stratigraphic data can be applied directly toward reconstructing ecosystem response to acid precipitation. Unfortunately, such undisturbed systems are rare in most regions. However, relatively unambiguous reconstructions can still be made provided that good historical data on watershed disturbance are available.

The PIRLA project involves stratigraphic analysis of several characteristics of sediment cores from approximately 35 lakes in four low-alkalinity regions in North America that currently receive acid deposition. The research is conducted by more than 30 investigators representing eight academic institutions.

The primary objective of the PIRLA project is to determine the extent to which lake acidification has occurred in these areas. The study is also designed to assess the relative importance of major potential causes of acidification.

The research is funded primarily by the Electric Power Research Institute. Support for the chironomid research (A. Uutala and M. Mitchell) is provided by the U.S. Environmental Protection Agency. Also, some investigators have obtained limited funds from their local institutions. Scheduled duration of the project is June 1983 to December 1986. This paper describes the purpose, approach, organization, and goals of the PIRLA project.

Study regions

Data for this study are being obtained from existing sources and analyses of short sediment cores taken from carefully selected lakes from four geographic regions (Fig. 1): (1) the Adirondack Mountains of New York; (2) northern New England (Maine, New Hampshire, and Vermont); (3) the Northern Great Lakes States (NGLS) (northern Minnesota including the Boundary Waters Canoe Area, northern Wisconsin, and the Upper Peninsula of Michigan); and (4) northern Florida (U. of Florida Ordway Preserve, Ocala National Forest, Appalachicola National Forest). These regions were chosen because they have bedrock and soil characteristics that make them potentially sensitive to acid precipitation (e.g., Galloway & Cowling, 1978; Norton *et al.,* 1982) and because they receive acid deposition (National Research Council, 1986). Each region receives different atmospheric deposition of strong acids, loading being greatest in the Adirondacks and northern New England (Barrie & Hales, 1984). There are no significant local emission sources that can affect the study lakes. Lakes in northern Minnesota receive precipitation with the highest pH (about 5.0) and thus serve as control lakes for the project, even though they do receive higher than background SO_4 and NO_3 loading.

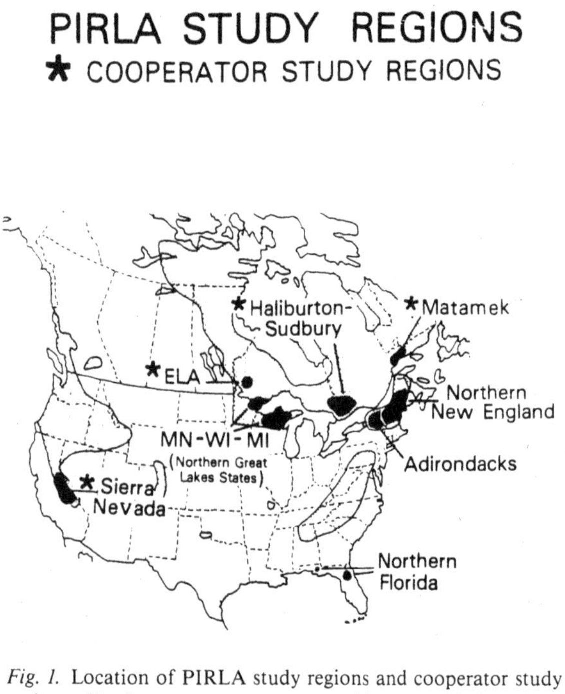

Fig. 1. Location of PIRLA study regions and cooperator study regions. Shading represents areas sensitive to acid deposition (adapted from Galloway & Cowling, 1984).

Study lakes (Table 1) were selected using the following criteria: (1) watersheds undisturbed, or minimal disturbance with good quality data on the nature and timing of disturbances; (2) no other lakes in the watershed; (3) morphometry and hydrology relatively simple; (4) water depth >8 m; (5) surface area 10 to 100 ha; and (6) watershed area to lake surface area ratio no more than 25:1. The lake set for each region included the following lake types wherever possible: (1) one relatively alkaline lake (pH > 7); (2) one clear, acidic lake (pH ≈ 4.5); (3) one acidic lake with high dissolved organic carbon content; (4) remaining lakes with pH in the range of 4.8 to 5.5 (potentially acidified or very sensitive). Exceptions were made if there were other compelling reasons for selecting a given lake. Also, some lakes initially selected were later rejected because of sediment dating or other problems. Watersheds of all lakes are forested and contain no, or minimal, cultural features (e.g., roads, houses). More specific information on PIRLA study lakes and preliminary project results are in Battarbee & Charles (1986) and Charles & Whitehead (1986).

Research projects similar to PIRLA are underway in other regions. We have made arrangements to exchange information with four of these projects, two in eastern Canada and two in the Sierra Nevada Mountains (PIRLA Cooperator Regions). The investigators in the projects in Canada are Mark Taylor and Hamish Duthie at the University of Waterloo and Geoff Davidson and David Schindler at the Fresh Water Institute of

Table 1. Characteristics of PIRLA study lakes cored for stratigraphic analysis. Mean and ranges of values (in parentheses) are presented. Number of lakes in the diatom-water chemistry calibration data set are listed in parentheses following the region name.

Region		Number of lakes	Surface area ha	Maximum depth m	pH Air-equilibrated	Alkalinity $\mu eq \cdot l^{-1}$	Ca $\mu eq \cdot l^{-1}$	SO_4 $\mu eq \cdot l^{-1}$
Adirondacks	(50)	15	51	11	5.3	41	111	125
			(0.4 – 515)	(1 – 24)	(4.3 – 7.0)	(– 48 to 410)	(30 – 430)	(65 – 154)
Northern New		5	11	19	5.5	5	62	86
England	(46)		(1 – 16)	(12 – 25)	(4.6 – 6.4)	(– 34 to 48)	(27 – 114)	(58 – 105)
Northern Great		9	42	11	5.7	19	51	80
Lakes States	(36)		(11 – 109)	(5 – 25)	(4.4 – 6.8)	(– 38 to 80)	(32 – 68)	(53 – 151)
Northern Florida	(32)	6	42	7	5.0	– 10	69	87
			(11 – 39)	(4 – 10)	(4.6 – 5.8)	(– 42 to 52)	(34 – 136)	(28 – 143)

Winnipeg, Canada. In the Sierras they are Robert Holmes and Mark Whiting (funded by the California Air Resources Board), and Mark Whiting, Donald Whitehead, and Donald Charles (funded by Southern California Edison).

Research plan

Overall organization of the PIRLA project is depicted in Fig. 2. Project investigators, research activities, and sediment characteristics to be analyzed are listed in Table 2.

Several analyses are being done on each core to provide independent lines of evidence and cross-checks to help interpret and verify changes which may have occurred in the lakes. This reduces the likelihood of incorrect interpretations and allows for a more thorough understanding of acidification causes and consequences.

Cores are being dated using Pb-210 (Appleby & Oldfield, 1978; Oldfield *et al.*, 1980; Davis *et al.*, 1984) and chronostratigraphic markers (e.g., *Ambrosia* rise, charcoal peaks associated with known local fires). Diatoms and chrysophytes are being analyzed to reconstruct lakewater pH and alkalinity (e.g., Battarbee, 1984, Charles, 1984, 1985; Charles & Norton, 1986; van Dam *et al.*, 1981; Davis *et al.*, 1982; Davis & Anderson, 1985; Renberg & Hellberg, 1982; Whitehead *et al.*, 1986; Smol *et al.*, 1984a, 1984b, 1986). Investigation of chironomid and chydorid remains also provides information on acidification trends. Analysis of pollen and charcoal provides both dating of events and reconstruction of watershed vegetational and disturbance history (e.g., Swain, 1973). Analysis of polycyclic aromatic hydrocarbons (PAH) (Hites *et al.*, 1977; Gschwend & Hites, 1981), selected trace metals, and soot particles permits the reconstruction of atmospheric deposition patterns associated with the combustion of fossil fuels. Investigation of metals allows assessment of: (1) rates of atmospheric and erosional inputs, (2) the extent to which various watershed cations are mobilized by acid precipitation, and (3) exchange of cations between the sediment and the water column induced by lake acidification (e.g., Norton, 1984; Charles & Norton, 1986). Further insights into lakewater changes are provided by studies of sulfur and sulfur isotopes (Mitchell *et al.*, 1984; Holdren *et al.*, 1984; Nriagu & Soon, 1985).

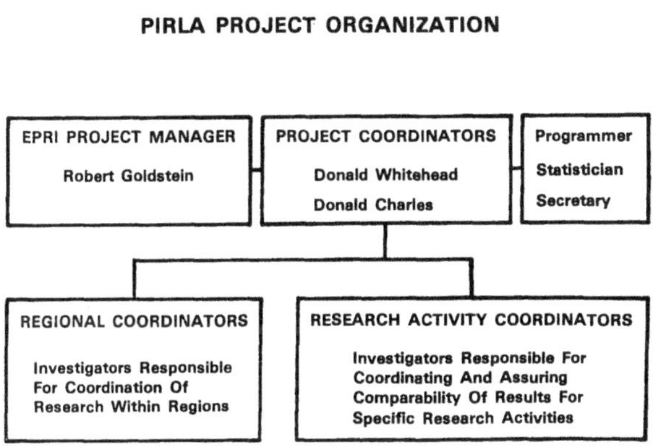

Fig. 2. PIRLA project organization.

Methods used in all regions (Charles & Whitehead, 1986) are consistent with each other, so data are comparable among regions. Quality control/quality assurance measures have been used to help ensure comparability.

Approach

By analyzing differences among lakes within regions and among regions, it will be possible

Table 2. PIRLA project investigators (Team PIRLA). Current university affiliations are indicated following investigators' names. Sediment analyses followed by an (I) are being done only on cores from a selected set of lakes (Intensive Study Lakes).

Project Coordinators
Donald Whitehead, IU
Donald Charles, IU
EPRI Project Manager
Robert Goldstein
Regional Coordinators
Adirondacks – Donald Charles, IU
– Donald Whitehead, IU
N. New England – Ronald Davis, UM
– Stephen Norton, UM
N. Great Lakes States – John Kingston, UMD
N. Florida – Raymond Bienert, UF
– Thomas Crisman, UF
*Ontario, Quebec – Mark Taylor, UW
– Hamish Duthie, UW
*Ontario – Geoff Davidson, FWI
– David Schindler, FWI
Computer Programmer
Mirian Tawata, IU
Michael Ford
Statistics Consultant
David Parkhurst, IU
Administrative Assistant
Rhea Baker, IU
Pollen/Charcoal – Donald Whitehead, IU
ADKS – Donald Whitehead, IU
– Robert Wise, IU
NNE – Linda Bacon, UM
NGLS – Linda Shane, UMM
NFLA – Robert Snodgrass, UF
Dating (Pb-210) – Stephen Norton, UM
ADKS, NNE, NGLS – Stephen Norton, UM
– Geneva Blake, UM
NFLA – Michael Binford, UF
Total Metals – Stephen Norton, UM
ADKS, NNE, NGLS – Stephen Norton, UM
– Marilyn Morrison, UM
NFLA – Raymond Bienert, UF
Sequentially Extracted Metals (I) –
Jeffrey White (AR), IU
Total Sulfur, Carbon, Nitrogen and
Hydrogen (I) – Myron Mitchell (AR)
Sulfur Isotopes – Brian Fry (AR), IU
Polycyclic Aromatic Hydrocarbons (I) –
Ronald Hites, Linda Roll (AR), IU
Variability Study – Russell Kreis, Jr.
(AR), UMD

Research Activity Coordinators
and Regional Investigators
Sediment Coring – Ronald Davis, UM
All Regions – RC
Watershed Analysis – Ronald Davis, UM
All Regions – RC
Sediment Subsampling – Stephen Norton, UM
Diatoms – John Kingston, UMD
ADKS – Donald Charles, IU
NNE – Dennis Anderson, UM
– Mark Whiting, UM
– Ronald Davis, UM
– Jesse Ford, CU
NGLS – John Kingston, UMD
– Keith Camburn, UMD
– Russell Kreis, Jr., UMD
NFLA – Roger Sweets, UMD
Chrysophytes (I) – John Smol (AR), QU
Chironomids (I) – Allen Uutala
ADKS – Allen Uutala, ESF
NFLA – Thomas Crisman, UF
Chydorids – Michael Patterson, DH
3 ADK lakes only
Water Chemistry – Robert Cook, IUP
– Donald Charles, IU
ADKS – Donald Charles, IU
NNE – Stephen Norton, UM
– Stephen Kahl, UM
– Ronald Davis, UM
– Linda Bacon, UM
NGLS – Robert Cook, IUP
NFLA – Robert Garren, UF
– Stephen Kahl, UM
Water Chemistry Quality Control – Robert Cook (AR), IUP
Data Management – Donald Charles, IU

Cooperator regions are indicated by an asterisk.
ADKS = Adirondacks; NNE = northern New England; NGLS = northern Great Lakes states; NFLA = northern Florida; AR = all regions; EPRI = Electric Power Research Institute; IU = Indiana University; UM = University of Maine at Orono; UMD = University of Minnesota at Duluth; UF = University of Florida; FWI = Fresh Water Institute (Winnipeg, Manitoba, Canada); CU = Cornell University; QU = Queens University (Kingston, Ontario); ESF = SUNY College of Environmental Science and Forestry; DH = Dalhousie University; IUP = Indiana University of Pennsylvania; UMM = University of Minnesota, Minneapolis; RC = Regional Coordinators.

to understand and predict more accurately the ecological effects of acid precipitation.

The fundamental questions being addressed by the PIRLA project are: Have lakes in acid-sensitive regions that receive low-pH precipitation acidified in recent years? If so, at what point in time did pH-related changes begin to occur? What was the rate and magnitude of change?

If lake acidification is documented, what was the primary cause(s)? Three major alternative hypotheses are being evaluated. These are presented, along with their logical consequences, in Fig. 3. The possibility of acidification being caused by local watershed changes (e.g., Rosenqvist, 1978) can be eliminated for lakes with no major recent watershed disturbances and can be evaluated for watersheds with known disturbance by examining the timing of diatom-inferred pH changes with respect to sediment indicators of watershed disturbance. Local emission sources or other local factors (e.g., extensive recent beaver activity) are not important causes of acidification in any of the lakes. Available historical climatic data (e.g., Bradley, 1986) will be evaluated for regional correlations between rainfall and temperature and lake acidification.

One of the greatest strengths of the PIRLA project is the ability to make regional comparisons and draw conclusions applicable at a national and international scale. This is possible because: (1) Information will be available from several lakes in four widely separated regions. (2) Information from the regions is obtained using the same procedures and is expressed in the same format, allowing it to be pooled and analyzed as one unified, comprehensive data base.

An advantage of PIRLA over smaller-scale projects is that many sediment characteristics are being analyzed by several investigators. No one researcher could provide all of the necessary types of analyses. Also, interaction among the investigators has served to stimulate ideas and generate constructive criticism within the project. The investigators (regional and research activity coordinators) are already familiar with lake/watershed systems in their regions, or with their analytical specialty, and have been doing research closely related to that involved with PIRLA project. Close contact is maintained by all collaborators to ensure comparability of data, exchange findings, and discuss progress. All investigators and diatomists meet twice each year.

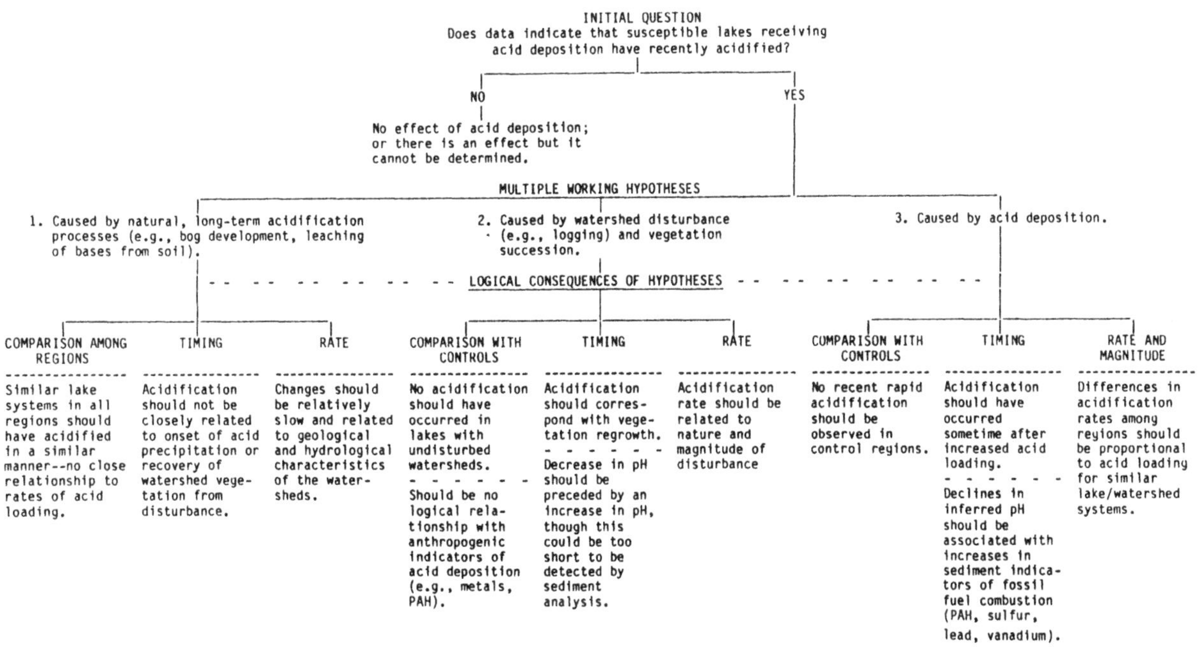

Fig. 3. Questions, multiple (alternative) working hypotheses, and logical consequences of the hypotheses being evaluated in the PIRLA project. Hypotheses will be examined by comparing data among regions, but can also be used to assess sediment data for individual lakes. All causes may be important and it may not be possible to discern a single primary cause for an individual lake or region.

Diatom- and chrysophyte-inferred lakewater chemistry reconstructions

Diatom and chrysophyte assemblages will be used to reconstruct past lakewater chemistry. Calibration data sets of water chemistry and diatom and chrysophyte assemblages are being developed for each region (Table 1). Water chemistry data (air equilibrated and non-air equilibrated pH, Gran plot alkalinity, conductivity, all major anions and cations, Si, Al, P, N, chlorophyll a, and other characteristics) have been measured for surface samples (0.5 m) for two summers in all regions except Florida. Analyses have been performed so that data from all regions is consistent and comparable.

The challenge of obtaining consistency in taxonomy among eight diatomists has been met by (1) preparing and maintaining a single PIRLA diatom taxa list, (2) meeting twice a year to discuss taxonomy, and (3) preparation of the PIRLA Diatom Iconograph, a compilation of photographic plates of taxa encountered during the project. The Diatom Herbarium at the Academy of Natural Sciences of Philadelphia is visited regularly. Several new taxa have been described by project diatomists (Camburn & Kingston, 1986; Charles, 1986).

The water chemistry and diatom/chrysophyte data for all PIRLA regions and Cooperator Regions will be assembled in a single data base. It will include data for over 200 lakes from several geographic regions, the largest such data base ever assembled. These data are being used to derive improved procedures for reconstructing lakewater chemistry. Initial calibration of diatom-pH predictive relationships will be based on approaches pioneered by Hustedt (1939), Nygaard (1956), and Meriläinen (1967), and recent refinements of Davis *et al.* (1982), Davis & Anderson (1985), Charles (1985), and Renberg & Hellberg (1982). Predictions based on these methods will be improved because information in the data base will allow more accurate assignment to pH categories, a task which is often difficult due to a lack of good ecological data.
ecological data.

Transfer functions, like those used to calculate past climatic conditions (Sachs *et al.*, 1977; Imbrie & Webb, 1981), will be developed. These functions will be based on the percentage of occurrence of individual taxa instead of percentge of diatoms in only three to five pH categories and will require a large data base. They should provide improved accuracy and flexibility in reconstructions.

New relationships will be developed to predict water chemistry characteristics other than pH. Aluminum and humic compounds are high priority because of their important role in the acidification process (e.g., Davis *et al.*, 1985).

Variability study

An important component of the PIRLA project is the statistical assessment of within- and between-core variability associated with replicate cores collected from the profundal zone of a single lake. The purpose of this study component is to determine how well one core represents the sediment record of an entire lake. Analyses are being based on evaluation of three cores taken from each of eight PIRLA study lakes.

Data management

All project data are maintained at Indiana University in the PIRLA Data Base Management System which uses the Scientific Information Retrieval (SIR) computer programs. Data initially are entered and verified at participating institutions, and much data compilation, analysis, and interpretation of data for each region is done by individual investigators. Higher level analysis, including comparative statistical and multivariate analysis of data from all regions, will be done using the PIRLA/DBMS.

Summary

The PIRLA project is a broadly interdisciplinary, multi-investigator study designed to assess the impact of acid deposition on watershed-lake systems in four acid-sensitive areas of North America. Dating of events will be provided by Pb-210 and chronostratigraphic markers; pH changes will be inferred from diatom and chrysophyte assemblages; deposition of by-products of fossil fuel combustion will be reconstructed from polycyclic aromatic hydrocarbons, soot particles, sulfur, and selected metals; watershed/lake changes related to acidification will be analyzed through studies of metals, metal speciation, sulfur, and sulfur isotopes; watershed disturbances will be determined from historical records and pollen and charcoal. As the methods applied to all cores are consistent, intra- and interregional comparisons will be possible.

References

Appleby, P. G. & F. Oldfield, 1978. The calculation of lead-210 dates assuming a constant rate of supply of unsupported ^{210}Pb to the sediment. Catena 5: 1–8.

Altshuller, A. P. & R. A. Linthurst (eds), 1984. The acidic deposition phenomenon and its effects: critical assessment review papers (EPA-600/8-83-016AF). U.S. Environmental Protection Agency, Washington, D.C.

Barrie, L. A. & J. M. Hales, 1984. The spatial distributions of precipitation acidity and major ion wet deposition in North America during 1980. Tellus 36B: 333–355.

Battarbee, R. W. & D. F. Charles, 1986. Diatom-based pH reconstruction studies of acid lakes in Europe and North America: A synthesis. Water, Air, and Soil Pollut. (in press).

Bradley, R. S., 1986. Uncertainties in trends in acid deposition: the role of climatic fluctuations. In Acid Deposition: Long-term Trends, chapter 3, pp. 93–108. National Academy Press, Washington, D.C.

Camburn, K. E. & J. C. Kingston, 1985. The genus *Melosira* from soft-water lakes with special reference to northern Michigan, Wisconsin, and Minnesota. In J. P. Smol, R. Battarbee, R. B. Davis & J. Meriläinen (eds), Diatoms and Lake Acidity, pp. 17–34. W. Junk, The Hague, The Netherlands.

Charles, D. F., 1984. Recent pH history of Big Moose Lake (Adirondack Mountains, New York, USA) inferred from sediment diatom assemblages. Verh. int. Verein. Limnol. 22: 559–566.

Charles, D. F., 1985. Relationships between surface sediment diatom assemblages and lakewater characteristics in Adirondack lakes. Ecology 66: 994–1011.

Charles, D. F., 1986. A new diatom species, *Fragilaria acidobiontica*, from acidic lakes in northeastern North America. In J. P. Smol, R. Battarbee, R. B. Davis & J. Meriläinen (eds), Diatoms and Lake Acidity, pp. 35–44. W. Junk, The Hague, The Netherlands.

Charles, D. F. & S. A. Norton, 1986. Paleolimnological evidence for trends in atmospheric deposition of acids and metals. In Acid Deposition: Long-term Trends, chapter 9, pp. 335–506. National Academy Press, Washington, D.C.

Charles, D. F. & D. R. Whitehead, 1986. Paleoecological investigation of recent lake acidification (PIRLA): Methods and project description. Electric Power Research Institute, Palo Alto, CA (in press).

Charles, D. F. & D. R. Whitehead, *et al.*, 1986. The PIRLA project (Paleoecological Investigation of Recent Lake Acidification): Preliminary results for the Adirondacks, New England, N. Great Lakes States and N. Florida. Water, Air, and Soil Poll. (in press).

Davis, R. B. & D. S. Anderson, 1985. Methods of pH calibration of sedimentary diatom remains for reconstructing history of pH in lakes. Hydrobiologia 12: 69–87.

Davis, R. B., D. S. Anderson & F. Berge, 1985. Palaeolimnological evidence that lake acidification is accompanied by loss of organic matter. Nature 316: 436–438.

Davis, R. B., C. T. Hess, S. A. Norton, D. W. Hanson, K. D. Hoagland & D. S. Anderson, 1984. ^{137}Cs and ^{210}Pb dating of sediments from soft-water lakes in New England (USA) and Scandinavia, a failure of ^{137}Cs dating. Chem. Geol. 44: 151–185.

Davis, R. B., S. A. Norton, C. T. Hess & D. Brakke, 1982. Paleolimnological reconstruction of the effects of atmospheric deposition of acids and heavy metals on the chemistry and biology of lakes in New England and Norway. Hydrobiologia 103: 113–123.

Drabløs, D. & A. Tollan (eds), 1980. Ecological Impact of Acid Precipitation. Proceedings of an international conference, Sandefjord, Norway, March 11–14, 1980. SNSF Project. Oslo, Norway.

Galloway, J. N. & E. B. Cowling, 1978. The effects of precipitation on aquatic and terrestrial ecosystems – a proposed precipitation chemistry network. Journal of the Air Pollution Control Association 28: 229–235.

Gschwend & R. A. Hites, 1981. Fluxes of polycyclic aromatic hydrocarbons to marine and lacustrine sediments in the northeastern United States. Geochim. Cosmochim. Acta 45: 2359–2367.

Hites, R. A., R. E. Laflamme & J. W. Farrington, 1977. Polycyclic aromatic hydrocarbons in recent sediments: the historical record. Science 198: 820–831.

Holdren, Jr., G. R., T. M. Brunelle, G. Matisoff & M. Wahlen, 1984. Timing the increase in atmospheric sulphur deposition in the Adirondack Mountains. Nature 311: 245–248.

Hustedt, F., 1939. Systematische and ökologische Untersuchungen über die Diatomeenflora von Java, Bali und Sumatra nach dem Material der Deutschen Limnologischen Sunda-Expedition III. Die ökologischen Factoren und ihr Einfluss auf die Diatomeenflora. Arch. Hydrobiol. Suppl. 16: 274–394.

Imbrie, J. & T. Webb III, 1981. Transfer functions: calibrating micropaleontological data in climatic terms. In A. Berger (ed), Climatic Variation and Variability: Facts and Theories, pp. 125–134. D. Reidel Publishing Company.

Meriläinen, J., 1967. The diatom flora and the hydrogen-ion concentration of the water. Ann. bot. fenn. 4: 51–58.

Mitchell, M. J., M. B. David & A. J. Uutala, 1984. Sulfur distribution in lake sediment profiles as an index of historical depositional patterns. Hydrobiologia 121: 121–127.

National Research Council, 1981. Atmosphere-biosphere interactions: toward a better understanding of the ecological consequences of fossil fuel combustion. National Academy Press, Washington, D.C., 273 pp.

National Research Council, 1986. Acid deposition, Long-Term Trends. National Academy Press, Washington, D.C., 506 pp. Washington, D.C. (in press).

Norton, S. A., J. J. Akielaszek, T. A. Haines, K. L. Stromborg & J. R. Longcore, 1982. Bedrock geologic control of sensitivity of aquatic ecosystems in the United States to acidic deposition. National Atmospheric Deposition Program (NC-141). Natural Resource Ecology Laboratory, Colorado State University, Fort Collins, Colorado, 50 pp.

Norton, S. A., organizer, 1984. Proceedings of a workshop on paleolimnological studies of the history and effects of acidic precipitation (Rockland, Maine, May 23–25, 1984). U.S. Environmental Protection Agency, Corvallis, Oregon, 425 pp.

Nriagu, J. O. & Y. K. Soon, 1985. Distribution and isotopic composition of sulfur in lake sediments of northern Ontario. Geochim. Cosmochim. Acta 49: 823–834.

Nygaard, G., 1956. Ancient and recent flora of diatoms and Chrysophyceae in Lake Gribsø. In K. Berg & I. C. Peterson (eds), Studies on the Humic, Acid Lake Gribsø, pp. 32–94,

pl. 1–12. Folia limnol. scand. 8: 1–273.
Oldfield, F., P. G. Appleby & D. Petit, 1980. A re-evaluation of lead-210 chronology and the history of the total lead influx in a small South Belgian pond. Ambio 9: 97–99.
Renberg, I. & T. Hellberg, 1982. The pH history of lakes in southwestern Sweden, as calculated from the subfossil diatom flora of the sediments. Ambio 11: 30–33.
Rosenqvist, I. Th., 1978. Alternative sources for acidification of river water in Norway. The Science of the Total Environment 10: 39–49.
Sachs, H. M., T. Webb III & D. R. Clark, 1977. Paleoecological transfer functions. Annual Review of Earth and Planetary Sciences 5: 159–178.
Smol, J. P., R. Battarbee, R. B. Davis & J. Meriläinen, 1986. Diatoms and lake acidity. W. Junk, The Hague, Netherlands, 307 pp.
Smol, J. P., D. F. Charles & D. R. Whitehead, 1984a. Mallomonadacean (Chrysophyceae) assemblages and their relationships with limnological characteristics in 38 Adirondack (New York) lakes. Can. J. Bot. 62: 911–923.
Smol, J. P., D. F. Charles & D. R. Whitehead, 1984b. Mallomonadacean microfossils provide evidence of recent lake acidification. Nature 307: 628–630.
Swain, A. M., 1973. A history of fire and vegetation in northeastern Minnesota as recorded in lake sediments. Quat. Res. 3: 383–396.
van Dam, H. G. Surrmond & C. J. F. ter Braak, 1981. Impact of acidification on diatoms and chemistry of Dutch Mooreland pools. Hydrobiologia 83: 425–459.
Whitehead, D. R., D. F. Charles, S. E. Reed, S. T. Jackson & M. C. Sheehan, 1986. Late-glacial and Holocene acidification history of Adirondack (N.Y.) lakes. In J. P. Smol, R. Battarbee, R. B. Davis & J. Meriläinen (eds), Diatoms and Lake Acidity, pp. 251–274. W. Junk, The Hague, The Netherlands.

Accepted 18 March 1986.

Hydrobiologia 143: 21–27, (1986)

^{210}Pb dating by low background gamma counting

P. G. Appleby[1], P. J. Nolan[2], D. W. Gifford[2], M. J. Godfrey[2], F. Oldfield[3], N. J. Anderson[4] & R. W. Battarbee[4]
[1]*Department of Applied Mathematics & Theoretical Physics,* [2]*Oliver Lodge Laboratory, Department of Physics,* [3]*Department of Geography, University of Liverpool, P. O.Box 147, Liverpool L69 3BX, England,* [4]*Palaeoecology Research Unit, Department of Geography, University College, London, England*

Keywords: ^{210}Pb and ^{226}Ra measurements, sediments

Abstract

Lead-210 and radium-226 measurements by direct gamma assay can now provide a record of changing concentrations in lake sedimemts sufficiently reliable and precise to form a suitable basis for age/depth and dry-sedimentation-rate calculations. There are additional benefits in terms of non destructive sample preparation and simultaneous assay for other environmentally significant gamma-emitting radioisotopes (e.g. ^{137}Cs and ^{241}Am). Results from L. Fleet, S. W. Scotland illustrate the value of this approach especially in lakes with disturbed catchments where variable input of supported ^{210}Pb has occurred.

Introduction

Lead-210 based sediment dating, first proposed by Goldberg (1963), has, over the last few years, been applied with increasing frequency to lakes within which sedimentation has been affected by catchment disturbance. This development reflects several trends. There has, for example been a growing interest in the use of sediment-based studies in the reconstruction of erosion rates and their relation to land changes (cf. Oldfield & Appleby 1984a) and also a need to test alternative hypotheses of lake water acidification from the fossil diatom content of sediments in acidified lakes with both disturbed and undisturbed catchments (Flower & Battarbee 1983; Battarbee 1984). Oldfield & Appleby (1984 a, b) have reviewed some of the implications of these developments for ^{210}Pb dating, in terms of analytical requirements and empirical testing

Most ^{210}Pb dating laboratories use alpha counting for determining lead-210 activity. The method involves sample-destructive chemical preparation followed by ingrowth of the alpha-emitting granddaughter isotope polonium-210. Additional procedures are required for counting the radium-supported activity of a sample (cf Eakins & Morrison, 1977) and in most cases, dated profiles have either completely lacked radium determinations (e.g. Kahl *et al.*, 1984) or relied on only a small number of measurements (e.g. Brush *et al.* 1982). Recent studies have shown that lack of adequate radium determinations seriously limits the accuracy of sediment chronologies in disturbed catchments where changes in sediment type often give rise to significant changes in radium concentrations (Oldfield *et al.* 1980).

Gäggeler *et al.* (1976) illustrated the use of direct gamma counting in ^{210}Pb dating but the standard errors associated with their measurements were unacceptably large. More recently, Durham & Oliver (1983) have published a dated profile from Lake Ontario for which the ^{210}Pb measurements have been done by direct gamma assay using an Aptec hyper-pure planar germanium detector. In their case, an alternative coaxial gamma detector was used for radium-266 and caesium-137 measurements. The combination of techniques led to acceptable analytical precision and the published results prompted the developments outlined below. In the present paper, we describe the approach to

^{210}Pb dating developed in the University of Liverpool using direct gamma assay, and we illustrate this approach by reference to work completed on the sediments of Loch Fleet, an acidified lake in Galloway, S.W. Scotland.

At the outset, the potential advantages of direct gamma assay include non-destructive sample preparation, a single analysis for both ^{210}Pb and ^{226}Ra assay on all samples, and simultaneous measurement of other gamma emitting radioisotopes of environmental interest (e.g. ^{137}Cs; ^{241}Am; ^{238}U; ^{40}K).

The Liverpool system

The radiometric data presented in this paper have been acquired using an Ortec HPGe GWL-80210-S well-type coaxial low background intrinsic germanium detector. This detector has a cylindral p-type germanium crystal of 50 mm diameter × 60 mm long. The well is along the axis of the cylinder and is 10 mm diameter × 40 mm long. Background suppression was achieved by means of a 100mm thick lead castle, a 305 mm diameter × 305 mm long sodium iodide (NaI(Tl)) escape suppression shield (supplied by John Caunt Scientific Ltd), and a 3 mm thick copper cylinder fitting over the end of the detector and sample. The configuration is shown in Fig. 1. The lead and copper shields reduce the background at the germanium detector due to its surrounding environment.

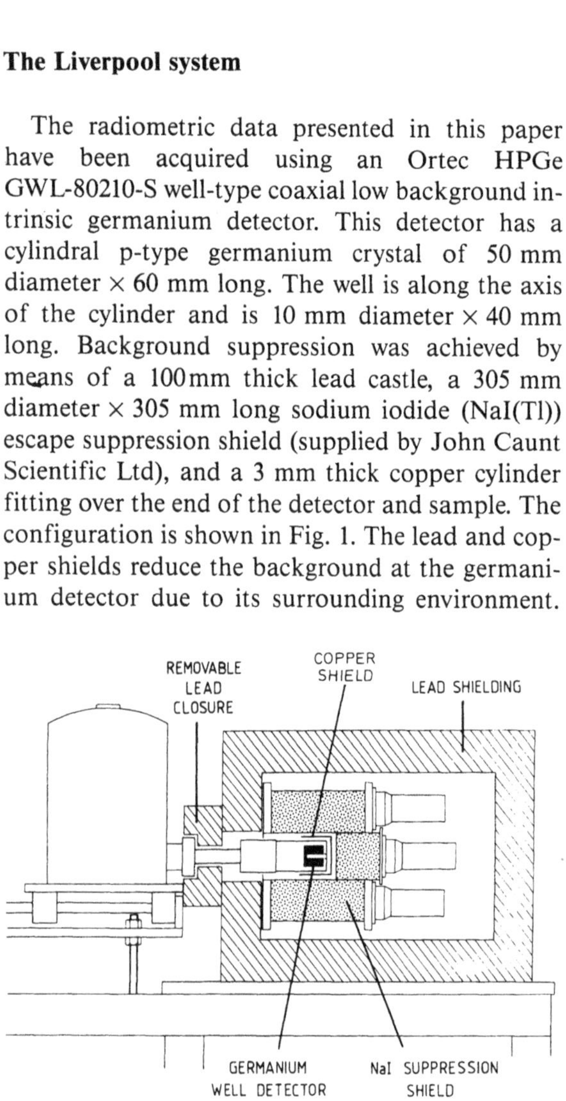

Fig. 1. The configuration of the Liverpool gamma based ^{210}Pb dating system.

The NaI(Tl) shield has two roles. It acts as a passive shield but also can be used to detect gamma-rays that Compton scatter out of the germanium detector and which also contribute to the background. This active suppression reduces the background by a further factor of 2–3. In this mode the system can however be used only for nuclides that emit a single gamma-ray, or for decay paths that only involve a single gamma-ray per nuclide. Samples (with masses from 0.5–3 g) are placed in specially fabricated plastic sample holders designed to fit inside the well recess of the detector.

The background characteristics of the system were optimised by a series of experiments using various combinations of shielding options. Table 1 summarises data from a series of background counts, each of 216000 sec duration, carried out over a period of 10 months. The low and constant background counts in the energy ranges of interest are essential to the achievement of acceptable precision and counting times.

Data were acquired using an ADC (analogue to digital computer) interfaced to a BBC microcomputer. This system was used to provide a dispersion of 0.5 keV per channel over a gamma ray energy range of 20 keV to 2 MeV. The system typically achieved a resolution of 1.2 keV at 100 keV and 2 keV at 1 MeV. The efficiency of the system for samples of the above configuration was measured using a series of calibrated radiometric sources and sediment samples whose activity was known. The absolute efficiency ranges from 50% at 47 keV (^{210}Pb) to 11% at 352 keV (^{226}Ra). A correction for self absorption was made for gamma rays of low energy. Files were transferred to the University mainframe IBM 3083 computer and analysed for a range of isotopes. Fig. 2 shows a typical spectrum, and Table 2 lists the concentrations of some of the isotopes. The analysis program automatically places the ^{210}Pb and ^{226}Ra concentrations in data files in a form which allows immediate sample by sample updating of chronologies using a range of models (c.f. Appleby & Oldfield 1983).

Loch Fleet ^{210}Pb chronology

a) Site characteristics and core location

Loch Fleet is a small (17 ha) upland (340 m) loch situated on the Cairnsmore of Fleet granitic intru-

Table 1. Background counts for selected isotopes.

Date	Counts per minute (over 3600 mins)		
	^{210}Pb (at 46.5keV)	^{214}Pb (at 352keV)	^{137}Cs (at 662keV)
31/12/84	0.131 ± 0.011	0.0251 ± 0.0061	0.0747 ± 0.0061
21/ 1/85	0.108 ± 0.011	0.0182 ± 0.0061	0.0689 ± 0.0058
28/ 1/85	0.100 ± 0.011	0.0160 ± 0.0061	0.0722 ± 0.0059
11/ 2/85	0.096 ± 0.011	0.0297 ± 0.0066	0.0705 ± 0.0063
29/ 3/85	0.119 ± 0.011	0.0376 ± 0.0063	0.0560 ± 0.0056
29/ 3/85	0.126 ± 0.012	0.0300 ± 0.0064	0.0581 ± 0.0056
5/ 7/85	0.109 ± 0.011	0.0353 ± 0.0070	0.0452 ± 0.0058
16/ 9/85	0.099 ± 0.011	0.0328 ± 0.0070	0.0603 ± 0.0058
Mean values	0.110 ± 0.013	0.0281 ± 0.0078	0.0632 ± 0.0101

sion in Galloway, South-West Scotland (Fig. 3). The lake has a pH of 4.4 and preliminary diatom data (Anderson & Battarbee unpubl.) show that recent acidification has occurred. In 1963 approximately 22% of the catchment was ploughed and drained prior to afforestation. Currently an integrated lake-catchment water quality study is being carried out at the site with a view to re-establishing a viable brown trout population.

As part of this study and with a view to reconstructing the recent history of the lake, the stratigraphy and distribution of limnic sediment is being examined. An initial attempt to use a core from the deepest part of the lake for dating was unsuccessful, since it produced an inconclusive and confusing ^{210}Pb timescale (cf. Battarbee *et al.* 1985). Subsequent fieldwork involved a surface sediment mapping and multi-core strategy to find more suitable core locations. The results indicated that most of the post-glacial organic sediments had ac-

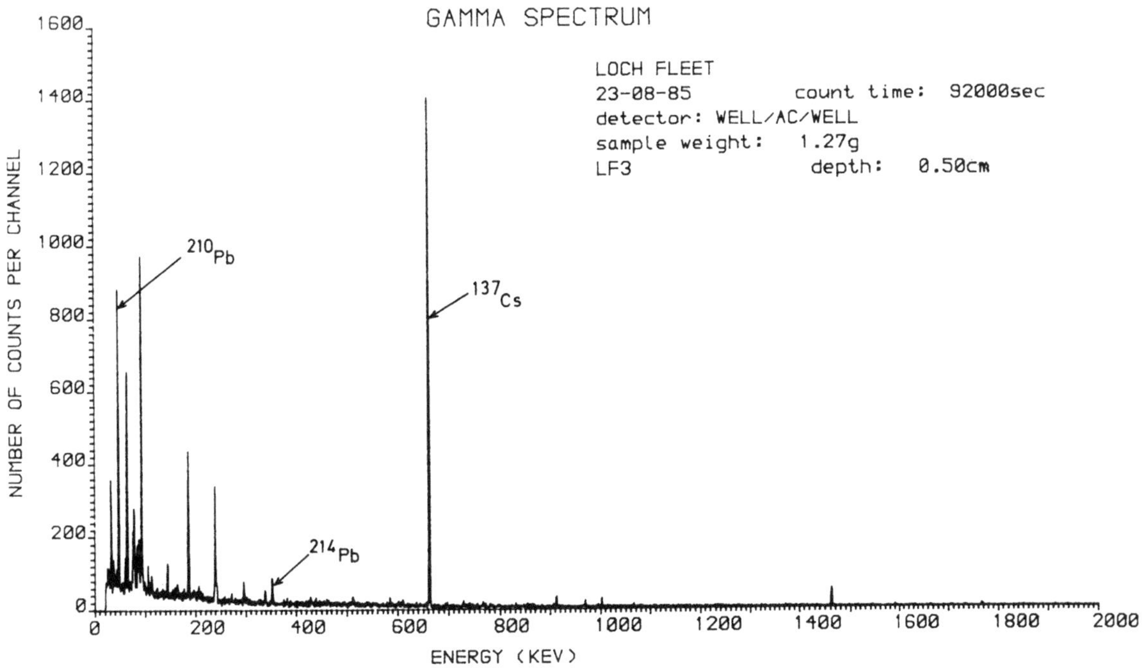

Fig. 2. A gamma spectrum from a Loch Fleet sediment sample.

Table 2. Radioisotopic concentrations in a sediment sample from Loch Fleet. Count Time: 92000 Sec. Loch Fleet LFL3. Date: 23.08.85. Weight: 1.270 g. Depth: 0.5 cm. Detector: WELL/AC/WELL.

Isotope	Energy keV	Gross counts	Net counts	FWHM keV	Conc PCI/G
PB-210	47.0	2696.0 ± 71.2	2526.0	1.4	25.54 ± 0.73
RA-226	352.7	230.0 ± 28.9	191.2	1.7	0.97 ± 0.15
	295.9	179.2 ± 27.8	165.0	2.0	
	1765.2	22.6 ± 7.4	20.3	2.7	
U-238	64.4	1923.0 ± 64.5	1706.2	1.4	17.69 ± 0.69
U-235	186.4	1471.0 ± 51.4	1054.9	1.8	1.92 ± 0.10
	144.3	259.0 ± 36.5	196.2	1.4	
	164.3	92.0 ± 32.1	78.2	1.6	
AM-241	60.2	152.5 ± 44.1	152.5	1.4	0.16 ± 0.05
CS-137	662.5	5417.5 ± 77.9	5320.6	1.9	20.76 ± 0.31
	32.8	999.0 ± 56.7	997.4	1.7	
	37.0	223.5 ± 44.3	221.6	1.4	
K-40	1462.1	231.2 ± 17.5	187.0	2.0	12.25 ± 1.23

cumulated on the slopes of the basin to the north of the area of deepest water (Anderson & Battarbee 1985), and, judging from the loss on ignition profiles, showed that catchment ploughing prior to afforestation had caused significant erosion and inwash of peaty soils into the lake. A series of Livingstone cores (L1, L2 and L3) were obtained from the northern part of the lake. Of these, L3, taken in May 1985 in 10 m water depth, had the longest post-glacial record, and, compared with the LOI records of all the various cores taken in the lake, had the most rapid post-ploughing accumulation rate. (Fig. 4). Because of this it is assumed that the core is located close to the centre of recent sediment accumulation in the lake despite its position some 200 m north of the maximum water depth.

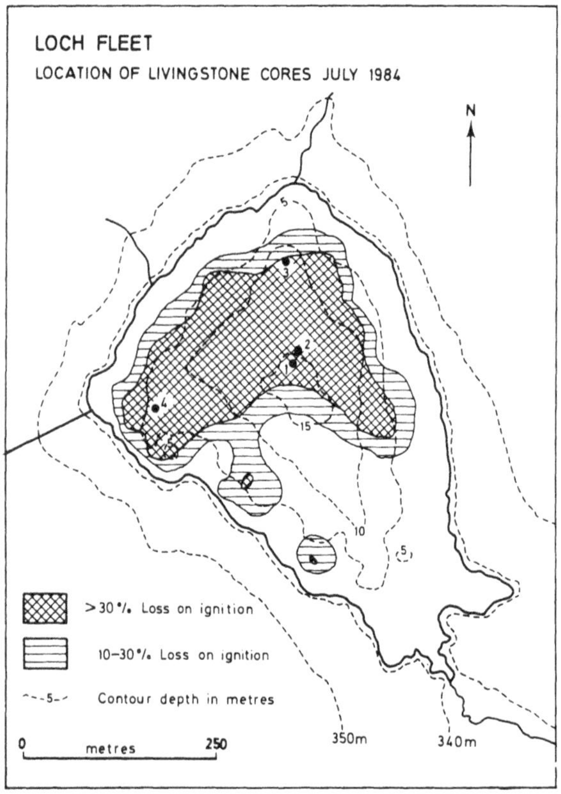

Fig. 3. Map showing location of Loch Fleet and core LFL3 used in this study.

b) ^{210}Pb chronology and land use history

Figure 4(a) shows the total ^{210}Pb concentrations versus depth for core LFL3 from Loch Fleet, and also the ^{226}Ra concentrations, both measured by gamma assay. Counting times ranged from 8×10^4s for near surface samples to 22×10^4s for samples near equilibrium. It is seen from the diagram that, except for near surface sediments, variations in the total ^{210}Pb concentration are largely governed by variations in the ^{226}Ra concentrations. The sudden reduction in the ^{226}Ra concentration above c100 cm is presumably associated with peat inwash from the catchment as a result of ploughing prior to afforestation in 1961.

The unsupported ^{210}Pb concentrations are plotted on Fig. 5(b). Because of the non-monotonic nature of this curve, calculations of a ^{210}Pb chronology by any model other than the CRS (Appleby & Oldfield 1978) model is not feasible. The CRS derived chronology is shown in Fig. 6. The total cu-

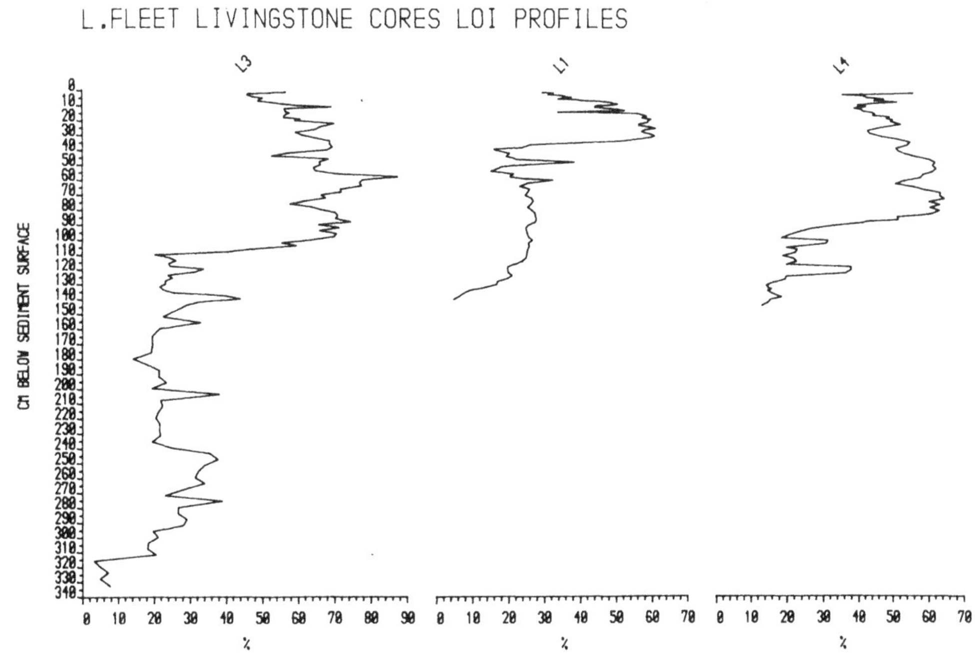

Fig. 4. Loss-on-ignition profiles from three 'Livingstone' cores taken from Loch Fleet. Coring sites are located on Fig. 3.

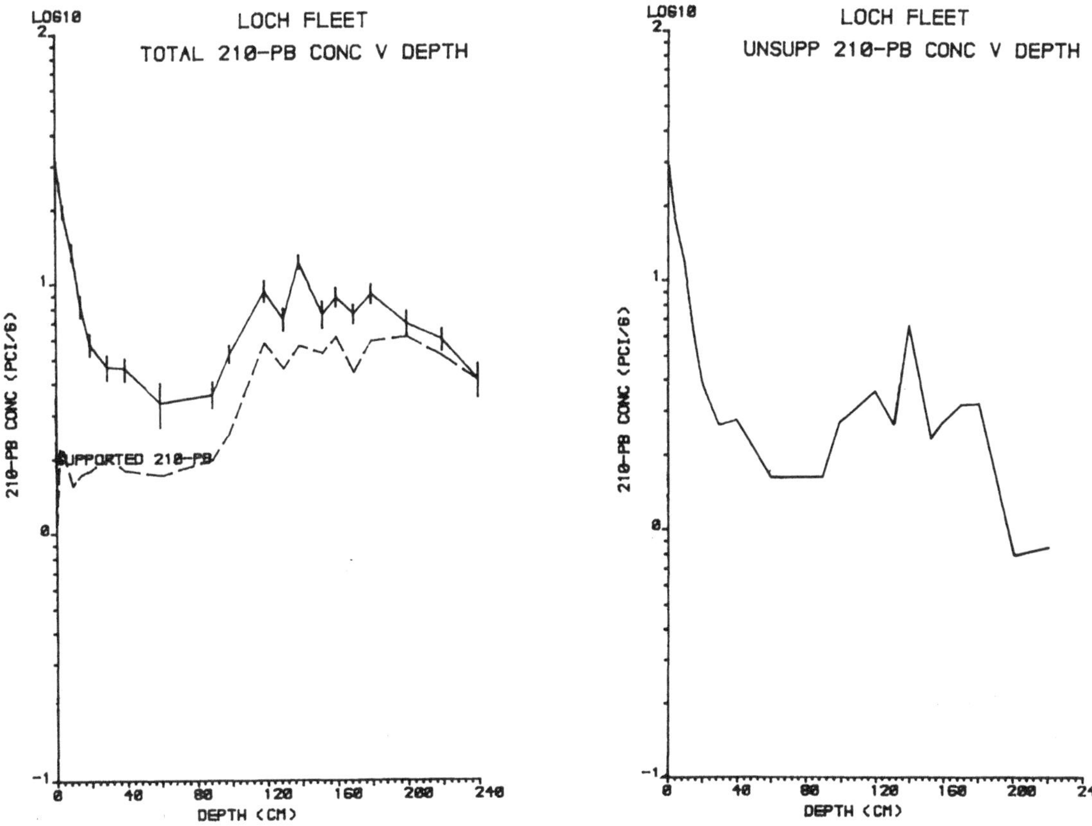

Fig. 5. Loch Fleet ^{210}Pb profiles; concentrations vs. depth. (a) Total ^{210}Pb and ^{226}Ra supported values, (b) unsupported ^{210}Pb.

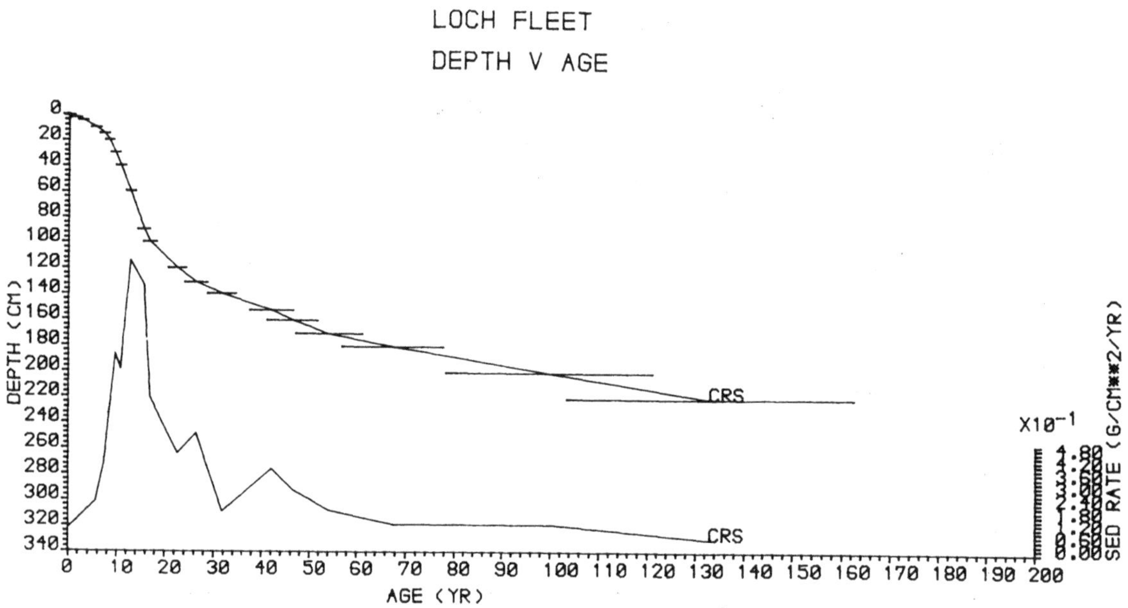

Fig. 6. Loch Fleet Core LFL3; depth v age and dry mass sedimentation rate.

mulative residual unsupported ^{210}Pb content of the core is abnormally high, and represents a mean ^{210}Pb flux of 2.0 ± 0.3 pCi/cm^2 yr. This is well in excess of the expected atmospheric flux (Appleby & Oldfield 1983), and the CRS model will be applicable only if the supply of ^{210}Pb from other sources, for example sediment focusing, has remained reasonable uniform. In support of this, we note that the CRS derived chronology is generally consistent with results from other afforested catchments in this region (Battarbee *et al.* 1985) in that they show a rapid increase in the sediment accumulation rate c1960. This, like the reduction in ^{226}Ra concentration at the same depth, is in all likelihood associated with the ploughing of the catchment in 1961 prior to afforestation. The sedimentation rate before 1960 parallels that at nearby Loch Grannoch (Battarbee *et al.* 1985).

Conclusions

From the results presented we conclude that direct gamma assay provides an excellent basis for ^{210}Pb dating especially at sites where changes in sediment type or source have occurred during the last 150 years. Given a sufficiently low and stable background, especially at low energies, acceptable analytical precision can be combined with a major reduction in the uncontrolled errors associated with uncertainties about variations in ^{226}Ra supported activity. A low background well-detector in the favourable environment provided by a 100 mm lead castle, NaI(Tl) escape suppression shield and copper screen, can give the required combination of characteristics along with major additional benefits in terms of the range of isotopes counted.

References

Anderson, N. J. & R. W. Battarbee, 1985. Loch Fleet, bathymetry and sediment distribution. *University College London Working Papers,* 10.

Appleby, P. G. & F. Oldfield, 1983. The assessment of ^{210}Pb data from sites with varying sediment accumulation rates. *Hydrobiologia* 103: 29–35.

Battarbee, R. W., 1984. Diatom analysis and the acidification of lakes. *Phil. Trans. R. Soc. Lond.* B. 305: 451–477.

Battarbee, R. W., P. G. Appleby, K. Odell & R. J. Flower, 1985. ^{210}Pb dating of Scottish lake sediments, afforestation and accelerated soil erosion. *Earth Surf. Processes and Landforms,* 10: 137–142.

Brush, G. L., E. A. Martin, R. S. DeFries & C. A. Rice, 1982. Comparison of ^{210}Pb and pollen methods for determining rates of estuarine sediment accumulation. *Quat. Res.* 18: 196–217.

Durham, R. & B. G. Oliver, 1983. History of Lake Ontario contamination from the Niagara River by sediment radiodating and chlorinated hydrocarbon analysis I. *Great Lakes Res.* 9: 160–168.

Eakins, J. D. & R. T. Morrison, 1977. A new procedure for the determination of lead-210 in lake and marine sediments. *NERC Report* 8475, H.M.S.O., London.

Flower, R. J. & R. W. Battarbee, 1983. Diatom evidence for recent acidification of two Scottish Lochs. *Nature* 305: 130–133.

Gäggeler, H., H. R. Von Gunten & W. Nyffeler, 1976. Determination of ^{210}Pb in lake sediments and air samples by direct gamma-ray measurements. *Earth Planet. Sci. Lett.* 33: 119–121.

Goldberg, E. D., 1963. Geochronology with ^{210}Pb. In: *Radioactive Dating.* Int. Atom. Energy Ag. Vienna: 121–131.

Kahl, J. S., S. A. Norton & J. S. Williams, 1984. Chronology, magnitude and palaeolimnology record of changing metal fluxes related to atmospheric deposition of acids and metals in New England. In O. P. Brichner (Ed.) *Geological Aspects of Acid Deposition,* pp. 23–35.

Oldfield, F. & P. G. Appleby, 1984a. A combined radiometric and mineral magnetic approach to recent geochronology in lakes affected by catchment disturbance and sediment redistribution. *Chem. Geol.* 44: 87–93.

Oldfield, F. & P. G. Appleby, 1984b. Empirical testing of ^{210}Pb-dating models for lake sediments. In E. Y. Haworth and J. W. G. Lund (Eds.), *Lake sediments and environmental history.* Leicester Univ. Press, Leicester, pp. 93–124.

Oldfield, F., P. G. Appleby & R. Thompson, 1980. Palaeoecological studies of lakes in the Highlands of Papua New Guinea. I. The Chronology of sedimentation. *J. Ecol.* 68: 457–477.

Accepted 18 March 1986.

Hydrobiologia 143: 29–35, (1986)

^{137}Cs as a tracer of recent sedimentary processes in Lake Michigan

Richard A. Cahill & John D. Steele
Illinois State Geological Survey, 615 East Peabody Drive, Champaign, IL 61801, USA

Keywords: Lake Michigan, geochemistry, ^{137}Cs, trace metals, sediment movement

Abstract

To determine recent sediment movement, we measured the levels of ^{137}Cs (an artificial radionuclide produced during nuclear weapons testing) of 118 southern Lake Michigan samples and 27 in Green Bay. These samples, taken from 286 grab samples of the upper 3 cm of sediment, were collected in 1975 as part of a systematic study of Lake Michigan sediment. ^{137}Cs levels correlated well with concentrations of organic carbon, lead, and other anthropogenic trace metals in the sediment. ^{137}Cs had a higher correlation with silt-sized than with clay-sized sediment (0.55 and 0.46, respectively). Atmospherically derived ^{137}Cs and trace metals are being redistributed by sedimentary processes in Lake Michigan after being incorporated in suspended sediment. We determined a distribution pattern of ^{137}Cs that represents areas of southern Lake Michigan where sediment deposition is occurring.

Introduction

The long radioactive half life (30.7 yr) and distinct pattern of ^{137}Cs introduction into the environment make it a very useful tracer of recent atmospheric, hydrologic, and sedimentologic processes. ^{137}Cs, a radioisotope produced by the atmospheric testing of nuclear weapons, began to be deposited in significant quantities in 1964. About 90 percent of the total flux of ^{137}Cs to the air/water interface of Lake Michigan occurred between 1955 and 1966, prior to the signing of the Limited Nuclear Test Ban Treaty of 1963 (Alberts & Wahlgren, 1981). Despite sporadic inputs in recent years, the amount of ^{137}Cs in the atmosphere has decreased since 1966 to near zero.

The geochemical behavior of ^{137}Cs would be expected to be similar to that of K, with adsorption and ion exchange as the dominant mechanisms for its removal from solution. Sibley *et al.* (1982) measured the effect of pH on the distribution coefficient of ^{137}Cs in a natural sediment/water system. They found that adsorption occurs rapidly; equilibrium is approached within the first day. In addition, pH over the range expected in natural waters did not affect ^{137}Cs adsorption.

^{137}Cs and ^{210}Pb are the radionuclides most widely used to determine rates of sedimentary accumulation in a variety of recent (within the last 20–100 years) depositional environments. The two isotopes are commonly used in combination, and, generally, yield comparable results. Sedimentation rates in Lake Michigan have been reported by Robbins & Edgington (1975a), Edgington & Robbins (1976), and Christensen & Chien (1981) using a combination of the two techniques. Sedimentation rates reported in the literature range from 0.1 to 1.2 mm yr^{-1} in 5 cores from southern Lake Michigan (Robbins & Edgington, 1975a). Christensen & Chien (1981) obtained accumulation rates of 0.01 to 0.44 mm yr^{-1} in three cores from central Lake Michigan. These modern accumulation rates agree with long term rates estimated by pollen analysis of 0.8 to 1.0 mm yr^{-1} in the areas of thick sediment in southern Lake Michigan and 0.5 mm yr^{-1} in the center of the southern basin (King, Lineback, & Gross, 1976).

Durham & Joshi (1984) determined that the levels of ^{137}Cs in Lake Michigan had decreased from 49.5 pCi m^{-3} in 1973 to 33.5 pCi m^{-3} in 1981. They attributed the decrease to radioactive decay, reduced input, and physical, chemical, and biological processes. Sediment contains 93 percent and 96 percent of the total ^{137}Cs in Lakes Superior and Huron respectively (Tracy & Prantl, 1983). Tracy & Prantl (1983) proposed that ^{137}Cs was removed from the water column within one to two years. Wahlgren & Nelson (1973) showed that ^{137}Cs was removed from Lake Michigan water in that time span. Because there was more ^{137}Cs in the water than could be accounted for by atmospheric input alone, Tracy & Prantl (1983) felt that ^{137}Cs was reintroduced to the water from bottom sediments. Wahlgren & Nelson (1973) proposed that particle size of the fallout controlled the rate of ^{137}Cs removal from Lake Michigan water. Submicron particles would be expected to be removed rapidly (1.3 yr) by incorporation into colloidal clay minerals while larger, biologically active particles remain in the water column on the order of 3–4 years.

The contribution of land runoff to the levels of ^{137}Cs in Lake Michigan has been estimated as negligible by Robbins & Edgington (1975b); whereas Wahlgren & Nelson (1975) estimated its contribution as 20 percent. The behavior of ^{137}Cs during sedimentary erosion and deposition has been studied in watersheds subject to different land use practices. Ritchie, McHenry & Gill (1975) and Ritchie & McHenry (1978) found that most ^{137}Cs fallout was retained within the watersheds where it was originally deposited. Moreover, they found that only between 6 percent and 19 percent of the original fallout moved via runoff from the primary deposit site. The principal cause of ^{137}Cs redistribution in watersheds is the high erosion potential of cultivated areas. They also found a high correlation between the clay and organic matter content of soils and levels of ^{137}Cs present. Muller, Sprugel & Kohn (1978) found that ^{137}Cs depended on the types of clay minerals present in the soil as well as the cation-exchange characteristics of the soil. ^{137}Cs uptake was also found to depend partially on the pH level.

Depositional processes that control the distribution of ^{137}Cs in Lake Michigan were investigated by Alberts & Muller (1979) and Robbins & Edgington (1975b). Alberts & Muller found that ^{137}Cs activity was constant with particle mass and that there was no simple relationship between isotope concentration and particle size. Robbins & Edgington (1975b) suggested that ^{137}Cs scavenging by river-borne sediment accounted for high levels of ^{137}Cs in lake sediments found north of major river mouths. However, it was not clear if the watershed contribution of ^{137}Cs were responsible for the high ^{137}Cs levels or if the high clay and organic content of the suspended river sediments effectively remove ^{137}Cs from the lake water column.

Materials and methods

In 1975 an extensive sediment survey of Lake Michigan was conducted in cooperation with the Canada Centre for Inland Waters. Grab samples were collected on a 12-by-12-km grid over most of the lake and collected on a more detailed 7-by-7-km grid in Green Bay (Fig. 1). The results of this work are published in Cahill (1981) and Cahill & Shimp (1984). These studies found that Lake Michigan could be classified into a number of sub-basins based on trace element signatures and grain size distributions. Recently the ^{137}Cs content of these samples has been measured both to define areas in the lake where active sedimentation is occurring and to identify the factors controlling ^{137}Cs distribution.

^{137}Cs activity is determined by counting the gamma activity of 10 g of dried sediment with a 15 percent efficient Ge(Li) detector for a minimum of 24 hours. The 661 keV photon activity in sediment samples is compared to the activity of the National Bureau of Standards Environmental Radioactivity Standard (NBS 4350B). The precision of counting statistics is typically less than ±20 percent. The ^{137}Cs measurements are checked by analysis of duplicate samples and reference samples.

Results

The samples used in this study were part of the first truly systematic and comprehensive sampling of the recent sediments of Lake Michigan. The samples used in the study are splits of the same samples previously described in Cahill (1981), which included analysis for 48 chemical elements

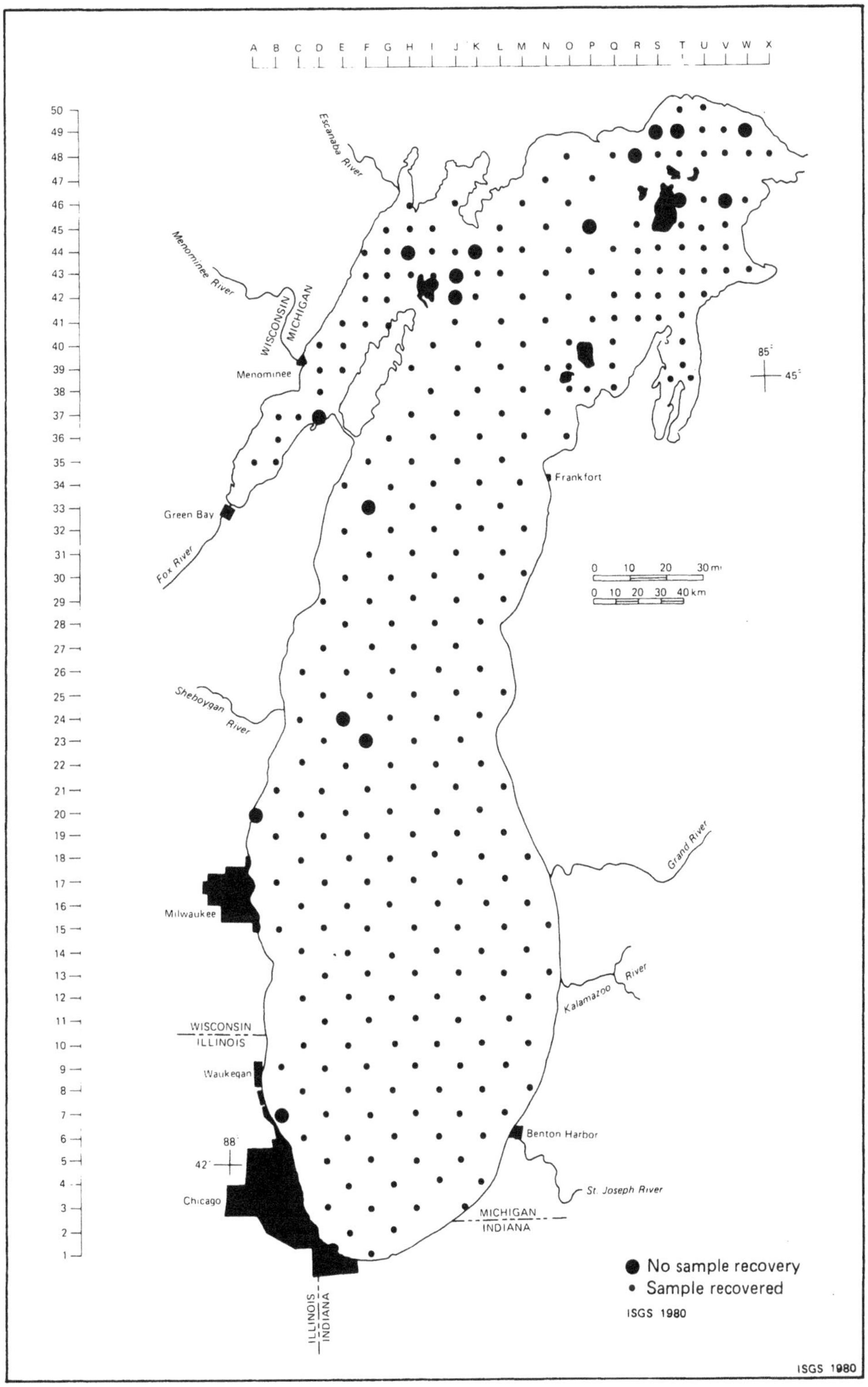

Fig. 1. Sample location grid for the 1975 cruise of the CSS LIMNOS in Lake Michigan.

and grain size. The two sets of information can be readily combined to discuss the influences on the ^{137}Cs distribution pattern.

The distribution of ^{137}Cs in Fig. 2 shows a pattern similar to that observed by Robbins & Edgington (1975b) with high levels of ^{137}Cs occurring near river mouths on the eastern side of Lake Michigan. The maximum levels of ^{137}Cs occur at grid locations D-21 (0.55 Bq g^{-1}) and J-23 (0.564 Bq g^{-1}, which are part of the Milwaukee and Grand Haven sub-basins respectively rather than the southern depositional basin. Neither of these areas in the lake was covered by Robbins & Edgington (1975b).

The ^{137}Cs results for southern Lake Michigan can be summarized as follows:

- The distribution of ^{137}Cs indicates that, over most of the lake, the levels are low (95/118 samples <0.012 Bq g^{-1}).
- The highest level of ^{137}Cs occurs predominantly on the east side of the lake in small zones of deposition (23/118 samples >0.012 Bq g^{-1}).
- The overall mean for ^{137}Cs is 0.075 Bq g^{-1}; the range is <0.003 Bq g^{-1} to 0.564 Bq g^{-1}.

The distribution of ^{137}Cs in Green Bay (Fig. 3) is similar to that of southern Lake Michigan:

- The distribution of ^{137}Cs indicates a low level throughout most of the bay (23/27 samples <0.006 Bq g^{-1}).
- Only 4/27 samples have a high level of ^{137}Cs (>0.014 Bq g^{-1}).
- The mean for ^{137}Cs in Green Bay is 0.046 Bq g^{-1}; the range is <0.004 Bq g^{-1} to 0.292 Bq g^{-1}.

Discussion

The interpretation of the results depends on certain assumptions: the sedimentation rates in Lake Michigan are seldom greater than 1.0 mm yr^{-1}, there is little or no downward movement of ^{137}Cs through the sediment column, the residence time for ^{137}Cs in the water column is short, and that little ^{137}Cs is re-entering the water column from the sediment. The first assumption implies that the upper 3 cm of sediment collected in 1975 contains a record back to at least 1945. If this statement is true, then the upper 3 cm of sediment includes 90 percent of the fallout ^{137}Cs that was deposited between 1954 to 1964. This rate is true for most of the lake except the mouths of the Fox and Menominee Rivers in Green Bay and the Grand River in southern Lake Michigan, where higher sedimentation rates occur. In these locations we would understimate the amount of ^{137}Cs present.

The second assumption of downward mobility has been discussed often in the literature, especially in regard to differences in sedimentation rates as determined by ^{210}Pb and ^{137}Cs. Robbins & Edgington (1975b) have noted that the variation of ^{137}Cs mobility from core to core may be due in part to either vertical mixing or recent sediment slumping. Mixing and downward movement of ^{137}Cs would therefore be important in areas of high bioturbation or shallow slope areas. At these locations we are probably underestimating the amount of ^{137}Cs present by using only a grab sample. Comparison of ^{137}Cs activity in a grab sample with the activity of a core collected at the same location would resolve this question.

The third assumption concerns the resuspension and movement of the uppermost layer of sediment. Chambers & Eadie (1981) have discussed the importance of the nepheloid layer in the chemical dynamics of southeastern Lake Michigan. Redistribution of sediments is a critical factor if patterns are to be compared from samples or cores collected over several years. Since all of our samples were collected within a few weeks of each other, we have a good picture of what the sediment distribution was in a small part of 1975.

A number of researchers have applied statistical techniques to clarify relationships between chemical and physical measurements in lake sediments. Linear correlation regression and cluster analysis were used to determine the controlling physical and chemical factors for ^{137}Cs distribution. The correlation is higher between silt-sized particles (0.55) than clay-sized particles (0.46) at the 99 percent confidence level, although neither correlation is very high. Glaciolacustrine clay-sized sediments, which occur between nearshore deposits and muds in depositional basins (Cahill, 1981) such as those sampled in Lake Michigan, are not of recent origin. For this reason, they would not be expected to have high levels either of ^{137}Cs or of metals derived largely from man-made activities. This fact may be part of the reason why the correlation of ^{137}Cs is lower with clay-sized sediments than with silt-sized sediments.

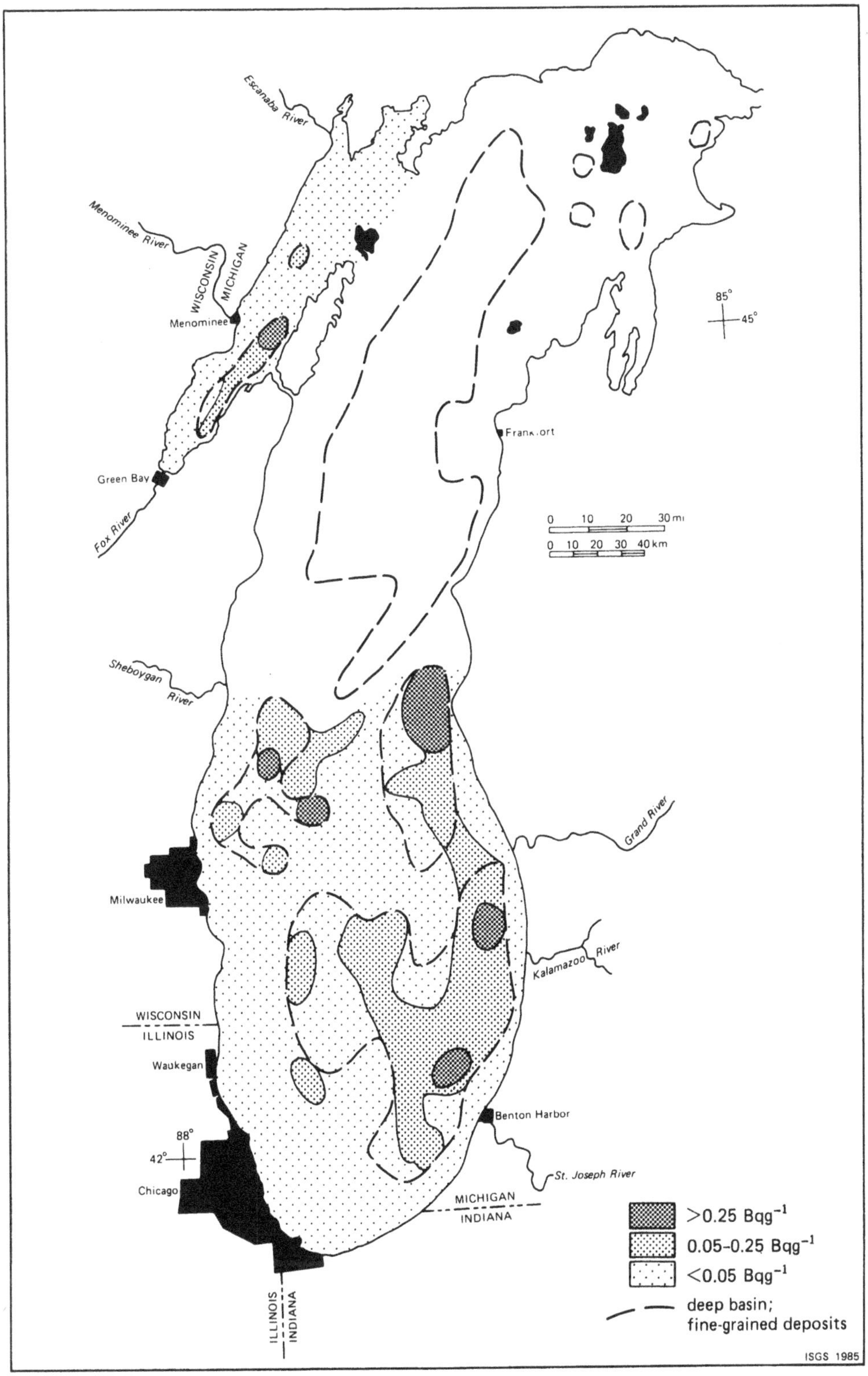

Fig. 2. ^{137}Cs distribution in Lake Michigan.

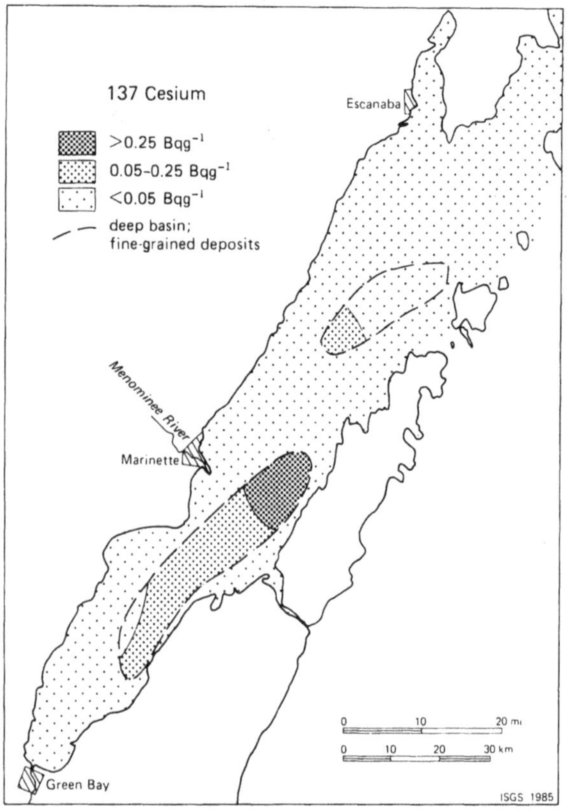

Fig. 3. ^{137}Cs distribution in Green Bay.

The correlation between organic carbon and ^{137}Cs is also high (0.58) due, in part, to the high correlation (0.74) between organic carbon and silt-sized sediments. It is difficult to directly determine which mechanism is the controlling factor in the incorporation of ^{137}Cs in the fine-grained sediments of Lake Michigan. Organic carbon coatings on silt- and clay-sized sediments could be more important than ion exchange with clay minerals. The evidence provided by statistical analyses of the data cannot directly answer this question. The highest correlation is between ^{137}Cs and Pb (0.74), which is also derived primarily from the atmosphere. The poor correlation with depth (0.26) is consistent with the results of Robbins & Edgington (1975b), who noted an inverse relationship between ^{137}Cs and water depth in southern Lake Michigan. The negative correlation (−0.56) between sand-sized sediment and ^{137}Cs is probably the result of the fact that sand-sized sediments do not provide a site for adsorption of ^{137}Cs. Cluster analysis placed ^{137}Cs in a group with Pb, silt, and organic carbon. The second cluster was clay content, grain size, and water depth.

Factor analysis is a statistical tool that summarizes relationships between variables in a matrix of factors. The loading values that emerge from the calculation specify what proportion of the total concentration of each element can be attributed to each factor. Seven factors were found to account for 87 percent of the variance in the surficial sediments of Lake Michigan (Cahill, 1981). The factor scores for each sample were calculated and then plotted using trend surface analysis. The resulting plots indicate where in the lake the common associations occur. The distribution pattern of ^{137}Cs corresponds to a trend surface plot of organic carbon, Hg, Pb, Sb, Zn, and Cd loadings from factor analysis scores reported in Cahill & Shimp (1984). This factor also contained a high proportion of silt-sized material. In contrast, the ^{137}Cs pattern does not correspond to a plot of a second factor, which represents 22 elements associated with clay-sized sediments.

The ^{137}Cs distribution pattern, which could be expected to correspond to the depositional sub-basins defined by Cahill (1981) because they are areas of rapid sediment accumulation, does in fact correspond well for the Grand Haven and Milwaukee sub-basins and Green Bay. In southern Lake Michigan, however, the ^{137}Cs is concentrated on the eastern side of the basin rather than uniformly distributed over the entire area. The influence of suspended river sediment, circulation patterns, and sediment resuspension all influence the distribution of ^{137}Cs in southern Lake Michigan.

A mass balance for ^{137}Cs was not attempted for Lake Michigan. The amount of ^{137}Cs in the sediments of Lake Michigan is highly variable (as shown in Fig. 2). A summation of the measurements of the ^{137}Cs content of the upper 3 cm of sediment may underestimate the total amount present. Although the atmospheric contribution to Lake Michigan is known, the contribution from the watershed is poorly understood and has not been measured. It would appear from the distribution pattern that river-suspended matter is an important source of ^{137}Cs. Perhaps a mass balance would be more feasible in northern Lake Michigan where runoff is less pronounced and there are fewer rivers entering the lake, but it is not clear that a mass balance contributes to the understanding of the distribution of ^{137}Cs.

References

Alberts, J. J. & R. N. Muller, 1979. The distribution of ^{239}Pu and ^{137}Cs in various particle size classes of Lake Michigan sediments. J. envir. Qual. 8: 20–22.

Alberts, J. J. & M. A. Wahlgren, 1981. Concentrations of ^{239}Pu, ^{137}Cs, and ^{90}Sr in the water of the Laurentian Great Lakes. Comparison of 1973 and 1976 values. Envir. Sci. Technol. 15: 94–98.

Cahill, R. A., 1981. Geochemistry of recent Lake Michigan sediments. Ill. Geol. Survey. Circ. 517: 94 pp.

Cahill, R. A. & N. F. Shimp, 1984. Inorganic contaminants in Lake Michigan sediments. In J. O. Nriagu & M. S. Simmons (eds), Toxic Contaminants in the Great Lakes. John Wiley & Sons, N.Y.: 393–423.

Chambers, R. L. & B. J. Eadie, 1981. Nepheloid and suspended particulate matter in south-eastern Lake Michigan. Sediment. 28. 439–447.

Christensen, E. R. & N. K. Chien, 1981. Fluxes of arsenic, lead, zinc, and cadmium to Green Bay and Lake Michigan sediments. Envir. Sci. Technol. 15: 553–558.

Durham, R. W. & S. R. Joshi, 1984. Dose equivalent commitments from fallout radionuclides in the open water of the Great Lakes, 1973–1981. Envir. Mont. 4: 405–417.

Edgington, D. N. & J. A. Robbins, 1976. Records of lead deposition in Lake Michigan sediments since 1800. Envir. Sci. Technol. 10: 266–274.

King, J. E., J. A. Lineback & D. L. Gross, 1976. Palynology and sedimentology of Holocene deposits in southern Lake Michigan. Ill. Geol. Surv. Circ. 496: 24 pp.

Muller, R. N., D. G. Sprugel & B. Kohn, 1978. Erosional transport and depostion of plutonium and cesium in two small midwestern watersheds. J. envir. Qual. 7: 171–174.

Ritchie, J. C., J. R. McHenry & A. C. Gill, 1974. Fallout ^{137}Cs in the soil and sediments of three small watersheds. Ecology 55: 887–890.

Ritchie, J. C. & J. R. McHenry, 1978. Fallout ^{137}Cs in cultivated and noncultivated North Central United States watersheds: J. envir. Qual. 7: 40–44.

Robbins, J. A. & D. N. Edgington, 1975a. Determination of recent sedimentation rates in Lake Michigan using ^{210}Pb and ^{137}Cs. Geochim. Cosmochim. Acta 39: 285–304.

Robbins, J. A. & D. N. Edgington, 1975b. Depositional processes and the determination of recent sedimentation rates in Lake Michigan. Fed. Conf. Great Lakes. Proc. No. 2: 378–390.

Sibley, T. H., A. L. Sanchez, E. A. Wurtz & W. R. Schell, 1982. Distribution coefficients for radionuclides in aquatic environments. U.S. Nuclear Reg. Comm. NUREG/CR-1853. 4: 52 pp.

Tracy, B. L. & F. A. Prantl, 1983. 25 years of fission product input to Lakes Superior and Huron. Wat. Air Soil Pollut. 19: 15–27.

Wahlgren, M. A. & D. M. Nelson, 1973. Residence times for ^{239}Pu and ^{137}Cs in Lake Michigan water. Argonne Nat. Lab. Rad. & Env. Res. Div. An. Rep., Jan.-Dec. 1973: 85–89.

Wahlgren, M. A. & D. M. Nelson, 1975. Plutonium in the Laurentian Great Lakes: comparison of surface waters. Verh. int. Ver. Limnol. 19: 317–322.

Accepted 18 March 1986.

Hydrobiologia 143: 37–44, (1986)

The chemistry of lake sediments in time and space

Daniel R. Engstrom[1] & Edward B. Swain[2]

Limnological Research Center 220 Pillsbury Hall, University of Minnesota Minneapolis, MN 55455, USA[2]Gray Freshwater Biological Institute Navarre, MN 55392, USA

Keywords: paleolimnology, multiple-cores, geochemistry, ^{210}Pb, pigments, myxoxanthophyll, Lake Minnetonka

Abstract

Five short cores (1.0–1.5 m) representing different depositional zones of an isolated bay of Lake Minnetonka, Minnesota (USA), were independently dated by ^{210}Pb and pollen analysis and were analyzed stratigraphically for elemental chemistry (following sediment fractionation) and sedimentary pigments (including myxoxanthophyll and oscillaxanthin). Because of good dating control, short-interval time-stratigraphic units could be traced across the basin, and lake-wide accumulation rates and concentrations could be estimated. The accumulation and concentration of each component at a core-site relative to basin-wide averages provide new indices, 'relative accumulation' and 'relative concentration', that are used to evaluate three major processes controlling sediment deposition: (1) the basin-wide flux, (2) the physical redistribution of bulk sediment, and (3) the selective bio-geochemical transport and fossilization of individual sedimentary components within the lake.

Introduction

Studies of the chemistry of lake sediments are typically confined to one of two possible dimensions within a basin, either stratigraphically at a single core site – the traditional time dimension of the paleolimnologist – or spatially across the lake bottom but with analyses limited to surficial deposits. This circumscribed view of lacustrine sediments is usually dictated by reasons of economy and because of the difficulty of securing reliable stratigraphic correlation between sites. Nevertheless, lake sediments are known to be spatially heterogenous in composition (e.g. Håkanson 1977, Hilton & Gibbs, 1984). Moreover, stratigraphic interpretations of some chemical variables such as Fe and Mn are actually dependent on changing deposition patterns within the basin (Mackereth, 1966). Thus the reconstruction of environmental history from single cores can be a fairly ambiguous enterprise, and in recent years a number of studies have appeared that utilize multiple cores to examine sediment deposition in both time and space. Such investigations have shown great potential for quantifying basin-wide sediment flux from sources such as erosion (e.g. Dearing *et al.*, 1981) and for revealing the shifting and not-so-predictible pattern of sediment deposition within a lake (e.g. Battarbee, 1978; Davis & Ford, 1982).

Our particular contribution to this growing literature is the analysis of multiple cores for algal pigments and elemental geochemistry, and the integration of these data along fine-scale time-stratigraphic units within a single basin. By comparing accumulation rates and concentrations among cores within these units we identify independent changes in three major processes controlling the pattern of sediment deposition: (1) the basin-wide flux, (2) the redistribution of bulk sediment, and (3) the selective bio-geochemical transport and fossilization of individual sedimentary components within the lake.

*Limnological Research Center Contribution No. 320.

Study site

Sediment cores for this study were obtained from the west basin of Crystal Bay in Lake Minnetonka, located 25 km west of Minneapolis, Minnesota (USA). The lake itself is composed of 16 discrete basins, some of which like Crystal Bay are essentially isolated lakes within the L. Minnetonka complex (Fig. 1). The west basin of Crystal Bay has a surface area of 140 ha and is bathymetrically conical with a maximum depth of 25 m. Most water flows into the bay via short channels from upper basins to the north and west and passes to the east across a shallow shelf; the direct catchment of Crystal Bay, excluding this drainage, is extremely small. All bays of Lake Minnetonka are in various stages of eutrophy (Megard, 1972) resulting from residential shoreline development of what are now the westernmost suburbs of Minneapolis. In 1982–83 mean-summer total-P in the epilimnion of Crystal Bay was 50 $\mu g\ L^{-1}$, chlorophyll averaged 25 $\mu g\ L^{-1}$, alkalinity measured 2.5 meq L^{-1}, and pH varied from 8.2 to 9.0 (Gries, unpublished data). Profundal waters are anoxic below the thermocline (7 m) by the end of May and contain high concentrations of P in both late summer (700 $\mu g\ L^{-1}$) and mid-winter (1600 $\mu g\ L^{-1}$). Like most North American lakes, Minnetonka was essentially oligotrophic before European settlement in the region, ca. 1853 (Bradbury, 1975).

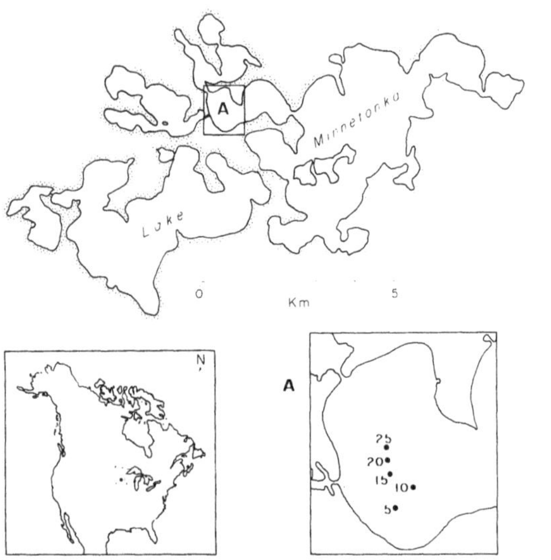

Fig. 1. Location of core sites in the west basin of Crystal Bay (insert A), Lake Minnetonka; depth in meters. The depositional environment represented by each core is approximated by bathymetric contours intermediate between sites.

Methods

Five short cores (10 cm diameter × 1.0–1.5 m length) were obtained at lake depths of 25 m to 5 m, at 5-m intervals along a roughly north-south transect (Fig. 1). Such a sampling regime was deemed suitable to characterize the depositional environment of this bathymetrically simple basin in which particulate inputs from discrete influent streams is negligible. Each sediment core was dated using ^{210}Pb techniques and the c.r.s. (constant rate of supply) model of Appleby & Oldfield (1983) – except the 5-m core which was dated by fitting a constant sediment flux to the ^{210}Pb profile (Murchie, 1985). Two additional stratigraphic markers were provided by pollen analysis: an 1870-horizon marking the regional onset of European agriculture (the rise in *Ambrosia,* chenopods, and ruderals), and an earlier vegetational change from oak-savanna to mesophytic hardwoods (the 'Big Woods' composed of *Ulmus, Tilia, Ostrya,* and *Acer*) that occurred about 1700 AD during the Little Ice Age (Grimm, 1983). Sediments from each core were analyzed at 2–4 cm intervals for a suite of 12 major elements by D.C.-argon-plasma spectroscopy following sequential extraction of authigenic (hydroxylamine hydrochloride/acetic acid) and allogenic (clastic silicate) fractions (Engstrom & Wright, 1984; Engstrom *et al.,* 1985). Fossil pigments (chlorophyll derivatives, total carotenoids, myxoxanthophyll, and oscillaxanthin) were extracted from these same samples using techniques developed by Sanger and Gorham (1972) and Swain (1985).

Calculations

Accumulation rates are often the preferred representation of stratigraphic data (as compared to concentration) because they remove the confounding influence of independent changes in the flux of the sediment matrix (Engstrom & Wright, 1984, and references therein). However, changes in accumulation at a particular core-site can result from several independent processes, each representing a different environmental signal. Firstly, the basin-wide flux for a particular sedimentary component may change, e.g. increased erosion of clastics or increased silica deposition by diatoms, thereby causing a simultaneous unidirectional change in accumulation at all potential core-sites. Secondly, the pattern of transport and deposition of the bulk sediment within the lake may change, e.g. the intensity of sediment focusing *sensu* Lehman (1975) may decrease (e.g. Davis & Ford, 1982), such that disproportional changes in accumulation will occur in different parts of the basin. And thirdly, the selective sorting and burial of individual sedimentary components may change spatially within the lake, e.g. hypolimnetic anoxia may increase the mobilization of redox-sensitive elements, the dissolution of authigenic carbonates, or the preservation of algal pigments, thereby causing site-specific changes in rates of accumulation.

Multiple-core data are used in this paper to disentangle these diverse effects by comparing the accumulation and concentration of a component at each core-site to basin-wide averages for individual time-stratigraphic units. We propose here two new indicies, 'relative accumulation' and 'relative concentration' to quantify these comparisons. The relative accumulation of a particular component at core-site 'x' (RA_x) is given by the equation:

$$RA_x = A_x / \bar{A} \quad (1)$$

where A_x is the core-specific accumulation of the component and $\bar{A}$ is its lake-wide average (for a given time-stratigraphic unit) calculated as:

$$\bar{A} = \sum_{x=1}^{n} A_x W_x \quad (2)$$

where W_x is the proportion of the depositional basin represented by each of n-cores.

Because relative accumulation is normalized to the basin-wide average, this ratio removes the influence of changes in basin-wide flux and so reveals variation in core-specific accumulation rates due to changes in deposition pattern within the lake. The deposition pattern for a given component, as previously stated, may be influenced by either the transport of the bulk sediment or the selective burial of the component. These latter two processes may be distinguished by the second index, relative concentration (RC_x), which is calculated as the ratio of core-specific concentration (C_x) to the basin-wide average concentration (for each time-stratigraphic unit):

$$RC_x = C_x / \bar{C} \quad (3)$$

$$\text{where: } \bar{C} = \sum_{x=1}^{n} C_x W_x \quad (4)$$

Variations in relative concentration reveal changes in the selective transport and burial of an individual component, because this index is not directly affected by changes in bulk-sediment transport or changes in basin-wide flux. The movement of bulk sediment alone will not change the concentration of a component at a core-site, and basin-wide flux without changes in sorting will affect the concentration of all cores equally.

Relative concentration may also be calculated as the ratio of the relative accumulation of the component to the relative accumulation of the bulk sediment, and should be calculated as such when a hiatus (where concentration, but not accumulation rate, is undefined) is present in any core.

The proportion of the depositional basin represented by an individual core (W_x) is difficult to assess when only a few cores are analyzed for a given lake, as in this study. We have assumed that sedimentation in Crystal Bay is largely depth dependent – a reasonable assumption for a basin of regular morphometry without deltaic deposits from major influent streams – and so have selected the depth contours intermediate between cores as delineating the depositional environment of each core (Fig. 1).

We suggest that the delineation of time-stratigraphic units should not exceed the reliability of core chronology and dating horizons for cross-correlation. In this study we have subdivided recent sediments (1980–1930 AD) into 10-year increments, because sedimentation rates are high (0.5–2.0 cm yr^{-1}) and ^{210}Pb dates are most precise for this period. For the interval 1930–1850 AD, 20-yr time-stratigraphic units are used, and a single 150-yr unit is defined between 1850 AD and the 1700 AD 'Big Woods' pollen horizon.

Results and discussion

Geochemical and pigment stratigraphy from the Crystal Bay cores can be divided into three distinct groups according to the patterns displayed in relative accumulation and relative concentration. Exemplary profiles are shown for each group along with those for total dry sediment. It should be noted that a depositional hiatus occurred at the 5-m core-site prior to 1960.

Total dry sediment

Accumulation profiles for total sediment show a monotonic increase up-core at all five sites such that mean basin-wide accumulation climbs by more than an order of magnitude from a pre-settlement (before 1870) rate of 50 g m^{-2} yr^{-1} to 1000 g m^{-2} yr^{-1} within the last decade (Fig. 2a). This marked increase reflects the impact of changing land-use within the catchment of Lake Minnetonka from virgin hardwood forests to agricultural cropland

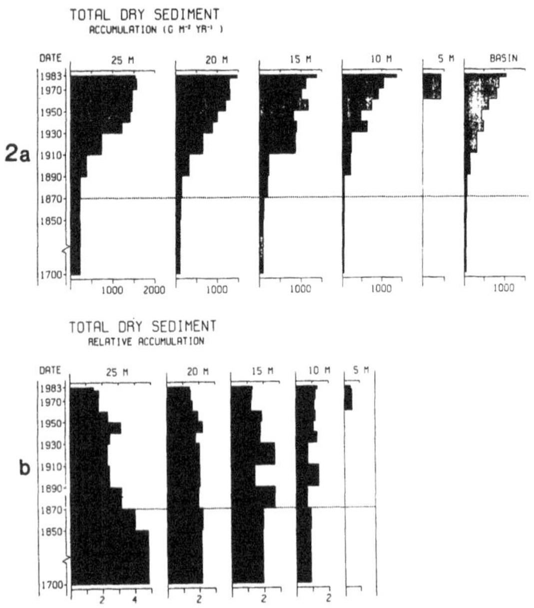

Fig. 2. Stratigraphic profiles for total dry sediment, calculated for 5 cores taken at the indicated water depths, plus the basin-wide mean accumulation as described in the text. The horizontal line at 1870 AD corresponds to the pollen horizon marking the regional onset of European agriculture. Note the change in vertical scale between 1700 and 1850 AD and the depositional hiatus at 5 m prior to 1960. (a) Accumulation rate (g m^{-2} yr^{-1}). (b) Relative Accumulation (the ratio of core-site to mean basin-wide accumulation rate, for each time-stratigraphic unit).

and eventually residential suburbs. With multiple-core data one can ask the question whether this increased sediment load was distributed about the basin in the same manner as the much lower (and compositionally different) pre-settlement flux. Profiles for relative accumulation (the ratio of core-site to average basin-wide accumulation) decrease up-core for the profundal sites (25 m and 20 m), are nearly flat at 15 m, and increase up-core at 10 m and 5 m. These patterns indicate that a greater proportion of the basin-wide sediment flux is settling at lake depths shallower than 15 m now than in the past (Fig. 2b). This is a fundamentally important observation if one is to make even qualitative interpretations of the magnitude of environmental change from a single core. Accumulation rates at the deep-water core-sites underestimate, and those in shallow water overestimate the relative impact of land-use change for the lake as a whole.

It may be further observed that total sediment accumulation is always greater at profundal sites than in shallower water, indicating that sediments have and continue to be preferentially focused into more protected depositional environments. However, the difference in accumulation rates among core-sites (i.e. the degree of focusing) has diminished in recent time in concert with the greater basin-wide sediment flux. While various hypotheses may be advanced to explain this changing pattern of sediment deposition, e.g. higher lake levels or changes in sediment composition, we agree with Murchie's (1985) conclusion that the overall increase in sediment flux is the actual cause. If sediment deposition in shallow water exceeds that which can be transported downslope by wave and current action, then sediments will begin to accumulate at that site (e.g. post-1960 at 5 m), and if the amount that can be transported is more or less fixed by the available energy of the basin, then as the sediment load increases a greater proportion will be permanently buried in littoral environments.

Authigenic Ca (and other major constituents)

Calcium as fine-grained $CaCO_3$ forms from 5 to 50% of the total dry sediment in Crystal Bay, increasing in concentration up-core. A sharp rise in %-Carbonate (from <10% to >20%) clearly marks the cultural horizon in all cores (except 5 m). Thus the accumulation of authigenic Ca increases continuously at all core-sites following settlement

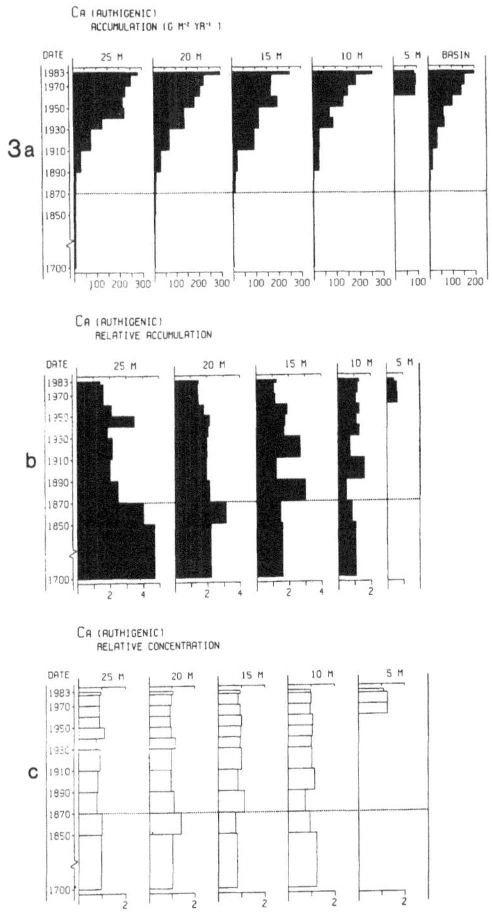

Fig. 3. Stratigraphic profiles for calcium (authigenic; primarily $CaCO_3$). The graphs are constructed as described in Fig. 2. (a) Accumulation rate (g m^{-2} yr^{-1}). (b) Relative Accumulation (the ratio of core-site to mean basin-wide accumulation rate, for each time-stratigraphic unit). (c) Relative Concentration (the ratio of core-site to mean basin-wide concentration, for each time-stratigraphic unit).

but at a rate of change far exceeding that for total dry sediment (from 1.4 g m^{-2} yr^{-1} prior to 1870 to 200 g m^{-2} yr^{-1} in 1980) (Fig. 3a). This change reflects the cultural eutrophication of Lake Minnetonka and the associated increase in primary production that shifted the carbonate equilibrium through greater CO_2 uptake (Kelts & Hsü 1978) and enhanced the midsummer precipitation of $CaCO_3$.

The relative accumulation of Ca is generally similar to that for total dry sediment, decreasing up-core at deep-water sites (particularly 25 m) and increasing up-core at shallow water sites (Fig. 3b). Because the intensity of sediment focusing has decreased post-settlement, at least part of this trend is due to the changing deposition pattern of the sediment load as a whole. However, carbonate solubility is enhanced by hypolimnetic anoxia (through increased dissolved CO_2), also associated with cultural eutrophication, so that selective mobilization of authigenic Ca from profundal regions might be expected in more recent sediments. This effect should be evident in the profiles for relative concentration (the ratio of core-site to mean basin-wide concentration) as an up-core decline at the profundal core-sites. However, the relative concentration curves are relatively flat for all cores (Fig. 3c) indicating that the deposition of authigenic carbonates is largely controlled by physical processes acting on the sediment matrix as a whole; hypolimnetic dissolution is apparently a minor factor in net carbonate sedimentation.

Relative concentration profiles were similarly flat at all sites for allogenic Al (and all measured allogenic components: K, Mg, Ba, Fe, Ti, Si) as well as for organic matter and biogenic-Si. Thus the deposition of fine-grained clastics, which comprise from 16% to 50% of Crystal Bay cores, organic matter, which constitutes from 17% to 42%, and diatom silica, which forms 3% to 25%, follow a similar pattern to that for carbonates and hence for the total dry sediment. While these diverse materials are clearly transported differentially within the basin (as evidenced by compositional differences among cores), this pattern of sorting has not noticeably shifted with the increased sediment load. Thus changes in overall sediment composition could not have caused the observed shift in bulk-sediment focusing.

Authigenic P (and other redox-sensitive elements)

Phosphorus accumulation increases following settlement at all but the 25-m core site, recording the lake-wide increase in P loading that has led to the recent eutrophication of Lake Minnetonka (Fig. 4a). Relative accumulation profiles (Fig. 4b) show an up-core decline in the proportion of the total phosphorus flux reaching profundal regions. Although due in part to a decreased focusing of the bulk sediment, this trend also results from the selective mobilization of phosphorus from deep-water sediments and/or its selective deposition in shallower parts of the basin, which has become

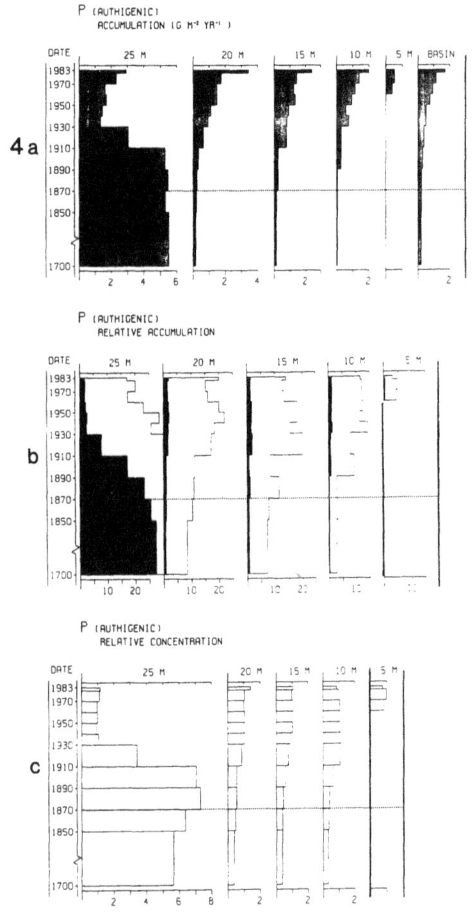

Fig. 4. Stratigraphic profiles for phosphorus (authigenic; primarily associated with iron). The graphs are constructed as described for Fig. 2. (a) Accumulation rate (g m^{-2} yr^{-1}); (b) Relative Accumulation (open curves represent 10X exaggeration); (c) Relative Concentration.

more pronounced over time. This conclusion is derived from the up-core decline in relative concentration shown in the profundal core at 25 m (Fig. 4c). Such trends are consistent with our understanding of phosphorus mobilization under the lowered redox potential of an anoxic hypolimnion (Mortimer, 1941, 1942).

Similar patterns in relative accumulation and relative concentration for authigenic Fe and Mn attest to the development of hypolimnetic anoxia associated with cultural eutrophication and the preferential transport of redox-sensitive elements from profundal environments (c.f. Mackereth, 1966; Davison et al., 1982, Engstrom & Wright, 1984). Such results clearly imply that the selection of core-sites has important bearing on interpretations that are made from the stratigraphy of these elements. For example, the reconstruction of basin-wide phosphorus loading from a single core would be very different using accumulation rates at 25 m and 20 m; P accumulation decreases up-core at the deeper core-site and increases up-core at the shallower one.

Myxoxanthophyll (and other pigments)

The interpretation of sedimentary pigments, like that of the redox-sensitive elements, can be complicated by spatial and temporal variation in factors controlling preservation within the sediments (Swain, 1985, and references therein). Most algal pigments are readily degraded to colorless constituents by photo- or biochemical-oxidation, so that rapid burial under an anoxic hypolimnion can greatly enhance their fossilization. Post-settlement increases in the accumulation of myxoxanthophyll at all core-sites document the progressive eutrophication of Crystal Bay and the development of a phytoplankton dominated by blue-green algae (Fig. 5a). Part of the increased basin-wide accumulation is due to greater primary production and part to better preservation.

Relative accumulation (Fig. 5b) shows an initially curious trend in the 25-m core of a decline between 1870 and 1910 and then an increase after 1950. This pattern is readily explained by reference to the relative concentration profiles (Fig. 5c) which show a relatively continuous up-core increase in the proportion of the basin-wide myxoxanthophyll concentration found in the deep-water sediments. While myxoxanthophyll preservation probably improved throughout the lake following eutrophication, the effect has been most strongly felt in deeper parts of the hypolimnion. Thus the selective preservation of myxoxanthophyll in profundal sediments coupled with a decreased intensity of sediment focusing can account for the observed trends in relative accumulation; the up-core decrease at 25 m until ca. 1910 is due to weaker focusing, and the subsequent increase is due to differential preservation.

Very similar stratigraphic profiles were found for oscillaxanthin as well as for chlorophyll derivatives and total carotenoids. However, the latter two pigment groups, being less susceptible to degrada-

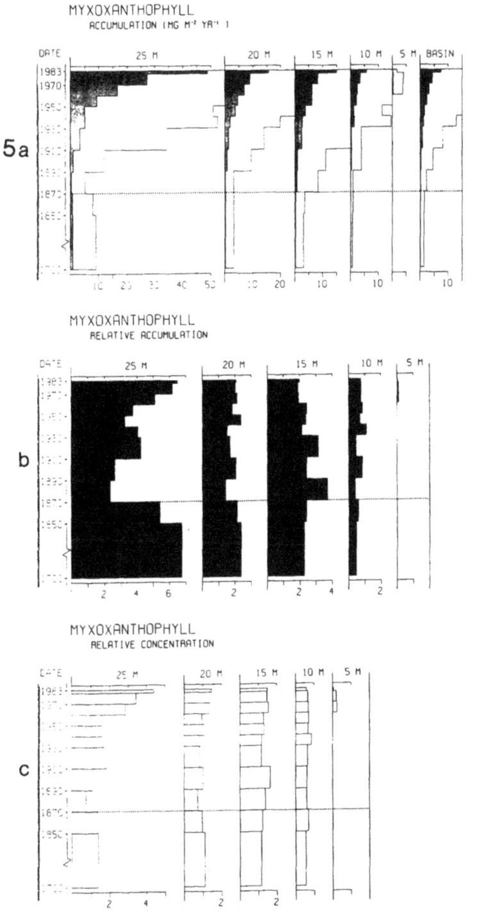

Fig. 5. Stratigraphic profiles for myxoxanthophyll (a carotenoid unique to blue-green algae); graphs as in Fig. 2. (a) Accumulation rate (mg m^{-2} yr^{-1}); open curves represent 10X exaggeration; (b) Relative Accumulation; (c) Relative Concentration.

tion than myxoxanthophyll or oscillaxathin (Swain, 1985), show a much smaller post-settlement increase in relative concentration in the profundal core. These results indicate that changes in trophic status can be greatly magnified in the pigment stratigraphy of deep-water sediments by the development of hypolimnetic anoxia. The most accurate history of eutrophication in Crystal Bay, if made from a single core, might well be obtained by determining pigment accumulation rates from a core taken in epilimnetic sediments where net accumulation is not so severely exaggerated by enhanced preservation.

Conclusions

Multiple cores, by providing a spatial dimension to the chemical stratigraphy of lake sediments, can help elucidate the various processes that control sediment deposition and thereby contribute to a more accurate reconstruction of environmental history. The ratios of accumulation and concentration at individual core-sites to basin-wide averages along time-stratigraphic units (relative accumulation and relative concentration) provide distinct signals for variations in basin-wide flux, bulk-sediment transport and diagenetic sorting. The elemental and pigment stratigraphy in five short cores from Crystal Bay, Lake Minnetonka, illustrate how anthropogenic increases in sediment loading from erosion, primary production and authigenic precipitation, concurrent with changes in sediment focusing, redox cycling and differential preservation, can be effectively disentangled and unambiguously interpreted from multiple-core data.

Acknowledgements

We thank S.L. Murchie for ^{210}Pb dating and L.C.K. Shane for pollen analysis. The Freshwater Foundation, Navarre, Minnesota provided funding for this project.

References

Appleby, P. G. & F. Oldfield, 1983. The assessment of ^{210}Pb data from sites with varying sediment accumulation rates. Hydrobiologia. 103: 29–35.

Battarbee, R. W., 1978. Biostratigraphical evidence for variations in the recent pattern of sediment accumulation in Lough Neagh, N. Ireland. Verh. int. Ver. Limnol. 20: 624–629.

Bradbury, J. P., 1975. Diatom stratigraphy and human settlement in Minnesota. Geological Society of America. Special Paper 171. 74 pp.

Davis, M. B. & M. S. Ford, 1982. Sediment focusing in Mirror Lake, New Hampshire. Limnol. Oceanogr. 27: 137–150.

Davison, W., C. Woof & E. Rigg, 1982. The dynamics of iron and manganese in a seasonally anoxic lake; direct measurement of fluxes using sediment traps. Limnol. Oceanogr. 27: 987–1003.

Dearing, J. A., J. K. Elner & C. M. Happey-Wood, 1981. Recent sediment flux and erosional processes in a Welsh upland lake-catchment based on magnetic susceptibility measurements. Quat. Res. 16: 356–372.

Engstrom, D. R., E. B. Swain & J. C. Kingston, 1985. A paleolimnological record of human disturbance from Harvey's Lake, Vermont: geochemistry, pigments and diatoms. Freshwat. Biol. 15: 261–288.

Engstrom, D. R. & H. E. Wright, Jr., 1984. Chemical stratigraphy of lake sediments as a record of environmental change. In E. Y. Haworth & J. W. G. Lund (eds), Lake Sediments and Environmental History. Leicester University Press. Leicester: 11–67.

Grimm, E. C., 1983. Chronology and dynamics of vegetation change in prairie-woodland region of southern Minnesota, U.S.A. New Phytol. 93: 311–350.

Håkanson, L., 1977. The influence of wind, fetch and water depth in the distribution of sediments in Lake Vänern, Sweden. Can. J. Earth Sci. 14: 397–412.

Hilton, J. & M. M. Gibbs, 1984. The horizontal distribution of major elements and organic matter in the sediment of Esthwaite Water, England. Chem. Geol. 47: 57–83.

Kelts, K. & K. J. Hsü, 1978. Freshwater carbonate sedimentation. In A. Lerman, (ed.), Lakes: Chemistry, Geology, Physics. Springer-Verlag, N.Y.: 295–323.

Lehman, J. T., 1975. Reconstructing the rate of accumulation of lake sediment: the effect of sediment focusing. Quat. Res. 5: 541–550.

Mackereth, F. G. H., 1966. Some chemical observations on post-glacial lake sediments. Phil. Trans. r. Soc. London B 250: 165–213.

Megard, R. O., 1972. Phytoplankton, photosynthesis, and phosphorus in Lake Minnetonka. Limnol. Oceanogr. 17: 68–87.

Mortimer, C. H., 1941. The exchange of dissolved substances between mud and water in lakes. J. Ecol. 29: 208–329.

Mortimer, C. H., 1942. The exchange of dissolved substances between mud and water in lakes. J. Ecol. 30: 147–201.

Murchie, S. L., 1985. ^{210}Pb dating and the recent geologic history of Crystal Bay, Lake Minnetonka, Minnesota. Limnol. Oceanogr. 30: 1154–1170.

Sanger, J. E. & E. Gorham, 1972. Stratigraphy of fossil pigments as a guide to the postglacial history of Kirchner Marsh, Minnesota. Limnol. Oceanogr. 17: 840–854.

Swain, E. B., 1985. Measurement and interpretation of sedimentary pigments. Freshwat. Biol. 15: 53–75.

Accepted 18 March 1986.

Hydrobiologia 143: 45–47, (1986)

Calculating net annual accumulation rates of sediment components exemplified by pollen

Ulf Segerström & Ingemar Renberg
Department of Ecological Botany, Umeå University, S-901 87 Umeå, Sweden

Keywords: absolute pollen counts, accumulation rates, net accumulation, pollen, sediment, sub-sampling

Abstract

The net annual accumulation rate of a sediment component can easily be estimated by determining the total content of the specific component (x) in a sediment sample, which represents a known number of years (yr) and certain area of the lake bottom (a), using the formula: annual accumulation $= x/(\mathrm{yr} \cdot \mathrm{a})$. To illustrate and test this procedure we have calculated mean net annual accumulation rates for sediment dry matter, mineral matter, organic matter and pollen grains of Scots Pine (*Pinus sylvestris* L.) for one short sequence of a varved lake sediment core. This paper includes descriptions of a simple method for sub-sampling fresh cores of consolidated sediments and an alternative method for making absolute pollen counts. The test results clearly show that in varved sediments quantitative sediment samples can be cut out with very high precision, i.e. the variable a and in this case also yr can be very accurately determined. They also show that dry sediment, mineral and organic matter and pine pollen grain contents (i.e. x) and thus, mean net annual accumulation values could be estimated with a good precision ($C < 3\%$). This procedure for estimating net annual accumulation rates is, of course, applicable in non-varved sediments and for many other sediment compounds provided that a reliable chronology can be determined.

Introduction

In palaeolimnological investigations it is often valuable to determine net annual accumulation rates (influx values) of various sediment components as a complement to concentration values or relative frequencies. Probably the most common method involves series of measurements and calculations on individual sediment samples (density; water content; amount (x) per gram dry weight, per gram wet weight and per volume), and finally amount (x) per area and unit-time after dating and estimation of the annual growth of the sediment (see e.g. Bengtsson & Enell, 1986; Digerfeldt, 1979). Calculations of accumulation rates of sub-fossils such as pollen for which the analysis starts from a fresh sample often comprise the addition of some marker grains (see e.g. Berglund & Ralska-Jasiewicz-nowa, 1986).

A variant of this common scheme to estimate accumulation values is to determine the total content of a specific component (x) in a sediment sample, which represents a known number of years (yr) and a certain area of the lake bottom (a), and to use the formula: annual accumulation $= x/(yr \cdot a)$. When this is done for a microfossil taxon such as a particular pollen species, annual accumulation rates of co-occurring microfossil types can easily be calculated from their percent frequencies.

To illustrate and test this procedure mean net annual accumulation rates for sediment dry matter, mineral matter, organic matter and for pollen grains of Scots pine (*Pinus sylvestris* L.) have been estimated for one short sequence of a varved lake sediment core. This work includes descriptions of a simple method for sub-sampling fresh cores of consolidated sediments and an alternative method for absolute pollen counting.

Methods

The sediment core was taken from Lake Kassjön in winter using a large Russian peat corer (length = 100 cm, width = 8 cm). For the analyses a slice comprising 12 years (from the time about 2920 B.C.) was cut out with a sharp thin-bladed knife (Fig. 1A). From this semicircular slice, ten pieces equal in size were cut out with a specially designed sediment cutter which consists of an aluminium frame, 18 × 20 cm, with two adjustable parallel running piano-wires, here set 1 cm apart (Figs 1B, C). Thus, each sediment piece represented 1 cm^2 of the lake bottom and the sediment deposited during 12 years.

Nine of these pieces were dried (at 105 °C for 24 h) to determine their dry matter content, and combusted (at 550 °C for 4 h) to determine their organic and mineral matter content. The tenth piece, taken for pollen analysis, was boiled in 5% potassium hydroxide, then treated with hydrofluoric acid and thereafter with an acetolysis mixture. After the acetolysis, which followed standard methods (e.g. Moore & Webb, 1979), the residue with the pollen grains was diluted with water and weighed (final sample weight = 11.794 g). This suspension was carefully homogenized by shaking and then ten sub-samples were transferred to plankton chambers and weighed (sub-sample weights between 0.531–0.620 g). The chambers (volume = 25 ml) were filled up with water and left to sediment for more than 18 h. In each chamber (sub-sample) all the pollen grains of Scots Pine (they are easy to identify) were counted with an inverted microscope (×100) and on the basis of each count the total number of Scots Pine pollen grains in the sediment piece was estimated. The values for the dry matter content, the organic and mineral matter content, as well as Scots Pine pollen content, were finally divided by 12 years and 1 cm^2 to obtain mean net annual accumulation values per cm^2.

Results and discussion

Nine replicate values for net annual accumulation of sediment (dry matter) show very little variation. Among replicate determinations of organic and mineral matter the variation also is small, although the coefficient of variation (C) is slightly

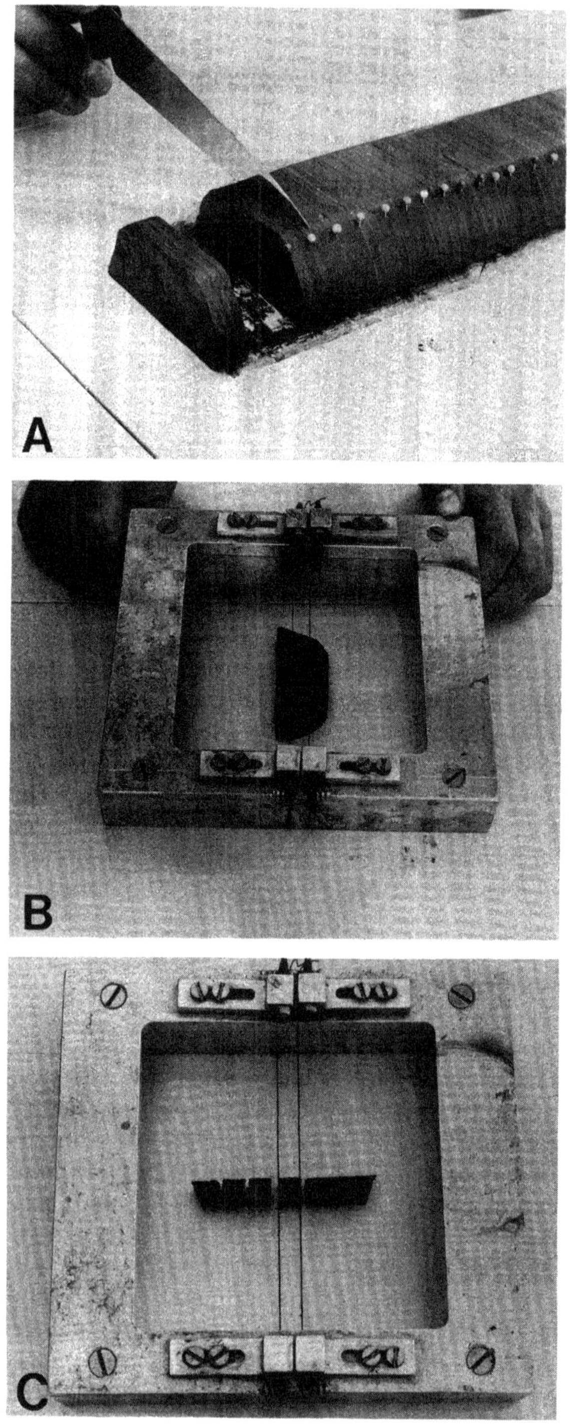

Fig. 1. A) After varve counting, the fresh sediment core is cut into slices (which comprise a desired number of years) using a sharp thin-bladed knife. B) and C) From each sediment slice, pieces representing a certain area of the lake bottom (here 1 cm^2) can be cut out with the sediment cutter (an aluminium frame, 18 × 20 cm, with two adjustable, parallel running piano-wires).

Table 1. The net annual accumulation of Scots pine (*Pinus sylvestris* L.) pollen grains, dry sediment, mineral matter and organic matter. The results are expressed as mean value ($\bar{X}$), minimum and maximum values (Range) and coefficient of variation ($C = SD \frac{100}{\bar{X}}$).

Net accumulation (cm^{-2} $year^{-1}$)	n	$\bar{X}$	Range	C
Scots pine pollen (no)	10	1615	1544 – 1689	2.7%
Dry sediment (mg)	9	28.032	27.675 – 28.408	0.97%
Mineral matter (mg)	9	23.161	22.925 – 23.458	0.84%
Organic matter (mg)	9	4.871	4.608 – 4.967	2.5%

higher for organic matter. The ten estimates of net annual accumulation of Scots Pine pollen grains likewise fall within fairly narrow ranges and the coefficient of variation is small (Table 1).

These tests show that it is quite possible to obtain estimates of mean net annual accumulation rates of dry matter, mineral matter and organic matter with very high precision. They also prove that it is possible to obtain representative sub-samples from a well homogenized water-pollen mixture, to count pollen grains by the inverted microscope method, and thus to make fairly accurate calculations of mean net annual accumulation rates of pollen grains within a sediment.

This method for estimating accumulation rates of pollen may be a valuable alternative to the existing methods. When pollen counts have been made in the manner described above, use of exotic spores (see e.g. Berglund & Ralska-Jasiewiczowa, 1986) is unnecessary. After standard pollen counts of ordinary microscopic slides, prepared from the remaining part of the acetolyzed material, the net accumulation of all other species found in the sample can be calculated simply from their relative frequencies and the relative frequency and net accumulation value for Scots Pine (or any other species that may have been chosen for absolute influx determination).

In any attempts to estimate annual accumulation values, errors due to the sampling and the analysis procedure can not be completely excluded. Sampling, sub-sampling, dating and determining the variable x all contribute to the total uncertainty of the results. In a case where varves are not available, dating probably will be most critical. If the sediment is unconsolidated or extremely firm, errors du to sub-sampling may also become of increased importance.

Acknowledgement

We thank the Carl Trygger's Foundation for Scientific Research for the financial support of this work.

References

Bengtsson, L. & M. Enell, 1986. Chemical analysis. In B. E. Berglund (ed.), Handbook of holocene palaeoecology and palaeohydrology. Chichester: 423 – 451.

Berglund, B. E. & M. Ralska-Jasiewiczowa, 1986. Pollen analysis and pollen diagrams. In B. E. Berglund (ed.), Handbook of holocene palaeoecology and palaeohydrology. Chichester: 455 – 484.

Digerfeldt, G., 1979. Guideline for lake investigations. In B. Berglund (ed.), Palaeohydrological changes in the temperate zone in the last 15000 years. IGCP 158 B. Lake and mire environments. Project Guide 1, general project description. Lund: 31 – 48.

Moore, P. D. & J. A. Webb, 1978. An illustrated guide to pollen analysis. Hodder and Stoughton, London., 133 pp.

Accepted 18 March 1986.

Hydrobiologia 143: 49–53, (1986)

Sedimentary records of accelerated nutrient loading in Florida lakes

Edward S. Deevey, Michael W. Binford, Mark Brenner & Thomas J. Whitmore
Department of Natural Sciences, Florida State Museum, University of Florida, Gainesville, Florida 32611, USA

Keywords: paleolimnology, eutrophication, trophic state, ^{210}Pb, transfer functions, Florida lakes

Abstract

Transfer functions relating trophic state (Carlson's $TSI_{chlorophyll\text{-}a}$) to present day accumulation rate of (1) nutrients, (2) cations, and (3) organic sediment, are computed using Binford's ^{210}Pb-dilution method. As computed from surficial sediments of 27 lakes, former trophic states are reconstructed for recent (^{210}Pb-dated) sedimentary histories of 14 lakes. Of the three kinds of models potentially available, model (3) (TSI vs. organic accumulation) is the weakest statistically, and may be unduly influenced by exceptional deposition and/or preservation of allochthonous organic matter. At present, however, it is the only model applicable to all 14 of the histories tested. Results are encouraging in that model accurately predicts observed TSI's in several mesotrophic and eutrophic lakes. Clearly significant increases (accelerations) are inferred only for 2 of the most eutrophic lakes of the set, while the model consistently overpredicts TSI's of the 7 most oligotrophic lakes. As Whitmore's diatom-assemblage index is a better predictor of TSI than is model (3) in the one eutrophic lake in which it has been tested, we expect more persuasive results when models (1) and (2) can be tested within a more complete set of analytical data. We were surprised to find 3 severely disturbed lakes among the 12 that show little or no acceleration in rate of eutrophication in recent decades, but we defer attempts at explanation until former nutrient loading can be tested by model (1).

Introduction

In our view (Binford *et al.*, 1983; Deevey, 1984) 'the objective of general ecosystem theory is explanatory description and prediction of a system's trajectory in time'. Among the sedimentary trajectories of lacustrine ecosystems described by paleolimnologists, there are some (Eocene Lake Uinta, for example) in which human influence has played no role; but we know of no Holocene trajectory that is unaffected by cultural disturbance. Since the Pallanza symposium on deep waters of lakes was informed of Neolithic influence on the Längsee and Windermere trajectories (Frey 1955; Tutin 1955), and of more recent changes in Swiss lakes (Thomas 1955; Zullig 1955), much effort has been directed to measuring human impact on lakes. We have participated in this effort because we find cultural disturbance to be a rich, quantitatively graded array of perturbing stresses, which have remarkably informative effects on dynamic structure of lacustrine ecosystems.

Notable advances have recently been made in study of one kind of impact, eutrophication. The process is normally quantified in terms of between-lake differences in waterborne nutrient (P and N) loading (Vollenweider 1968; Dillon & Rigler 1975; textbook account in Chapra & Reckhow 1983). As historical ecologists, however, we have to notice that nutrient-loading models are static descriptions, not of trajectories, but of contemporary differences in *amounts* (concentrations) of phytoplankton biomass and nutrients. Such models have no explicit time-dimension, and yield no inferences about the *rates* at which conditions have been attained, or are likely to change in the future.

Working first in the Maya area of Guatemala (Deevey *et al.* 1979; Deevey & Rice 1980; Rice *et al.* 1983) and more recently in peninsular Florida, a similar but more deeply buried karst district, we have used stratigraphic methods to evaluate *rates of change* of trophic state and nutrient and ionic loading. Without such measurements, the first few of which are presented in this paper, we can imagine no way of making ecosystem models predictive.

Methods

Outline of methodology

In this study we use synoptic and diachronic methods in conjunction. Trophic states and nutrient and other solute loads have been measured synoptically in hundreds of Florida lakes (Baker *et al.* 1981; Shannon & Brezonik 1972a, 1972b; Brezonik & Shannon 1971; Brezonik *et al.* 1969; Huber *et al.* 1982; Canfield 1981); for measurement of trophic state we use one or more components of Carlson's trophic-state index (TSI) (Carlson 1977), usually $TSI_{chlorophyll}$. As eutrophication can sometimes be negative, even when P-loading is increasing (Brenner 1983), we reasoned that rates of eutrophication are governed only in part by nutrient export from land. Our first step was therefore to select ca. 30 basin-lake systems that provide a wide range, not only of disturbance and trophic state, but of solute loading, degree of anoxia, and other capacity factors that might either buffer or amplify a given nutrient income. Only 14 lakes have been chosen and cored so far.

The second step was to estimate transfer functions, relating water-column properties to sedimentary concentrations of nutrients, major cations, and biological fossils, in surficial sediments of 91 lakes (Brenner & Binford, in prep.). We found it possible, using the ^{210}Pb-dilution method (Binford & Brenner 1986) to express transfer functions as net (instantaneous) accumulation rates. Examples are shown by Brenner & Binford (this volume). For the present, we restrict the data set to the 34 lakes in which ^{210}Pb data are available for surficial sediments.

The third step, with which this paper is concerned, was to project transfer coefficients diachronically, to ^{210}Pb-dated levels in short cores.

The fourth step, synoptic comparison of sedimentary histories across the lake district, is required for full-scale interpretation. Our first 14 histories of rates of change of trophic state, reconstructed from incomplete analytical data, comprise a rather biased sample and only provisional interpretations are possible.

Field and analytical methods

Surficial sediments were sampled with an Ekman dredge, while cores were obtained by piston corers. Organic matter was assessed from loss on ignition at 550 °C. ^{210}Pb concentration in surface muds and cores was measured by the content of the ^{210}Po granddaughter according to Eakins and Morrison (1978) with modification (Binford & Brenner 1986). Chlorophyll-a data were provided by several investigators (Huber *et al.* 1982; R. A. Garren, C. L. Clarkson & R. W. Bienert, pers. comm.). Downcore ^{210}Pb dates are calculated from the unmodified constant-rate-of-supply (CRS) model (Goldberg 1963; Appleby & Oldfield 1978, 1983). For Lake Parker only, diatom assemblages yield independent estimates of former trophic state, derived from the diatom-index:TSI relationship developed by Whitmore (1985).

Results

The transfer coefficient used in this paper relates TSI_{chlor} to organic-matter accumulation rates in 27 of the 34 survey lakes. Core-top data from 7 lakes were excluded. Of all the empirical water-column:sediment relationships, the organic matter-TSI regression (Fig. 1) is one of the weakest ($r^2 = 0.26$). However, it is the only one available so far for 7 of the 14 cores shown in Fig. 2. Net accumulation rate of sedimentary organic matter is at least a familiar measure of trophic state, and percent organic matter is a variable reported by most paleolimnologists. Some of the wide scatter in Fig. 1 probably reflects various increments of allochthonous organic matter, even though blackwater lakes were excluded from this study. When cores are more completely analyzed, accumulation rates of N, S, and P are sure to give better estimates of TSI, while some of the strongest relations, though unexpected, are between TSI and cation accumulation rates (Brenner & Binford, this volume).

Application of the transfer down-core gives the

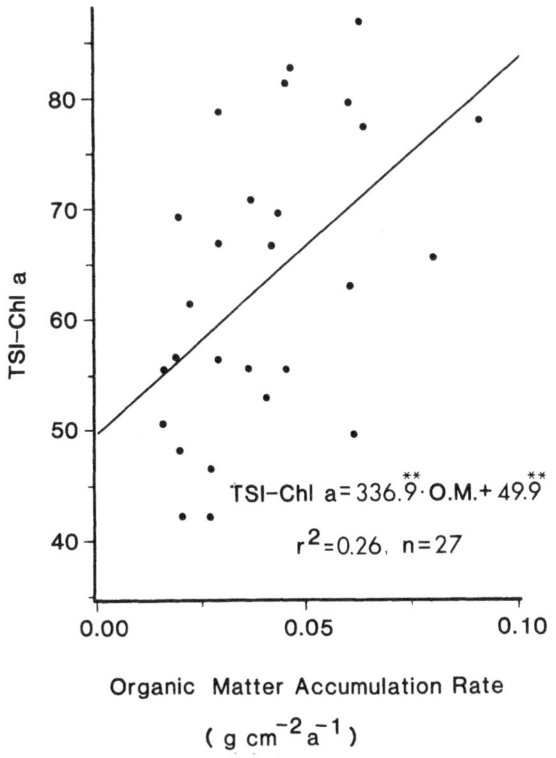

Fig. 1. Organic matter accumulation rate vs. TSI-Chl *a* for 27 Florida lakes. ** = $P < 0.01$.

historical TSI_{chlor} inferences shown in Fig. 2. Only two lakes, Parker and Hawthorne, show clearly significant accelerations in recent decades. Very recent accelerations in several lakes, notably Lake Wauberg, are probably artifacts of the CRS dating model. The transfer-function model overestimates modern TSI in undisturbed oligotrophic systems (all of the last 8 are undisturbed except for Lake Geneva), probably for statistical reasons. Underestimates in the two most eutrophic lakes are plainly not due to admixture of allochthonous organic matter, but probably arise either from faulty curve-fitting or from weakness of TSI_{chlor} as a measure of trophic state.

In Lake Hawthorne, accelerated eutrophication is doubtless due to the usual causes (growth of the town of Hawthorne) but hypereutrophy of Lake Parker is probably attributable to drainage from

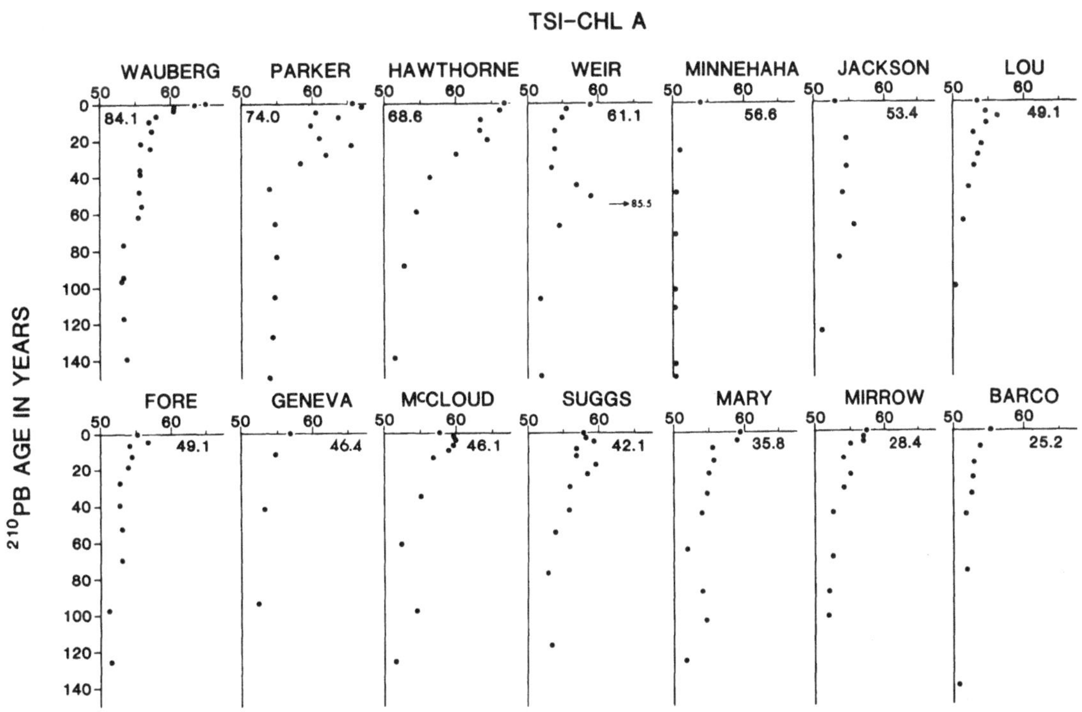

Fig.2. Historical TSI-Chl *a* inferences for 14 Florida lakes. Inferred values are based on the transfer function relating organic matter accumulation rate to TSI-Chl *a*. Modern Carlson's TSI-Chl *a* values are indicated for each lake.

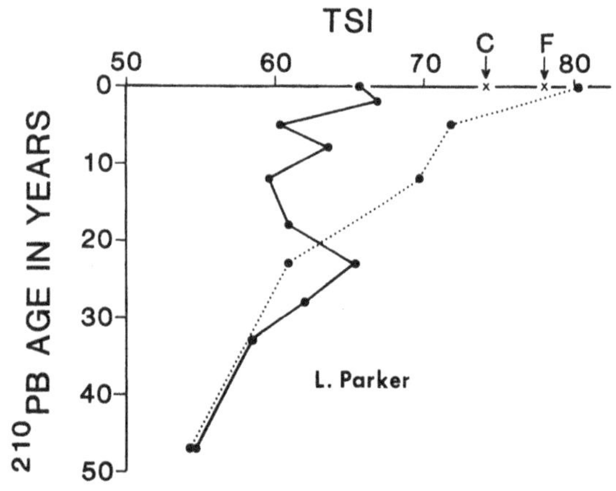

Fig. 3. Historical TSI inferences for Lake Parker based on organic matter accumulation rate (solid) and a diatom index (dotted). C and F mark, respectively, the modern Carlson's TSI-Chl *a* and Average Florida TSI values for the lake.

phosphate strip-mines in addition to urban population growth. The diatom index (Fig. 3) gives an independent view of changing trophic state in this lake. The diatom-assemblage index uses 'Average Florida TSI', a modified version of the Carlson TSI (Whitmore 1985). The diatom-assemblage index (dotted line) is a better predictor of modern trophic state than is the rate of accumulation of organic matter (solid line).

Discussion

Our first attempt at measuring rate of change of trophic state in a series of lakes has given encouraging results, but the sample of 14 lakes is biased toward oligotrophic sandhill lakes, and district-wide comparisons are premature. Among sources of variance over which we need better control, we can list the following:

1. Our trophic-state index, TSI_{chlor}, is at best a measure of chlorophyll concentration, scaled logarithmically, and while concentrations are correlated with primary productivity, they are not equivalent to rates of C-assimilation. Perhaps primary productivity would be a better correlate of the organic accumulation rate.
2. Sedimentary organic matter is a mixture of allochthonous and autochthonous material, inseparable by our methodology.
3. Decomposition of organic matter (diagenesis) proceeds at variable rates, both across a lake district and from surficial to deeper sediments in a single lake. Incomplete diagenesis at the mud-water interface may explain the core-top increases of organic matter accumulation seen in Lakes Wauberg and Weir. Eventually, we hope to evaluate diagenesis by comparing rates of accumulation of diatoms and chitinous fossils with apparent rates of change of more labile compounds.
4. Single, mid-lake sediment cores are subject to sediment focusing, and may give misleading estimates of lake-wide accumulation rates.

Looking past these difficulties, we consider the most surprising result of this investigation: rates of eutrophication have not been sharply accelerated, even in such disturbed lakes as Geneva, Minnehaha, and Weir. Without estimates of changing P- and N-loading, we are not sure how this fact should be explained. Among factors that might depress rates of eutrophication, we have provisionally rejected only one. Siltation had this effect in Guatemala, during Maya times, but Florida soils seem much less easily erodable than Peten soils, and the topography is less conducive to massive colluviation. Anyway, rates of accumulation of organic and inorganic matter are positively correlated in the 14 cores shown in Fig. 2, whereas they were negatively correlated in the Peten Lakes.

Acknowledgements

We thank E. A. Fisher and M. Sid Flannery for help in the field, and Ray Bienert for loss-on-ignition determination of the Barco core. This research was supported in part by NSF Grants DAR 79-24812 and DEB 82-11380.

References

Appleby, P. G. & F. Oldfield, 1978. The calculation of lead-210 dates assuming a constant rate of supply of unsupported ^{210}Pb to the sediment. Catena 5: 1–8.

Appleby, P. G. & F. Oldfield, 1983. The assessment of ^{210}Pb data from sites with varying sediment accumulation rates. Hydrobiologia 103: 29–35.

Baker, L. A., P. L. Brezonik & C. R. Kratzer, 1981. Nutrient loading-trophic state relationships in Florida lakes. Publication 56, Water Resources Research Center, Univ. of FL. Gainesville, FL.

Binford, M. W., E. S. Deevey & T. L. Crisman, 1983. Paleolimnology: an historical perspective on lacustrine ecosystems. Ann. Rev. Ecol. Syst. 14: 255–286.

Binford, M. W. & M. Brenner, 1986. Dilution of ^{210}Pb by organic sedimentation in lakes of different trophic states, and application to studies of sediment-water interactions. Limnol. Oceanogr. 31: 584–595.

Brenner, M., 1983. Paleolimnology of the Maya region. Ph.D. thesis, University of Florida, Gainesville, FL. 249 pp.

Brenner, M. & M. W. Binford, This volume. Material transfer from water to sediment in Florida lakes. Hydrobiologia 143: 55–61.

Brenner, M. & M. W. Binford, In prep. Relationships between concentrations of sedimentary variables and trophic state in Florida lakes. Unpublished manuscript.

Brezonik, P. L., W. H. Morgan, E. E. Shannon & A. D. Putman, 1969. Eutrophication factors in north central Florida lakes. Publication No. 5, Water Resources Research Center, Univ. of FL., Gainesville, FL.

Brezonik, P. L. & E. E. Shannon, 1971. Trophic state of lakes in north-central Florida. Publication No. 13, Water Resources Research Center, Univ. of FL., Gainesville, FL.

Canfield, D. E., Jr, 1981. Chemical and trophic state characteristics of Florida lakes in relation to regional geology. Final report to Cooperative Fish and Wildlife Research Unit, Univ. of FL., Gainesville, FL., contract No. 14-16-0009-79-064. 444 pp.

Carlson, R. E., 1977. A trophic state index for lakes. Limnol. Oceanogr. 22: 361–369.

Chapra, S. C. & K. H. Reckhow, 1983. Engineering approaches for lake management. Volume 2: Mechanistic modeling. Butterworth Publishers, Boston, MA. 492 pp.

Deevey, E. S., 1984. Stress, strain, and stability of lacustrine ecosystems. In E. Y. Haworth & J. W. G. Lund (eds.), Lake Sediments and Environmental History. University of Minnesota Press, Minneapolis: 203–229.

Deevey, E. S. & D. S. Rice, 1980. Coluviacion y retencion de nutrientes en el distrito lacustre del Peten, Guatemala. Biotica 5: 129–144.

Deevey, E. S., D. S. Rice, P. M. Rice, H. H. Vaughan, M. Brenner & M. S. Flannery, 1979. Mayan urbanism; impact on a tropical karst environment. Science 206: 298–306.

Dillon, P. J. & F. H. Rigler, 1975. A simple method for predicting the capacity of a lake for development based on lake trophic status. J. Fish. Res. Bd Can. 32: 1519–1531.

Eakins, J. D. & M. T. Morrison, 1978. A new procedure for the determination of lead-210 in lake and marine sediments. Int. J. appl. Radiat. Isot. 29: 531–536.

Frey, D. G., 1955. Langsee: a history of meromixis. Mem. Ist. ital. Idrobiol. Suppl. 8: 141–164.

Goldberg, E. D., 1963. Geochronology with lead-210. In Radioactive Dating, I.A.E.A., Vienna, pp. 121–131.

Huber, W. C., P. L. Brezonik, J. P. Heaney, R. E. Dickinson, S. D. Preston, D. S. Dwornik & M. A. DeMaio, 1982. A classification of Florida lakes. Final report ENV-05-82-1 to the Florida department of Environmental Regulation, Tallahassee, FL.

Rice, D. S., P. M. Rice & E. S. Deevey, 1983. El impacto de los Mayas en el ambiente tropical de la cuenca de los Lagos Yaxha y Sacnab, El Peten, Guatemala. America Indigena 43 No. 2: 261–297.

Shannon, E. E. & P. L. Brezonik, 1972a. Eutrophication analysis: a multivariate approach. Journal of the Sanitary Engineering Division, Proceedings of the American Society of Civil Engineers vol. 98, No. SA 1: 37–57.

Shannon, E. E. & P. L. Brezonik, 1972b. Relationships between lake trophic state and nitrogen and phosphorus loading rates. Envir. Sci. Tech. 6: 719–725.

Thomas, E. A., 1955. Stoffhaushalt und sedimentation im oligotrophen Aegerisee und im eutrophen Pfaffiker und Greifensee. Mem. Ist. ital. Idrobiol. Suppl. 8: 357–465.

Tutin, W., 1955. Preliminary observations on a year's cycle of sedimentation in Windermere, England. Mem. Ist. ital. Idrobiol. Suppl. 8: 467–484.

Vollenweider, R. A., 1968. Recherches sur l'amenagement de l'eau; (OECD: Paris. DAS/CSI/68.27).

Whitmore, T. J., 1985. Diatom transfer functions for assessing the cultural eutrophication of Florida lakes. M.S. thesis, Univ. of Florida, Gainesville, FL. 143 pp.

Zullig, H., 1955. Sedimente als Ausdruck des Zustandes eines Gewassers. Mem. Ist. ital. Idrobiol. Suppl. 8: 485–530.

Accepted 18 March 1986.

Hydrobiologia 143: 55–61, (1986)

Material transfer from water to sediment in Florida lakes

Mark Brenner & Michael W. Binford
Department of Natural Sciences, Florida State Museum, Gainesville, Florida 32611, USA

Keywords: paleolimnology, Florida lakes, trophic state, sediment chemistry, accumulation rate, ^{210}Pb

Abstract

We describe statistical relationships between chemical aspects of surficial sediments from 34 Florida lakes, and trophic state of the overlying waters. Trophic state is expressed by Carlson's TSI-Chl *a*. The objectives of the study are two-fold: 1) to understand processes that govern the transfer of material across the mud-water interface, and 2) to develop transfer functions for inferring historical TSI measures in chemically analyzed, ^{210}Pb-dated cores. Simple regression of organic matter content or nutrient (C, N, P) concentration in surface sediments vs. TSI yields nonsignificant or weak positive correlations. However, using a novel application of ^{210}Pb assay, net accumulation rates of the materials are estimated and show a better correlation with trophic state. Cation (Ca, Mg, Fe, K) and sulfur concentrations in surface muds are poorly related to corresponding TSI's. Net accumulation rates of these elements (Ca, Mg, Fe, K, S) are positively correlated ($P < 0.01$) with TSI-Chl *a*. Chemical data from the Florida surficial sediment survey suggest that inferred levels of past trophic state should be based on net accumulation rates of chemical constituents rather than their simple sediment concentrations.

Introduction

Florida is a lake-rich state with more than 7000 named water bodies (Canfield 1981). Recent, rapid population growth in the state stresses Florida's aquatic systems, as they are increasingly exploited for private, agricultural, industrial and recreational uses. By 1980, Florida was ranked the 7th most populous state, with $9.75 \cdot 10^6$ inhabitants. Projections indicate that by the year 2000, Florida will be home to $14.25 \cdot 10^6$ people (Morris, 1983).

Despite the abundance of lakes in the state, systematic, long-term limnological study of basins on the peninisula did not begin until about 1970. Many aquatic systems had experienced severe cultural impact for many years prior to limnological measurements. A number of recent studies have evaluated the modern trophic status of Florida's lakes (Baker *et al.*, 1981; Brezonik *et al.*, 1969; Canfield, 1981, 1983; Shannon & Brezonik, 1972a; Shannon & Brezonik, 1972b). However, there is not a single Florida lake for which sufficient long-term limnological data exist to document the magnitude or rate of change of cultural eutrophication.

We are currently studying the vulnerability of Florida lakes to various kinds of catchment disturbances. We will employ paleolimnological techniques to reconstruct the trophic trajectories of some 25 carefully selected lakes (Binford *et al.*, 1983; Deevey, 1984). Our project has several long-term goals:

(1) We will analyze ^{210}Pb-dated sediment cores from 25 lakes to assess quantitatively historical (last 150 yrs) trophic state changes
(2) We will use multivariate models to relate historical land-use information from the watersheds to inferred trophic state changes in the 25 sampled lakes

(3) We will construct a model that predicts lacustrine trophic response for proposed land use practices in a drainage basin.

To realize these goals, we seek sedimentary chemical variables that accurately reflect the trophic state of the overlying waters. Few studies quantitatively relate surface sediment characteristics to lake waters for many basins of a region (Gorham, 1960; Dean & Gorham, 1976; Rybak, 1969; Gorham *et al.*, 1974; Guilizzoni *et al.* 1983). A preliminary study in Florida by Flannery *et al.* (1982) indicated that a survey approach was justified. Therefore, we assessed chemical content of surficial sediments from a suite of Florida lakes that vary widely in trophic state in search of constituents that vary systematically as a function of trophic state.

We used simple linear regression to assess the relationship between concentrations of sedimentary variables (organic matter, C, N, P, pigments) and trophic state of 91 lakes. Direct evaluations of productivity are scarce for all but a few Florida lakes, and we rely on Carlson's (1977) TSI-Chl *a* as a quantitative measure of trophic state. Sedimentary concentrations of variables proved to be uncorrelated or showed weak positive correlation with TSI (Brenner & Binford, in prep.). Several factors account for the poor relationships. Most important is the variable dilution of materials of interest by inorganics (primarily sand) across our suite of study lakes.

We next employed a novel application of ^{210}Pb techniques in which unsupported (fallout) ^{210}Pb in surficial sediments was used as a dilution marker to measure modern (instantaneous) net accumulation rates of materials on the lake bottom. We applied the methodology to a 34-lake subset of our original lake list (Fig. 1) and calculated net accumulation rates of organic matter, C, N, P, and pigments (see Binford & Brenner (1986) for a detailed discussion of the method). Net accumulations of materials proved to be better correlated with trophic state than did simple sedimentary concentrations. In the present study we examine the feasibility of using sedimentary cations (Fe, Ca, Mg, K) and sulfur as indicators of trophic state.

Methods

Surficial sediment samples were collected near the center of each lake using an Ekman dredge (n = 27 lakes) or a mud-water interface corer (n = 7 lakes). The top-most 1–2 cm of sediment were removed for analyses. ^{210}Pb-dated cores from 18 Florida lakes suggest that this interval represents the last 2–10 years of sedimentation. All samples were assessed for wet density, % dry weight, and % loss on ignition at 550 °C. Dredge samples were also combusted at 990 °C to measure carbonate-CO_2 loss (Dean, 1974). Core-top samples of dry mud for P and cation analyses were digested with HNO_3-$HClO_4$ according to Standard Methods (APHA 1975). Dredge samples were combusted at 550 °C and the ash was digested in 1N HCl according to Andersen (1976). Cation concentrations in the digestate were read by atomic absorption. Sulfur was read turbidimetrically (APHA, 1975) on dredge-sample digestates. ^{210}Po was assessed in dry

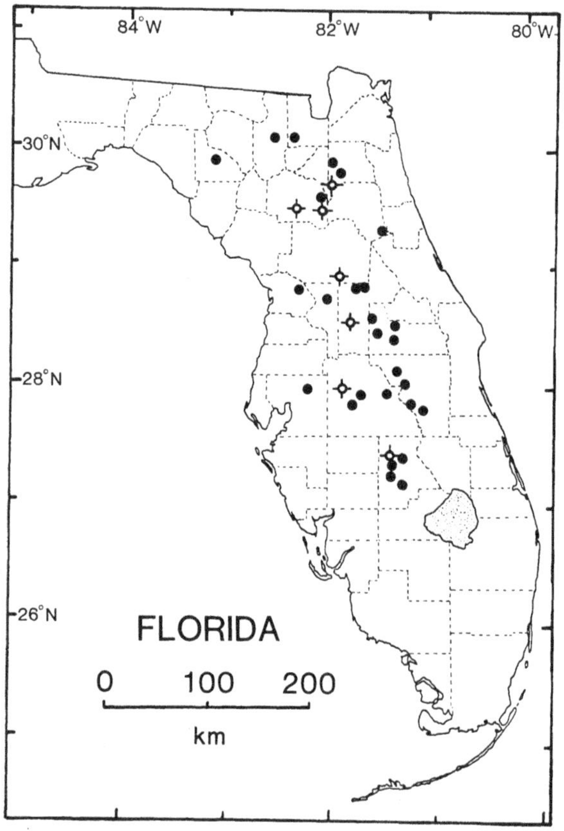

Fig. 1. Peninsular Florida, showing the location of the 34 lakes in the surface sediment study. ● = lakes sampled with an Ekman dredge. -ф- = lakes sampled with a mud-water interface corer.

material after waiting 14 months for the granddaughter isotope to come in to equilibrium with ^{210}Pb (Eakins & Morrison, 1978). Median TSI-Chl *a* values for each lake were calculated using water column chlorophyll data from a data set maintained by the Water Resources Research Center at the University of Florida. A summary of the limnological data for the lakes is presented by Huber *et al.* (1982).

Results

Table 1 shows organic matter, Fe, Ca, Mg, K, and S concentrations in surficial sediments of the Florida lake data set. L.O.I. at 990 °C was < 4.0% in all 27 dredge samples, reflecting the low carbonate content of most Florida muds.

Table 1. Organic matter, sulfur, and cation concentrations in surface sediments of Florida lakes.

Lake	Organic matter (%)	Fe	Ca	Mg	K	S
		(mg g⁻¹)				
1. Apopka	64.1	3.7	58.5	5.8	0.9	8.5
2. Butler	1.6	0.4	0.3	0.1	0.1	0.2
3. Clay	48.4	4.6	5.4	1.8	0.8	4.7
4. Cypress	31.5	15.8	7.3	2.4	1.1	3.8
5. Eagle	41.0	6.2	6.9	2.8	1.1	4.0
6. Fairview	8.2	2.4	2.9	0.5	0.4	1.6
7. Geneva	32.4	5.9	2.7	1.5	1.4	–
8. Griffin	62.0	5.3	42.7	2.3	0.6	9.1
9. Hawthorne	81.5	5.6	9.7	0.8	0.8	–
10. Howard	40.7	6.1	10.7	1.6	1.4	4.8
11. Jackson	32.6	9.2	3.7	1.8	0.8	–
12. Jessamine	35.9	8.5	12.6	2.3	1.5	5.7
13. Josephine	5.8	0.9	0.6	0.2	0.1	0.3
14. June-in-Winter	7.7	2.3	0.7	0.2	0.1	1.1
15. Kingsley	39.2	7.8	4.4	1.0	0.8	3.9
16. Kissimmee	0.8	0.7	0.1	0.0	0.0	0.2
17. Little Red Water	39.8	8.4	2.4	0.7	0.5	1.9
18. Marian	55.2	10.4	9.1	1.4	0.7	5.3
19. Minnehaha	6.1	3.6	0.8	0.4	0.2	–
20. Ocean Pond	39.0	10.1	1.2	0.6	0.5	1.8
21. Okahumpka	67.1	21.6	12.3	1.3	0.4	8.9
22. Parker	64.1	8.3	21.8	2.3	2.0	–
23. Pierce	62.3	6.9	10.5	1.3	0.7	5.0
24. Santa Fe	36.2	5.4	2.6	1.3	1.1	1.3
25. Sheelar	24.3	0.8	0.2	0.0	0.4	0.6
26. Stella	14.0	2.7	2.8	0.7	0.5	2.2
27. Thonotosassa	40.0	8.8	14.5	2.1	1.4	8.0
28. Tohopekaliga	8.1	3.0	2.1	0.4	0.1	1.8
29. Townshend Pond	79.6	3.1	4.7	0.5	0.2	2.2
30. Tsala Apopka	82.3	4.6	19.4	1.3	0.9	5.6
31. Watertown	60.3	13.6	15.5	2.0	0.9	7.8
32. Wauberg	72.4	5.9	6.6	1.2	1.2	–
33. Weir	63.3	4.3	5.1	1.3	0.1	–
34. Yale	52.7	6.3	7.5	1.8	0.5	5.8

Linear regression analysis was used to examine the relationship between concentrations of sedimentary cations and sulfur vs. TSI-Chl *a* (Figs. 2A-E). TSI is plotted as the dependent variable because we plan to use sediment core chemical data to infer past levels of TSI. Sedimentary concentrations of Fe and K are uncorrelated with TSI, while Ca, Mg, and S concentrations show weak positive correlations with trophic state. Earlier results for this 34-lake data set showed that for nutrient concentrations, only N was significantly correlated with TSI-Chl *a* ($r = 0.38$, $P < 0.05$). When net accumulation rates of cations and sulfur were computed and regressed against TSI-Chl *a*, all models were found to be significant at $P < 0.01$ (Figs. 2F-J).

Discussion

We have shown that surface sediment concentrations of many biologically important elements (C, N, P, S, Ca, Mg, Fe, K) are uncorrelated or only weakly positively correlated with trophic state. However, a statistically significant positive relationship exists between net accumulation rates of these elements to the sediment, and TSI-Chl *a*.

Sedimentary C and N were shown to be bound exclusively in organic matter (Brenner & Binford, in prep.). Higher autochthonous production and concomitant increased sedimentation would be expected to result in higher delivery rates of C and N to sediments. Sedimentary phosphorus may be bound in both the organic and inorganic phase, but the proportion of P in each phase varies basin-to-basin. Phosphate district lakes accumulate P rapidly, much of it deposited as large-particle inorganics (Binford & Brenner, in press., Prepas & Vickery, 1984).

Table 2 shows several within-sediment correlations. Positive correlation between all elemental concentrations and % organic matter suggests that the elements are associated with the organic fraction, but these results can be misleading. In most of our surface sediment samples, the inorganic fraction appears to be comprised primarily of fine and coarse sands, devoid of extractable ions. Across the suite of lakes, inorganic matter constitutes 17.7%-99.2% of the sediment dry weight. Because of the variable volumetric displacement by coarse inorganics, cation and S concentrations would be expected to be positively related to % organic matter, even if the elements are not bound to or contained within organics.

High rates of inorganic sedimentation dilute final concentrations of organic matter, S, Fe, Mg, and K in surface muds. Sulfur, Ca, and Mg appear to reach the lake bottom primarily bound to or within organic material, as their respective accumulation rates are all positively correlated with organic accumulation ($P < 0.01$). Lack of correlation be-

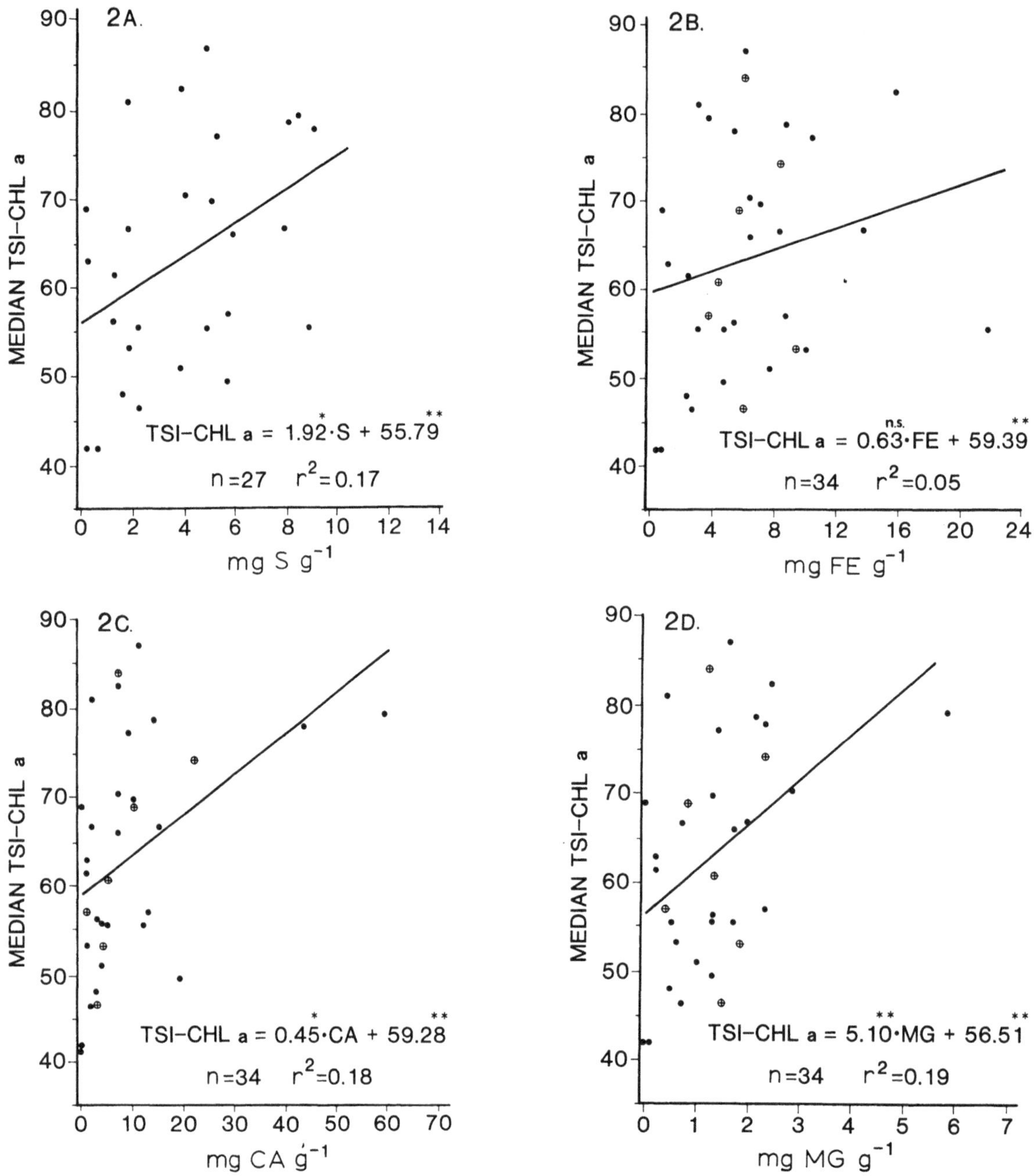

tween organic accumulation rate and Fe or K accumulation rates hints that these elements may in part be associated with sedimentary inorganics, the small amount of silt and clay present. This claim remains speculative until future studies partition the inorganic fraction by particle size.

The variance in the accumulation rate vs. TSI plots is attributable to at least two principal factors: 1) The inability to discern sedimented materials of autochthonous and allochthonous origin. For example, high rates of net material accumulation will be measured in oligotrophic (low TSI) systems that receive large allochthonous loads. 2) Variable basin-to-basin diagenetic changes in recent muds,

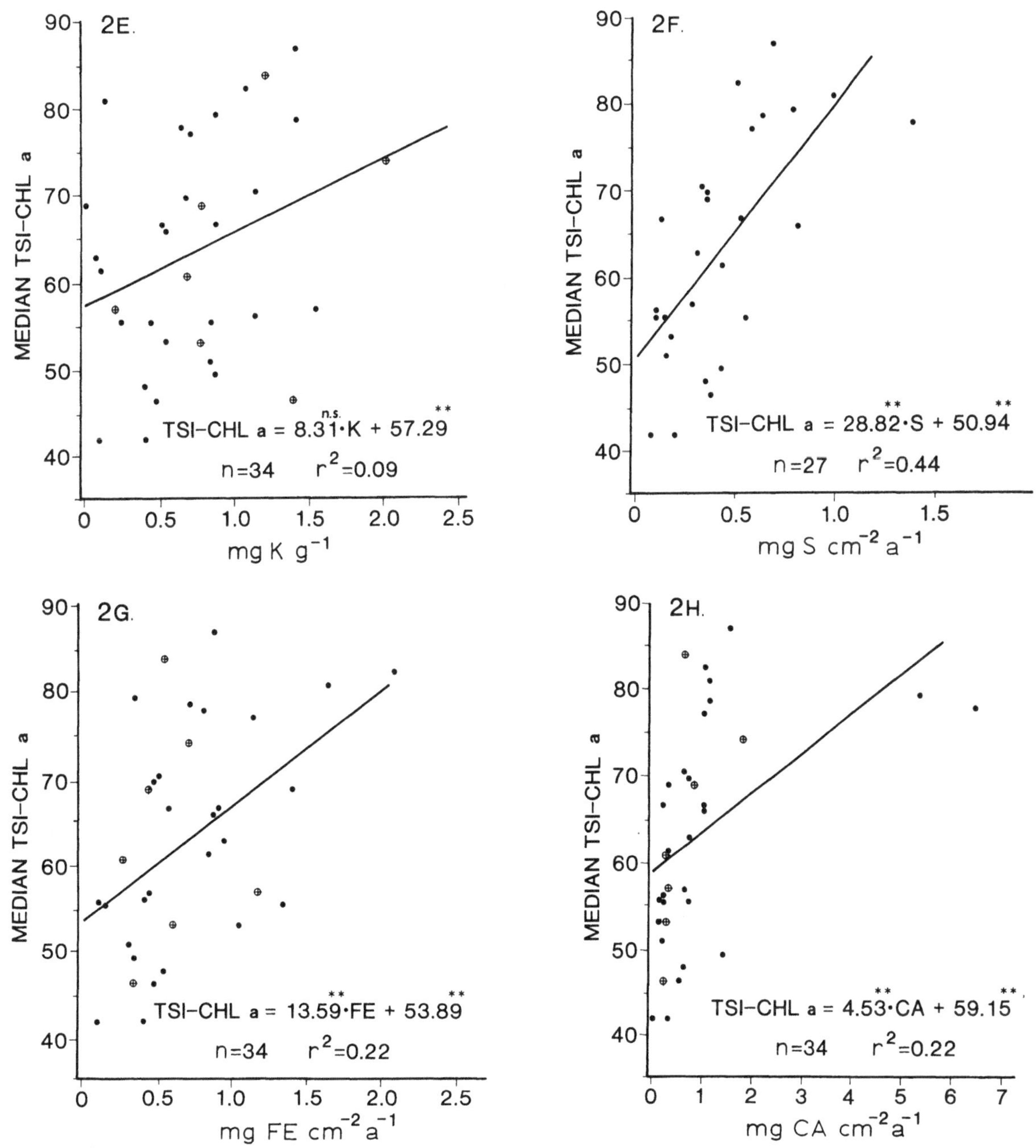

Table 2. Correlations between surface sediment variables of the surveyed Florida lakes. Only statistically significant correlations are shown.

			% organic matter	organic accumulation rate	inorganic accumulation rate
concentration	($mg\ g^{-1}$)	S	0.72**		-0.51**
		Fe	0.40*		-0.44**
		Ca	0.50**	0.40*	
		Mg	0.45**		-0.42*
		K	0.43*		-0.52**
accumulation rate	($mg\ cm^{-2}\ a^{-1}$)	S		0.56**	
		Fe			
		Ca		0.52**	
		Mg		0.48**	
		K			
% organic matter				0.35*	-0.61**
organic accumulation rate					
inorganic accumulation rate					

** = P<0.01.
* = P<0.05.

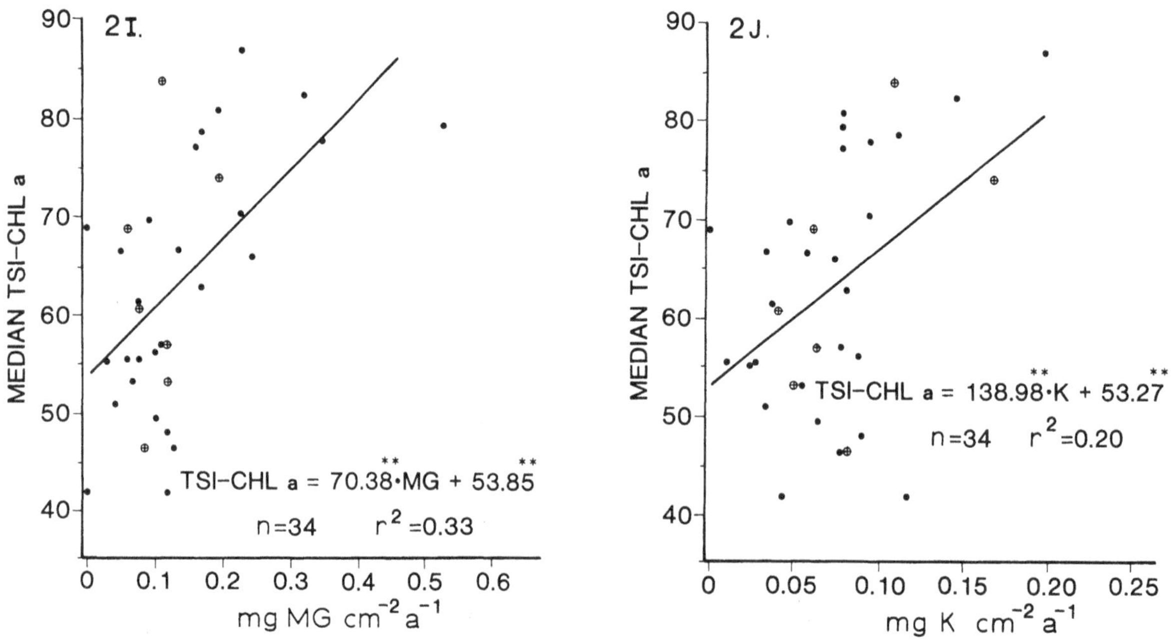

Fig. 2. Figs. 2A–E show regressions of elemental concentrations in surface sediments ($mg\ g^{-1}$ dry sediment) of Florida lakes vs. TSI-Chl *a* of overlying waters. Figs. 2F–J show regressions of net accumulation rates of the elements ($mg\ cm^{-2}\ a^{-1}$) vs. TSI-Chl *a* of the overlying waters. ● = lakes sampled with an Ekman dredge. ⊕ = lakes sampled with a mud-water interface corer.

causing inequality between actual and measured net accumulation rates.

Changing material concentrations in cores from a single lake basin may permit accurate interpretation of altered trophic conditions. Nevertheless, we recommend that downcore, historical TSI inferences in Florida basins be based on net accumulation rates of biologically important elements. Inferences should be checked against alternative lines of evidence (e.g. diatoms, cladocera, and midge remains) that are perhaps less prone to diagenetic change.

Acknowledgements

We thank Elizabeth Fisher and Michael S. Flannery for help with field work and lab analyses. The University of Florida Institute of Food and Agricultural Sciences Soils Lab provided assistance with cation analyses. Edward S. Deevey gave us helpful comments on the manuscript. Vonceil Levine, and Romona Barnes typed the various drafts of the paper. This research was supported by NSF grants DAR 79-24812, EAR 82-14308 and DEB 82-11380.

References

American Public Health Association, 1975. Standard methods for the examination of water and wastewater, 14th ed.

Andersen, J. M., 1976. An ignition method for determination of total phosphorus in lake sediments. Wat. Res. 10: 329–331.

Baker, L. A., P. L. Brezonik & C. R. Kratzer, 1981. Nutrient loading – trophic state relationships in Florida lakes. Publication 56, Water Resources Research Center, Univ. of Florida, Gainesville.

Binford, M. W. & M. Brenner, 1986. Dilution of ^{210}Pb by organic sedimentation in lakes of different trophic states and application to studies of sediment–water interactions. Limnol. Oceanogr. 31: 584–595.

Binford, M. W., E. S. Deevey & T. L. Crisman, 1983. Paleolimnology: an historical perspective on lacustrine ecosystems. Ann. Rev. Ecol. Syst. 14: 255–286.

Brenner, M. & M. W. Binford, in prep. Relationships between concentrations of sedimentary variables and trophic state in Florida lakes.

Brezonik, P. L., W. H. Morgan, E. E. Shannon & A. D. Putnam, 1969. Eutrophication factors in north central Florida lakes. Publication No. 5, Water Resources Research Center, Univ. of Florida., Gainesville.

Canfield, D. E., Jr., 1981. Chemical and trophic state characteristics of Florida lakes in relation to regional geology. Final Report to Cooperative Fish and Wildlife Research Unit, Univ. of Florida, Gainesville, Contract No. 14-16-0009-79-064. 444 pp.

Canfield, D. E., Jr., 1983. Prediction of chlorophyll a concentrations in Florida lakes: the importance of phosphorus and nitrogen. Wat. Res. Bull. 19: 255–262.

Carlson, R. E., 1977. A trophic state index for lakes. Limnol. Oceanogr. 22: 361–369.

Dean, W. E., Jr., 1974. Determination of carbonate and organic matter in calcareous sediments and sedimentary rocks by loss on ignition: Comparison with other methods. J. Sed. Pet. 44: 242–248.

Dean, W. E., Jr. & E. Gorham, 1976. Major chemical and mineral components of profundal surface sediments in Minnesota lakes. Limnol. Oceanogr. 21: 259–284.

Deevey, E. S., Jr., 1984. Stress, strain, and stability of lacustrine ecosystems. p. 203–229. In E. Y. Haworth & J. W. G. Lund (eds.), Lake sediments and environmental history. University of Minnesota Press, Minneapolis, 411 pp.

Eakins, J. D. & R. T. Morrison, 1978. A new procedure for the determination of Lead-210 in lake and marine sediments. Int. J. appl. Radiat. Isotopes. 29: 531–536.

Flannery, M. S., R. D. Snodgrass & T. J. Whitmore, 1982. Deepwater sediments and trophic conditions in Florida lakes. Hydrobiologia 92: 597–602.

Gorham, E., 1960. Chlorophyll derivatives in surface muds from the English lakes. Limnol. Oceanogr. 5: 29–33.

Gorham, E., W. G. Lund, J. E. Sanger & W. E. Dean, Jr., 1974. Some relationships between algal standing crop, water chemistry, and sediment chemistry in the English Lakes. Limnol. Oceanogr. 19: 601–617.

Guilizzoni, P., G. Bonomi, G. Galanti & D. Ruggiu, 1983. Relationship between sedimentary pigments and primary production: evidence from core analyses of twelve Italian lakes. Hydrobiologia 103: 103–106.

Huber, W. C., P. L. Brezonik, J. P. Heaney, R. E. Dickinson, S. D. Preston, D. S. Dwornik & M. A. DeMaio, 1982. A classification of Florida Lakes. Final Report ENV-05-82-1 to the Florida Department of Environmental Regulation, Tallahassee, Florida.

Morris, A., 1983. The Florida handbook 1983–1984, 19th ed. The Peninsular Publishing Company, Tallahassee, Fl. 693 pp.

Prepas, E. E. & J. Vickery, 1984. Contribution of particulate phosphorus (>250 μm) to the total phosphorus pool in lake water. Can. J. Fish. aquat. Sci. 41: 351–363.

Rybak, J. I., 1969. Bottom sediments of the lakes of various trophic type. Ekol. Pol. A: 611–662.

Shannon, E. E. & P. L. Brezonik, 1972a. Relationships between lake trophic state and nitrogen and phosphorus loading rates. Envir. Sci. Tech. 6: 719–725.

Shannon, E. E. & P. L. Brezonik, 1972b. Eutrophication analysis: a multivariate appraoch. Journal of the Sanitary Engineering Division, Proc. am. Soc. civ. Engineers 98, SA1: 37–57.

Accepted 7 April 1986.

Hydrobiologia 143: 63–69, (1986)

Pollen as a chronometer and sediment tracer, Burrinjuck Reservoir, Australia

Robin L. Clark
Division of Water and Land Resources, CSIRO, G.P.O. Box 1666, Canberra, A.C.T. 2601, Australia

Keywords: pollen, sediment dating, sediment tracer, reservoirs, erosion

Abstract

Pollen analysis is widely used to reconstruct vegetation and land use histories, but can also provide sedimentological information. At Burrinjuck Reservoir, in south-eastern Australia, annual grass pollen peaks are used to distinguish each year's sediment, even when there are no visible laminations. In conjunction with other dating methods, this allows the determination of year by year influxes of all sediment components. Pollen grains in the Burrinjuck sediments are shown to be predominantly waterborne so that they can be used to trace sediment to its source in particular vegetation stands. Pollen concentration and the proportion of damaged pollen might also distinguish sediment eroded from topsoils and that from subsoils. Pollen analysis can thus be used to locate specific erosion events in both time and space.

Introduction

Pollen analysis is one of many techniques being applied in a large project aimed at both describing and explaining changes in the location, type and rate of soil erosion over the past hundred years in different parts of Australia. The time-scale of decades and the historical period covered are the most appropriate for providing a sound basis for decisions about how water and land resources are to be managed for the future. Reservoir sediments were chosen because Australia has relatively few permanent lakes and because both trap efficiency and sedimentation rate in reservoirs are often high, allowing fine resolution of the sedimentary record. Reservoir sediments offer a wealth of opportunities for process studies, either on their own or in conjunction with environmental monitoring, and, with the fine resolution that is often possible, are ideal for studying the recent past of aquatic and terrestrial ecosystems and the effects of pollutants (Clark & Wasson, 1986).

The purpose of most palynological studies is to reconstruct vegetation and land use histories on a variety of time scales. Pollen analysis of Australian reservoir sediments is being used to provide such general vegetation and land use records on a time scale of decades, but also to study the effects on vegetation and soil erosion of short-term climatic changes, to determine specific vegetation responses to disturbances such as fire or the sudden reduction in rabbit populations in the 1950s, and to reconstruct the details of changes in species composition in arid and semi-arid grazing lands.

In sedimentological studies, pollen grains can provide much more information. Preliminary results from an Australian reservoir show that pollen can be used in new ways as a chronometer and as a sediment tracer.

The reservoir

Burrinjuck reservoir is a large irrigation water storage in the south-east of New South Wales, Australia (35°00′S, 148°35′E; Fig. 1). Construction began in 1907 and water storage in 1912, but the reservoir was not filled until a record flood in 1925

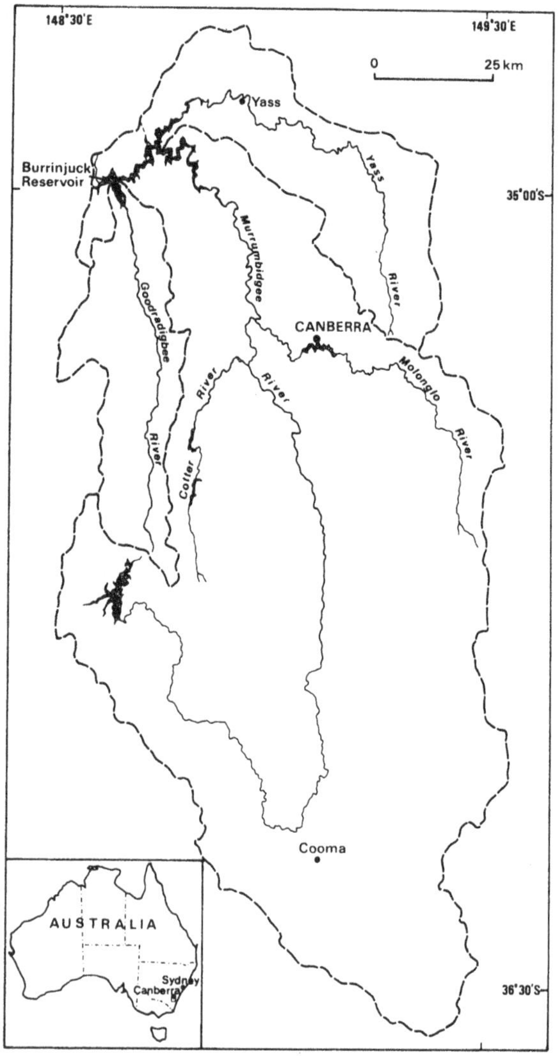

Fig. 1. The location of Burrinjuck Reservoir and its catchment, showing the three major subcatchments of the Yass, Murrumbidgee and Goodradigbee Rivers.

that overtopped the wall and threatened to destroy the dam. The wall was raised and capacity increased in the 1950s. When the reservoir is full, the water level is 364 m ASL (Australian Height Datum), the surface area is 57 km^2, the volume of water stored is 1.03×10^9 m^3 and maximum depth is 67 m. The total catchment area of 13 500 km^2 is divided into three main subcatchments, each feeding into separate arms of the reservoir (Fig. 1). The Yass River catchment, of 1 680 km^2, was largely cleared last century for pastures and crops and is now mostly used for sheep grazing. The Goodradigbee River catchment, of 1 260 km^2, and two small catchments (total area 200 km^2) around the main basin of the reservoir are almost entirely native forests dominated by trees of the genus *Eucalyptus.* The large Murrumbidgee River catchment (10 320 km^2) contains both pastures and native forests, as well as plantations of introduced *Pinus* species and the Canberra urban area.

Rainfall over the catchment varies from 500 to 1 000 mm p.a., evenly distributed throughout the year and falling as snow on the higher peaks and plateaus in winter. Rainfall is highly variable from year to year and the main inflow to the reservoir comes from the spring snow melt. The reservoir is monomictic with summer stratification and turnover in autumn, but this is often disrupted by droughts or major summer inflows (Australian Government, Department of Housing and Construction 1981).

Coring of sediments in Burrinjuck Reservoir began in November, 1982, when 11 cores were taken from the main basin, within a 30 m radius, about 2 km upstream from the dam wall. Visible stratigraphic units (A to O, Fig. 2) were matched between cores and these correlations were later checked by chemical analyses. Canonical variate analysis showed that the agreement between the stratigraphic classification and the numerical classification based on 12 chemical elements and wet density ranged from 60% to 90% for the 5 units A to E and was 100% for all 10 units F to O (I. R. Willett, pers. comm.). Later sediment profiling by seismic sounding showed a uniform layer of reservoir sediment over the old river terraces, interrupted only by the infilled former river channel. A further series of cores was taken in 1984 along a transect across the reservoir, including the original coring site. The same sedimentary units could be traced along this transect and work is now in progress to determine the variability of properties, including pollen content, within units across the reservoir. Cores have been taken and sediment sampled at many other sites within the reservoir, but this paper is concerned only with results from the original (1982) set of cores.

Pollen as a chronometer

Dating such recent sediments presents problems,

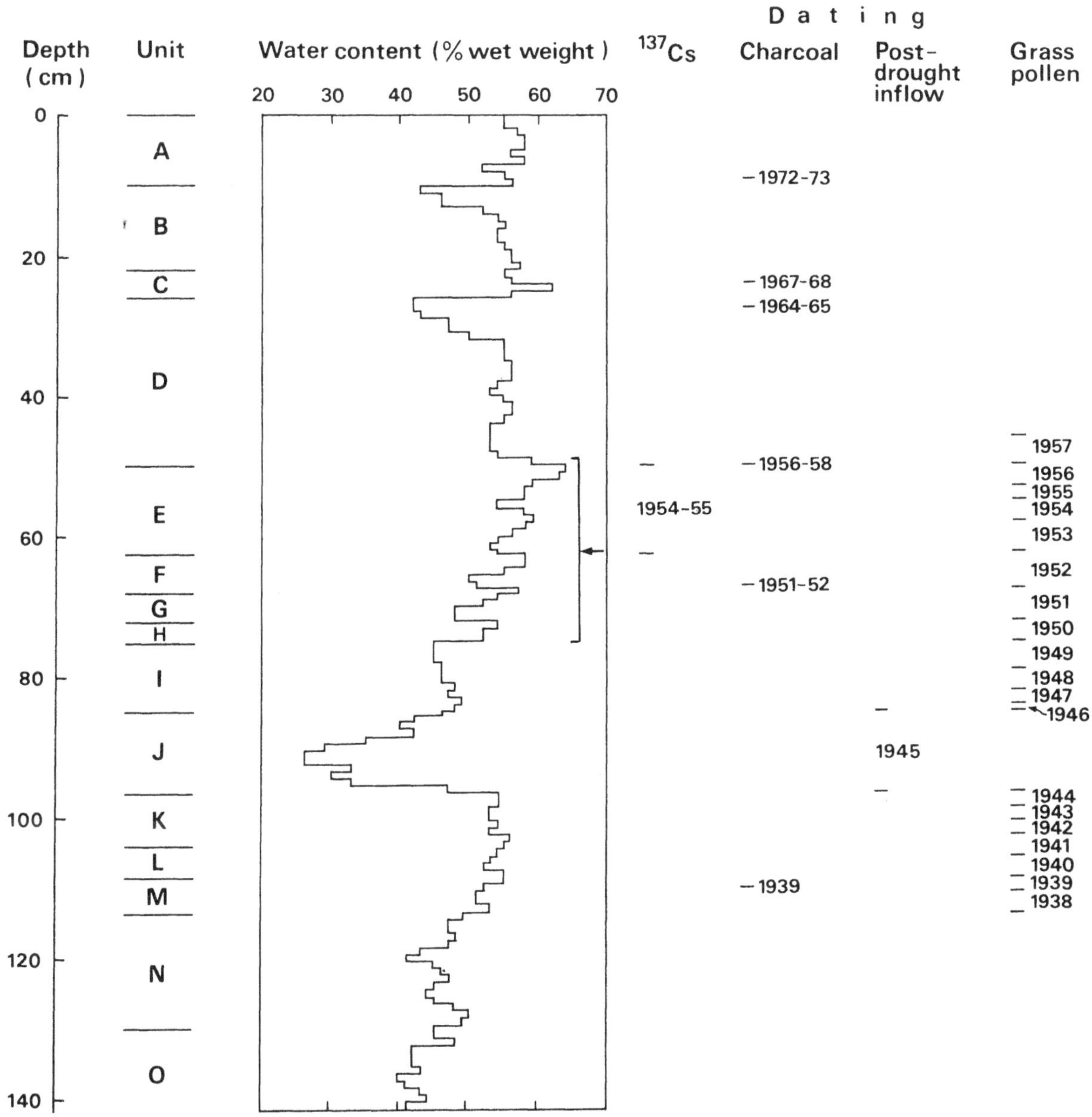

Fig. 2. Dating the 1982 cores from the main basin of Burrinjuck Reservoir. Visible stratigraphic units are labelled A to O; the arrow on the water content curve shows part of the sequence that suggests annual cycles.

but a chronology has been built up using a variety of methods (Fig. 2). Lead-210 was not useful for dating because the atmospheric component forms only a small fraction of the total flux and sedimentation in the reservoir has been neither continuous nor at a constant rate.

The first appearance of caesium-137, produced by nuclear weapons testing, is in unit E and, as calculations suggested this radionuclide would not have diffused vertically more than 4 cm in these sediments, it is assumed that sediments deposited in 1954 are in unit E. Visible charcoal bands were matched with major regional bushfire seasons to give further dates; all charcoal bands and all fire

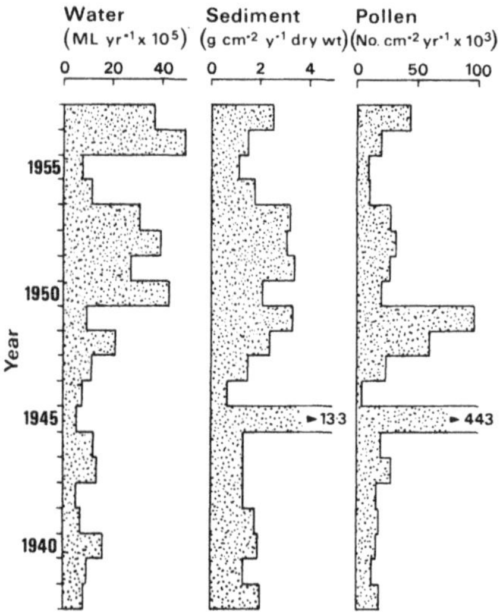

Fig. 3. Annual water inflow to Burrinjuck Reservoir and annual deposition of sediment and pollen over the years 1938–57.

years were accounted for, except 1944 when there was little rainfall to wash charcoal into the reservoir. A thick layer of coarser-grained sediment was attributed to an intense rainstorm at the end of a drought in 1945. Subsequent coring has shown a 60 cm thick layer of homogeneous sediments at the base of the sequence, thought to have been deposited in the 1925 flood, and a layer deposited in 1983, at the end of a severe drought, similar to that of 1945.

In parts of the cores, changes in sediment water content that matched visible stratigraphic units suggested annual deposition cycles (Fig. 2). Pollen analysis of contiguous samples taken at 1 cm intervals showed peaks in grass pollen at the same stage of successive cycles. Assuming that all peaks in grass pollen mark annual spring to early summer grass flowering, then each year's sediment can be identified. Pollen analyses have not been completed for the whole sequence, but each year's deposition has been determined in both laminated and unlaminated sediment for the years 1938–1957.

There are likely to be other sites where fine resolution pollen analysis of sediments can be used to distinguish annual layers even where there are no visible laminations. Sedimentation rate must be high enough for several samples to be taken from each year of deposition and pollen concentration sufficient to allow unequivocal interpretation.

Pollen as a sediment tracer

Palynologists have usually chosen sites where they could assume that most pollen grains in lake sediments were carried there through the air and deposited on the lake surface. As a consequence, much time and effort has been devoted to the study of aerial transport of pollen (see Prentice 1985). More recent studies of waterborne pollen grains have revealed that very large numbers are carried by streams (e.g., Crowder & Cuddy, 1973; Starling & Crowder, 1981; Brown, 1985). The excellent pioneering work of Peck (1973) showed that waterborne pollen grains in the sediments of two small reservoirs far outnumbered airborne pollen grains, and similar results were found in lakes by Bonny (1976, 1978).

Plants produce large quantities of pollen and, while many are carried through the atmosphere, most are deposited on the ground surface either directly or through the agency of animals. This might be underneath the parent plant, singly or within flowers, or at some distance after being carried by animals or wind, in the latter case after falling from the air or being washed out by rainfall. Surface soils and organic litter often contain large numbers of pollen grains (Dimbleby, 1957; Peck, 1973) that are subject to physical and chemical decay and microbiological destruction. Pollen grains at the soil surface are just another category of silt-sized particles that can be washed into streams and lakes with any other particulate sediment component.

Whether airborne or waterborne pollen grains predominate in the sediments of a lake or reservoir depends on relationships between the size of the catchment, the surface area of water, the topography and vegetation of the catchment, the timing and intensity of rainfall and the timing, direction and strength of winds. Burrinjuck Reservoir has a surface area of 57 km^2 and a large catchment of 13 500 km^2, so it is likely that most pollen grains in its sediments have been carried there by streams. If this is true, then pollen grains can be used as sediment tracers at Burrinjuck.

Fig. 3 shows the annual inflow of water to the reservoir for the years 1938–1957, along with the annual influx of sediment (dry weight) and of pollen grains. The last two have been obtained by adding all samples from sediments deposited in each year. Water inflow is used as it integrates rainfall and runoff over the entire catchment and gives a better indication than rainfall of sediment transport capacity. For the 20 samples there is no correlation between water inflow and sediment ($r = 0.11$) or pollen ($r = 0.04$) but an excellent correlation ($r = 0.90$) between pollen and sediment influxes when variables are logarithmically transformed to correct for the extreme outlying values of 1945.

The size distribution of sediment particles in the core samples was determined by the pipette method (Black 1965) in three size classes: clay (<2 μm), silt (2–63 μm) and sand (>63 μm). The amount of sand in most samples was insignificant. For the logarithmically transformed data over 20 years the correlation between annual pollen influx and annual clay influx was good ($r = 0.82$), but that between pollen and silt influxes ($r = 0.92$; Fig. 4) was better than that between pollen and either clay or total sediment deposition. Pollen grains range in size from 5 μm to 200 μm diameter, with most between 15 μm and 40 μm, but as their density is low they can be expected to move in air and water with smaller mineral particles. Matsushita (1985) found that pollen grains of various sizes in suspended sediment precipitated with particles of 11–44 μm. As pollen grains are among the smaller sediment particles in the silt size range their relative number increases as the mass of silt increases, thus explaining the power function indicated by the relationship between the logarithmically transformed pollen and silt influxes.

The correlation between pollen and silt suggests that the pollen in the Burrinjuck sediments is overwhelmingly waterborne and that pollen can be used as a sediment tracer to locate areas within the catchment from which the sediment came. Further, as the concentration of pollen increases, so does the number of damaged pollen grains (Fig. 5). This is only a general trend, not a good correlation, but it is likely that sediments with a high proportion of damaged pollen were derived from eroded topsoils containing degraded pollen grains. If they were from re-eroded older sediments then the general trends of vegetation change as indicated by the pollen would not be so consistent, responses of individual species, such as regeneration of *Callitris* after the reduction of rabbit populations in the 1950s, would not appear so clear-cut and annual grass pollen peaks might not be distinguishable.

All the points argued above are being tested by independent means, but there is much that can be

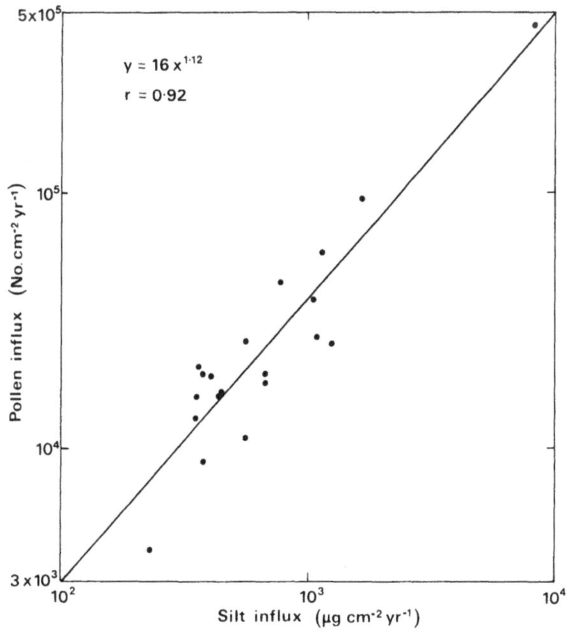

Fig. 4. Annual pollen influx to the Burrinjuck sediments plotted against annual silt influx, both on logarithmic scales.

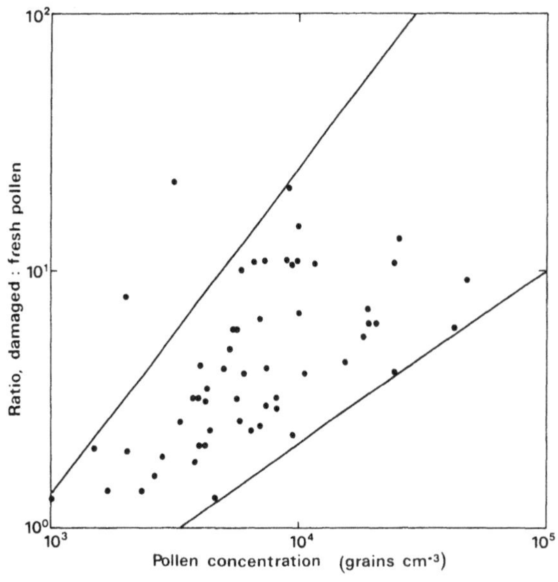

Fig. 5. The ratio of damaged to fresh pollen plotted against pollen concentration, both on logarithmic scales, for all samples analyzed from the Burrinjuck sediments.

deduced about the origin of sediments in Burrinjuck Reservoir for the years 1938–1957, for which pollen analyses are complete. Changes in abundance of the major pollen types indicate which parts of the catchment sediment came from, and high pollen numbers per unit of weight of silt, with low clay content and a high proportion of damaged pollen grains, indicates erosion from topsoils. As there are no data on size distribution within the silt range, clay content is used as an indicator of whether high pollen concentrations are due to a predominance of smaller silt-sized particles or to larger numbers of pollen grains in the sediment source.

Year by year changes in the major pollen types in the Burrinjuck sediments are shown in Fig. 6. Most Myrtaceae pollen, that includes the genera *Eucalyptus, Leptospermum* and *Melaleuca,* comes from native forests, most Poaceae pollen from pastures and almost all *Pinus* pollen from pine plantations. It is evident from Fig. 6 that most erosion in 1938–9, 1942–3 and 1955 was in the native forests of the Goodradigbee and Murrumbidgee catchments; in 1945–46 there was severe erosion of pastures, but as there was no erosion in the pine forests of the Murrumbidgee catchment in these years, the most intense rainfall may have been limited to the Yass catchment and the northern part of the Murrumbidgee; in 1953 the pastures were again eroded; and in 1956 there were serious floods that eroded the pastures and pine plantations of the

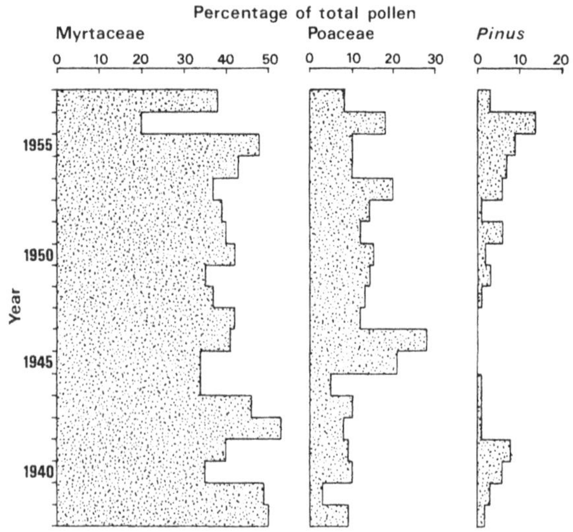

Fig. 6. Mean percentages of the three major pollen types in the Burrinjuck sediments for the years 1938–57.

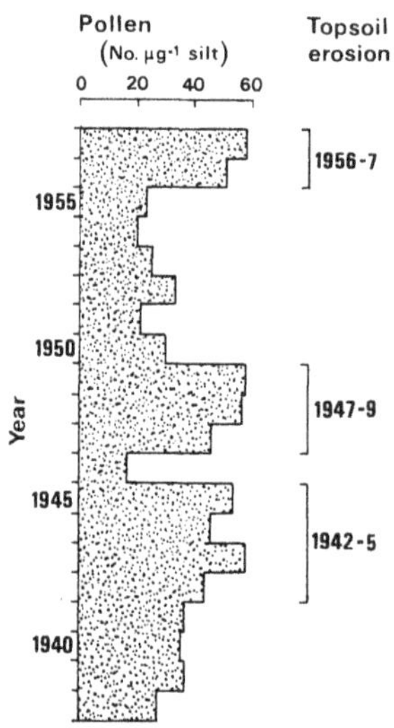

Fig. 7. Mean pollen concentration per unit weight of silt in the Burrinjuck sediments for the years 1938–57, with years of likely topsoil erosion marked.

Yass and Murrumbidgee catchments, but not the native forests.

Fig. 7 shows the number of pollen grains per unit weight of silt for the same years, 1938–57. The most severe erosion from topsoils occurred in the drought years 1942–45, in 1947–49, that were dry, and in the floods of 1956–57.

With this information it is possible to look for the recorded rainfall events responsible for mobilizing sediment in particular years and limited areas of the Burrinjuck catchment. Once these events are located in both time and space then the antecedent conditions can be determined and related to land use. Answers can be sought to questions such as why there was erosion from specific topsoils in the native forests in 1942–43 or from the pastures of the Yass catchment in 1945, why the 1956 floods affected pastures and pine forests more than native forests, or why there was little topsoil erosion in the years 1950–55.

Conclusion

Pollen in some sediments can provide more information than has usually been considered possible. In conjunction with other chronological methods, pollen can be used to distinguish annual layers, even in unlaminated sediment, so yearly influxes of all sediment components can be calculated. Where pollen has predominantly been transported by stream flow, then the source of sediments can be located in particular vegetation types. Pollen can also be used to distinguish between sediments eroded from topsoils and those from subsoils, such as gullies or stream banks. The results presented here indicate that pollen can be extremely valuable in sedimentological studies, providing information not otherwise obtainable about sedimentation rates and sources of sediments.

Acknowledgements

I am most grateful to all those who have been involved in the Burrinjuck project, but particularly, in this facet of the study, to Bob Wasson, Ian Willett and Ian Francis.

References

Australian Government, Department of Housing & Construction, 1981. Water quality monitoring, Murrumbidgee and Molonglo Rivers, Burrinjuck Reservoir, September 1977 to April 1981. Canberra, 77 pp.

Black, C. A. (Ed.-in-Chief), 1965. Methods of soil analysis, Part 1, American Society of Agronomy, Madison, 770 pp.

Bonny, A. P., 1976. Recruitment of pollen to the seston and sediment of some Lake District lakes. J. Ecol. 64: 859–887.

Bonny, A. P., 1978. The effect of pollen recruitment processes on pollen distribution over the sediment surface of a small lake in Cumbria. J. Ecol. 66: 385–416.

Brown, A. G., 1985. The potential use of pollen in the identification of suspended sediment sources. Earth Surf. Proc. Landforms 10: 27–32.

Clark, R. L. & R. J. Wasson, 1986. Reservoir sediments. In: P. de Dekker & W. D. Williams (eds), Limnology in Australia, CSIRO, Melbourne, Dr W. Junk, Dordrecht: 497–507.

Crowder, A. A. & D. G. Cuddy, 1973. Pollen in a small river basin: Wilton Creek, Ontario. In: H. J. B. Birks & R. G. West (eds), Quaternary Plant Ecology. Blackwell, Oxford: 61–77.

Dimbleby, G. W., 1957. Pollen analysis of terrestrial soils. New Phytol. 56: 12–28.

Matsushita, M., 1985. The behaviour of streamborne pollen in the Kako River, Hyogo Prefecture, Western Japan. Quat. Res. (Tokyo) 24: 57–61.

Peck, R. M., 1973. Pollen budget studies in a small Yorkshire catchment. In: H. J. B. Birks & R. G. West (eds), Quaternary Plant Ecology. Blackwell, Oxford: 43–60.

Prentice, I. C., 1985. Pollen representation, source area, and basin size: toward a unified theory of pollen analysis. Quat. Res. 23: 76–86.

Starling, R. N. & A. Crowder, 1981. Pollen in the Salmon River system, Ontario, Canada. Rev. Palaeobot. Palynol. 31: 311–334.

Accepted 18 March 1986.

Hydrobiologia 143: 71–77, (1986)

Sedimentary pigments as an index of the trophic status of Lake Mead

Michael S. Adams[1] & Richard T. Prentki[2]
[1]*Department of Botany and the Institute for Environmental Studies, University of Wisconsin, Madison, WI 53706, USA,* [2]*Formerly with the Lake Mead Limnological Research Center, University of Nevada, Las Vegas (Present address is Minerals Management Service, USDI, Alaska OCS Region, P.O. Box 101159, Anchorage, AK 99510-1159, USA)*

Keywords: Lake Mead, sedimentary pigments

Abstract

Sedimentary pigment data were analysed from 6 cores from Lake Mead, the reservoir resulting from the impoundment of the Colorado River by Boulder (now Hoover) Dam in 1935. Surficial sedimentary pigments correlated well with contemporaneously measured primary productivity of the water column at the 6 coring stations. For the coring station nearest the sewerage inflow from the City of Las Vegas, the historical record of pigments closely paralleled the increase in human population, sewerage loading to the reservoir, and the associated phosphorus loadings. For three of the cores, pigments were closely correlated with total nitrogen from the same levels in the cores. Pigment concentrations in cores from different parts of the reservoir corroborate findings of other studies which indicate major differences in productivity of the upper and lower basins of Lake Mead (the lower basin having higher primary productivity). The pigment data from the cores provide some indications of the effect of the damming of the Colorado River above the Grand Canyon in 1963 to form Lake Powell, which significantly reduced phosphorus loading downstream to Lake Mead, and which may have adversely affected the largemouth bass fishery.

Introduction

Lake Mead, impounded in 1935 by the damming of the Colorado River in Black Canyon on the Nevada-Arizona border (Prentki & Paulson, 1983) provides an ideal site to evaluate the results of sedimentary pigment degradation – product analyses as indicators of change in trophic status over time in a large reservoir. Part of Lake Mead has been impacted by the growth of the city of Las Vegas, which has discharged wastewater through Las Vegas Wash into Las Vegas Bay. The construction of the Glen Canyon Dam in 1963 significantly affected Lake Mead, in that the formation of Lake Powell behind the Glen Canyon Dam greatly altered characteristics of the Colorado River inflow to Lake Mead (Paulson & Baker, 1980; Paulson, 1981; Paulson *et al.*, 1980; Prentki & Paulson, 1983). For example, suspended sediment load decreased by 70-80%, and total phosphorus load by 96%. Currently, the Las Vegas Wash provides 60% of the phosphorus inputs to its reservoir (Paulson & Baker, 1980).

The Lake Mead Limnological Center of the University of Nevada, Las Vegas conducted an intensive study of the Lake Mead sediments, first coring the sediments in 1979 (Fig. 1). The purpose of this study was to evaluate the historical record in the sediments, in relation to the history of the reservoir. Specifically, one of the questions asked is whether an hypothesized decrease in upper basin primary productivity since the construction of Lake Powell has occurred. If so, it could be a cause of the decline of the largemouth bass fishery in Lake Mead (Prentki & Paulson, 1983).

One of us (Adams) arranged with the Center to obtain sediment samples in order to analyze the sedimentary pigment degradation products in a laboratory at the University of Wisconsin in Madison. The Center was involved in an intensive study

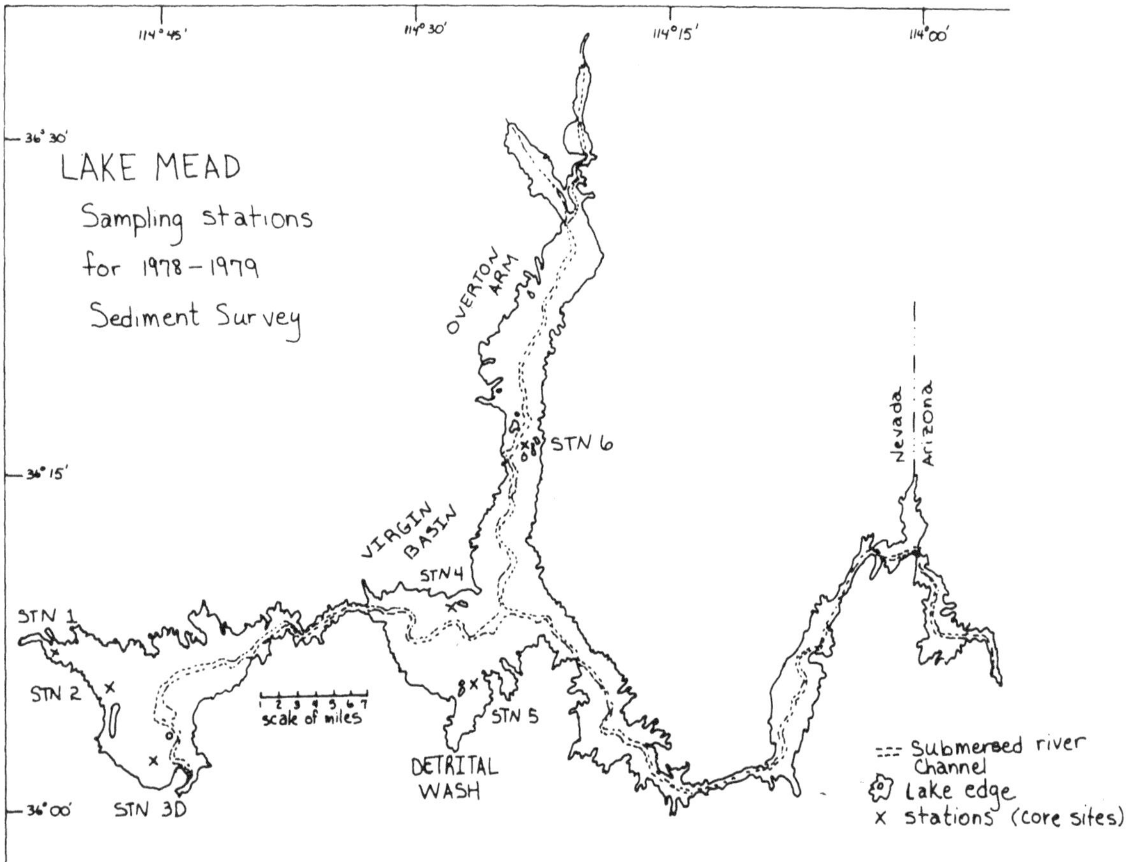

Fig. 1. Map of Lake Mead, showing sampling (core) stations.

of the chemical and physical characteristics of the sediment, and we desired to examine sedimentary pigment products as part of the overall project. We expected that the record of pigment concentrations in the sediment should provide an indication of the changes in lake primary productivity with respect to the effects from (1) the increase in population of Las Vegas, and (2) the construction of Lake Powell and its consequent role in decreasing suspended sediment and nutrient loads to the Lake Mead Reservoir. In addition, we analyzed pigment degradation products in surficial sediments from the upper portions of sediment cores to determine what corrrelation exists between contemporaneously measured phytoplankton productivity and the most recently deposited pigments. Several levels in the core have been dated by the Limnological Center, and the procedures followed are reported in Prentki, Paulson & Baker (1981).

Methods

Sediment samples which were collected from Lake Mead by coring were shipped to Madison, WI for spectrophotometric pigment analyses. There, sediments were analyzed after extraction with the two-solvent system (95% acetone for chlorophyll derivatives, and 30°–60° petroleum ether for carotenoids) used by Gorham & Sanger (1976). Wavelength absorption for chlorophyll derivatives was read at the 665 nm peak after scanning the acetone extract from 520 nm to 800 nm, and correcting for background absorbance. Carotenoids were read in the petroleum extract at the 450 nm peak after scanning from 380 nm to 550 nm. Details of the methodology were provided to us directly from Jon Sanger (personal communication). Calculation of pigment concentrations were based on sedimentary pigment degradation units (SPDU value): absorbance due to the respective pigment in 100 ml sample of the respective solvent, extracted from one gram of organic material.

We calculated the organic weights based on organic carbon analyses provided to us by the Lake Mead Limnological Center.

Results

I. Sedimentary pigments and contemporary phytoplankton primary productivity

Surficial sediment indices of derived plant pigments (both phaeophytin and carotenoids) correlated well with the in-lake measured phytoplankton productivity values published by Paulson, Baker and Deacon (1980) (Table 1). The published phytoplankton production values were determined from monthly measurements at each of the six sampling stations for one year during 1977–1978, which were summed to provide an annualized primary production value at each station. The correlations of primary productivity with phaeophytin and carotenoids were r = 0.996 and r = 0.972, respectively. Both correlations were significant at the level $P \leq 0.001$, indicating that the two measurements of sedimentary pigments were good indications of contemporary primary productivity.

Table 1. Surficial sediment indices of derived plant pigments (phaeophytin and carotenoids), and phytoplankton productivity for 6 stations in Lake Mead. Phytoplankton productivity values* are from Paulson, Baker and Deacon, 1980. Sedimentary pigments are expressed as SPDU values, which are 'sedimentary pigment degradation units' measured spectrophotometrically and expressed per unit of sediment organic weight (calculated from organic carbon).

Core number	Phaeophytin in sediment at sediment/ water interface	Carotenoids in sediment at sediment/ water interface	Phytoplankton productivity* $g\ C \cdot m^{-2} \cdot year^{-1}$
1	221	398	1257
2	152	200	817
3	79	98	475
4	40	48	159
5	40	68	152
6	38	73	189

Correlation of phaeophytin with phytoplankton productivity, R = 0.996, P = 0.001.
Correlation of carotenoids with phytoplankton productivity, R = 0.972, P = 0.001.

II. Sedimentary pigment indices in Las Vegas Bay and Boulder Basin

Comparison of the data presented on Clark County population growth, and associated sewerage discharge parameters (Fig. 2) with the data on sedimentary pigments for Station 1 in Las Vegas Bay (Fig. 3, 4) graphically demonstrate the rela-

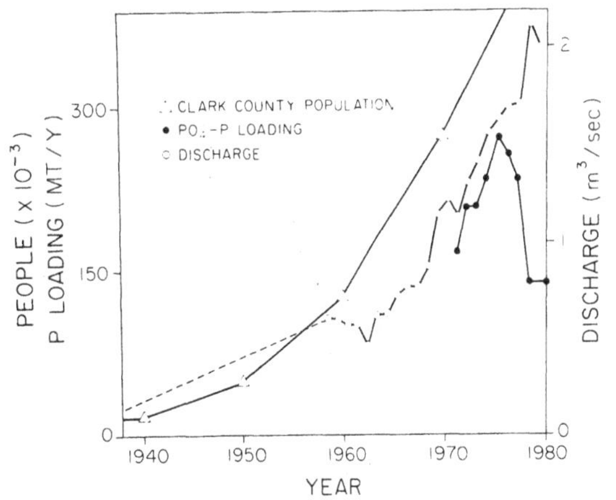

Fig. 2. Clark County population change, phosphorus loading to Lake Mead through Las Vegas Wash, and total discharge to Lake Mead through Las Vegas Wash.

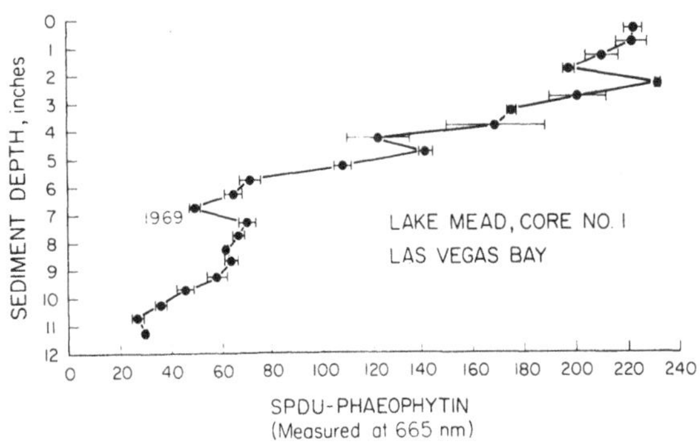

Fig. 3. Core 1, Las Vegas Bay (phaeophytin).

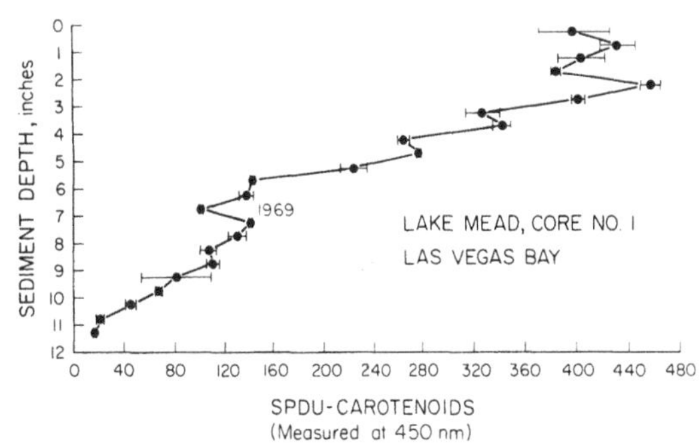

Fig. 4. Core 1, Las Vegas Bay (carotenoids).

tionship observable between the two sets of parameters. Quite clearly, there is an apparent increase in sedimentary pigments which parallels the growth in population of Clark County over a period of years, and the increase in sewerage-water discharge to the Bay. We did not consider it appropriate to attempt to calculate statistical relationships between the data sets, due to the small number of dates that are available with certainty for the different levels in the core. However, it is evident that increases in sedimentary pigment indices with time approximate the changes in population and loading statistics. Visual inspection of the curves provides considerable circumstantial evidence for the effect of the city's population growth on primary productivity of Las Vegas Bay. It should be noted that the decrease in phosphorus loading to the Bay through Las Vegas Wash subsequent to tertiary treatment in recent years may be related to the change in the trend line of sedimentary pigments.

Core number 1, from Las Vegas Bay, shows increasing pigment concentrations (Figs. 3, 4) in the lower 9 inches of sediment. The upper two inches of sediment recorded by the core show a stabilization with respect to changes in pigment concentration over time. About 4 inches of sediment were laid down between the level identified as 1969, and the beginning of the period of stabilization of pigment concentration. Thus, our data suggest an increase in primary productivity of the Las Vegas Bay portion of the lower Lake Mead Basin including the time since the construction of the Powell Reservoir. Our findings corroborate those of Prentki and Paulson (1983) whose estimates of primary production changes based on organic matter deposition indicate increase in primary productivity of the Lower Basin, but differ from the data of Paulson (1981) who presents evidence of decreasing chlorophyll concentrations in the inner Las Vegas Bay. A significant point to be made regarding the data from Core 1 from Las Vegas Bay is that this core has pigment concentrations remarkably higher than from cores from the Virgin Basin (the latter reflecting effects on productivity from upstream inputs to Lake Mead), and higher than for any of the other cores including nearly Boulder Basin (Figs. 5–8).

The Boulder Basin core data appear as intermediate between extremes of indications of very high productivity in the Las Vegas Bay area and very low productivity in the upper Lake Mead Ba-

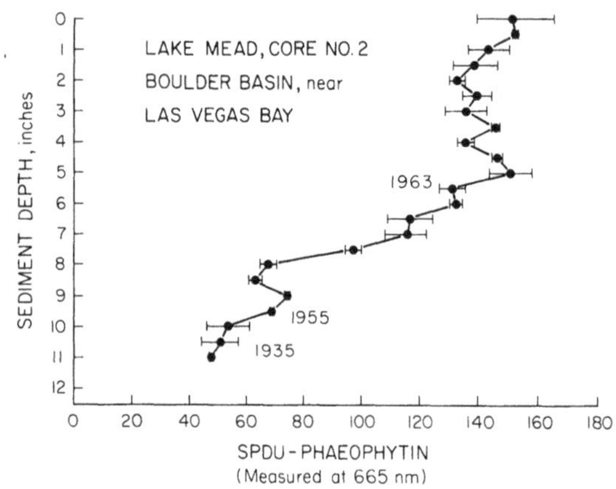

Fig. 5. Core 2, Bouler Basin near Las Vegas Bay (phaeophytin).

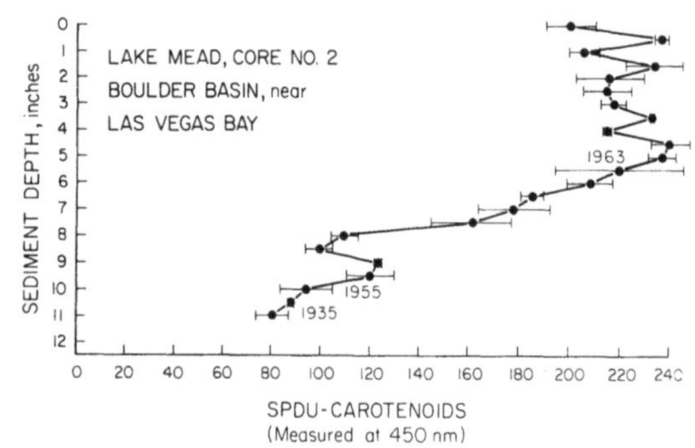

Fig. 6. Core 2, Boulder Basin near Las Vegas Bay (carotenoids).

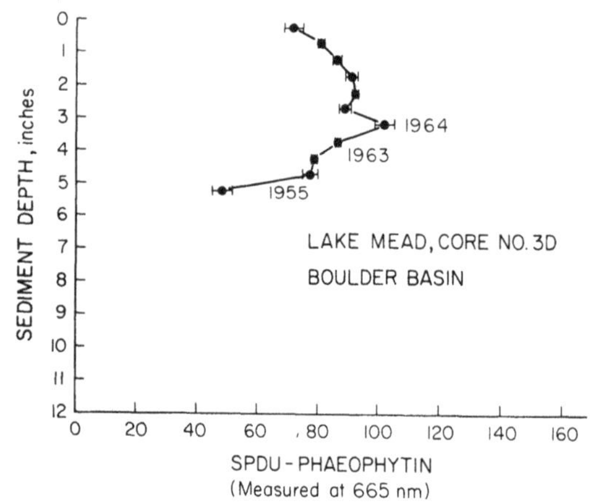

Fig. 7. Core 3D, Boulder Basin (phaeophytin).

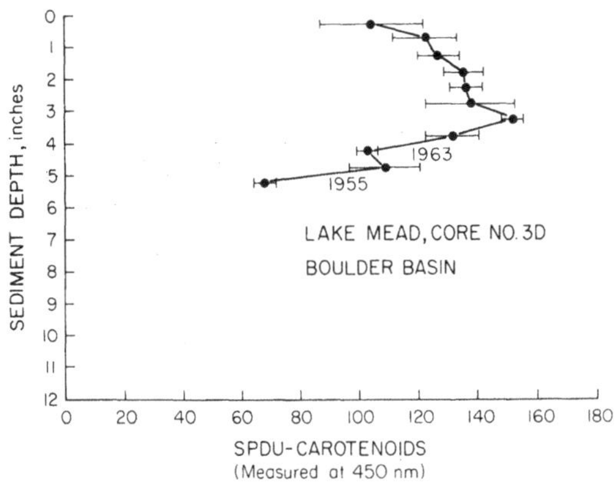

Fig. 8. Core 3D, Boulder Basin (carotenoids).

sin as evidenced by the data from the Virgin Basin Cores. The Boulder Basin core number 2 reflects stabilization in primary productivity occurring during the time period starting immediately subsequent to the construction of the Powell reservoir. Core 3D suggests declining primary productivity following the construction of the upstream reservoir.

III.

The most significant point to be made regarding the pigment data from the two cores from the Virgin Basin area (Figs. 9, 10, 11) is that the pigment concentrations were markedly lower than from most other sediments investigated, and they particularly contrasted with the data from Las Vegas Bay. Trends within the Virgin Basin cores seemed much less important than the overall low concentration of pigments, suggesting that primary production in the Virgin Basin area has been low throughout

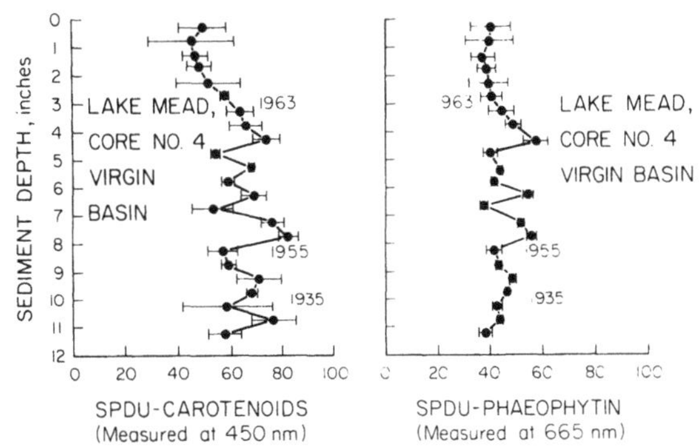

Fig. 9. Core 4, Virgin Basin (carotenoids and phaeophytin).

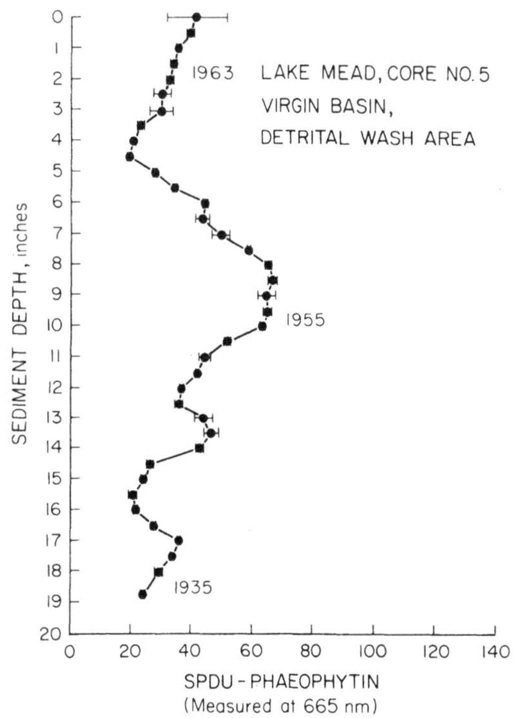

Fig. 10. Core 5, Virgin Basin detrital wash (phaeophytin).

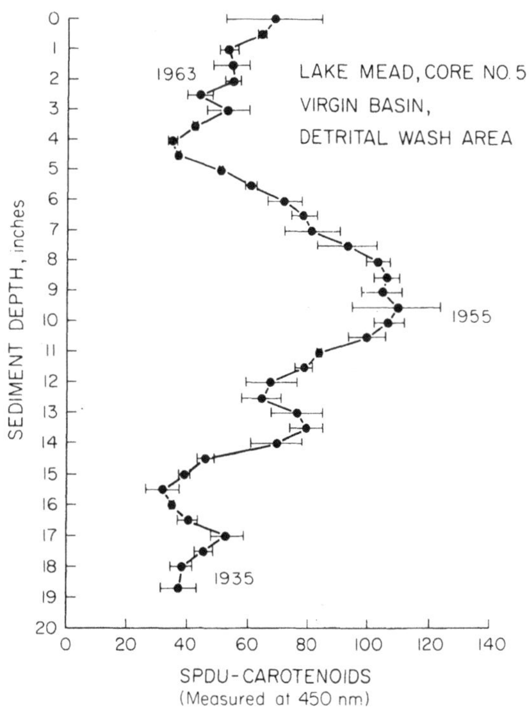

Fig. 11. Core 5, Virgin Basin detrital wash (carotenoids).

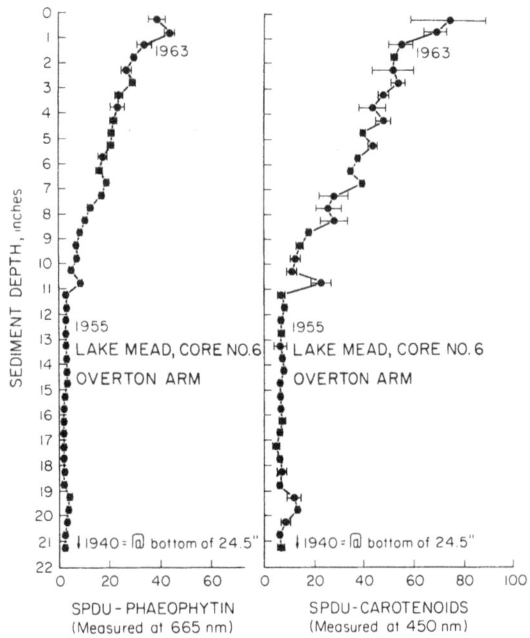

Fig. 12. Core 6, Overton Arm (phaeophytin and carotenoids).

much of the history of the reservoir. The alternative explanation would be that sedimentation of phytoplankton remains has been less in this part of the lake, or since much of the organic matter is allochthonous, the Virgin Basin cores may be low in pigment content due to extensive dilution of the material by the allochthonous inputs of organic matter.

The data from Overton Arm (Fig. 12) also suggest low primary production in that area of the reservoir, but the difference here is that while sedimentary pigments were extremely low in the older parts of the core, a steady increase has occurred since about the late 1950's.

IV. Relationship of pigments and nitrogen content of the cores

Visual inspection of the N content of the cores from the six stations, in relation to sedimentary pigment concentrations, suggested parallel changes in three of the six cores: 1, 2, and 6. It was appropriate to conduct a simple linear regression on these data, since the pigments were analyzed from subsamples of the same sample set for which nitrogen analyses were conducted at Las Vegas. The results of the linear regression (Table 2) showed that sedimentary N as % of dry wt. was correlated with phaeophytin and carotenoids (each calculated separately) at a level of significane $P \leq 0.001$.

Discussion

Prentki & Paulson (1983) stated that 'Las Vegas Wash inflow is confined to the Lower Basin of Lake Mead where historically it has elevated phytoplankton productivity.' They report that the high loading and high productivity have effected lowered N concentrations. Las Vegas Bay, and parts of Boulder Basin, have been N limited since 1972 (Paulson, 1981, as cited by Prentki & Paulson, 1983). The formation of Lake Powell has thus resulted in a situation in which the Upper Basin of Lake Mead has become more P limited and the Lower Basin more N limited (Prentki & Paulson, 1983). Decreases in reservoir-wide productivity presumably have resulted from the changes in nutrient loading (Paulson & Baker, 1980; Prentki & Paulson, 1983).

Our data on the sedimentary pigment degradation products analyzed from the core from Site #1 in Las Vegas Bay suggest that phytoplanktonic primary productivity in the Las Vegas Bay has experienced a steady increase since initial submergence, but that stabilization in primary production

Table 2. Correlation of sedimentary pigments with nitrogen in the sediment cores.

	Core 1 Las Vegas Bay	Core 2 Boulder Basin	Core 6 Overton Arm
Number of paired samples	23	23	36
Range of nitrogen (N as % of dry wt.)	0.053 – 0.199	0.018 – 0.190	0.066 – 0.132
Correlation with phaeophytin	0.94	0.94	0.60
Significance level	0.001	0.001	0.001
Correlation with carotenoids	0.93	0.93	0.53
Significance level	0.001	0.001	0.001

has occurred, beginning at approximately the time of the start of tertiary nutrient treatment of the effluent that reaches the Lake via Las Vegas Wash.

The high concentration of sedimentary pigments in the core from Las Vegas Bay (station 1) support the conclusions that can be reached from the organic carbon deposition rate studies conducted on the same core (Prentki, Paulson & Baker, 1981). Prentki *et al.* found organic carbon deposition of about 120 g $C \cdot m^{-2} \cdot y^{-1}$ averaged for the period beginning in 1969, and such a high level of deposition seems in concert with the results from the pigment studies.

The observed relationship supports the utility of sedimentary pigment analyses as indicators of trophic status for Lake Mead, because Prentki & Paulson (1983) demonstrated in their paper (their Fig. 6) a very good correlation between autochthonous organic carbon sedimentation and phytoplankton production.

The very highly significant correlation of contemporary primary productivity measured in six different parts of Lake Mead with surficial concentrations of sedimentary pigment degradation products provides direct evidence of the accuracy of pigment analysis as indicators of primary productivity in Lake Mead.

The very highly significant correlation of sedimentary pigment indices with sedimented nitrogen throughout the cores number 1, 2 and 6 (Figs. 3, 4, 5, 6, 12) would support a contention that Las Vegas Bay, part of Boulder Basin, and the central part of Overton Arm have been N-limited over most of the course of the history of Lake Mead.

Sedimentary pigment data from Core #2, in Boulder Basin, indicates stabilization in primary production in that area of Lake Mead beginning not long after the construction of the Powell Reservoir. It seems possible that this stabilization may be due, in part, to the tertiary treatment of sewerage effluent entering the Las Vegas Wash.

Sedimentary pigment data from core #3D, from the part of Boulder Basin near the Thalweg suggests decrease in primary production after construction of Powell Reservoir.

Sedimentary pigment data from core 4 and 5 from the Virgin Basin (near, and distant from the Thalweg, respectively) indicates probable dilution of pigments by sediment load prior to the construction of Powell Reservoir.

Sedimentary pigment data from Overton Arm suggests that primary productivity prior to about 1955 was extremely low, relative to total deposition. On the basis of the sedimentary pigment data, there appears to have been a steady but gradual increase in primary productivity from 1955 to 1963.

Sedimentary pigment data must be interpreted with caution, due to problems originating from differential rates of pigment degradation in the water column and sediment and from dilution, but we suggest that for the system studied (Lake Mead), the data provide information useful to our understanding of the probable history of change in trophic status. The data should be interpreted in conjunction with information derived from more traditional methods.

References

Gorham, E. & J. E. Sanger, 1976. Fossilized pigments as stratigraphic indicators of cultural eutrophication in Shagawa Lake, northeastern Minnesota. Geol. Soc. Am. Bull. 87: 1638–1642.

Lari, J. M. & J. I. Sanders, 1970. The 1963–64 Lake Mead Survey. U.S. Bureau of Reclamation Report No. REC-OCE-20-21. 169 pp.

Paulson, L. J., 1981. Nutrient management with hydroelecric dams on the Colorado River system. Lake Mead Limnological Res. Ctr. Tech. Rept. No. 8. Univ. Nev., Las Vegas. 39 pp.

Paulson, L. J. & J. R. Baker, 1980. Nutrient interactions among reservoirs on the Colorado River. Pages 1647–1656 in H. G. Stefan, ed. Symposium on surface water impoundments. ASCE June 2–5, 1980. Minneapolis, Minn.

Paulson, L. J., J. R. Baker & J. E. Deacon, 1980. The limnological status of Lake Mead and Lake Mohave under present and future powerplant operations of Hoover Dam. Lake Mead Limnological Res. Ctr. Tech. Rept. No. 1. Univ. Nev., Las Vegas. 229 pp.

Prentki, R. T. & L. J. Paulson, 1983. Historical patterns of phytoplankton productivity in Lake Mead. In: M. Adams & V. Lamarra, (eds.), *Aquatic Resources Management of the Colorado River Ecosystem*, Chapter 8, pp. 105–123.

Prentki, R. T., L. J. Paulson & J. R. Baker, 1981. Chemical and Biological Structure of Lake Mead Sediments. Final report to the U.S. Bureau of Water and Power Resources on Chemical and Biological Analysis of Lake Mead Sediments (Contract No. 14-06-300-2718). James F. Deacon, Principal Investigator.

Accepted 18 March 1986.

Hydrobiologia 143: 79–86, (1986)

Biogenic silica and phosphorus accumulation in sediments as indices of eutrophication in the Laurentian Great Lakes

C. L. Schelske, D. J. Conley, E. F. Stoermer, T. L. Newberry & C. D. Campbell
Great Lakes Research Division, University of Michigan, Ann Arbor, Michigan 48109, USA

Keywords: paleolimnology, diatoms, nutrients, phytoplankton production

Abstract

Biogenic silica (BSi), total phosphorus (TP), and biologically available phosphorus (AVP) were measured in short cores from Lake Michigan, Lake Erie, and Lake Ontario. Peaks in BSi concentration and peaks in BSi:TP or BSi:AVP ratios provided stratigraphic signals of water column silica (Si) depletion as a response of increased diatom production to P enrichment and decreased diatom production resulting from silica depletion. By contrast the stratigraphic record of P accumulation provided very weak signals of the historical nutrient enrichment in the water column. These results indicate that system P recycling has a higher rate constant than Si recycling and, as a consequence, that relatively small levels of P enrichment can increase diatom production and sedimentation eventually causing Si depletion and Si-limited diatom production in the water mass.

Introduction

The Laurentian Great Lakes provide an ideal system in which to assess the utility of using biogenic silica (BSi) accumulation in sediments as an index of eutrophication. Diatoms, historically, have been important components of the phytoplankton assemblage and, excluding the influence of silica-limited waters, are important components in the present day flora (Stoermer, 1984). Phytoplankton growth in these systems is controlled by phosphorus (P) supplies and P enrichment increases diatom production if silica (Si) supplies are not limiting (Schelske *et al.*, 1986).

Little information is available on historical changes in the nutrient chemistry of Great Lakes waters. Historical data on total phosphorus (TP) concentrations are not available or are very limited (Beeton, 1969), but historical TP concentrations have been estimated from computer simulations (Chapra, 1977). The only useful historical data on Si concentrations are those from Lake Michigan (Schelske, 1985) where the onset of epilimnetic Si depletion occurred as recently as the late 1960s (Schelske & Stoermer, 1971). The BSi stratigraphy of cores provides evidence that Si depletion in Lake Erie and Lake Ontario waters occurred by the late 1800s (Schelske *et al.*, 1983; Stoermer *et al.*, 1985a) before there are useful records of water chemistry. Here we will show: (1) that BSi accumulation in sediments provides a stronger signal of initial effects of phosphorus enrichment than accumulation of either TP or biologically available phosphorus (AVP) and (2) that ratios of either BSi:TP or BSi:AVP in the sediment stratigraphy can be used to infer the onset of Si limitation for diatom production and to demonstrate that BSi accumulation provides a better signal of P enrichment than P accumulation.

Methods

Sediment cores were collected with a Benthos gravity corer (BEN) from stations in Lake Michigan, Lake Ontario, and Lake Erie. Cores were either collected by us or splits of Lake Erie cores were obtained from cores collected by J. A. Robbins as part of the HI-SED project. Cores were sectioned in the field, freeze dried, and analyzed for nutrients. ^{210}Pb data for Lake Erie cores were obtained from J. A. Robbins.

Samples for BSi (ca. 30 mg) were weighed and leached with Na_2CO_3 at 85 °C. After 2, 3, and 4 hr of leaching, samples were withdrawn for analysis of soluble silica on a Technicon Auto Analyzer with the hetropoly blue method. A linear regression of these time course data extrapolated to zero time was used to correct for mineral interference (Krausse *et al.,* 1983).

Samples for AVP (ca. 50 mg) were leached for 17 hr in 45 ml NaOH at 25 °C in a shaker bath. Samples were neutralized with 0.9 N HCl, centrifuged, and analyzed for soluble reactive phosphorus. Samples for TP (ca. 50 mg) were digested in 20 ml 5.0% H_2SO_4 and 10 ml 5.0% K_2SO_8 in an autoclave for 20 min. Samples were neutralized with 1.8 N NaOH, centrifuged, and analyzed for soluble reactive phosphorus. Soluble reactive phosphorus was measured with an Auto Analyzer.

Atomic ratios of BSi:TP and BSi:AVP were calculated to determine effects of Si limitation on diatom production. Ratios were utilized because the results are independent of sediment dating.

Results

BSi accumulation and Si depletion

The stratigraphy of BSi in cores from Grand Traverse Bay, Lake Michigan, conforms with the hypothetical model proposed earlier (Schelske *et al.,* 1983) which we interpret as evidence for epilimnetic Si depletion. The model predicts an increase in BSi accumulation in the sediments as the result of P enrichment until the reservoir of Si in the water mass is utilized and Si supplies become limiting for diatom production. In these cores, the largest increase in BSi concentrations occurred after 1940 and a decrease in BSi concentration signalling epilimnetic Si depletion occurred about 1970 (Fig. 1). The timing of the decrease in water column Si concentrations and the associated increase in BSi accumulation in the sediments corresponds to a period of increased P loading beginning in the late 1940s (Fig. 2). In addition, the Si depletion signal

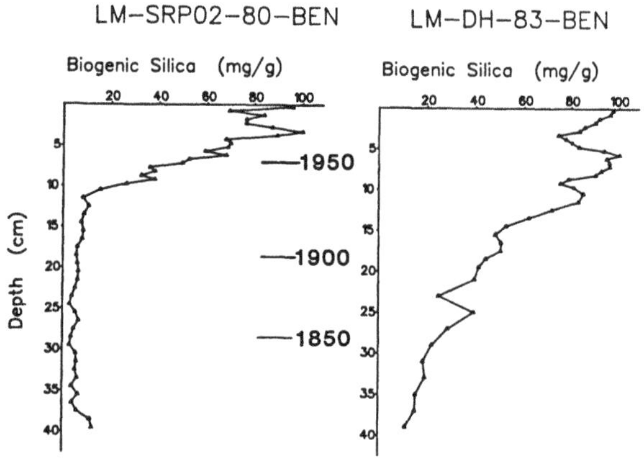

Fig. 1. Biogenic silica concentration vs. depth in cores from Grand Traverse Bay, Lake Michigan. The BSi peak at 6 cm in DH-83 has been dated at 1970 by stratigraphic correlation with core SRP02 (see Schelske *et al.,* 1983).

in the sediments occurred about 1970 when it is known from experimental evidence that Si became limiting for diatom production (Schelske & Stoermer, 1971).

Previous analyses of the BSi stratigraphy of single cores from Lake Ontario and Lake Erie

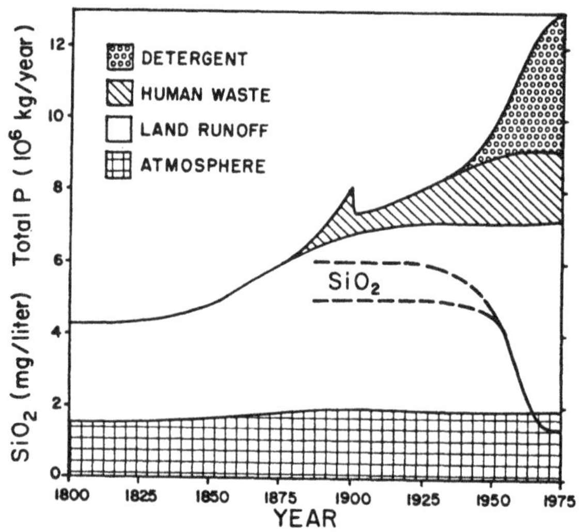

Fig. 2. Computer simulated total phosphorus loads to Lake Michigan from Chapra (1977). The curves for silica concentration represent winter maximum concentrations. The higher winter maximum (6.0 mg l^{-1}) is based on assumptions presented by Schelske & Stoermer (1971) and the lower value is based on data reviewed by Schelske (1985). The onset of summer epilimnetic silica limitation coincides with the reduction in winter maximum concentration to 1.4 mg l^{-1}.

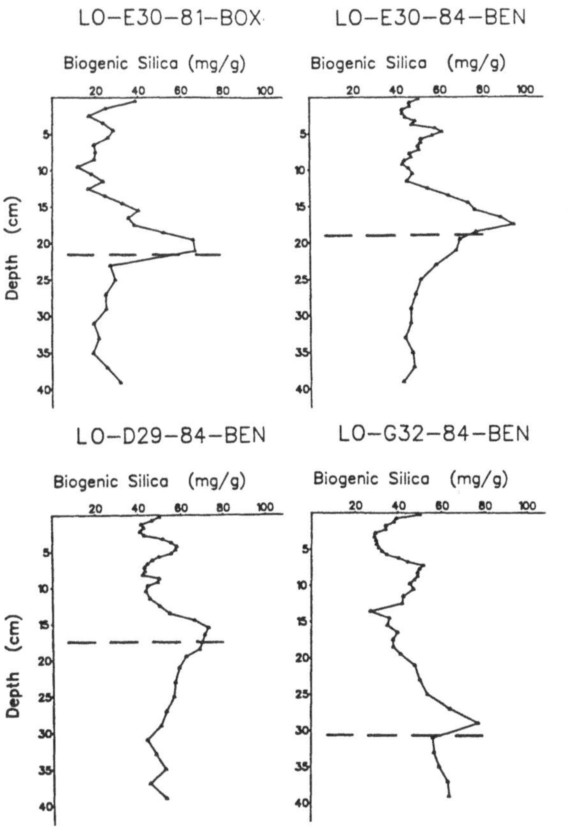

Fig. 3. Biogenic silica concentration vs. depth in cores from Lake Ontario. The post-settlement BSi peak shown by the dashed line was correlated at 1850 with the BSi peak in two ^{210}Pb dated cores (Schelske *et al.*, in prep.).

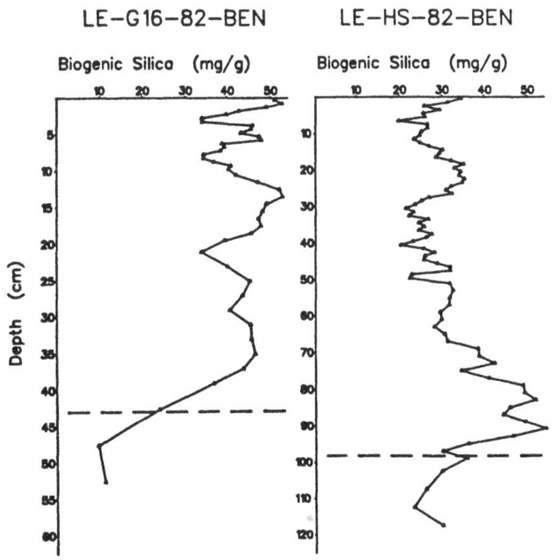

Fig. 4. Biogenic silica concentration vs. depth in cores from Lake Erie. The post-settlement rise in biogenic silica shown by the dashed line was dated at 1880 with ^{210}Pb. Mass sedimentation rates calculated from ^{210}Pb data for cores G-16-82 and Hs-82 were 0.108 and 0.474 mg cm^{-2} yr^{-1}, respectively.

(Schelske *et al.*, 1983) provide evidence that Si depletion occurred about 1850 and 1880, respectively, during the initial stages of European settlement. These pulses of BSi were found in other cores from these lakes (Figs. 3 and 4) and occurred before simulated TP concentrations increased to 10 μg P l^{-1} (Chapra, 1977). We believe that these early signals represent epilimnetic Si depletion. It is also apparent from closer examination of these cores that BSi concentrations increased and then decreased after 1950. We believe this secondary peak in BSi provides the stratigraphic record of events that led to severe Si depletion in Lake Ontario and the eastern and central basins of Lake Erie. Severe Si depletion reduces Si to 0.4 mg SiO_2 l^{-1} or less before the lake is stratified thermally (Stoermer *et al.*, 1985a; Schelske *et al.*, 1986).

P accumulation

Much greater increases in P enrichment have occurred historically in the lower lakes than in the upper Great Lakes. Modelled TP concentrations in the lower lakes (excluding the western basin of Lake Erie) increased 5- to 6-fold, whereas in the upper lakes concentrations increased only 2-fold at most (Chapra, 1977). Larger increases in P accumulation therefore would be expected in the sediments of Lake Erie and Lake Ontario than in Lake Michigan.

Increases in P accumulation in Lake Ontario and Lake Erie, however, were smaller than the 5- to 6-fold increase in modelled TP concentrations. In Lake Erie, TP and AVP concentrations increased <2-fold in core G16 from the central basin (Fig. 5). In two cores from Lake Ontario, TP concentrations increased <2-fold for 10-yr intervals between 1700 and the present while AVP concentrations increased 3-fold at most (Schelske *et al.*, in prep). Kemp *et al.* (1972) also reported that TP concentrations in cores increased about 2-fold in Lake Ontario and 2- to 3-fold in Lake Erie.

Increases in P accumulation in Lake Michigan

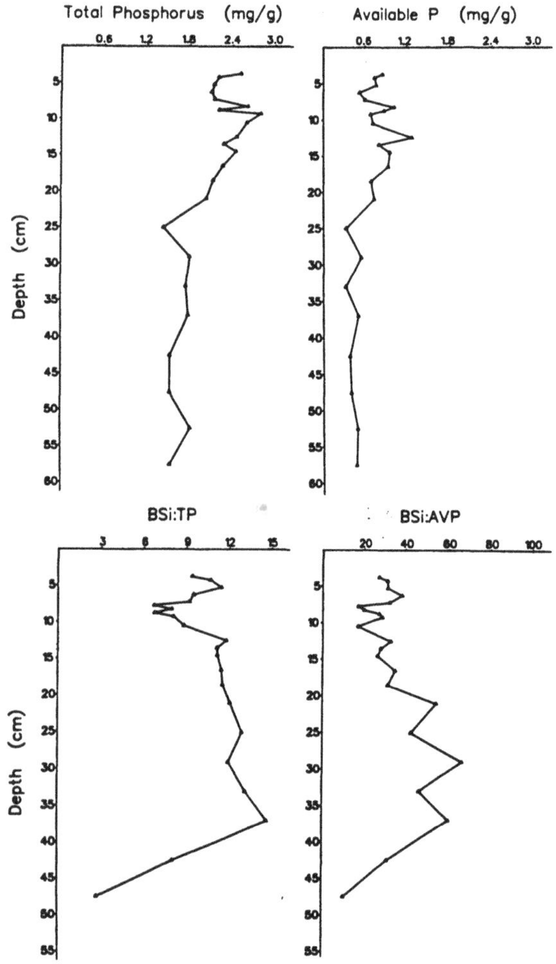

Fig. 5. Total phosphorus and biologically available phosphorus concentrations and ratios of these forms of phosphorus to biogenic silica vs. depth in core G16-82 from the central basin of Lake Erie. (see Fig. 4 also.)

cores were not apparent from concentration data except for core C37 from Green Bay (Fig. 6). But even in this core the maximum concentration was less than twice as large as concentrations found below 10 cm and, without the spikes in concentration of AVP and TP centered at 5 to 6 cm, the trend for increased accumulation would be much smaller. By contrast, no obvious trend in TP or AVP concentration was found in core I26 or K8. However, TP and AVP concentrations above 10 cm were slightly greater than those below 10 cm in core K8 indicating that P enrichment had occurred in this core. Small increases in TP and AVP concentrations found in recent sediments of cores C37 and K34 might be expected because the largest change in simulated TP loadings occurred after 1940 (Fig. 2). Greater increases in P accumulation normalized to annual fluxes would be obtained if an accurate estimate of pre-settlement sedimentation rates could be obtained for Lake Michigan.

Large spikes in P concentration common in Lake Michigan cores appeared to be the result of changes in accumulation of AVP because the spikes were found for AVP and also for TP which includes the AVP fraction. These spikes indicate that AVP storage in the sediments may be associated with episodic events. It has been suggested that sediment resuspension associated with severe storms affects the timing of permanent burial of sedimented materials (Robbins & Edgington, 1975) and that decreases in water column TP concentrations result from winters with greater than average ice cover (Rockwell *et al.,* 1980). Either of these processes could affect P storage and produce a large AVP spike in the sedimentary record. This result agrees with data from Williams *et al.* (1976) who reported that concentrations of apatite P were fairly constant in the stratigraphy of Lake Erie cores and that recent increases in concentration of TP resulted from inputs of nonapatite inorganic P.

In inferring accumulation based on nutrient concentration it has been assumed that sedimentation rates were constant. Although this assumption may be valid in obtaining an average post-settlement sedimentation rate over the past 100 to 130 yr for ^{210}Pb dated cores, it may not be valid in comparing pre- and post-settlement sedimentation rates. If post-settlement rates in the lower Great Lakes were three times larger than pre-settlement rates and accumulation rates for P are adjusted accordingly, resulting post-settlement increases in TP or AVP accumulation in Lake Erie would be on the order of the 6-fold increase expected from computer simulations by Chapra (1977). Annual BSi fluxes like P fluxes for all lakes would undoubtedly be greater than the range in concentrations if these fluxes were calculated over a period which included pre-settlement stratigraphy (see Schelske *et al.,* 1985).

BSi:TP ratios

It is apparent that the Si depletion signal based on patterns of BSi accumulation in cores from

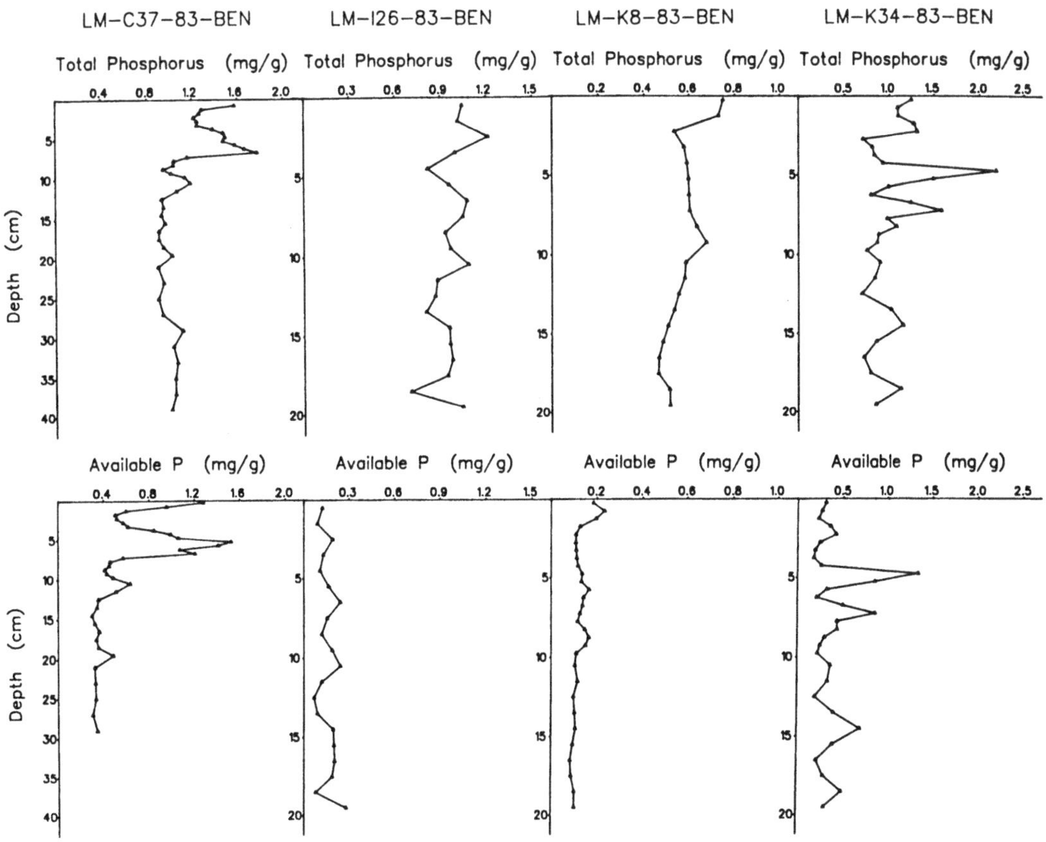

Fig. 6. Total phosphorus and biologically available phosphorus concentrations vs. depth in Lake Michigan cores. The mass sedimentation rate for core K34 was obtained from ^{210}Pb measurements made in our laboratory.

Grand Traverse Bay (Fig. 1) was more pronounced than in cores from open Lake Michigan (Fig. 7). Atomic ratios of BSi:TP and BSi:AVP in the stratigraphy of open Lake Michigan cores, however, provide evidence for Si depletion in recent sediments. Ratios which are independent of sediment dating show that the relative increase in BSi accumulation was greater than that for TP or AVP. In general maximum ratios were found in recent sediments or at depths dated between 1967 and 1970 in K34, which coincides with the approximate time for Si depletion (Fig. 2). This pattern agrees with our hypothetical model in that ratios should decrease after Si became limiting for diatom production and P utilization by non-siliceous phytoplankton increased. However, the earlier increase in both ratios as P enrichment increased is not explained by this simple model.

Table 1. Ratios of BSi:AVP and BSi:TP in sediment cores from the Great Lakes.

	BSi:AVP	BSi:TP
Green Bay[a]		
Recent	100–200	60–90
Maximum	350–400	150
Lake Michigan[a]		
Maximum	350–400	60–100
Lake Ontario[b]		
Recent	10–20	5–10
Maximum	50–100	20–30
Lake Erie[a]		
Recent	30–40	10–15
Maximum	50–60	10–15

[a] This paper.
[b] Schelske et al., in prep.

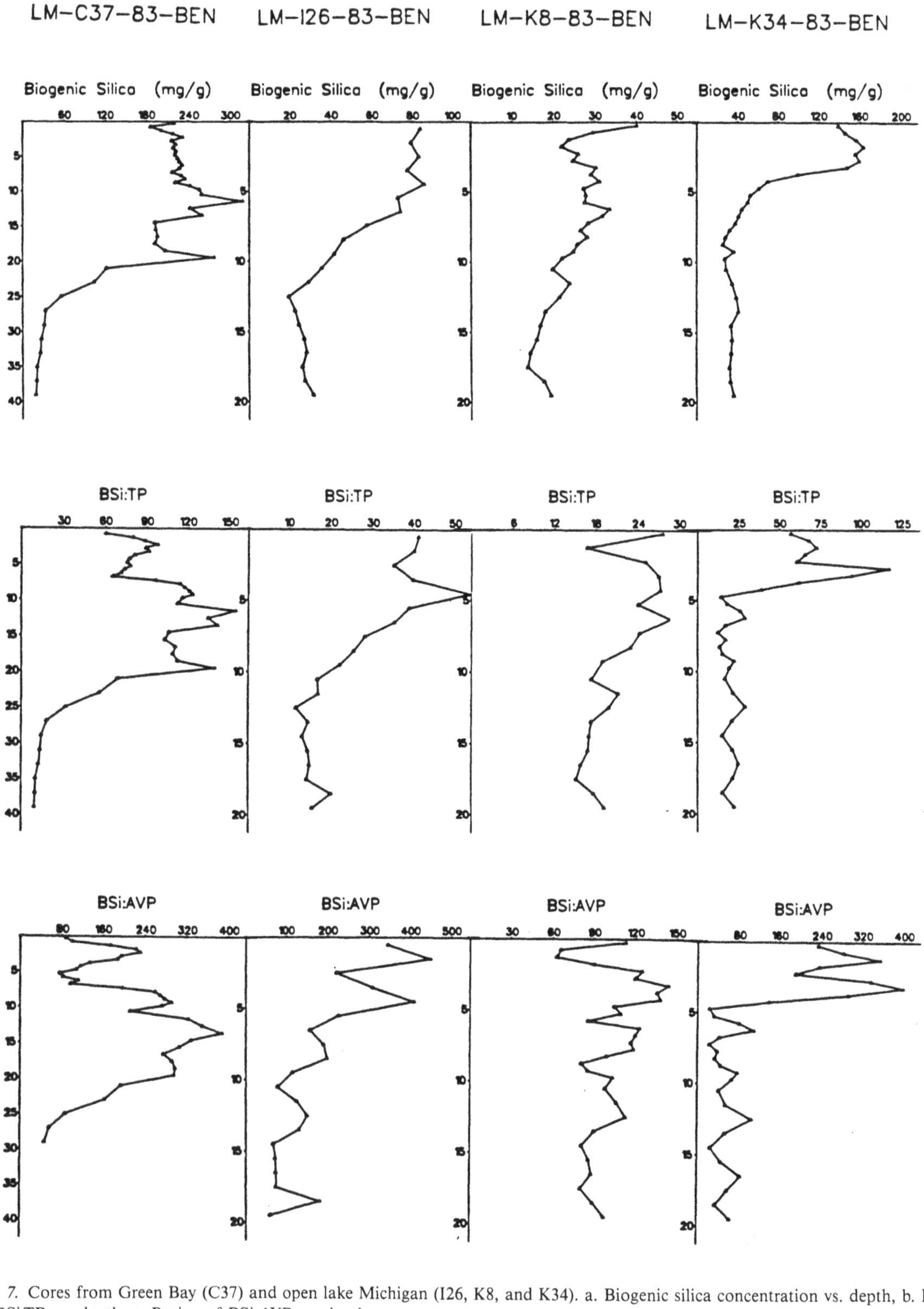

Fig. 7. Cores from Green Bay (C37) and open lake Michigan (I26, K8, and K34). a. Biogenic silica concentration vs. depth, b. Ratios of BSi:TP vs. depth, c. Ratios of BSi:AVP vs. depth.

In the lower lakes, ratios also changed in response to P enrichment. In core G16 from central Lake Erie, BSi:TP ratios increased soon after settlement (1850) and BSi:AVP ratios apparently peaked somewhat later (Fig. 5). Both ratios decreased in post-settlement sediments to minimum levels between 7 and 10 cm which according to ^{210}Pb dates occurred between 1965 and 1972. The general decrease in the BSi:AVP ratio after post-settlement indicates that smaller fractions of AVP supplies were being utilized for diatom production and that diatom production was Si limited. In Lake Ontario cores, ratios of BSi:TP and BSi:AVP also increased before the period of presumed epilimnetic Si depletion in the mid 1800s (Schelske *et al.*, in prep.). In these cores, the increase in ratios was followed by a period in which ratios decreased rapidly so peaks in ratios corresponding in time to peaks in BSi concentration were produced. We believe that the earliest peak was associated with epilimnetic Si depletion and that either a peak in BSi concentration or in BSi:TP and BSi:AVP ratios can be used as evidence for epilimnetic Si depletion in Lake Michigan, Lake Ontario, and possibly in Lake Erie.

Discussion

The peak in BSi accumulation in the sediments which has been termed the Si depletion signal is produced because diatom production can be increased in response to P enrichment if Si supplies are available and then is decreased when the reservoir of Si in the water mass is depleted and Si becomes limiting for diatom production (Schelske *et al.*, 1983). However, patterns of BSi accumulation will vary among the Great Lakes because of differences in morphometry, hydraulic residence time, and nutrient supplies. It follows that a deep lake would have a larger reservoir of Si than a shallower lake and that renewal of Si supplies from external sources would be more important in a shallow lake or shallow embayment than in deeper waters. Lake Erie is the shallowest of the Great Lakes and has the shortest hydraulic residence time. Phosphorus concentrations in Lake Erie, therefore, increased rapidly in response to nutrient loading (Chapra, 1977) and the resulting increases in diatom production depleted the Si reservoir in the water column even though Si supplies were renewed continually by inputs from Lake Huron which has relatively large Si concentrations (Schelske *et al.*, 1983). As a consequence, post-settlement BSi production was maintained at high levels in the central basin of Lake Erie for about 100 years (Fig. 4) because a major fraction of the Si input could be utilized in diatom production, especially when P enrichment was adequate to produce severe Si limitation. Likewise, the period of increased BSi accumulation in the highly enriched waters of Green Bay was longer than in the less enriched waters of open Lake Michigan (Fig. 7) because Si supplies in Green Bay could be renewed from tributary inputs and by exchange of water with Lake Michigan.

Although changes in BSi accumulation in the sediments are caused by P enrichment in the water column, changes in the accumulation of P in the sediment stratigraphy were relatively small compared to those for BSi. The largest changes were found in Lake Ontario where TP or AVP concentrations only increased 2- to 3-fold over the length of the core or a relative change comparable to the increase in BSi accumulation associated with epilimnetic Si depletion in the mid 1800s (Schelske *et al.*, in prep.). The important difference in accumulation of the two nutrients is the greatest increase in BSi concentration occurred at a time when the increase in TP or AVP concentrations in the sediment stratigraphy was much less. Likewise, a very small increase in TP or AVP concentration or a larger increase in BSi:TP and BSi:AVP ratios was associated with the increase in BSi concentration in Lake Michigan cores.

That BSi accumulation in the sediments increased more relative to TP or AVP than either form of P was evident from an examination of BSi:TP or BSi:AVP ratios. In Lake Ontario cores, a peak in BSi:TP and BSi:AVP ratios was found at the same depth as the peak in BSi accumulation associated with epilimnetic Si depletion (Schelske *et al.*, in prep.). A peak or increase in these ratios in Lake Michigan cores also marked the time we believe that epilimnetic Si depletion occurred. The peak in ratios and increase in ratios at the onset of Si depletion provide evidence that BSi accumulation is a better index of eutrophication than P accumulation in sediments. These results are important because they demonstrate that recycling of P which is primarily a biologically mediated process has a higher rate constant than the recycling of Si which is controlled by physical and chemical processes (Schelske *et al.*, 1986) and that BSi accumulation can affect the geochemistry of Si even if 90% or more of the BSi production is recycled annually (Schelske, 1985). These results also suggest that the efficiency of P utilization for diatom production increases with nutrient enrichment during the period in which water column silica reserves are being depleted. Shifts in diatom species composition and the provenance of diatom production or other physical, chemical, or biological factors may promote greater system efficiency in the utilization of available supplies of P for BSi production and sedimentation.

In summary, the record of BSi accumulation in the sediments of the Great Lakes provides a sensitive and general index that can be used to assess the historical impacts of P enrichment. BSi accumulation provides a more sensitive index of small levels of P enrichment than P accumulation in the sediments. It is sensitive enough to record effects of nutrient enrichment in Lake Superior where TP concentrations have not exceeded 4–5 μg TP l^{-1} (Stoermer *et al.*, 1985b). It provides a signal which can be used to identify epilimnetic Si depletion in Lake Michigan, Lake Erie, and Lake Ontario that resulted from P enrichment to levels <10 μg TP l^{-1}. It also provides a secondary signal which we have associated with severe Si depletion characterized by water column Si concentrations no greater than 0.4 mg SiO_2 l^{-1} during most of the year in Lake Erie and Lake Ontario. Severe Si depletion probably results from P enrichment to 15 or 20 μg TP l^{-1}. Identification of these stages in the trophic history of the Great Lakes has added significance because the historical chemical changes have corresponding analogues in the siliceous microfossil stratigraphy (Stoermer *et al.*, 1985a; Stoermer *et al.*, 1985b).

Acknowledgments

We thank John Robbins for providing splits of samples from cores taken as part of the Hi-SED project and for providing ^{210}Pb data and Wayne Gardner for supplying phosphorus data for Lake Ontario cores. We also acknowledge grants from the National Science Foundation (OCE-8216588), Environmental Protection Agency (R810396-01), and the Michigan Sea Grant College Program (NA80-AA-D-0072) which were used to develop techniques for assay of bioavailable phosphorus. Contribution No. 458, Great Lakes Research Division, University of Michigan.

References

Beeton, A. M., 1969. Changes in the environment and biota of the Great Lakes, pp. 150–187. In Eutrophication: causes, consequences, correctives. Nat. Acad. Sci., Washington, D.C.

Chapra, S. C., 1977. Total phosphorus model for the Great Lakes. J. Envir. eng. Div. 103: 147–161.

Kemp, A. L. W., C. B. J. Gray & A. Mudrochova, 1972. Changes in C, N, P, and S in the last 140 years in three cores from Lakes Ontario, Erie, and Huron. In: H. E. Allen & J. R. Kramer (eds), Nutrients in Natural Waters, pp. 251–279.

Krausse, G. L., C. L. Schelske & C. O. Davis, 1983. Comparison of three wet-alkaline methods of digestion of biogenic silica in water. Freshwat. Biol. 13: 73–81.

Robbins, J. A. & D. N. Edgington, 1975. Determination of recent sedimentation rates in Lake Michigan using ^{210}Pb and ^{137}Cs. Geochim. Cosmochim. Acta 39: 285–304.

Rockwell, D. C., D. S. Devault, M. F. Palmer, C. V. Marion & R. J. Bowden, 1980. Lake Michigan intensive survey 1976–1977. U.S. Environmental Protection Agency Report No. EPA-905/4-80-003-A.

Schelske, C. L., 1985. Biogeochemical silica mass balances in Lake Michigan and Lake Superior. Biogeochemistry. 1: 197–218.

Schelske, C. L. & E. F. Stoermer, 1971. Eutrophication, silica, and predicted changes in algal quality in Lake Michigan. Science 173: 423–424.

Schelske, C. L., E. F. Stoermer, D. J. Conley, J. A. Robbins & R. M. Glover, 1983. Early eutrophication in the lower Great Lakes: new evidence from biogenic silica in sediments. Science 222: 320–322.

Schelske, C. L., D. J. Conley & W. F. Warwick, 1985. Historical relationships between phosphorus loading and biogenic silica accumulation in Bay of Quinte sediments. Can. J. Fish. aquat. Sci. 42: 1401–1409.

Schelske, C. L., E. F. Stoermer, G. L. Fahnenstiel & M. Haibach, 1986. Phosphorus enrichment, silica utilization, and biogeochemical depletion in the Great Lakes. Can. J. Fish. aquat. Sci. 43: 407–415.

Schelske, C. L., D. J. Conley, W. S. Gardner & R. A. Bourbonniere, in prep. Biogenic silica, phosphorus, and carbon in Lake Ontario sediments.

Stoermer, E. F., 1984. Research on Great Lakes algal communities: Problems from the past, lessons for the future. J. Great Lakes Res. 10: 143–155.

Stoermer, E. F., J. A. Wolin, C. L. Schelske & D. J. Conley, 1985a. An assessment of changes during the recent history of Lake Ontario based on siliceous microfossils preserved in the sediments. J. Phycol. 21: 257–276.

Stoermer, E. F., J. P. Kociolek, C. L. Schelske & D. J. Conley, 1985b. Siliceous microfossil succession in the recent history of Lake Superior. Proc. Acad. Nat. Sci., Philadelphia 137: 106–118.

Williams, J. D. H., T. P. Murphy & T. Mayer, 1976. Rates of accumulation of phosphorus forms in Lake Erie. J. Fish. Res. Bd Can. 33: 430–439.

Accepted 18 March 1986.

Hydrobiologia 143: 87–91, (1986)

Mallomonadacean (Chrysophyceae) Scales: Early biotic paleoindicators of lake acidification

Helga Hartmann & Christian Steinberg
Bayerisches Landesamt für Wasserwirtschaft, Lazarettstr. 67, D-8000 Munich 19, FRG

Keywords: sediment, *Mallomonas, Synura,* scales, acidification, paleoindication

Abstract

17 subfossil mallomonadacean species have been detected in the acidified Kleiner Arbersee. The most abundant one is *M. crassisquama.* In the top most layers, it is replaced by acid tolerant *Synura* spp. and several acidophilic and acidobiontic *Mallomonas* spp. Mallomonadaceae respond to acidification (and alkalinization) of lakes prior to benthic diatoms. Mallomonadaceae live euplanktically and mainly in spring, when most of the annual acid loading enters the lakes.

Introduction

The siliceous scales of mallomonadacean species are well preserved (cf. Battarbee *et al.,* 1980; Smol, 1980) in lake sediments. The identification even on the species level is possible. Thus, the examination of subfossil mallomonadacean assemblages in lake sediments allows reconstruction of paleolimnological lake conditions, eutrophication (Kristiansen, 1981; Smol *et al.,* 1983) as well as acidification (Smol, 1980; Smol *et al.,* 1984a, b).

The aim of this study is to compare the paleoindication of mallomonadaceans with that of diatoms, as diatom-inferred pH is as well established tool in reconstructing former pH-conditions in lakes (for instances, Renberg & Hellberg, 1982; Battarbee, 1984; and for central Europe, Arzet & Steinberg, 1984).

Study site and methods

The lake under study is Kleiner Arbersee, a small (9.4 ha surface, 9.7 m max. depth and 2.7 m mean depth) cirque lake in the Bavarian Wood located at 918.4 m altitude. The surrounding bedrock consists of gneiss, covered with spodosol soils. The catchment vegetation consists of a natural mountain forest, mainly of spruce (*Picea abies*) and beech (*Fagus sylvaticus*) with small moieties of fir (*Abies alba*) and mountain maple (*Acer pseudoplatanus*), the coniferous trees show severe symptoms of the central European forest decay. Present water chemistry is given by Arzet *et al.* (1986).

Sediment cores were taken by scuba divers in 1982. The cores were immediately sliced into 1 cm sections. Prior to preparation for diatom analysis as described by Arzet *et al.,* (1986), the subsamples were stored frozen. For SEM-investigation the cleaned and washed samples were dried on glass cover slips and then affixed onto SEM stubs. These were coated with gold using a sputter coater (Polaron E 5000) and the material was examined in a Hitachi S 450 operating at 20 KV. In each sediment subsample at least 200 scales were identified and counted.

Results

In the sediments of Kleiner Arbersee, 17 mallomonadacean species could be detected: *Mallomonas actinomola* Asmund & Takahashi var. *maramuresensis* Peterfi & Momeu, *M. allantoides*

Harris, *M. annulata* (Harris & Bradley) Harris, *M. calceolus* Bradley, *M. canina* Kristiansen, *M. caudata* Iwanoff em. Krieger, *M. crassisquama* (Asmund) Fott, *M. flora* Harris & Bradley, *M. lelymene* Harris & Bradley, *M. paludosa* Fott, *M. papillosa* Harris & Bradley var. *monilifer* Harris, *M. pillula* Harris, *M. pugio* Bradley, *M. transsylvanica* Peterfi & Momeu, *Synura echinulata* Korsh., *S. petersenii* Korsh. and *S. sphagnicola* Korsh.

Only 6 species were quantitatively significant. SEM micrographs of subfossil scales of these 6 species are presented in Fig. 2. The relative abundances of these species are given in Fig. 1.

M. crassisquama which shares up to 95% of all scale counts is the most abundant species in the deeper sediment layers. It is followed by *M. caudata* which makes up to 17% in the 31 cm layer. Above 14 cm depth, the numbers of *M. crassisquama* decline rapidly to almost zero at the sediment top. However, it spikes again in 5 cm depth. As *M. crassisquama* scales are very characteristic, a misidentification can be excluded here. *M. crassisquama* is obviously replaced by *Synura echinulata* (finally ca 80% of scale counts) and to a lesser degree by *S. sphagnicola*. Besides this displacement, *M. canina* and *M. calceolus* spiked subsurficially, each sharing not more than 10% each. It is noteworthy, several other species have been found only in the upper 6 cm of the sediment like *M. allantoides, M. pugio, M. transsylvanica, M. flora, M. paludosa* and *M. papillosa* var. *monilifer*.

Discussion

Mallomonas crassisquama is distributed worldwide (e.g. Battarbee *et al.*, 1980; Cronberg & Kristiansen, 1980; Smol *et al.,* 1984a, b). It seems to have wide ecological tolerances. However, *M. crassisquama* appears to avoid the most acid localities (Cronberg & Kristiansen, 1980). Smol *et al.* (1984a, b) report on threshold acidities of pH 5.0. Thus, the rapid decline of this species indicates obviously that acidities of Kleiner Arbersee have declined below that threshold pH during the growing season of *M. crassisquama*. The reoccurrence of *M. crassiscuama* may probably be attributed to constructions of a road and recreational facilities in the catchment some decades ago. The input of neutralizing material is documented by higher concentration of base cations, e.g. Na, K, Mg (Steinberg *et al.,* 1984a). Diatom-inferred pH also shows increasing pH values, however, this increase lies well within the range of standard deviation.

Also *M. caudata* is thought to be one of the most

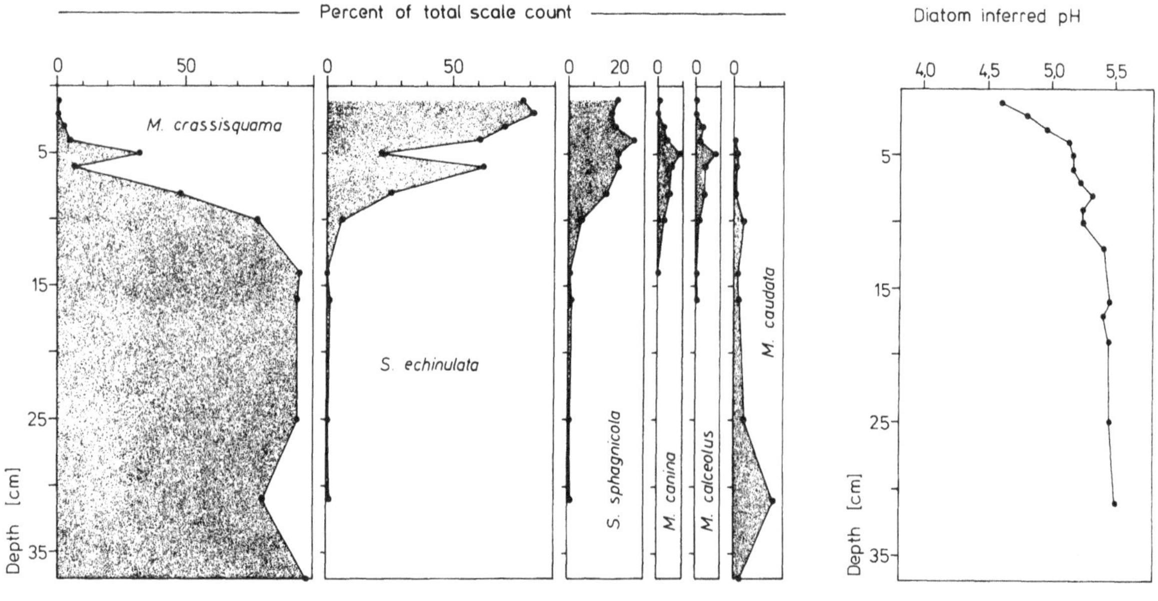

Fig. 1. Relative frequency of mallomonadacean scales in the sediments of Kleiner Arbersee and diatom inferred pH.

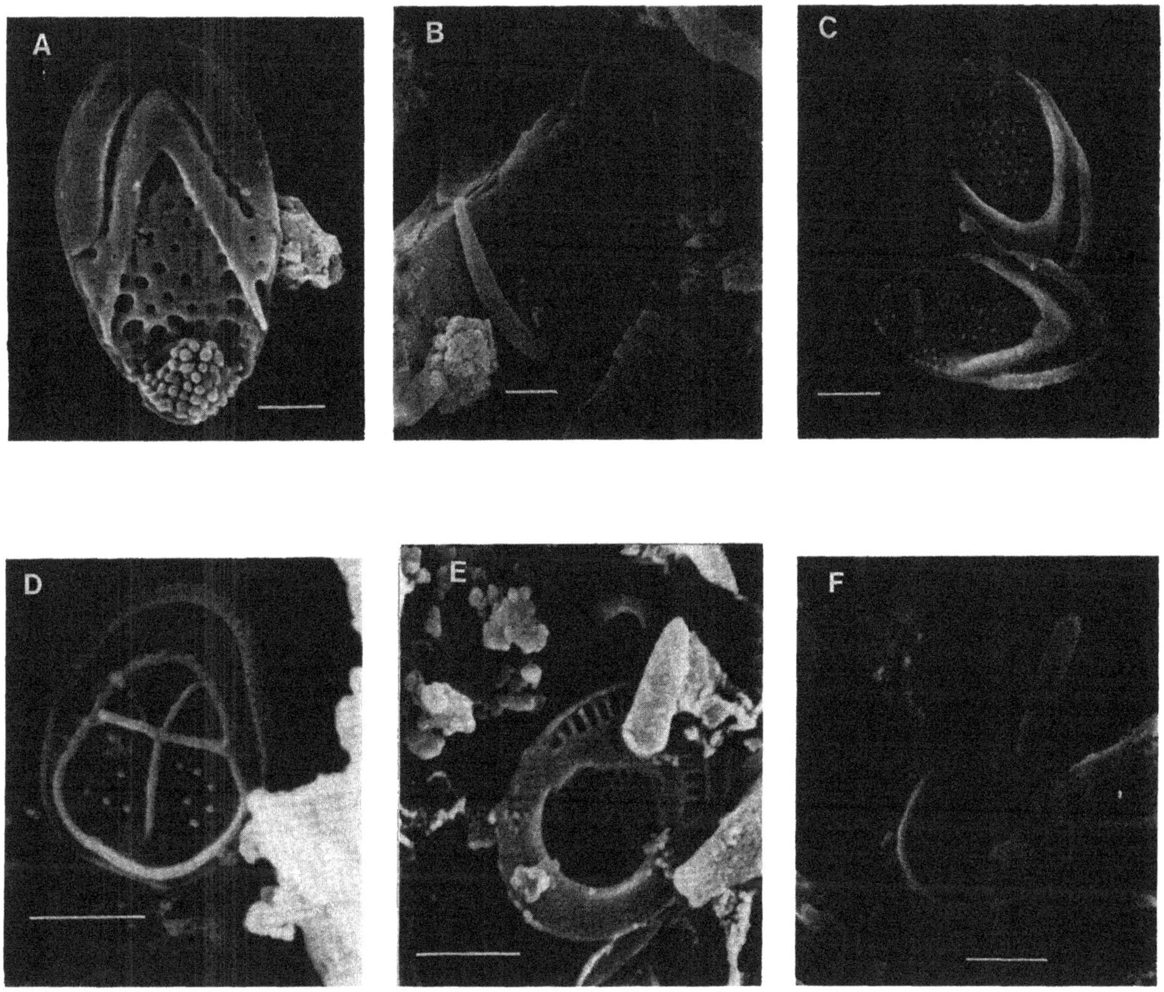

Fig. 2. SEM micrographs of scales of most abundant mallomonadacean species in the sediments of Kleiner Arbersee.
A: Mallomonas crassisquama. – B: Mallomonas caudata. – C: Mallomonas calceolus. – D: Mallomonas canina. – E: Synura echinulata. – F: Synura sphagnicola. White bar indicates one μm.

adaptable mallomonadacean species. In the Adirondack lakes, however, it tended to be most successful in more alkaline water (Smol *et al.*, 1984b). In Kleiner Arbersee, the early decline of this species is coincident with that statement. It is evident that the paleodistribution of both species shows increasing acidity in the lake.

M. crassisquama and to a lesser degree *M. caudata* are replaced by the *Synura* species mentioned above. *S. echinulata* is known from a wide variety of lakes. *S. sphagnicola* is considered to be characteristic for acid waters. Smol *et al.* (1984b), however, found it also in slightly alkaline Adirondack lakes. Nevertheless, the highest abundances of both species were recorded in acid Adirondack waters. Thus, these species appear to be acid tolerant rather than acidophilic or even acidobiontic. The ecological data were taken from Harris & Bradley (1960), Kristiansen (1975, 1978, 1979, 1982), Peterfi & Momeu (1976), Cronberg & Kristiansen (1980), Kling & Kristiansen (1983), Donaldson & Stein (1984), and Smol *et al.* (1984). The pH-classification is done according to Hustedt (1939).

M. hamata and *M. hindonii*, which are characteristic for recent lake acidification in numerous Adirondack lakes (Smol *et al.* 1984a, b), did not occur in Kleiner Arbersee. In this lake, their position is obviously occupied by the two *Synura* species and, in layers around 5 cm depths, by *M. canina* and *M. calceolus*. *M. calceolus* is known to prefer acid to circumneutral conditions (Bradley, 1964; Kristiansen, 1978; Cronberg & Kristiansen, 1980; Dürrschmidt, 1982). In Kleiner Arbersee, its occurrence seems to reflect increasing acidity: it occurred, when the lake is slightly acidified and it disappeared, when acidification continues. Moreover, *M. canina* also gives evidence for increasing acidity in Kleiner Arbersee. This species has so far only been found in Lake Hundsø (Kristiansen, 1982), an acid (pH 4.4 to 4.9) clear-water lake in Jutland, Denmark. It is yet unknown why the number of this species declines in the upper layers.

The occurrence of acidophilics and acidobiontics like *M. allantoides, M. pugio* and *M. transsylvanica* as well as *M. flora, M. paludosa* and *M. papillosa* var. *monilifer* presents also a clear evidence of recent lake acidification.

In the sediment of Kleiner Arbersee, mallomonadacean assemblages indicate that acidity in this lake is somewhat below pH 5.0. This is at least valid for the growing seasons of the Mallomonadaceae (mainly spring). It is not yet feasible to infer pH-values from qualitative and quantitative scale composition. It is almost impracticable to transform scale counts into numbers of individuals of the different species. This restriction is not valid for diatoms where a frustule quals an individual. Diatom-inferred pH by Index B (Renberg & Hellberg, 1982; Fig. 2, right) shows a decline from about 5.5 to 4.6 (Steinberg *et al.,* 1984b). This decrease runs more or less parallel to that of *M. crassisquama*. The slight decrease in the pH-values inferred from diatoms below 10 cm depth again lies well within the range of the standard deviation. A significant pH-decrease indicated by diatoms does not occur before 5 cm depth. However, abundances of scales of *M. crassisquama* start to decrease significantly already above the 14 cm-layer. That means a significant and considerable delay in the time of diatom response. It suggests that mallomonadacean spp. respond more rapidly than diatoms. In other words, surface water chemistry is affecting Mallomonadaceae prior to diatoms with respect to acidification (and probably alkalinization). The differences in response may be explained by the different living conditions of these two algal groups. *Mallomonas* and *Synura* spp. are exclusively euplanktic, while in acid and acidified lakes, diatoms prefer the benthic habitats (except a very few species, such as *Asterionella ralfsii*). The sediments, however, provide at least two buffer mechanisms against acid loading: cation exchange by humic substances and dissolution of carbonates. Furthermore, Mallomonadaceae grow mostly during spring, when most of the annual acid loading enters the lake during snow melt.

It may be concluded that mallomonadacean species in acidified lakes respond to the lowest seasonal pH for short term as well as for long term acidification. They form very useful paleoindicators of lake acidification.

Acknowledgements

The authors would like to thank Drs. G. Cronberg and M. Dürrschmidt for providing valuable reprints with SEM micrographs. Thanks are also due to Drs. P. Smol for identifying rare subfossil scales and to E. Haworth for correcting the manuscript. This work was supported by the Umweltbundesamt under UFO-Pl. no. 102 04 333.

References

Arzet, K. & C. Steinberg, 1984. In Ermangelung von langjährigen Meßreihen: Paläolimnologie. In: Gewässerversauerung in der Bundesrepublik Deutschland. Umweltbundesamt Materialien 1/84: 169–183.

Arzet, K., D. Krause-Dellin & C. Steinberg, 1986: Acidification of four lakes in the Federal Republic of Germany, as reflected by diatom assemblages, cladoceran remains and sediment chemistry. In: Smol, J. P. *et al.* (eds.): Diatoms and lakes acidification. Developments in Hydrobiology, Dr. W. Junk, The Hague.

Battarbee, R. W., 1984: Diatom analysis and the acidification of lakes. Phil. Trans. r. Soc. Lond. B. 305: 451–477.

Battarbee, R. W., G. Cronberg & S. Lowry, 1980: Observations on the occurrence of scales and bristles of *Mallomonas* spp. (Chrysophyceae) in the microlaminated sediments of a small lake in Finnish North Karelia. Hydrobiologia 71: 225–232.

Bradley, D. E., 1964. A study of the *Mallomonas, Synura* and *Chrysosphaerella* of Northern Iceland. J. gen. Microbiol. 37: 321–333.

Cronberg, G. & J. Kristiansen, 1980. Synuraceae and other Chrysophyceae from central Småland, Sweden. Bot. Notiser 133: 595–618.

Donaldson, D. & J. Stein, 1984. Identification of planktonic Mallomonadaceae and other Chrysophyceae from selected lakes in the lower Fraser Valley, British Columbia (Canada). Can. J. Bot. 62: 525–539.

Dürrschmidt, M., 1982. Studies on the Chrysophyceae from South Chilean inland waters by means of scanning and transmission electron microscopy, II. Arch. Hydrobiol., Suppl. 63: 121–163.

Harris, K. & D. E. Bradley, 1960. A taxonomic study of *Mallomonas*. J. gen. Microbiol. 22: 750–777.

Hustedt, F., 1939. Ökologische und systematische Untersuchung über die Diatomeenflora von Java, Bali und Sumatra. Arch. Hydrobiol., Suppl. 16: 1–155, 74–394.

Kling, H. J. & J. Kristiansen, 1983. Scale bearing Chrysophyceae (Mallomonadaceae) from Central and Northern Canada. Nord. J. Bot. 3: 269–290.

Kristiansen, J., 1975. Studies on the Chrysophyceae of Bornholm. Bot. Tidsskr. 70: 126–142.

Kristiansen, J., 1978. Studies on the Chrysophyceae of Bornholm. II. Bot. Tidsskr. 73, 71–85.

Kristiansen, J., 1979. Observations on some Chrysophyceae from North Wales. Br. phycol. J. 14: 231–241.

Kristiansen, J., 1981. Distribution problems of Synuraceae (Chrysophyceae). Verh. Int. Ver. Limnol. 21: 1444–1448.

Kristiansen, J., 1982. *Mallomonas canina* sp. nov. (Chrysophyceae), a new member of sect. Heterospinae. Nord. J. Bot. 2: 293–296.

Peterfi, L. S. & L. Momeu, 1976. Romanian *Mallomonas* species studied in light and electron microscopes. Nova Hedwigia 27: 353–392.

Renberg, I. & T. Hellberg, 1982. The pH history of lakes in south-western Sweden, as calculated from the subfossil diatom flora of the sediments. Ambio 11: 30–33.

Smol, J. P., 1980. Fossil synuracean (Chrysophyceae) scales in lake sediments: a new group of poleoindicators. Can. J. Bot. 58: 458–465.

Smol, J. P., S. R. Brown & R. N. McNeely, 1983. Cultural disturbances and trophic history of a small meromictic lake from central Canada. Hydrobiologia 103: 125–130.

Smol, J. P., D. F. Charles & D. R. Whitehead, 1984a. Mallomonadacean microfossils provide evidence of recent lake acidification. Nature 307: 628–630.

Smol, J. P., D. F. Charles & D. R. Whitehead, 1984b. Mallomonadacean (Chrysophyceae) assemblages and their relationships with limnological characteristics in 38 Adirondack (New York) lakes. Can. J. Bot. 62: 911–923.

Steinberg, C., R. Meijer, U. Palm & H. Nirschl, 1984a. Chemisch-stratigrafische Untersuchungen an Sedimenten der Bayerwaldseen. In: Gewässerversauerung in der Bundesrepublik Deutschland. Umweltbundesamt Materialien 1/84: 304–314.

Steinberg, C., K. Arzet & D. Krause-Dellin, 1984b. Gewässerversauerung in der Bundesrepublik Deutschland im Lichte paläolimnologischer Studien. Naturwissenschaften 71: 631–633.

Accepted 18 March 1986.

Hydrobiologia 143: 93–103, (1986)

The relationship between surface sediment diatom assemblages and pH in 33 Galloway lakes: Some regression models for reconstructing pH and their application to sediment cores

R. J. Flower
Palaeoecology Research Unit, Department of Geography, University College, 26, Bedford Way, London, WC1H 0AP, UK

Keywords: Galloway lakes, pH reconstructions, diatoms

Abstract

The species composition of surface sediment diatom assemblages in 33 Galloway lakes, pH range c. 4.5 to 7.4, is related in a statistically significant manner to water acidity (or to factors closely associated with pH). Predictive models of summer mean pH, using simple regression equations of Index B (Renberg & Hellberg, 1982) with both Scandinavian and Galloway data sets, and multiple regression equations using diatom pH preference groups and individual species, are described and applied to fossil diatoms in six sediment cores. Although multiple regression of individual species gives the highest correlation coefficient ($r^2=0.87$) in the modern data set this method is least appropriate for reconstructing pH values from fossil material where predictor species are often absent. Of the four methods examined here it is suggested that multiple regression of diatom preference groups is probably the most suitable for pH reconstructions from sediment cores.

Introduction

Lake acidification can be recorded by changes in the species of diatoms deposited in sediment but quantification of trends, in terms of pH, requires use of transfer functions developed from modern diatom-lake water pH data sets. There are several ways in which diatom assemblages can be manipulated to produce estimates of water acidity. The index approach pioneered by Nygaard (1956) and refined by Meriläinen (1967) and Renberg & Hellberg (1982) is based on diatom index-pH relationships. More recently alternative models predicting pH have been developed using multiple regression of species, pH preference groups, or species clusters with pH, or by relating the first principal component scores from PCA of modern diatoms to measured pH (Davis & Anderson, 1985; Charles 1985; Huttunen & Meriläinen, this volume).

Despite the cosmopolitan distribution of most diatom taxa the precise relationship between the species composition of diatom assemblages and water pH is likely to vary geographically. Such factors as climate, soil, vegetation and anthropological impacts are all likely to influence the composition of local diatom communities. Until these effects are clarified it is desirable to reconstruct historical pH values in lake sediment cores using a predictive model derived from locally collected data. In this paper surface sediment diatom assemblages from 33 Galloway lakes are compared with measured lake water pH and results are used to reconstruct palaeo-pH values for six lake sediment cores from the same region. Various methods for predicting pH, based on simple or multiple regression of diatoms on measured pH, are examined.

Sites and methods

The sample sites are located in Galloway, SW Scotland and comprise of 33 acid to circumneutral lakes (Fig. 1). The group includes a wide range of

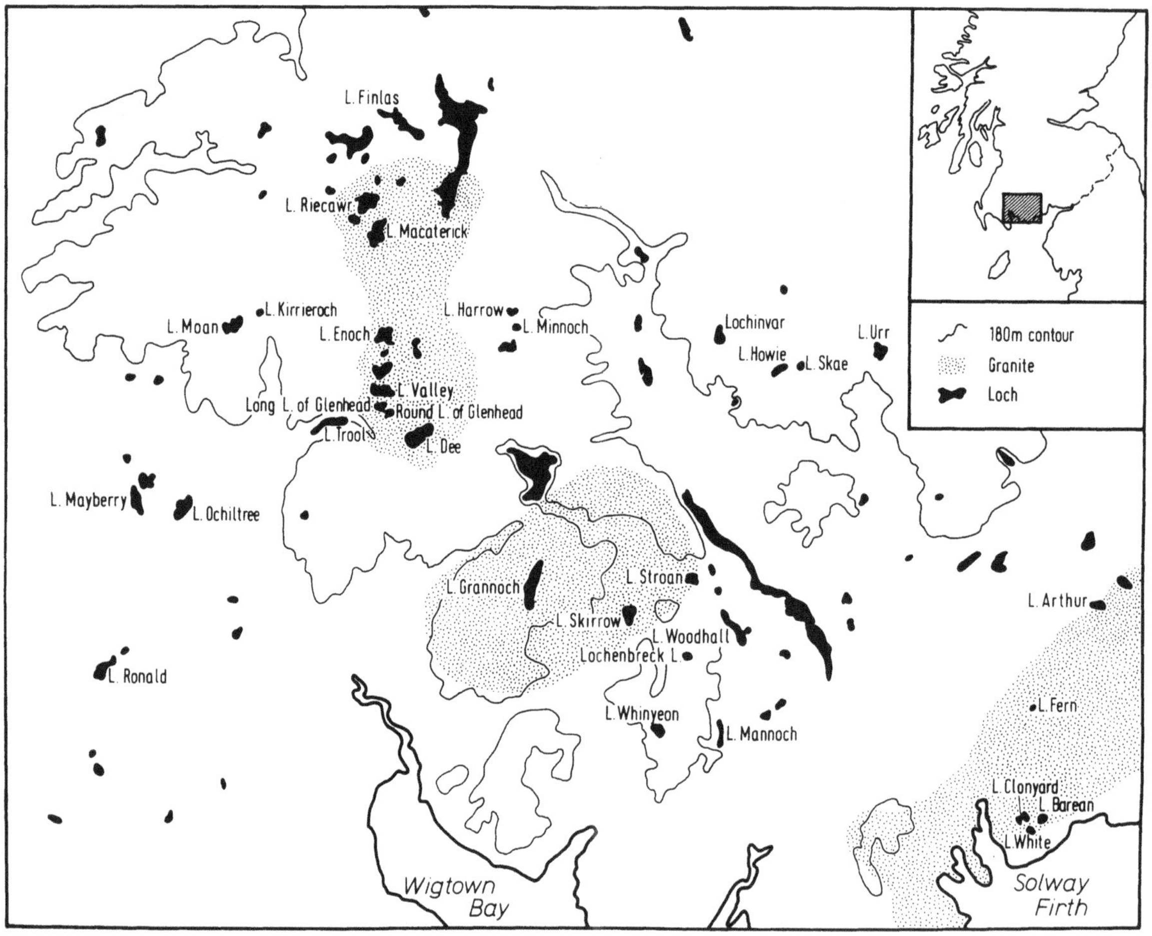

Fig. 1. The distribution of the 33 Galloway sample lakes with inset showing the region of interest in southwest Scotland (UK).

lakes differing strongly in size, depth, altitude, macrophyte development and water colour (Flower, in prep. and see Table 3). Solid geology of the sites is either granite or Silurian/Ordovician sedimentary and metamorphic rocks. Catchment soils below about 100 m altitude where drainage is good are mainly acid brown-earths but above this level blanket peat predominates. Additionally, the catchments of many of the lakes have been afforested with conifers within the past 30 years.

Surface sediment samples for diatom analysis were collected on several occasions between 1981 and 1984 from the deepest point in each lake using a Kajak corer. Sediment cores ca. 80 cm in length were taken from six lakes using a mini-Mackereth corer (Mackereth, 1969). The sediment characteristics and accumulations rates for all six cores are given in Flower & Batterbee (1983) and Battarbee *et al.* (1985). Diatom analysis was carried out at × 1000 magnification using a Zeiss microscope on sediment samples oxidized with 30% hydrogen peroxide and mounted in Mikrops. Between 500 and 600 diatom valves were counted in each sample but where planktonic taxa predominated a count of at least 300 periphytic valves was made to ensure good representation of the periphyton component.

Water samples for measurement of pH and conductivity were collected on several occasions. Determination of pH was made using a Orion meter with Ross electrodes and always within several hours of collection. Fuller chemical analysis of the samples for November 1984 was carried out by the Solway River Purification Board, Scotland, using standard methods.

Results

i) Modern pH measurements: Results of the pH determinations made in this study are presented with those of other workers in Table 1. The replicability of pH from year to year is particularly good for the most acid sites. There are, however, some clear temporal differences in the pH data and most of the pH determinations made in winter 1984 are consistently lower than those made in the summers of 1983/84 and 1985. Also those lakes in which the pH is most variable occur in the poorly buffered pH range 5.0–6.0.

Mean pH for each site was calculated geometrically combining the data in several ways, by using all the pH data in Table 1, by excluding the low pH values recorded in November 1984, and from the summer measurements only (Table 2). The lakes are ordered in Tables 1 and 2 according to mean summer pH, where Loch Enoch is the most acid site (pH 4.5) and Loch Arthur is the least acid (pH = 7). Fuller water chemistry results are available for the November 1984 samples and Table 3 generally indicates increasing ionic concentrations with pH but there is no clear correspondence between water colour and pH. Calcium con-

Table 1. Measured pH values for the 33 Galloway sample lakes 1978 – 1985. The 1978 – 82 and 1981 – 82 pH data columns are geometrical mean values calculated from 8 measurements.

Site	April 1979[1]	1978 – 82[2]	1981 – 82[3]	1983 – 84 (Summer)	1984 (Winter)	1985 (Summer)
1. Loch Enoch	4.54	4.42	4.5	4.5	4.5	4.5
2. Loch Valley	4.64	4.60	4.7	4.7	4.6	4.7
3. Loch Grannoch	4.53	4.55	4.6	4.8	4.4	4.6
4. Long Loch of Glenhead	4.72	4.70	–	4.7	4.7	4.8
5. Round Loch of Glenhead	4.67	4.65	4.7	4.7	4.7	4.7
6. Loch Macaterick	5.10	5.10	–	5.1	4.8	5.2
7. Loch Trool	4.93	4.90	–	5.3	4.9	5.1
8. Loch Kirrieroch	–	–	–	5.1	5.0	5.3
9. Loch Dee*	5.18	5.00	5.3	6.7	5.5	6.6
10. Loch Moan	5.08	–	–	5.9	4.7	5.1
11. Loch Stroan	5.02	–	–	5.6	4.4	5.2
12. Loch Harrow	4.95	5.10	–	5.5	4.7	5.4
13. Loch Riecawr	5.60	5.70	–	5.3	4.9	5.8
14. Loch Howie	5.73	–	–	5.4	5.6	6.1
15. Loch Muck	5.76	–	–	5.7	5.0	5.8
16. Loch Skirrow	5.25	5.05	–	5.5	4.7	5.9
17. Loch Minnoch	4.86	4.90	–	6.0	5.0	5.6
18. Loch Skae	–	–	–	5.5	6.1	6.5
19. Loch Finlas	5.91	–	–	5.7	5.5	6.2
20. Loch Mayberry	5.48	–	–	7.1	4.8	5.6
21. Loch Ochiltree	5.80	–	–	6.4	6.0	6.3
22. Loch Urr	–	–	–	6.0	6.4	7.0
23. Loch Mannoch	6.90	–	–	6.5	6.1	6.7
24. Loch Barean	6.99	–	–	–	6.8	6.8
25. Lochenbreck	–	–	–	7.0	6.3	6.8
26. Loch Ronald	–	–	–	7.3	6.2	6.7
27. Loch Invar	6.59	–	–	7.1	6.3	6.7
28. Loch Woodhall	6.80	–	–	6.9	6.6	6.9
29. Loch Whinyeon	–	–	–	6.7	7.0	7.1
30. Loch Fern	6.66	–	–	7.1	6.7	6.9
31. Loch Clonyard	7.03	–	–	7.1	6.8	6.9
32. Loch White	7.09	–	–	7.1	7.0	7.0
33. Loch Arthur	7.12	–	–	7.6	6.8	7.2

* Extensively limed since 1981.
[1] Wright & Henriksen, 1979.
[2] Harriman pers comm.
[3] Flower & Battarbee, 1983.

Table 2. Various mean lake water pH values calculated geometrically using all 1978 – 1985 measurements, 1978 – 1985 measurements but excluding winter (November) 1984 data, and summer 1983 – 1985 measurements only.

	Mean summer pH (1983 – 85)	Mean pH (1978 – 85 ex Nov. 1984)	Mean pH (1978 – 85)
1. Loch Enoch	4.5	4.5	4.5
2. Loch Valley	4.7	4.7	4.7
3. Loch Grannoch	4.7	4.6	4.6
4. Long Loch of Glenhead	4.7	4.7	4.7
5. Round Loch of Glenhead	4.7	4.7	4.7
6. Loch Macaterick	5.2	5.1	5.0
7. Loch Trool	5.2	5.0	5.0
8. Loch Kerrieroch	5.2	5.2	4.8
9. Loch Dee*	5.3	5.1	5.1
10. Loch Moan	5.3	5.2	5.2
11. Loch Stroan	5.4	5.2	4.8
12. Loch Harrow	5.5	5.2	4.8
13. Loch Riecawr	5.5	5.6	5.3
14. Loch Howie	5.5	5.6	5.3
15. Loch Muck	5.7	5.8	5.4
16. Loch Skirrow	5.7	5.3	5.1
17. Loch Minnoch	5.7	5.1	5.1
18. Loch Skae	5.8	5.8	5.9
19. Loch Finlas	5.9	5.9	5.8
20. Loch Mayberry	5.9	5.8	5.3
21. Loch Ochiltree	6.3	6.1	5.9
22. Loch Urr	6.3	6.3	6.3
23. Loch Mannoch	6.6	6.7	6.4
24. Loch Barean	6.8	6.9	6.6
25. Lochenbreck	6.9	6.9	6.7
26. Loch Ronald	6.9	6.9	6.5
27. Lochinvar	6.9	6.7	6.7
28. Loch Woodhall	6.9	6.9	6.8
29. Loch Whinyeon	6.9	6.9	6.9
30. Loch Fern	7.0	6.9	6.8
31. Loch Clonyard	7.0	7.0	6.9
32. Loch White	7.0	7.1	7.0
33. Loch Arthur	7.4	7.2	7.1

* Calculated from pre-1983 pH data.

centrations are notably low at all sites, especially in the low pH lakes (pH < 5), and reflect the catchment geologies.

ii) Modern diatom assemblages: Percentage frequencies of the common diatoms present in each surface sediment sample are shown in Fig. 2 ordered according to lake water pH (cf. Table 2). The acidobiontic species *Tabellaria quadriseptata* Knud. and *T. binalis* (Ehr?) Grun. are restricted to the most acid sites (pH < 5.0). *Eunotia veneris* (Kutz.) Mull. and *T. flocculosa* (Rabh.) Kutz. are also common in the more acid lakes between about pH 5 and 5.5. On the other hand *Anomoeoneis vitrea* (Kutz.) Ross and several *Fragilaria* species are more widely distributed but show some preference for acid water in the range pH 5 to 6.5. *Achnanthes minutissima* Kutz. and planktonic *Cyclotella* spp., however, show increased abundance in the circumneutral waters (pH = 6) and *Stephanodiscus* spp. occur in low abundance in the least acid lakes.

iii) Diatoms as predictors of pH: In the absence of continuous pH measurements the 'mean' pH of a lake can be calculated in several ways, means of all the data available, means of summer values, or means of winter values (cf. Table 2). Using simple regressions of log Index B (cf. Renberg & Hellberg, 1982) against various estimates of lake mean pH, a best fit regression occurred using mean summer pH

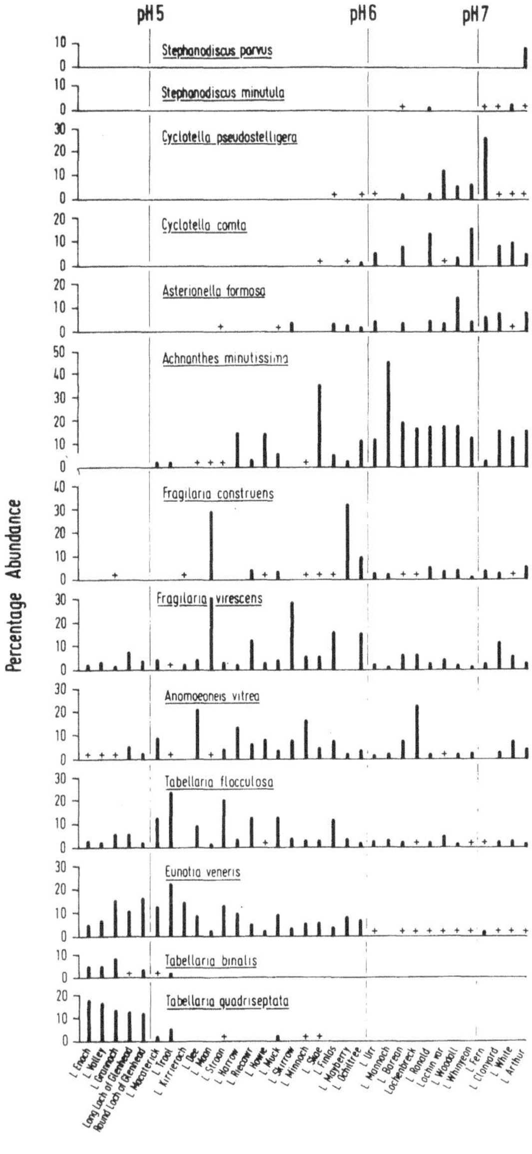

Fig. 2. Percentage frequency distributions of the common surface sediment diatom taxa occurring in the 33 Galloway lakes.

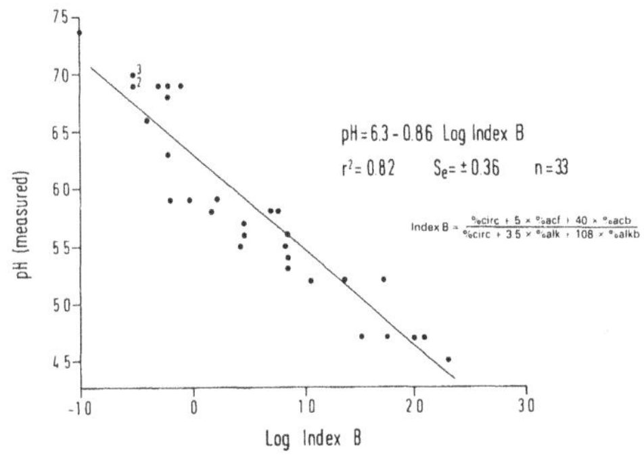

Fig. 3. Simple regression of log Index B values calculated from the surface sediment diatom assemblages with measured summer mean pH.

values (Table 4). The regression of pH on log Index B using the Galloway data (Fig. 3) generates marginally different 'a' and 'b' coefficients and a lower r^2 value than the pH-log Index B regression equation of Renberg & Hellberg (1982) where a Scandinavian data set was used.

Since Index B calculations use predetermined weightings for the various pH preference groups, multiple regression (MR) of these groups against measured summer mean pH was carried out (Table 4). To avoid assigning individual taxa to pH preference groups, MR of individual species abundances was also performed. Nine predictor species were picked out by step-wise MR and identified as those having a significant effect (=1.0%) on the r^2 value (Table 4). Lake water pH values are back-calculated using four methods, Index B with coefficients from the Scandinavian data set (Renberg & Hellberg, 1982), Index B using the Galloway coefficients, by MR of species pH preference groups and by MR of individual species (Fig. 4). The results can be compared with the measured summer mean pH values (Fig. 5). Good agreement occurs between pH values calculated according to the various methods, especially between the Index B methods and MR of the diatom pH preference groups. MR of individual species shows some divergence from the others for several lakes at the middle and top of the pH range, sites 13–15 and the eight least acid sites. Compared with measured pH the calculated values are overestimated for sites 12–15 by all methods and for 18 to 20 by Index B and MR of preference. The greatest point errors in predicted pH are +0.8 and +0.7 of a unit by MR of preference groups and Index B Scandinavian coefficients respectively. The order of regression significance of the various methods, according to the simple correlation coefficients of the predicted pH with measured pH, is MR of species > MR of preference groups > Index B Scandinavian coefficients > Index B Galloway coefficients. Normalizing the percentage diatom data by arc-cosine transformation failed to produce r values that were significantly better than those produced by the raw data.

Table 3. Chemical analysis of lake water samples collected in November 1984. All concentrations are in mg l^{-1}.

	Na	K	Ca	Mg	SO_4	Cl	Total alkalinity (as Ca CO_3)	Colour (Absorbance at at 250 nm)	pH	Conductivity $\mu S\ cm^{-1}$ 20 °C
1. Loch Enoch	4.0	0.2	0.3	0.5	2.6	10.2	0	0.04	4.5	45
2. Loch Valley	4.3	0.3	0.5	0.6	4.0	11.2	0	0.08	4.5	45
3. Loch Grannoch	5.5	0.2	0.9	0.7	5.3	11.6	0	0.17	4.4	62
4. Long Loch of Glenhead	4.5	0.3	0.7	0.6	3.5	9.2	0.4	0.14	4.7	45
5. Round Loch of Glenhead	4.6	0.3	0.6	0.6	3.6	10.8	0.2	0.09	4.7	47
6. Loch Macaterick	5.0	0.4	0.8	0.7	3.0	9.8	1.0	0.29	4.8	45
7. Loch Trool	4.7	0.4	0.7	0.6	4.0	8.4	0.8	0.10	4.9	48
8. Loch Kerrieroch	5.7	0.3	0.7	0.8	4.2	13.2	1.1	0.15	5.0	54
9. Loch Dee*	4.2	0.3	1.0	0.7	3.7	10.0	1.5	0.11	5.5	41
10. Loch Moan	7.3	0.2	1.5	1.2	8.1	15.6	0.9	0.31	4.8	99
11. Loch Stroan	7.8	0.3	1.5	1.0	7.0	15.0	0	0.25	4.4	80
12. Loch Harrow	4.1	0.2	0.7	0.5	3.4	7.6	0.5	0.08	4.7	40
13. Loch Riecawr	4.9	0.3	0.9	0.8	4.5	10.4	2.4	0.24	4.9	50
14. Loch Howie	6.7	0.3	2.5	1.3	7.7	15.8	1.7	0.04	5.6	74
15. Loch Muck	6.1	0.2	1.3	1.1	5.6	11.2	2.2	–	5.0	59
16. Loch Skirrow	7.2	0.3	1.8	1.1	7.2	15.2	1.2	0.22	4.7	68
17. Loch Minnoch	4.4	0.3	0.8	0.6	4.5	7.8	–	0.09	5.0	44
18. Loch Skae	6.2	0.2	2.7	1.3	7.8	12.2	3.4	0.10	6.1	66
19. Loch Finlas	5.7	0.3	1.4	0.9	5.5	11.0	3.0	–	5.5	59
20. Loch Mayberry	7.9	0.4	2.1	1.9	9.0	–	0.5	–	4.8	99
21. Loch Ochiltree	7.5	0.6	2.7	1.9	9.8	19.2	2.2	–	6.0	91
22. Loch Urr	5.2	0.5	3.0	1.7	8.4	12.6	4.8	0.37	6.4	67
23. Loch Mannoch	10.1	0.6	3.9	2.1	9.9	19.6	5.5	0.25	6.1	100
24. Loch Barean	–	1.3	6.2	3.8	16.4	32.0	14.8	0.21	6.8	104
25. Lochenbreck	5.2	0.2	3.3	1.5	9.9	16.6	3.5	0.10	6.5	84
26. Loch Ronald	9.0	1.0	4.5	2.1	8.6	22.4	5.3	–	6.2	116
27. Lochinvar	5.5	0.4	3.3	1.8	11.1	11.6	3.4	0.26	6.3	74
28. Loch Woodhall	8.9	0.7	5.8	2.8	11.8	16.8	11.5	0.28	6.6	103
29. Loch Whinyeon	6.7	0.3	5.9	1.5	6.7	13.4	14.9	0.17	7.0	90
30. Loch Fern	10.6	2.2	6.4	3.0	16.2	21.2	7.9	0.45	6.7	136
31. Loch Clonyard	14.8	1.7	7.2	4.0	17.2	29.2	18.9	0.22	6.8	158
32. Loch White	17.6	1.4	8.0	4.1	15.1	32.8	22.2	0.22	7.0	158
33. Loch Arthur	6.7	1.1	6.7	2.0	11.7	16.0	15.8	0.09	6.8	171

* Extensively limed since 1981.

Application of regression coefficients to diatoms in sediment cores

The four predictive equations of lake water pH are applied to sediment core diatom assemblages in six lakes (Fig. 6). The pH reconstruction methods demonstrate similar trends and at five of the six sites show lower pH values in the most recent sediments. The Loch Grannoch core is unusual in showing pH minimum around 29 cm depth caused by a peak in the frequency distribution of acidobiontic *T. quadriseptata* (cf. Battarbee & Flower, 1984). The various methods disagree by up to 0.6 of a pH unit (see 50 cm depth in the Loch Skirrow core) but values for the most recent sediments, where pH values are typically lower, are usually in closer agreement. Exceptions occur at three sites, Loch Enoch, Loch Dee and Loch Skirrow where pH calculated using MR of individual species is consistently different to pH values calculated by the other methods. At the former site species MR values are consistently higher and at the latter two sites they are consistently lower. Additionally, at Loch Grannoch pH values calculated according to MR of the preference groups are consistently lower. In the Loch Dee core three predicted pH curves shows pH declining towards the sediment surface but that of MR of individual species shows no trend change.

Only the pH values derived from the core tops can be compared directly with modern measurements of water pH and Table 5 shows that all the

Table 4. Regression equations relating measured pH and diatoms in the 33 Galloway lakes.

Index B regression equations

1. Regression of Index B with mean summer pH values:
   pH $= 6.30 - 0.86$ log Index B
   where, $r^2 = 0.82$ Se $\pm$ 0.36 P = < 0.001.

2. Regression of log Index B with mean pH of entire data set excluding winter 1984 values.
   pH $= 6.22 - 0.87$ log Index B
   where, $r^2 = 0.77$ Se = $\pm$ 0.42 P = < 0.001.

3. Regression of log Index B with mean all pH measurements
   pH $= 6.06 - 0.84$ log Index B
   where, $r^2 = 0.75$ Se = $\pm$ 0.44 P = < 0.001.

Multiple regression equations

1. Multiple regression of diatom pH preference groups with mean summer pH
   pH = 7.82 − 0.035 ·% acidophilous − 0.037 ·% acidobiontic
   + 0.100 ·% alkalibiontic − 0.013 ·% circumneutrals
   0.015 ·% alkaliphilous
   where, $r^2 = 0.82$ Se = $\pm$ 0.36 P = < 0.001.

2. Multiple regression of individual species with mean summer pH
   pH = 5.37 − 0.010 ·% *E. veneris* 0.044 ·% *T. quadriseptata*
   + 0.051 ·% *C. comta* + 0.041 ·% *C. pseudostelligera*
   + 0.026 ·% *Ach. minutissima* + 0.060 ·% *Ast. formosa*
   + 0.237 ·% *S. minutula* + 0.021 ·% *Anom. vitrea*
   + 0.134 ·% *F. elliptica*
   where, $r^2 = 0.87$ Se = $\pm$ 0.31 = < 0.001.

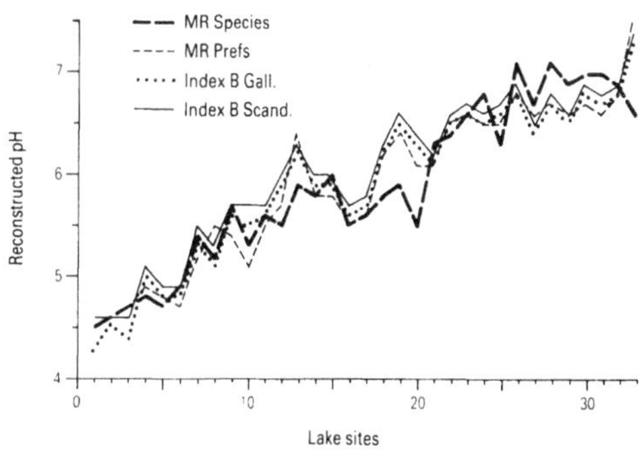

Fig. 4. The pH values calculated from the surface sediment diatom assemblages for each of the 33 lakes according to the Index B and multiple regression methods. The methods are Index B using Scandinavian Regression coefficients (Renberg & Hellberg, 1982), Index B using Galloway regression coefficients, multiple regression of diatom pH preference groups, and multiple regression of individual species.

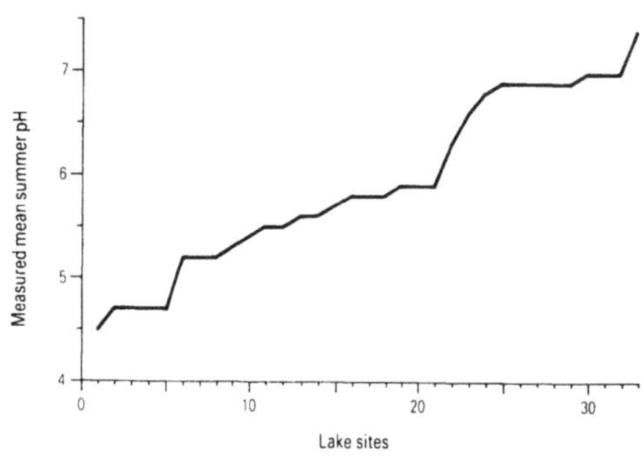

Fig. 5. Mean summer pH values for each lake with sites sequenced according to decreasing acidity.

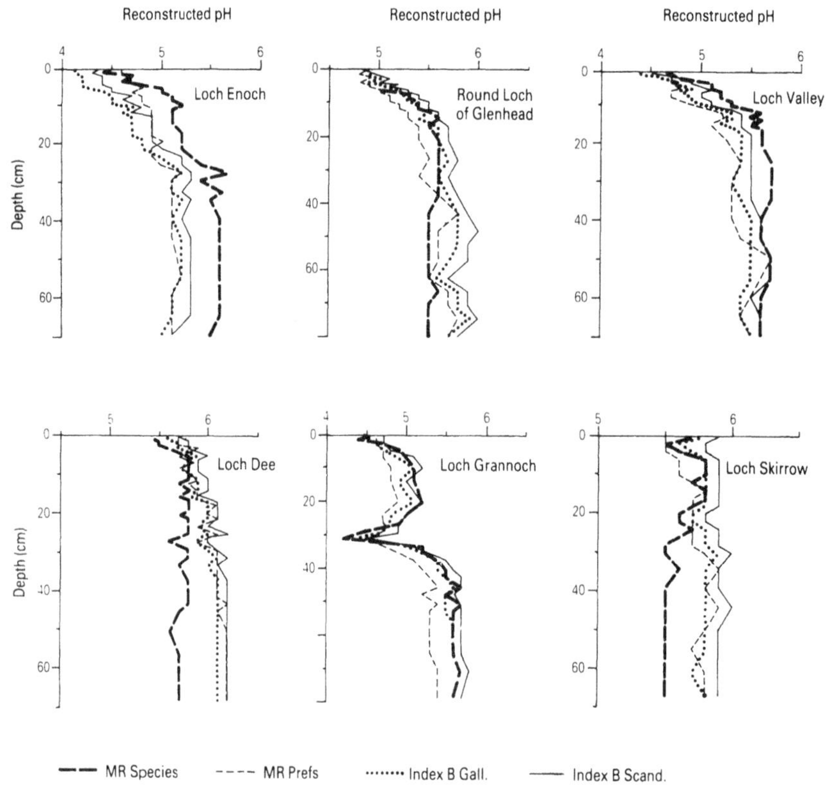

Fig. 6. pH values reconstructed from the fossil diatoms in six Galloway lake sediment cores according to the Index B and multiple regression methods.

Table 5. A comparison of the measured mean summer pH with pH values calculated using the Index B and multiple regression methods. Underlined values differ from the measured by 0.3 pH units or more.

Site	Mean measured pH	Index B Scand.[1]	Index B Gall.[2]	MR[3] of pH reference groups	MR species
Loch Enoch	4.5	4.4	<u>4.2</u>	4.6	4.5
Loch Valley	4.7	4.6	4.5	4.6	4.6
Round Loch of Glenhead	4.7	4.9	4.8	4.8	4.6
Loch Dee	5.3	<u>5.7</u>	<u>5.6</u>	5.4	<u>5.7</u>
Loch Grannoch	4.7	4.7	4.6	4.6	4.7
Loch Skirrow	5.7	<u>6.0</u>	5.9	5.8	<u>6.0</u>

[1] Scandinavian data set (Renberg & Hellberg 1982).
[2] Galloway data set.
[3] MR = Multiple regression.

methods give results that are within 0.3 pH unit or less of the measured value. The calculated pH values of 4.2 for the Loch Enoch core top (by Index B Galloway) and 5.7 pH for the Loch Dee core top (by Index B Scandinavia and MR of species) are the least accurate estimates. No consistent differences between the various methods are apparent in Table 5.

Discussion

The surface sediment diatoms in the Galloway lakes are shown to be distributed according to water acidity or to factors closely covarying with pH. Significant regression equations between diatoms and pH are developed using Index B, diatom pH preference groups, and diatom species. The highest regression coefficient resulted from multiple regression of abundances of nine predictor species against mean summer pH. However all the regressions with diatoms and summer mean pH have a regression coefficient > 0.8. The overall standard error of regression is about ± 0.3 of a pH unit for all the methods but for individual pH estimates error will be about twice this value (cf. Ryan *et al.*, 1976; Sutcliffe, pers. comm.). However the data indicate that the real degree of error associated with a particular calculated pH will vary according to the range in which that estimate occurs (cf. Flower, 1982). Predicted estimates appear to be most accurate for the sites at the upper and lower ends of the pH range where measured pH values (see Table 1) are temporally stable (cf. Huttunen & Meriläinen, 1982). On the other hand lakes in the pH 5 to 6 range are poorly buffered and small seasonal changes in alkalinity can cause large shifts in pH making estimation of a mean pH difficult without many measurements.

There are several merits and criticisms of both the Index B and multiple regression approaches to predicting pH. If Index B is applied to very acid floras the index can become very sensitive to the frequency of circumneutral forms. For example an ecologically insignificant increase from 2 to 4% in these forms will cause a 50% change in the value of the index when alkaliphilous taxa are absent. Moreover the weightings of the pH preference groups used in the Index may vary according to the data set. Problems associated with weightings in the Index B equation are avoided by MR of diatom pH preference groups and of individual species. The former, as in Index B, requires diatoms to be grouped according to pH preference but occasionally the allocation may be unknown or in doubt. The species MR avoids this ecological categorization but derives its significance from only a selected number of species. For the Galloway data set this number is nine so information from the remaining 280 taxa is ignored.

All methods, whether based on simple or multivariate statistics, of predicting pH from sedimentary diatom assemblages depend on the relationship between modern pH and modern diatom communities. There are, however, several inherent problems in this approach. The data set must be large enough to both satisfy statistical rules and to provide significant correlation. Charles (1985) suggests that the modern data set should be limited to sites below pH 6. However, the ecological range sampled will almost certainly overlap to a large extent the range experienced at a particular lake over the past few hundred years.

In addition, the integrity of any mathematical expression relating diatoms and pH clearly depends on accurate correspondence between the two. This correspondence can be weakened by non-pH factors, such as abundance of macrophytes, substrate variations, and depth, affecting the abundance of diatom taxa. For example in the surface sediment sample from site 8 (Loch Kirrieroch) a > 20% frequency of *Navicula sohrensis* occurred. This diatom is rare elsewhere and its abundance here is possibly associated with light penetration to sampled surface sediment. Inwashed catchment diatoms which do not reflect the lake water pH may also affect the accuracy with which pH can be predicted (Battarbee & Flower, 1984).

Inadequacy of the measured pH data can also disrupt the pH – diatom relationship, particularly for sites where only a few pH measurements are available. As is the case for the Adirondack pH – diatom data (Charles, 1985), our results indicate that the diatom distribution in the Galloway lakes is most closely related to a mean of the summer pH values when diatoms are actively growing, rather than to the usually lower winter pH values. Overestimation of pH at several Galloway sites (see Fig. 4) by all four predictive methods may reflect inaccurate mean pH values based on too few pH determinations. More measurements would probably allow the sites to be resequenced in a more accurate order of mean pH. Despite individual weaknesses in the different approaches, and problems with atypical sites, the overall strength of the diatom – pH relationship permits reliable historical reconstructions of pH to be undertaken. However it is probably most accurate in demonstrating relative trends and changes in trends of lake water acidity.

It is of importance to know which of the various regression equations gives the most reliable reconstructed pH results. However, the lack of historical data for acidified lakes prevents comparisons between predicted and measured pH through time. Inferences from the modern data set indicate that the best predictor of pH is species MR but there are several reasons why this method may be inappropriate. The significance of the species MR relationship with pH for the Galloway data depends upon 9 predictor species picked out of the modern data set. In core material several of these species are often numerically unimportant or are absent all together if the core site occurs beyond the pH range of a particular taxon. For example only three of the nine predictor species occur in the Loch Enoch core and in the Loch Dee core the two taxa that decline strongly in abundance (*C. kützingiana* and *A. microcephala*) are not represented in the MR equation. Additionally the species MR method is most sensitive to large abundances of one or more of the predictor species and to species not in the modern data set which leads to erratic results (Davis & Anderson, 1985). Closer agreement is found between results of the Index B methods and MR of preference groups based on pH classification of diatom assemblages. These methods use the same diatom counts but for Galloway data, MR of the pH preference groups is preferred to Index B methods as arbitrary weightings are avoided. Also MR of preference groups is less sensitive to variation in the circumneutral forms. Other promising predictive methods for reconstructing pH values from diatoms include ordination techniques (Hill, 1979), as used by Huttunen & Meriläinen (1983, 1986, and this volume), Charles (1985), and Davis & Anderson (1985) and possibly categorical methods (Wrigley, 1985). Evaluation of these approaches is planned for future research using a UK pH – diatom data set.

Acknowledgements

I express my gratitude to Rick Battarbee for originally recognising Galloway as a prime area in which to study the recent history of lake acidification and for making a considerable contribution to the organisation of this paper. I thank David Tervert of the Solway River Purification Board for freely carrying out chemical analysis of the water samples and Paul Raven, John Anderson, Giti Flower and Viv Jones for help in the field. I thank also Paul Schooling for advice on computational methods. The work was carried out under a grant from the Department of the Environment, UK.

References

Battarbee, R. W. & R. J. Flower, 1984. The inwash of catchment diatoms as a source of error in the sediment-based reconstruction of pH in an acid lake. Limnol. Oceanogr. 29: 1325 – 1329.

Battarbee, R. W., P. G. Applebly, K. Odel & R. J. Flower, 1985. 210-Pb dating of Scottish lake sediments, afforestation and accelerated soil erosion. Earth Surface Processes and Landforms 10: 137 – 142.

Charles, D. F., 1985. Relationships between surface sediment diatom assemblages and lake water characteristics. Ecology 66: 994 – 1011.

Davis, R. B. & D. S. Anderson, 1985. Methods of pH calibration of sedimentary diatom remains for reconstructing history of pH in lakes. Hydrobiologia 120: 69 – 87.

Flower, R. J., 1982. A comparative limnological survey of Rathlin Island, Co. Antrim, with special reference to the diatoms. Proc. R. Ir. Acad. 80B: 1 – 20.

Flower, R. J. & R. W. Battarbee, 1983. Acid lakes in the Galloway uplands, South West Scotland: catchments, water quality, and sediment characteristics. Working paper No. 6, Palaeoecology Research Unit, University College London, 56 pp.

Hill, O. M., 1979. Twinspan – a fortran program for detrended correspondence analysis. Cornell University, Ithaca, N.Y.

Huttunen, P. & J. Meriläinen, 1983. Interpretation of lake quality from contemporary diatom assemblages. Hydrobiologia 103: 91 – 98.

Huttunen, P. & J. Meriläinen, 1986. Applications of multivariate techniques to infer limnological conditions from diatom assemblages. In J. Smol *et al.* (Eds), Diatoms and lake acidity. Junk. In press.

Mackereth, F. J. H., 1969. A short core sampler for subaqueous deposits. Limnol. Oceanogr. 14: 145 – 151.

Meriläinen, J., 1967. The diatom flora and the hydrogen-ion concentration of the water. Ann. Bot. Fenn. 4: 51 – 58.

Nygaard, G., 1956. The ancient and modern flora of diatoms and chrysophyceae in lake Gribsø. Fol. Limnol. Scand. 8: 32 – 94.

Renberg, I. & T. Hellberg, 1982. The pH history of lakes in southwestern Sweden, as calculated from the subfossil diatom flora of the sediments. Ambio 11: 30 – 33.

Ryan, T. A., L. B. Joiner & B. F. Ryan, 1976. Minitab student handbook. Massachusetts, Duxbury Press. 341 pp.

Wright, R. F. & A. Henriksen, 1980. Regional survey of lakes and streams in southwestern Scotland, April 1979. SNSF Project IR 72/80, Oslo.

Wrigley, N., 1985. Categorical data analysis for geographers and environmental scientists. UK, Longman. 576 pp.

Hydrobiologia 143: 105–112, (1986)

Diatom biostratigraphy and comparative core correlation within a small lake basin

N. J. Anderson
Palaeoecology Research Unit, University College London, 26, Bedford Way, London WC1H 0AP, U.K.

Keywords: diatoms, magnetic minerals, multiple cores, core correlation, slotseq

Abstract

A number of short cores from Lough Augher, Co. Tyrone, N. Ireland, with water depths ranging from 3–12 m, were independently correlated using various parameters: diatom biostratigraphy, magnetic mineral stratigraphy (χ, SIRM, S-Ratio) and loss-on-ignition (LOI). Correlating between deep and shallow water sites using LOI proved problematical, although this was possible with both bio- and magneto-stratigraphy. Agreement between these techniques was generally good, but decreased with increasing sediment depth. Although correlation errors are considerably less than those derived from dating, they do suggest that exclusive use of unsubstantiated correlations based on magnetic parameters may lead to erroneous estimates of dry mass or micro fossil accumulation rates, particularly when a whole basin approach is taken.

Introduction

The complexity of sediment accumulation patterns in lake basins means that one core is unlikely to be representative of a whole lake. Consequently, multi-core approaches have become increasingly common (Bloemendal *et al.*, 1979; Davis & Ford, 1982; Evans & Rigler, 1982). Since the radiometric dating of individual cores is prohibitively expensive cores are correlated by various methods (cf. Dearing, 1982) to a centrally located, dated master core. Following the pioneering study of Thompson *et al.* (1975) magnetic susceptibility and related magnetic parameters (Oldfield *et al.*, 1978) have been widely used in whole basin projects (Bloemendal *et al.*, 1979; Dearing, *et al.*, 1981; Foster *et al.*, 1985), largely because of its speed compared to, for example, biostratigraphy. Thompson *et al.* (1975) found good support for the magnetic based correlations from the diatom biostratigraphy. Since then, however, there have been few comparative studies to test the validity of the derived correlations, although the study of Thompson & Edwards (1982) is a notable exception.

Despite the application of quantitative methods for stratigraphic correlation in geology (e.g. Cubitt & Reyment, 1982) such correlation in palaeolimnological studies is still largely subjective, based upon visual identification and matching of assumed synchronous horizons. However, the program SLOTSEQ (Birks 1979) has become widely available, and although developed for between site comparison of fossil pollen assemblages is suitable for within basin comparison of diatom biostratigraphy.

The present study attempts to assess the accuracy of the various methods by offering comparative correlations based upon bio- and lithostratigraphic and mineral magnetic parameters for a number of accurately located cores from a small eutrophic lake. Lough Augher, Co. Tyrone, Northern Ireland is a simple single basin of maximum depth 14.5 m and an area of 9.25 ha (Fig. 1). Until recently it directly received untreated creamery effluent, which has had a profound influence on its ecology.

Methods

Mini-cores (Mackereth, 1969) were collected and

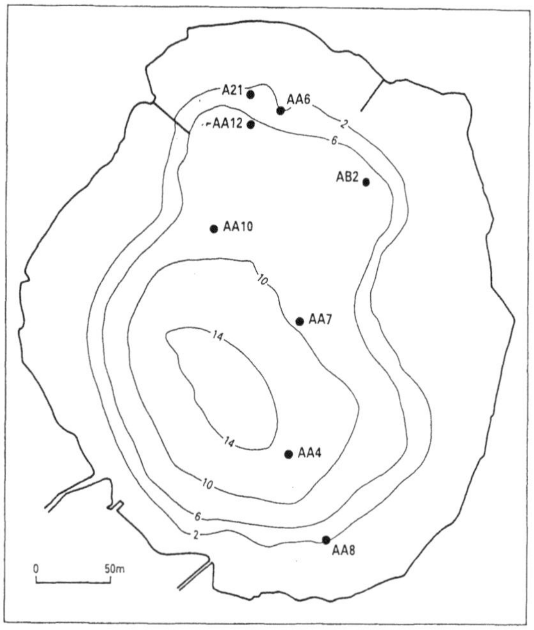

Fig. 1. Location of cores referred to in text. Prefixes for date of coring: 'A': 1981; 'AB': 1983. Contour depth in meters.

accurately located (Battarbee *et al.*, 1983) in 1980, 1981 and 1983. The locations of cores referred to in this paper are presented in Fig. 1. Sample preparation follows standard methodologies: loss-on-ignition (Dean, 1974), diatom analysis (Battarbee, 1985). Magnetic susceptibility (χ) was measured on a Bartington Instruments sensor; SIRM and reverse field parameters were grown in a pulse magnetiser and measured with a Minispin Fluxgate magnetometer (cf. Thompson *et al.*, 1980). By comparison of χ and SIRM profiles, 12–14 assumed synchronous features were identified in each core as a basis for magnetic correlations.

Diatom biostratigraphic zones were identified visually, using features recognisable in all cores, and these were largely confirmed by use of a statistical zonation programme (Birks, 1979) on a selected 15 taxa. The same taxa were used for SLOTSEQ between core comparisons (Birks 1979). SLOTSEQ attempts to slot two biostratigraphic profiles so as to reduce the statistical dissimilarity between pollen/diatom percentages in two core sequences. The resultant goodness-of-fit is indicated by the statistic Psi and because the programme will always produce correlations between cores the interpretation of this value is very important (Gordon & Birks, 1974). However, there are few guidelines, except that with lower values a better fit is implied. Further subjectivity is involved in identifying and ignoring blocked levels (Edwards & Thompson, 1984). Blocking occurs when numerous levels in one core are not disassociated and a correlation of

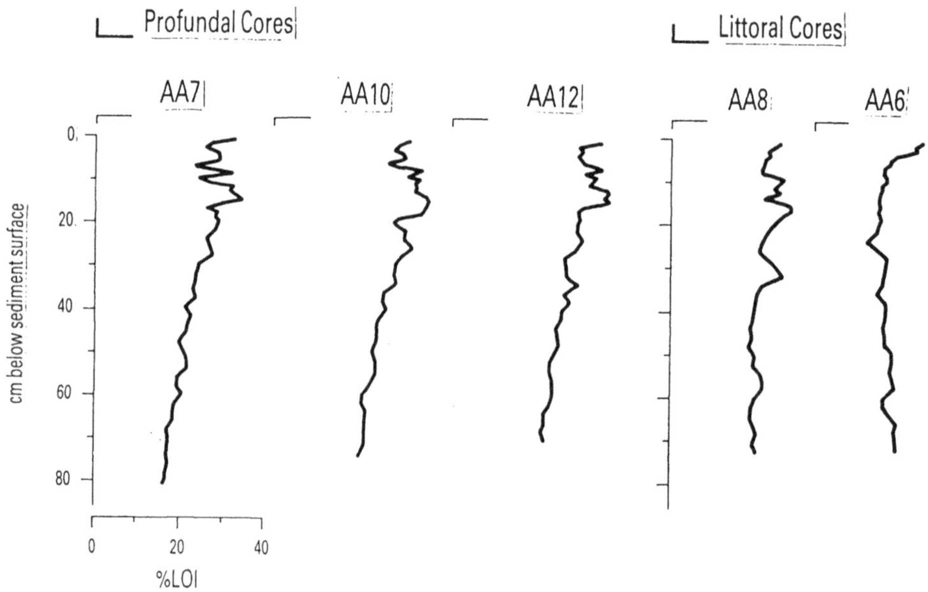

Fig. 2. Comparative loss-on-ignition profiles for selected cores.

this group as a whole with two corresponding levels in the second core is suggested. Arbitrary decisions then have to be made at what level (i.e. > 2) blocked slots should be disregarded, and where blocking is excessive the number of slots is greatly reduced.

Results

1. Correlation from LOI profiles (Fig. 2)

LOI profiles from water depths > 6 m are very similar, but provide few correlatable points below the subsurface peaks. Shallow water cores are more heterogeneous and are difficult to relate to those from deeper water (Fig. 2). These atypical LOI profiles perhaps reflect localised concentration of organic matter from macrophyte decay at the littoral sites.

2. Correlation from diatom biostratigraphy (Fig. 3)

Core AA7 has biostratigraphy representative of the profundal sediments (Fig. 3) and shows the clear response of the planktonic diatoms to the increased nutrient loading from the creamery. There is a succession from *Melosira ambigua* (Grun.) O. Muller, through *Fragilaria crotonensis* Kitton, *Diatoma elongatum* (Lyngbye) Ag. and *Asterionella formosa* Hassall to dominance by a small *Stephanodiscus* complex (Anderson, 1984). There are also clearly defined stratigraphies of benthic and epiphytic forms (e.g. *Cocconeis placentula* Ehr.) although there is greater variability in these profiles (Anderson, in prep.). Despite variation in the percentage occurrence and profile form in e.g. *Stephanodiscus parvus* Stoermer & Hakansson (Fig. 6) the clear response of the diatom communities to the eutrophication of the lake has been found in all cores studied (Anderson, in prep.) and so the biostratigraphic zones defined (Figs. 3 & 4) could be used to correlate between cores. Also, isolated but clearly repeatable occurrences of individual taxa (e.g. *Cyclostephanos dubius* (Fricke) Round) were used for correlation.

3. Correlation using SLOTSEQ (Table 1)

Results from the program SLOTSEQ gave low Psi values (0.77–1.27, Table 1), presumably indicative of goodness of fit between the cores (Gordon & Birks, 1974), but blocking (cf. Edwards & Thompson, 1984) reduces the number of slots between cores. This, and the fact that the program always forces profiles to correlate, causes some aberrant results (e.g. A21 v AB2, A21 v AA4 in Fig. 7) when compared to the other methods utilised. However, both these slots do have the highest Psi values (Table 1).

Table 1. Slotseq Psi Values.

AA10					
AA12	0.92				
AA7	0.80	1.00			
AA4	0.87	1.19	0.81		
A21	0.91	0.77	1.20	1.27	
AB2	0.80	1.14	1.04	1.16	1.22
	AA10	AA12	AA7	AA4	A21

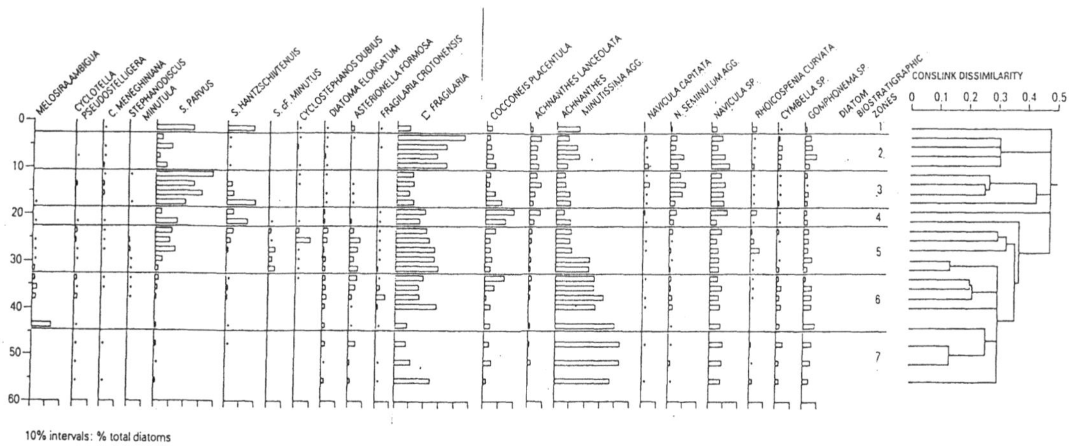

Fig. 3. AA7 relative frequency diatom biostratigraphy for selected taxa. Σ *Fragilaria* excludes *F. crotonensis*. Zone boundaries were determined by visual examination of taxa.

4. Correlation using magnetic parameters (Fig. 4)

SIRM profiles show good similarity (Fig. 4). All cores have a surface peak and immediately subsurface minimum values, but fine detail and maximum values are quite variable, making it difficult to identify a given feature with any degree of reliability. S-ratio (IRM-100mT/IRMsat) is a concentration independent parameter and values have a clear decline in all cores at c. 30–40 cm, and another, secondary, decline which matches well with the AU-3/4 diatom zone boundary (Fig. 5).

5. Comparison of methods (Figs. 6 & 7)

The suggested correlation frameworks for the different parameters were developed independently of each other and then compared to see if similar results were obtained. Despite the coarser sampling interval used in the biostratigraphic analysis there is good agreement between the less ambiguous features of both magnetic and diatom biostratigraphy, particularly in the upper half of the cores (Fig. 6). For instance, the secondary S-ratio decline agrees clearly with the shoulder of SIRM curves and the increase in *S. parvus* percentages; which is one of the criteria used to delimit the AU 3/4 boundary (Fig. 6). Below the *S. parvus* increase however, SIRM profiles are less repeatable (Fig. 4), and biostratigraphic resolution is less clear prior to the *Melosira ambigua* increase (Fig. 3), and hence there is more uncertainty in the correlations. This is sug-

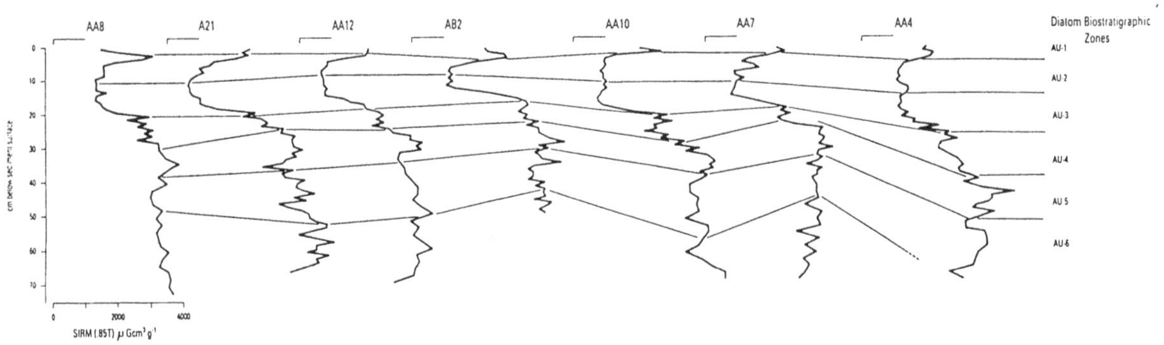

Fig. 4. SIRM (.85T) profiles for selected cores. Diatom biostratigraphic zones determined as for Fig. 3 and as described in text.

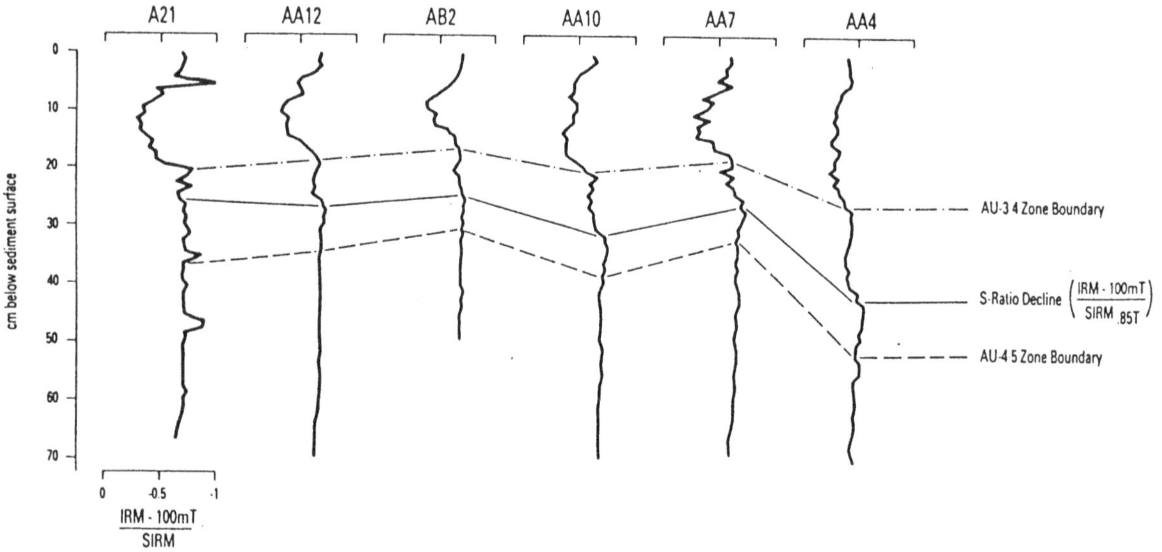

Fig. 5. S-Ratio (IRM-100mT/IRMsat) for selected cores.

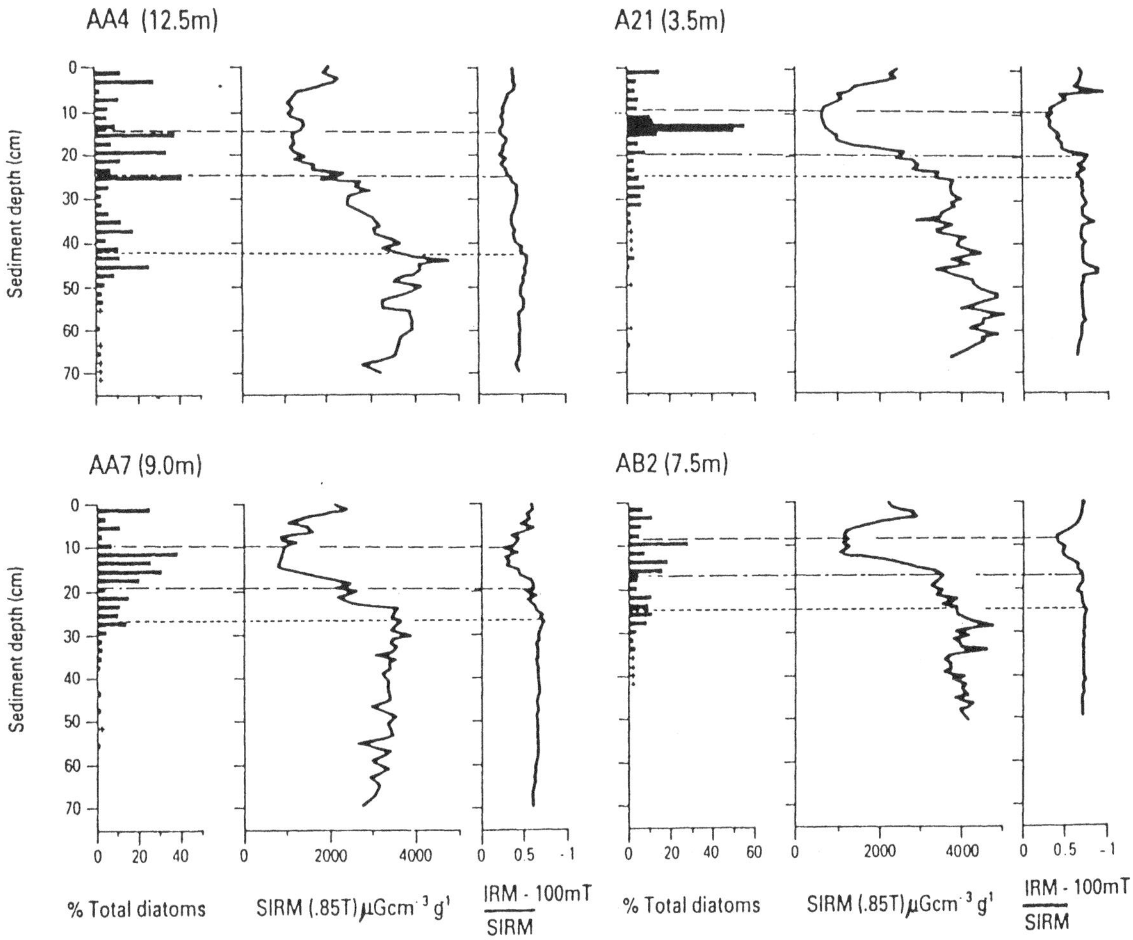

Fig. 6. Comparative stratigraphy for selected cores; profiles from left to right: *Stephanodiscus parvus*, SIRM, and S-Ratio. Lines represent assumed synchronous features for each stratigraphic profiles in all cores:
----- initial S-Ratio decline
—·—·— SIRM shoulder
– – – – – AU-2/3 diatom biostratigraphic zone boundary.

gested in the Shaw diagrams (Fig. 7).

In Shaw diagrams (Shaw, 1964) uniform accumulation rates are presented by a 1:1 straight line (y = x). Changes of slope indicate between core differences of accumulation rate, whilst deviations from a straight line (i.e. dog-legging) suggest relative changes of accumulation rate for a given time period. Where more than one method of correlation has been used, the degree of scatter of the assumed synchronous points can suggest possible errors of correlation. In Fig. 7 the scatter of points is often less than ± 2 cm in terms of sediment depth, particularly in the upper c. 30 cm (e.g. AA4 v AA7) suggesting that reliable correlations have been made in this case. With increasing depth down core there is often a greater lateral spread and bifurcation in the plots (e.g. A21 v AA4), demonstrating conflicting and increasing errors of correlation. The similarity of the slope of these scatter plots also shows the evenness of linear accumulation rates within the basin as a whole and the independence, to some extent, of accumulation rate and water depth.

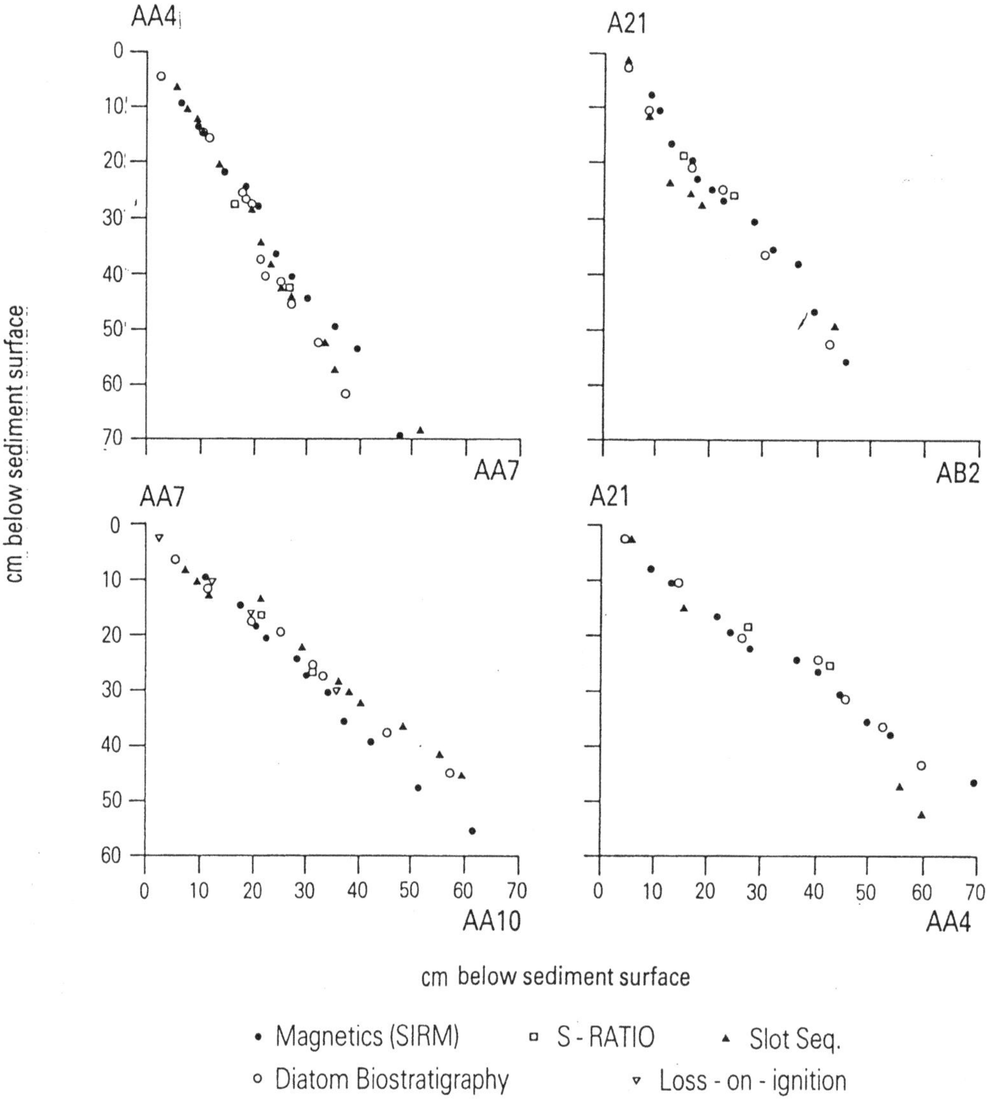

Fig. 7. Shaw diagrams for selected pairs of cores. Each point represents an assumed synchronous feature for both cores, derived using various correlation parameters.

Discussion

Sediment-water interface mixing from both biological and physical processes will destroy temporal resolution in the stratigraphic record (Berner, 1980). The extent of mixing may vary spatially with differing water depth (i.e. increased resuspension in shallow water, and more pronounced seasonal anoxia and reduced bioturbation in deep water) and may therefore cause problems for correlation between profundal and littoral sites.

The increased sediment-water interface mixing in shallow water sites is assumed to cause the differences in biostratigraphy in Fig. 6. The shallow water diatom profiles have poorer resolution compared with deeper water sites (e.g. A21 v AA4), and as a result some taxa with lower percentages than those of *S. parvus* have ambiguous profiles when seen in isolation. However, despite these individual profile differences between profundal and littoral sites it was still possible to place the zone boundaries with some reliability and therefore facilitate biostratigraphic correlation. This was because all the cores, regardless of water depth show the same response to the lake's eutrophication, and because the profiles of numerous taxa were considered in the placement of a zone boundary.

The single high percentage spike of *S. parvus* (cf. A21 in Fig. 6) was found in all cores with water depths less than c. 6 m, and while this can be resolved within the visually defined zonation scheme, it proves to be problematical for SLOTSEQ. Besides being unable to cope with these assumed mixing induced differences, SLOTSEQ involves subjective decisions about which blocked levels should be disregarded, and the interpretation of the Psi values. Taxa with low percentages, and/or sporadic presence (e.g. *Cyclostephanos dubius, Asterionella formosa*, Fig. 3) may have little effect on statistical zonations and SLOTSEQ, but can be used in visual comparisons. However, Clark *et al.* (1986) have demonstrated how chord-distance dissimilarity coefficients can be used to minimise differences between high and low percentage occurrence taxa in pollen assemblages and reduce statistical noise generated by infrequently occurring taxa.

Correlations using magnetic parameters can be complicated by the relationship between magnetic grain size and sediment particle size distributions, and their effect on χ, SIRM (Bjork *et al.*, 1981; Bradshaw & Thompson, 1985; Thompson & Morton, 1979), as much as sediment interface mixing. This is particularly acute where there are particle size gradients between littoral and profundal sediments (e.g. Jones, 1980). Despite the apparent uniformity of the SIRM profiles, the finescale variability suggests that sediment transfer processes, even within a small lake such as Augher, do not result in uniform magnetic mineral sediment distributions and concentrations. This results in problems of identifying individual features in all cases. The concentration independent S-ratio profiles show much better uniformity and would suggest that the composition of the magnetic assemblages is very similar at the different sites. In Augher, SIRM profiles are very repeatable, especially when compared with some published studies (e.g. Dearing *et al.*, 1981). However, in comparison with the diatom biostratigraphy SIRM profiles are more ambiguous and errors are indicated. That errors have been suggested in the Augher SIRM correlations would indicate that correlations from more variable data sets, such as that of Dearing *et al.* (1981), may be tentative at best, and some circumspection required in their interpretation.

In Augher the eutrophication of the Lough by the creamery is clearly recorded in the sediments, and consequently in the upper 40 cm correlations are explicit, regardless of the methods used. However, with increasing depth in the sediments features become more infrequent and ambiguous; implying that in less disturbed systems core correlation may be more difficult than in those sites where anthropogenic influences (e.g. cultural eutrophication, Battarbee (1978); acidification, Flower & Battarbee (1983)) have not yet left a clear sediment record.

The Shaw diagrams demonstrate the variability of the various methods used in this study, and suggest that because of the errors involved not so much emphasis should be placed on small inferred changes in accumulation rate. Together with the associated dating errors, this would suggest that excessive zonation and correlation of cores (e.g. Dearing *et al.*, 1981) may generate noise rather than valid estimates of microfossil or dry mass accumulation rates, particularly in a whole basin approach.

Conclusion

Considering other errors involved (e.g. dating) the differences in correlations in this study are minimal and do suggest that reliable correlations can be obtained by a variety of methods. However, the comparisons do suggest those that should be checked by further sampling. More reliable and objective quantitative methods are required for core correlation and there is an increased need for comparative studies, perhaps with careful consideration of fewer cores and corroboration of the more rapid techniques by biostratigraphy.

Acknowledgements

Simone Gatsky and Alick Newman kindly drew the diagrams for this paper and Dr H. A. Viles typed and commented on the manuscript. Prof. F. Oldfield and research staff at the University of

Liverpool permitted use of magnetic equipment and provided helpful discussion. Dr R. W. Battarbee reviewed the manuscript and has shown continued interest in the work over the years. The majority of this work was undertaken while the author was in receipt of an NERC studentship.

References

Anderson, N. J., 1984. The palaeoecology and taxonomy of some *Stephanodiscus* species from the recent sediments of a eutrophic lake. Br. phycol. J. 19: 189 (Abstract).

Anderson, N. J., in prep. Recent sediment accumulation in a small lake basin, with special reference to diatoms: thesis University of London.

Battarbee, R. W., 1978. Observations on the recent history of Lough Neagh and its drainage basin. Phil. Trans. R. Soc., Lond. B 281: 303–345.

Battarbee, R. W., 1985. Diatom Analysis. In B. E. Berglund (ed.), Handbook of Holocene Palaeoecology and Palaeohydrology. J. Wiley, Lond.: 527–570.

Battarbee, R. W., C. Titcombe, K. Donelly & J. Anderson, 1983. An automated technique for the accurate positioning of sediment core sites and the bathymetric mapping of lake basins. Hydrobiologia 103: 71–74.

Berner, R. A., 1980. Early diagenesis: a theoretical approach. Princeton University Press, Princeton.

Birks, H. J. B., 1979. Guideline for numerical treatment of data. In B. E. Berglund (ed.), Palaeohydrological changes in the temperate zone in the last 15000 years. IGCP 158B Lake and Mire Environments Project Guide, Vol. 1. Department of Quaternary Geology, Lund.

Bjork, S., J. A. Dearing & A. Jonsson, 1982. Magnetic susceptibility of Late Weichselian deposits in southeastern Sweden. Boreas 11: 99–111.

Bloemendal, J., F. Oldfield & R. Thompson, 1979. Magnetic measurements used to assess sediment influx of Llyn Goddionduon. Nature 280: 50–53.

Bradshaw, R. & R. Thompson, 1985. The use of magnetic measurements to investigate the mineralogy of Icelandic lake sediments and to study catchment processes. Boreas 14: 203–215.

Clark, J. S., J. T. Overpeck, T. Webb & W. A. Patterson, 1986. Pollen stratigraphic correlation and dating of barrier-beach peat sections. Rev. Palaeobot. Palynol. 47: 145–168.

Cubitt, J. M. & R. A. Reyment (eds.), 1982. Quantitative Stratigraphic Correlation. J. Wiley, Chichester.

Davis, M. B. & M. S. Ford, 1982. Sediment focussing in Mirror Lake, New Hampshire. Limnol. Oceanogr. 27: 137–150.

Dean, W. E. Jr., 1974. Determination of carbonate and organic matter in calcareous sediments and sedimentary rocks by loss on ignition: comparison with other methods. J. sedim. Petrol. 44: 242–248.

Dearing, J. A., 1982. Core correlation and total sediment influx. In B. E. Berglund (ed.), Palaeohydrological changes in the temperate zone in the last 15000 years. IGCP 158B. Lake and Mire Environments Project Guide, Vol. III, Department of Quaternary Geology, Lund.

Dearing, J. A., J. K. Elner & C. M. Happey-Wood, 1981. Recent sediment flux and erosional processes in a Welsh upland lake-catchment based on magnetic susceptibility measurements. Quat. Res. 16: 356–372.

Edwards, K. J. & R. Thompson, 1984. Magnetic, palynology and radiocarbon correlation and dating comparisons in long cores from a Northern Irish lake. Catena 11: 83–89.

Evans, R. D. & F. H. Rigler, 1980. Measurement of whole lake sediment accumulation and phosphorus retention using lead-210 dating. Can. J. Fish. Aquat. Sci. 37: 817–822.

Flower, R. J. & R. W. Battarbee, 1983. Diatom evidence for recent acidification of two Scottish lochs. Nature 305: 130–133.

Foster, I. D. L., J. A. Dearing, A. Simpson & A. D. Carter, 1985. Lake catchment based studies of erosion and denudation in the Merevale catchment, Warwickshire, U.K. Earth Surf. processes & Landfs 10: 45–68.

Gordon, A. D. & H. J. B. Birks, 1974. Numerical methods in Quaternary palaeoecology. II. Comparison of pollen diagrams. New Phytol. 73: 221–249.

Jones, J. G., 1980. Some differences in the microbiology of profundal and littoral lake sediments. J. Gen. Microbiol. 117: 285–292.

Mackereth, F. J. H., 1969. A short core sampler for subaqueous deposits. Limnol. Oceanogr. 14: 145–151.

Oldfield, F., J. A. Dearing, R. Thompson & S. Garret-Jones, 1978. Some magnetic properties of lake sediments and their possible links with erosion rates. Pol. Arch. Hydrobiol. 25: 321–331.

Shaw, A. B., 1964. Time in Stratigraphy. McGraw-Hill, N.Y.

Thompson, R., R. W. Battarbee, F. Oldfield & P. E. O'Sullivan, 1975. Magnetic susceptibility of lake sediments. Limnol. Oceanogr. 20: 687–698.

Thompson, R. & K. J. Edwards, 1982. A Holocene palaeomagnetic record and a geomagnetic mastercurve from Ireland. Boreas 11: 335–349.

Thompson, R. & D. J. Morton, 1979. Magnetic susceptibility and particle-size distribution in recent sediments of the Loch Lomond drainage basin, Scotland. J. sedim. Petrol. 49: 801–812.

Thompson, R., J. Bloemendal, J. A. Dearing, F. Oldfield, T. A. Rummery, J. C. Stober & G. M. Turner, 1980. Environmental applications of magnetic measurements. Science 207; 481–486.

Accepted 18 March 1986.

Hydrobiologia 143: 113–118, (1986)

Interpretation of bryozoan microfossils in lacustrine sediment cores

Thomas L. Crisman[1], Ulrike A. M. Crisman[1] & Michael W. Binford[2]
[1]*Department of Environmental Engineering Sciences, University of Florida, Gainesville, FL 32611, USA*
[2]*Florida State Museum, University of Florida, Gainesville, FL 32611, USA*

Keywords: Florida lakes, bryozoan statoblasts, sediment

Abstract

Bryozoan statoblasts were isolated from the surficial sediments of 30 Florida lakes representing a gradient of trophic conditions. A prerequisite for lake selection was a good data base for water chemistry that was collected within the past 5–10 years. A novel technique of employing 210-Pb as a dilution tracer was used to estimate sedimentation rates for each lake. The net annual accumulation rate for bryozoan statoblasts was calculated for each lake and correlated with twelve physical, chemical, and biological variables. The results suggest that bryozoan distributions are strongly controlled by the extent of the littoral zone and phytoplankton biomass. It appears that bryozoan abundance in excessively weeded lakes may be limited by phytoplankton in spite of the richly developed habitat. Separation of the importance of food requirements from habitat requirements was not possible from this initial survey.

Introduction

Frey (1964) noted that while bryozoan statoblasts are common subfossils in lacustrine sediments and are relatively easy to identify to species, knowledge of extant populations and controlling factors for statoblast formation is scant. He further suggested that once such information is available, bryozoans may prove to be a valuable paleolimnological tool.

Two classes of bryozoans (Ectoprocta) are found in freshwater. Most species are members of the Phylactolaemata, a class restricted to freshwater and represented in the paleolimnological record by statoblasts. The second class, Gymnolaemata, produces winter buds of sclerotized incipient zooids (hibernacula) rather than statoblasts (Bushnell & Rao, 1974). These are not as likely to be preserved in sediments.

Statoblasts are two-valved chitinous 'overwintering' reproductive structures containing cells and yolky nutrient and are formed from the funiculus of the polypide (Bushnell, 1966). They are produced by all four families of the Phylactolaemata. Four major types of statoblasts are recognized based on external morphology and mode of dispersal (Rogick, 1943; Bushnell, 1966). Both sessoblasts and piptoblasts lack peripheral float cells and are thus poorly dispersed from the producing colony. Sessoblasts are always attached to a substrate. Spinoblasts are the largest statoblasts (0.7–1.75 mm) and are characterized by peripheral float cells fringed with hooks or spines. The final statoblast type, floatoblasts, also have peripheral float cells but lack marginal armament. These are by far the most common type of statoblasts found in sediments and reach their greatest development in the Plumatellidae.

Pennak (1953) noted that bryozoans are most common along the shoreline in water less than 1 meter deep but are frequently encountered up to a depth of 5–10 meters. He suggested that colonies generally avoid bright light, choosing instead to attach to the underside of rocks, logs, and macrophytes.

While it is tempting to assume that the presence of bryozoan statoblasts in lacustrine cores can be used to trace historical changes in the structure and extent of littoral zones, few quantitative data exist on the modern ecology of the group. In part this is a reflection of the fact that colonies are dispersed sporadically both in space (within the littoral zone) and time. It is difficult to find a population that can be monitored quantitatively for an extended period.

Bushnell (1966) collected bryozoans from 122 freshwater habitats in Michigan over a four-year period. He noted that while some species appeared to display some specificity in their choice of macrophytes as a substrate, others did not. These conclusions were supported by addition survey work in Colorado (Bushnell, 1968). In both studies Bushnell also qualitatively related the distribution of individual species to pH, alkalinity, color, and trophic state but did not attempt to quantitatively separate the importance of water chemistry, trophic state and habitat structure as factors controlling the observed species distribution patterns.

The present study is an initial attempt at quantifying the relationship between the annual accumulation rates of bryozoan statoblasts in surficial sediments and individual lacustrine parameters. Particular emphasis has been placed on separating the importance of basin morphometry, water chemistry, trophic state, and extent of the vegetated littoral zone as controlling factors for bryozoan distributions.

Methods

Surface sediments were collected during 1982 from 30 Florida lakes representing a gradient of trophic state and water chemistry. Lake selection was restricted to those basins for which detailed limnological data had been collected within the past 10 years. Sediments were collected from the deepest portion of each basin by means of an Ekman grab. Each Ekman sample was visually inspected for signs of disturbance of the sediment-water interface, after which the top 2 cm of sediment were collected and placed in separate containers for subsequent dating and bryozoan analyses.

The sediments to be dated were left undisturbed for 14 months to allow ingrowth of the polonium granddaughter, after which 210-Pb was assayed by a modification of the method of Eakins and Morrison (1978). Methodological details, rationale for the technique, and data for the lakes included in the present study are presented by Binford and Brenner (1985) and need not be elaborated upon here.

Each sediment sample (16–27 cm^3) to be used for bryozoan analyses was cooked in 10% KOH for 30 minutes after which it was filtered through an 80 μm nitex screen using a water rinse. The entire 80 μm fraction was placed in a plankton counting wheel and examind under a dissecting microscope. Bryozoan statoblasts were removed individually using an Eppendorf micropipet and placed in a plexiglass vial, the bottom of which was removed and replaced with a 40 μm nitex screen. The vial was always set upright in a shallow container filled with water sufficient to cover 1/3 of the vial volume. After examining the entire 80 μm fraction, the vial containing the statoblasts was rinsed with ethyl alcohol to remove all water and placed in a shallow container with TBA sufficient to cover the nitex screening on the vial bottom.

Permanent microscope slides were prepared by initially placing a drop of silicon oil onto a heated glass slide. With a wide bore Eppendorf pipet, statoblasts were transferred from the vial containing TBA to the silicon oil. At the end of the operation, the pipet was rinsed repeatedly with TBA into the vial, and the vial was examined under a dissecting microscope to insure that all statoblasts had been transferred to the slide. Afterwards, a coverslip was added to the cooled slide and ringed with clear nail polish to make a permanent mount.

The entire area of each slide was examined at 200× under a binocular microscope. Although statoblasts were identified to species using taxonomic keys provided in Pennak (1953), Edmondson (1959), and Bushnell (1965a, b, c), the present paper considers only the relationship between the annual accumulation rates of total bryozoan statoblasts in surface sediments and selected limnological parameters. The relationship displayed by individual species will be presented elsewhere (Bushnell & Crisman, in preparation).

Annual accumulation rates of total bryozoan statoblasts were calculated for each lake and correlated with recent physical, chemical and biological data collected by various investigators and compiled by Huber *et al.* (1982). Approximately 80% of all limnological data were collected post 1970. Given the fact that the upper 2 cm of sediment likely represent 2–10 years of accumulation, the bryozoan and limnological data are considered contemporaneous.

Results and discussion

The lakes of the present study represent a broad range of trophic conditions from ultraoligotrophic to hypertrophic (total phosphorus 0.005–0.69 $\mu g\ l^{-1}$) within a color range of 5–118 mg Pt l^{-1}. The number of bryozoan statoblasts per cm^3 of wet sediment ranged from 0–5.8 and averaged .65. Annual accumulation rates ($\#/cm^2 yr^{-1}$) ranged from 0–5.21 with a survey mean (N=30) of .87 $cm^{-2}\ yr^{-1}$.

Bryozoans are generally thought of as being filter feeders whose greatest abundance is in the littoral zone. If we accept that statoblast formation is directly proportional to population size, then the annual accumulation of statoblasts in profundal sediments may be a reflection of, singly or in combination, three general factors: 1) basin morphometry, 2) extent and structure of the littoral zone, and 3) trophic state. Correlations between statoblast accumulation rates and selected limnological parameters have been performed as a means of separating the importance of each of these factors on observed statoblast accumulations (Table 1).

The influence of basin morphometry on statoblast accumulation rates was assessed in two ways. Because bryozoans are assumed to be generally restricted to the shallow littoral, statoblast accumulation rates at the center of a lake may be inversely proportional to lake area as a reflection of long distance transport from the producing population. In spite of the wide range of lake sizes in the present study (7–14449 ha), statoblast accumulations were not significantly correlated with lake area (r^2=.10).

The lack of a relationship between statoblast accumulation rates and lake area could reflect in part the fact that Florida lakes are generally shallow thus providing bryozoan habitat over the entire bottom. Although data were not available for all lakes, mean depth for a subset of lakes (N=9) was 2.89 m (range 1.29–5.99 m). There was a trend toward decreasing statoblast accumulations with increasing mean lake depth (r^2= –.36), but the relationship was not significant. Mean depths for the lakes analyzed were all within the habitat range in which bryozoans are commonly encountered (Pennak, 1953). The lack of a significant relationship between statoblast accumulations and both lake area and mean depth is felt to be a reflection of the shallow nature of Florida lakes and does not preclude the possibility of significant relationships in other lake regions displaying greater morphometric diversity.

Extensive surveys in Michigan (Bushnell, 1966) and Colorado (Bushnell, 1968) have demonstrated the importance of macrophytes as bryozoan substrates. Huber *et al.* (1982) summarized the results of a state funded macrophyte survey that included 21 of the lakes of the present study. Statoblast ac-

Table 1. Results of correlation analysis between net annual accumulation rates of bryozoan statoblasts in surficial sediments ($\# cm^{-2} yr^{-1}$) and selected limnological variables.

	N	r^2	Significance
Lake area (ha)	23	0.10	ns
Mean lake depth (m)	9	-0.36	ns
Total area colonized by macrophytes (ha)	21	-0.003	ns
Percent lake bottom colonized by macrophytes	21	0.17	ns
Color	26	-0.19	ns
Conductivity	26	0.18	ns
pH	26	0.24	ns
Alkalinity	25	0.23	ns
Total phosphorus	28	-0.07	ns
Total nitrogen	28	0.07	ns
Total Kjeldahl nitrogen	20	0.30	ns
Chlorophyll *a*			
All lakes	28	0.07	ns
Without lakes of > 50% macrophytes	22	0.52	0.05

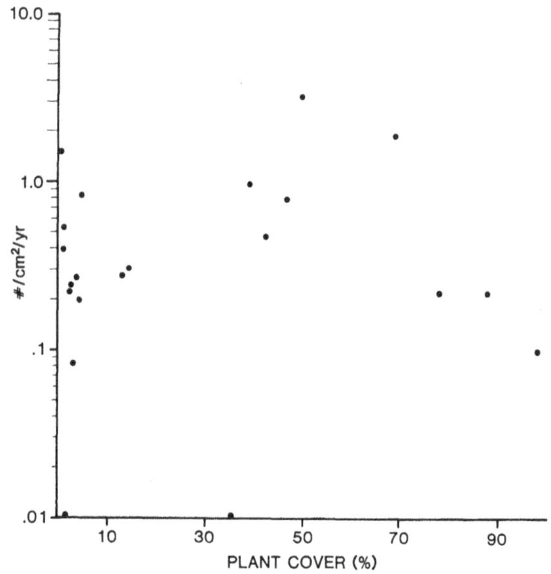

Fig. 1. Net annual accumulation rates (# cm^{-2} yr^{-1}) of bryozoan statoblasts in surficial sediments of 21 Florida lakes relative to the percent of lake bottom colonized by macrophytes.

cumulation rates were not significantly correlated (Table 1) with either the total hectares or the percent of bottom area colonized by macrophytes (emergents, submergents, free-floating, and floating leaved combined).

Closer examination of the data suggests a curvilinear relationship between statoblast accumulation rates and percent of the lake bottom colonized by macrophytes (Fig. 1). Statoblast accumulations increase progressively with increasing importance of the littoral zone up to approximately 50% area coverage by macrophytes above which accumulations steadily decline. Such a pattern is consistent with a shift in the dominant controlling factor for bryozoan populations from substrate availability below 50% plant cover to food availability above.

Support for such an interpretation is twofold. First, the four lakes displaying high statoblast accumulation rates ($>.4/cm^2/year$) coincident with extremely low macrophyte cover ($<4\%$) are all hypertrophic systems with abundant plankton for filter feeders such as bryozoans. Second, Canfield *et al.* (1984) monitored macrophyte abundance and water column chlorophyll *a* concentrations in 32 Florida lakes as well as a lake undergoing progressive macrophyte removal as part of a restoration effort. Their data suggested that water column chlorophyll *a* values in Florida lakes are significantly lower than predicted from phosphorus:chlorophyll equations only when the percent of total lake volume occupied by macrophytes exceeds 30%. They attributed this deviation to competition between macrophytes and phytoplankton for nutrients. Such evidence for significantly reduced phytoplankton abundance in macrophyte dominated lakes coupled with the progressive decline in statoblast accumulation rates in lakes with greater than 50% bottom coverage by macrophytes suggest that filter feeding bryozoans may be food limited in macrophyte dominated Florida lakes.

Color, conductivity, pH and alkalinity have often been proposed to have some direct influence on bryozoan distributions (Bushnell, 1966, 1968). Color ranged from 5–118 mg Pt l^{-1} in the present survey with highly colored lakes being intentionally excluded. This narrow range of conditions may partially explain the lack of a significant relationship ($r^2 = -.19$) for statoblast accumulation rates (Table 1). Although Bushnell (1966, 1968) noted some degree of pH and alkalinity preference for *Fredericella sultana, Plumatella fruticosa,* and *Plumatella casmiana,* total bryozoan statoblast accumulation rates were not significantly correlated with conductivity (range 23–395 $\mu S\ cm^{-1}$), pH (range 4.1–8.9) and alkalinity (range .5–120 mg l^{-1} as $CaCO_3$) in the present study.

The failure of the total bryozoan community to display strong relationships with general water chemistry is not surprising. When completed, data for individual species may yield clearer results (Bushnell and Crisman, in preparation). As with most heterotrophs, the effect of water chemistry on bryozoan species is likely to be indirect through its influence on macrophyte and phytoplankton communities rather than a direct physiological response.

Total phosphorus, total nitrogen, total Kjeldahl nitrogen, and chlorophyll *a* were used as variable of trophic state in the present study. No significant relationship was noted between statoblast accumulation rates and either nitrogen or phosphorus concentrations (Table 1).

Chlorophyll *a* values were used to calculate the trophic state index (TSI_{CHL}) of Carlson (1977). Statoblast accumulation rates were not significantly correlated with TSI_{CHL} when all lakes were con-

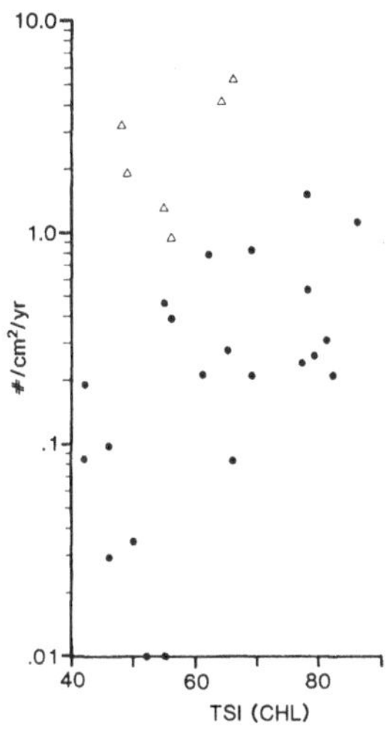

Fig. 2. Net annual accumulation rates (# cm^{-2} yr^{-1}) of bryozoan statoblasts in surficial sediments of 28 Florida lakes relative to the trophic state index (TSI_{CHL}) of Carlson (1977) based on chlorophyll. Triangles denote lakes with greater than 50% bottom coverage by macrophytes.

sidered, but removal of the six lakes having greater than 50% bottom coverage by macrophytes (Fig. 2) produced a statistically significant relationship ($r^2 = .52$, $p < .05$). Such an approach is justified as Canfield *et al.* (1984) have demonstrated that chlorophyll *a* concentrations underestimate trophic state in Florida lakes when the percent of total lake volume occupied by macrophytes exceeds 30%.

While the significant positive relationship between statoblast accumulations and TSI_{CHL} for lakes with <50% macrophyte coverage is suggestive of the controlling influence of phytoplankton biomass on population levels of filter-feeding bryozoans, the order of magnitude greater accumulation rates in lakes of comparable chlorophyll values but >50% macrophyte coverage stresses the important interaction of habitat (macrophytes) and food (phytoplankton) as controlling factors (Fig. 2). Separation of the importance of these two factors is beyond the scope of this initial survey.

The results of the current study have shown that the annual accumulation rates of bryozoan statoblasts in lacustrine sediments are a reflection of both the extent of the littoral zone and the trophic state of the lake. At present, paleolimnological interpretation should be made cautiously. It is not possible, for example, to determine whether enhanced statoblast accumulations are the result of an expansion of the littoral zone with no change in phytoplankton biomass or a contraction in littoral associated with an expansion of phytoplankton. We can not clearly separate the contribution of littoral versus planktonic autotrophic communities to total lake production using bryozoan statoblasts alone.

Paleolimnological interpretation of the statoblast record is further complicated by a number of problems. The controlling mechanism for both statoblast formation and germination are not clearly defined. The production of statoblasts displays a great deal of interspecific variability ranging from 1–2 statoblasts/polypide for *Fredericella sultana* to 15–20/polypide for *Plumatella repens* (Bushnell, 1973). Additionally, some species produce more than one type of statoblast. Quantification of the importance of the individual species to the total bryozoan assemblage of a lake must await calculation of statoblast production coefficients for each species present. Such models must be constructed for individual lake regions because as pointed out by Bushnell (1973) only 4 of the 45–50 species of freshwater bryozoans worldwide are considered cosmopolitan with species endemism in many regions, especially the tropics, approaching 40%. Finally, the need for a large volume of sediment for processing may preclude the use of statoblasts as a paleolimnological tool where only small diameter cores were collected.

While this initial investigation has shown the potential importance of statoblasts as a paleolimnological tool, it has also demonstrated the validity of the paleolimnological approach to understanding the ecology of extant bryozoans. Within a single lake, bryozoan distributions are often patchy in both space and time. It therefore is difficult to obtain quantitative data on individual populations. The use of statoblast accumulation rates is a novel way of integrating both space and time and may be of value in future ecological research on freshwater bryozoans.

Acknowledgements

We wish to thank M. Brenner and E. Fisher for assistance in the field and W. E. Bolch for giving free access to his laboratory for our 210-Pb analyses. A special thanks goes to J. H. Bushnell of the

University of Colorado and F. J. Maturo of the University of Florida for familiarizing us with the bryozoan literature and generally providing encouragement throughout. As always, S. O. and L. Crisman assisted in the preparation of the final manuscript. This project was funded by National Science Foundation grants DAR 79-24812 awarded to E. S. Deevey, T. L. Crisman, and P. L. Brezonik and DEB 82-11380 awarded to E. S. Deevey and T. L. Crisman.

References

Binford, M. W. & M. Brenner, 1985. Dilution of 210-Pb by organic sedimentation in lakes of different trophic states, and application to studies of sediment-water interactions. Limnol. Oceanogr. (In Press).

Bushnell, J. H., 1965a. On the taxonomy and distribution of freshwater Ectoprocta in Michigan. Part I. Trans. am. microsc. Soc. 84: 231–244.

Bushnell, J. H., 1965b. On the taxonomy and distribution of freshwater Ectoprocta in Michigan. Part II. Trans. am. microsc. Soc. 84: 339–358.

Bushnell, J. H., 1965c. On the taxonomy and distribution of freshwater Ectoprocta in Michigan. Part III. Trans. am. microsc. Soc. 84: 529–548.

Bushnell, J. H., 1966. Environmental relations of Michigan Ectoprocta, and dynamics of natural populations of *Plumatella repens*. Ecol. Monogr. 36: 95–123.

Bushnell, J. H., 1968. Aspects of architecture, ecology, and zoogeography of freshwater Ectoprocta. Atti. Soc. ital. Sci. nat. Museo civ. St. nat. Milano 108: 129–151.

Bushnell, J. H., 1973. The freshwater Ectoprocta: a zoogeographical discussion. In G. P. Larwood (ed.), Living and Fossil Bryozoa. Academic Press, London: 503–520.

Bushnell, J. H. & K. S. Rao, 1974. Dormant or quiescent stages and structures among the Ectoprocta: physical and chemical factors affecting viability and germination of statoblasts. Trans. am. microsc. Soc. 93: 524–543.

Canfield, D. E. Jr, J. V. Shireman, D. E. Colle, W. T. Haller, C. E. Watkins, II & M. J. Maceina, 1984. Prediction of chlorophyll *a* concentrations in Florida lakes: Importance of aquatic macrophytes. Can. J. Fish. aquat. Sci. 41: 497–501.

Carlson, R. E., 1977. A trophic state index for lakes. Limnol. Oceanogr. 22: 361–369.

Eakins, J. D. & R. T. Morrison, 1978. A new procedure for the determination of lead-210 in lake and marine sediments. Int. J. Appl. Radiation Isotopes 29: 531–536.

Edmondson, W. T. (ed.), 1959. Freshwater biology. John Wiley and Sons, New York, 1248 pp.

Frey, D. G., 1964. Remains of animals in quaternary lake and bog sediments and their interpretation. Ergeb. Limnol. 2: 1–114.

Huber, W. C., P. L. Brezonik, J. P. Heaney, R. E. Dickinson, S. D. Preston, D. S. Dwornik & M. A. DeMaio, 1982. A classification of Florida lakes. Publication 72. Water Resources Research Center, University of Florida, Gainesville, 491 pp.

Pennak, R. W., 1953. Fresh-water invertebrates of the United States. Ronald Press, New York, 769 pp.

Rogick, M. D., 1943. Studies on fresh-water bryozoa. XIV. The occurrence of *Stolella indica* in North America. Ann. N.Y. Acad. Sci. 45: 163–178.

Accepted 18 March 1986.

Hydrobiologia 143: 119–121, (1986)

On the relationship between the *Bosmina* taxa *coregoni* and *thersites* (Cladocera), as indicated by subfossil remains

Wolfgang Hofmann
Max-Planck-Institut für Limnologie, Abt. Mikrobenökologie, Postfach 165, D-2320 Plön, F.R.G.

Keywords: lake sediments, *Bosmina coregoni* f. *thersites,* long term morphological variation, polymorphism, taxonomy

Abstract

A succession of morphs from *Bosmina coregoni* f. *coregoni* to *B. c.* f. *thersites,* which vary in shell and antennule characters, occurred in an undated 176 cm long sediment core from the Neversdorfer See (N. Germany).

Introduction

Shells and head shields with first antenna of the planktonic cladoceran *Bosmina* are preserved in lake sediments. On the basis of these remains direct observation of long term morphological variation is possible. This may document the evolution of the various morphs in the subgenus *Eubosmina* and may clarify their taxonomic rank (Frey, 1964).

The simplest explanation of the results relating to *Bosmina coregoni* Baird obtained from North German lakes was that morphological transition from *kessleri* to *coregoni* morphs occurred by mucro reduction (Hofmann, 1977, 1978, 1984).

The mucroless *B. coregoni* f. *coregoni* is the most common species of the subgenus *Eubosmina* in eutrophic lakes in North Germany (Schleswig-Holstein). However, in some highly eutrophic lakes an easily recognized form with a backward-arched shell occurs. This is considered a subspecies: *B. coregoni thersites* Poppe (Flössner, 1972; Lieder, 1983). Analysis of *Bosmina* remains in a sediment core from a *thersites* lake has been carried out to obtain further information on the relationship between the taxa *coregoni* and *thersites.*

Material and methods

The Neversdorfer See is a highly eutrophic lake (84 ha, maximum depth 10 m, mean depth about 4 m) about 60 km southwest of Kiel, N. Germany.

Two sediment cores of 150 and 176 cm, respectively, were taken from sites, approximately 2 km apart (5 m water depth) using a 'Klein-Rammkolbenlot' (Züllig, 1969). The cores were cut into 5 cm slices. Preparation of the sediments followed Hofmann (1978).

Bosmina shells which represent the range of variation in a sample were drawn by means of a camera lucida in a scale of 160:1. Lengths of the first antennae were measured and numbers of the segments were counted (both distal to the triangular scale). The relative abundance of chydorid and *Bosmina* species were determined.

Results and discussion

Results according to *Bosmina* morphology and to the percentages of the cladoceran taxa were alike in both cores. The following data refer to core A from the eastern lake basin.

In the cladoceran assemblage *Chydorus sphaericus* predominated (40–64%) throughout the core, followed in abundance by *Bosmina longirostris* (mostly between 20 and 30%), whereas *Bosmina coregoni* mostly ranged between 4 and 11%. The remaining chydorids had frequencies below 10%, the most abundant species being *Alona rectangula,*

Alona quadrangularis, and *Leydigia quadrangularis.* The cladoceran assemblage as a whole distinctly reflects the eutrophic conditions (Whiteside, 1970; Flössner, 1972). Although the lake was rather shallow the planktonic elements (*Bosmina, Chydorus sphaericus*) clearly predominated, accounting for more than 90% of the cladoceran remains and indicating a high planktonic/littoral ratio (Alhonen, 1970).

A plot of the shapes of the *Bosmina coregoni* shells against sediment depth shows that only the form *coregoni* occurred in the lowermost layers while at 112 cm high-arched *gibbera* morphs and at 67 cm back-arched *thersites* morphs appeared, producing a high degree of polymorphism, because the *coregoni* forms were still present (Fig. 1). In the uppermost sediments *Bosmina* was only represented by typical *thersites* specimens. Hence a monomorphic *coregoni* phase was followed by a phase of polymorphism and finally by a uniform *thersites* stage.

A steady increase in the number of the segments of the first antenna was discernible: In the zone 172–122 cm the number mainly ranged between 12 and 16 (mean about 14). At 112 cm there was a significant shift to higher values (mostly 16–22, mean: 17–18) and in the 17 cm sample a third increase occurred resulting in values largely above 20 segments (maximum: 28, mean: 22–23) (Fig. 2).

Similarly, there was an enormous increase in antennule length: 172–122 cm: means 157–194 µm; 117–57 cm: means 205–274 µm; 52–2 cm: means 322–416 µm (Fig. 3). As with the number of segments this change was due not only to increasing range of variation but also to a displacement of the range, i.e. by the disappearance of short antennules with low segment numbers.

There was no overlap between individuals from the lowermost and the uppermost sediment layers with respect to antennule length and number of segments (Fig. 4).

The morphological changes in both the shell and the first antenna clearly show a development from *coregoni* to *thersites* morphs, the latter being characterized by back-arched shells and long first antennae, with a high number of segments (Flössner, 1972; Lieder, 1957).

It is however not clear whether polymorphism and subsequent transition to *thersites* morphs oc-

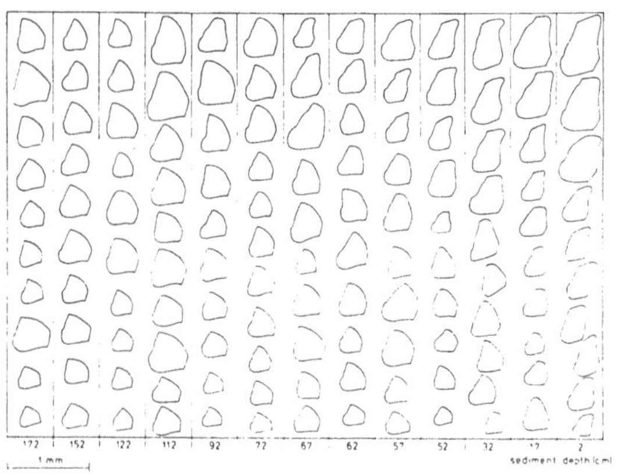

Fig. 1. Neversdorfer See, *Bosmina coregoni:* shape of the shells of 10 representative specimens from different sediment layers.

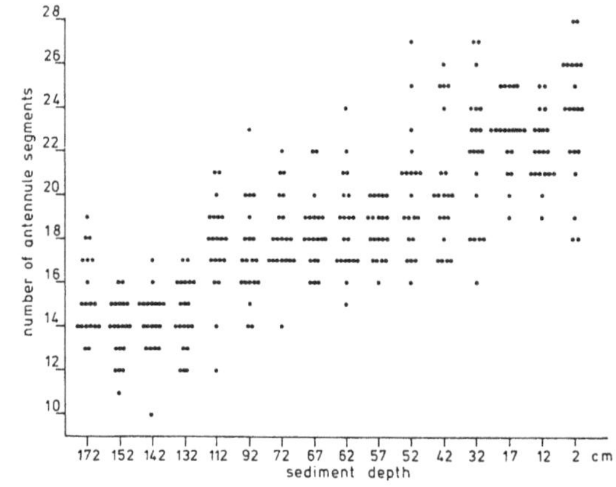

Fig. 2. Neversdorfer See, *Bosmina coregoni:* number of antennule segments in relation to sediment depth.

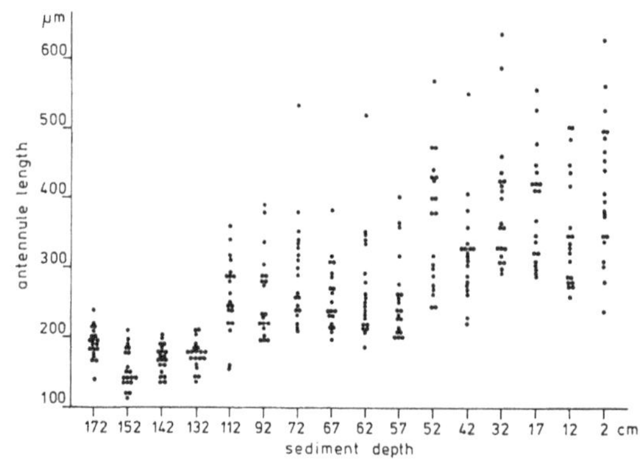

Fig. 3. Neversdorfer See, *Bosmina coregoni:* antennule length in relation to sediment depth.

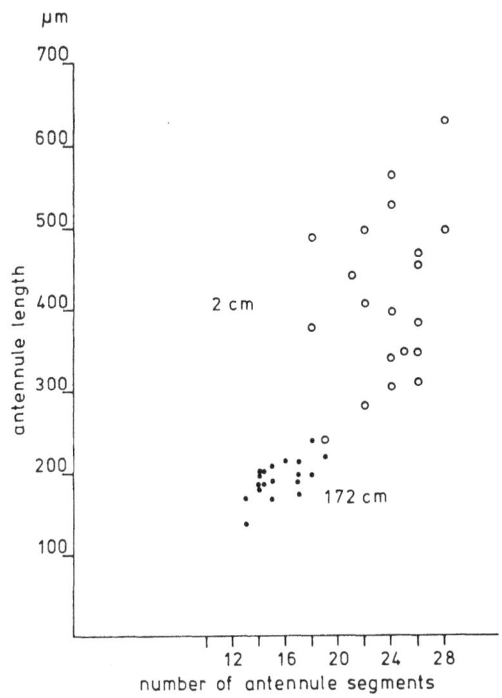

Fig. 4. Neversdorfer See, *Bosmina coregoni,* specimens from 2 cm and 172 cm sediment depth: number of antennule segments against antennule length.

curred within the original population or if it was induced by immigration of a second population. The most significant change occurred between 120 and 115 cm sediment depth, as indicated by the appearance of high-arched shells and an increase in antennule length and segment numbers.

Three different pathways for this development are possible:

1. The morphological changes occurred within a single *Bosmina coregoni* population, induced by changing environmental conditions. This implies that the morphological forms observed were morphs of *B. coregoni,* which suggests a close relationship between ecological conditions and *Bosmina* morphology.

2. Morphological variation was due to immigration of another population and so reflects species succession. This implies a species rank of *B. coregoni* relative to *gibbera* and *thersites.*

3. Immigration of *gibbera* morphs and subsequent hybridization with the original population led to polymorphism and, under the influence of changing environmental conditions, to the predominance of *thersites* morphs. This supports the subspecific status of the taxa involved.

Closer observation of the transitional phases and data from other *thersites* lakes are required to answer these questions. Further analysis should help to elucidate the taxonomical relationships between the *coregoni* taxa as well as the influence of environmental conditions on the *Bosmina* morphology.

References

Alhonen, P., 1970. On the significance of the planktonic/littoral ratio in the cladoceran stratigraphy of lake sediments. Comment. biol. 35: 1–9.

Flössner, D., 1972. Kiemen- und Blattfüßer, Branchiopoda, Fischläuse, Branchiura. In: M. Dahl & F. Peus (eds.), Die Tierwelt Deutschlands, 60: 1–499.

Frey, D. G., 1964. Remains of animals in Quaternary lake and bog sediments and their interpretation. Arch. Hydrobiol. Beih. Ergebn. Limnol. 2: 1–114.

Hofmann, W., 1977. *Bosmina (Eubosmina)* populations of the Großer Segeberger See during late glacial and postglacial times. Arch. Hydrobiol. 80: 349–359.

Hofmann, W., 1978. *Bosmina (Eubosmina)* populations of Großer Plöner See and Schöhsee lakes during late-glacial and postglacial times. Pol. Arch. Hydrobiol. 25: 167–176.

Hofmann, W., 1984. Postglacial morphological variation in *Bosmina longispina* Leydig (Crustacea, Cladocera) from the Großer Plöner See (north Germany) and its taxonomic implications. Z. zool. Syst. Evolut.-forsch. 22: 294–301.

Lieder, U., 1957. Beiträge zur Kenntnis des Genus *Bosmina* (Crustacea, Cladocera). IV. Versuch einer Monographie der Untergattung *Eubosmina* Seligo 1900. Diss. Berlin.

Lieder, U., 1983. Revision of the genus *Bosmina* Baird, 1845 (Crustacea, Cladocera). Int. Rev. Hydrobiol. 68: 121–139.

Whiteside, M. C., 1970. Danish chydorid Cladocera: modern ecology and core studies. Ecol. Monogr. 40: 79–118.

Züllig, H., 1969. Ein Klein-Rammkolbenlot zur Gewinnung ungestörter Sedimentprofile. Schweiz. Z. Hydrol. 31: 128–131.

Accepted 18 March 1986.

Hydrobiologia 143: 123–128, (1986)

Ecological correlates of net accumulation rates of Cladocera remains in lake sediments

Michael W. Binford
Florida State Museum, University of Florida, Gainesville, FL 32611, USA
Present address: Graduate School of Design, Harvard University, 48 Quincy St., Cambridge, MA 02138, USA

Keywords: paleolimnology, Florida, lake sediments, accumulation rates, Bosminidae, Chydoridae, lake trophic state, lake acidity

Abstract

Paleolimnological study of plant and animal microfossils is based on either relative abundances or net accumulation rates of various taxa. Interpretation of the latter is difficult because ecological factors that influence accumulation rates have been unknown. I use a novel method of ^{210}Pb assay to estimate net accumulation rates of bosminids and chydorids at the sediment-water interface of 15 Florida (U.S.A.) lakes which represent a wide range of trophic state, water chemistry, and morphometry. Thirty-two species of Chydoridae, and two genera of Bosminidae were recovered from the lake sediments. Accumulation rates of 6 of the 11 more abundant chydorid species and both bosminid taxa were positively correlated with trophic state as measured by chlorophyll *a* concentration. Lake size was positively correlated with accumulation rates of 4 chydorid species and both bosminids. Weedbed area was positively correlated with accumulation rates of 2 chydorids, but neither bosminid. Lakewater pH was positively correlated with only one chydorid species' accumulation rate. Multiple regression models explain statistically nearly all of the variance in each of the limnological variables, and can be used as transfer functions for paleolimnological inference.

Introduction

David G. Frey published the first, and only, clear demonstration that lake sediments were better collectors of the chydorid Cladocera than 30 years of netting in the littoral (Frey, 1960). Furthermore, his data suggested a positive relationship between the number of remains in the sediments and the standing crop of living plankton in the lake. In another paper, Frey (1962) presented evidence that the ecology of the family has been unchanged at least for the last 100000 years. Frey's major contribution in these two papers was unequivocal demonstration of the close correspondence between the fossil and living assemblages of chydorids, which set the stage for uniformitarian studies in paleolimnology.

Subsequent work has depended on measurements of the concentration of Cladocera remains in lake sediments, or examination of relative abundances of different species of the assemblages. The former measurement may not be a good indicator because deeper sediments are compacted as they are buried, which increases the concentration per unit volume of cladoceran remains. Interpretation of the latter measurement, and community-type indices (e.g. diversity, equitability, ordination axes, etc.) derived from it, is vague for several reasons. First, statistics derived from relative abundances are not robust because the data represent a closed universe. Ad hoc solutions such as omitting rare species are adopted to conform to the assumptions of classical statistical methods. Second, the ecological meaning of diversity indices are if not unknown, then at least controversial. Finally, ordination methods, originally derived for vegetation studies, are usually interpreted by analysis with statistical methods which are affected by the interdependence of variables.

The objective of this paper is to examine the relationship between limnological factors and the net accumulation rates of chydorid and bosminid remains. Accumulation rates of sedimentary constituents in surficial sediments are measured with a new method (Binford & Brenner, 1986) in 15 lakes on the Florida (U.S.A.) peninsula. I demonstrate correlations between accumulation rates of several species and measures of trophic state, lake size, and extent of aquatic weedbeds. Lakewater acidity is a correlate of only one species' accumulation rate. The correlations are the bases of transfer functions useful for application to paleolimnological studies.

Methods

Fifteen lakes in peninsular Florida were selected to represent a gradient of trophic state. The study lakes are a subset of the same lakes reported by Deevey *et al.* (this volume), Brenner & Binford (this volume), and Crisman & Crisman (this volume). Physical, chemical, and biological measurements were made by others over the past 10–20 years. The data are maintained at the University of Florida as described by Huber *et al.* (1982). Limnological variables are expressed as median values of all observations ever recorded for a particular lake.

I collected 2-cm deep surficial sediments from the center of each lake with an Ekman grab, and retained only those samples with an undisturbed mud-water interface.

^{210}Pb activity of the surface sediment was measured by alpha-spectrometry of the granddaughter ^{210}Po after extraction by the method of Eakins & Morrison (1978). Accumulation rate of dry sediment was calculated by the method of Binford & Brenner (1986), which is the fallout rate of ^{210}Pb (mBq cm^{-2} yr^{-1}) divided by the measured ^{210}Pb activity (mBq g^{-1}).

Quantitative counts of cladocerans were made by deflocculating 10 to 20 ml wet sediment in hot 10% KOH for several hours, then filtering through 102 μm and 20 μm netting. The filtrate was washed with water and tertiary butanol and suspended in a known volume of tertiary butanol, usually 5–20 ml. A variable volume of the suspension, from 0.20 ml to 0.40 ml, was mounted in silicone oil on a microscope slide. Usually more than 200 chydorid remains were counted, but remains were too sparse in a few samples. The number of remains was adjusted to the number of individuals by considering only the most abundant remain for each species.

Correlation and regression statistics were calculated with the Statistical Analysis System (SAS) in use at the Northeast Regional Data Center of the University of Florida.

Results

I found a total of 32 species of Chydoridae, and 2 genera of Bosminidae. Frey (1982: Table 9) lists 58 species of Chydoridae known to occur in Forida. Frey's sampling has covered many lakes over several years. My samples are small and quantitative, and would not be expected to contain all species. Of the 32 chydorids, 11 of them occurred in more than 5 of the lakes. I concentrate on the 11 abundant species for the remaining analyses. *Eubosmina* spp. were found in all the samples, while *Bosmina longirostris* was missing in only one lake.

The statistical correlation between accumulation rates of the animals and several limnological variables is shown in Table 1. The variables were chosen because most paleolimnological work has been concerned with them. Chlorophyll *a* concentration is correlated with primary productivity, and data

Table 1. Correlation coefficients describing the relationship between chydorid and bosminid accumulation rates and selected limnological variables (* $p < 0.05$, ** $p < 0.01$). Only taxa with more than 5 occurrences are included.

Taxon	TSI_{chl}	Chl-a	pH	Weed Area	Lake Area
Alona circumfimbriata	0.58*	0.68**	0.75**	–	0.73**
Alona quadrangularis	–	0.52*	–	–	–
Alona karua	–	–	–	0.90**	–
Alona rustica	0.58*	0.73**	–	–	0.73*
Alona verrucosa	−0.53*	–	–	–	–
Camptocercus sp.	–	–	–	–	–
Chydorus piger	–	–	–	–	–
Chydorus cf. *sphaericus*	–	0.65**	–	0.90**	–
Ephemeroporus hybridus	–	0.62*	–	–	0.86**
Leydigia leydigi	–	0.62*	–	–	0.87**
Rhyncotalona falcata	–	–	–	–	–
Total Chydoridae	0.58*	0.75**	–	–	0.96**
Bosmina longirostris	0.66**	0.74**	–	–	0.60*
Eubosmina spp.	0.74**	0.88**	–	–	0.94**
Total Bosminidae	0.77**	0.90**	–	–	0.92**

are available for the entire suite of lakes, whereas measurements of productivity are not. Both Carlson's (1977) TSI-chl and chlorophyll concentration are included to examine the effect of a scaled, log transformation on the correlation.

The accumulation rate of only one species, *Alona verrucosa*, is negatively correlated with any of the variables. Its negative correlation with TSI-chl *a* is also the only one not improved by using chlorophyll *a* concentration. Likewise, the accumulation rate of only one species, *Alona circumfimbriata*, is correlated with pH. Each of the other limnological variables is correlated with accumulation rates of several chydorids.

Accumulation rates of both bosminids are correlated with all the variables except pH and weedbed area. As with the chydorids, correlation coefficients between bosminids and TSI-chl are improved by using chlorophyll *a* concentration.

The 21 chydorids not shown on the table have a similar pattern of correlations. Accumulation rate of one additional species was correlated with TSI-chl *a*, none with pH, three with chlorophyll *a* concentration, six with weedbed area, and two with lake area.

When accumulation rates of all 32 chydorids or both bosminids are added together, the total accumulation rates are correlated with every variable except weedbed area and pH.

As examples of the accumulation rate-limnological variable relationships, Figs. 1 and 2 show the total chydorid and bosminid accumulation rates as a function of chlorophyll *a* concentration and pH. These variables were not strongly correlated, but there is a definite trend for higher accumulation rates of cladocerans at higher pH and higher chlorophyll *a* concentration.

Transfer functions for paleolimnological studies can take many forms. My choice here is multiple regression of a limnological characteristic as a dependent variable, and the accumulation rates of individuals of many species of cladocerans as independent variables.

Table 2 presents the 'best' multiple regression

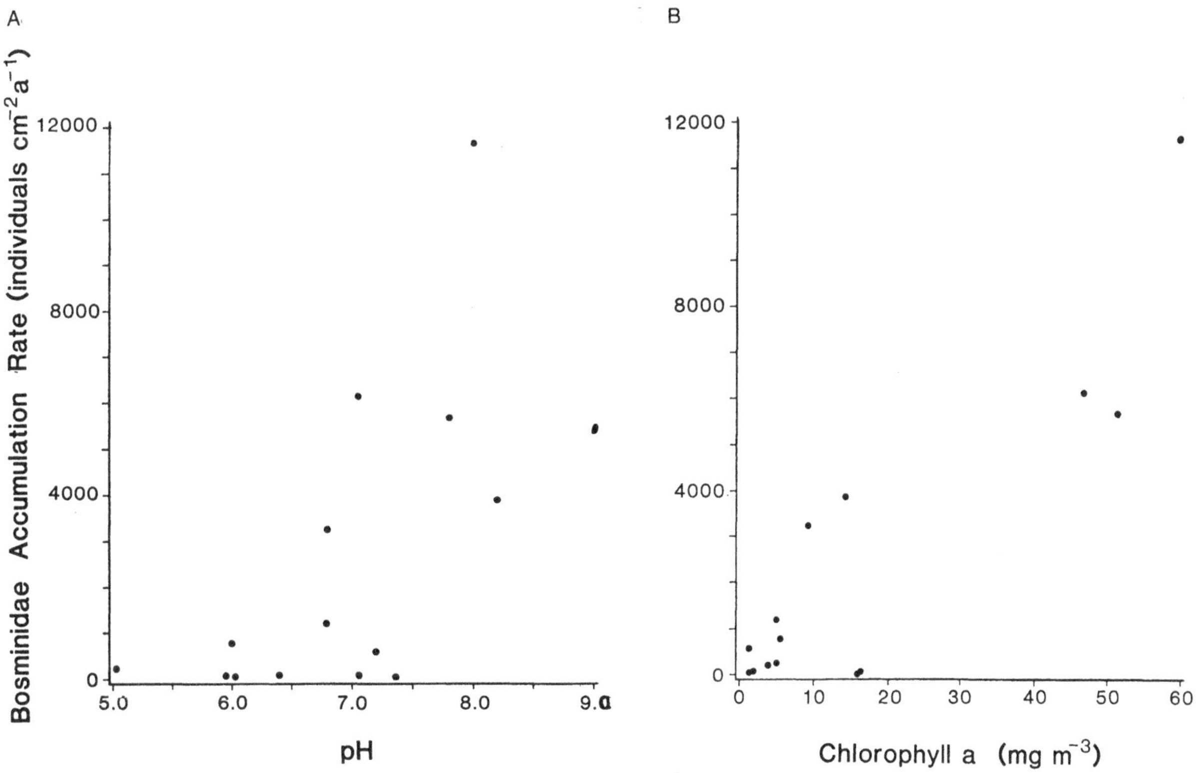

Fig. 1. Bosminidae accumulation rates as a function of a) pH and b) Chlorophyll *a* concentration. Total accumulation rate is the sum of the accumulation rates of both bosminid taxa.

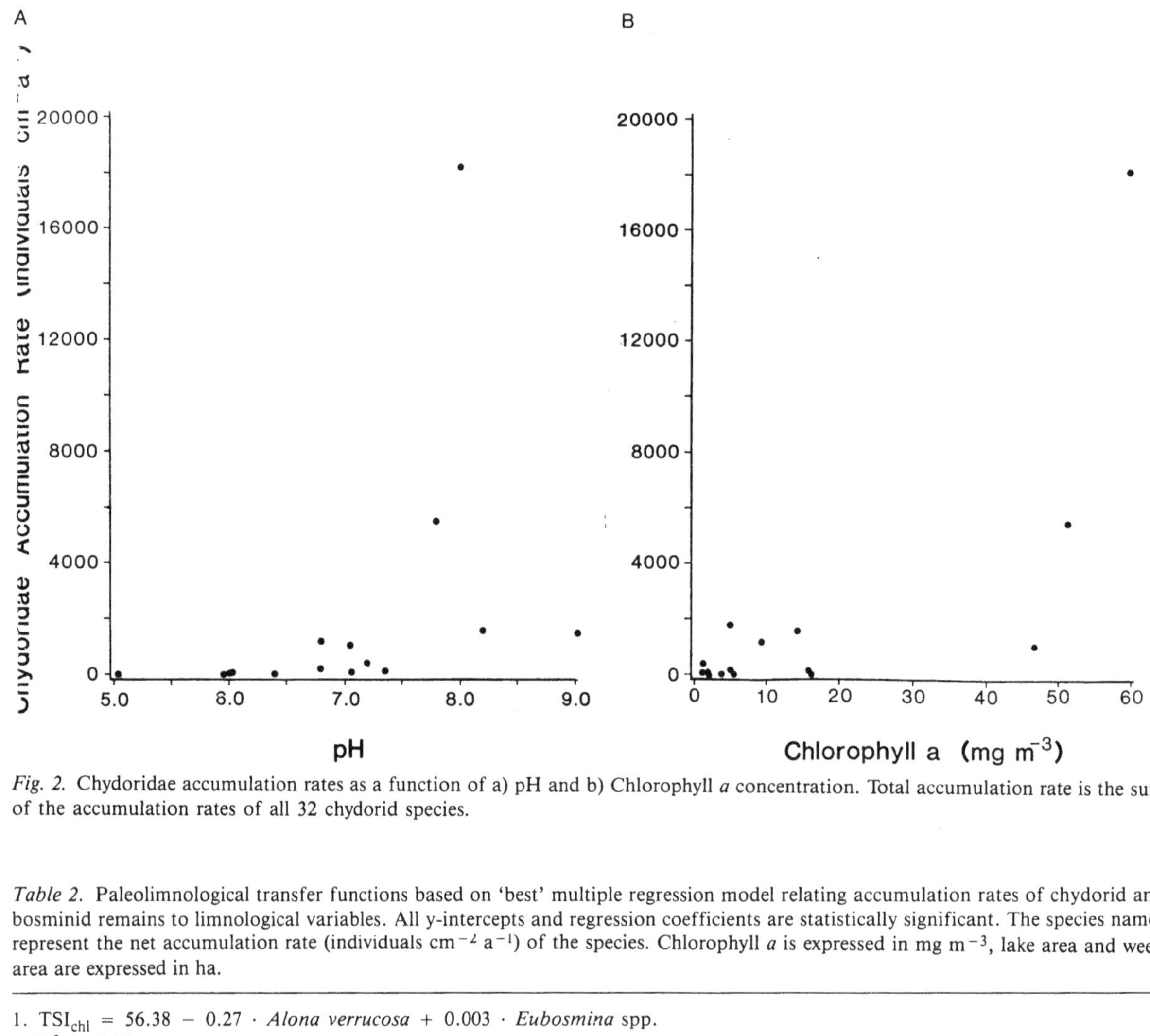

Fig. 2. Chydoridae accumulation rates as a function of a) pH and b) Chlorophyll *a* concentration. Total accumulation rate is the sum of the accumulation rates of all 32 chydorid species.

Table 2. Paleolimnological transfer functions based on 'best' multiple regression model relating accumulation rates of chydorid and bosminid remains to limnological variables. All y-intercepts and regression coefficients are statistically significant. The species names represent the net accumulation rate (individuals cm^{-2} a^{-1}) of the species. Chlorophyll *a* is expressed in mg m^{-3}, lake area and weed area are expressed in ha.

1. TSI_{chl} = 56.38 − 0.27 · *Alona verrucosa* + 0.003 · *Eubosmina* spp.
   $R^2 = 0.69$

2. Chl-*a* = 7.07 − 0.71 · *Alona karua* + 0.91 · *Alona quadrangularis*
   − 0.08 · *Alona rustica* − 0.20 · *Alona verrucosa* +
   0.22 · *Ephemeroporus hybridus* + 0.004 · *Eubosmina* spp.
   $R^2 = 0.95$

3. pH = 6.66 + 0.004 · *Alona circumfimbriata* + 0.16 · *Alona karua*
   + 0.43 · *Alona quadrangularis* − 0.04 · *Alona rustica*
   − 0.02 · *Alona verrucosa* + 0.08 · unidentified *Alona*
   + 0.01 · *Leydigia leydigi* − 0.29 · *Monospilus dispar*
   − 0.004 · *Eubosmina* spp.
   $R^2 = 0.99$

4. Weed area = 86.6 + 85.0 · *Alona karua* − 7.6 · *Ephemeroporus hybridus* + 0.86 · *Leydigia leydigi*
   $R^2 = 0.99$

5. Lake area = 1325.7 − 249.9 · *Alona karua* + 10.2 · *Chydorus* cf. *sphaericus* − 8.1 · *Leydigia leydigi* + 1.7 · *Eubosmina* spp.
   $R^2 = 0.98$

models for each of the variables of interest. Best models were selected by stepwise addition of independent variables until there was no significant improvement of R^2, the coefficient of determination. In no case was the total accumulation rate of chydorids or bosminids a better model than that of individual species. The proportion of variance in the dependent variable explained by the regression models (R^2) is surprisingly high in each case. As seen above in the correlation results, chlorophyll *a* concentration is more closely related to cladoceran accumulation rates than is TSI-chl *a*. Although accumulation rates of only one species of chydorid, and neither of the bosminids were correlated with pH, a regression model containing 9 taxa explains 99% of the variation of pH in the 15 lakes studied.

In part, this result is an artifact of multiple regression. If one includes enough independent variables, all the variance of any quantitative measure can be explained statistically, even if there is no cause-effect relationship between the variables. Fortunately, the pH regression is the only one that uses more than 6 independent variables. Furthermore, the utility of a regression model is not reduced if there is no true causal relationship. Predictions of the dependent variable are still valid.

Discussion

Paleolimnologists have assumed that more animals in the sediment meant more productive lakes (e.g. Deevey 1942, 1955), but as Livingstone (1955) and Whiteside (1983) pointed out, concentrations of materials can be misleading if accumulation rates of other sedimentary constituents have varied through time. At least for a few lakes in Florida, I have demonstrated the feasibility of using accumulation rates of some animal microfossils as predictors of limnological variables, and therefore for uniformitarian studies in paleolimnology. Accumulation rates of aquatic animals are strongly correlated with trophic state, lake size, and in some cases the area of a lake covered by weedbeds. Increased accumulation rates of the organisms do indeed mean a greater mass of chlorophyll *a* in lakewater, or a larger extent of weedbeds. The correlation with lake size may be the result of a general relationship between lake size, productivity, and alkalinity. Larger Florida lakes are shallower, more alkaline, and more productive than smaller lakes (Brenner & Binford, in prep.).

Accumulation rates of Cladocera are not well correlated with water chemistry. Cladocera are, with a few exceptions, primary consumers. Their populations respond more to higher-level biotic interactions such as food and habitat availability, predation, and competitive interactions than to strictly chemical factors within tolerance ranges.

Many paleolimnologists are studying the long-term effects of human activity with ^{210}Pb-dated cores. Some models for ^{210}Pb dating can derive instantaneous sediment accumulation rates, from which the accumulation rate of any constituent can be calculated. Changes of trophic state and of lakewater acidity are two of the most important effects of recent disturbance. The accumulation rates of chydorids and bosminids can be used to infer historical measures of chlorophyll *a* concentrations, pH, and other variables.

To extend my multiple-regression models to other lake districts would be unwise. The ^{210}Pb dilution method has not been demonstrated outside the Florida lake district. Chydorids and bosminids may have different relationships with limnological variables. Even if the direction of the effect is the same, the magnitude may be completely different. I recommend repeating this work for each lake district before attempting to apply transfer functions down-core.

Acknowledgements

I thank Elizabeth Fisher for help in the field, and Mark Brenner and E. S. Deevey for helpful discussions and editorial comments. This research was supported by NSF grants DAR 79-24812 and DEB 82-11380.

References

Binford, M. W. & M. Brenner, 1986. Dilution of ^{210}Pb by organic sedimentation in lakes of different trophic states, and application to studies of sediment-water interactions. Limnol. Oceanogr. 31: 584–595.
Brenner, M. & M. W. Binford, 1986. Material transfer from water to sediment in Florida lakes. Hydrobiologia 143: 55–61.
Carlson, R. E., 1977. A trophic state index for lakes. Limnol. Oceanogr. 22: 361–369.

Crisman, T. L., U. L. Crisman & M. W. Binford, 1986. Interpretation of Bryozoa microfossils in lacustrine sediments. Hydrobiologia 143: 113–118.

Deevey, E. S., 1942. Studies on Connecticut lake sediments. III. The biostratonomy of Linsley Pond. Am. J. Science. 240: 233–264.

Deevey, E. S., 1955. Paleolimnology of the Upper Swamp deposit, Pyramid Valley. Rec. Canterbury Mus. (New Zealand) 6: 291–344.

Deevey, E. S., M. W. Binford, M. Brenner & T. J. Whitmore, 1986. Sedimentary records of accelerated nutrient loading in Florida lakes. Hydrobiologia 143: 49–53.

Eakins, J. D. & R. T. Morrison, 1978. A new procedure for the determination of Lead-210 in lake and marine sediments. Internat. Journal of Applied Radiation and Isotopes. 29: 531–536.

Frey, D. G., 1960. The ecological significance of cladoceran remains in lake sediments. Ecology 41: 684–699.

Frey, D. G., 1962. Cladocera from the Eemian interglacial of Denmark. J. Paleont. 36: 1133–1154.

Huber, W. C., P. L. Brezonik, J. P. Heaney, R. E. Dickinson, S. D. Preston, D. S. Dwornik & M. A. DeMaio, 1982. A classification of Florida lakes. Final Report ENV-05-82-1 to the Florida Dept. Environmental Regulation, Tallahassee, Florida.

Livingstone, D. A., 1957. On the sigmoid growth phase in the history of Linsley Pond. Am. J. Science 255: 364–373.

Whiteside, M. C., 1983. The mythical concept of eutrophication. Hydrobiologia 103: 107–111.

Accepted 18 March 1986.

Hydrobiologia 143: 129–134, (1986)

Cladoceran remains as indicators of lake acidification

Dieter Krause-Dellin & Christian Steinberg
Bayerisches Landesamt für Wasserwirtschaft, Lazarettstr. 67, D-8000 München 19, FRG

Keywords: paleolimnology, chydorid cladocerans, lake acidification

Abstract

Wide gaps in the records of limnological parameters from the past 30 to 50 years prevent a documentation of lake acidification. To solve this problem, methods must be found to infer former pH-conditions from subfossils. Therefore cladoceran remains were studied in undisturbed sediment cores of selected, poorly buffered lakes in the F.R.G. As a result, a correlation was found between the lakewater pH and the species composition of the assemblage of chydorid cladocerans. This correlation is highly significant and points out the distinct influence of low pH values (<7) on this group of organisms. Reconstruction of the pH-history of three studied lakes evidences that acidification of poorly buffered lakes in the F.R.G. has taken place, even though it was not noticed previously.

Introduction

Diatom analysis of lake sediments has become quite common when dealing with acid lakes and trying to clear up patterns of lake acidification. In contrast, the remains of other groups of organisms have served somewhat as accompanying data only. This is true especially for cladoceran remains. In this study we want to give some evidence for cladoceran remains as sensitive paleoindicators of lake water acidity.

Material and methods

Sediment cores were collected from twenty-three soft water lakes in the Federal Republic of Germany and three high alpine lakes in Austria (Fig. 1). All lakes are situated on poorly buffered ground and spodsolic soils, hence sensitive to acid precipitation. Sediment cores were sampled by a gravity-sampler (after Kajak) at the deepest part of each lake and subsampled into 1 cm slices immediately after sampling. Cladoceran microfossils were recovered from sediments according to recommendations of DeCosta (1964). Each count of cladoceran remains included – if possible – at least 200 fragments of the most abundant species. Numbers of the most abundant skeletal part of each species were taken for further calculations.

Results and discussion

The group of chydorid cladocerans was chosen for investigating pH dependency because of its high diversity, its closely related species and good preservation. To detect the influence of acidity on chydorid cladocerans, counts of the topmost sediment layers were correlated with electrochemically measured pH values of lake surface water.

Evidence is given that there are distinct differences in chydorid communities according to acidity as well as altitude of the lake. Significant reduction in species number and diversity occurs when pH decreases and altitude increases. Altitude-dependency does not refer to high alpine lakes alone. Similar results are reported by Fryer (1984)

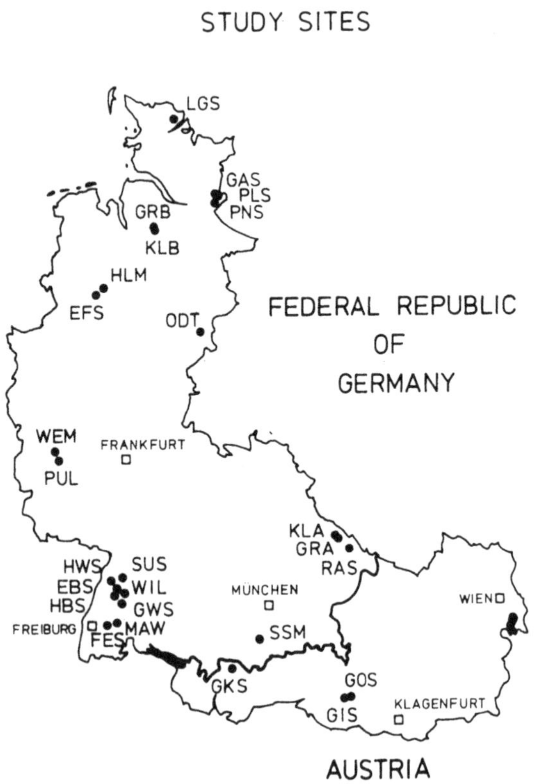

Fig. 1. Lakes studied. LGS Lüngsee, GAS Garrensee, PLS Plötschersee, PNS Pinnsee, GRB Grosser Bullensee, KLB Kleiner Bullensee, HLM Heiliges Meer, EFS Erdfallsee, ODT Oderteich, WEM Weinfelder Maar, PUL Pulvermaar, HWS Herrenwiesersee, SUS Schurmsee, EBS Elbachsee, WIL Wildsee, HBS Huzenbachersee, GWS Glaswaldsee, MAW Mathisleweiher, FES Feldsee, SSM Schwarzsee, GRA Grosser Arbersee, KLA Kleiner Arbersee, RAS Rachelsee. GKS Gossenköllesee GIS Gippersee GOS Goaßelesee

for Yorkshire lakes and by Brakke *et al.* (1984) for New England lakes. Fryer summarized that, within Yorkshire, many crustaceans, belonging to several taxonomic groups, are markedly restricted to lowland, generally alkaline habitats, while others are similarly restricted to upland, acidic situations or to such and to acidic lowland heath.

However, the presence or absence of single species cannot serve as a significant pH-indicator. Only comparison of the percentage composition of chydorid assemblages at different pH-values makes it evident that at certain pH levels there are chydorid communities composed of characteristic amounts of certain species. In other words, many species show a maximum abundance at certain pH-levels. This observation is comparable to the pH-classification of diatoms pioneered by Hustedt (1939). We therefore tried to find a procedure for chydorids similar to the pH-inference of diatoms. In order to establish the pH groups of chydorids independently of our own data, we used data from Fryer (1968, 1980, 1984), Fryer & Forshaw (1979), Flössner (1972) and Whiteside (1970). Table 1 shows all species present in the studied lakes grouped into pH-categories according to their maximum abundances.

In order to prove the applicability to our own data, cluster analyses were performed. The set of species was restricted to those present in at least six of the studied lakes. Indifferent species were ex-

Table 1. Chydorid species present in the studied lakes grouped into five pH groups according to their maximum abundances.

acidobiontic (acb)	
Alonella excisa	maximum abundance at pH < 5.5
acidophilous (acf)	
Alona rustica	
Alona guttata	
Alona intermedia	maximum abundance at pH < 7
Acroperus elongatus	
Rhynchotalona falcata	
Chydorus ovalis	
circumneutral (cir)	
Alona affinis	
Eurycercus lamellatus	
Graptoleberis testudinaria	maximum abundance at pH ≈ 7
Pleuroxus truncatus	
Chydorus piger	
Chydorus latus	
alkaliphilous (alkf)	
Alona quadrangularis	
Alona rectangula	
Pleuroxus uncinatus	
Pleuroxus laevis	maximum abundance at pH > 7
Pleuroxus adunctus	
Camptocercus rectirostris	
Monospilus dispar	
Disparalona rostrata	
indifferent	
Chydorus sphaericus	
Alonella nana	no clearcut pH preference
Acroperus harpae	
Pleuroxus trigonellus	

cluded. The percentage shares of each pair of species in the studied lakes were compared by linear correlation. Data matrices were composed of the resulting correlation coefficients. The UPGMA (unweighted pair-group method using arithmetic averages) was used as the clustering method with the sum of squares providing a measure of resemblance. The first cluster was analysed to detect species at various levels of acidity. The resulting plot (Fig. 2) shows two clusters and one species, *Alonella excisa,* being isolated. The first cluster contains all acidophilous and all circumneutral species. Additionally, one alkaliphilous species is enclosed in the first cluster. Similarly, the second cluster is composed of all alkliphilous species plus one acidophilous and one circumneutral species. It is obvious, that pH-grouping fits well to the data of our study, even though acidophilous and circumneutral species could not be separated by this procedure.

A second cluster analysis tried to group lakes according to their chydorid assemblage. Two main clusters occur (Fig. 3). One cluster contains all lakes with pH values below 5 except one (pH = 5.8). Vice versa the other cluster shows all lakes above pH values of about 6 except one.

Furthermore there is a striking increase in sum of squares distance from low to high pH, meaning that the low pH lakes are very similar in chydorid composition whereas lakes above 5.8 differ much more. A quite similar restriction of chydorid communities at low pH levels was also observed by Fryer (1980) in Great Britain.

To evaluate the dependency of chydorid assemblage on mean water pH percentages of acidobiontic, acidophilous and circumneutral species were correlated to turnover water pH by multiple linear correlation. Alkaliphilous species were involved indirectly by completing each set of data to 100%. The regression equation reads:

$$y = 6.674 - 0.031 \times \%acb - 0.035 \times \%acf + 0.003 \times \%cir$$

This regression is highly significant ($p < 0.001$). It shows an $r^2 = 0.92$ and a standard deviation of 0.33. As a formal simplification, percentages of the pH groups including their coefficients are combined to a single term, called index C. The y-axis intercept 6.674 symbolises a theoretical pH value for the exclusive presence of alkaliphilous species. Index C represents the shares of other pH-groups at lower pH values.

$$y = 6.674 - \text{index C}$$

Fig. 4 shows a plot of index C against electrochemically measured pH. It points out the extreme

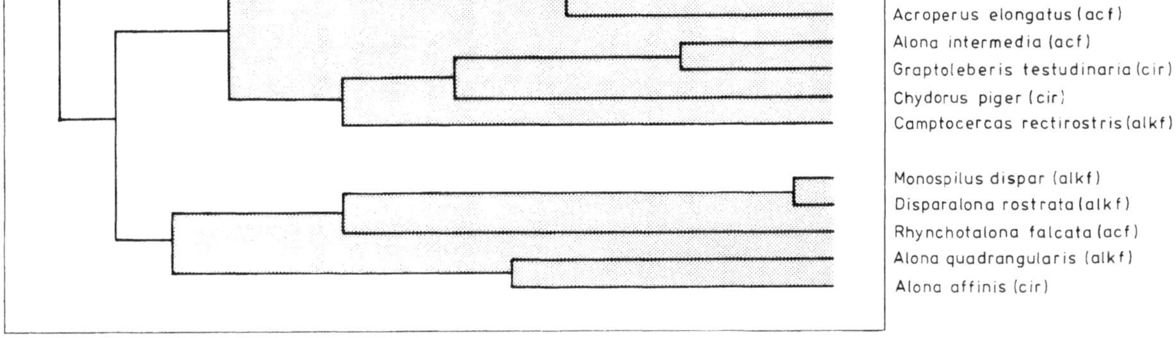

Fig. 2. Cluster dendrogram of chydorid species based on abundances at various pH conditions.

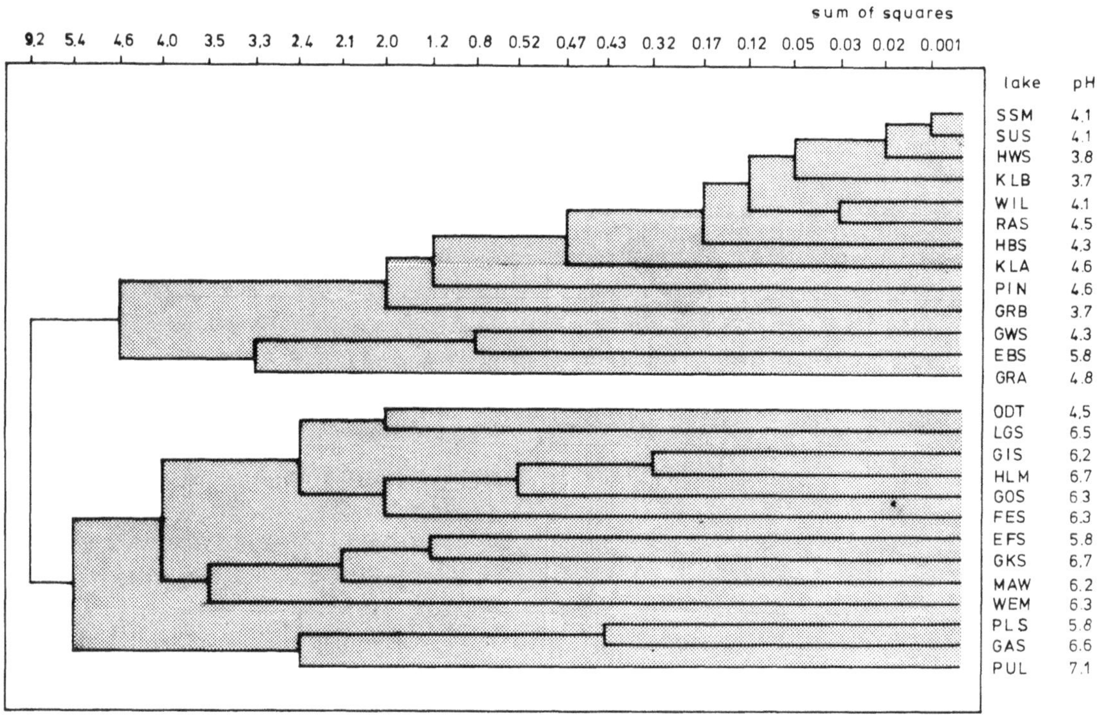

Fig. 3. Cluster dendrogram of studied lakes based on their chydorid assemblage.

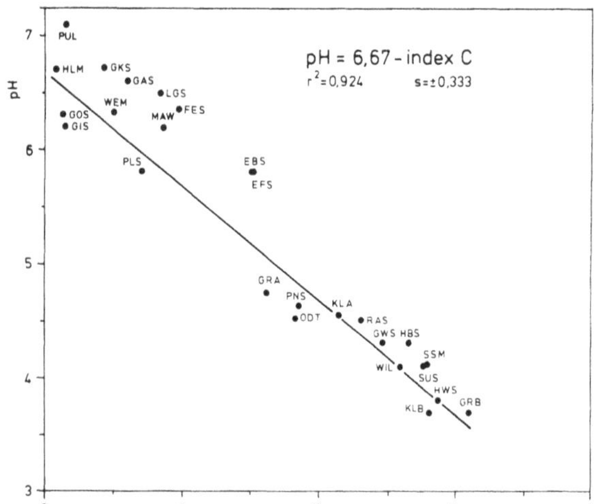

Fig. 4. Index C inferred from chydorid assemblage of 26 lakes plotted against electrochemically measured lake water pH.

uniformity of chydorid communities at a pH below 5, whereas above 5 the communities are more variable, producing a wider range of discrepancy. Within the range of index C < 1.0 the equation often leads to a slight underestimation of surface water pH.

Paleoindication of lake acidification

Index C represents a reliable tool to indicate lake water pH. The pH history of some of the lakes studied has therefore been reconstructed and compared with diatom-based reconstructions (Index B of Renberg & Hellberg, 1982) in order to evaluate their significance.

Fig. 5 shows the pH reconstruction of Großer Arbersee within the uppermost 40 cm of sediment. The date of 1963 was derived from ^{137}Cs dating. Chydorid analysis detects only a weak trend of acidification in the uppermost sediment layers. As the pH decline lies well within the standard deviation of index C, this change is not significant. Chydorid inferred pH values do not correspond well with diatom inferrence. Differences up to 0.40 pH units between index B and index C in the downcore samples may give room for speculation about other influences than acidity on the species composition. However, no significant differences occur in Pinnsee (Fig. 6). Chydorid assemblage indicates a severe drop of pH from about 6 to values of about 4.8 at the surface. These differences also occur

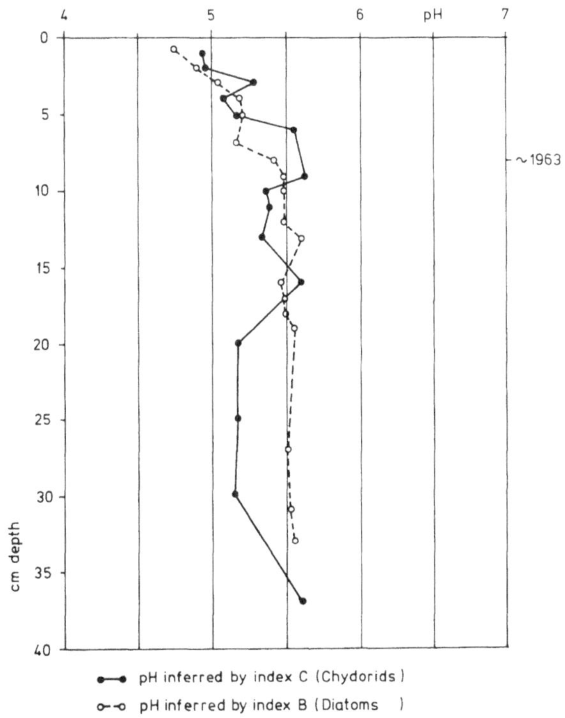

Fig. 5. Recent pH history of Grosser Arbersee inferred from chydorid and diatom assemblage. Diatom inferred pH from Steinberg *et al.*, 1984.

within diatom inferrences. Moreover, almost every short-term trend in pH fluctuations is documented simultaneously.

The same is true for Herrenwiesersee (Fig. 7). Chydorid inferred pH values prove the lake to have low pH conditions already for a long time, however being stable in the earlier period. At a sediment depth of about 13 cm, after a short rise to 4.7, pH begins to decrease continuously up to the surface. Except of the 5 cm layer, chydorid and diatom inferred reconstructions do not differ by more than 0.3 units. Summarizing the results of pH reconstruction by chydorid cladocera there is strong evidence for a good applicability of index C. A comparison with methods using diatoms frustules shows a good correspondence. So pH reconstruction by chydorid cladocerans seems to be a promising approach.

Furthermore, the results from both groups of organisms (chydorids and diatoms) support the supposition that lake acidification has recently taken place in poorly buffered lakes in the F.R.G.

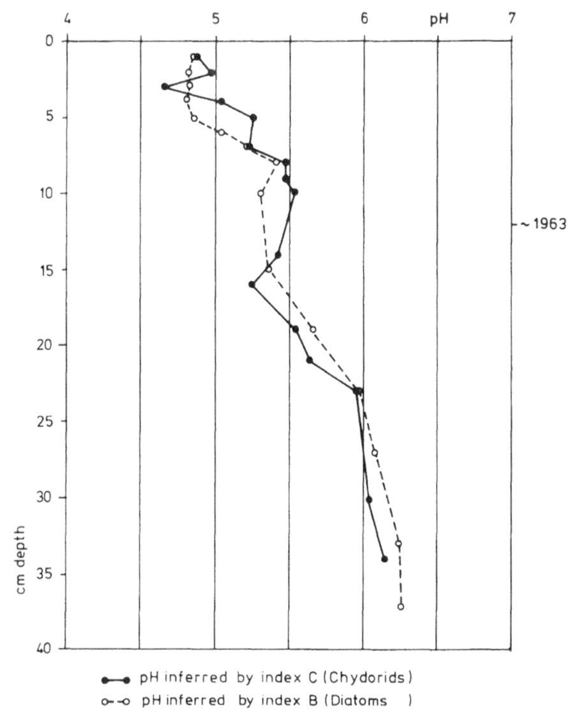

Fig. 6. Recent pH history of Pinnsee inferred from chydorid and diatom assemblage. Diatom inferred pH from Steinberg *et al.*, 1984.

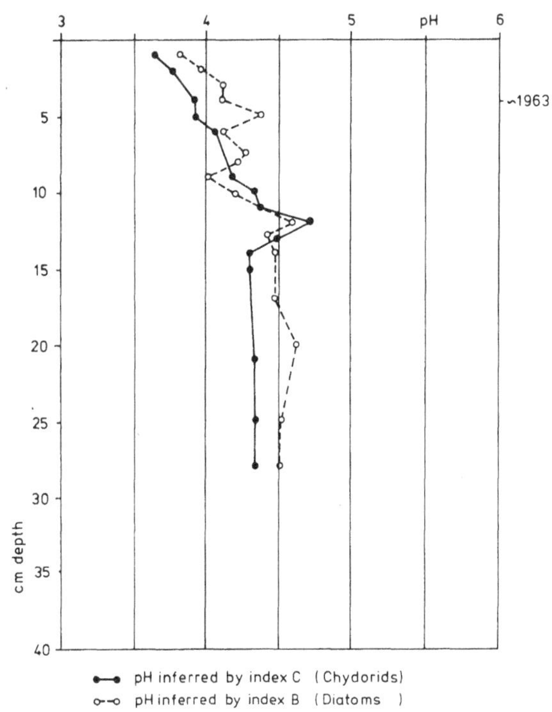

Fig. 7. Recent pH history of Herrenwiesersee inferred from chydorid and diatom assemblage. Diatom inferred pH from Steinberg *et al.*, 1984.

Acknowledgement

This study is supported by the Umweltbundesamt, Berlin, FRG, UFO-Plan. Nr. Wasser 10204333.

References

Brakke, D. F., R. B. Davis & K. H. Kenlan, 1984. Acidification and Changes over Time in the Chydorid Cladocera Assemblage of New England Lakes. Acid Precip. Ser. 6, Cap. 5: 85–104.

DeCosta, J. J., 1964. Latitudinal distribution of chydorid Cladocera in the Mississippi Valley, based on their remains in surficial lake sediments. Invest. Indiana Lakes Streams 6: 65–101.

Flössner, D., 1972. Die Tierwelt Deutschlands, 60. Teil, Kiemen- und Blattfüsser, Brachiopoda, Fischläuse, Branchiura. Gustav Fischer-Verlag, Jena: 501 pp.

Fryer, G., 1968. Evolution and adaptive radiation in the chydoridae (Crustacea, Cladocera): A study in comparative functional morphology and ecology. Phil. Trans. r. Soc., Lond., Ser B 254: 221–385.

Fryer, G., 1980. Acidity and species diversity in freshwater crustacean faunas. Freshwat. Biol. 10: 41–45.

Fryer, G., 1984. The ecology and distribution of Chydorus ovalis (Cladocera: Anomopoda) and its relevance to the recent history of the southern Pennines. Naturalist 109: 97–102.

Fryer, G. & O. Forshaw, 1979. The freshwater cladocera of Rhum (Inner Hebrides) – a faunistic and ecological survey. Biol. J. linn. Soc. 11: 333–367.

Hustedt, F., 1939. Ökologische und systematische Untersuchung über die Diatomeenflora von Java, Bali und Sumatra. Arch. Hydrobiol., Suppl. 16: 1–155, 74–394.

Renberg, I. & T. Hellberg, 1982. The pH-history of lakes in southwestern Sweden as calculated from the subfossil diatom flora of the sediment. Ambio 11: 30–33.

Steinberg, C., K. Arzet & D. Krause-Dellin, 1984. Gewässerversauerung in der Bundesrepublik Deutschland im Lichte paläolimnologischer Studien. Naturwissenschaften 71: 631–633.

Whiteside, M. C., 1970. Danish chydorid Cladocera: Modern ecology and core studies. Ecol. Monogr. 40: 79–118.

Accepted 18 March 1986.

Hydrobiologia 143: 135–142, (1986)

Magnesium and strontium in non-marine ostracod shells as indicators of palaeosalinity and palaeotemperature

A. R. Chivas[1], P. De Deckker[2] & J. M. G. Shelley[1]
[1]*Research School of Earth Sciences, The Australian National University, G.P.O. Box 4, Canberra A.C.T. 2601, Australia*
[2]*Department of Geography, Monash University, Clayton Vic. 3168, Australia*

Keywords: palaeolimnology, trace elements, magnesium, strontium, salinity, temperature, calcite, water chemistry, palaeochemistry

Abstract

Measurements of the Ca, Sr, and Mg contents of individual calcitic shells of non-marine ostracods and their host waters, both in lakes and controlled aquaria, permit the calculation of the distribution coefficients of Sr/Ca and Mg/Ca partitioning in ostracod shells. We report new $K_D[Sr]$ for seven genera of non-marine ostracods and $K_D[Mg]$ for *Cyprideis* at 25° C.

Strontium partitioning is virtually temperature-independent, and is related to the Sr/Ca of the host water, and in Ca^{2+}-saturated waters, to the salinity of the water. Magnesium partitioning is dependent on both temperature and Mg/Ca of the host water.

For simple closed-basin lakes (crater lakes are ideal), the Sr content of ostracods is a sensitive indicator of salinity and thus evaporation/precipitation changes, which in turn, indicate variations in continental climate. A 10000-year continuous palaeosalinity record established by Sr and Mg contents of fossil ostracods for Lake Keilambete, southeastern Australia, is in close agreement with an independent palaeosalinity estimate based on sediment textures.

We suggest rules that allow Sr and Mg analyses of suites of individual fossil ostracod shells from lacustrine sediments to be interpreted in terms of palaeosalinity and palaeotemperature variations.

Introduction

Non-marine ecosystems, like those in all environments, continuously evolve and therefore undergo changes. Because there have always been progressive environmental changes as well as recurrent seasonal ones, the aquatic biota are adapted to cope with the ever-changing physico-chemical aspects of the environment they inhabit. In response to environmental changes, elements of the biota may appear or disappear, and in some cases will barely survive or thrive in the ever-changing waterbodies.

To document past conditions within waterbodies, a number of approaches can be taken. One is to examine the remains of the biota which lived in the waterbody, then relate these to the biota's particular ecological requirements, and so reconstruct the nature of the aquatic environment. Another approach is to study the sediments which deposited from the waterbody and relate those to the physico-chemical parameters which controlled their formation and deposition. The latter approach is perhaps the most comprehensive because sediments form a *continuous* register of the conditions which affect them. The biota, on the other hand, being adapted to survive variations (sometimes only slight) of the aquatic environment without obvious change as a community, may provide only general information about their host environment. A third approach is to identify the physico-chemical parameters of the aquatic environment which affect the morphology and chemical composition of fossilized parts of the organism.

Although some communities may not apparently alter during a change of environmental conditions, a chemical signature may be registered within the hard parts of the organism (e.g. varying uptake of trace elements in a carbonate or silica shell) or slight changes in the morphology of the animals may occur (e.g. appearance or disappearance of spines, nodes, deformities).

The aim of this paper is to demonstrate that the chemical composition of the calcitic shells of ostracods varies in response to the chemistry and temperature of the host water, and that this can be used successfully to reconstruct past conditions in aquatic environments.

Ostracods

Ostracods are bivalved microcrustaceans that occur in nearly all types of aquatic environments (marine, non-marine, including non-marine saline) and, because of the calcitic composition of their bivalved shells, are readily preserved as fossils. Ostracods moult their bivalve shells up to nine times before reaching adulthood, building new shells from elements in the host water obtained *only* at the time of formation (Turpen & Angell, 1971).

A number of authors (Bodergat & Andreani (1980), Bodergat (1983, 1985), Chivas *et al.* (1983, 1985, 1986) for non-marine ostracods, and Chave (1954), Cadot *et al.* (1972) and Cadot & Kaesler (1977) for marine species) have attempted to determine the relationships that exist between host water and ostracod shell composition. *In vitro* experiments have established a relationship between temperature and Mg uptake by ostracods' low-Mg calcite shells, and demonstrated the relationship between Mg and Sr concentration in the host water, and the Mg and Sr contents in ostracod shells (Chivas *et al.*, 1983, 1986).

Several authors (Carbonel & Peypouquet (1983), Peypouquet *et al.* (1983) and Tölderer-Farmer & Carbonel (in prep.) have documented modifications in the architectural morphology of ostracod shells and attempted to identify the environmental parameters which might produce these morphologies. So far, no controlled laboratory experiments have been set up to demonstrate these morphological changes. However, once morphological changes are thoroughly explained, an extremely valuable tool in palaeolimnology will become available, especially if combined with the work based on trace-element analyses in ostracod shells.

Magnesium and strontium in living ostracods

It is now common knowledge that Mg and Sr, among other elements, can be incorporated in traces within the calcium carbonate lattice during calcite formation (see Lippmann (1983) for review). This phenomenon has been studied in detail for authigenically precipitated calcite and for biogenic calcite. For biogenic calcite, molluscs, cephalopods and foraminifers have been the most studied organisms. Data on these are summarized in Renard (1985). Cadot & Kaesler (1977) and Bodergat (1983) have assessed the Mg content of some ostracod shells, but it is the work of Chivas *et al.* (1983, 1985, 1986) which experimentally established the relationship between Mg and Sr uptake in ostracod shells and the properties of the host water controlling this uptake.

In the laboratory, ostracods were allowed to moult and build new shells under controlled conditions of temperature and water ionic composition (namely, well defined Mg/Ca, Sr/Ca and salinity). From the compositions of the new shells the coefficients of partitioning for Mg and Sr in living ostracods were established. Experimental conditions are described in Chivas *et al.* (1983). This K_D (presented in molar ratios) is established as follows:

$$K_D{}^m[Me]_{(T)} = \frac{{}^m(Me/Ca)_{\text{ostracod shell}}}{{}^m(Me/Ca)_{\text{host water}}}$$

where Me is either Mg or Sr, and T is the temperature of the experiment.

Analysis of the Mg/Ca or Sr/Ca of fossil ostracod shells, coupled with the experimentally determined K_D values for individual ostracod species, enables determination of Mg/Ca or Sr/Ca of the waters in which the ostracods lived. If temperature has a pronounced effect on trace-element uptake, information on palaeotemperature can be obtained after experimental determination of the thermodependence of the K_D of the trace elements in the ostracod shells.

Chivas *et al.* (1983) have demonstrated the strong positive dependence on temperature of Mg uptake in ostracods. For the large species *Mytilocypris henricae*, these authors indicate a shell Mg-content at 15° C of 6720 ± 1160 ppm (1σ) and at 25° C of 9320 ± 1410 ppm (1σ) for several water compositions.

In addition, the same authors showed that Mg is preferentially incorporated at the expense of Ca in the ostracod shell during the early stages of shell calcification. During continued growth, as the shell weight increases, Ca is taken up in proportionately greater quantities and after some days of growth (which is temperature dependent*), the Mg/Ca of the shell becomes constant with additional calcification, provided environmental conditions remain unchanged. The enhanced early uptake of Mg compared to that for Ca, appears similar to that documented by Kitano *et al.* (1971) for the abnormal uptake of Sr and Ba during the early stages of inorganic calcium carbonate precipitation. This phenomenon was thought by Kitano *et al.* (1971) to be caused by the different nature of the calcite precipitates at the early stages of the formation of calcite crystals which are much finer-grained than the stable calcite crystals formed at a later stage. It is possible that the same phenomenon applies to the early stage of ostracod shell formation where abnormally high Mg contents are recorded. Consequently, well calcified specimens of fossil ostracod shells must be chosen for chemical analysis.

Temperature has an important effect on the partitioning of Mg during low-Mg calcite formation because calcite and $MgCO_3$ have similar trigonal crystallographic structures. On the other hand, because $SrCO_3$ has an orthorhombic structure, Sr partitioning is not strongly affected by temperature during the formation of calcite; the opposite behaviour for Mg and Sr uptake occurs during the formation of aragonite which is orthorhombic. This is consistent with the findings of Chivas *et al.* (1983, 1985) which showed that Sr uptake by ostracod shells depends on the Sr/Ca of the host water and, in comparison to Mg, is little influenced by ambient temperature.

The K_D values of some ostracods have been established (Table 1). Similar K_D are found for species grouped within the same genus. For example, three species of *Diacypris* have the same K_D[Sr], as do some closely related genera (e.g. *Australocypris* and *Mytilocypris*).

Data on K_D[Sr] can be readily obtained from field collections since water temperature has little effect on Sr uptake. Accurate K_D[Mg] values can only be obtained through laboratory cultures grown at a series of constant temperatures.

An additional difficulty with Mg measurements on ostracod shells grown *in vitro* is that valves formed in the laboratory have to be fully calcified otherwise Mg/Ca values are likely to be too high and be unrepresentative of fully calcified shells. Unfortunately, ostracods appear to be stressed in aquaria, perhaps because of limited food availability; usually a small portion of aquatic grass (e.g. *Ruppia* which has been thoroughly washed) is considered sufficient food, and ensures chemical changes that might occur within the experimental water are minimal. These experiments have so far shown that ostracods rarely reach the weight of specimens collected from their natural habitat, and therefore data on Mg/Ca from ostracods grown in the laboratory have to be treated with caution, unless shells are fully calcified. A comparison of the weights of laboratory-grown shells and those from the natural environment is perhaps the easiest way to assess the extent of laboratory calcification, since it is almost impossible to detect full calcification with a binocular microscope.

*The time taken for *M. henricae* to reach a weight in excess of 120 μg/valve is 13 $\pm$ 4 (1σ) days at 15° C and 7 $\pm$ 3 days at 25° C.

Significance of study site

The choice of a site from which fossil ostracods are to be collected for the study of the trace elements of their shells is extremely important. First of all, it is necessary to choose a lake which always benefited from the same drainage system so as to be able to assume with some confidence that chemical changes within the lake water only resulted from changes of conditions within the lake's drainage area. Crater lakes (either of volcanic or meteoritic impact origin) with well defined and restricted catchments, are ideal sites. Those crater lakes known to be little affected by the regional groundwater (like those sealed by a hard clay layer formed during pedogenesis) are preferred because they act as giant rain gauges; changes in the salinity of the lake water should result only from variations in evaporation and precipitation affecting the lake and its catchment.

The selection of a crater lake which still retains water is preferred as a study site because this allows direct measurement of K_D[Me] values for modern ostracods living in the lake water. In that case, modern Mg/Ca values can also be related to modern water salinity, and if the lake water apparently remained $CaCO_3$-saturated through time (as indicated by a continuous record of authigenic carbonate sediment), a change of Mg/Ca in the fossil ostracods would permit direct calculation of past water salinities.

Mg and Sr in fossil ostracod shells

The volcanic maar Lake Keilambete in southwestern Victoria in southern Australia was selected to study the effects of past salinity changes on the chemical composition (Mg/Ca and Sr/Ca) of fossil ostracod shells. This lake has a well documented

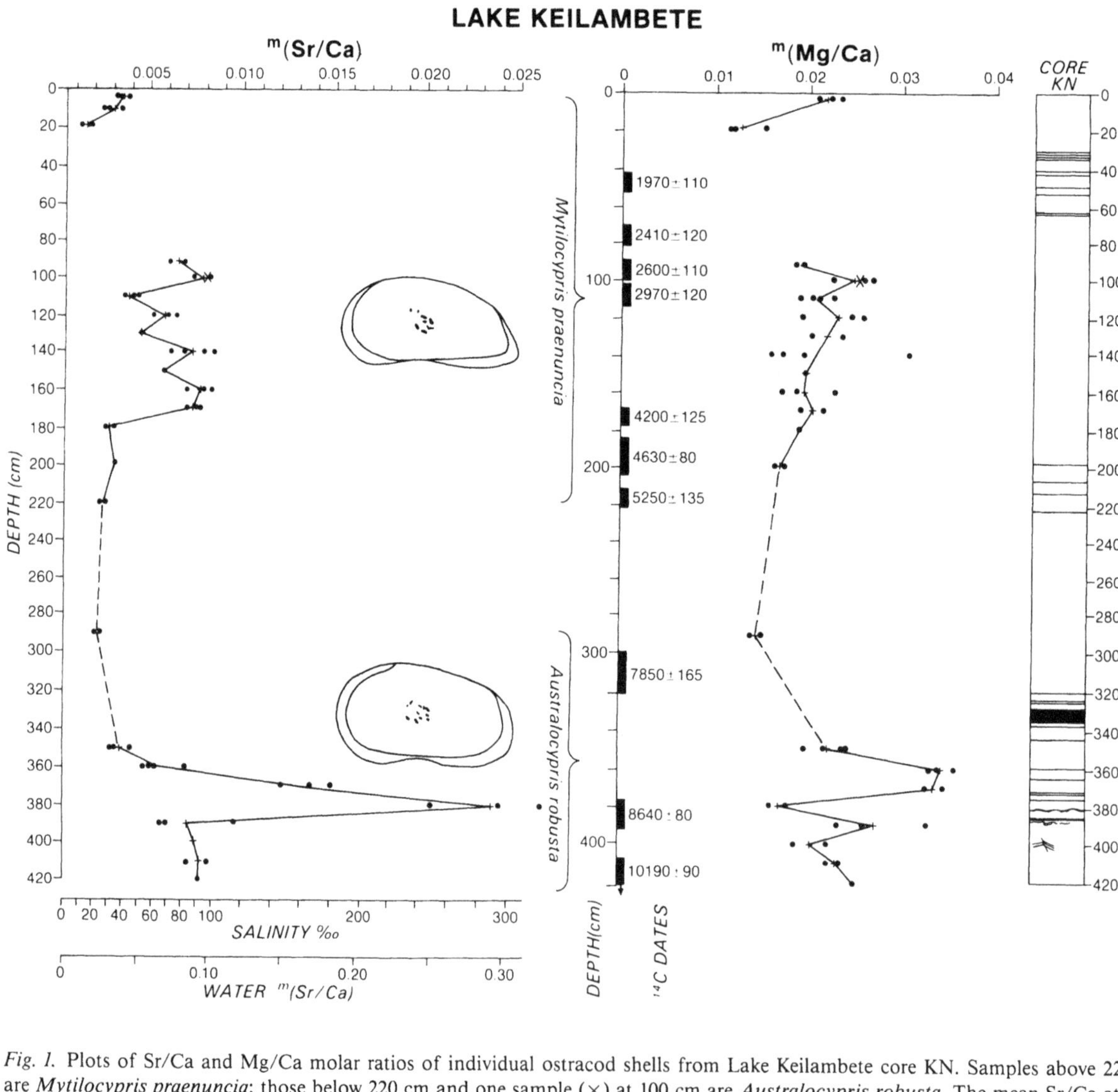

Fig. 1. Plots of Sr/Ca and Mg/Ca molar ratios of individual ostracod shells from Lake Keilambete core KN. Samples above 220 cm are *Mytilocypris praenuncia*; those below 220 cm and one sample (×) at 100 cm are *Australocypris robusta*. The mean Sr/Ca at each horizon is marked by a cross (+). The palaeosalinity and palaeo-Sr/Ca ratios of the lake (lower abscissae) are constructed using a K_D[Sr] of ~0.082 measured for modern *Australocypris* and *Mytilocypris*. The text presents further interpretation of the high values below 360 cm. Radiocarbon data are assigned by interpolation with determinations from other cores from the lake that have been correlated using the carbonate laminae shown in the right-hand column.

Holocene sedimentary record, with well defined palaeosalinity changes (Bowler, 1981), a well preserved ostracod fauna (De Deckker, 1982) and a good chronology (summarized by De Deckker, 1982).

Chivas *et al.* (1985) constructed a salinity curve for the lake's last 10000 years from Sr/Ca of ostracod shells from core KN from Lake Keilambete. This curve matched the one constructed by Bowler (1981) from sediment-size analyses by relating modern sediment particle size with present water depths, and consequently water salinity (Fig. 2). The study of Chivas *et al.* (1985) is further developed here and Mg/Ca analyses reported on the same ostracods from which Sr/Ca data were obtained. Fig. 1 shows plots of Sr/Ca and Mg/Ca analyses for individual ostracod valves or fragments of valves. Only specimens of the congeners *Mytilocypris praenuncia* and *Australocypris robusta* were analysed here. Since the salinity tolerance of these 2 species is quite different from one another (De Deckker, 1982, 1983), they rarely occur together in the natural environment. However, in one sample (depth 100 cm in core KN), individuals of both species were available for analysis, allowing us to determine that both species when growing in

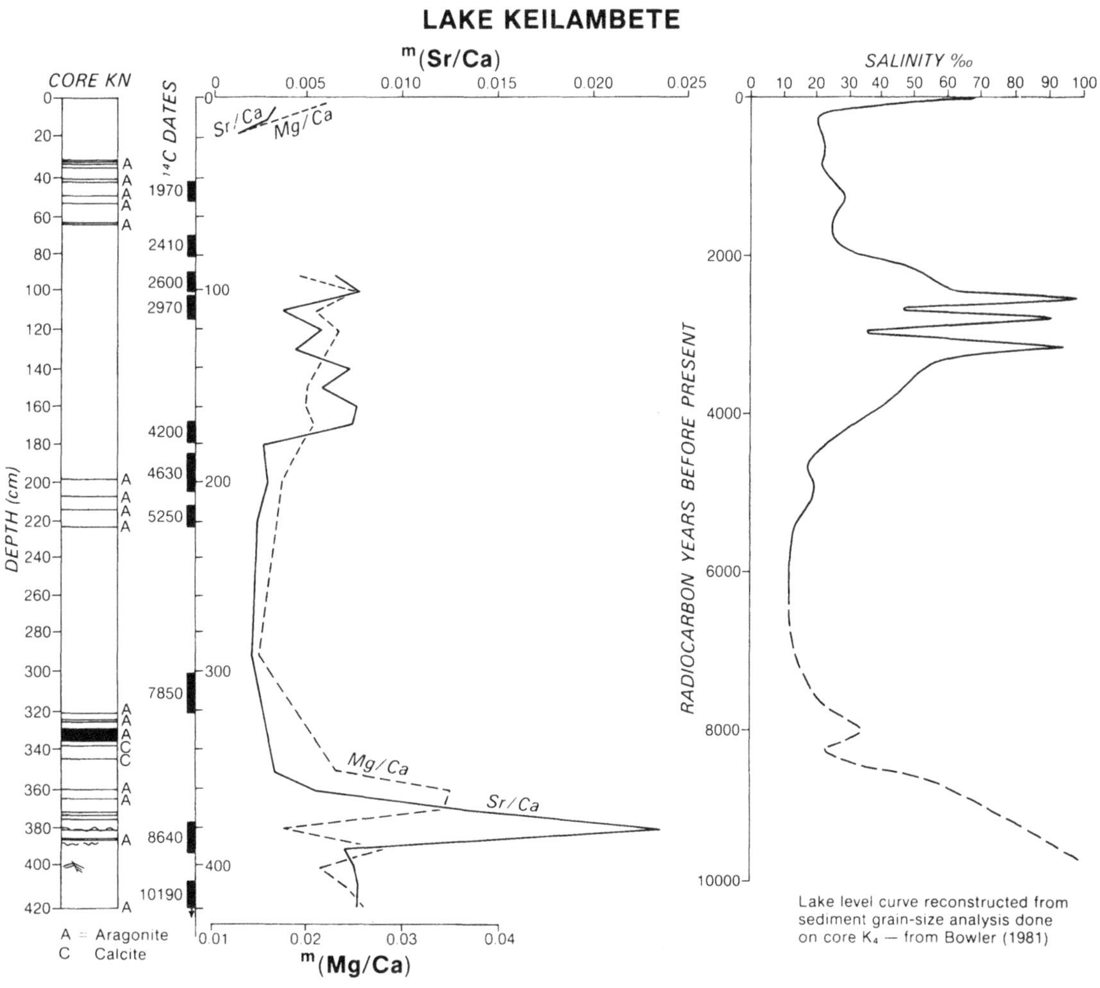

Fig. 2. Palaeo-Sr/Ca and palaeo-Mg/Ca curves for core KN from Lake Keilambete compared to the smoothed palaeosalinity curve (based on 40 samples) of Bowler, 1981. The palaeosalinity curve is derived from a water-level curve that is based on sediment grain-size analysis (Bowler, 1981).

Lake Keilambete waters had a similar K_D[Sr] of ~0.082. This value was also found by Chivas *et al.* (1985) from modern *Australocypris* individuals found living in the lake. However, this value is extremely low compared to the value of 0.22 obtained for a number of species belonging to *Australocypris* and *Mytilocypris* which have been collected in numerous saline lakes in Victoria and South Australia (Table 1). This anomalous value of 0.082 is thought to result from the unusually high Sr/Ca of Lake Keilambete water (0.075–0.078) compared to that of other Australian lakes (the seawater value is ~0.0082). Lake Keilambete's Sr/Ca value is the second highest out of 85 measurements made on Australian lake waters and we suggest that this value may be the reason for the unusual K_D[Sr] of the 2 ostracod species in the lake. This phenomenon appears to have functioned for at least the last 8000 years in the lake (= ~ the upper 360 cm of sediments in core KN). Below that level, it appears that the K_D[Sr] of ~0.082 for *Australocypris* is not applicable because the mean water salinity value (~300‰) calculated from the Sr/Ca ostracod data* appears incorrect since *A. robusta* is not known to live in such a high salinity. The salinity range of *A. robusta* from modern lakes is 7–145‰ (De Deckker, 1982). In addition, there is evidence of disturbance in the laminations at level 380 cm in core KN and below (see stratigraphical column in Fig. 1) suggesting a very low water level, with perhaps even a short period of dry lake around level 400 cm. This phenomenon had already been found by De Deckker (1982) in core KG.

At present we are unable to calibrate the Mg/Ca data from ostracod shells in terms of water salinity because of the strong thermodependence of the K_D[Mg] in calcitic shells. Laboratory experiments carried out under controlled temperatures need to be performed with *Australocypris* specimens that are allowed to moult in water with the composition of Lake Keilambete water. However, it is clear that the Mg/Ca curve for core KN (Figs. 1 & 2) generally follows the same trend of the Sr/Ca curve which was interpreted in terms of salinity by Chivas *et al.* (1985). Where the Mg/Ca curve is not parallel to the Sr/Ca curve, we ascribe this to variations in palaeotemperature.

As pointed out by Chivas *et al.* (1986), temperature changes can also be detected through analyses of $\delta^{18}O$ of ostracod shells combined with Sr and Mg analyses. For example, an increase in water temperature and/or salinity produces an increase in Mg/Ca in ostracod shells, whereas an increase in

*Note that the Sr/Ca measurements for that lower portion of the core are different from those presented in Chivas *et al.* (1985) which resulted from a plotting error in the original paper. This new plot does not alter the original interpretation of the data nor the correlation with Bowler's (1981) data. The grey clay encountered below 390 cm in core KN was not found in Bowler's (1981) core K4 and therefore could not have been interpreted in term of lake level and salinity.

Table 1. Calculated distribution coefficients K_D[Sr] and K_D[Mg] for several genera of non-marine ostracods from both natural lakes and laboratory cultures.

Ostracod genus	Strontium K_D[Sr]		Magnesium K_D[Mg]
Cyprideis	0.474±0.061	F, n=37	
	0.475±0.057	L, n=32 (25 °C)	0.0046±0.001 L, n=15 (25 °C)
Australocypris *Mytilocypris*	0.208±0.048	F, n=89	
Australocypris (Lake Keilambete)	0.824±0.0051	F, n=16	
Limnocythere	0.350±0.058	L, n=16 (20 °C)	
(L. mowbrayensis)	0.350±0.050	F, n=10	
Diacypris (3 species combined)	0.212±0.020	F, n=13	
Reticypris	0.237±0.027	F, n=7	
Cyprinotus (C. edwardi)	0.176±0.014	F, n=10	

F = field collection, L = laboratory culture, n = number of analyses of individual ostracod valves. Errors in K_D are 1σ values.

temperature produces a decrease in $\delta^{18}O$ and an increase in salinity causes an increase in $\delta^{18}O$ and Sr/Ca. A combination of these three measures (Mg/Ca, Sr/Ca, and $\delta^{18}O$) should provide a good definition of past salinity and temperature changes in lacustrine systems.

Conclusions

The following statements are relevant to the study of (palaeo) lacustrine environments using Sr and Mg in ostracod shells.

1. Sr/Ca analyses of ostracods of a single species from a single environment are similar for both adult and juvenile ostracod shells and also for poorly calcified shells or those which suffered some leaching by acid waters. Mg/Ca measurements are only indicative of the environment if restricted to fully calcified adult specimens or those at the last pre-adult moult stage. Mg analyses of partly leached specimens are to be avoided.

2. Past water salinities can be reconstructed for a site with a well defined catchment, and which is little affected by the groundwater, provided the K_D[Sr] of the ostracod genera found as fossils can be obtained, and with knowledge of the amount of Sr in the modern lake water and its salinity. Even if the K_D[Sr] values are not available, as will be the case for extinct species, a detailed relative palaeosalinity record can be obtained using Sr/Ca of ostracods.

3. If at a given single stratigraphic horizon, the Sr/Ca values of several individual ostracod shells of the same species are significantly variable, it is possible to say that the salinity varied substantially, but no information on temperature can be obtained because Sr/Ca values are little affected by temperature changes.

4. If at a given single stratigraphic horizon, Mg/Ca values of individual ostracod shells of the same species are similar, salinity and temperature variations more than likely must have been minimal during the lives of the ostracods.

5. If at a given stratigraphic horizon, Mg/Ca values of individual ostracod shells of the same species differ from one another, salinity and/or temperature of the lake water must have varied during the lives of the ostracods. If Sr/Ca measurements of the same ostracods are similar, then only temperature fluctuations occurred. If in the latter case the Mg/Ca values have decreased (or increased) compared with those of the ostracods from the underlying stratigraphic levels, a temperature decrease (or increase, respectively) can be postulated. This would be confirmed with $\delta^{18}O$ analyses of the ostracods.

6. With adequate knowledge of K_D[Sr], K_D[Mg], and the thermodependence of K_D[Mg] for the same ostracods, and knowledge of modern salinity and data on Sr and Mg in the water from the lake which is cored, it would be possible to reconstruct past lacustrine-salinity and -temperature changes.

References

Bodergat, A.-M., 1983. Les ostracodes, témoins de leur environnement: approche chimique et écologique en milieu lagunaire et océanique. Docum. Lab. Géol. Lyon 88: 1–246.

Bodergat, A.-M., 1985. Composition chimique des carapaces d'ostracodes. Paramètres du milieu de vie. Bull. Centres Rech. Explor.. Elf-Aquitaine, Mém. 9: 379–386.

Bodergat, A.-M. & A.-M. Andreani, 1980. Mise en évidence de la réponse adaptive d'une espèce euryhaline, *Cyprideis torosa* (Jones, 1850) à des conditions écologiques difficiles par l'analyse multi-élémentaire en spectrométrie de masse à étincelle. In J. Martinell (ed.), Concept and method in Paleontology. Departamento Paleontologia, Univ. Barcelona: 135–139.

Bowler, J. M., 1981. Australian salt lakes. A palaeohydrological approach. Hydrobiologia 82: 431–444.

Cadot, H. M. & R. L. Kaesler, 1977. Magnesium content of calcite in carapaces of benthic marine Ostracoda. Paleont. Contr. Univ. Kansas 87: 1–23.

Cadot, H. M., W. R. van Schmus & R. L. Kaesler, 1972. Magnesium in calcite of marine Ostracoda. Geol. Soc. Am. Bull. 83: 3519–3522.

Carbonel, P. & J. P. Peypouquet, 1983. Ostracoda as indicators of ionic concentrations and dynamic variations: methodology (Lake Bogoria, Kenya). In R. F. Maddocks (ed.), Applications of Ostracoda, Univ. Houston. Geosci., Houston: 264–276.

Chave, K. E., 1954. Aspects of the biogeochemistry of magnesium. 1. Calcareous marine organisms. J. Geol. 62: 266–283.

Chivas, A. R., P. De Deckker & J. M. G. Shelley, 1983. Magnesium, strontium, and barium partitioning in nonmarine ostracode shells and their use in paleoenvironmental reconstructions – a preliminary study. In R. F. Maddocks (ed.), Applications of Ostracoda, Univ. Houston, Geosci., Houston: 238–249.

Chivas, A. R., P. De Deckker & J. M. G. Shelley, 1985. Strontium content of ostracods indicate lacustrine palaeosalinity. Nature 316: 251–253.

Chivas, A. R., P. De Deckker & J. M. G. Shelley, 1986. Magnesium content of non-marine ostracod shells: a new palaeosalinometer and palaeothermometer. Paleogeogr. Palaeoclimatol. Palaeoecol. 54: 43–61.

De Deckker, P., 1982. Holocene ostracods, other invertebrates and fish remains from cores of four maar lakes in southeastern Australia. Proc. R. Soc. Vict. 94: 183–220.

De Deckker, P., 1983. Notes on the ecology and distribution of non-marine ostracods in Australia. Hydrobiologia 106: 223–234.

Kitano, Y., N. Kanamori & T. Oomori, 1971. Measurements of distribution coefficients of strontium and barium between carbonate precipitate and solution – abnormally high values of distribution coefficients measured at early stages of carbonate formation. Geochem. J. 4: 183–206.

Lippmann, F., 1973. Sedimentary carbonate minerals. Springer-Verlag, Berlin, pp. 228.

Peypouquet, J. P., P. Carbonel, M. Taieb, J. J. Tiercelin & G. Perinet, 1983. Ostracoda and evolution process of paleohydrologic environments in the Hadar Formation (the Afar Depression, Ethiopia). In R. F. Maddocks (ed.), Applications of Ostracoda, Univ. Houston, Geosci., Houston: 277–285.

Renard, M., 1985. Géochimie des carbonates pélagiques. Docum. Bur. Recherches. Géol. Min. (France) 85: 1–650.

Accepted 18 March 1986.

Hydrobiologia 143: 143–157, (1986)

On the preservation of carapaces of some limnic ostracods: An exercise in actuopalaeontology

D. L. Danielopol[1], L. M. Casale[2] & R. Olteanu[3]
[1]*Limnologisches Institut d. ÖAW, A5310 Mondsee, Austria*
[2]Instituto de Pesca do Interior, Seceo de Limnologia, Sao Paulo, Brazil (present address, see 1)
[3]*Institutul Geologic, Str. Caransebes 1, Bucharest, Romania*

Keywords: taphonomy, limnic Ostracoda, actuopalaeontology, Palaeogene and Neogene fossils

Abstract

The way in which the ostracod carapace decomposes through biological and physico-chemical processes is illustrated with Recent and Neogene non-marine ostracods (candonines, herpetocyprines and cytherids) from lacustrine and groundwater habitats. The degree of carapace preservation for Recent Candoninae, in the Mondsee (Austria) is shown to be dependent on the sedimentary microenvironment. The carapaces of herpetocyprines, because of their special morphological structure, decay very easily. This could explain the rare occurrrence of this ostracod group in Palaeogene deposits in Europe.

Introduction

Limnic Ostracoda have a very long fossil record, ranging from Carboniferous to Recent, the consequence of the preservation of their calcareous carapaces in sedimentary deposits (McKenzie, 1971). To date, no systematic study has considered the taphonomy of limnic Ostracoda. At first glance two problems are of paramount interest in basic and applied micropalaeontological research: 1. The relationship between the degree of carapace preservation and the sediment microenvironment. 2. The causes of selective destruction of the carapaces of some ostracod groups.

For marine ostracods, Oertli (1971, 1975) documented the way in which the degree of carapace preservation can be used for palaeoecological reconstructions of the environment. Kontrovitz (1966) and more recently Whatley *et al.* (1982) showed experimentally how ostracod shells break due to physical agencies. Valuable data on the selective destruction of Holocene ostracod shells by biological and chemical agents are also reported by Frydl (1982).

In the present contribution we describe how the carapace of limnic ostracods separate from the soft parts during moulting or after death. We also show how carapaces and/or valves are corroded, first by microorganisms (bacteria and/or fungi) and later by chemical agencies (Fig. 1).

Material and methods

We studied the state of preservation of carapaces and valves from living animals, as well as carapaces and valves which are deposited on/within sediment after the moulting or death of the animal, and which are separated from the soft parts of the body. In the latter case, both Recent and fossil material were examined. In the former the material comes mainly from drinking water wells and was obtained during a survey of the groundwater ostracods of Romania carried out by the senior author during 1970–1972. *Herpetocypris* aff. *brevicaudata* Kaufm. was found in a shallow well, 4 m deep from the village of Gogosu, near Turnu-Severin; *Pseudocandona serbani* Dan. was sampled in a well from the village of Jiblea, near Rimnicu-Vilcea; *Mixtacandona* sp. aff. *chappuisi* Klie originated from a well in the village of Cuhea, in the Iza Valley (see also Danielopol, 1980, 1982); *Microdar-*

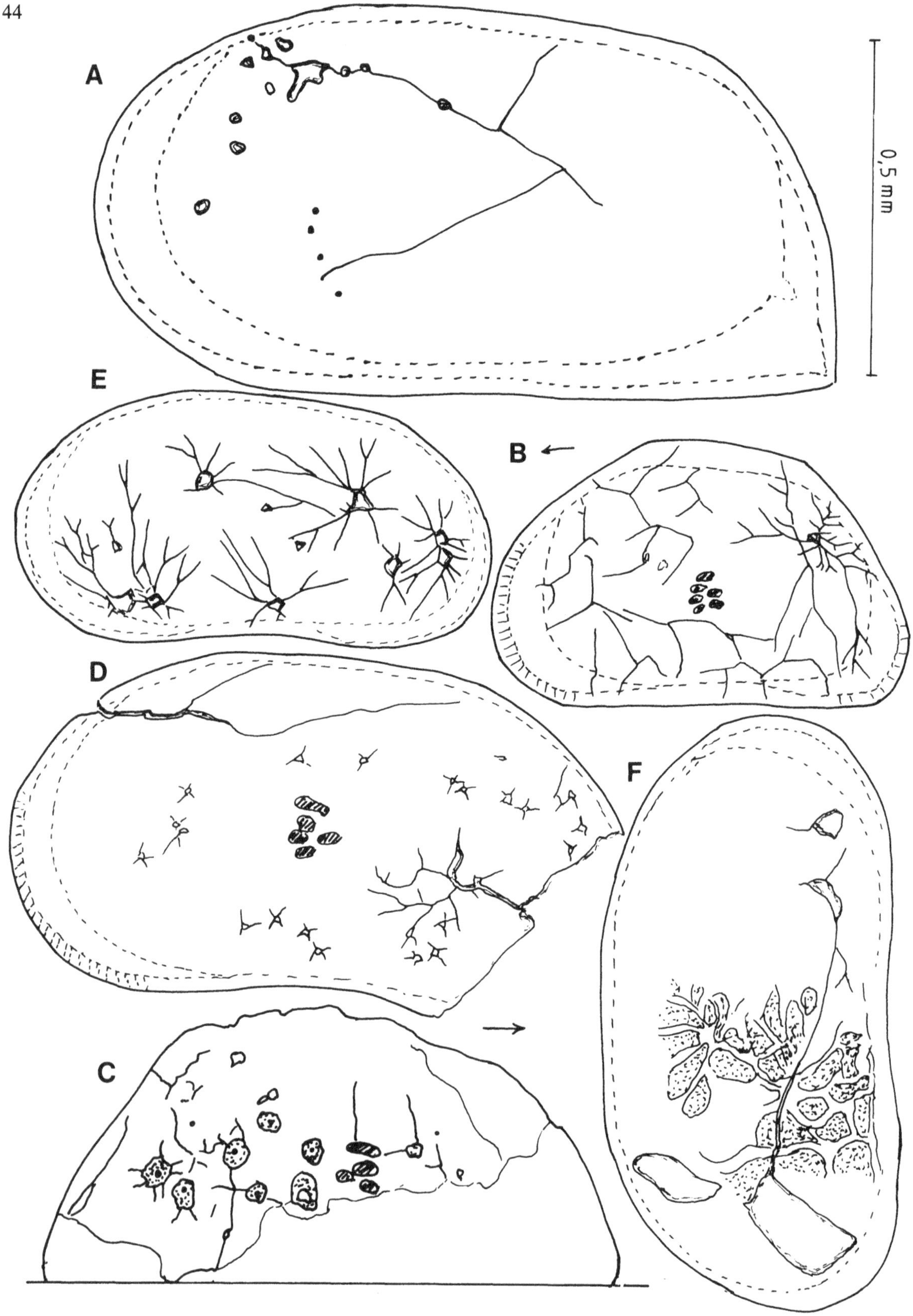
A
0,5 mm
E
B
D
F
C

winula zimmeri (Menz.) was collected on a floating fen at Lake Caldarusani, near Bucharest.

Living and subfossil ostracods also used in this study come from the Mondsee (i.e. *Cytherissa lacustris* Sars and *Leucocythere mirabilis* Kaufm.) from site MO-7, 18 m in depth (see Danielopol et al., 1985).

The fossil material used here was collected by one of us (R.O.) mainly from outcrops in Upper Pannonian sediments at the localities of Grosi and Sintesti (between Deva and Lugoj, Banat, Romania). From these localities we studied *'Candona'* sp. aff. *lobata* (Zalany), *'Candona'* sp. aff. *lunata* (Mehes), *Bakunella dorsoarcuata* Zalany. The ostracod assemblages and the stratigraphy of the Grosi outcrops, are given in Olteanu (1971). The sediment and fauna are typical of a lacustrine phase (shallow depth and low salinity) of the Pannonian Basin (for a palaeogeographical reconstruction of the Paratethys see Rögl & Steininger, 1983). We also studied one valve of *Bakunella dorsoarcuata* collected by Dr. E. Hanganu (Bucharest) from Pliocene sediments of the Dacic Basin (Eastern Paratethys) in the Fore-Carpathians (Jud. Prahova, Romania).

Investigation of the structure of carapaces and valves was done using transmitted and stereo-light microscopes and a scanning electron microscope. The ultrastructure of the hard integument was investigated on valves which had been intentionally fractured with dissecting needles.

Observations on shell ultrastructure and on the separation of the soft body from the (hard) carapace

Bate & East (1972), Dépèche (1979, 1982), Okada (1981), inter alia, described in detail the fine structure of the carapace integument of several limnic and marine ostracods. In some of the Candoninae and Cypridinae which we studied (e.g. *Mixtacandona* sp., Fig. 2, A, B) the epicuticle appears to be undulating (thin strips of the epicuticle are not attached to the underlined calcified layer). The procuticle is built of small calcite crystals (Fig. 4, A, B, E), which are enclosed by a matrix of organic membranes. The outer lamella is traversed by pore channels belonging to the normal setae (sensilae). The nervous cell of these setae are located in an rounded space within the calcified procuticle (Fig. 4, A, B). In the case of the Herpetocypridini we noticed that vacuolar voids occur within the procuticle (Fig. 4, A) which in transmitted light resemble small dots. Some of these voids are pyriform (Fig. 4A), others are rounded (Dépèche, pers. comm.). In the case of *Herpetocypris* aff. *brevicaudata, Candona neglecta* and *Microdarwinula zimmeri,* where the valves are transparent, the crystals of the outer lamella are tightly packed and oriented with the flat side parallel to the lateral side of the valve. Those areas where the wall is spongeous (i.e. where voids occur between the calcitic layer) appear opaque in transmitted light. For example see the central muscle scars of *Microdarwinula zimmeri* (Fig. 2, C, D) or the vacuolar voids in the case of the previously mentioned herpetocyprines.

Most podocopid ostracods moult eight times. During each moult the calcified carapace separates from the soft body. Those carapaces with poorly developed hinge structures, such as the Cypridacea, Darwinulacea and most limnic Cytheracea, readily disarticulate after each moult, which explains why mainly separate juvenile valves of these groups are found. Only in the case of a catastrophic mortality will the number of juvenile carapaces be higher than those of the separate valves. We call this latter situation the 'Whatley effect' after the author who first discussed the taphonomical implications of this process (Whatley, 1983). The 'Whatley effect' can be observed in the case of a sudden desiccation of shallow water bodies (we observed this among ostracods in karstic pools and channels in several caves in France and Romania, as well as in tem-

Fig. 1. Various types of preservation of fossil and subfossil ostracod valves. A – *'Candona'* sp., aff. *lobata* (Zalany), adult (Sintesti, Upper Pannonian), left valve with holes and fractures; B – *'Candona'* sp., aff. *lunata* (Méhès), juvenile (Grosi, Upper Pannonian), left valve with superficial microborings and holes; C – *'Candona'* sp., adult? (Sintesti, Upper Pannonian), right valve (broken) with holes, fractures, superficial microborings and opaque areas around the pore canals; D–F–*Candona neglecta* Sars, juveniles (Mondsee, 47 m depth, Recent); D – left valve (broken) with holes, and with superficial microborings; F – right valve with holes and fractures, etched in glycerine (pH-5.3).

porary pools, in the wetlands, formed during the spring).

Danielopol (1980) showed, that during the last part of the life of an adult ostracod (i.e. when the sexual activity usually stops), the animal moves slowly, especially with its limbs being hardly retracted within the carapace, even when physical and/or biological disturbances occur. At the end of

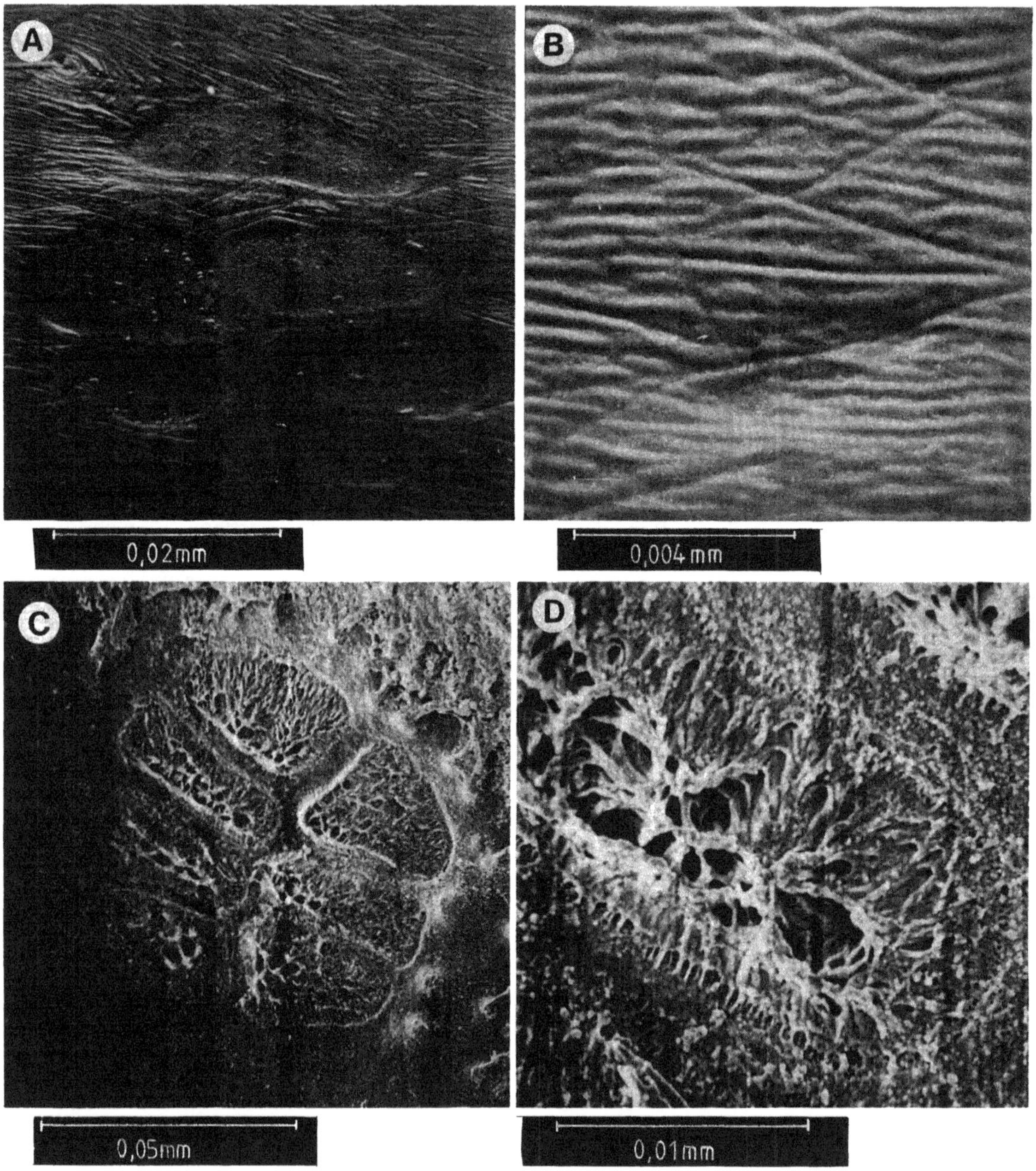

Fig. 2. Details of the outer and inner sides of the valve surface A, B – *Mixtacandona* sp., aff. *chappuisi* Klie, adult, living (Cuhea, Recent), right valve, outer surface in the central muscle scar area; A – The adductor muscle scars and the surrounding undulated epicuticle; B – detail of the epicuticle shape; C, D – *Microdarwinula zimmeri* (Menz), adult, living specimen (Caldarusani, Recent), right valve, inner surface in the central muscle scars area; C – the central adductor muscle scars with their inner spongeous structure, general view, D – detail of a muscle scars from C.

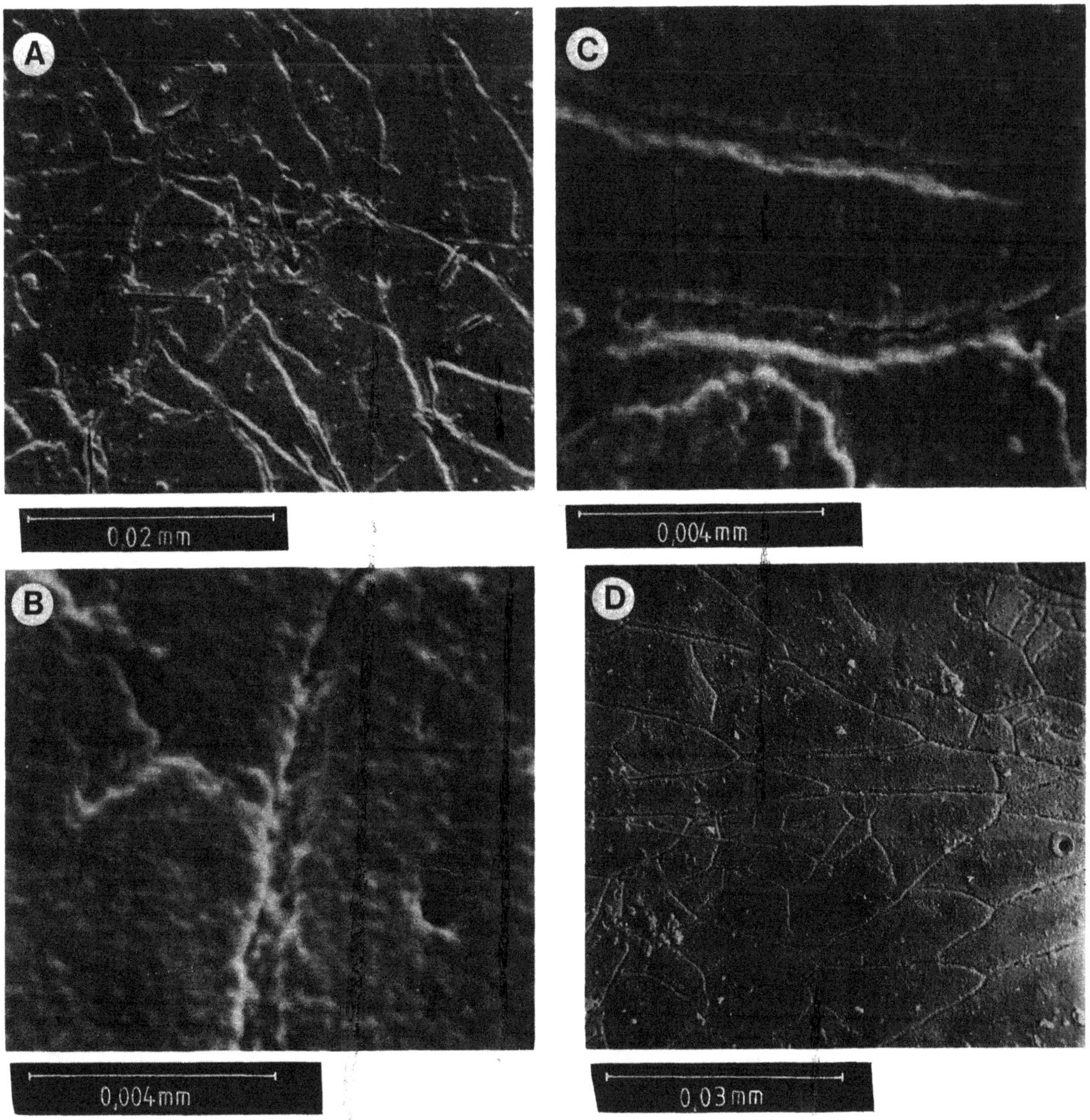

Fig. 3. Details of the superficial microborings on the outer side of the carapace; A – C – *Herpetocypris* sp., aff. *brevicaudata* Kaufm. (Gogosu, Recent), adult left valve; D – *Pseudocandona serbani* Dan. (Jiblea, Recent), juvenile, right valve.

this senile phase, the animal dies with, in most cases, its carapace open. Also, when living in waters of low oxygen or on substrates rich in organic matter and with high densities of epizoic microorganisms (bacteria, fungi, Protozoa) the ostracod will also die with its carapace open. During adverse life conditions, some ostracods can enter a torpid phase (i.e. they remain for longer periods of time with the carapace closed). Such situations occur during low temperature (Delorme & Donald, 1969) and/or poor food conditions, or during the partial desiccation of the sediment. In these latter instances, ostracods will die from this case with the valves remaining closed.

The soft body of those ostracods which die with the carapace open eventually decomposes due to

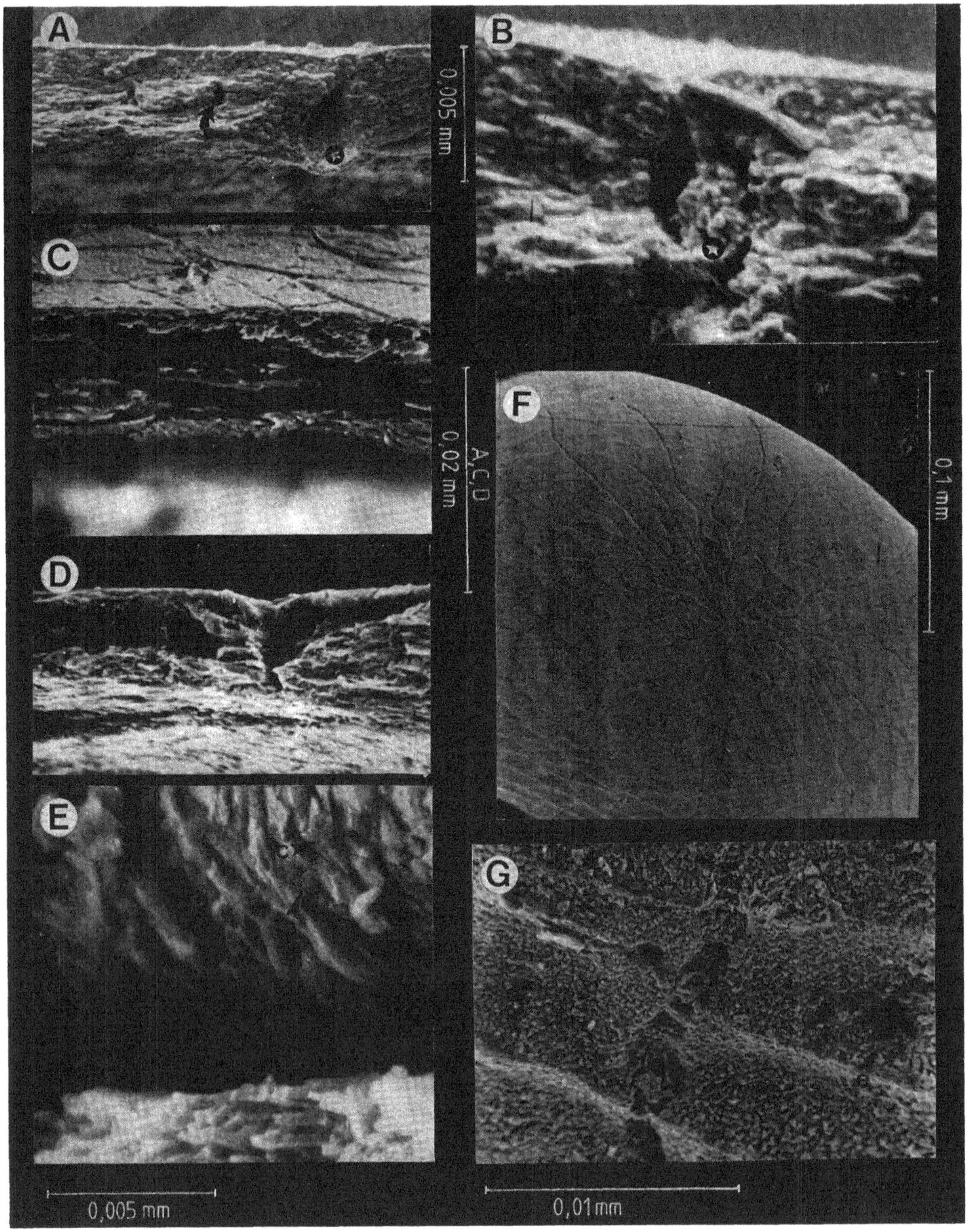
A
B
C
D
E
F
G
0,005 mm
A,C,D
0,02 mm
0,1 mm
0,005 mm
0,01mm

microbial activity. In addition, carnivorous organisms can contribute to the separation of the soft body from the carapace. For instance the ostracods *Heterocypris incongruens* and *Cytherissa lacustris* very efficiently remove the soft body of a dead ostracod from its carapace. As a result of these processes, small particles of sediments may infill the (open) carapace. During high sediment accumulation rates the carapace can be rapidly filled with sediment, and can subsequently be closed up due to superincumbent loading. In agitated environments, or those starved of sediment, the opposite will occur, i.e. the carapace will disarticulate easily into two valves.

Processes of biological and physico-chemical degradation of carapaces

We shall describe here the formation of microborings, holes and fractures on/in the carapace of several limnic ostracods from Recent and Neogene environments:

a. Microborings

In many cases, a network of thin shallow channels is noticeable on the surface of 'dead' valves and carapaces of limnic ostracods (Fig. 1, 3, 4, C, F, G, 5, 6). These channels are produced by organisms referred to here as microborers. Because in most of the cases it is difficult to identify these organisms only the results of their activities is mentioned; they are called microborings.

In the case of the Recent and Neogene limnic ostracods studied here, microborings occur mainly in the epicuticle and the upper part of the procuticle (Fig. 1, 3, 4, F, G, 5, A–C). They form a dendritic pattern, which in many cases is dichotomously ramified (Fig. 1, B, D, E, 3, 4, F, 5, A–C, 6, C). The channels are very straight, normally less than 1 μm. Some of these channels converge in larger holes (Fig. 1, 5, A, C). Because of the shape and size of the studied microborings and because of fact that some of them come from aphotic habitats (groundwater), we believe that they are not produced by algae but by fungi and/or bacteria (Drs Philippon and Le Campion-Alsumard pers. comm.). Chytridiales fungi especially (det. Dr. K. Messner) and/or filamentous bacteria like Actinomycetes (det. Prof. M. Moser) could produce the microborings figured in the present paper. The channels are probably produced by the thin hyphae and the larger holes by the sporangia like structures. Both fungi and Actinomycetes are probably attracted by the nutrient source of the organic epicuticle and the procuticle matrix. These two groups of organisms are known to decompose complex organic molecules like chitin with enzymes produced by the hyphae or sporangia (Jagnow, 1957, Okafor, 1966). Fig. 6, D shows an microborer on a 'dead' valve of *Leucocythere mirabilis.* In Fig. 6, A it is possible to see how the microborer destroyed the areas rich in organic matter around a normal pore canal and seta; from there the hyphae radiated centrifugaly. Fig. 6, C shows that the microboring channels remain within the superficial layer of the outer lamella. Fig. 3, C shows that the microborings on *Herpetocypris* aff. *brevicaudata* penetrate into the procuticle. A deep microboring is visible in *Cytherissa lacustris* (Fig. 7, C). The penetration into the calcified procuticle is only possible by dissolution of the calcitic crystals. Veldkamp (1955) showed that many chitin decomposing bacteria like Actinomycetes beside decomposing organic molecules, also produce lactic acid which can significantly lower the pH of the microenvironment. This seems also to be the case of the microborers which attack ostracod valves. The production of an acidic microenvironment immediately surrounding the hyphae or sporangia allow the localised dissolu-

Fig. 4. Details of the outer calcareous lamella of fossil and Recent ostracods; A – E – *Herpetocypris* sp., aff. *brevicaudata* Kaufm. (Gogosu, Recent), adult; A, B – right valve (broken) of a living specimen; A – inner structure of the 'normal' pore voids (white star) and the vacuolar voids (black star); B – detail of the nervous cell (white star) of the 'normal' seta; C, D – 'dead' valve, biologically and chemically corroded (note on the outer surface the microborings and within the calcified procuticle, the large voids are caused by chemical dissolution of the calcitic crystals; the white star indicates the place of the 'normal' pore void; E – details of the chemically corroded procuticle (c= calcitic crystals); F, G – *Bakunella dorsoarcuata* (Zalany), (Fore Carpathians, Jud. Prahova, Pliocene), adult, right valve; F – superficial microborings; G – details of a microboring and of small epicuticle (e) rests.

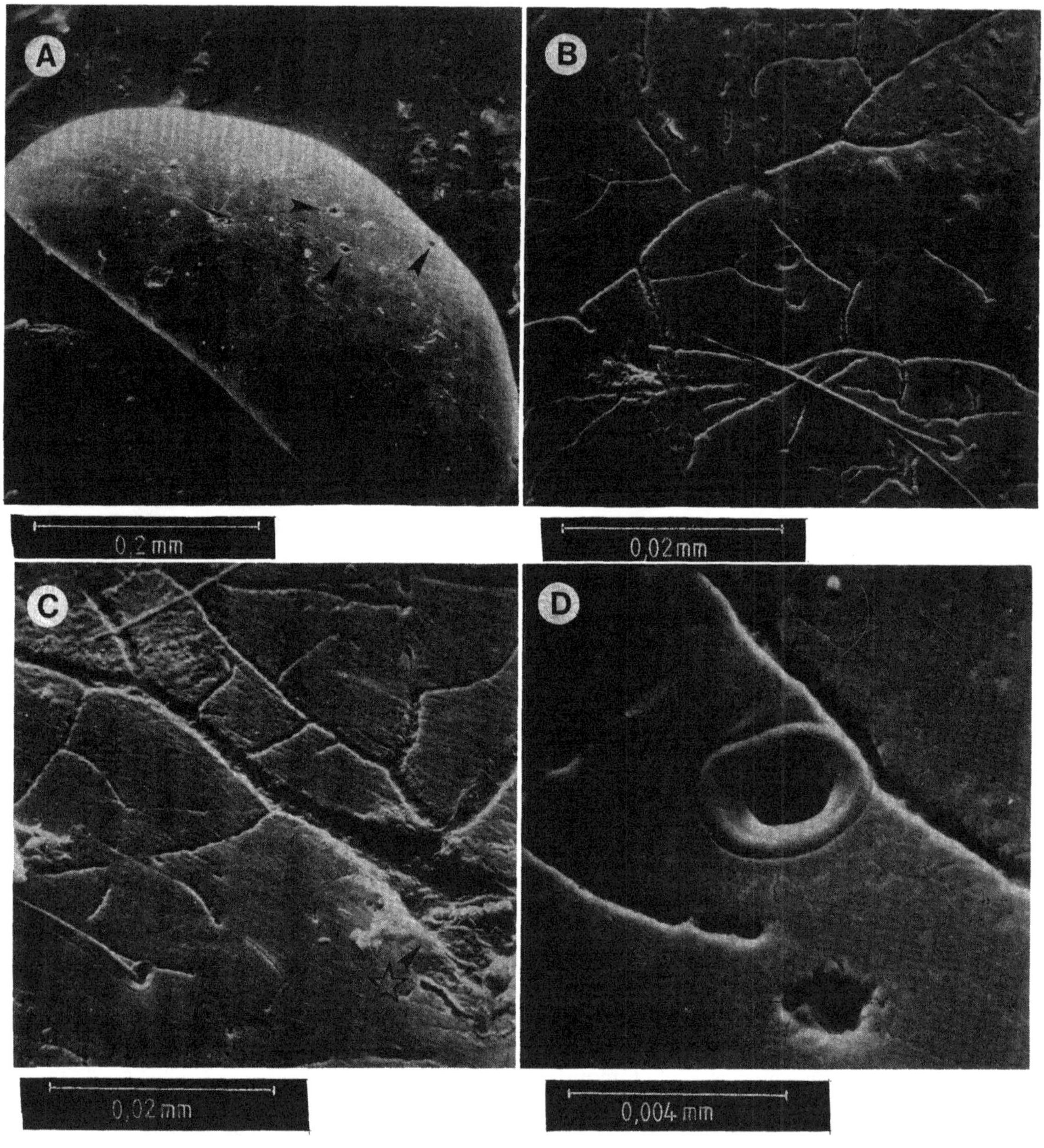

Fig. 5. Details of the superficial microborings and holes (arrows) on the outer side of a valve; *Mixtacandona* sp., aff. *chappuisi* Klie (Cuhea, Recent), adult, left 'dead' valve; A – general view; B – D – details of the microborings and holes around a normal pore opening.

tion of the calcitic layer and, therefore, the penetration of the microorganisms into the inner part of the valve.

Microborings such as those we describe from Recent ostracods (i.e. *Mixtacandona* sp., *Herpetocypris* aff. *brevicaudata, Candona neglecta, Leucocythere mirabilis*) can be also seen in fossil Neogene ostracods (*Bakunella dorsoarcuata* (Fig. 4, F, G), *'Candona'* aff. *lobata* (Fig. 1, A) and *'Candona'* aff. *lunata* (Fig. 1, B) from the Upper

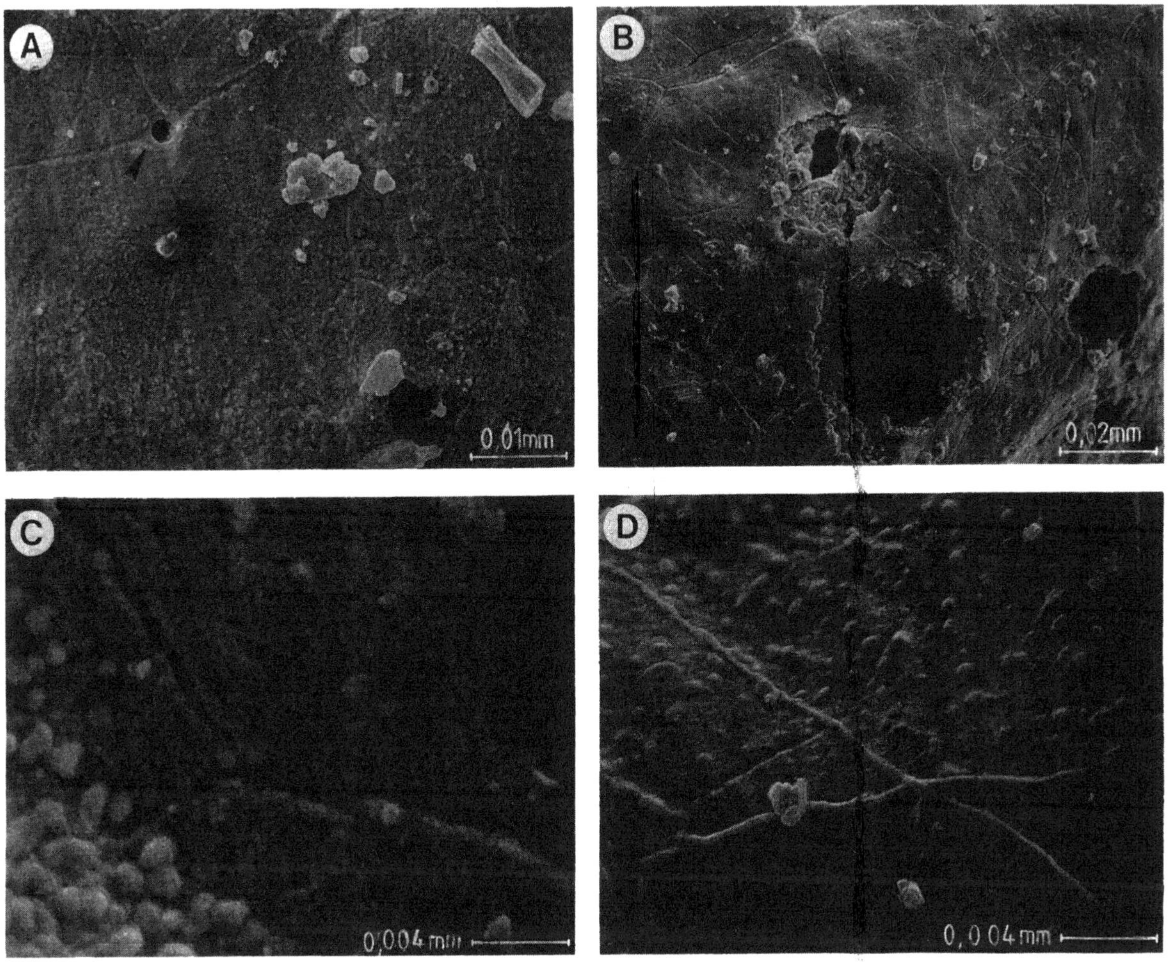

Fig. 6. Details of a 'dead' valve with microborer (bacteria or fungi), microborings and holes; *Leucocythere mirabilis* Kaufm. (Mondsee, Recent), adult, right valve. A – details of microborings around a 'normal' pore canal: B, C – details of microborings and holes; D – hyphae of a microborer.

Pannonian (Miocene) and the Pliocene of the Paratethys). Freels (1980) figured Candoninae valves from Pliocene limnic sediments of Turkey with microborings. All these data suggest that the biological corrosion of limnic ostracods by fungi and/or filamentous bacteria like Actinomycetes is a very common phenomenon and which has a history of at least 11 My years (our oldest material is Upper Miocene). It is worth mentioning that the microborings of marine ostracods differ in shape and structure (e.g. see the microborings of *Agrenocythere pliocenica* figured by Benson, 1972).

It is interesting to note that microborings have never been noticed on carapaces of living ostracods, despite the fact that fungi and/or filamentous bacteria can cover the surface of the shell (Fig. 7, E). However, Okafor (1966) found that the chitinoclastic microorganisms are inhibited by phenolic compounds. In the case of the ostracods studied here, it is likely that, only after death of the organism, a chemical alteration of the molecular structure of the epicuticle occurs, thus allowing the activity of microborers to start. Veldkamp (1955) and Warnes and Randles (1977) noted that the activity of chitinoclastic bacteria was strongly diminished under anaerobic conditions. Further, Schrader *et al.* (1983:225) noted that 'carbonate derived from Foraminifera and calcareous nan-

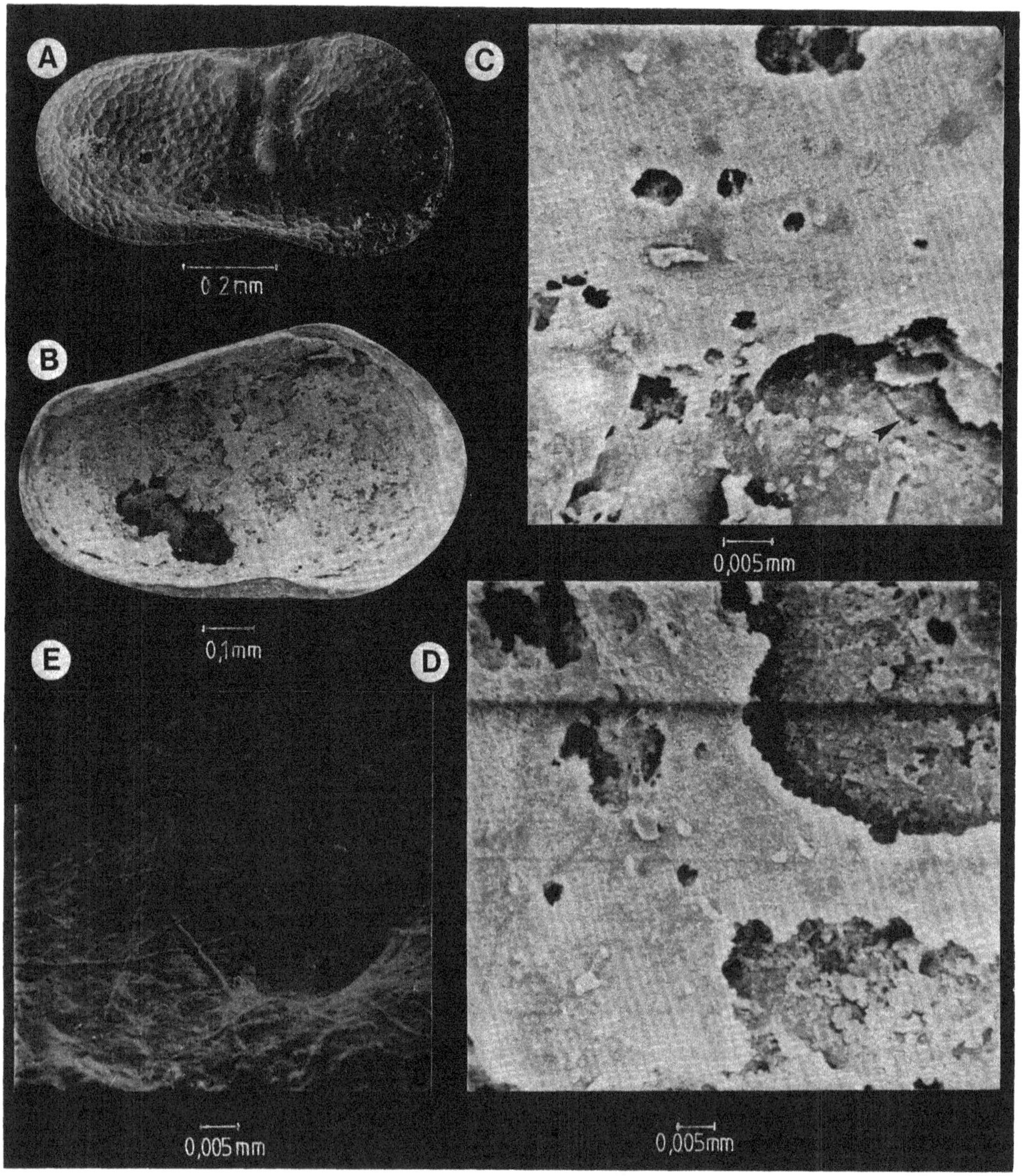

Fig. 7. 'Dead' valves after biological and chemical corrosion; A – *Leucocythere mirabilis* Kaufm. (Mondsee, Recent), adult, right valve, general view; B – E – *Cytherissa lacustris* Sars, adult, left valve (B – D); B – general view inner side; C, D – details of the corroded (calcite) procuticle; arrow indicates a deep microboring tube within the calcite layer; E – outer surface of the carapace covered with filamentous bacteria or fungi.

noplankton can occasionally be better preserved in oxygen depleted depositional environments'. In the case of the ostracods, one should expect that for the valves which remain for only a short time in contact with an oxic environment (e.g. they are rapidly burried in habitats of high sedimentation and/or the sediment becomes anoxic, the activity of microborers on them will be lower than in oxic environments and/or in habitats with a low sedimentation rate (see below).

b. Holes

Our material shows several types of holes which are caused by microborers and also which result from partial dissolution by acid water. Already noted are shallow holes most probably initiated by the sporangia of microborers. Their shape is triangular or polygonal. Other holes could be initiated by the mechanical destruction of the epicuticle which, in some places, does not tightly adhere to the procuticle (Fig. 5, D). The dissolution of the inner calcified procuticle will deepen those holes down to the perforation of the valve (Fig. 6, B). A second type of hole is caused by the enlargement of the normal pore canals and/or the vacuolar voids, through the destruction of the organic matrix and the dissolution of the calcitic structure (Fig. 1, C, 4, D, 6, A, B). The activity of 'agressive' water further dissolves the carbonate and the inner structure of the procuticle form a kind of micro-karst topography (Fig. 4, C, D). This is seen in *Herpetocypris* aff. *brevicaudata* studied here. Simple chemical corrosion can start from the inner side of the procuticle as in *Cytherissa lacustris* (Fig. 7, B – D). The presence of large inner and unstructured voids causes those areas to appear opaque and chalky; this is best seen under transmited light (e.g. Fig. 1, C). We produced similar holes by etching a valve of *Candona neglecta* (Fig. 1, C) which was kept in glycerine (pH 5,3) for a week. The shallow channels of the microborers became especially enlarged and the surrounding area became opaque. Flessa & Brown (1983) obtained similar results through experiments on bivalves. Thin valves with a spongeous central muscle scars area, like in *Leucocythere mirabilis,* are particularly exposed to perforation and hole formation.

c. Fractures

Biologically and chemically corroded carapaces/or valves become biomechanically weakened and fracture more easily (Fig. 1, A, D, F). However, we did not observe a strict correlation between the network of microborings and holes with the directions of fractures. As Whatley *et al.* (1982) observed experimentally, the fracturing of a valve or carapace is the result of the mechanical stress which forms on the whole calcified wall. A fracture can start from holes or can traverse the tracks of microborings which become paths of weakness (Fig. 1, A, D, F). The valves of limnic ostracods in high energy environments, as in littoral habitats, are already corroded and, therefore, particularly prone to perforation and fracture (see following section). Similar observations were made by Philippon and Plaziat (1975) in their study of gastropod shells from an African mangrove area.

5. Relationships between the degree of carapace and/or valves preservation and sediment accumulation rates

Oertli (1971: 139) pointed out the interest petroleum geologists have in finding paleoecological indicators for environments with high rates of sediment accumulation and of low oxygen. He stated (op.cit.): 'A relatively rapid sedimentation rate must in effect increase the chances for hydrocarbon formation... The higher the sedimentation rate... the greater chances of a sufficiently rapid burial so that the organic matter will not be mineralized and thus retain the potential for a later, and partial, transformations into hydrocarbons.'

In order to check such a relationship between ostracod preservation and the degree of the sediment accumulation, we investigated the rate of sedimentation in the Mondsee for the last 20 – 30 years (Irlweck & Danielopol, 1985). Secondly, we looked in a semiquantitative fashion at the degree of ostracod preservation (of whole assemblages) at sites with various types of sediment accumulation (low rates in sub-littoral habitats to high rates in profundal deposits (D.L.D. unpubl.). Finally, one of us (L.M.C.) quantified for a single species, *Candona neglecta,* the degree of preservation at various sites of the Mondsee where the history of the sediment accumulation was already known.

Until 30 years ago the Mondsee remained a mesooligotrophic lake with well oxygenated water in the whole lake. At the time such species as *Cytherissa lacustris* were more common than today (Danielopol *et al.,* 1985). During the late fifties, and early sixties, during the construction of the adjacent Vienna-Salzburg highway, about 900000 m^3 of soil was discharged into the lake, thus dramatically increasing the sedimentation rates at least in the central part. Subsequently from the end of sixties, the lake experienced annual algal blooms as a result of increasing tourist activity, and sewage discharge into the lake. All these factors increased the rate of autochthonous sediment deposition within the lake (Schmidt *et al.,* 1985). Irlweck & Danielopol (1985) showed by using the Cs-137 and Pb-210 dating methods that in the profundal zone of the northern and central part of the Mondsee, the mean annual sedimentation rates varied between 4 and 7 mm/yr. These areas were subjected during the last years to major allochthonous and autochtonous sediment accumulation. In those areas, during the late summer and the autumn, oxygen content becomes greatly depleted. In the southern part of the lake, and on the slopes of the sublittoral zones (18–20 m depth), sedimentation rates are much lower i.e. 0.5–2.5 mm/yr. In the latter areas, good levels of oxygenation occur and delicate species such as *Cytherissa lacustris* still form abundant populations. We studied samples from site Mo-7 at a depth of 16 and 18 m, 2–3 cm below the water sediment interface. At that site, Irlweck & Danielopol (1985) measured annual sedimentation rates which vary between 0.5 and 2.5 mm for the last 20–30 years. For comparative purposes, samples from the northern part of the lake (Mo-9, 47 m depth) were also studied. There annual sedimentation rate of 6.2 mm/y were found in the upper 15 cm sediment (see Irlweck & Danielopol, 1985 and Danielopol *et al.,* 1985).

Examination of the sediment from site Mo-9 reveals that the stratigraphical sequence (Fig. 9, A) can be divided into three zones. Between depths of 30 cm and 16 the sediment is grey and more or less homogeneous; the silt fraction dominates with also a small sand fraction; between 16 and 9 cm the sediment is slightly light grey and contains a significant fraction of gravel (Fig. 9, B). This sediment most probably represents the event of discharge of the large amount of refuse during the period of the highway construction. The upper part of the sediment column is strongly laminated, with black and

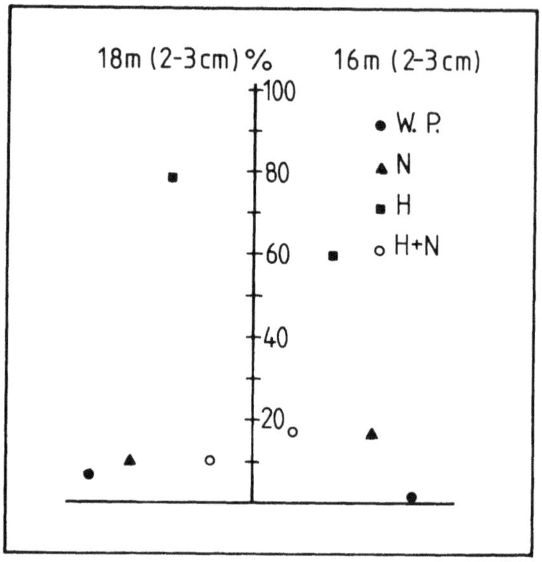

Fig. 8. Distribution of various types of valves in the sediments of the Mondsee (site MO-7, 18 m depth, 2–3 cm below the water sediment interface); WP – valves well preserved; N – valves with a network of microborings; H – valves with holes; NH – valves with microborings and holes; for each sample 100 valves examined.

light coloured lamellae; this has been produced since 1968 and is the result of sedimentation of autochthonous algae which seasonally decay (Schmidt *et al.,* 1985). The black layers are due to the pyrite formation under anoxic conditions.

Fig. 8 shows that for *Candona neglecta* the percentage of well preserved (WP) valves is very low at site Mo-7 at 16 and 18 m depths, and the percentage of the valves with microborings and with holes (also including fractures) is high.

Investigation of the preservation of *Candona neglecta* (Fig. 9, C, D), shows that in the lower part of the sedimentary column (which was deposited during a period of low sedimentation rates when compared with the upper half, viz. from 15–16 cm) the degree of well preserved valves is low, and a fairly high percentage of valves with a network of microborings occurs. The opposite is visible in the upper part of the column, where the percentage of well preserved valves increases substantially and the corroded ones decrease. In this latter zone, a higher rate of sedimentation is noticeable for the last decades; in the upper part, the sediment is also periodically anoxic. We believe that the

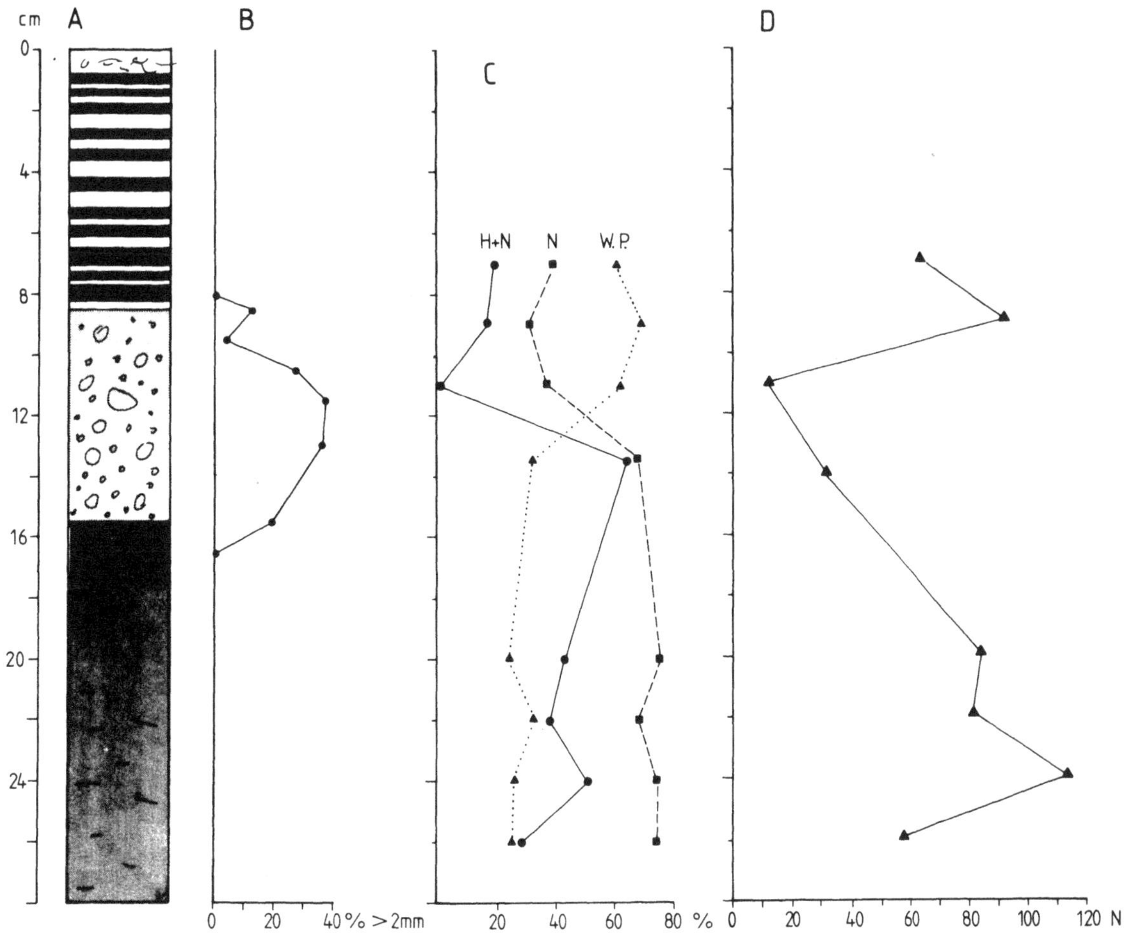

Fig. 9. Sediment core from the site MO-9, at a depth of 47 m (Mondsee) and the distribution of various types of valves of *Candona neglecta* Sars; A – Stratigraphic column; B – distribution of the gravel size fraction in the core; C – ostracod distribution; well preserved valves (WP); (N) – valves with a network of superficial microborings, (H) – valves with holes; (D) – the number of valves which have been examined for the degree of preservation.

better degree of preservation of *Candona neglecta* valves, in the latter part of the core, is due to the faster burial processes and to the shorter time during which the ostracod shells remain in contact within an oxic medium. This scenario is very similar to the one observed by Schrader *et al.* (1983) for the foraminifers and nannoplankton as mentioned previously. In conclusion, we believe that the degree of ostracod preservation can be also related to the type of sedimentation. For example, within a quiet environment and with high sedimentation rate and rapid formation of anoxic sediments, carapaces and/or valves of limnic ostracods become isolated from extensive biological and/or physico-chemical degradation. This observation surely would be of interest to the petroleum industry. Therefore, more investigations on fossil material are needed in order to test this model derived on a Recent ostracod species.

6. Why are there so few herpetocyprines in the Palaeogene of Europe?

The main genera of the tribe Herpetocypridini (Cyprinidae) can easily be recognized from the morphological structures of the limbs and of the carapace (Danielopol & McKenzie, 1977). The representatives of the genera *Herpetocypris* Sars and *Psychrodromus* Dan. & McKenz. are widely distributed in cool temperate waters. On the other hand, representatives of the genera *Stenocypris* Sars, *Acocypris* Vavra, *Chrissia* Hartm., *Parastenocypris* Hartm. mainly occur in warm temperate and subtropical regions (e.g. in Africa, Asia and Central America (McKenzie, 1971). Identification of *Stenocypris* is especially easy by the structure of the carapace alone because of the radial septa which unite the external and the inner calcareous lamellae. Triebel (1953), for example, found a *Stenocypris* sp. in the Chattian (Upper Oligocene) lacustrine sediments of the Mainz Basin in Europe.

The palaeoenvironment during the Eocene and the Oligocene in Europe was subtropical to warm temperate (Pearson, 1978). Despite this similarity and the fact that lacustrine ostracods are well known from several Eocene and Oligocene deposits of England, France and Germany (see review in Keen, 1975) herpetocyprines have been seldom found (Triebel, 1953). Triebels sample consists of 30 broken valves of *Stenocypris*. We believe that the scarcity of thermophilous herpetocyprines in the fossil record in Europe could be due to the poor preservation of the carapaces and/or valves after the death of these ostracods. Following the model of the degration of the *Herpetocypris* aff. *brevicaudata* presented above, it is very likely that the thermophilous herpetocyprines, like the *Stenocypris* species, which have a high density of vacuolar voids in the procuticle, suffer corrosion under high temperatures and 'agressive' water more easily than other limnic cypridids and cytherids which benefit from a compact calcitic layer. This should explain why in so few Palaeogene herpetocyprines have been found so far.

Acknowledgements

Dr. H. J. Oertli and the Research Center ELF-Aquitaine (Pau) stimulated and partly supported the work of the senior author in 1972/73. Prof. M. Moser (Innsbruck), Drs. K. Messner (Vienna), T. Le Campion-Alsumard (Marseille), J. Philippon (Alger), helped to identify the ostracod microborings. Prof. H. Löffler (Vienna), Dr. R. H. Benson (Washington), Dr. P. De Deckker (Canberra), Dr. F. Dépèche (Paris), Prof. J. Schneider (Göttingen), Dr. R. Whatley (Aberystwyth) provided useful information and/or commented on various drafts of this paper. Dr. E. Hanganu (Bucharest) offered material for study. The following colleagues helped during the field and laboratory work: D. Dancau (Bucharest), W. Geiger and K. Maier (Mondsee), C. Lété (Bordeaux), A Loser (Vienna), Fr. Saffon and F. Passini (Pau). The SEM microphotographs were taken in the laboratories of ELF-Aquitaine and Zoological Institute, University of Vienna. This contribution was partly funded by the 'Fonds zur Förderung de wissenschaftlichen Forschung', Project P 5614 attributed to the senior author. I. Gradl (Mondsee) typed the manuscript.

References

Bate, R. H. & B. A. East, 1972. The structure of the Ostracoda carapace. Lethaia 5: 177–194.

Benson, R. H., 1972. The Bradleya problem, with descriptions of two new psychrospheric Ostracoda genera, *Agrenocythere* and *Poseidonamicus* (Ostracoda: Crustacea). Smithson. Contr. Paleobiol. 12, 138 pp.

Danielopol, D. L., 1982. Nouvelles données sur les Candoninae (Ostracoda) hypogés de Roumanie et Yougoslavie. Bull. Mus. natn. Hist. nat. Paris, 4ser. 4, A: 369–396.

Danielopol, D. L. & K. G. McKenzie, 1977. *Psychrodromus* gen.n. (Crustacea, Ostracoda) with redescription of the cypridid genera Prionocypris and Ilyodromus. Zool. Scr. 6: 301–322.

Danielopol, D. L., W. Geiger, M. Tölderer-Farmer, C. P. Orellana & M.-N. Terrat, 1985. The Ostracoda of Mondsee: spatial and temporal changes during the last fifty years. In D. Danielopol, R. Schmidt & E. Schultze (eds.). Contributions to the paleolimnology of the Trumer Lakes (Salzburg) and the lakes Mondsee, Attersee and Traunsee (Upper Austria). Limnol. Inst. ÖAW, Mondsee: 99–121.

Delorme, D. & D. Donald, 1969. Torpidity of freshwater ostracodes. Can. J. Zool. 47: 997–999.

Depeche, F., 1979. Ultrastructure de la paroi externe des 'micro-Ostracodes' du Jurassique moyen normand. N. Jb. Geol. Paläont. Mh. 1979/6: 340–348.

Depeche, F., 1982. Ultrastructure of the wall of two living ostracods, *Herpetocypris chevreuxi* (Sars) and *Pontocythere elongate* (Brady), in comparison with fossil ostracods from the Middle Jurassic of Normandy. In R. H. Bate, E. Robinson & L. M. Sheppards (eds.). Fossil and Recent ostracods. Ellis Horwood Ltd., London: 61–75.

Flessa, K. W. & T. J. Brown, 1983. Selective solution of macro-invertebrate calcareous hardparts: a laboratory study. Lethaia, 16: 193–205.

Freels, D., 1980. Limnische Ostrakoden aus Jungtertiär und Quartär der Türkei. Geol. Jahrb., 39: 3–169.

Frydl, P. M., 1982. Holocene ostracods in the southern Boso

Peninsula. In: T. Hanai (ed), Studies on Japanese Ostracoda. Univ. Mus. Univ. Tokyo Bull., 20: 61–140.
Irlweck, K. & D. L. Danielopol, 1985. Caesium-137 and lead-210 dating of recent sediments from Mondsee (Austria). Hydrobiologia 128: 175–185.
Jagnow, G., 1957. Beiträge zur Ökologie der Streptomyceten. Arch. Mikrobiol., 26: 175–191.
Keen, M. C., 1975. The palaeobiology of some upper Palaeogene fresh-water ostracodes. Bull. Am. Paleont. 65: 271–283.
Kontrovitz, M., 1966. An investigation of ostracode preservation. Florida Acad. Sci. Quart. J., 29: 171–177.
McKenzie, K. G., 1971. Paleozoogeography of freshwater Ostracoda. In H. J. Oertli (ed), Paleoecologie Ostracodes. Bull. Centre Rech. Pau – SNPA, 5 suppl.: 207–238.
Oertli, H. J., 1971. The aspect of ostracode faunas – A possible new tool in petroleum sedimentology. In H. J. Oertli (ed), Paleoecologie Ostracodes. Bull. Centre Rech. Pau – SNPA, 5 suppl.: 137–151.
Oertli, H. J., 1951. The conservation of ostracode tests – Observations made under the scanning electron microscope. In F. M. Swain, L. S. Kornicker & R. F. Lundin (eds), Biology and Paleobiology of Ostracoda. Bull. Am. Paleont. 65: 549–576.
Okada, Y., 1982. Structure and cuticle formation of the reticulated carapace of the ostracode *Bicornucythere bisanensis*. Lethaia, 15: 85–101.
Okafor, N., 1966. The ecology of micro-organisms and the decomposition of insect wings in the soil. Plant Soil, 25: 211–236.
Olteanu, R., 1971. Study of ostracods within the Upper Pannonian deposits (zone E) from Grosi Locality (Banat). D. S. Inst. Geol. Bucuresti, 57: 85–101 (in romanian, english abstract).
Pearson, R., 1978. Climate and evolution. Academic Press, London, 274 pp.
Philippon, J. & J.-Cl. Plaziat, 1975. Roles respectifs de la corrosion et des Cryptrogames perforantes dans la destruction des coquilles de Mollusques des mangroves. Consequence sur la fossilisation. C.r. Acad. Sci. Paris D 281: 617–620.
Rögl, F. & F. F. Steininger, 1983. Vom Zerfall der Tethys zu Mediterran und Paratethys. Die neogene Paläogeographie und Palinspastik des zirkum-mediterranen Raumes. Ann. Naturhist. Mus. Wien 85/A: 135–163.
Schmidt, R., J. Müller & J. Froh, 1985. Laminated sediments as a record of increasing eutrophication of the Mondsee. In D. Danielopol, R. Schmidt & E. Schultze (eds), Contributions to the paleolimnology of the Trumer Lakes (Salzburg) and the lakes Mondsee, Attersee and Traunsee (Upper Austria). Limnol. Inst. ÖAW, Mondsee: 122–131.
Schrader, H., G. Cheng & R. Mahood, 1983. Preservation and dissolution of foraminiferal carbonate in an anoxic slope environment, southern gulf of California. In J.E. Meulenkamp (ed) Reconstruction of marine paleoenvironments. Utrecht Micropaleontological Bulletins: 205–227.
Triebel, E., 1953. Genotypus und Schalenmerkmale der Ostracoden-Gattung *Stenocypris*. Senckenbergiana, 34: 5–14.
Veldkamp, H., 1955. A study of the aerobic decomposition of chitin by microorganisms. Med. Landbouw. Wageningen 55: 127–174.
Warnes, C. E. & C. I. Randles, 1977. Preliminary studies on chitin decomposition in lake Erie sediments. Ohio J. Sci., 77: 224–230.
Whatley, R., 1983. The application of Ostracoda to palaeoenvironmental analysis. In R. Maddocks (ed), Applications of Ostracoda to economic and scientific problems. Univ. Houston, Houston: 51–77.
Whatley, R. C., K. Trier & P. M. Dingwall, 1982. Some preliminary observations on certain mechanical and biophysical properties of the ostracod carapace. In R. H. Bate, E. Robinson & L. M. Sheppard (eds), Fossil and Recent ostracods. Ellis Horwood Ltd., London: 76–104.

Accepted 18 March 1986.

Hydrobiologia 143: 159–165, (1986)

Information value of chironomid remains in the uppermost sediment layers of a complex lake basin

Kansanen, Pekka H.
Vesitekniikka Ltd, Terveystie 2, SF-15870 Salpakangas, Finland

Keywords: Chironomidae, subfossil remains, pollution, redeposition, sediment, palaeolimnology

Abstract

The subfossil remains of chironomids were analyzed from four short (35–50 cm) sediment cores in eutrophicated Lake Vanajavesi, southern Finland. The chironomid analysis was found to be a useful palaeolimnological technique for indicating the following aspects: (1) The pollution history of the lake during the last 150 years. A succession from an oligotrophic *Micropsectra-Monodiamesa* community, through eutrophic *Chironomus* communities, to the disappearance of the chironomid fauna, was detected. (2) The bottom dynamic conditions at the sampling site. (3) The water level changes of the lake. (4) The significance of bioturbation in the core chronology and stratigraphy. The disappearance of the chironomid remains in the cores was correlated with the occurrence of annually laminated sediment.

The most important factor disturbing the interpretation of the cores was the redeposition of head capsules by erosion and transportation. Detailed identification of the remains is necessary to avoid the misinterpretations caused by redeposition. Bioturbation reduces the time resolution of the cores. It is necessary to use dating methods which take the effect of bioturbation into account.

Introduction

Palaeolimnological methods can be used not only in the study of lake development in the past, but also in the monitoring of recent environmental changes. Regular monitoring of Finnish lakes began in the 1960s, when eutrophication and pollution had progressed for several decades and in many places had reached alarming dimensions. Often no reliable knowledge of the natural state and the early stages of change in these lakes exists. This kind of information is invaluable when evaluating the significance of the observed changes and future prospects of the lake related to various levels of waste water loading.

The value of a long time series of zoobenthos investigations in the monitoring of lakes has been demonstrated, e.g. by Kansanen & Aho (1981). One of the most common groups of benthic animals which leave identifiable remains in lake sediments is Chironomidae (Diptera). The aim of this study was to test the feasibility of using the palaeolimnological analysis of chironomid remains in Lake Vanajavesi, southern Finland, (1) in the assessment of the lake's recent pollution history and (2) as an indicator of other environmental conditions (e.g. water level lowering).

Study site, material and methods

The study area (Fig. 1) forms a part of the drainage basin of the River Kokemäenjoki, southern Finland (roughly 61°N, 24°E). Lake Vanajavesi is one of the most polluted of large inland lakes in Finland.

The bottom topography of the lake basin is rather complicated (Fig. 1). It comprises several sub-basins. The largest sub-basin, Vanajanselkä, has an area of 119.2 km^2, mean depth of 7.9 m, maxi-

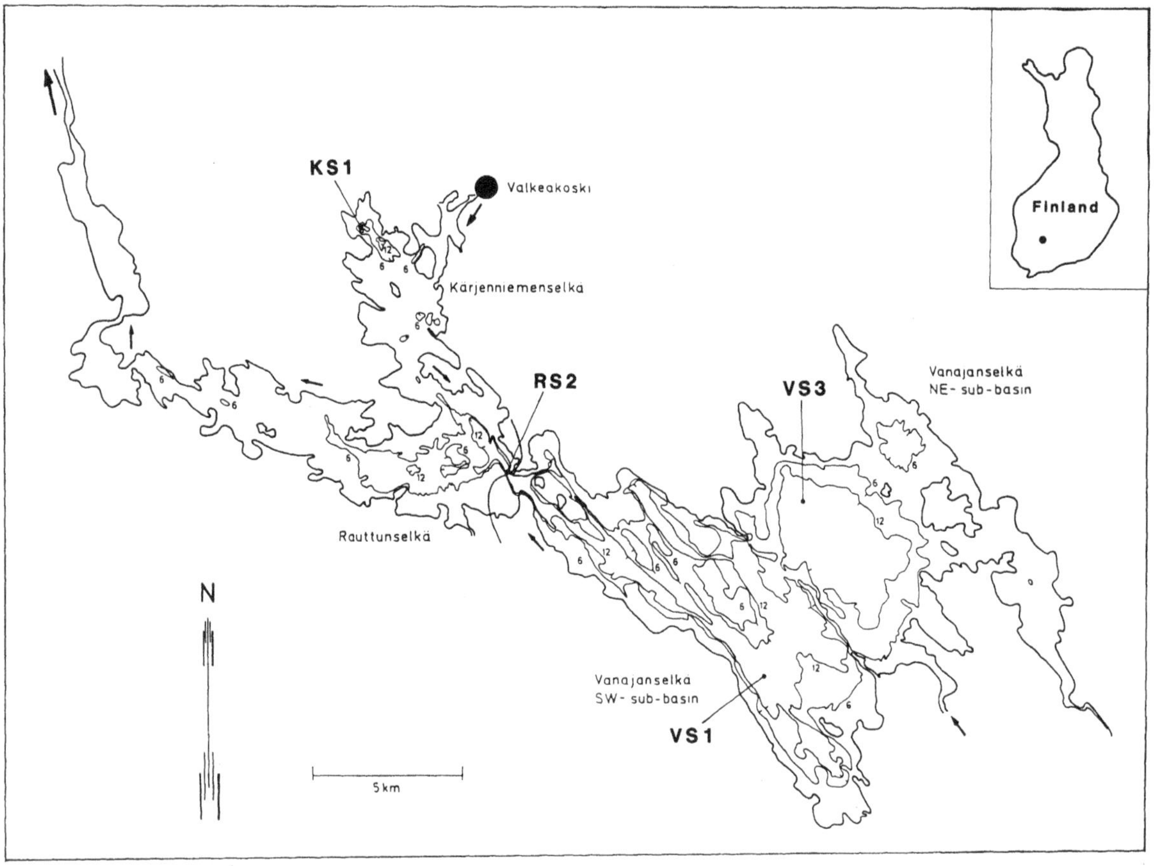

Fig. 1. Location and bathymetric map of the study area (contour lines of 6 and 12 m). The coring sites are indicated.

mum depth of 24 m and mean volume of 936.7×10^6 m^3

The whole study area is nowadays highly eutrophic and some parts are heavily polluted. Algal blooms and hypolimnetic oxygen depletion are common phenomena in this previously more or less oligotrophic lake. The Valkeakoski industrial centre is the most important source of waste waters in the study area (Fig. 1). About 90% of the total BOD_7 load and 33% of the phosphorus load come from the wood-processing and chemical industry and the town of Valkeakoski. A specific pollutant in these waste waters is zinc. This is due to the effluents of a rayon fibre factory, which started its production in 1943. In 1980 the zinc load was over 1000 kg d^{-1}. A detailed description of the study area is given e.g. by Kansanen & Aho (1981) and Kansanen & Jaakkola (1985).

Four short profiles (length 35–50.5 cm) were taken from the study area (Fig. 1). Profile KS1 was taken from Kärjenniemenselkä, the most polluted sub-basin, which receives the waste waters of Valkeakoski. Profile RS2 was taken from the sub-basin of Rauttunselkä, which is less polluted than Kärjenniemenselkä. Two profiles, VS1 and VS3, were taken from the main basin of Vanajavesi, which is eutrophicated but still in better condition than the other parts of the lake. The zinc content of the surficial sediments indicates well the degree of pollution (Fig. 2).

A Kajak-type gravity corer was used in sampling. At site KS1 the loose uppermost layers with annually laminated sediment were frozen *in situ* using a sampler of the type described by Huttunen & Meriläinen (1978). The cores were sectioned into 0.5–2.3 cm thick slices. Subsamples of known volume (10 cm^3) were taken from the fresh sediment with a cut sampling syringe. In the case of frozen

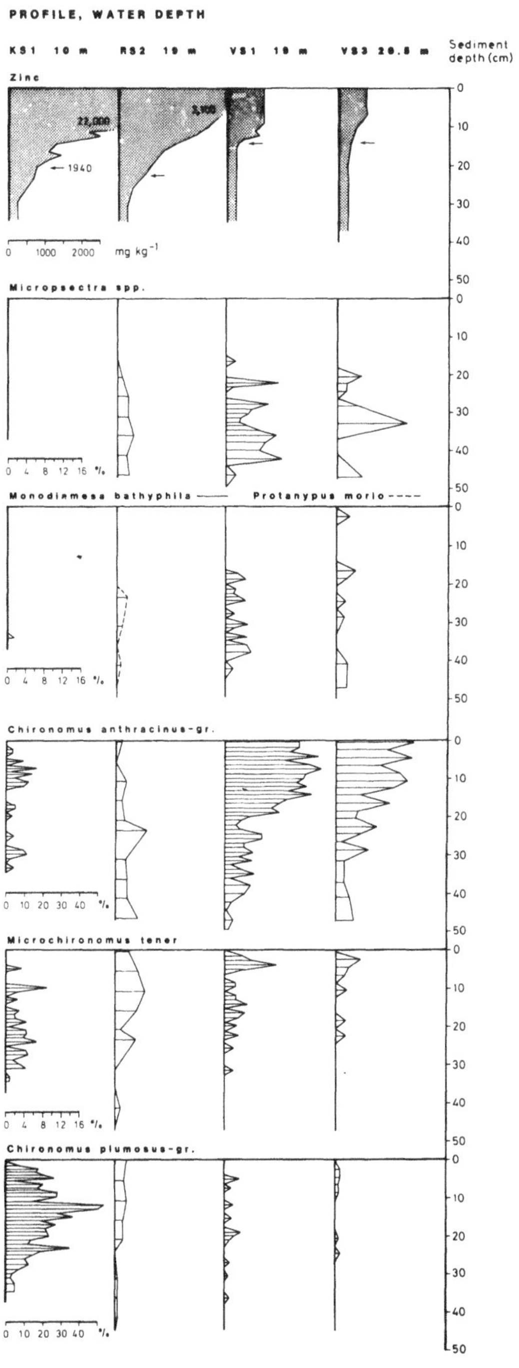

Fig. 2. The zinc concentration of the sediment and the proportion (%) of selected taxa in the subfossil fauna in four profiles of Lake Vanajavesi. The approximate level of the year 1940 is indicated by an arrow on the zinc profiles.

sediment (KS1), sections cut for biostratigraphic analyses were dried and weighed. In the analysis of subsamples methods presented by Hofmann (1971) and Warwick (1980) were applied. The procedure is described in detail by Kansanen (1985).

The total material consists of 4,327 subfossil remains (3.927 head capsules, 5 adult hypopygia and 1 pupal exuvia). All the material was mounted permanently on glass slides in polyvinyl lactophenol. The core chronology was mainly based on ^{210}Pb, ^{137}Cs and $^{239,240}Pu$ dating and in the case of profile KS1 on the analysis of annually laminated sediment (see Kansanen & Jaakkola, 1985).

Results and discussion

The stratigraphy of chironomid fauna indicates clearly pollution succession in all four profiles. The changes in the subfossil fauna of the profile KS1 differ from other profiles, lacking strictly profundal elements of the fauna because of the lesser water depth at that site (10 m, other profiles 19–20.5 m).

Micropsectra spp. are now totally vanished faunal elements in Vanajavesi (Fig. 2). Two subfossil hypopygia were identified as *M. insignilobus* Kieffer or *M. lindegergi* Säwedal, which are typical profundal species of oligotrophic lakes. *Monodiamesa bathyphila* (Kieffer) and probably also *Protanypus morio* (Zett.) belong still to the present fauna, but have clearly decreased during this century. Both species prefer oligotrophic conditions. *Chironomus anthracinus*-gr. consists of two species: in the profundal zone *C. anthracinus* Zett. and *C. neocorax* Wülker & Butler. An increase of this group is especially evident in profiles VS1 and VS3. *Microchironomus tener* (Kieffer) also is primarily an increasing species in Lake Vanajavesi (Fig. 2). It is known that this species prefers eutrophy (Kansanen *et al.*, 1984). *C. plumosus*-gr., which at least in the present fauna consists of only one species in the profundal, *C. plumosus* (L.), at first strongly increased in profiles KS1 and RS2, but a clear decrease can be seen in the uppermost 10 cm of profile KS1. *C. plumosus* indicates strongly eutrophic conditions.

To summarise, the following succession of chironomid fauna seems to have taken place in the profundal of Lake Vanajavesi during this century.

1. Oligotrophic *Micropsectra-Monodiamesa* community (whole lake).

2. Moderately eutrophic *C. anthracinus* community (earlier Rauttunselkä, present Vanajanselkä).

3. Strongly eutrophic *C. plumosus* community (present Rauttunselkä, earlier Kärjenniemenselkä).

4. Barren profundal with occasional *C. plumosus* populations (present Kärjenniemenselkä). For further details of eutrophication succession, see Kansanen (1985).

The dynamic bottom conditions (erosion, transportation, accumulation) in the vicinity of the sampling site have a significant effect both on the qualitative composition and the quantity of the subfossil fauna. All coring sites were selected from accumulation bottoms on the basis of mapping of surficial sediments by means of granulometric analysis (Kansanen & Jaakkola, 1985). The short distance from the shore and the steep slope of bottom between the shore and the sampling site clearly increases the influx of total chironomid remains to sediments (Fig. 3).

The effects of redeposition can also be seen in the proportion of the strictly littoral elements in the fauna. The average contribution by littoral species (for definition, see Kansanen, 1985) was in KS1 16.7%, in RS2 28.6%, in VS1 12.6% and in VS3 16.8% (Fig. 3). The much greater influx of chironomid remains in profile RS2 is probably due to occasional slumping of sediment owing to the steep bottom gradient (Kansanen & Jaakkola, 1985).

The water level of Lake Vanajavesi was lowered artificially in the 18th century and again in 1857. The latter lowering was more significant, causing a drop of 2 m in the water level. In both cases a large amount of the near-shore sediment was washed back into the lake before the soil became bound by the invading vegetation. The effects of the water level lowering can clearly be seen in the profile VS1 (Fig. 4).

The proportion of strictly littoral taxa, especially of *Limnophyes* spp., is significantly higher below the level of 28 cm than in the uppermost layers.

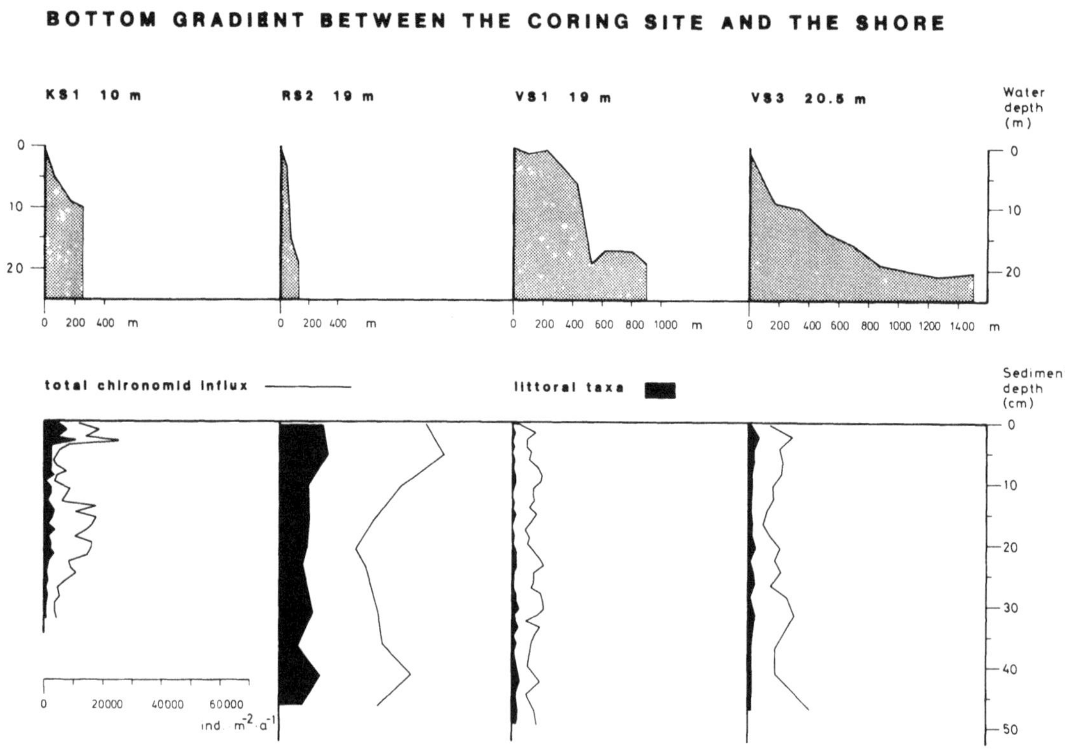

Fig. 3. The bottom topography between the coring sites and the nearest shoreline, and the influx of littoral and of total chironomid remains at each site.

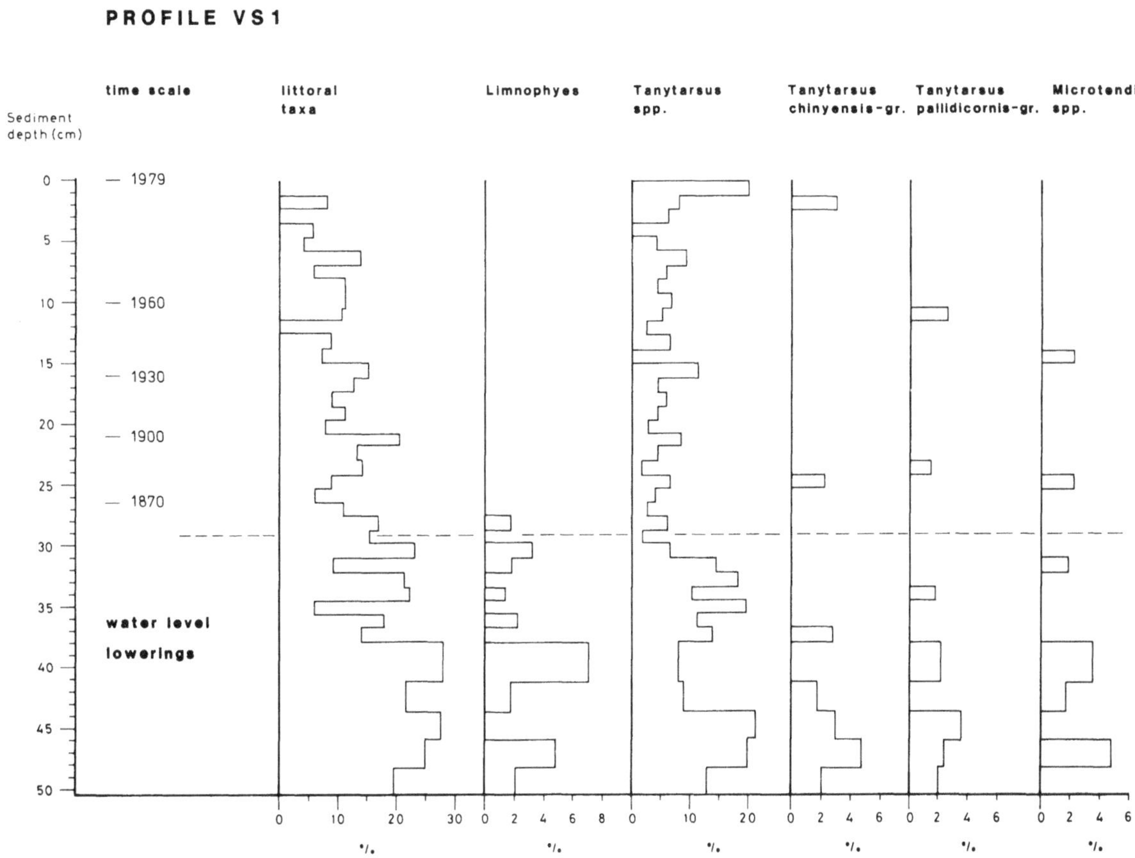

Fig. 4. The proportion (%) of littoral species and of some selected taxa in profile VS1.

Limnophyes spp. in lake sediments are derived from the terrestrial fauna of shoreline biotopes. The genus *Tanytarsus* includes both littoral and profundal species. Due to taxonomic difficulties it was not possible to distinguish between all species or species groups of *Tanytarsus*. The high proportion of *Tanytarsus* below 28 cm indicates that most of *Tanytarsus* head capsules originate from the littoral zone. *T. pallidicornis*-gr. and *T. chinyensis*-gr. belong to the littoral fauna.

The effects of bioturbation, which is caused by the benthic fauna, can also be seen in the chironomid stratigraphy. In Lake Vanajavesi, where Chironomidae is the most important benthic group, the highest population densities of the macrozoobenthos are at present over 4000 ind. m^{-2} in Vanajanselkä (Kansanen & Aho, 1981). The density of the meiozoobenthos can be over 600000 ind. m^{-2} (Kansanen, 1981). In profile KS1 there was a sudden decrease in 1957 in the influx of head capsules of the total Chironomidae, especially of species belonging to that part of the subfossil fauna which is not redeposited from the littoral zone (Fig. 5). At this time the formation of annually laminated sediment began. The zoobenthos living at the site was then destroyed by the highly increased waste water loading, and during the subsequent period 1957–1981 the subfossil fauna mainly reflects conditions in the littoral zone.

Conclusions

Redeposition of chironomid remains by erosion and transportation of sediment is one of the most important disturbing drawbacks to simple interpretation of eutrophication successions in sediment profiles from large and complex lake basins. It is

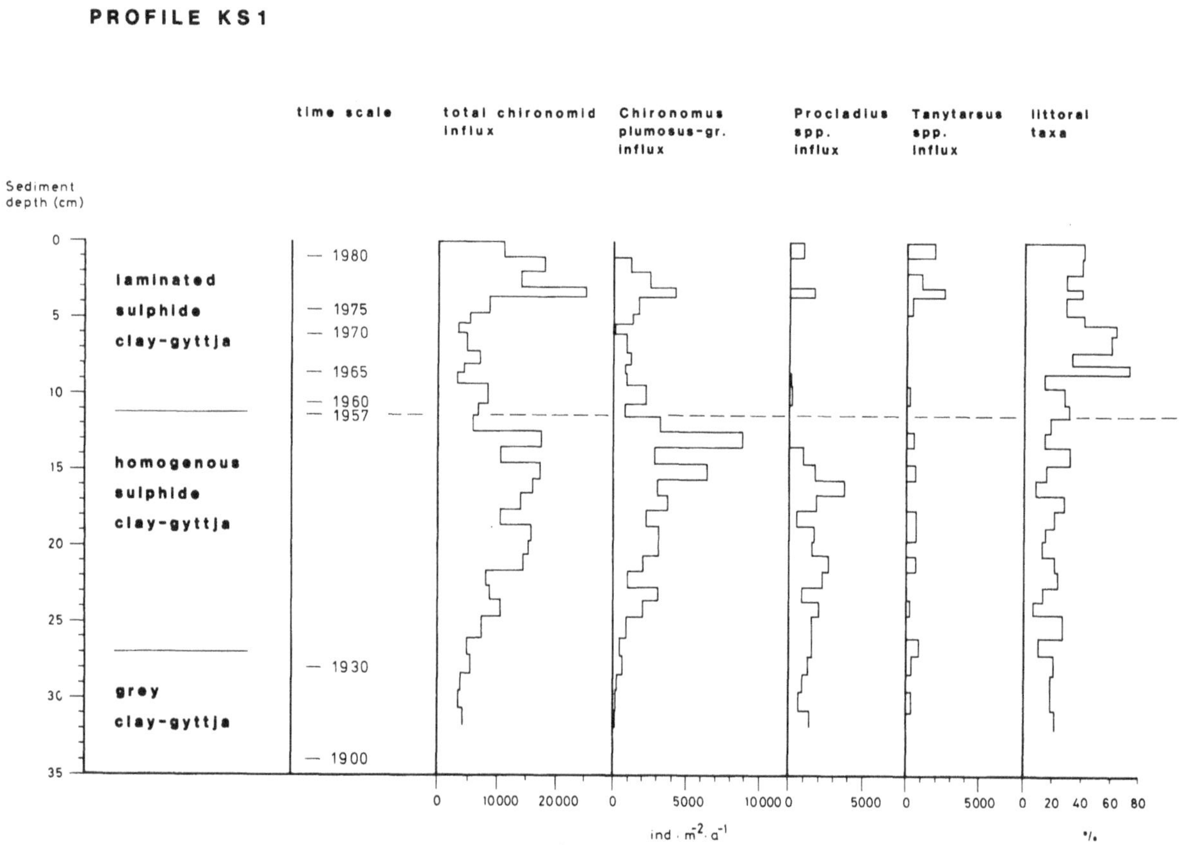

Fig. 5. The influx of total chironomids and of some taxa in profile KS1. The proportion (%) of littoral species also is given.

necessary to select the coring sites carefully and to compare the stratigraphy of several cores. Detailed identification of chironomid species or at least species groups is the only way to avoid the misinterpretations caused by redeposition. Several taxonomic problems still exist in some genera, e.g. *Tanytarsus.*

Despite difficulties, a palaeolimnological study of chironomid remains can produce valuable reference data for a monitoring programme based on zoobenthos investigations. The structure of the subfossil community indicates also changes in the water level of the basin. Bioturbation is often an important disturbing factor in the core stratigraphy. Chironomid analysis indicates also the occurrence of bioturbation activity in the profile. To avoid problems caused by bioturbation, it is necessary to use dating methods which take the effects of mixing into account (cf. Kansanen & Jaakkola, 1985).

Acknowledgements

Financial aid from the Maj and Tor Nessling Foundation, The Finnish Entomological Society and the Finnish Cultural Foundation is gratefully acknowledged.

References

Hofmann, W., 1971. Die postglaziale Entwicklung der Chironomiden- und Chaoborus-Fauna (Dipt.) des Schöhsees. Arch. Hydrobiol. Suppl. 40: 1–74.

Huttunen, P. & J. Meriläinen, 1978. New freezing device providing large unmixed sediment samples from lakes. Ann. bot. fenn. 15: 128–130.

Kansanen, P. H., 1981. Effects of heavy pollution on the zoobenthos in Lake Vanajavesi, southern Finland, with special reference to the meiozoobenthos. Ann. zool. fenn. 18: 243–251.

Kansanen, P. H., 1985. Assessment of pollution history from recent sediments in Lake Vanajavesi, southern Finland. II. Changes in the Chironomidae, Chaoboridae and Ceratopogonidae (Diptera) fauna. Ann. zool. fenn. 22: 57–90.

Kansanen, P. H. & J. Aho, 1981. Changes in the macrozoobenthos associations of polluted Lake Vanajavesi, southern Finland, over a period of 50 years. Ann. zool. fenn. 18: 73–101.

Kansanen, P. H., J. Aho & L. Paasivirta, 1984. Testing the benthic lake type concept based on chironomid associations in some Finnish lakes using multivariate statistical methods. Ann. zool. fenn. 21: 55-76.

Kansanen, P. H. & T. Jaakkola, 1985. Assessment of pollution history from recent sediments in Lake Vanajavesi, southern Finland. I. Selection of representative profiles, their dating and chemostratigraphy. Ann. zool. fenn. 22: 13–55.

Warwick, W. F., 1980. Palaeolimnological of the Bay of Quinte, Lake Ontario: 2800 years of cultural influence. Can. Bull. Fish. aquat. Sci. 206: 1–117.

Accepted 18 March 1986.

Hydrobiologia 143: 167–174, (1986)

Paleolimnology of mid-elevation lakes in the Kenya Rift Valley

J. L. Richardson & R. A. Dussinger
Department of Biology, Franklin and Marshall College, Lancaster, PA 17604, USA

Keywords: diatoms, ostracodes, paleoclimate, Kenya, Pleistocene, Holocene

Abstract

Three small lakes of the Kenya Highlands have been subject to intermittent paleoenvironmental investigation for more than 50 years. In this report, the stratigraphic, core-based history of Lake Naivasha (elev. 1890 m) is extended backward to 20000 B.P. and is related to less complete 30000-year core stratigraphies from nearby lakes Nakuru and Elmenteita. The paleolimnological reconstructions, primarily diatom-based, include geochemical parameters and the first ostracode stratigraphies from tropical Africa. The analyses from all three lakes converge on a history of low and fluctuating water levels over approximately the past 4500 years, preceded by a major Holocene high stand (ca. 10000–6000 B.P.) and a brief earlier rise centered on 12500 B.P. The lakes were mostly very low during the late Pleistocene (13000–30000 B.P.), though modestly elevated levels may have occurred early in this time period. The ostracode record, comprising at least 15 species, was relatively continuous in the Nakuru core but was restricted to sharply demarcated regions in cores from the other lakes.

Introduction

The floor of the East African Rift Valley arches broadly upward as it crosses the equator, reaching elevations of 1800–2400 m in the Kenya Highlands (Fig. 1), where the valley is flanked to east and west by the still higher Aberdare Range (3900 m) and Mau Escarpment (3000 m). The three lakes discussed in the present paper occupy this elevated region of the valley and are separated from each other by less than 50 km. At times of high water within the last 8000 years their maximum separation was probably less than 10 km. Because of their proximity to each other, and also because of their limnological contrasts (two are closed, alkaline lakes today, while one is fresh and apparently drained by underground seepage), their paleolimnological records present unusual opportunities for comparison.

Exposed lake sediments and strandlines indicate that the two alkaline lakes, Nakuru and Elmenteita, once were joined as a single large lake nearly 200 m deep and 800 km^2 in area (Washbourn-Kamau, 1971; Butzer *et al.*, 1972). Today their depths fluctuate between 0 and 5 m, and Nakuru (the larger of the two) covers an area of no more than 40 km^2 – often less. The freshwater Lake Naivasha presently is the largest of the trio, its area in recent decades ranging between about 110 and 175 km^2 and the depth of its main basin between about 3 and 8 m. Naivasha, too, formerly was much larger, though because its basin has a lower surface outlet this lake never became as large or deep as the combined Lake Nakuru-Elmenteita (Washbourn-Kamau, 1975). Naivasha, which lies closer to the eastern wall of the rift than the other lakes, receives considerably less direct rainfall than they but is fed by larger, more reliable streams.

This paper summarizes our core-derived knowledge of water-level and biological change in these lakes since 30000 B.P. A detailed, 9000-year record from Lake Naivasha (Richardson & Richardson,

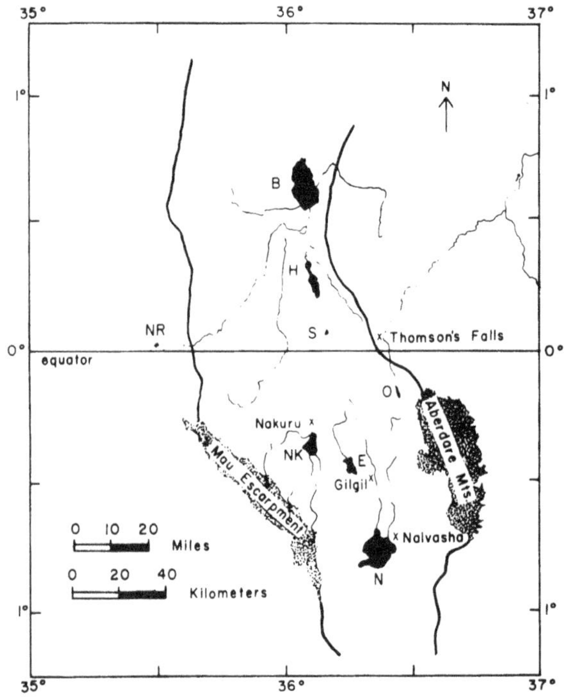

Fig. 1. Location map for Rift Valley lakes in the Kenya Highlands. Main towns and boundaries of the Rift Valley are also shown. Abbreviations for lakes: B = Baringo, E = Elmenteita, H = Hannington (Bogoria), N = Naivasha, NK = Nakuru, NR = Narasha, O = Ol Bolossat, S = Solai.

1972) and an account of the ostracode stratigraphy of Lake Nakuru (Cohen *et al.,* 1983) have appeared already, and additional detailed stratigraphies for all three lakes are in preparation. For more than 50 years the landscapes surrounding the present lakes have attracted the attention of geomorphologists, stratigraphers and paleoclimatologists, because of their abundant evidence of higher lake levels in the past (Nilsson, 1931, 1940; Flint, 1959; Washbourn-Kamau, 1971, 1975). Also, for at least as long a period, archeologists and paleoanthropologists have investigated the rich record of early man in these basins (cf. Leakey, 1931; Posnansky, 1967; Isaac *et al.,* 1972; Ambrose, 1984). One aim of our core-based paleolimnologic studies is to develop sufficiently precise paleoclimatic records to provide a detailed ecological context for prehistoric cultural and habitation changes in this region. We dedicate this paper to the memory of Glynn Isaac; friend, field companion and initiator of this joint endeavor linking paleolimnology and paleoanthropology.

Materials and methods

A 1960–61 expedition led by D. A. Livingstone of Duke University obtained an initial core from the Crescent Island Crater on the eastern edge of Lake Naivasha, using a lightweight piston sampler. In 1969–70 four additional cores were collected, using heavier piston-coring equipment of Livingstone's from (1) the Crescent Island Crater, (2) the main basin portion of Naivasha, (3) Lake Nakuru and (4) Lake Elmenteita. The present account concentrates on cores 2–4. The Naivasha cores provide more complete continuous records than those from Elmenteita and Nakuru, but the latter extend further back in time.

Laboratory analyses were made of several mineralogic and paleochemical parameters, mostly to be reported elsewhere; of fossil diatoms, on which the paleolimnologic reconstructions are primarily based; and of fossil ostracodes. Diatoms were cleaned by the peroxide-dichromate method of Van der Werff (1953), mounted permanently in Hyrax, and counted at 1000X; except where their densities were very low, at least 300 were counted per level, usually at 20 cm stratigraphic intervals. For ostracode analysis, wet mounts of 0.04 cc sediment samples, for the most part also at 20 cm stratigraphic intervals, were counted at 20–40X using a dissecting microscope.

Results and interpretation

1. Diatoms, inferred chemistry and water levels

Figs. 2 and 3 present summaries of the diatom stratigraphies of Lakes Elmenteita, Nakuru and the main basin of Naivasha, together with interpretations of the diatom zones in terms of changing water chemistry. In Figs. 3 and 4 these inferences are translated into probable water level changes over the periods recorded by these cores; Richardson and Richardson's (1972) record of water level change from Naivasha's Crescent Island Crater is added to Fig. 4.

The first third (30000–20000 B.P.) of the time period for which we have records is represented only by the Nakuru and Elmenteita cores, and there imperfectly, by sediments composed largely of volcanic ash and pumice. The fossil diatoms, though

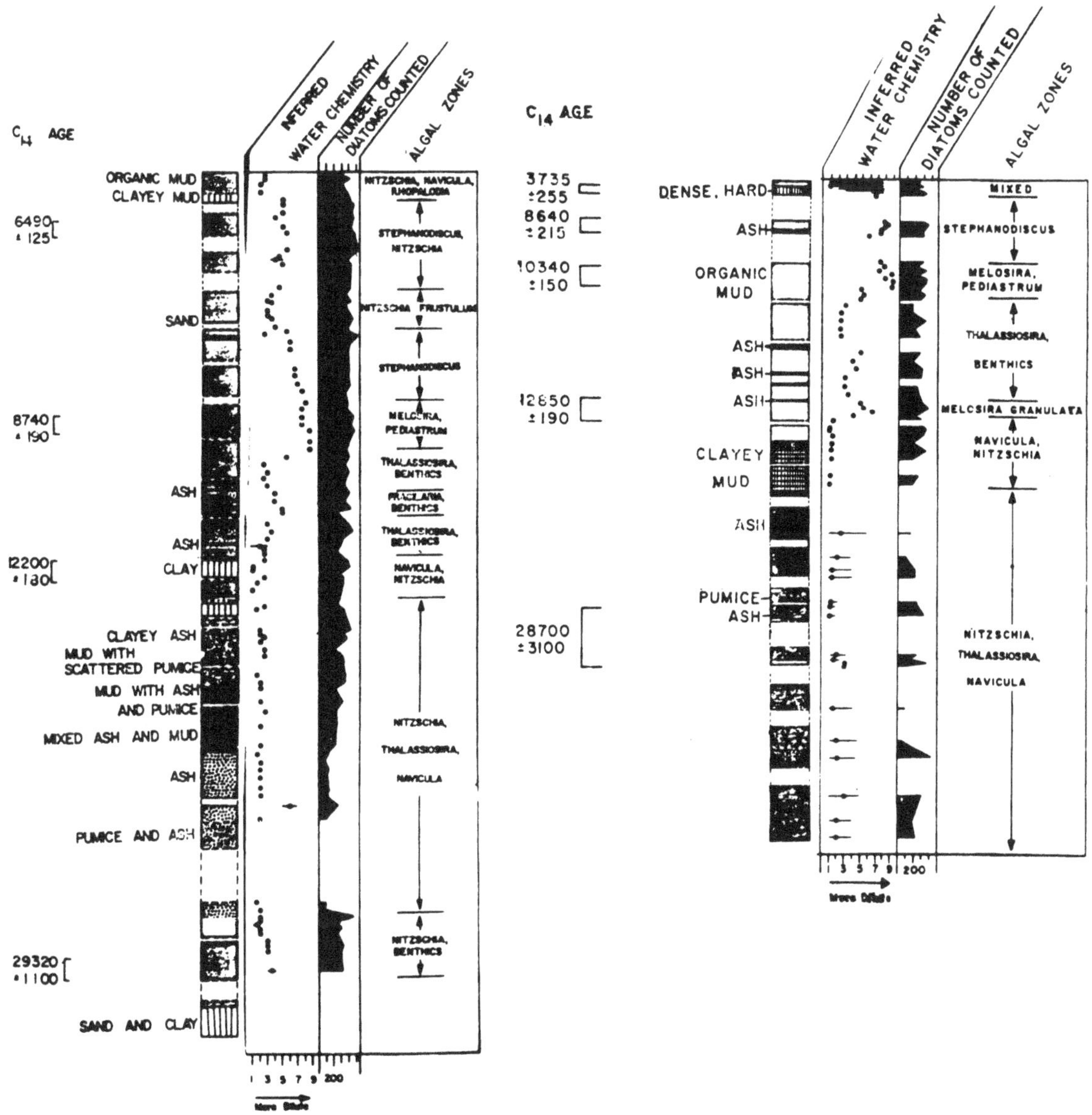

Fig. 2. Core stratigraphies from Lake Elmenteita (left) and Lake Nakuru (right). The diatom records, presented here in terms of zonal dominance patterns, are the basis of the inferred records of total solute content. The dilution scale is relative; both lakes varied over approximately a hundred-fold range of dilution and concentration.

relatively sparse, indicate that these two lakes certainly were separated during the early period, and were low and quite probably intermittently dry. The diatom flora was dominated by *Nitzschia frustulum, Nitzschia palea* (?), *Navicula el kab* and *Thalassiosira rudolfi.*

In the second phase of our time sequence (20000–10000 B.P.), the Nakuru and Elmenteita records are joined by that of Naivasha's main lake (Fig. 3). All three lakes remained low through most of this period, but Naivasha was relatively fresh (alkalinity 1–15 meq/l) whereas the other two lakes were generally highly concentrated (alkalinities 50–150? meq/l). In its most dilute phases Naivasha was apparently a small, weedy pond characterised by *Eunotia* spp., *Pinnularia* spp., *Navicula zanoni* and *Cocconeis placentula* v. *euglypta;* but more characteristically it was a shallow, more open and more alkaline habitat dominated by a variety of planktonic and tychopelagic diatoms.

These conditions terminated between 13000 and 11500 B.P., apparently rather asynchronously

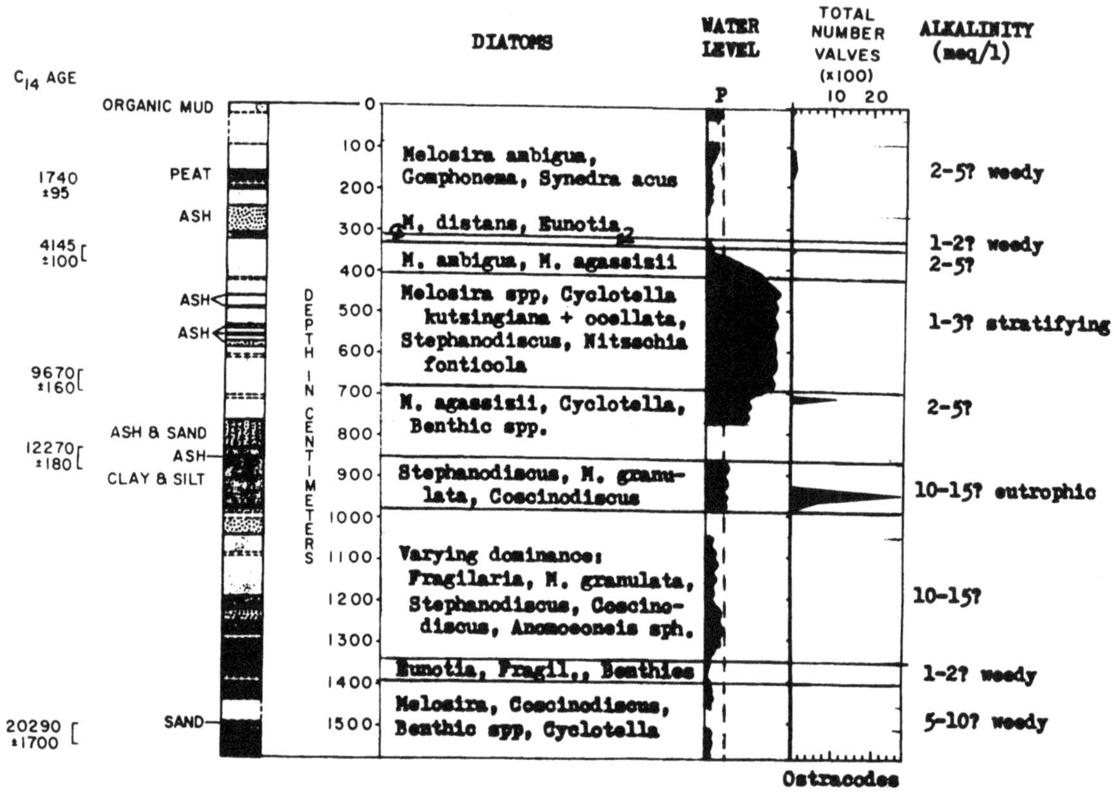

Fig. 3. Core stratigraphy from the main basin of Lake Naivasha, showing diatom zones, regions of ostracode occurrence, and inferred alkalinity and lake level changes.

among the three lakes. The Nakuru core records the earliest (c. 12800 B.P.) and most abrupt move toward a more dilute condition (= a higher water level) in the form of a brief but major pulse of *Melosira granulata* var. *angustissima* and *Fragilaria brevistriata.* Lake Nakuru then apparently receded to a more alkaline condition characterized by *Thalassiosira rudolfi* and/or *Coscinodiscus rothii, Rhopalodia gibberula* and several benthic taxa, remaining in this state until the lake's major freshening phase began (c. 10500 B.P.).

The early freshening of Lake Elmenteita was apparently a much weaker episode than that of neighboring Lake Nakuru, and the radiocarbon date suggests that in Elmenteita this freshening occurred closer to 12000 B.P. than to 12800. Elmenteita then receded, acquiring a flora similar to that of Nakuru, until joining its neighbor in the major freshening that began c. 10500 B.P. Lake Naivasha also expanded circa 12500 B.P., to levels probably close to those of the present lake. However, judging from the character of its sediment and its dominant diatoms (*Stephanodiscus rotula* v. *minutula, Melosira granulata, Thalassiosira rudolfi*), the Naivasha of 12500 B.P. was quite different ecologically from today – more fertile, more alkaline, and richer in suspended clay and silt. In Naivasha this phase persisted, without the subsequent recession noted for Nakuru and Elmenteita, until all three lakes, apparently contemporaneously, began to expand to maximum size and freshness.

The most recent 10000 years of our combined record is divisible into two major phases of roughly equal duration. As this period began, levels of all three lakes were rising (or had just risen) to high, fresh levels – the greatly expanded lakes alluded to in the introduction. In both Nakuru and Elmenteita, which probably became confluent at about this time, an early dominance of *Melosira agassizii* (accompanied by the chlorophyte *Pediastrum simplex*) gave way to dominance by *Stephanodiscus rotula.* One can interpret this change as a transition from maximally fresh to somewhat less dilute conditions as the lakes stopped rising; probably, too, Nakuru-Elmenteita now became stratified at least seasonally, with silica depletion in the epilimnion favoring the ascendancy of *Stephanodiscus* (cf. Kilham, 1971). At times during this large-lake phase Elmenteita-Nakuru probably lost water via a surface outlet to the north, but this outflow doubtless was relatively minor. Naivasha, with heavier riverine inflows and a lower overflow channel from its basin, became much more truly an open, throughflow lake during this high-water phase than did the other two lakes, and developed a more varied set of dominant planktonic diatoms.

The Elmenteita core provides a more complete record than does the Nakuru core of the later stages of the large lake in their joint basin. A phase of somewhat lower, more alkaline conditions circa 7500 B.P. is signalled by the emergence of *Nitzschia frustulum* and, briefly, *Thalassiosiria rudolfi* as dominants; thereafter, although *Stephanodiscus* returned strongly (probably signalling a rise in water level), *N. frustulum* remained as an important element, suggesting that the lake remained more concentrated ionically than it had been in the period 8000–10000 B.P. Naivasha's core record, by contrast, indicates relatively little limnological change and continued high water levels throughout this time period (circa 10000–5500 B.P.). About 5000 years ago all three lakes apparently began falling rapidly. In Naivasha, where the decline is signalled much more definitely (its core record for this period is much more complete), dominance passed quickly from *Melosira agassizii* (at 4145 B.P.) to *M. ambigua* and then to *M. distans* and *Eunotia* spp. in a peaty layer signalling shallow, weedy, dilute conditions. The lake probably dried completely for a brief time thereafter (the Crescent Island Crater core records 3000 B.P. as being the date of lowest water level), and sometime after 1700 B.P. Naivasha assumed its present character as a shallow, open, relatively fresh lake dominated by *Melosira ambigua* and *Synedra acus* with *Gomphonema* common in the extensive macrophytic growths.

We believe that lakes Nakuru and Elmenteita may have become confined to their own sub-basins even before their precipitous decline circa 5000 years ago. For much of the period since that decline they apparently have been dry. During shallow phases the upper sediments of the dilute, pre-5000 B.P. lake phase seem to have been mixed with more recent, alkaline-lake sediments. In addition, some sediments probably were removed by deflation when the lakes dried completely. The present highly alkaline lakes Nakuru and Elmenteita are at most times heavily populated by blue-green algae, with *Nitzschia frustulum* and *Navicula el kab* the most characteristic diatom associates.

2. Ostracodes

When Dussinger (1973) first determined the ostracode stratigraphy of these cores, too little was known of the ecology of East African species to allow detailed use of ostracodes in our paleolimnological reconstructions. Rather, we envisioned that from the diatom stratigraphy we would be able to make inferences about the ecology of stratigraphically co-occurring ostracodes, thereby building a stronger foundation for the use of African ostracodes as direct paleolimnological indicators in future African studies.

Within the past decade, in fact, studies primarily by other investigators (Peypouquet *et al.*, 1979; Cohen *et al.*, 1983) have begun to use ostracodes directly in paleolimnological reconstructions. Indeed, an interpretation of our Nakuru core record using ostracodes as the primary indicators has now been published, and compared (Cohen *et al.*, 1983) to the diatom-based interpretation presented in this paper. There is very good agreement between the two separately-prepared interpretations, the chief difference being an indication from the ostracodes of a somewhat more dilute (= larger?) Lake Nakuru circa 25000 B.P. than is indicated by the diatoms (cf Fig. 4).

The ostracode taxa identified from the sediment of all three lakes are listed in Table 1. The Nakuru ostracode record (see Cohen *et al.*, 1983) is more extensive than that of Elmenteita and Naivasha, although a rich ostracode fauna (containing all the species found in the Nakuru record) also characterized Elmenteita between 12000 and 7000 B.P. (Ostracodes were absent during this time span only during the *Melosira/Pediastrum* phase.) Elsewhere in the Elmenteita core ostracodes were very scarce; in opposition to Lake Nakuru, they do not seem to have been common in Elmenteita during its low-water phases. In the core from Naivasha's main ba-

Table 1. Occurrence of fossil ostracode taxa in sediments of lakes Nakuru, Elmenteita and Naivasha.

Taxon	Ecological notes
Limnocythere sp. A	Mid- to highly alkaline phases of all three lakes
L. sp. B	Most generally abundant species in Elmenteita and Naivasha, and widely occurring in Nakuru core; tolerant of wide alkalinity range.
L. sp. C	Mid to lower alkalinity phases; scattered occurrence in all lakes.
L. michaelsonii	High to intermediate or quite low alkalinity levels; in all lakes
Gomphocythere obtusata	Scattered in all cores; apparently tolerant of wide alkalinity range
Cypridopsis sp. A	Restricted occurrence in all lakes; mid- to low alkalinities?
C. sp. B	Similar to *C.* sp. A
C. sp. C	Similar to *C.* sp. A
C. sp. D	Mid- to high alkalinities in all lakes
Zonocypris calcarata	Restricted occurrence, in Elmenteita & Nakuru only. Low alkalinities.
Oncocypris sp.	Low to mid- high alkalinities; restricted occurrence in all lakes.
Cypria sp.	Intermediate alkalinities; all lakes
Cypretta cf. *C. murati*	Scattered; rel. dilute phases in Nakuru and Elmenteita; absent from Naivasha.
Mecynocypria cf. *M. reniformis*	Mid- to mid-low alkalinities, all lakes
Potamocypris arcuata	Only in Naivasha; restricted here to mid-alkalinity, open water conditions.

sin ostracode occurrence was even more sharply restricted (Fig. 3), and although the narrow zones of occurrence were rich in species, ostracodes by themselves yield a very incomplete reconstruction of Naivasha's history.

Discussion

The striking high stand recorded for all three lakes circa 10000–5000 B.P. represents a major climatic phenomenon already known from other studies in many parts of Africa (cf. Butzer *et al.*, 1972; Street & Grove, 1976). Information that relates to the nature, timing and duration of the drier phases, and to lesser fluctuations of climate and lake conditions within the major wet phase, may be a more significant contribution from these cores. For example, the diatoms in these stratigraphies give only faint, equivocal evidence – and the ostracodes none at all – of the higher water levels about 16000–19500 B.P. for which Holdship (1976) found evidence at Lake Manyara, an Eastern Rift Lake some 400 km south of our study area. Harvey, 1976; see Richardson *et al.*, 1978) also found evidence for elevated levels of Lake Mobutu Sese Seko (650 km NW) at about this time. An earlier freshening phase, circa 25000–27000 B.P., that also was recorded by both Holdship and Harvey, seems better supported in our cores by ostracode evidence (Fig. 4) than by diatoms. These early wetter episodes must have been relatively minor at our sites.

The proximity of Lakes Nakuru, Elmenteita and Naivasha to each other makes the contrasts among their core histories (Fig. 4) particularly instructive.

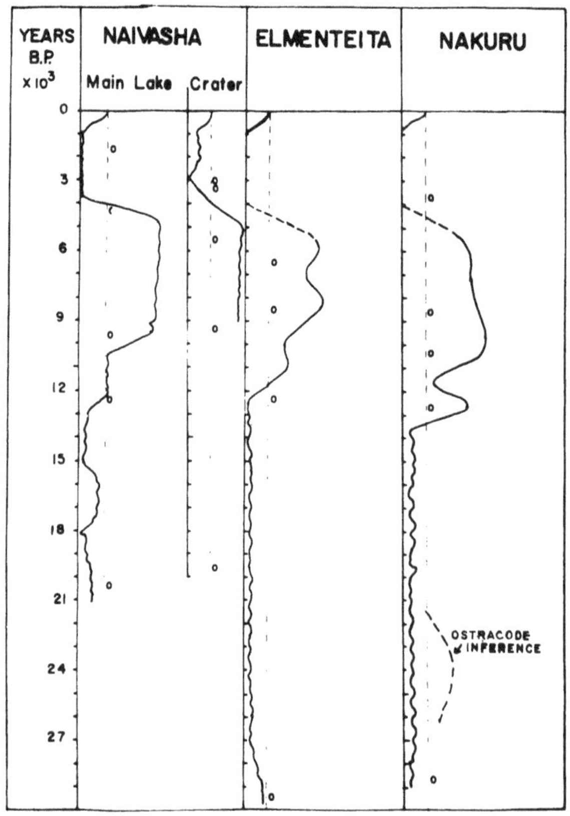

Fig. 4. Inferred water level changes for all lakes, including the crescent Island Crater of Naivasha (analyzed by Richardson & Richardson, 1972). The inferences are diatom-based; an alternative interpretation based on ostracodes is shown for the lower part of the Nakuru core. In other parts of this core diatom- and ostracode-based stratigraphies generally agree closely. Carbon-dated strata are indicated by circles, and average modern water levels by dashed lines.

There are noteworthy differences, such as the fluctuation of Nakuru about 12800 B.P. that seemingly had no parallel of equivalent magnitude in the other two lakes. Some of these contrasts may be resolved by closer analysis, or may be artifacts of interruptions in the record of one or another of the lakes due to incomplete core recovery. However, it is clear that hydrological differences between Naivasha and the other two lakes underlie some of these contrasts, and these differences have produced records that complement each other in the degree of resolution they provide. When these lakes were high, Naivasha's biota and chemistry seem to have remained relatively stable, owing no doubt to the copious surface inflow and outflow at this time. Conversely, the deep Elmenteita-Nakuru lake of the major high-water period retained much of the ecological climatic sensitivity of a closed-basin lake; its inflow rivers and outflow channel were smaller and the lake only overflowed at times of very highest water. It is in the records from Elmenteita-Nakuru, rather than Naivasha, that biotic changes preserve evidence of a fluctuating water chemistry and climate during the high-water period 5000 to 10000 years ago, rather than uniformly moist conditions. But while the Elmenteita-Nakuru record is more fine-tuned during the high-water portion of the lakes' history, at low-water times Naivasha's record is more sensitive. Since this lake seldom shrank to dryness, its accumulating sediments preserved evidence of limnological and climatic fluctuations at times when the other two lakes were dry. Only from Naivasha, for example, has it been possible to ascertain that the time of *maximum* Holocene aridity occurred about 3000 B.P. in these basins. As we have indicated, ostracodes are a consistent and interpretively valuable component of the stratigraphy of Lake Nakuru, and to a lesser extent of Lake Elmenteita, but in the Naivasha main lake core their occurrence is highly localized. As ecological knowledge of East African ostracodes continues to accumulate, their usefulness (and use) in paleolimnological interpretation will increase but our results suggest that ostracode analysis will be fruitful for only some lake stratigraphies, not all.

A number of workers (Hustedt, 1949; Richardson, 1968, 1969; Hecky & Kilham, 1973; Richardson *et al.*, 1978; Gasse *et al.*, 1983) have directed their attention to the water chemistry relationships of modern East African diatoms. Consequently, the paleoecological meaning of many diatom species and assemblages is quite well understood, and we believe that the diatom-based interpretations we have made here are sound. But both questions and opportunities remain with respect to the diatom stratigraphies. For example, why have we been unable to infer with greater precision the dilution status of both late and early portions of the Nakuru history (Fig. 2)? Is this uncertainty caused by post-depositional mixing of sediments representing more than one stage of a chemically fluctuating lake? Or, alternatively, did the diatom species recorded together in these regions of the core actually coexist? If the second suggestion is correct their chemical ecology may be more eurytopic than we presently suspect. Other questions relate to the *Melosira agassizii/Pediastrum* zone of Lakes Nakuru and Elmenteita: is this assemblage indicative, as we tend to think, of the phase of maximum dilution in these lakes? And if so, does it represent the time of highest lake level and maximum overflow, or instead, a prior phase of lower but rapidly rising levels? In the case of Naivasha, the diatom record raises similar questions, but also opportunities. Here the stratigraphies seem unusually conducive, for example, to untangling questions about ecological differences within the genus *Melosira.* At least four distinct taxa of *Melosira* alternated and sometimes coexisted as dominant elements in Naivasha through nearly the entire, changing, 20000-year history we now have from this lake. The record is sufficiently detailed to provide strong indications of the conditions favoring each taxon of *Melosira,* and more detailed analysis with this genus as the focus should be fruitful.

Acknowledgements

We thank all of the following: The Office of the President, the Kenya National Park Service for permission to work in Kenya and in Lake Nakuru National Park; Daniel Livingstone for the loan of coring equipment; Paul Buckland, Thomas Harvey, Glynn Isaac, Peter Kilham, Alice Richardson and Paul Waiswa for assistance in drilling operations; and Andrew Cohen, L. D. Delorme, Linda Dussinger, Timothy Morris, Dean Pappas and George Ruddy for assistance with laboratory analysis or interpretation. This research was supported by grants from the National Science Foundation to J. L. Richardson at Franklin and Marshall College, and to D. A. Livingstone at Duke University.

References

Ambrose, S. H., 1984. Holocene Environments and Human Adaptations in the Central Rift Valley, Kenya. Ph.D. Thesis, University of California (Berkeley).

Butzer, K. W., G. L. Isaac, J. L. Richardson & C. Washbourn-Kamau, 1972. Radiocarbon dating of East African lake levels. Science 175: 1067–1086.

Cohen, A. S., R. Dussinger & J. Richardson, 1983. Lacustrine paleochemical interpretations based on eastern and southern African Ostracodes. Palaeogr. Palaeoclimat., Palaeoecol. 43: 129–151.

Dussinger, R. A., 1973. The Environmental Reconstruction of African Rift Lakes Using Ostracodes as Paleoecological Indicators. Thesis, Franklin and Marshall College, Lancaster, PA.

Flint, R. F., 1959. Pleistocene climates in eastern and southern Africa. Geol. Soc. Amer. Bull. 70: 343–374.

Gasse, F., J. F. Talling & P. Kilham, 1983. Diatom assemblages in East Africa: classification, distribution and ecology. Rev. Hydrobiol. trop. 16: 3–34.

Harvey, T. J., 1976. The Paleolimnology of Lake Mobutu Sese Seko, Uganda-Zaire: The Last 28000 Years. Ph.D. Thesis, Duke University.

Hecky, R. E. & P. Kilham, 1973. Diatoms in alkaline, saline lakes: ecology and geochemical implications. Limnol. Oceanogr. 18: 53–71.

Holdship, S. A., 1976. The Paleolimnology of Lake Manyara, Tanzania: A Diatom Analysis of a 56 Meter Sediment Core. Ph.D. Thesis, Duke University.

Hustedt, F., 1949. Süsswasser-Diatomeen. Explor. du Parc National Albert. Mission H. Damas (1935–1936), No. 8, 1–199. Bruxelles, Inst. des Parcs Nat. du Congo Belge.

Isaac, G. L., H. V. Merrick & C. M. Nelson, 1972. Stratigraphic and archeological studies in the Lake Nakuru basin, Kenya. Paleoecology of Africa 6: 225–232.

Kilham, P., 1971. A hypothesis concerning silica and the freshwater planktonic diatoms. Limnol. Oceanogr. 16: 10–18.

Leakey, L. S. B., 1931. The Stone Age Cultures of Kenya Colony. Cambridge, Cambridge University Press. 288 pp.

Nilsson, E., 1931. Quaternary glaciations and pluvial lakes in British East Africa. Geogr. Ann. 13: 249–349.

Nilsson, E., 1940. Ancient changes of climate in British East Africa and Abyssinia. Geogr. Ann. 22: 1–79.

Peypouquet, J.-P., P. Carbonel & J. de Heinzelin, 1979. Les ostracodes indicateurs de l'évolution des environments lacustres de la fin du Cénozoïque dans la branche orientale du Rift africain. Bull. Inst. Géol. Bassin d'Aquitaine, Bordeaux 25: 201–219.

Posnansky, M., 1967. Excavations at Lanet, Kenya. Azania 2: 89–114.

Richardson, J. L., 1968. Diatoms and lake typology in East and Central Africa. Int. Revue ges. Hydrobiol. 53: 299–338.

Richardson, J. L., T. J. Harvey & S. A. Holdship, 1978. Diatoms in the history of shallow East African lakes. Pol. Arch. Hydrobiol. 25: 341–353.

Richardson, J. L. & A. E. Richardson, 1972. History of an African rift lake and its climatic implications. Ecol. Monogr. 42: 499–534.

Street, F. A. & A. T. Grove, 1976. Environmental and climatic implications of late Quaternary lake-level fluctuations in Africa. Nature 261: 385–390.

Van der Werff, A., 1953. A new method of concentrating and cleaning diatoms and other organisms. Verh. int. Ver. Limnol. 12: 276–277.

Washbourn-Kamau, C. K., 1971. Late Quaternary lakes in the Nakuru-Elmenteita basin, Kenya. Geograph. J. 137: 522–535.

Washbourn-Kamau, C. K., 1975. Late Quaternary Shorelines of Lake Naivasha, Kenya. Azania 10: 77–92.

Accepted 18 March 1986.

Hydrobiologia 143: 175–190, (1986)

Upper Pleistocene and Holocene lakes in the An Nafud, Saudi Arabia

Erhard Schulz[1] & John W. Whitney[2]
[1]*Geographisches Institut, Am Hubland, D 8700 Würzburg, FRG;* [2]*USGS, Box 25046, MS 913, Denver Federal Center, Denver, Colorado 80225, USA*

Keywords: Arabian peninsula, An Nafud, present environment, quaternary limnology, vegetation history

Summary

Two major lake periods were discovered in the sand sea of An Nafud and the surrounding areas. In Upper Pleistocene large lakes occurred around the dune area and in the interior of the sand sea. Their deposits were formed between 34000 and 24000 BP. The lakes were not conformous to the present dune relief. They had an extension of several km^2 and a depth of ca. 10 m. Holocene lakes (8400–5400 BP) were of minor extension and restricted to the interdune depressions. Lake sediments consist of cemented sand, calcareous crusts and diatomites. Upper Pleistocene lakes were fresh water lakes, the Holocene lakes were mostly swamps depending on rising and falling aquifers in the dunes. Plant remains as pollen and macro rests show that the environmental changes didn't exceed the system of semidesert comparable to the modern plant cover. However, the Upper Pleistocene lake deposits contain some more soudanian elements as the Holocene sediments in the pollen spectra. Climatically the lake formations are interpreted as depending on a stronger influence of the mediterranean cyclones or an interaction of them with monsoonal air masses.

Introduction

The environmental and climatic history of the northern Arabian peninsula is still almost unrevealed. Holocene records from Syria and northern Israel give some contradictory informations about the environmental development of theat region (Niklevsky & van Zeist, 1970; Tsukada ref. in Bottema, van Zeist, 1981). Investigations in the northern part of the Arabian peninsula may therefore provide solutions for these questions and allow comparisons with the environmental history of northern Africa. The southern margin of the mediterranean zone in North Africa has seen several periods of lake formations during the Quaternary (Pachur, 1984; Petit-Maire, 1982). Oxygen isotope data from Holocene aquifers indicate an increasing continentality of the climate from the west to the east (Sonntag *et al.,* 1980).

In order to compare the different hypotheses on the environmental history of this region and their climatic implications the sand sea of An Nafud was chosen as study area. Situated on the same latitude as the former lakes in northern Fezzan, the An Nafud is known as one of the most severe deserts and as an area of highly mobile sands (Fryherger, 1979). Reports of last century's voyagers (Euting, 1896), however, indicated, that there are lake sediments in the dune area which could provide informations about the environmental conditions during their formation and allow comparisons with the results of studies in the Rub al Khali (McClure, 1978). The investigations in the An Nafud and the adjacent sand seas included geomorphic observations as well as sedimentological and palynological analyses of the collected sediments. These methods provide results which may determine the type and extent of former lakes and the type of the regional vegetation in comparison with the modern pollen rain. Climatic conditions can be deduced from both of these sources.

Present conditions

The landscape of north central Arabia is dominated by peneplains and inselberg of the Arabian shield. Most of the inselbergs on the shield are remnants of granitic intrusions of the late Precambrium. Jebel Aja, situated between Hail and the An Nafud, is the largest inselberg in that area with an elevation of about 900 m rising ca. 500 m above the surrounding plain. The pediments arround the inselberg either expose the ingenious rock surface or are overlain by alluvial debris.

The eastern edge of the Arabian shield is surrounded by a series of cuestas in palaeozoic to mesozoic sandstones and limestones. Their escarpments are ca. 100 to 150 m high. Pediments in front of the escarpments are commonly terminated at topographic depressions containing playas and dunes of the Ad Dhana sand sea. Streambeds on the peneplain are incised only a few meters. They are mostly filled with silty sand and some gravel lenses. Most of the wadis on the peneplain are tributaries of the Wadi Rhima. Mountain wadis, however, are narrow valleys which receive more frequent local runoff.

The sand seas of the Arabian peninsula are of quaternary age. The An Nafud is situated just north of the Arabian shield and is composed mainly of 60–120 m high barchanoid and traverse dunes. Nafud Urayk is a small sand sea that lies completely on the Arabian shield and is confined by several inselbergs and by tributaries of Wadi Rhimah. Interdune depressions are common in all sand seas and remains of lake sediments occur in some of these depressions. These lake beds were deposited during different late Quaternary periods, giving evidence of more effective precipitation. Older lake beds range in age from about 34000 to 24000 BP and are overlain by the large principal dunes. The younger lake deposits are middle Holocene in age and they were formed in interdune depressions which still exist today (Whitney *et al.*, 1983).

Climatically the region is situated in the area of winter rainfall receiving 50 to 100 mm precipitation. On rare occasions a summer storm from the south indicated that a monsoon influence from the southern regions will reach the study area. Mean annual temperatures vary across the study area from 24° to 10 °C. Average summer temperatures in the An Nafud are 30 °C and winter temperatures are from 8 °C to 10 °C. The coincidence of rainfall and periods of low temperatures is partly responsible for the effectiveness of the low precipitation in the study area. The climate diagrams of Jeddah, Hail and Rhyad show the increasing continentality of climate from the West to the East (see Fig. 1).

Vegetation

The plant cover of the study area is mostly characterized as an extreme desert (Zohary, 1973; Barth, 1983). Zohary follows the system of Monod (1954) distinguishing deserts from semideserts or savannas by the mode of distribution of permanent vegetation. In a desert the permanent vegetation is contracted or linear, meaning restricted to wadis or depressions where groundwater or runoff is available. The annual rainfall is too low and irregular for the permanent vegetation. In a semidesert or savanna permanent vegetation occurs in a diffuse distribution depending on the annual rainfall.

Contrary to previous descriptions the study area presents a sparse but diffuse permanent vegetation (Schulz & Whitney, 1985). The region takes part of the large belt of semideserts that is situated south of the Mediterranean sea and is influenced both by Mediterranean flora and climate. In general, vegetation is built up of communities of the saharo-sindian, irano-turanian and mediterranean-presaharan zones including those of Acacia-Panicum of the large wadis, Anabasis-Salsola-Rhanterium of the plains or Artemisia-Calligonum-Stipagrostis of the dunes.

The different types of vegetations mainly depend on geomorphology and substratum according to its waterholding capacity. Typical relations between the different landscapes and vegetation units are schematically illustrated in Fig. 8 which shows the northern edge of the Arabian shield and An Nafud sand sea and typical inselbergs, pediments, wadis and peneplains on the shield.

Mountains bear a very thin shrub cover of *Rhus* or *Ephedra* together with tufted grasses like *Cymbopogon, Stipagrostis* or *Hyparrhenia,* whereas the wadis have shrub and tree communities of *Acacia, Ochradenus, Astragalus, Anabasis, Fagonia* a.o. The surrounding pediments are covered with a sparse Anabasis-Rhanterium vegetation. The wide

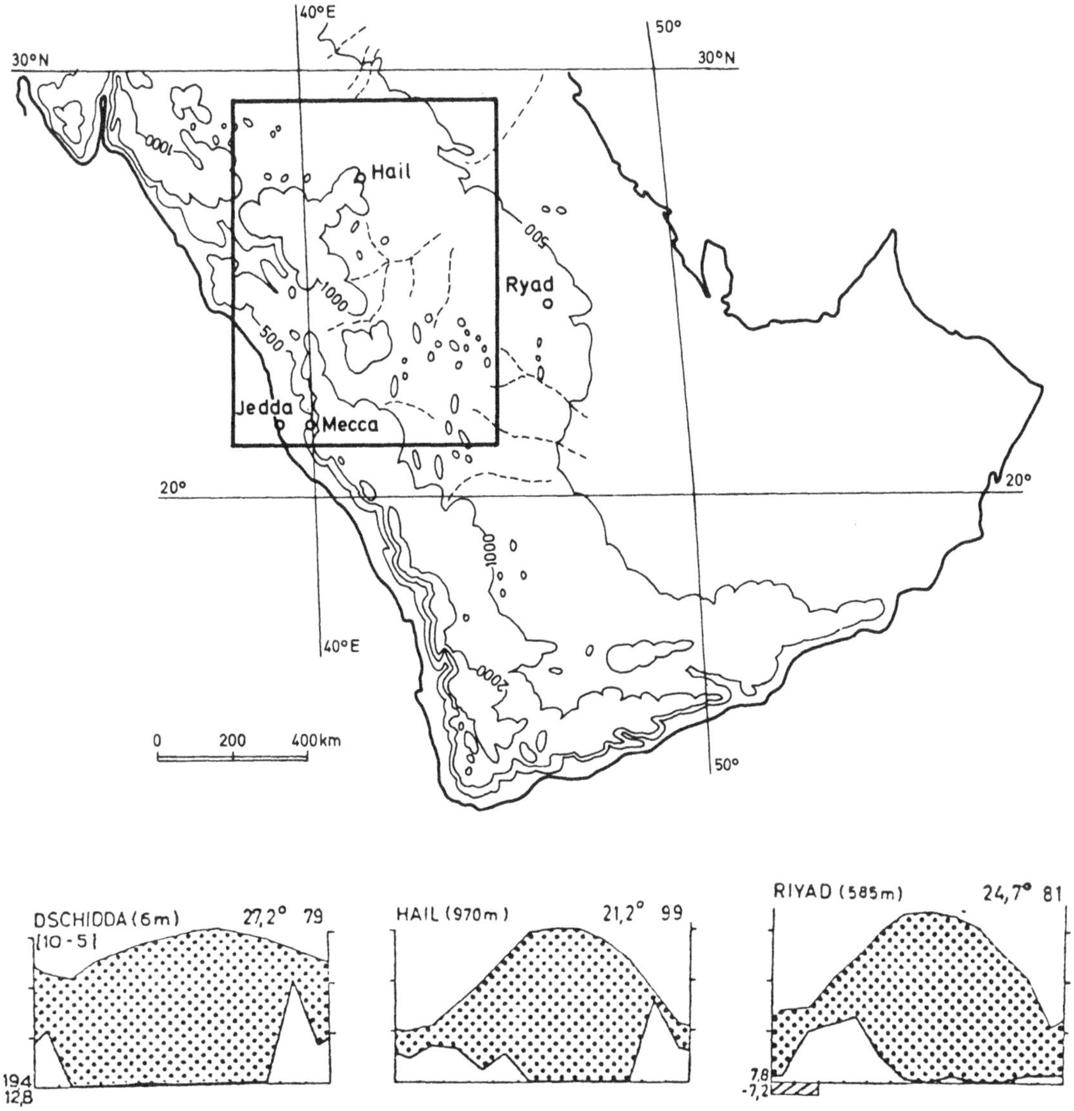

Fig. 1. Map of Arabian peninsula showing the location of study area and climate diagrams of Jeddah, Hail and Riyad (Walter & Lieth, 1965).

plains of the Arabian shield commonly expose a Rhanterium-Astragalus shrub cover on silty areas whereas *Anabasis, Hamada* or *Salsola* are characteristic for the sandy parts of the plains. In the southwestern edge of the study area the wadivegetation is dominated by Acacia-Panicum communities together with *Rhazya* and *Hyphaene.*

The dunes support a cover of shrubs (*Artemisia, Calligonum, Ephedra*) tufted grasses (*Stipagrostis*) and *Cyperus* together with various herbs. Achab floras forming short time vegetation plots of annual herbs and grasses, which show up after rainfalls, are very characteristic for the region.

Present pollen rain

In order to get a comparison for fossil pollen spectra the present pollen precipitation was measured with help of silicone oil soaked gauze filters exposed behind a vehicle for a certain distance to collect the upthrown dust. This method is largely approved in arid regions (Cour, 1974; Schulz, 1980). The filters were taken in the main vegetation types on the plains between Hail and Buraydah, in the dune area east of Unayza and in the plain between Unayza and Hanikah as well as in the large wadis east of Medinah (see Fig. 2).

Fig. 3 shows that the bulk of pollen comes from

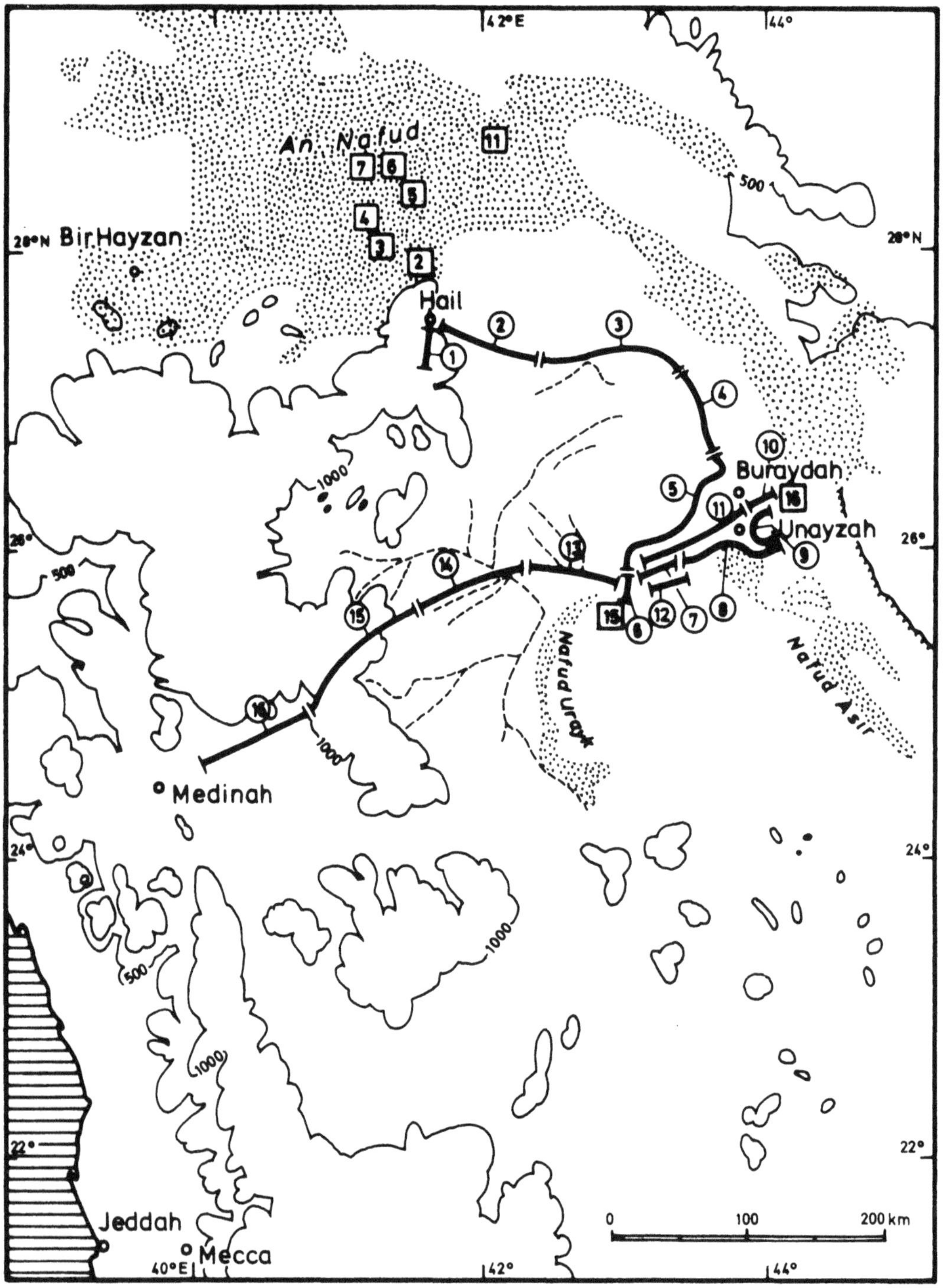

Fig. 2. Location of sample sites and filters in the study area.

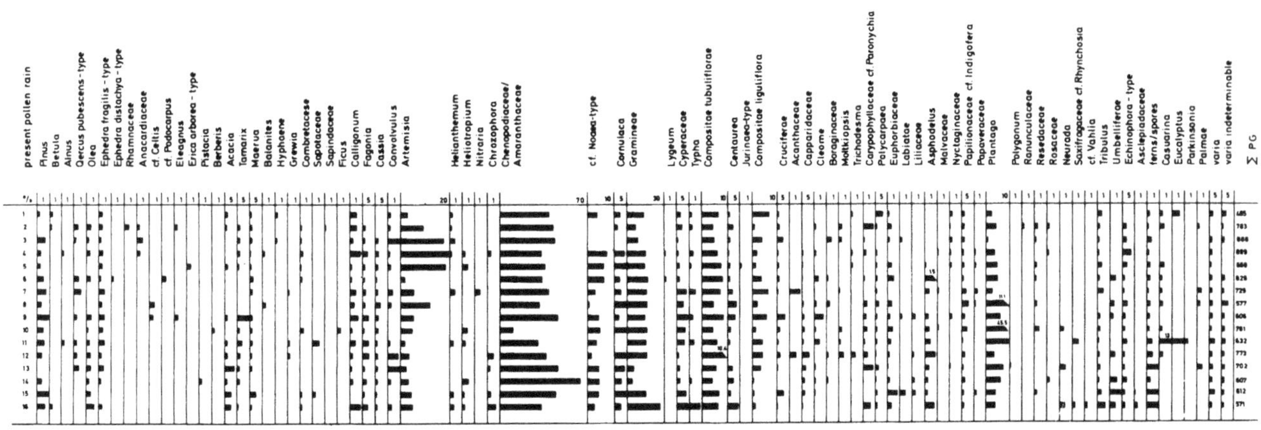

Fig. 3. Present pollen rain in northern Saudi Arabia.

the Chenopodiaceae/Amaranthaceae, Gramineae, *Calligonum, Artemisia,* Compositae and *Plantago.* The spectra are dominated by elements of the regional vegetation whereas the portion of elements transported over long distances to the area is relatively low. Pollen of *Pinus, Betula, Alnus, Quercus, Olea* or Sapotaceae, Sapindaceae and Combretaceae does not exceed one percent each.

Filters taken in the plains between Hail and Unayzah (1–7) expose the dominance of Cheno/Am, *Artemisia,* Compositae and *Plantago,* whereas Gramineae have lower values. The higher percentages of *Artemisia* and *Calligonum* may be explained by transport from the nearby sand sea of An Nafud.

In the dune area (8–12) the Cheno/Am have lower values in the pollen spectra and *Artemisia, Calligonum,* Compositae are dominating together with Gramineae and Chenopodiaceae. *Plantago* and *Moltkiopsis* are part of the typical achab-floras.

The southern part of the study area (13–16) is again characterized in the present pollen rain by a predominance of Chenopodiaceae/Amaranthaceae. Elements like *Artemisia, Calligonum,* Gramineae or *Plantago* are low again. In the last filters (15/16) there is a certain rise in the percentages of *Acacia,* Gramineae, *Tribulus* and *Maerua* reflecting the Acacia-Panicum vegetation of the larger wadis near Medinah.

In conclusion, the present pollen rain gives a fairly good image of the main vegetation types whereas the portion of pollen, which is transported over long distances, is relatively low. Northern elements are constantly present in this part but tropical elements like *Grewia,* Combretaceae or Sapotaceae occur only sporadically. Investigations on the present pollen rain in Syria and Lebanon showed similar trends especially in the semidesert areas (Bottema & Barkoudah, 1979).

Upper Pleistocene and Holocene lakes

One of the most striking phenomena in the sand seas of the Arabian peninsula is the presence of various lake beds in different extension. Principally one can distinguish two main types of sediments (see Fig. 4). The first type are largely extended calcareous crusts, marls and rarely diatomites. They appear in the external parts of the sand sea as well as in the upper parts of dune depressions. Their thickness may vary between 0.30 m and 2.30 m. A great number of these lake deposits are underlaying the dunes showing that they are not conformous to the present dune configuration. Several radiacarbon datings indicate, that these deposits were formed in the period between 34000 and 24000 BP (Whitney, 1982).

The second type of deposits one can find at the base of the dune depressions itself. They consist of cemented sands, thin layers of marls and rarely some diatomites. Slumping phenomena as well as root or rhizome remains are common. Different to the Upper Pleistocene lake beds these sediments are conformous to the present dune configuration. They were deposited in the already existing dune re-

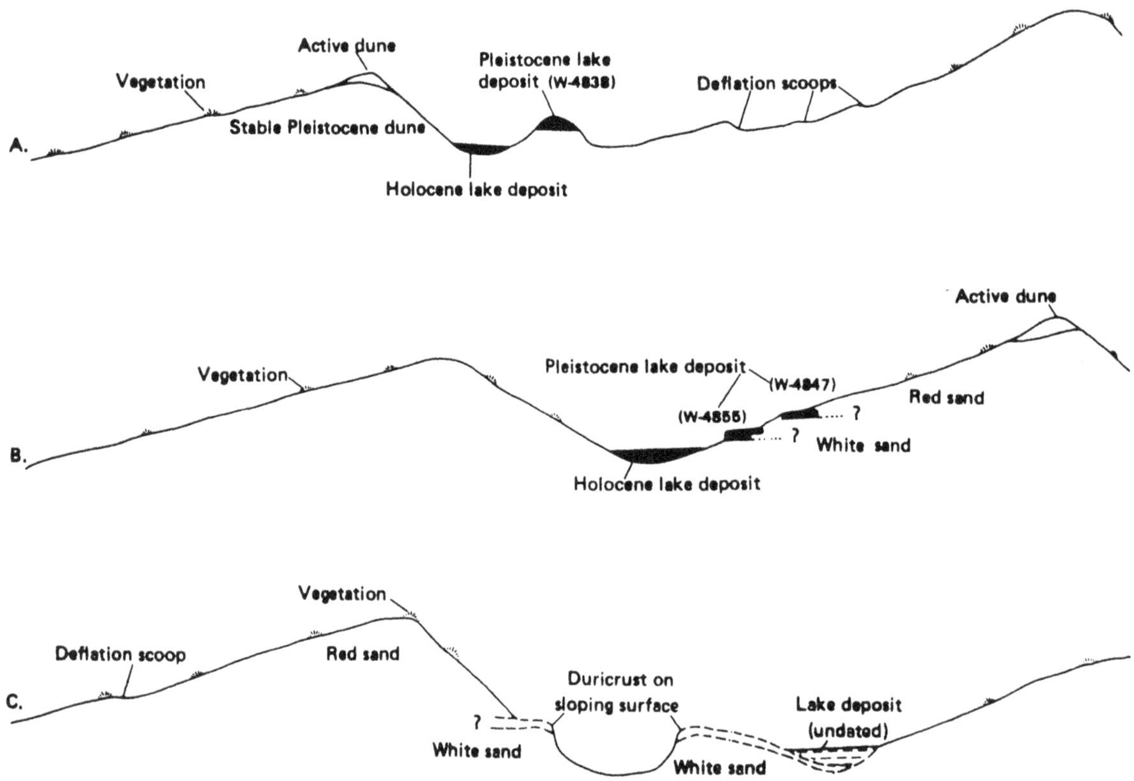

Fig. 4. Schematic cross sections in the An Nafud illustrating the relationship between Quaternary lake deposits and both active and stable sand dunes (Whitney *et al.,* 1983).

lief in the period between 8400 and 5400 BP (Whitney, 1983).

Both types of lake deposits show some similar traits. They are clearly stratified with interbedding sand layers and they show all kinds of discontinuities between them. This indicates that they do not represent continous lake phases but more or less short time water bodies with accentuated oscillations connected to the water table in the large dunes. Because of these facts one cannot expect continous sequences of fossil records but instead only short phases of different or similar environmental conditions.

Upper Pleistocene lakes

Sediments of these lakes were collected in the upper part of dune depression in southern and central An Nafud as well as from the great lake near Juba (see Fig. 2). Further informations derive from the Bi'r Hayzan diatomite in the southwest of the An Nafud (Whitney, 1982; Bradbury, pers. comm.). The lake beds consist of calcareous crusts and marls and show very often an interbedding of eolian sand. Sediments of profile 4 derive from the great lake near Juba in the external part of An Nafud which were deposited directly on the peneplain. The other sediments were collected in the dune depressions indicating by their presence that at least a certain part of the sand sea was already built up at that time.

Palaeolimnological information

Sediment types may provide indications of the type of Upper Pleistocene lakes, even they are more indirectly. With the exception of the Bi'r Hayzan diatomite they were no fossils giving informations on the environmental conditions of the lakes itself. The characteristic interbedding with eolian sand and the stratification in distinct layers demonstrate that the lakes in the dune area did change their extension and depth several times. Apparently they were formed by the changing water table in the dunes when the large sand bodies could serve as water collectors. Lake beds in the external parts of the An Nafud show that these lakes had an extension of several km^2.

The investigation of the diatomite of Bir Hayzan (Whitney, 1982; Bradbury, pers. comm.) provided some more precise informations on the type of these lakes (see Fig. 7). Most of the diatoms are part of the plankton (*Stephanodiscus, Cyclotella*). They are common in freshwater lakes in middle latitudes with a depth of about 10 m. The base of the sequence has indicators for stagnant water such as *Synedra ulna, Amphora ovalis* or *Cymbella cistula*. The lake could have been monomictic and not strongly affected by running water.

Palaeovegetation

In the pollenspectra derived from the lake beds there is a clear predominance fo Gramineae, Cyperaceae and shrubs like *Calligonum, Fagonia, Convolvulus* was well as herbs like Compositae, *Moltkiopsis,* Capparidaceae, *Polycarpaea* and *Plantago.* The dunes were stabilized by a grass and shrub vegetation certainly some more dense as nowadays. The lakes were surrounded by greater belts of *Typha* and *Phragmites.*

There is a certain portion of pollen transported over long distances such as *Pinus, Betula, Carpinus, Quercus* and even Combretaceae and *Commiphora.* Probably they were captured by the large water surface working as a great pollen trap. There is no evidence that the northern elements were present in the regional vegetation, because a compainging herb flora is lacking. *Salvadora,* however, was certainly present in the area because of its over-representation in profile 7. Probably *Acacia, Maerua* and *Balanites* will have been present in the mountain area. Pollen of *Commiphora* is very little represented in the modern pollen rain (Schulz, 1980), so there might be a chance that *Commiphora* was present at this period. Generally there is no sign for a denser tree vegetation neither in form of mediterranean woodland nor of tropical savanna. Compared with the modern pollen precipitation the fossil spectra represent the grass and shrub communities of a semidesert in Upper Pleistocene times.

These first pollen spectra cannot provide information on the density of the former vegetation. They only describe the main vegetation types of the region.

Holocene lakes

Lake beds of Holocene age (ca. 8400–5400 BP) appear in An Nafud and Nafud Asir at the bottom of interdune depressions. They consist of duricrusts, cemented sands and rarely of diatomites. Their thickness is faible and only the profile in Nafud Asir was more than 2 m thick. Comparable to the Upper Pleistocene lake beds these deposits do not represent permanent lake phases but alternating lacustrine periods between times of eolian sand deposition. Contrary to the Upper Pleistocene lakes the Holocene ones were deposited in an already existing dune relief and they were conformous to the present dune configuration.

Palaeolimnological informations

The presence of numerous root or rhizome remains as well as slumping phenomena demonstrate, that these lakes represent only shallow water bodies or swamps. One of the rare findings of mollusca in profile 6 consists of *Biomphalaria* and *Melanoides* indicating swampy areas or very shallow lakes. Again these sequences were depending on the oscillating water table in the dunes but in a minor scale in Upper Pleistocene times.

Palaeovegetation

Rhizome and root remains show the presence of *Phragmites* and *Typha* swamps in the depressions. The pollen spectra of the deposits in An Nafud and Nafud Asir indicate a clear predominance of Gramineae, Cyperaceae, shrubs like *Artemisia, Convolvulus, Cornulaca* and herbs like *Moltkiopsis,* Compositae, Polycarpaea as well as *Plantago.* Trees are only presented by *Acacia, Maerua Balanites* and *Hyphaene.* The portion of pollen transported over long distances is very low and consists only of *Pinus* and perhaps Anacardiaceae.

Compared to the modern pollen spectra the Holocene vegetation is to describe as a semidesert of grasses and some shrubs similar to the present one. The presence of numerous swamps will have provoked a denser herb cover. The portion of the

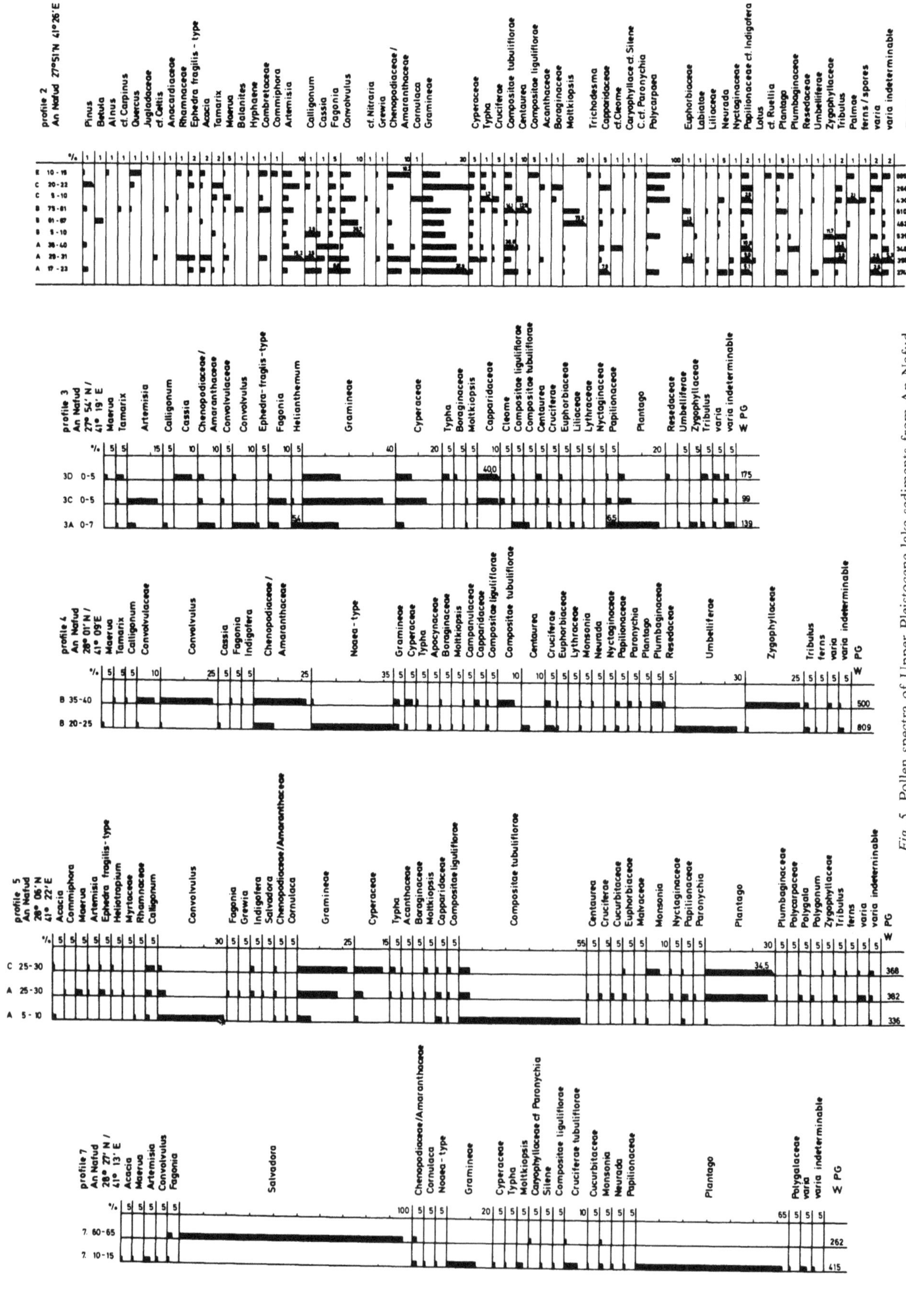

Fig. 5. Pollen spectra of Upper Pleistocene lake sediments from An Nafud.

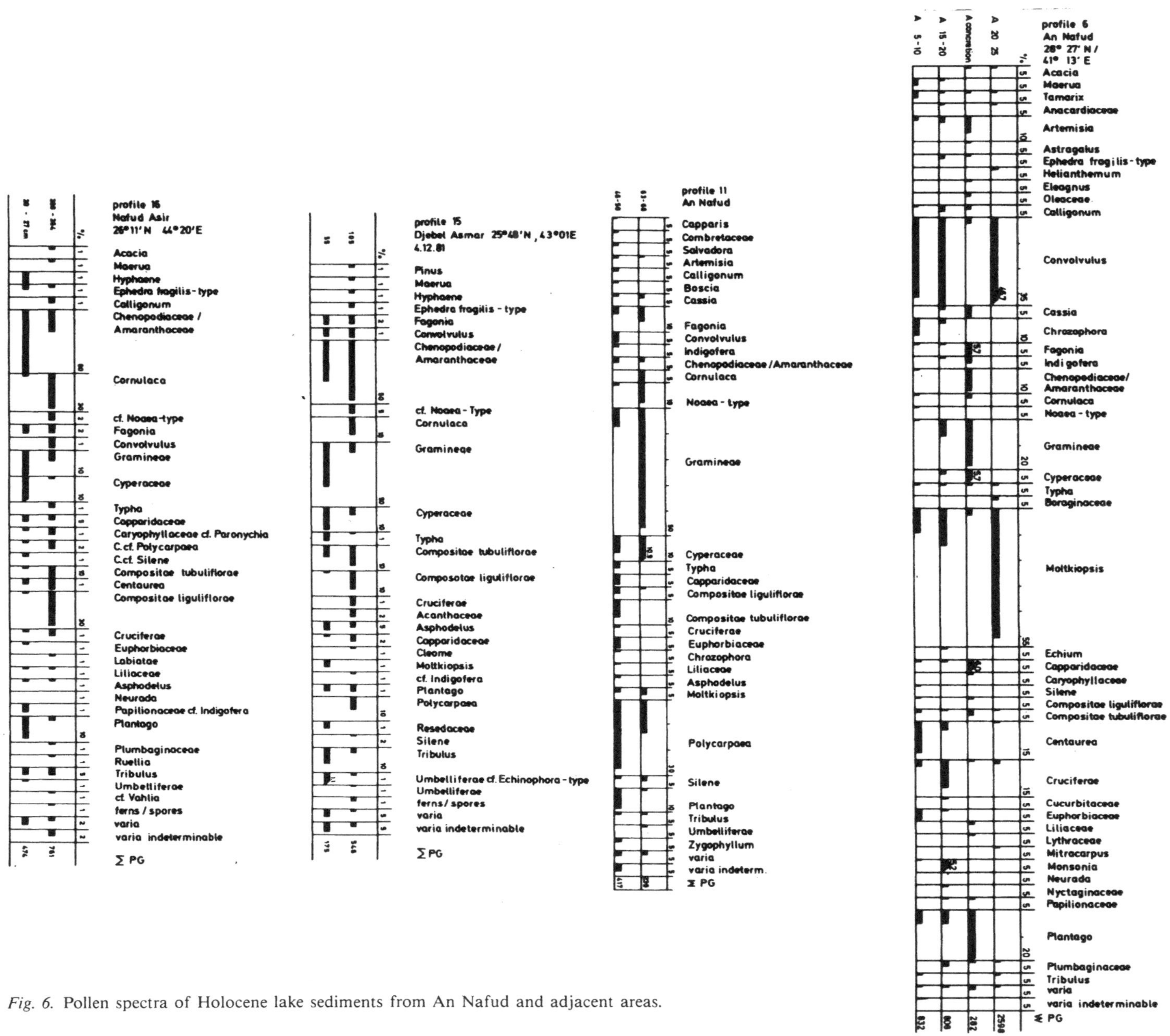

Fig. 6. Pollen spectra of Holocene lake sediments from An Nafud and adjacent areas.

long distance transport may be explained by the lack of water surfaces wich could act as pollen traps. Pollen spectra derived from a soil profile at the edge of Djebel Asmar near Nafud Urayk (profile 16) show the representants of the surrounding plains dominated by Chenopodiaceae, Compositae and Cruciferae. This underlines the dependence of vegetation on the type of substratum in the semi-desert. Certainly the number of achab floras was more frequent in time and space. They could also stabilize the dunes and reduce sand movements.

Conclusion and remarks on palaeoclimates

Palaeolakes are commonly taken as exellent tools to reconstruct former climates and even to read its parameters. Different lake levels have been compared worldwide and used to reconstruct the former precipitation (Street & Grove, 1979; Kutzbach, 1980). On the other hand this could lead to compare completely different types of lakes and their fluctuations not regarding the dynamic of those lakes.

The former lakes of An Nafud, Nafud Asir and Nafud Urayk provide informations about the local as well as the regional situation. Type and extension

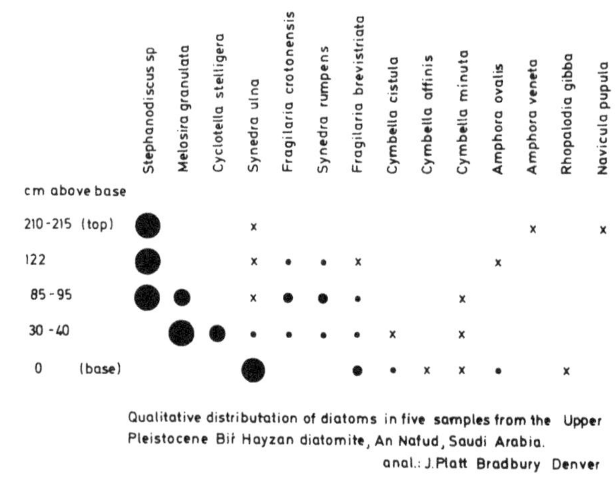

Fig. 7. Qualitative distribution of diatoms in the Upper Pleistocene Bi'r Hayzan diatomite.

is deduced from the geomorphic position, sedimentology and the content of fossils. Their age is dated directly or indirectly from the stratigraphical position. Vegetation type is reconstructed from pollen and some macro remains. In Upper Pleistocene lakes with an extension of some km^2 occured in a relief which was not comparable to the present one. Contrary to that, the Holocene lakes had only shal-

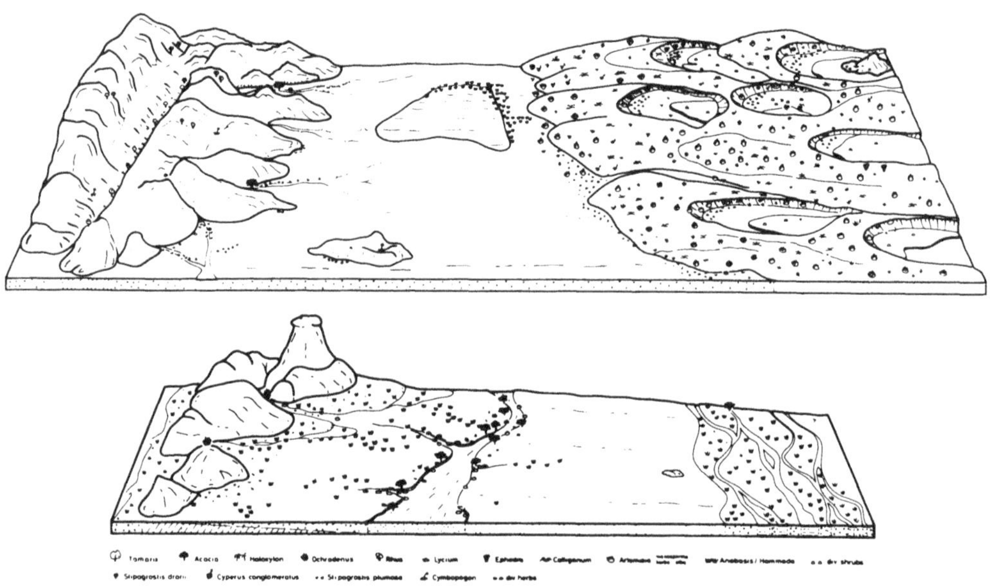

Fig. 8. Present landscape at the northern edge of the Arabian shield. Permanent diffuse shrub and grass vegetation in the dune area. Diffuse and permanent shrub in different types in the mountains and plains depending on substratum type. Semidesert.

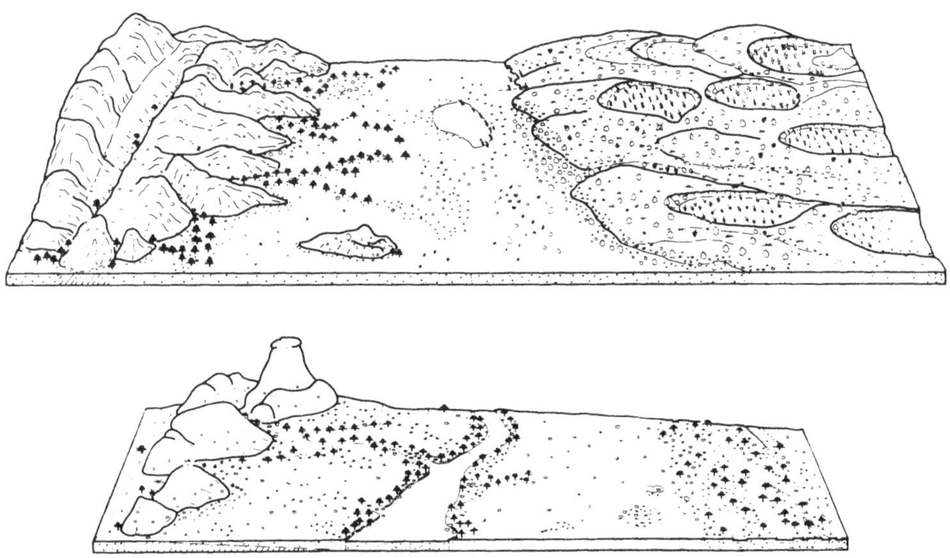

Fig. 9. Reconstruction of Holocene environment. Semidesert. Vegetation more dense and rich in floristic elements but comparable to the modern vegetation. Swamps and small shallow lakes in the dune depressions. Playas and wadis in the plains.

low water bodies or were even swamps and they were situated in the already existing dune configuration.

The main conclusion on the landscapes history in Upper Pleistocene and Holocene is that the sand seas as well as the peneplains were covered by a semidesert vegetation. The content of floristic elements as deduced from the pollen spectra was more or less the same as today. Only in Upper Pleistocene there occurred some more tropical elements in the region. This vegetation was able to stabilize the dunes by a certain root carpet even when the soil surface was not necessarily covered in total. Short or medium time variations in the plant cover are included in the picture of former landscapes which remained in the system of semideserts. Figs. 9, 10 give a reconstruction model of these landscapes in Holocene and Upper Pleistocene.

The former climate is only to reconstruct from indirect informations. Extension and stratigraphy of the former lakes demonstrate that there must have been periods of humidity long enough to build up the large water bodies. The obtained radiocarbon data reflect the already existing lakes, so the humid periods must have existed a certain period before. Lakes could be built up by precipitation in the region itself and/or by large sources fed by groundwater, which was collected in a greater area. Since groundwater is easily collected and stored in large sand bodies, dune areas may rapidly produce shallow lakes or swamps in their foor regions reflecting only the waterholding capacity of the dunes and not greater climatic fluctuations (Rognon, 1980). The existence of palaeolakes does not imply a pluvial period of largely augmented rainfall in every case, but it could be caused by the combination of lower temperatures, more developed cloud cover, storage of humidity in the sand sea, groundwater transport over long distances etc.

In the case of An Nafud there is no indication for groundwater movement over long distances. The collection and storage of groundwater and its seepage in front of an escarpment may have influenced only the lakes of Nafud Asir. The other lakes in the study area were alimentated either by runoff on the peneplains around the sand seas or by the water table in the sand bodies. This shows, that there was a higher precipitation in the region itself.

From the Bir Hayzan diatomite one can get some informations on the type of Upper Pleistocene lakes. The lake phases must have been long – about 900 years deduced from the stratigraphy of the deposit (Whitney, 1982). The spectra of diatoms indicate a temperate perhaps monomictic lake under a temperature regime comparable to the present

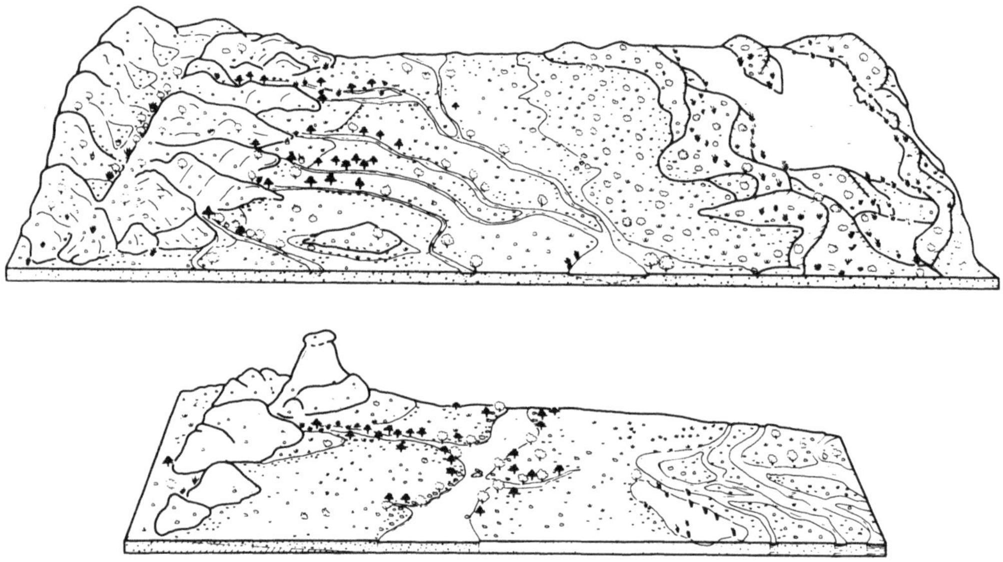

Fig. 10. Reconstruction of Upper Pleistocene environment. Dominance of longitudinal dunes in the sand sea. Large lakes in the interdunal depressions. Extended wadis in the plains together with shallow lakes. Semidesert vegetation. Relatively dense and rich in floristic elements compared with the present vegetation.

one – especially in winter times (Bradbury, pers. comm.).

Times of lake presence and periods of higher precipitation must not have been strictly comtemporaneous. It was necessary to have the long periods of rainfall before, and greater lakes could persist in the region when the rainfall was already reduced again. Higher precipitation could have been caused by more frequent cyclonic activities but also by the monsoonal summer rain or an interaction of both regimes.

The model of interaction of both rainfall regimes stands against the theory of shifting climatic belts either to the South or to the North. The slightly higher presence of tropical elements in Upper Pleistocene spectra must not necessary reflect a stronger tropical regime as it could have been trapped by greater water surfaces.

For Holocene times it is not necessary to presume climatic changes. Some more frequent invasions of cyclonic rainfalls would have been sufficient to build up the shallow water bodies in the dunes. An interaction with monsoonal rainfall is not excluded, but it is not indicated in the records. The amount of rainfall was certainly lower than in Upper Pleistocene and the fluctuations of lakes and swamps were more or less simultaneous to those of the rainfall.

For both periods we do not have further informations about the amount of rainfall per year, but it did'nt exceed the climatological amplitude of the semidesert system. Comparison to other lacustrine and fluvial deposists on the Arabian peninsula are not numerous. Former lakes on contemporaneous periods are documented for the Wadi Rhima area (Al Sayari & Zötl, 1978) or for the lakes of Rub al Khali and Mundafan basin (McClure, 1978). Lakes in Mundafan basin did exist for longer periods than those of the An Nafud probably due to the groundwater influence of Jebel Tuwayk escarpment.

Comparison with Upper Pleistocene or Holocene lakes from other regions as Turkey (Roberts, 1983), Syria (Kaiser, 1973), Egypt and Sudan (Gabriel & Kröpelin, 1983, Haynes, 1982, Mehringer *et al.,* 1979), Fezzan (Petit-Maire, 1982) or Ethiopia (Gasse, 1974) show that lakes did exist in those regions in more or less comparable times, but it would provoke a comparison of different lake systems formed under variable climatic conditions. This would exceed our first approach to reconstruct the former landscape and its environmental conditions leading later on to a discussion on climate together with the results of other disciplines.

Photo 1. Plain between Djebel Aja and An Nafud. Diffuse Anabasis-Artemisia shrub.

Photo 2. Southern margin of An Nafud north of Djebel Aja. Barchanoid dunes. Semidesert vegetation.

Photo 3. Upper Pleistocene lake beds near Juba.

Photo 4. Dune depression in southern An Nafud with Holocene lake deposits.

Photo 5. Holocene diatomite in southern An Nafud (P 6)

Photo 6. Section of Holocene lake deposits in Nafud Asir (P 16).

Aknowledgements

We are grateful to Mr. J. Platt Bradbury, Denver, for analysing the Bi'r Hayzan diatomite and to Mr. K. Weppler, Würzburg, for the preparation of maps and figures.

References

Barth, H. K., 1983. Die physisch-geographischen Grundlagen modernen Kulturlandschaftswandels. In Koszinowsky, L. (ed.), Saudi Arabien. Ölmacht und Entwicklungsland. Mitteilungen des Deutschen Orient Instituts 20: 7–60.

Al Sayari, S & J. G. Zötl, 1978. Quaternary Period in Saudi Arabia. Springer Verlag, Wien, 334 pp.

Bottema, S. & Y. Barkoudah, 1979. Modern pollen precipitation in Syria and Lebanon and its relation to vegetation. Pollen Spores 21: 427–480.

Bottema, S. & W. van Zeist, 1981. Palynological evidence for the climatic history of the Near East 50000–6000 B.P. Colloques inernationaux du CNRS, 598, Prehistoire du Levant: 111–132.

Cour, P., 1974. Nouvelle techniques de détection de flux et de retombées polliniques. Etude de la sédimentation des pollens et des spores a la surface du sol. Pollen Spores 16: 103–141.

Euting, J., 1896. Tagebuch einer Reise in Inner Arabien. Leiden, Teil I u. II, 554 pp.

Fryberger, S., 1979. Dune formes and wind regime. In McKee, E. D. (ed.), A study of global sand seas. U.S. Geol Survey, professional paper 1052: 137–169.

Gabriel, B. & S. Kröpelin, 1983. Jungquartäre limnische Akkumulationen in NW-Sudan. Z.f. Geomorph. N.F. Suppl. Bd. 48: 131–143.

Gasse, F., 1974. Les diatomées des sediments holocènes du bassin du lac Afrera (Giulielli), Afar septentrional (Ethiopie), Essai de réconstitution de l'evolvution du milieu. Int. Rev. ges. Hydrobiol 59: 95–122.

Haynes, C. U., 1982. Great sand sea and Selima sand sheet, Eastern Sahara: geochronology of desertification. Science 217: 629–633.

Jado, A. R. & J. G. Zötl, 1984. Quaternary period in Saudi Arabia vol. 2. Springer Verlag Wien, 360 pp.

Kaiser, K. H. *et al.,* 1973. Quartärstratigraphische Untersuchungen aus dem Damaskus-Becken und seiner Umgebung. Z. Geomorph, N. F. 17: 263–353.

Kutzbach, J. E., 1980. Estimation of past climate and palaeolake Chad, North Africa, based on a hydrological and energy balance model. Quat. Res. 14: 210–223.

McClure, A. A., 1978. Ar Rub al Khali. In Al Sayary, S. & Zötl, J. G. (eds.), Quaternary period in Saudi Arabia, Springer Verlag, Wien: 252–263.

Mehringer, P. J. *et al.,* 1979. A pollen record from Birket Qarun and the recent history of the Fayum, Egypt. Quat Res 11: 238–256.

Monod, T., 1954. Modes 'contracte et diffus' de la végétation saharienne. In Cloudsley Thompson, J. L. (ed.), Biology of deserts, Institute of Biology Tavistock House, London: 35–44.

Niklevsky, J. & W. van Zeist, 1970. A late quaternary pollen diagram from northwestern Syria. Acta bot. neerl. 29: 737–754.

Pachur, H. J., 1984. Die Bedeutung palaeoklimatischer Befunde aus den Flachbereichen der östlichen Sahara und des nördlichen Sudan. Z.F. Geomorph, N.F. Suppl. 50: 50–78.

Petit-Maire, N. (ed.), 1982. Le Shati, lac pleistocène du Fezzan Libya. CNRS, Paris, 118 pp.

Roberts, N., 1983. Age, palaeoenvironment and climate significance of late pleistocene Konyalake Turkey. Quat. Res. 19: 154–171.

Rognon, P., 1980. Pluvial and arid phases in the Sahara: the role of not climatic factors. Palaeoecology of Africa 12: 45–62.

Rognon, P., 1982. Modifications des climats et des environments en Afrique du Nord et au Moyen Orient depuis 20000 B.P. In Bintliff, J. L. & W. van Zeist (eds), Palaeoclimates, palaeoenvironments and human communities in the Eastern Mediterranean region in late prehistory. BAR Internat. series 133: 67–97.

Schulz, E., 1980. Zur Vegetation der östlichen zentralen Sahara und zu ihrer Entwicklung im Holozän. Würzburger Geographische Arbeiten 51, 194 pp.

Schulz, E. & J. Whitney, 1985. Vegetation on the Arabian shield and adjacent sand seas. U.S. Geol Survey open file report 85–116, 45 pp.

Sonntag, C. *et al.,* 1980. Isotopic identification of saharan groundwater, groundwater formation in the past. Palaeoecology of Africa 12: 159–171.

Street F. A. & A. J. Grove, 1979. Global maps of lake level fluctuations since 30000 y B.P. Quat Res 12: 83–118.

Walter, H. & H. Lieth, 1960. Klimadiagram Weltatlas. VEB Gustav Fischer Jena, 437 pp.

Whitney, J., 1982. Preliminary geological investigation of the Bi'r Hayzan diatomite deposit, Kingdom of Saudi Arabia. U.S. Geol. Survey, open report 02–7, 25 pp.

Whitney, J., 1983. Erosional history and surficial geology of western Saudi Arabia. U.S. Geol. Survey, Technical report, TR 041, 90 pp.

Whitney, J. *et al.,* 1983. The environmental history and present condition of the northern sand seas of Saudi Arabia. U.S. Geol. Survey, open file report 0395, 39 pp.

Zohary, M., 1973. Geobotanical foundations of the Middle East. 2 vols. Gustav Fischer Stuttgart, 799 pp.

Hydrobiologia 143: 191–197, (1986)

Late Quaternary palaeohydrology of Lake Huinaymarca (Bolivia)

Scenarios based on ostracods fauna

Ph. Mourguiart[1], P. Carbonel[1], J.-P. Peypouquet[1], D. Wirrmann[2] & C. Vargas[3]
[1]*IGBA Département de Géologie et Océanographie, Université de Bordeaux I, Avenue des Facultés F-33405 Talence Cedex, France*
[2]*Centre de Recherches OSTROM, 70/74 Route d'Aulnay, F-93140 Bondy, France*
[3]*Departamento de Geologia, Universidad, La Paz, Bolivia*

Keywords: Lake Huinaymarca, Lake Titicaca, ostracods, paleohydrology

Abstract

In lake Titicaca, the distribution of the modern ostracod fauna appears to be controlled by a combination of two factors: a) the equilibrium between carbonates and organic matter, b) the ionic composition of the lake water and its tributaries. Therefore, the *Limnocythere-Pampacythere* group generally predominates on an alkaline-carbonate lake floor. In parts of the lake receiving an NaCl input from tributaries, *Cyprideis* and *Cyprinotus* occur. These are generally absent elsewhere. The *Candonopsis* group is found in the deepest part of the lake.

Analysis of ostracod assemblages recovered from a number of cores permitted the reconstruction of the hydrological evolution of lake Huinaymarca for the last 10000 years. This included: low water levels, variations of oxygenation related to the position of the thermocline, and interconnections between different basins of the lake.

Introduction

Lake Titicaca is located on the high plateau of Peru and Bolivia (Altiplano, about 3800 m high) between latitude 15°45′–16°30′ S and longitude 68°30′–70° W. Its surface area is 8000 km^2 (Fig. 1). It is divided into two sub-basins connected by the Tiquina Strait (20 m deep). The northern basin or Great Lake has a mean depth of 135 m while the south-eastern lake or Small Lake or Lake Huinaymarca is 9 m deep. Lake level fluctuations relate to seasonal phenomena (Carmouze *et al.*, 1978, 1981).

Lake Huinaymarca, which is the subject of our study here, is characterized by three interconnected sub-basins (Fig. 1):

a the eastern, which is fairly deep (3–10 m),
b the western, on average quite deep (10–20 m),
c the northern, which forms a 40 m depression (Chua).

The present-day outflow of the lake Huinaymarca is the Rio Desaguadero, which serves as a link with the rest of the Altiplano (Lake Poopo, etc.) area.

Surface sediments are either organic-rich or carbonate-rich, and their distribution is controlled by depth (Boulange *et al.*, 1981). The carbonate sediments occupy the eastern part of the lake (Figs. 1–2). Water masses in the Small Lake are poorly mineralized (2 mg/l) and predominantly carbonate/sodic in nature (Carmouze *et al.*, 1981), except on the coast, near the outlet of the affluents.

Modern day ostracod distribution

The lake comprises a relatively rich benthonic fauna, principally ostracods. Presence or absence of the ostracods appears to be controlled by distribution of the sediments and the depth: the deepest parts of the lake with organic rich sediments are without ostracods; areas situated up to 10 m with carbonate-rich sediments and charophytes well developed are very rich in ostracods. The effects of the ionic composition of the lake water is marked

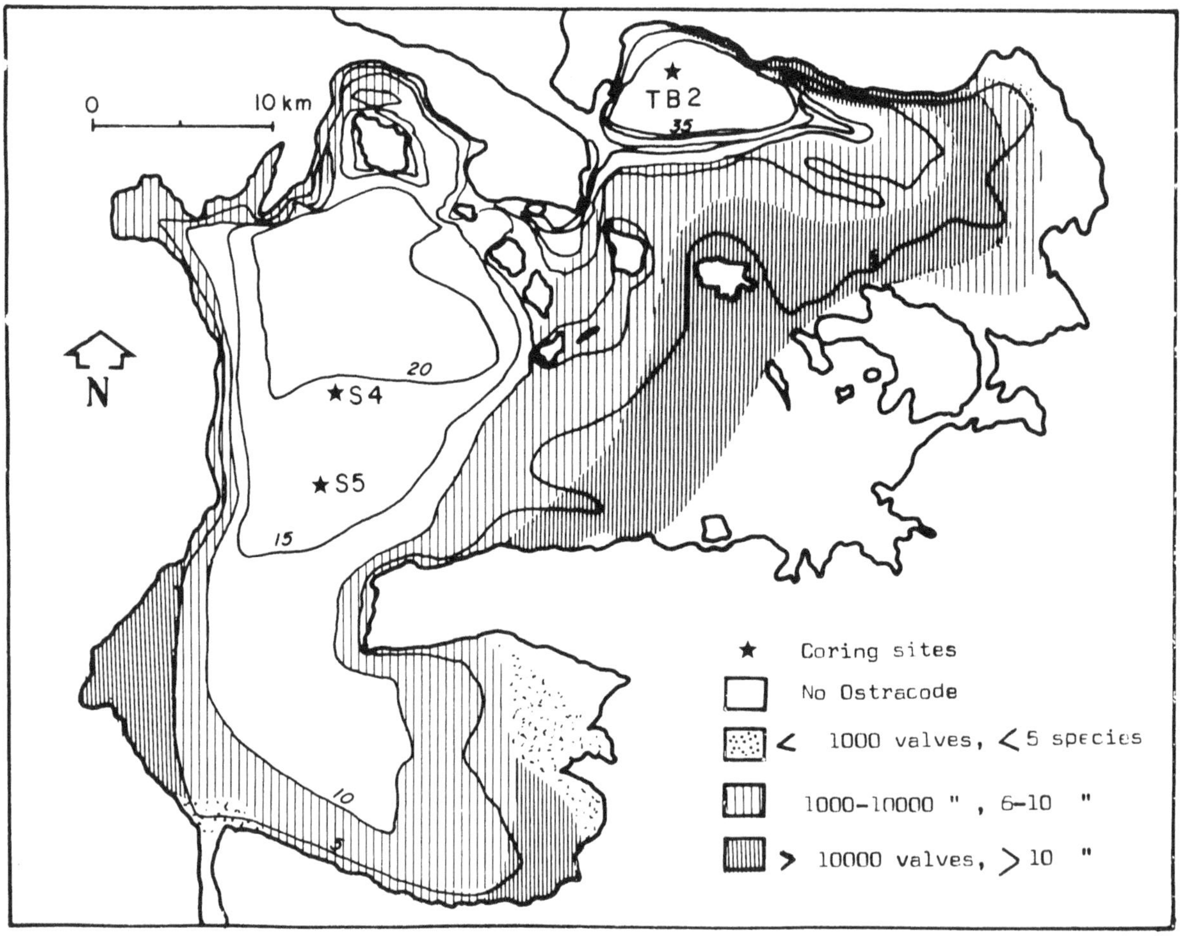

Fig. 1. Distribution of ostracod fauna and location of the cores.

specific composition of the ostracod faunas: the group *Limnocythere-Pampacythere* is generally dominant because the waters are mainly alkaline-carbonate (Carbonel *et al.*, 1983; 1983). In parts of the lake with waters rich in NaCl, *Cyprideis* and *Cyprinotus* generally occur (mouth of Rio Catari-Pallina). In deepest parts the *Candonopsis* group is dominant.

In sum, the ostracod distribution is closely related to development of macrophytes, especially charophytes, to chemical composition of the bottom waters, and to distribution of the sediments, according to the depth (Vargas, 1982, Carbonel *et al.*, 1983). These environments are the result of several processes which varied through time and which are herewith documented from the analysis of a range of cores taken from the lake (Fig. 1).

Analyses of the cores and interpretation of the results

Cores were recovered in the eastern basin (S4 and S5) within a depth range of 15–20 m, and in the Chua depression (TB 2) at the depth of 40 m. Additional sampling in the eastern part of the lake supports the following preliminary observations (A3, A4, B1, B2).

Core S4 (Fig. 3a)

Length: 90 cm and taken at 19 m. Three lithologic units are recognized (Vargas, 1982):

unit 1 (90–32 cm): biogenic-calcareous mud, rich in charophytes,

unit 2 (32–10 cm): organo-calcareous sediments,

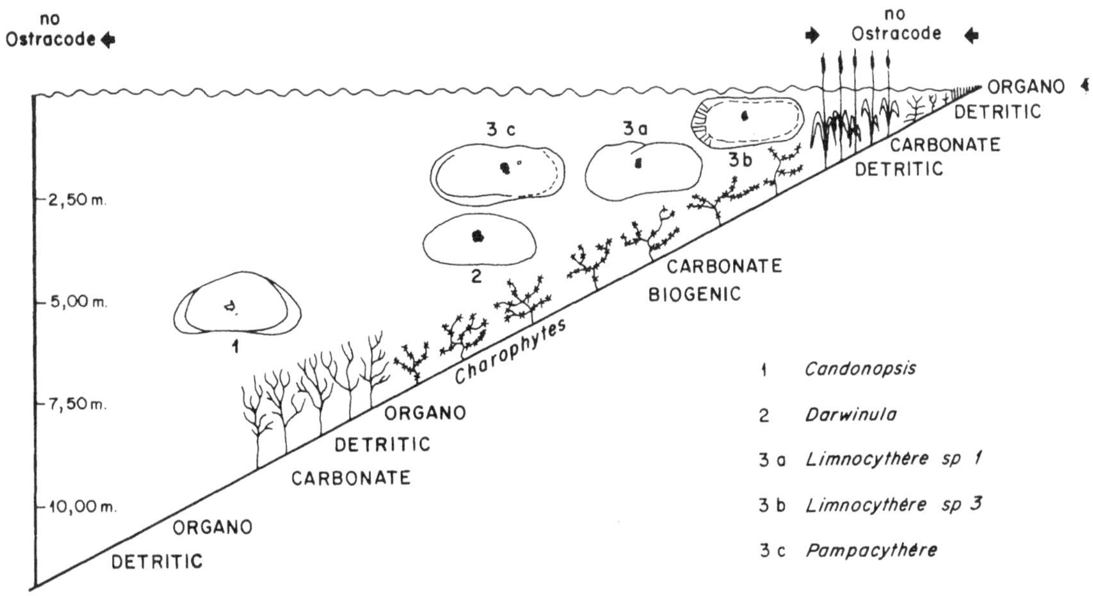

Fig. 2. Distribution of the main groups of ostracods related to water depth and sediment type.

unit 3 (10 cm – top): similar to modern-day muds.

Ostracods occur only in units 1 and 2. In unit 1, the fauna is rich and diversified, dominated by the *Candonopsis* group that characterizes the deepest environments colonized by the ostracods. In unit 2, there is a complete change in faunal structure and composition – density drops and diversity increases. The previously dominant species are replaced by faunas typical of Na, Cl waters: *Cyprideis* and *Cyprinotus*, and phytal forms. The passage from unit 2 to unit 3 occurs without any obvious transition.

Interpretation of the hydrochemical evolution of core S4: in the basal unit (1), there is evidence of an infralittoral calm environment with clear water masses with no obvious change. The overall faunal community points out to carbonate-sodic waters (Carbonel *et al.*, 1983). Unit 2, an important environmental change occurs; it probably relates to a lowering of lake level and waters become NaCl-rich and more saline. In unit 3, the absence of faunas, as well as the presence of organic sediment, imply a sudden deepening of the lake.

To sum up, core S4 testifies to a phase of low lake level, poorly saline waters but with distinct ionic composition, recorded by a change of ostracods assemblage (Carbonel *et al.*, 1983). This phase is followed by a period of lake level some 10 m below the present-day level.

Core TB 2 (Fig. 3b)

Length: 483 cm and taken in the Chua depression at a depth of 39 m. Four lithologic units were recognized:

unit 2 (480 – 360 cm): fine grained clay and azoic mud,

unit 2 (360 – 320 cm): clay vegetable debris and several shelley layers with polygonal agregates,

unit 3 (320 – 260 cm): clay-sand with few molluscs,

unit 4 (260 cm – top): organo-carbonate sediment rich in charophytes and gastropods.

Only unit 4 contains large numbers of ostracods. In this apparently homogeneous sequence, ostracofaunas undergo significant changes.

Interpretation of hydrochemical changes in summital unit of core TB 2: between 260 and 210 cm, the fauna is poor and poorly diversified. This low density and low diversity suggests the existence of an environment of less than 3 m deep with alkaline-carbonate waters. Between 210 and 160 cm, the microfauna diversifies and is more

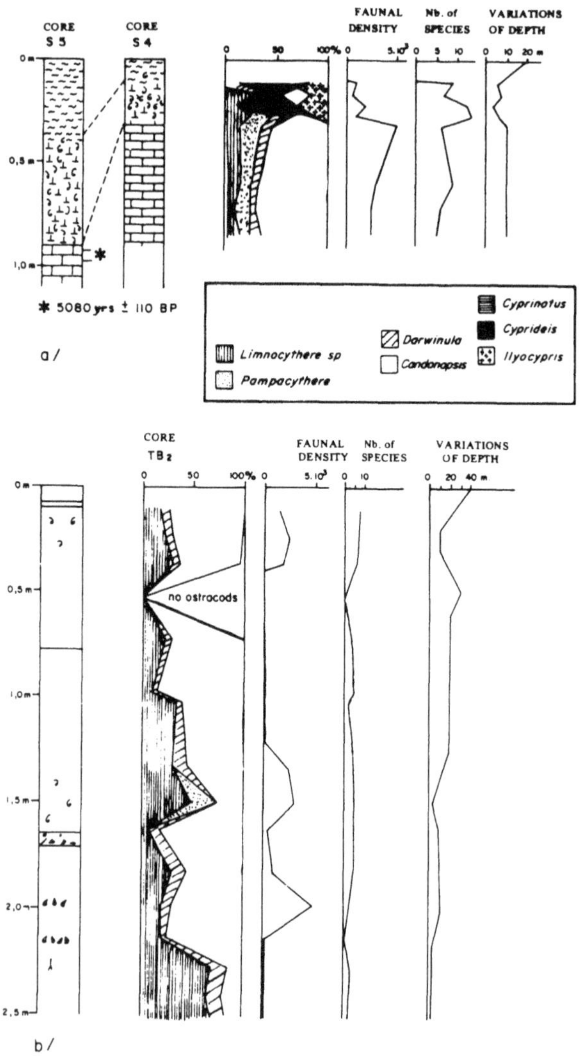

Fig. 3. Lithology and faunal distribution in two cores: a) core S 4 (western basin); b) core TB 2 (Chua depression).

abundant (*Candonopsis* dominance). It implies a significant deepening of the lake without any noticeable change in the water chemistry. Between 160 and 140 cm, the *Pampacythere-Limnocythere* group prevails over *Candonopsis* (< 30%). Furthermore, the biogenous fraction is made up of numerous molluscs, vegetable debris, diatoms and fishes, suggesting a shallow and probably well oxygenated environment, rich in nutrients. Between 140 and 60 cm, the ostracofauna is poor. The faunal assemblage is largely dominated by *Candonopsis.* The low number of individuals suggests a succession of phases with dominant stratification and phases with water mixing and circulation favouring colonization by benthic species. Between 60 and 40 cm, the disappearance of a microfauna reflects an increase in water depth, and consequently the development of a more marked thermocline. Between 40 and 20 cm, the ostracods are abundant and well diversified. It is similar to that at depth 210–160 cm, suggesting therefore a rapid lowering of the lake level. The top of the core is azoic and it corresponds to the thermal stratification conditions that prevail at the present day (Lazzaro, 1981).

Discussion

Evolutionary trends for the two cored sites vary in the following way: changes in water chemistry in the western basin at the water/sediment interface. In the Chua depression, the water chemistry remains the same; however, several hydrological variations were observed, suggesting the existence of a more or less ephemeral thermocline. These variations, in fact, result from lake level fluctuations.

In the western basin, the change in water chemistry at the water-sediment interface is probably caused by a disconnection of the Great Lake. This qualitative change cannot however, be accounted for only by the lowering of the water level. It is likely that the presence of NaCl-rich waters resulted from an inversion of the circulation in the southern part of the lake. This would have provoked the arrival of waters draining evaporitic sediments deposited previously (Wirrmann, in prep.).

In the Chua depression, the seasonal thermocline, observable today, became increasingly ephemeral with the lowering of the water level and it finally disappeared. This is well reflected in the quantitative and qualitative structure of the ostracod fauna (core TB 2).

It is therefore apparent that all the changes mentioned above were triggered by the water level variations. Circulation patterns in both basins can also be inferred from the water depth. Assuming a sequence of water level drops, the following scenario can be envisaged (Fig. 4):

a) Today, the western basin and the Chua depression receive water from the Great Lake. The thermocline in the Chua depression is very high and water depth clearly controls the absence of ostracods (Fig. 4a).
b) A 10 m level drop: circulation and communica-

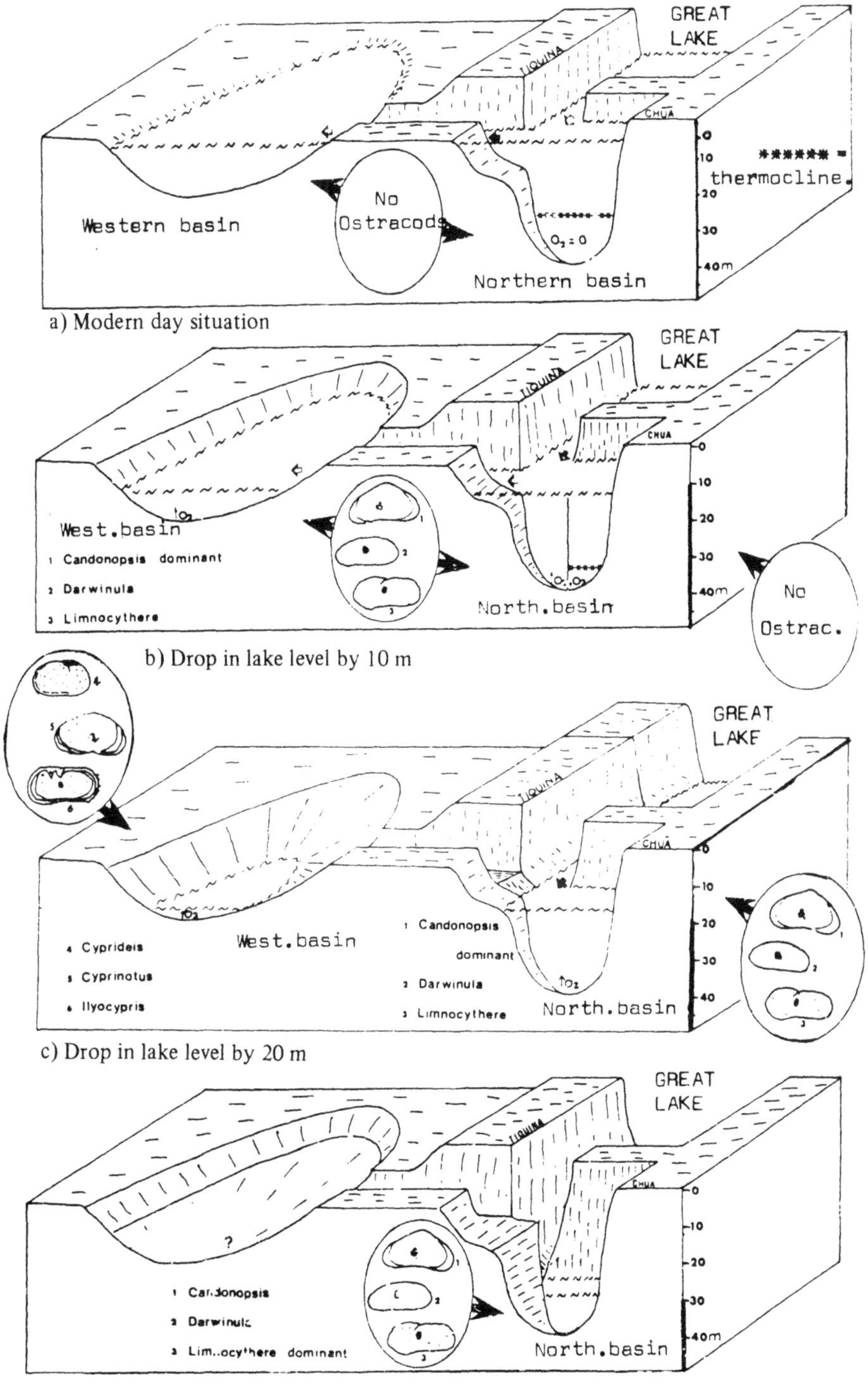

Fig. 4. Schematic diagram explaining the possible scenarios that would occur when the lowering of the lake is recorded, based on ostracods.

tion patterns are the same as above. As a result of this drop, reoxygenation at the water-sediment interface within the western basin is indicated by the fauna. The scarcity of the microfauna suggests a decrease in the thermocline of the Chua depression (Fig. 4b).

c) A 20 m level drop: circulation conditions within the Chua depression do not alter, but the thermocline disappears. The predominance of *Candonopsis* indicates a water depth of about 15 m. Communication with the western basin is, however interrupted (submerged ridges). Ionic concentration of waters increases, but more importantly, water chemistry changes (seen by the occurrence of *Cyprideis*), because of a probable reversal of the Desaguadero water and the arrival of Cl-rich waters from this particular region (Fig. 4c).

d) A 30 m level drop: communication with the Great Lake is minimal; the Chua waters just reach 10 m (*Limnocythere* predominance). The western basin is probably fully drained (Fig. 4d).

In sum, declining lake level leads to the isolation of basins and subsequently to a completely different chemistry at the water/sediment interface, probably due to a change in water quality. This change is well shown by the faunas (Table 1).

However, a question remains: how can lake level fluctuations be integrated in the evolution pattern since the last glacial maximum? For the time being, it is very difficult to propose a detailed chronology of the lake's fluctuating levels because of lack of dates. Nevertheless, results so far obtained suggest the timing of a low lacustrine level encountered in core S4: less than 5100 years B.P. and more than 3,650 years B.P. (Wirrmann *et al.*, 1983).

Conclusion

Preliminary results based on sedimentological analysis and the distribution of the ostracod fauna from two cores taken in the lake Huinaymarca show that significant water level fluctuations occurred in the lake after the last glacial maximum, i.e. at about 13000 years B.P. (Servant & Fontes, 1978). Environments determined by present-day faunas are also detectable in the past. Some of the faunal assemblages recorded in the sediments differ in structure from those represented at present day. They indicate environmental conditions and ecosystems for which no present day analogues exist, as well as different chemistry at the water-sediment interface and different water depths. These variations may, of course, be attributed to the level fluctuations and consequently to regional climatic variations. Nevertheless, one has to be aware that water chemistry and subsequently the river inputs are also controlled by the basin's tectonism.

Additional data, mainly from the polymorphism of the ornamentation of the shells analyses will provide further informations on the changes of water chemistry (Carbonel *et al.*, 1983), whereas sedimentological and mineralogical data will provide informations on the dynamic evolution of the lake (Servant & Fontes, 1984).

Finally, an accurate chronology of events should enable us to envisage possible 'scenarios', and thus reconstruct with more accuracy the history of the lake at the end of the Pleistocene.

Table 1. Scenarios of water circulation between Great Lake Titicaca and Lake Huinaymarca with lowering of the lake level by faunal recording.

	Faunas		Communications between	
	Western basin	Northern basin	Great lake/W. basin	G. lake/Chua
1 – Present level	No ostracods	No ostracods	+	+
2 – −10 m	*Candonopsis* Limnocythere	No ostracods or very rare Candonopsis	+	+
3 – −20 m	Cyprideis Cyprinotus	*Candonopsis* Limnocythere	−	+
4 – −30 m	No ostracods	*Limnocythere* Candonopsis	−	±

Acknowledgements

This work was carried out within the 'GEOCIT' research program part in cooperation between O.R.S.T.O.M. and the Department of Geology of the University of La Paz (Bolivia). We wish to thank P. de Deccker (Monash University, Australia) for revision of the text and Pr. Oldfield (University of Liverpool, U.K.) for his suggestions. We thank Mrs. D. Morel (I.G.B.A.) for translating our original text.

References

Boulangé, B., C. Vargas & L. A. Rodrigo, (1981). La sédimentation actuelle dans le lac Titicaca. Rev. Hydrobiol. trop. 14 (4): 299–309.

Carbonel, P., E. Grosdidier, J. P. Peypouquet & J. J. Tiercelin, (1983). Les ostracodes, témoins de l'évolution hydrologique d'un lac de rift. Exemple du lac Bogoria, rift Gregory, Kenya. Bull. Centres Rech. SNEA(P) 7 (1): 301–313.

Carbonel, P., J. P. Colin, D. L. Danielopol, H. Löffler & I. Y. Neustrueva, (1983). The paleoecology of limnic ostracods. A review of some major topics. Proc. 2er Congrès Paleoécol. Lyon, in press, 90 p.

Carmouze, J. P., C. Arze & J. Quintanilla, (1978). Circulacion de materia (agua-sales disueltas) através del sistema fluviolacustro del Altiplano: la regulacion hidrica é hydroquimica de los lagos Titicaca y Poopo. Cah. O.R.S.T.O.M., sér. Géol. X (1): 49–68.

Carmouze, J. P. E. Aquize-Jaen, (1981). La régulation hydrique du lac Titicaca et l'hydrologie de ses tributaires. Revue Hydrobiol. trop. 14 (4): 311–328.

Carmouze, J. P., C. Arze & J. Quintanilla, (1981). Régulation hydrochimique du lac Titicaca et l'hydrochimie de ses tributaires. Revue Hydrobiol. trop. 14 (4): 329–348.

Lazzaro, X., (1981). Biomasses, peuplements phytoplanctoniques et productivité primaire du lac Titicaca. Revue Hydrobiol. trop. 14 (4): 349–380.

Servant, M. & J. Ch. Fontes, (1978). Les lacs quaternaires des hauts-plateaux des Andes boliviennes. Premières interprétations paléoclimatiques. Cah. O.R.S.T.O.M., sér. Géol. X (1): 9–23.

Servant, M. & J. Ch. Fontes, (1984). Les basses terrasses fluviatiles du Quaternaire récent des Andes boliviennes. Datations par le 14C. Interprétation paléoclimatique. Cah. O.R.S.T.O.M., sér. Géol. XIV: 15–28.

Vargas, C., (1982). La sédimentation lacustre subactuelle d'un bassin intramontagneux: le lac Titicaca (partie lac Huinaymarca, Bolivie). Thèse 3ème cycle Bordeaux I, n°1763, 91 p.

Wirrmann, D. & J. Ch. Fontes, (1983). Oxygène 18, Carbone 13 et 14 dans quatre carottes du lac Huinaymarca (Bolivie). Premiers résultats. In press.

Accepted 13 May 1986.

Hydrobiologia 143: 199–208, (1986)

The Pleistocene lake deposits of the NE Baza Basin (Spain): salinity variations and ostracod succession

P. Anadón[1], P. De Deckker[2] & R. Julià[1]
[1]*Institut J. Almera (C.S.I.C.), c/. Martí i Franqués, s/n. 08028 Barcelona, Spain*
[2]*Department of Geography, Monash University, Clayton, Victoria 3168, Australia*

Keywords: paleolimnology, saline lake, ostracod paleoecology, non-marine foraminifera

Abstract

The Guadix-Baza Basin in Spain covers an area of approximately 3000 km^2 and yields a sedimentary sequence ranging from Lower Miocene to Pleistocene. Twenty five meters of Lower Pleistocene lacustrine sediments have been located in the NE part of the Basin at about 1000 meters in altitude. This sequence which overlies dolomitic mud flat deposits consists of limestones, calcareous and dolomitic mudstones, dolostones, silty clays, sands and gravels. Salinity fluctuations and short dry episodes, related to lake level oscillations, have been recorded by textural, mineralogical and faunal changes throughout the sequence. Ostracods, which are the most commonly encountered fossils, permit to detect recurrent changes in water salinity and regime, and solute composition. The faunal changes indicate an alternation of slightly saline and bicarbonate-rich water (when ostracods and gastropods occur) with a saline NaCl-dominant water (in which ostracods, *Cerastoderma* bivalves and non-marine foraminifers are found).

The frequent and recurrent hydrochemical changes in the Baza Basin in the Early Pleistocene point to a climate of high contrast like in the Mediterranean region today but with a greater availability of water within the system compared to the present situation in the area.

Introduction

Most of Spain is today under the influence of an semi-arid climate and consequently its lakes, even at altitudes over 1000 m in endorheic regions, retain ephemeral waters. However, it is known that large lakes did occur in Spain as evidenced by substantial lacustrine deposits. For example, in the southern portion of Spain, the Baza Basin (Fig. 1) is characterized by such deposits covering an area of approximately 900 km^2. There, a lacustrine sequence spanning since the Lower Pliocene is exposed, and it can reach over 100 m in thickness. A feature typical of some portions of the lacustrine deposits is the abundance of ostracods and foraminifers sometimes in thick and unconsolidated beds (up to 3 m thick), where these fossils constitute up to 90% of the sediment (e.g. 12000 ostracod valves per ml of sediment). Of interest in this region too is the recently investigated large and diversified mammal fauna recovered with sediments associated to this lacustrine sequence.

The aim of this paper is to present a detailed description of the Pleistocene portion of the lacustrine sequence associated to the fossil mammals and provide a palaeoenvironmental reconstruction of this site.

It is hoped that this type of investigation, based on the reconstruction of different facies, will permit a better understanding of these lake deposits and allow their chronological correlation in the future.

Geological setting

The Alpine Betiç Chain, in the south of the Iberian Peninsula, includes some intramontane ba-

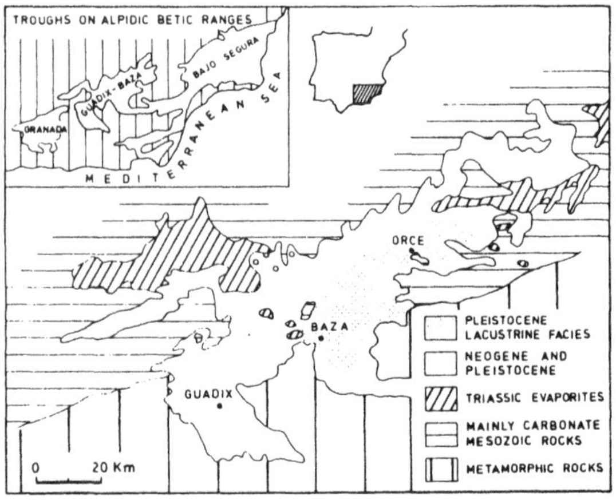

Fig. 1. Map showing the location of Guadix-Baza Basin and areal extent of the Pleistocene lacustrine deposits.

sins (Granada, Guadix-Baza, Bajo Segura, etc.). Substantial sedimentary sequences, ranging from Neogene to Pleistocene in age, are found in those depressions. The Guadix-Baza depression (37°30′ N, 2°30′ W and about 1000 m a.s.l.) is the largest and extends over 3000 km^2 (Fig. 1). This depression is surrounded by Mesozoic and Palaeozoic sequences of the Inner Zone of the Betic Chains. The southern limit of the Betic Zone consists of metamorphosed Palaeozoic rocks. Inside the depression and within its northern boundary, mainly Mesozoic carbonate and evaporitic rocks of the Subbetic and Prebetic Zones are found. Material filling up the Guadix-Baza ranges between Early Miocene (Aquitanian) to Pleistocene in age (Vera, 1970) and shows quite a variety of facies which relate to an alternation of marine and continental phases.

Climatic data

The Guadix-Baza Basin has a typically semi-arid, Mediterranean type of climate and is characterized by an annual rainfall of 300–500 mm. Rainfall is most common in spring and autumn; winter is cold and dry, and summer dry and warm. Mean annual temperature is 19° C, whereas the mean of maxima is 36° C in summer and the mean of minima is −2 to −4° C in winter. It is important to point out that climatic conditions of the mountains (up to 3000 m in altitude) surrounding the depression are completely different: annual rainfall there can reach more than 1000 mm.

Stratigraphy of the NE portion of the Baza depression

In, the Orce-Venta Micena region, Lower to Middle Pleistocene lacustrine deposits belong to the Baza Formation as defined by Vera (1970). Recently Vera *et al.* (1985) defined 2 members for this region: – a 'red detrital member' overlain by an upper 'silty-calcareous member'. These informal units are mentioned here as 'lower, alluvial unit' and 'upper, lacustrine unit' respectively. In some places of the basin the lower alluvial unit overlies lacustrine deposits of a Late Pliocene – Early Pleistocene age, and corresponds to a calcareous sequence, called 'lower member' in Vera *et al.* (1985).

The units in the Orce – Venta Micena region show lateral facies transitions to evaporitic gypseous deposits towards the centre of the lake basin.

Lower alluvial unit

This unit is well exposed in the Orce – Venta Micena region where it reaches up to 40 m. Towards the central part of the Basin (Baza area), a transition to lacustrine facies is noticeable.

The Lower alluvial unit is formed by calcareous and dolomitic mudstones with interbedding of sandstones and rare cross-bedded conglomerates. The mudstones are red to pink in colour and are characterized by abundant mottling and bioturbation. The poorly cemented sandstones are quartzose or in some places, litharenitic. Ostracods and reworked foraminifers are found in some of the sandstones beds.

These deposits correspond to the mud flat facies of a large lake during a major water level recession, with also sparce channels with a coarse-grained load. The presence of evaporite minerals (gypsum and minor halite) in some places further supports this interpretation.

This unit is not further discussed here.

Upper lacustrine unit

A variety of lithologies are encountered: calcare-

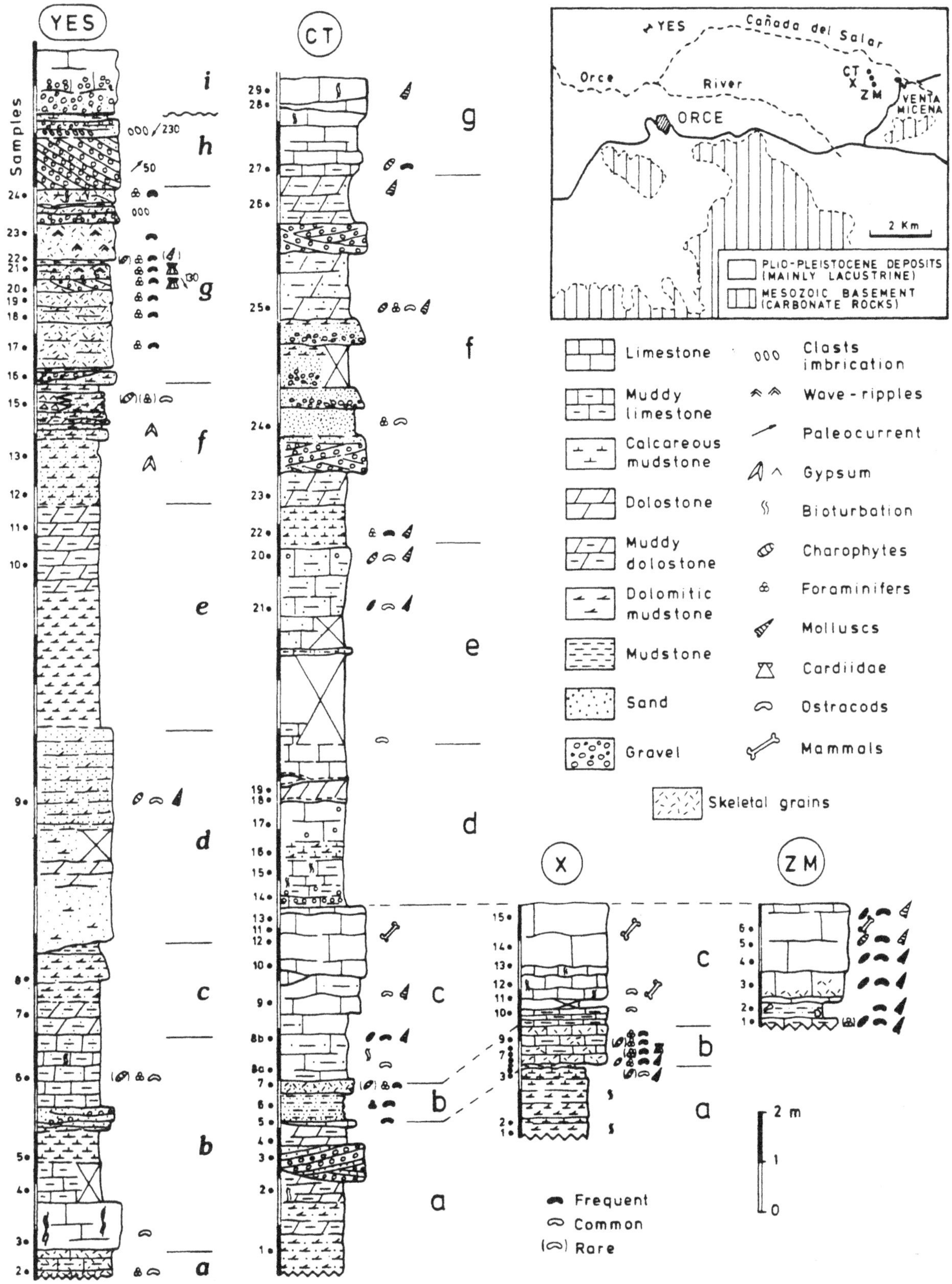

Fig. 2. Studied sections. Carbonate rocks have been differentiated according to the following values. Limestone and dolostone: 80–100% carbonate, muddy limestone and muddy dolostone: 60–79% carbonate, calcareous mudstone and dolomitic mudstone: 40–59% carbonate.

ous and dolomitic mudstones, sands, sandstones, gravels, limestones and dolostones altogether covering a thickness of 25 m in the area studied (Fig. 2).

These lithologies show a complex vertical succession and sequences examined in different localities demonstrate numerous lateral facies transitions with changes occurring over a few meters. Fig. 2 illustrates examples of such transitions within this unit and they are detailed below.

Yeseras section (YES). From base to top

a) 0.5 m: white sandy limestones and quartzose-skeletal sands. Abundant ostracods and foraminifers. This level overlies several meters of red to brown mudstones belonging to the lower alluvial unit. It represents the first recorded saline lake phase.
b) 4.5 m: white bioturbated limestones, muddy limestones and calcareous mudstones within which very coarse sandstones with gravel lenses occur. Only one level consisting of muddy limestones is fossiliferous (ostracods, foraminifers and rare charophytes). The lacustrine calcareous deposits can be related to a low salinity bicarbonate water phase with episodic detrital inputs.
c) 1.9 m: white dolomitic sandy mudstones devoid of fossils. They correspond to a mud flat environment linked to a shrinking phase of the lake.
d) 4.25 m: fine quartz sands with a dolomitic mudstone matrix. The lower part of this level shows evidence of channel erosion at the base and has a lenticular pattern. The upper part is laminated and contains frequent gastropods, ostracods and charophytes. This level can be interpreted as a fluvial channel deposit (lower portion) cut into the mud flat and affected by an expansion of the lake.
e) 4.5 m: white dolomitic mudstones and muddy dolostones devoid of fossils. This level corresponds to a retreating phase of the lake with mud flat deposits.
f) 2.5 m: sandy dolomitic mudstones interbedded with muddy dolostones. Rare ostracods and locally lenticular gypsum crystals (several cm in diameter). These deposits probably were formed in a transitional zone between a mud flat and a low salinity lacustrine environment.
g) 3.9 m: ostracod sands. The sand-size particles consist mainly of ostracod shells (*Cyprideis*) and a minor fraction of foraminiferal tests (*Ammonia*). Isolated shells of *Cerastoderma* cf. *glaucum* (identif. J. Porta) occur. In the lower part of the bed, quartz grains are abundant (up to 50%) but decrease towards the top. In the upper part, most of the grains are skeletal, but locally thin beds of quartz sands and small gravels are present. Wave and current ripples plus minor cross bedding and water escape structures occur in the upper part of this level. In the rippled layers, laminae show varying proportions of skeletal to quartz grains. The sedimentary features and biota suggest that these deposits were formed along the shore of a saline and permanent-water lake. The sands were reworked by current and wave action.
h) 1.5 m: gravels within tabular cross bedding. They consist of rounded to subangular quartz and metamorphic pebbles. These gravels are overlain by horizontally bedded gravels showing imbrication. This level may be related to a small delta mouth bar.
i) 1.25 m: cemented gravels topped by calcrete crust overlying unconformably the lacustrine sequence.

Venta Micena profiles (CT, X and ZM sections)

The most complete lacustrine succession in the Venta Micena area is shown in the CT section. There, the outcrops of the lower portion demonstrates sudden lateral lithological changes (compare with sections X and ZM). In this area the following beds can be observed from the base to the top:

a) 3.1 m: (minimum thickness): dolomitic mudstones and muddy dolostones, red to white in colour. Presence of bioturbation (in some cases attributed to roots) and mottling. Lenticular layers of cross bedded gravels and sands are also present. The sand grains consist mainly of angular quartz and the gravels originate principally from Mesozoic carbonate rocks. No evidence of biota.

   This bed belongs to the upper part of the underlying alluvial unit and is interpreted as a mud flat deposit. The sand and gravel layers probably correspond to deposits originated by tractive currents in small channels over the mud flat.
b) 0.9–1.0 m: white, fine to medium and thinly bedded sands interbedded with grey calcareous

mudstones and limestones. The sands, with a variable degree of cementation, consist of angular quartz grains and skeletal grains in varying proportions.

This bed is characteristic of a lacustrine sequence. The lower layers correspond to a freshwater lake (X-3, CT-5) evolving into a permanent saline lake. Level X-7 shows lumachelle accumulations of *Cerastoderma.*

c) 2.3 – 3.4 m: white limestones, up to 100% calcite. A rich mammalian fauna of Early Pleistocene age was recovered from this level (Moyà *et al.*, 1980). The presence of accumulations of vertebrate remains, sometimes linked to bedding planes, indicate the presence of sedimentation breaks and of reworking. Salinity was lower than for the underlying level b.
d) 3.4 m: white to greyish muddy limestones with green and yellow vertical mottling. Presence of one layer of white dolostones plus another of pale red calcareous and dolomitic mudstones also with mottling. No fossils, however the presence of bioturbation is attributed to roots. All this suggests a marginal lake-type deposit during a receding water level phase.
e) 3.7 m: not well exposed, grey sandy limestones and muddy limestones with sparse floating pebbles. Fossils here indicate a more open lacustrine environment than the underlying level, resulting from a lake expansion phase.
f) 7.4 m: calcareous and dolomitic mudstone and muddy-sandy dolostones with cross bedded sands and some gravel layers. This level probably corresponds to a saline lake with frequent tractive currents from fluviatile inputs.
g) 2 m: white limestones, with burrowing near the top. Ostracods and charophytes occur at the base (CT-27) and gastropods casts at the top.

Biota

Ostracods

Fifteen species of ostracods have been recovered from the sections. They are listed in Table 1 and their respective numbering and order referred to in that Table and Figs. 3 – 5 are followed here for simplicity. Most of the ecological information mentioned here for these species is taken from the excellent work of Diebel & Pietrzeniuk (1984) unless acknowledged otherwise. For illustrations of the species refer to the latter paper and previous ones by the same authors. Salinity data comes from De Deckker (1981).

1. *Ilyocypris bradyi* is a holarctic species which is oligothermophilous, and euryplastic; it is found in flowing waters as well as temporary ones and can tolerate salinities up to 4.5‰. Usually, it is found in bicarbonate-dominant waters and crawls on soft muddy substrates (De Deckker, unpubl.). Shell architecture is known to vary substantially.

2. *Ilyocypris gibba* is also holarctic but prefers warm and fresh waters and tolerates flowing ones. De Deckker (unpubl.) believes that this species is conspecific with *I. bradyi*, and because it lives in warmer waters, it is smaller and develops a different shell ornamentation (hence the obvious lateral protuberances characteristic of the 'taxon') resulting from different calcium carbonate saturation levels in the water.

3. *Candona angulata* commonly frequents slightly saline waters up to 14‰ salinity.

4. *Candona neglecta* also frequents slightly saline waters up to 15.7‰ salinity.

5. *Candona compressa* is found in lakes as well as ephemeral pools. It also tolerates slightly saline waters up to 5.7‰ salinity.

6 and 7. *Cyprideis torosa* is well known for its tolerance to saline waters. It has been found living in waters up to 60‰. Usually it is found in NaCl-rich waters. Noded valves (=7N), and often highly reticulated ones, are found in waters below 6 – 10‰ (Vesper, 1972) whereas smooth ones (=6S) are always found above these values, and then waters are NaCl-dominant.

C. torosa requires permanent water to reproduce because its eggs cannot withstand desiccation. It usually lives in among organic debris and is on top of the meiobenthos food chain in some estuaries (Heip, 1976). Herman *et al.* (1983) found that this species can reach extremely high numbers (up to + 2000 individuals/10 cm^2). This would explain the thick 'sandstones' bands (see profile YES for example) consisting almost entirely of *Cyprideis* valves. On those occasions, the *C. torosa* valves are always smooth.

8. *Heterocypris salina* is a common inhabitant of either ephemeral or permanent, slightly saline waters. It prefers NaCl-dominant waters (De Deck-

Table 1. List of ostracods recorded in the studied sections.

SECTION	SAMPLE	Ilyocypris bradyi	Ilyocypris gibba	Candona angulata	Candona neglecta	Candona compressa	Candona juvenile	Cyprideis torosa S = smooth N = nodose	Limnocythere sp	Heterocyrpis salina	Heterocypris incongruens	Herpetocypris chevreuxi	Prionocypris serrata	Eucypris virens	Ammonia becarii	Charophyte	REMARKS
Ostracod number (referred to in Figs. 3–5)		1	2	3	4	5		6S-7N		8	9	10	11	12			
YES	24							×× S							×		
	23							×××S						×			
	22							×× S							××	1	
	21							×××S							××		Cerastoderma
	20							×××S							××		
	19							×××S							×		
	18							×××S							××		
	17							×××S							×		
	15	×		×						×			×			1	
	9							××S+×N								×	
	6						×	1 juv.							××	1	+ 2 Potamocypris aff. producta
	2							× S							××		
CT	27	×						×							×	×	
	25						×									×	
	24							×							×		Cyprideis reworked?
	22		××				×			fragment					Elphidium	×	
	21		×	××												××	
	8b	×	×	×						×						××	
	7	×						××S+×N		×					×	1	
	6	×						×× S,N			×				×		
ZM	6			×					1	×		×				×××	
	5		1	1							××					×	1 Candonopsis kingslei 2 ? Candona pratensis
	4	×		×						×	×	×				×××	
	3		×	×		×				×						××	
	2		×				×			rare		1 fragment					
	1	×	×	×				rare S		×		×			×	××	
X	9							×××S,N							××		Cerastoderma
	7							×××S,N							××	×	Cerastoderma
	6							×× S,N									
	5	××		××	×			juv. N	2	1					×	×	
	3	×××			×					2						×	1 juv. Cyprididae

few; ×× common; ××× abundant.

ker, unpubl.) and is found in salinities up to 20‰ salinity or more.

9. *Heterocypris incongruens* is a cosmopolitan and euryplastic species mostly found in ephemeral pools. It also frequents slightly saline waters up to 20‰ salinity and prefers HCO_3^--rich waters (De Deckker, unpubl.) compared to its congerer *H. salina.*

10. *Herpetocypris chevreuxi* frequents slightly saline waters (K. Martens and K. Wouters, pers. comm.).

11. *Prionocypris serrata* is a freshwater species usually found in slowly flowing waters where aquatic plants occur. In Diebel and Pietrzeniuk (1984), it is refered to as *P. zenkeri.*

12. *Eucypris virens* is a cosmopolitan species, found commonly in temporary pools but it also occurs in permanent waters. It is euryplastic and can

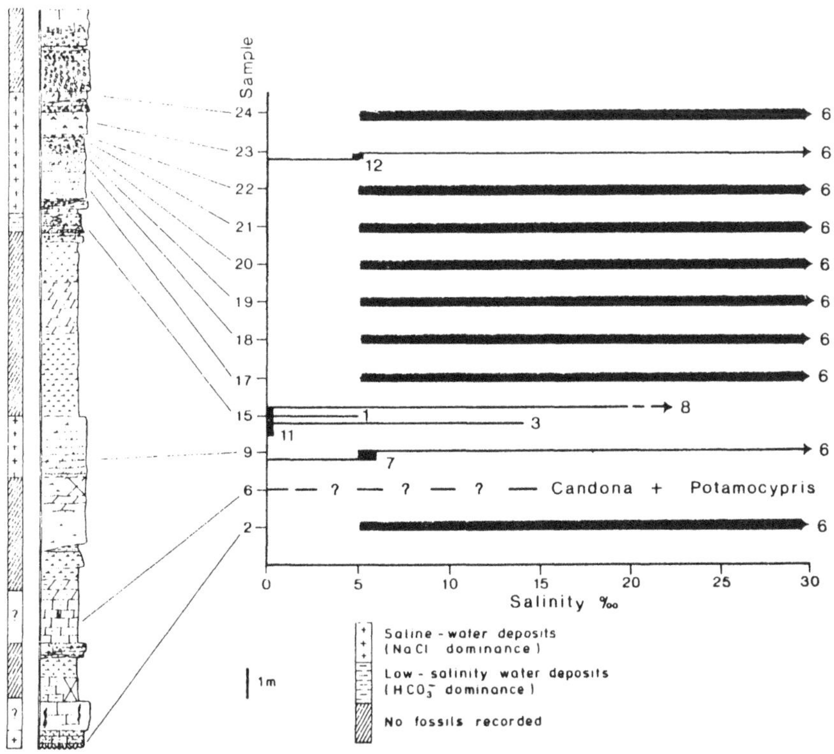

Fig. 3. Yeseras section. Ostracod species distribution and salinity range within which all species can occur together within the sample (shaded zones in the diagram). Numbers within the diagram refer to ostracods in Table 1.

withstand low salinities up to 5‰. It prefers HCO_3^--rich waters (De Deckker, unpubl.).

The other species mentioned in Table 1 are too rare and of no palaeoecological significance to be worthy of mention here. For more information on their ecology refer to Diebel and Pietrzeniuk (1984).

Foraminifera

Ammonia beccarii is an estuarine foraminifer capable of withstanding a wide range of salinities (7–67‰, Bradshaw, 1957). Cann & De Deckker (1981) discussed the occurrence of this foraminifer in non-marine inland (= athalassic) saline waters. It requires permanent water to reproduce and NaCl-dominant waters (De Deckker, unpubl.). In permanent inland saline lakes (with NaCl waters), it is commonly accompanied by *Cyprideis torosa.*

Elphidium sp. is represented by only a few specimens. It will not be discussed here except that it is no surprise to find this genus associated to inland salt lakes of NaCl chemistry. For more details on this refer to Cann & De Deckker (1981).

Mollusca

The bivalve *Cerastoderma glaucum* and numerous gastropods have been encountered in a number of samples. They are the subject of a separate study (material sent to J. C. Rosso). Nevertheless, it is no surprise to find *C. glaucum* in abundant numbers in some of the beds deposited under permanent saline conditions (conditions determined from examination of the ostracod fauna) as there is ample documentation on the presence of this species in similar deposits elsewhere (see volume on Le Shati edited by Petit-Maire, 1982).

Here, it is also commonly found with *Cyprideis torosa* and *Ammonia beccarii*, although the latter species is sometimes given different specific names by various authors. Carbonel & Pujos (1982) and De Deckker (unpubl.) believe that architectural differences seen between some *Ammonia* specimens which are referred by others as belonging to different species, e.g. *A. beccarii* and *A. tepida*, are simply the result of differences in water conditions (e.g. temperature fluctuations) and chemistry.

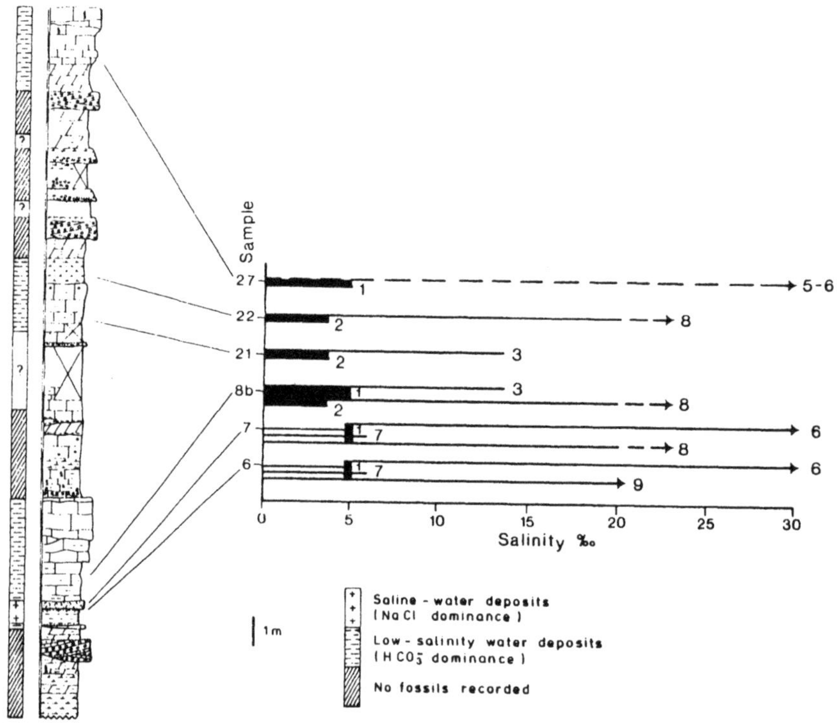

Fig. 4. Venta Micenla CT section. Ostracod species distribution and salinity range within which all species can occur together within the sample (shaded zones in the diagram). Numbers within the diagram refer to ostracods in Table 1.

Palaeoenvironmental reconstruction

Based on the knowledge of modern ostracod and foraminifer ecology, and also from the understanding of sedimentary facies, we present a palaeoenvironmental reconstruction of the studied sites.

Basically, two major lake phases can be distinguished. One corresponds to a saline water phase with NaCl dominance as evidenced by the smooth shelled *Cyprideis*. *Ammonia* is often present and usually other members of the biota are absent except for *Cerastoderma* and charophytes. The lake must have retained permanent water. The thick and consolidated *Cyprideis* bands represent a lake shore facies where these ostracods are accumulated after death. Cross stratification of these ostracod sands further confirms this interpretation. The second phase is characterized by lower salinities, usually well below 10‰ (for more details see Figs. 3, 4 and 5). Bicarbonate dominance is apparent from the ostracod data. *Cyprideis* specimens are coarsely reticulated and heavily noded. Some of the other ostracod species indicate that the lake could have been of an ephemeral nature. However, nearly all ostracods indicate that the lake water could have been slightly saline at all times. Exception perhaps occurs for level-sample 15 in profile YES (see Fig. 3) where *Prionocypris serrata* was found. This bicarbonate-rich water phase favoured a more important deposition of carbonates whereas the more saline NaCl phase coincided with a greater supply of clastic material. Based on field observations, further substantiated by fossil ostracod studies, it appears that the changes of hydrochemical nature of the lake were rather sudden and common.

An additional, significant type of environment has been recognized in the profiles. It represents a marginal facies to the lake during a receding water level phase. It is the site of sedimentation of muddy material towards the lake by overland or runoff flows during flood-like events from the surroundings. This facies is understandably devoid of fossils because of the high energy nature of its environment during deposition, dolomitization, and

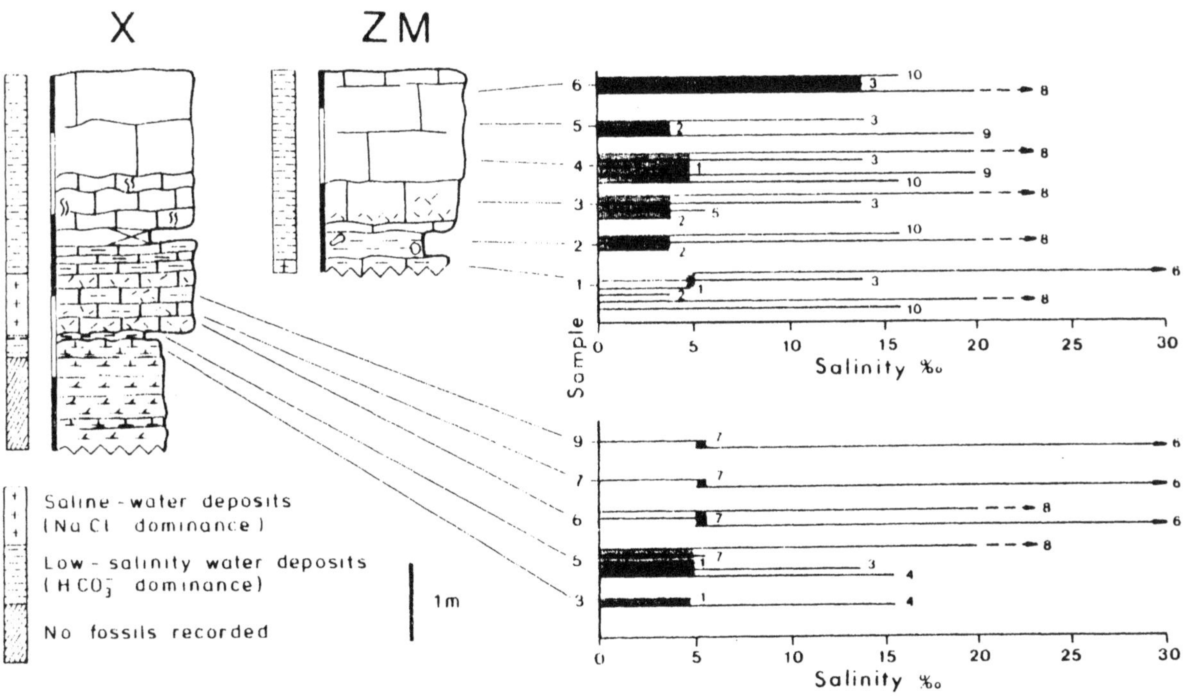

Fig. 5. Venta Micena, X and ZM sections. Ostracod species distribution and salinity range within which all species can occur together within the sample (shaded zones in the diagrams). Numbers within the diagrams refer to ostracods in Table 1.

its frequent aerial exposure as further evidenced by root casts. Pebbly layers within channels are encountered within this facies in association to some erosional features.

The interpretation of facies for profiles YES and CT easily shows that these could easily be synchroneous and be part of the same lake. This is further substantiated by the chronology based on mammal remains recovered at both sites. However, the continuous hydrochemical alternations recorded at both sites will render accurate lateral correlation rather difficult. So far, no unique and overall chemical or biological event could be detected which, otherwise, would have allowed correlation between both profiles.

These recurrent changes which occurred throughout the lacustrine sequences studied here indicate climatic conditions of high contrast, typical of the Mediterranean area today. However, compared to the conditions prevailing today in Southern Spain, there must have been more water available within the system during deposition at the four sites. We are unable at this stage to relate this phenomenon to either precipitation or evaporation.

References

Bradshaw, J. S., 1957. Laboratory studies on the rate of growth of the foraminifer 'Streblus beccarii (Linné) var. tepida (Cushman)'. J. Paleont. 31: 1138–1147.

Cann, J. H. & P. De Deckker, 1981. Fossil Quaternary and living Foraminifera from athalassic (non-marine) saline lakes, southern Australia. J. Paleont. 55: 660–670.

Carbonel, P. & M. Pujos, 1982. Les variations architecturales des microfaunes du lac de Tunis: relations avec l'environnement. Oceanol. Acta, sp. issue 1982: 79–85.

De Deckker, P., 1981. Ostracods of athalassic saline lakes. A review. Hydrobiologia 81: 131–144.

Diebel, K. & E. Pietrzeniuk, 1984. Jungpleistozäne Ostracoden aus Sedimenten der Parkhohlen von Weimar. Quartärpaläontologie 5: 285–319.

Heip, C., 1976. The life cycle of Cyprideis torosa (Crustacea, Ostracoda). Oecologia 24: 229–245.

Herman, P. M. J., C. Heip & G. Vranken, 1983. The production of Cyprideis torosa Jones 1850 (Crustacea, Ostracoda). Oecologia 58: 326–331.

Moyà-Solà, S., J. Agustí, J. Gibert & J. Pons-Moyà, 1981. El yacimiento cuaternario de Venta Micena (España) y su importancia dentro de las asociaciones faunísticas del Pleistoceno inferior europeo. Paleont. Evol. 16: 39–53.

Petit-Maire, N. (ed.), 1982. *Le Shati*. Lac pleistocène du Fezzan (Libye). Editions CNRS, Paris, 118 pp.

Vera, J. A., 1970. Estudio estratigráfico de la Depresión de Guadix-Baza. Bol. Geol. Min. 81: 429–462.

Vera, J. A., J. Fernández, A. C. López-Garrido & J. Rodriguez-Fernández, 1985. Geologia y estratigrafía de los materiales plio-pleistocenos del sector de Orce-Venta Micena (Prov. Granada). Paleont. Evol. 18: (in press).
Vesper, B., 1972. Zur Morphologie und Ökologie von Cyprideis torosa (Jones, 1850) (Crustacea, Ostracoda, Cytheridae) unter besonderer Beruck sichtigung seiner Biometrie. Mitt. hamb. zool. Mus. Inst. 68: 21–77.

Accepted 18 March 1986.

Hydrobiologia 143: 209–212, (1986)

Organic microscopic remains in Miocene lacustrine sediments near Libros (Teruel, Spain)

Maria Jose Gil Quilez
Catedra de Ciencias Naturales, Escuela Universitaria Profesorado EGB, Universidad de Zaragoza, San Juan Bosco no. 7, 50009 Zaragoza, Spain

Keywords: paleolimnology, sponges, Chrysophyceae, cysts, diatoms

Abstract

Siliceous remains from Miocene lacustrine sediments near Libros (Teruel, Spain) are studied. Most of them are sponge spicules and may be assigned to *Ephydatia fluviatilis.* Some chrysophycean cysts and several diatom genera (*Melosira, Cyclotella, Fragilaria, Navicula, Pinnularia,* and *Cymbella*) have also been found.

Introduction

The sections studied are of Miocene lacustrine sediments near to Libros (Teruel, Spain). The sediments, known for a long time to be highly fossiliferous, contain abundant remains of Arthropoda (insects, Arachnida), Mollusca (Gastropoda: *Lymnaea, Bithynia, Planorbis*). Among the Vertebrates, it is worth noting the remains of some mammals (mainly teeth), as well as of Amphibia (*Rana pueyoi, R. quellenbergi*) (Navas, 1922). No previous study of siliceous remains (sponge spicules, chrysophycean cysts and diatoms) has been made. We report here on the different kinds of sponge spicules present in the sections and associated with the remains of other organisms and we also give a first estimate of their changes in abundance, thus pointing to the possibility of a more detailed paleolimnological study of this area in the future.

Study area

The sediments analysed lie between Libros and Riodeva, about 40 km to the SE of Teruel. They are Miocene (Vindobonian-Pontian) in age. The basal levels of the stratigraphic column studied are located in a deserted mining village, now a "ghost town", on the road from Libros to Riodeva (see Fig. 1). Section A (levels 1 to 8) consists of marly clays, and some levels show many remains of Gastropoda and plants. Section B (levels 9 to 16), consists of gypsiferous marls. Most of them, like the earlier section, contain laminations, Gastropoda and plant remains. Section C (levels 17 to 25) contains alternatively gypsum and marls. In levels 18 and 19 sulphur nodules appear. Section D was hidden by slumping and in section E there are alternating gypsum and marl bands (see Fig. 2). The top of the column consists of a massive stratum of beige coloured sandstone that provides a direct basis of comparison with the column studied by Cortazar (1885).

Methods

Nitric acid was added to the sediment and the sample heated until disintegration. The material was then reduced with potassium dichromate and washed to remove the acid. The sediments were diluted and mounted in a highly refractive medium, Naphrax. Samples from each level were treated, except those which consisted of gypsum. Subsequently a large number of megascleres and all the gemmoscleres found along the column were measured.

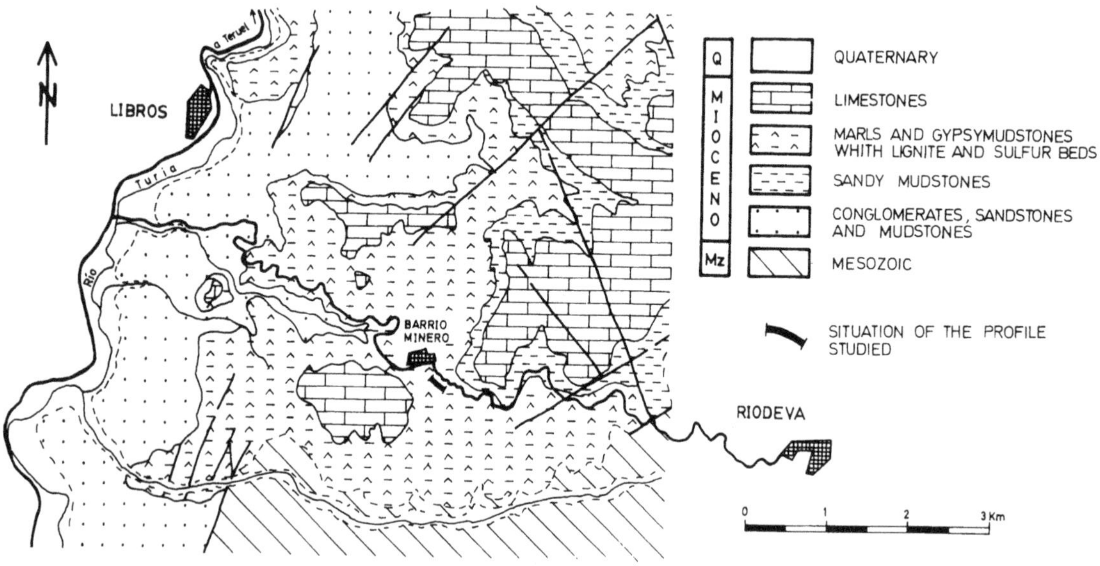

Fig. 1. Location of the profile studied and geological map of the surroundings.

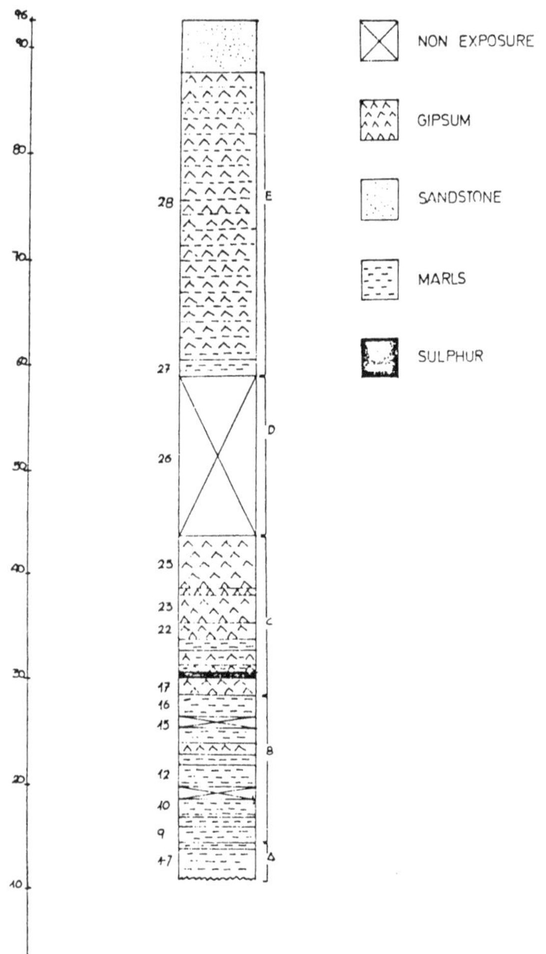

Fig. 2. Geological sections studied.

Results

In section A remains appear only in level 7 and (in small quantities) level 8. In section B, siliceous remains are found in all levels although in layers number 9 and 10 the numbers were small. In section C the remains appear only in levels 21, 18 and 27 and in the marls at level 28. Most of the spicules are megascleres of the amphioxis type and they are smooth or slightly curved. They range in length from 70 to 450 μm, 307 μm being the average length. The gemmoscleres appear more sporadically, all of them being of the amphiaster type with a mean aster diameter of 21 μm (range 12 to 31 μm). The mean width of the shaft is 6.7 μm (range 3 to 8 μm) and the gap between the two asters is 26–60 μm. The dimensions as well as the morphology of the gemmoscleres and megascleres coincide with the characteristics reported for *Ephydatia fluviatilis* (see Fig. 3).

The chrysophycean cysts that have been found can be grouped according to their morphology (see Fig. 4) into:

Simple spherical unsculptured cysts with or without collar
Spherical cysts with collar and projections
Spherical or ovoid cysts with collar and geometrical patterns formed by connecting ridges
Ovoid unsculptured cysts with or without collar
Ovoid cysts with patterns formed by non con-

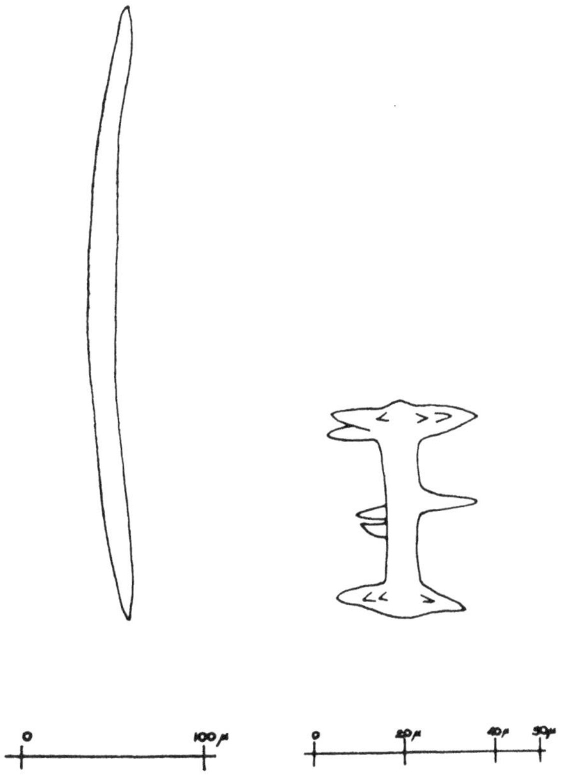

Fig. 3. Gemmo- and megasclere: on the basis of their morphometry they may be assigned to *Ephydatia fluviatilis*.

necting and usually irregular disposed ridges
Ovoid cysts with large hook-like projection near pore
Cysts with two pores.

The diatoms, like the cysts, vary considerably along the column. Species of the genera *Melosira, Cyclotella, Fragilaria, Navicula, Pinnularia,* and *Cymbella* were found.

Conclusion

The sediments were accumulated in anoxic, stratified or non turbulent water. Productivity was probably high. The variety of Chrysophyceae points to a considerable seasonality. The abundant remains of sponges and of benthic diatoms point to an important development of the littoral zone. A more intensive study of the sediments is recommended including some consideration of chemical fossils.

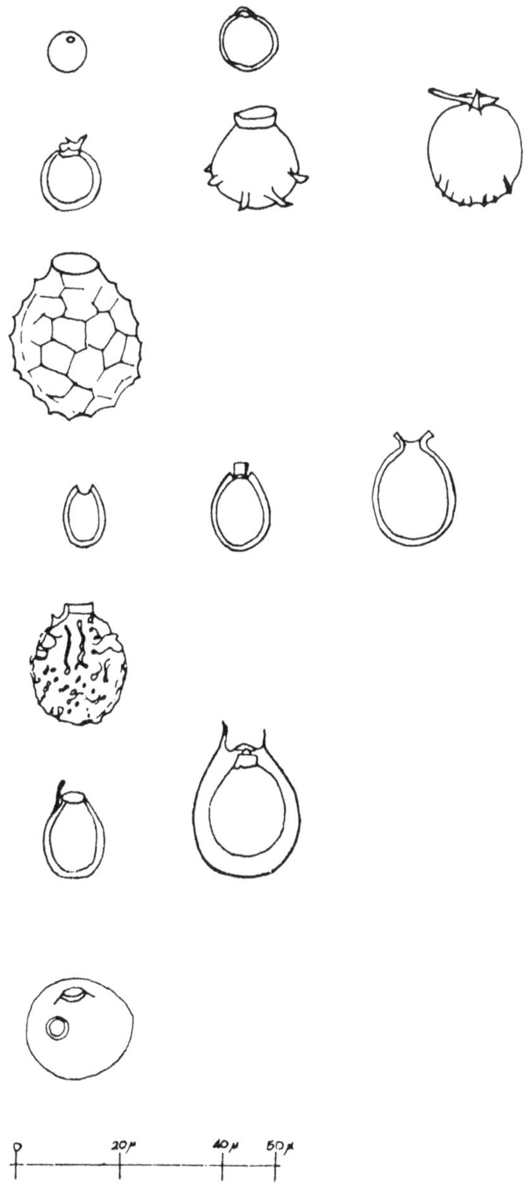

Fig. 4. Chrysophyscean cysts.

Acknowledgements

This research was supported by the Instituto de Estudios Turolenses.

References

Cortazar, D., 1885. Bosquejo fisico-geologico y minero de la prov. de Teruel. Bol. Com. Map. Geol. Esp. 12: 171–286.

Igme, 1978. Memoria explicativa de la hoja 612. ADEMUZ. MAGNA 1/50000, no. 612 (26–24).
Laubenfels, M. W., 1963. Porifera. In: R. C. Moore, (ed.), Treatise on Invertabrate Paleontology, Part E: Archaeocyatha and Porifera: E22–'E22–'113. Geological Society of America. University of Kansas Press, 113 pp.
Navas, L., 1922. Algunos fosiles de Libros (Teruel). Bol. Soc. iber. Cienc. nat. 21: 52–61.
Navas, L., 1922. Algunos fosiles de Libros (Teruel), adiciones y correcciones. Bol. Soc. iber. Cienc. nat. 21: 172–175.
Penney, J. T. & A. Racak, 1968. Comprehensive of a world wide collection of Freshwater Sponges (Porifera: Spongillidae). U.S. nat. Mus. Bull. 272: 1–175.
Roman, F., 1928. Sur quelques reste de Mammiferes decouvertes par le R. P. Longinos dans les argiles pontiques de Libros (Teruel, Espagne). Bol. Soc. iber. Cienc. nat. 27: 198–204.

Accepted 18 March 1986.

Hydrobiologia 143: 213–216, (1986)

Non-siliceous algae in a five meter core from Lake Kinneret (Israel)

U. Pollingher
Kinneret Limnological Laboratory, Israel Oceanographic and Limnological Research, Tel-Shikmona, P.O.B. 8030, Haifa 31080, Israel

Keywords: paleolimnology, Lake Kinneret, Chlorophyta, diatoms, eutrophication

Abstract

The composition and succession of non-siliceous algae, studied in a five meter core from Lake Kinneret (Israel), are described. Only Chlorophyta species were recorded, probably due to the standard palynological sample processing which was used. In the lower part of the core, from the bottom to 300 cm (interval 5500–2500 years B.P.), *Botryococcus braunii* was the only common alga. Relevant changes in algal diversity and abundance occur at 300 cm. Many species of green algae were recorded for the first time (*Pediastrum, Scenedesmus, Coelastrum*, etc.). These changes may be related to an increase in nutrient concentration as a consequence of cultural disturbance. In the interval 300–0 cm, a succession of *Pediastrum* species is followed. The recovered green algae are extant in the present plankton of Lake Kinneret. They also constitute an important part of the algae found in the profundal sediments today.

Introduction

Lake Kinneret lies 209 m below MSL in the northern part of the Syro-African Rift Valley. It is the only freshwater lake in Israel and one of the few natural bodies of fresh water in the Middle East.

The lake acquired its present configuration only 20000 years ago. Earlier, during the Wurm period, a part of the present lake was occupied by the Lisan Lake. The Lisan sediments, deposited about 60000 to 20000 years B.P., are poor in fossils and testify to brackish conditions (Begin *et al.*, 1974).

Microfossil studies in Lake Kinneret have been few and devoted only to pollen (Horowitz, 1971; Baruch, 1983) and diatoms (Ehrlich in Stiller *et al.*, 1984; and in Pollingher *et al.*, in press).

The composition and succession of non-siliceous algae in a five meter core are presented in this paper and contribute to the identification of the trophic changes which occurred in the lake in the last 5500 years, changes which are probably correlated with human activity.

Site description

Lake Kinneret (32°45′–32°53′ N and 35°30′–35°38′ E) has a surface area of 168 km^2 and a volume of 4300×10^6 m^3. Its mean depth is 25 m and maximum depth 43 m. The lake is of the warm monomictic type, with winter holomixis and a stratification period from May–June to December.

The Jordan River draining the northern watershed is the main inlet to the lake. The waters of the Jordan River belong to the calcium–hydrogen carbonate type and have a chloride content of 20 mg l^{-1}, whereas the lake water belongs to the sodium chloride type and has a chloride content of ~ 250 mg l^{-1}. The level of water nutrients is relatively low.

The phytoplankton biomass is dominated by Pyrrhophyta-Chlorophyta assemblages, with a very low contribution of diatoms and blue-green algae. A yearly water bloom of the dinoflagellate *Peridinium cinctum* fa. *westii* (Lemm.) Lefevre develops from February to June. It is accompanied by other dinoflagellates. During the summer, coenobial and nanoplanktonic green algae predominate, together with Cyanophyta and nanoplanktonic diatoms (*Cyclotella* spp.) (Pollingher, 1978, 1981; Pollingher *et al.*, in press).

The principal components of the phytobenthos are the diatoms (Round, 1978) followed by Cyanophyta (Dor, 1974), euglenoids and cryptomonads.

Material and methods

A 5 m long core (KIND-4) was taken in the southern part of the lake (Fig. 1) at 23 m water depth (Stiller *et al.*, 1984). Radiocarbon dating of the core shows that it represents about the last 5500 years of the lake's history (Thompson *et al.*, 1985). The core description and the sedimentation rate can be found respectively in Stiller *et al.* (1984) and Thompson *et al.* (1985).

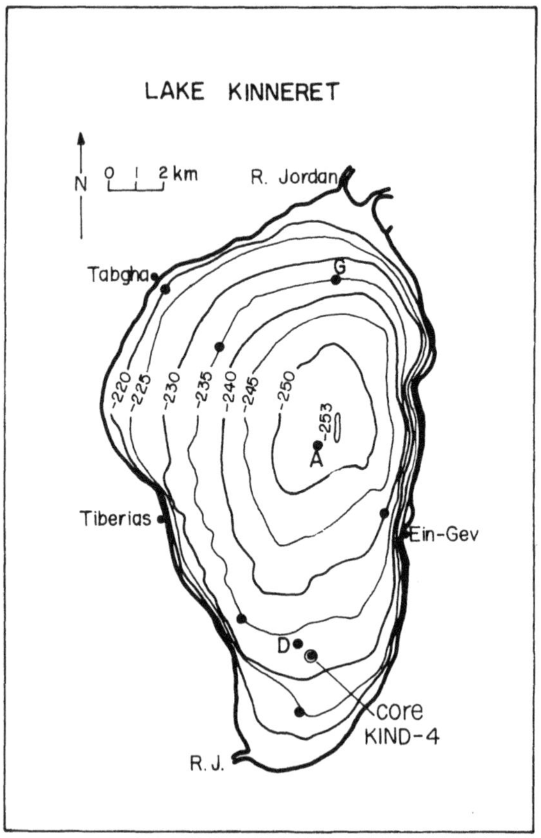

Fig. 1. Bathymetric map of Lake Kinneret with location of core KIND-4.

The isotope O^{18}/O^{16} ratios suggest a mild change in climate; the earlier period was somewhat warmer than the more recent one (Stiller *et al.*, 1984; Stiller & Kaufman, 1985).

Since we were looking for dinoflagellate cysts, standard palynological methods for sample processing were used (i.e., treatment with HCL, HF and acetolysis, Baruch, 1983). In these samples, however, no cysts or other remains of dinoflagellates were recorded. All the preserved algae belong to Chlorophyta. The absence of Cyanophyta and Pyrrhophyta may be correlated with the destructive effect of HF acetolysis.

Taxonomy for identification of Chlorophyta species was that of Komarek & Fott (1983).

Results and discussion

The composition and relative abundance of the non-siliceous remains present large variations.

From the bottom of the core to 300 cm (5500–2500 years B.P.), *Botryococcus braunii* Kutz was the only common form, accompanied by *Pediastrum simplex* Meyen var. *sturmii* in low numbers. Large coccoons of flatworms (?) were present, together with fungi. At 300 cm a notable change in algal diversity and abundance occurred. *Botryococcus* decreased, *P. simplex* var. *sturmii* became dominant, and the following algae were observed for the first time: *P. simplex* var. *echinulatum* Wittr., *P. boryanum* var. *brevicorne* (only coenobia composed of 4 cells), *P. boryanum* (Turp.) Menegh. var. *boryanum* (coenobia composed of a few cells), *Tetraedron minimum* (A.Br.) Hansg., *T. trigonum* (Nag.) Hansg., *Coelastrum reticulatum* (Dang.) Senn, and *Scenedesmus quadricauda* (Turp.) Breb. (small coenobia). In the following interval, from 300 to 180 cm (2500–1400 years B.P.), *P. simplex* var. *echinulatum* became very abundant and dominant, *P. simplex* var. *sturmii* decreased, and an increase of *Scenedesmus* and *Coelastrum* occurred. The remains of Cladocera (*Bosmina*) were more abundant and fragments of *Chironomus* were recorded.

P. simplex var. *echinulatum* continued to dominate the algal assemblages in the upper sediments. At 1 m depth, in addition to all recorded species,

Botryococcus reappeared in higher numbers and large coenobia of *P. boryanum* var. *boryanum* are recorded for the first time. At the same depth, samples prepared using only hydrochloric acid revealed also the presence of *Cosmarium laeve* Rbh. and *Staurastrum* sp.

The described succession of species and their abundance indicate differences in the environmental conditions, differences which are also confirmed by the composition and frequency of diatoms, studied by Aline Ehrlich (in Stiller *et al.*, 1984; and in Pollingher *et al.*, in press).

In the lower part of the core, where only *Botryococcus* is common, the diatom assemblages are dominated by large *Cyclotella* and *Stephanodiscus*, species known from oligotrophic waters.

Higher in the core (340–290 cm), *Cyclotella* spp. are replaced by large *Melosira agassizii* Ostenfeld and *Stephanodiscus* cf. *astrea* Ehr., i.e. an indication of an increase in nutrients, which may explain the drastic change in diversity and abundance of green algae in the interval 300–180 cm. The interval 280–240 cm is nearly barren of diatoms.

The increase and diversification of *Pediastrum* spp. above the 300 cm level likewise suggest an increase in trophic state (Fredskild, 1983; Prat & Daroca, 1983). In the phytoplankton of Lake Kinneret, *Pediastrum* spp. are more abundant in winter when the concentration of nutrients is higher (Pollingher, 1981). In 1974, when higher levels of nutrients were recorded, the abundance of Chlorophyta, and among them *Pediastrum* spp., increased significantly (Serruya & Pollingher, 1977).

The interval 280–180 cm (2300–1400 years B.P.) is also characterized by a sharp rise in *Olea* pollen, signifying large scale cultivation in the area around the lake (Baruch, 1983). This period corresponds to the Hellenite-Roman-Byzantine periods, when the region around the lake was more populated than today and olive oil was an important export product (Josephus [Flavius], 1959).

In the upper 180 cm of the core (1400 B.P. to present), the diatoms are only accessory elements in the total phytoplankton (a situation similar to that prevailing today in the lake). During this period the limnological conditions seem to have been quite similar to those of the lake today. All the species of diatoms and green algae from this part of the core (except coenobia of *P. boryanum* var. *boryanum* composed of 4 cells) are extant in the present lake, and all of the recovered taxa of Chlorophyta constitute an important part of the algae found in the present profundal sediments (Serruya *et al.*, 1974). *P. duplex* Meyen has not yet been recovered from the core.

The $^{13}C/^{12}C$ ratio in the carbonate fraction of the core, which indicates the past intensity of algal productivity (Stiller *et al.*, 1984), shows that lower values prevail at the bottom and higher ones at the top. This again indicates a general trend of the lake to become more eutrophic. The start of eutrophication, and as a consequence Chlorophyta development, may be correlated with the cultural disturbance of the basin which occurred during the Hellenite-Roman-Byzantine periods.

Acknowledgements

The author is grateful to Dr. M. Stiller for the initiation of the multidisciplinary study of the core and core sampling. The author is also much indebted to Mr. U. Baruch for the processed samples, and thanks are given to Ms. K. Diskin for typing the manuscript.

References

Baruch, U., 1983. The palynology of a late Holocene core from Lake Kinneret. M. A. Thesis, Hebrew University, Jerusalem, 29 pp.

Begin, Z. B., A. Ehrlich & Y. Nathan, 1974. Lake Lisan, the Pleistocene precursor of the Dead Sea. Isr. geol. Surv. Bull. 62: 30 pp.

Dor, I., 1974. Considerations about the composition of benthic algal flora in Lake Kinneret. Hydrobiologia 44: 255–264.

Fredskild, B., 1983. The Holocene development of some low and high Arctic Greenland lakes. Hydrobiologia 103: 217–224.

Horowitz, A., 1971. Climatic and vegetational developments in northeastern Israel during Upper Pleistocene – Holocene times. Pollen Spores 13: 255–278.

Josephus [Flavius], 1959. The Jewish War. Translated by G. A. Williamson. Penguin Books Ltd., Harmondsworth, Middlesex, 411 pp.

Komarek, J. & B. Fott, 1983. Das Phytoplankton des Susswassers. 7. Teil, 1. Halfte. Chlorophyceae, Ordnung Chlorococcales. Schweizerbart'sche Verlag., Stuttgart, 1044 pp.

Pollingher, U., 1978. The phytoplankton of Lake Kinneret. In C. Serruya (ed.), Lake Kinneret. Monographiae Biologicae 32. Junk, The Hague, pp. 229–242.

Pollingher, U., 1981. The structure and dynamics of the phytoplankton assemblages in Lake Kinneret, Israel. J. Plankton Res. 3: 93–105.

Pollingher, U., A. Ehrlich & S. Serruya, in press. The planktonic diatoms of Lake Kinneret, Israel, during the last 5000 years, their contribution to the algal biomass. Hedwigia.
Prat, N. & M. V. Daroca, 1983. Eutrophication processes in Spanish reservoirs as revealed by biological records in profundal sediments. Hydrobiologia 103: 153–158.
Round, F. E., 1978. The benthic algae. In C. Serruya (ed.), Lake Kinneret. Monographiae Biologicae 32. Junk, The Hague, pp. 323–328.
Seruya, C. & U. Pollingher, 1977. Lowering of water level and algal biomass in Lake Kinneret. Hydrobiologia 54: 73–80.
Serruya, C., M. Edelstein, U. Pollingher & S. Serruya, 1974. Lake Kinneret sediments: Nutrient composition of the pore water and mud-water exchanges. Limnol. Oceanogr. 19: 489–508.
Stiller, M. & A. Kaufman, 1985. Paleoclimatic trends revealed by the isotopic composition of carbonates in Lake Kinneret. Z. Gletscherkunde Glazialgeologie.
Stiller, M., A. Ehrlich, U. Pollingher, U. Baruch & A. Kaufman, 1984. The late Holocene sediments of Lake Kinneret (Israel) – multidisciplinary study of a five meter core. Geol. Surv. Isr. Curr. Res. 1983–1984: 83–88.
Thompson, R., G. M. Turner, M. Stiller & A. Kaufman, 1985. Near East paleomagnetic secular variation record in sediments from the Sea of Galilee (Lake Kinneret). Quat. Res. 23: 175–188.

Accepted 18 March 1986.

Hydrobiologia 143: 217–225, (1986)

Limnologic consequences of the decline in hemlock 4800 years ago in three Southern Ontario lakes

Mary M. Boucherle[1,3], John P. Smol[1], Tracie C. Oliver[1], S. R. Brown[1] & R. McNeely[2]
[1]*Department of Biology, Queen's University, Kingston, Ontario, Canada K7L 3N6*
[2]*Terrain Sciences Division, Geological Survey of Canada, 601 Booth Street, Ottawa, Ontario, Canada K1A 0E8*
[3]*Present address: Grassy Fork Fisheries, P.O. Box 154, Martinsville, Ind. 46151, U.S.A.*

Keywords: Cladocera, diatoms, eutrophication, hemlock, pigments, vegetational change

Abstract

Four thousand eight hundred years ago hemlock (*Tsuga canadensis*) populations were decimated throughout eastern North America. We have studied the effects of this loss from the terrestrial community on three Southern Ontario lakes: Little Round Lake, Sunfish Lake, and McKay Lake. This study includes the use of cladocerans, diatoms, chrysophytes, and bacterial pigments to assess the limnologic changes that occurred in these lakes. Each lake experienced a change in trophic status that coincided with the loss of hemlock from its catchment, but the change in the aquatic biota was different in each lake. The lakes' size may have been the most influential factor governing the response to this terrestrial disturbance.

Introduction

Only one event of widespread loss of a single tree species in eastern North America is known from Quaternary history (Davis, 1965). Hemlock (*Tsuga canadensis*) was decimated over a very short span (~60 years) 4800 years ago throughout Eastern North America (Davis, 1981). This event was recorded in the pollen assemblages of lake sediments from Upper Michigan to New Brunswick and from Canada to the southern Appalachian Mountains (Davis, 1965). Davis (1981) attributes this abrupt decline in hemlock to a pathogen because a single tree species was affected over a wide geographical area in a very short time span. This change in the terrestrial community, loss of hemlock and replacement by deciduous species, has been shown to influence neighboring aquatic communities (Whitehead *et al.*, 1973). They believe that increased nutrient input occurred in Berry Pond, Massachusetts at the time of the hemlock decline due to increased erosion and a larger input of deciduous leaf litter resulting from the replacement of hemlock by hardwoods and changes in spring and summer runoff. Likens & Davis (1975) also have evidence of increased nutrient input in Mirror Lake, New Hampshire at the time of the hemlock decline. We have investigated a number of paleolimnological lines of evidence to ascertain the effects of the hemlock decline on the biota of Little Round Lake, Sunfish Lake, and McKay Lake in Southern Ontario. Specifically, we wanted to investigate whether this disturbance in the terrestrial community coincided with trophic changes in the aquatic community.

Pollen assemblages from the sediments of the three lakes were analyzed stratigraphically to determine where the hemlock decline occurred. The period of the decline was then investigated by close interval contiguous samples that were analyzed for cladoceran and algal remains, and bacterial pigments. Cladocera provide information on lake trophic status and have proven to be reliable indicators of nutrient enrichment in a wide variety of lakes (Frey, 1955; Deevey, 1969; Brugam, 1978; Boucherle & Züllig, 1983). Analysis of siliceous algal microfossils, diatoms and chrysophytes, also supplies data on trophic changes in the lakes (Brugam, 1978; Munch, 1980; Smol *et al.*, 1983).

Carotenoids and chlorophylls of the photosynthetic bacteria provide ecological evidence of past populations (Brown *et al.*, 1984; Lotter & Boucherle, 1984). Purple sulfur bacteria (Chromatiaceae) and brown-coloured varieties of the green sulfur bacteria (Chlorobiaceae) have absolute requirements for both light and sulfide, but the Chlorobiaceae tolerate both lower light intensities and higher concentrations of sulfide. They are arranged along vertical gradients of light intensity (decreasing with depth), and concentrations of sulfide (increasing with depth). The photosynthetic pigments of the two groups are distinctive and group specific. These pigments form quantitatively measurable biochemical fossils that reflect the relative abundance of the organisms that produced them, and thus contribute important information about the light regime under which they grew.

Interpretations of the postglacial histories of each of these lakes appear elsewhere: Little Round Lake, Brown *et al.*, 1977; Daley *et al.*, 1977; Smol & Boucherle, 1985; Sunfish Lake, Smol, 1982; McIntosh, 1983; McKay Lake, Oliver, 1986. These lakes were chosen for study because they are or have been meromictic, have laminated sediments, and provide good habitats for sulfur bacteria. Detailed analysis of the recent history of Little Round Lake has been presented by Smol *et al.*, 1983, and Brown *et al.*, 1984. The lakes have been of interest to other paleolimnologists because of their laminated sediments (Little Round Lake and McKay Lake, Tippett, 1964) and the presence of meromixis (Sunfish Lake, Duthie & Carter, 1970; Sreenivasa & Duthie, 1973; Adams & Duthie, 1976).

Materials and methods

A 4.35-m long sediment core was extracted from Little Round Lake using a Brown piston corer (1956) for the first meter and a Livingstone corer (Wright *et al.*, 1965) for the remainder of the core. The same method was used to obtain a 4.85-m long sediment core from Sunfish Lake and a 4.62-m long sediment core from McKay Lake. All of these cores were collected from the deepest part of the lakes' basin.

The cores were extruded in the laboratory, split lengthwise, photographed, and sediment color, texture, and structure were described, then they were sectioned at 1-cm intervals. Sediment at and adjacent to the Hemlock Decline Zone was analyzed at 1-cm intervals. Pollen preparation followed standard techniques (Faegri & Iverson, 1975). At least 400 pollen grains were identified per sample. Our cladoceran and algal microsfossil preparation methods are outlined in Smol (1983) and Smol & Boucherle (1985). Cladocera remains were extracted and analyzed; either 200 microfossils or 10 coverslips were counted, whichever came first. Siliceous algal microfossils were prepared and at least 600 diatom valves and mallomonadacean scales were identified and enumerated for each sample. The carotenoids, okenone and isorenieratene, were extracted and analyzed using the methods of Brown *et al.* (1984). Densitometric scanning of reversed-phase thin-layer chromatograms was used for pigment quantification.

Results

All three study lakes are located in Southern Ontario; Little Round Lake (44°45′N, 76°43′W); Sunfish Lake (43°28′N, 80°35′W); and McKay Lake (45°27′N, 75°40′W). Some physical and chemical characteristics of the lakes are presented in Table 1. Although the lakes are of similar size today, we have evidence that McKay Lake was approximately three times larger at the time of the hemlock decline (McNeely, unpublished). Each of these lakes lies in a different type of bedrock ranging from Precambrian gneiss and marble overlain by sandy tills at Little Round Lake to shale, salt, and gypsum surrounded by sandy kames at Sunfish Lake, to Ordovician rocks covered by Pleistocene marine clays at McKay Lake. These lakes have similar water chemistries and all three are calcareous (Table 1). Each lake has been meromictic and during meromictic periods the sediments have been annually laminated (Tippett, 1964; Smol, 1982; Smol & Boucherle, 1985). During the hemlock decline all three lakes are considered to have been meromictic. The lakes are presented in order of the most affected to the least affected, that is Little Round Lake, Sunfish Lake, McKay Lake.

Little Round Lake

In Little Round Lake preservation of both the biochemical and morphological fossils was good. Hemlock declined from over 30% of the pollen sum at 168 cm to 4% at 162 cm. The concentration

Table 1. Some physical and chemical characteristics of Little Round Lake, Sunfish Lake and McKay Lake.

Lake	Area (ha)	Maximum depth (m)	Catchment area (km^2)	Alkalinity (mg/l CaCO)	Calcium (mg/l)	Conductivity (μS)
Little Round	7.4	16.8	1.09	~120	~50	~400
Sunfish	8.3	20.0	2.2	~125	~75	~600
McKay	8.0	10.5	1.4	~100	~400	~400
McKay*	22.8	18.6	1.24			

* McKay Lake at the time of the hemlock decline.

of the planktonic cladocerans decreased slightly at the time of the hemlock decline. Planktonic cladocerans were comprised of 90% *Bosmina longirostris* (O. F. Müller) and less than 10% *Daphnia longispina* (O. F. Müller) and *Daphnia pulex* (de Geer) together. The relative abundance of planktonic species was unaffected at the time of the hemlock decline, whereas the littoral species composition was very much affected; *Chydorus faviformis* Birge attains its highest percentage (50%) and largest concentration for the entire postglacial history at the 165 cm level, and 9 out of 13 species that had been present disappeared (Fig. 1).

The Chlorobiaceae and Chromatiaceae populations both declined following the hemlock decline, as evidenced by the decreased concentrations of isorenieratene and okenone (Fig. 2). The Chromatiaceae recovered and populations were larger than before the decline. The Cholorbiaceae also recovered but the population was not as large as before the hemlock decline.

The diatoms were dominated by *Cyclotella stelligera* (Cl. & Grun) V.H., although *Cyclotella michiganiana* Skv., *Cyclotella comta* (Ehr.) Kutz., *Fragilaria crotonensis* Kitt., and *Asterionella formosa* Hass. were also common (Fig. 3). In the sample at 165 cm, *Stephanodiscus hantzschii* Grun. appeared and dominated the flora. This peak in *S. hantzschii* correlates precisely to the sudden increase in the cladoceran *Chydorus faviformis*. *Mallomonas pseudocoronata* (Prescott)/*crassisquama* (Asmund) Fott composed over 95% of the fossil mallomonadaceae scales throughout this zone. The concentration of the diatoms and chrysophytes decreased at the time of the hemlock decline (Fig. 2).

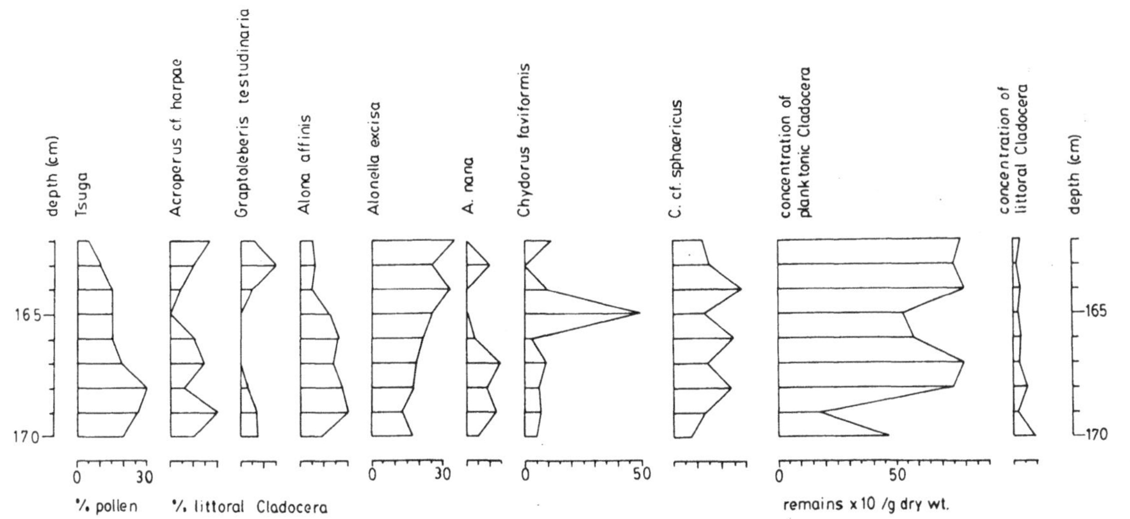

Fig. 1. Percent *Tsuga*, relative abundance of littoral Cladocera and the concentration of littoral and planktonic Cladocera in Little Round Lake during the hemlock decline.

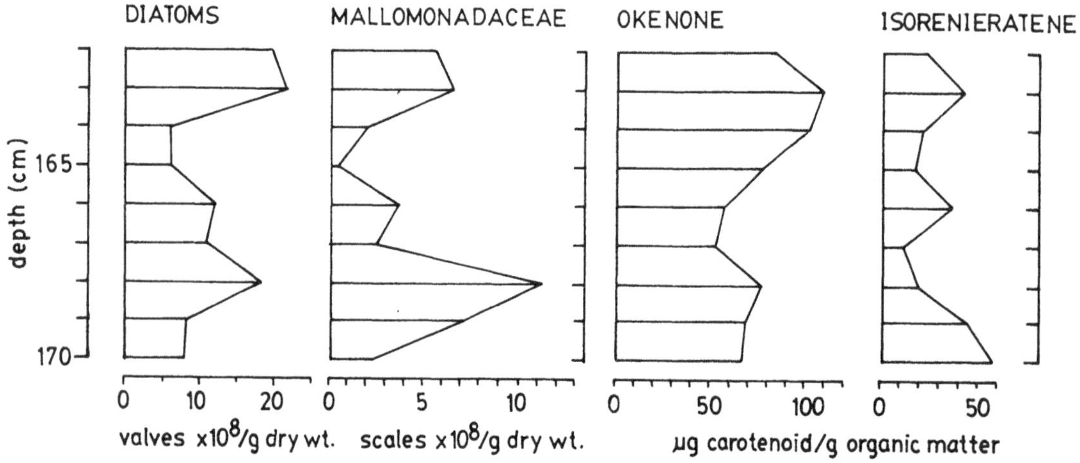

Fig. 2. Concentrations of diatoms, Mallomonadaceae, okenone and isorenieratene in Little Round lake during the hemlock decline.

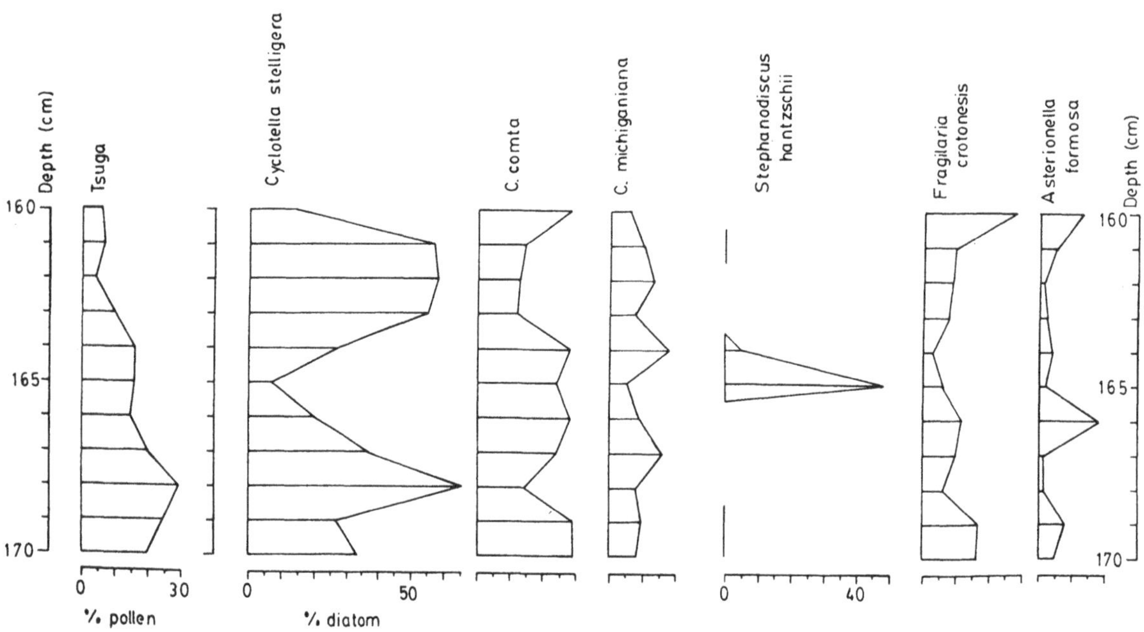

Fig. 3. Percent *Tsuga* and relative abundance of diatoms during the hemlock decline in Little Round Lake.

Sunfish Lake

In Sunfish Lake the hemlock decline occurred between 255 and 250 cm levels where the hemlock percentage fell from 30% to 10% and then continued to decline to less than 4% at 240 cm (Fig. 4). The concentration of planktonic cladocerans increased following the hemlock decline (Fig. 4). *Daphnia pulex* was dominant following the hemlock decline except at 250 cm where *Bosmina longirostris* peaked. This period was one of two periods during the lake's entire postglacial history when *Daphnia pulex* was present in appreciable numbers and reached its highest relative abundance at this time. Changes in the littoral cladocerans were unremarkable except that at 250 cm *Chydorus faviformis* was recorded for the first time in the lake's history (Fig. 5). By the 245 cm sample the cladoceran populations had decreased but the species composition remained the same. The Chromatiaceae populations, as reflected by their pigment okenone, showed a noticeable decrease during the hemlock decline period (McIntosh, 1983). Chlorobiaceae populations, as measured by their pigment isorenieratene, showed a slight increase at this time. Siliceous algal microfossils were not preserved in the portion of the core containing the hemlock decline, precluding a detailed investigation.

McKay Lake

In McKay Lake hemlock began declining from 28% at 341 cm to 19% at 340 cm and then decreased to 6% at 330 cm (Fig. 6). The planktonic cladocerans were represented by only one species, *Bosmina longirostris,* whereas the littoral cladocerans were more diverse with as many as 15 species present. Species changes were slight during the hemlock decline but a large increase in concentration of littoral species was recorded at 335 cm.

The diatom flora was dominated by *Cyclotella stelligera* throughout this period (Fig. 7). Diatom concentrations peaked twice during this period, once at 339 cm following the decrease in hemlock and again at 333 cm following a small increase in *Stephanodiscus hantzschii.* This species previously recorded in only trace amounts increased slightly at 336 cm just before the increase in littoral cladocerans at 335 cm. These changes were followed by a gradual decrease in *C. stelligera* and an increase in *C. michiganiana.*

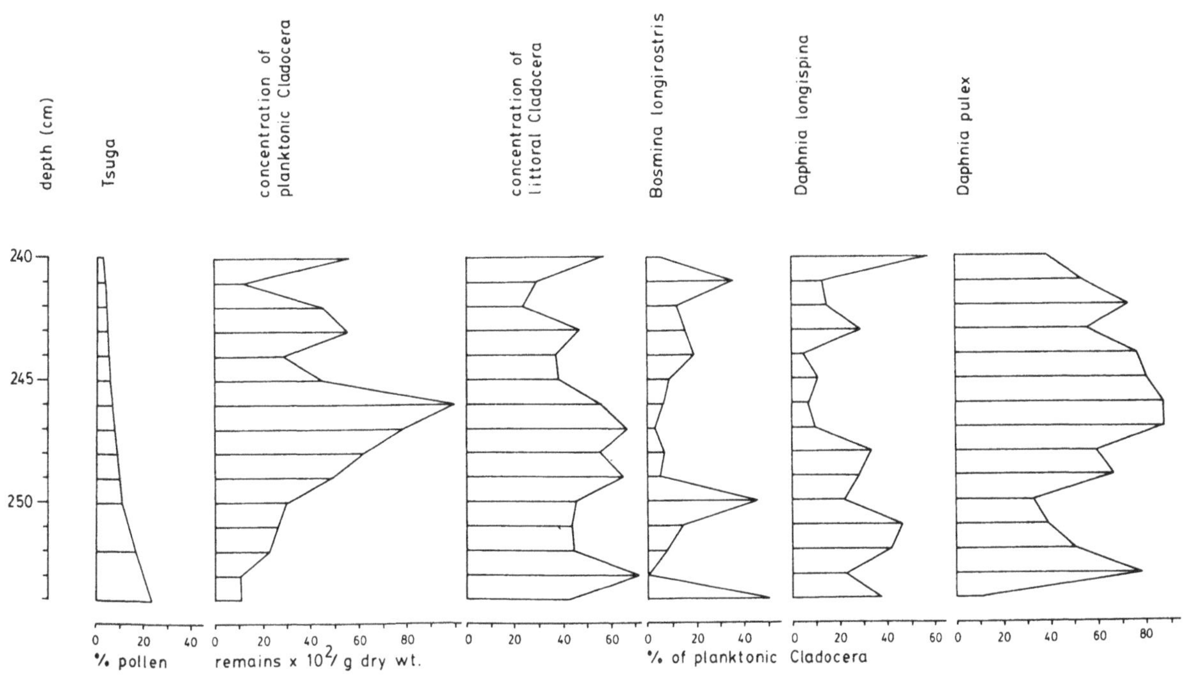

Fig. 4. Percent *Tsuga*, concentrations of planktonic and littoral Cladocera and relative abundance of planktonic Cladocera in Sunfish Lake during the hemlock decline.

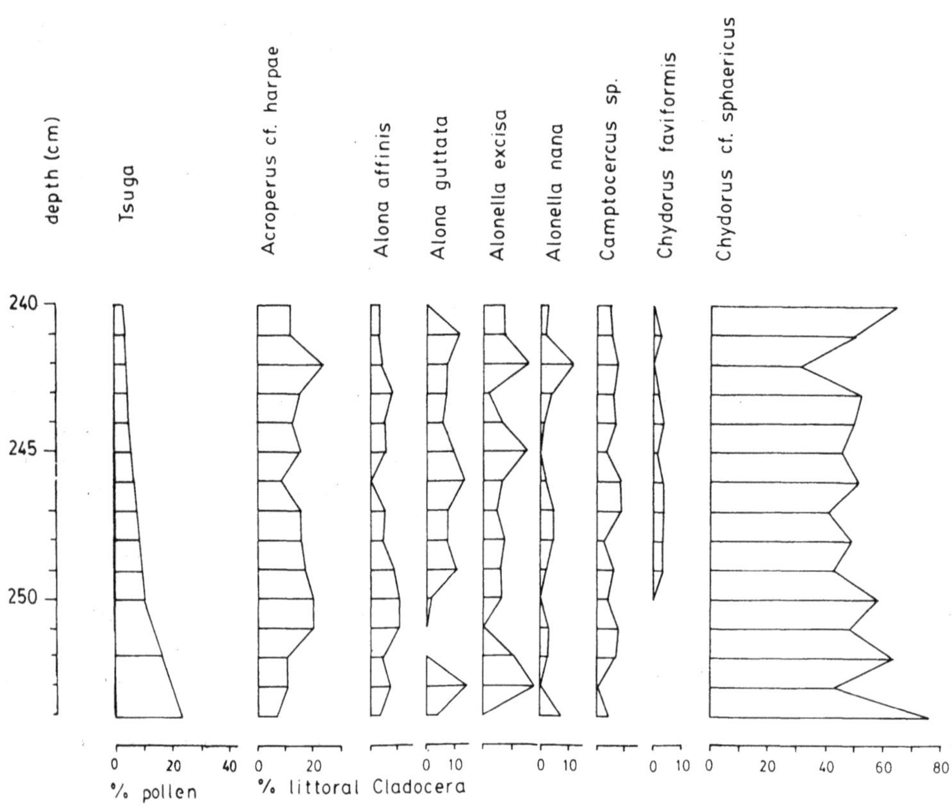

Fig. 5. Percent *Tsuga* and relative abundance of littoral Cladocera in Sunfish Lake during the hemlock decline.

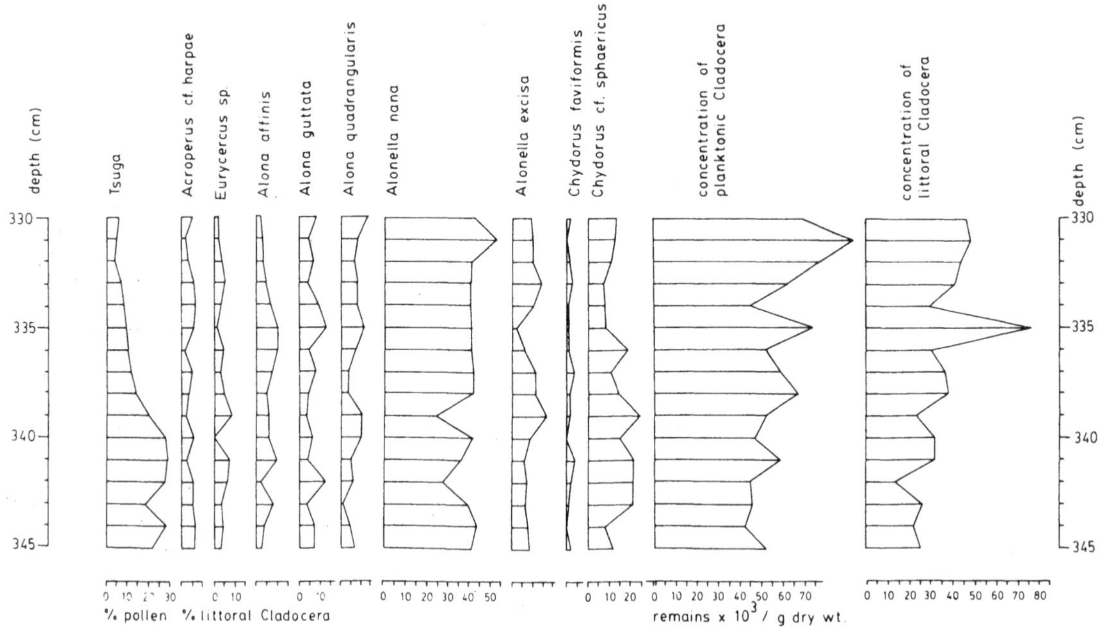

Fig. 6. Percent *Tsuga,* relative abundance of littoral Cladocera, and concentrations of planktonic and littoral Cladocera in McKay Lake during the hemlock decline.

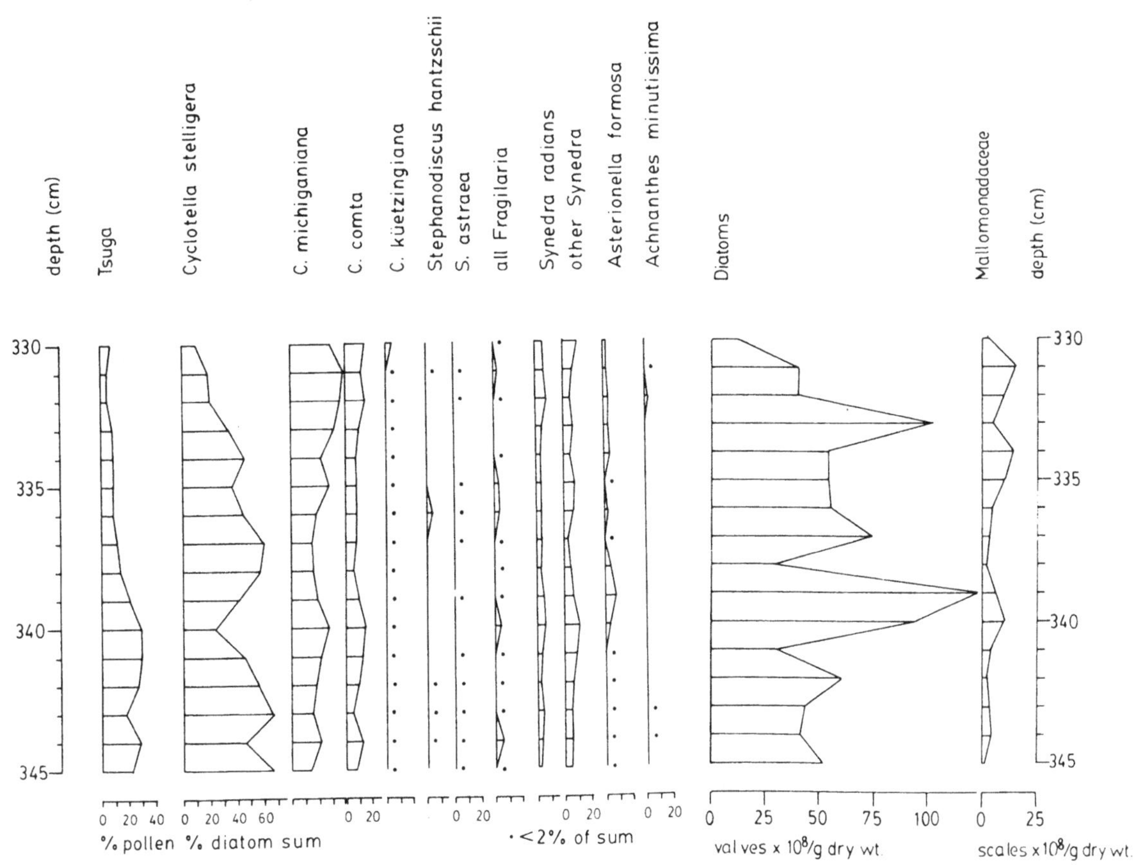

Fig. 7. Percent *Tsuga,* relative abundance of diatoms and concentrations of diatoms and Mallomonadaceae in McKay Lake during the hemlock decline.

Discussion

There is a tendency among North American limnologists to regard the eutrophication of lakes solely as a recent phenomenon, that is the result of human activities. Within the last century, increasing density of human populations has led to higher nutrient budgets of lakes, with a subsequent increase in algal production (Edmondson & Lehman, 1981; Züllig, 1982). This increased production began with colonial settlement and land clearance and has been further accelerated by the addition of domestic and agricultural effluents. Lakes in undisturbed watersheds appear to eutrophy naturally at a slower rate because their watersheds supply only small amounts of essential plant nutrients. Our study provides evidence of accelerated eutrophication brought about by natural events occurring in the terrestrial community.

Little Round Lake experienced a eutrophic phase following the hemlock decline. Changes in the cladoceran and algal communities and the bacterial populations indicate a switch from oligotrophic conditions at 168 cm to eutrophic conditions at 165 cm, while the relative abundance of hemlock decreased 15% in the pollen assemblage. The algal community responded by the appearance and sudden increase (to 40%) of *Stephanodiscus hantzschii,* a species characteristic of eutrophic lakes (Brugam, 1979). At the same time an oligotrophic species, *Cyclotella stelligera* was markedly reduced. This species shift is similar to the one recorded in Little Round Lake's recent sediments that coincides with cultural activities in the catchment (Smol et al., 1983). The concentration of mallomonadacean scales decreased during the hemlock decline. This decrease is interpreted as further evidence of nutrient enrichment, because Mallomonadaceae are usually replaced by other groups of algae under eutrophic conditions (Smol *et al.*, 1983). This type of response also was seen in the littoral cladocerans as *Chydorus faviformis* peaked at the same time and many other species disappeared. This eutrophic phase was acute, but short-lived as the fauna and flora had begun to return to mesotrophic assemblages by the next sediment increment.

Populations of Chlorobiaceae and Chromatiaceae waned following the hemlock decline in Little Round Lake. This response is attributed to the increased shading caused by higher algal production in the overlying water column. The impact of light reduction would be greatest and would occur first on the Chlorobiaceae, by virtue of their lower position, but with continued eutrophication the Chromatiaceae would also be reduced. Recovery of the bacterial populations indicates an improvement in light regime as the algal production again declined; and additionally, as sulfide moved upward in the water column as a result of mineralization of algal debris.

Sunfish Lake experienced a phase of moderate nutrient enrichment following the hemlock decline based on changes in the cladoceran communities. The appearance of *Chydorus faviformis*, the dominance of *Daphnia pulex* and the increase in concentration of the planktonic cladocerans are all symptomatic of nutrient enrichment.

The initial nutrient enrichment in Sunfish Lake was relatively short-lived. By the 245 cm sample the concentration of planktonic species had diminished and by the 240 cm sample *Daphnia pulex* was no longer dominant. The lake appears to have had a slight nutrient enrichment that persisted for a considerable time (900 years) during the hemlock minimum when deciduous trees dominated. This idea is supported by the continued presence of *Chydorus faviformis* and *D. pulex* and research on the chemistry of the Sunfish Lake sediments (Adams & Duthie, 1976).

McKay Lake appears to have experienced less nutrient enrichment at the time of the hemlock decline than either Little Round Lake or Sunfish Lake. *Stephanodiscus hantzschii,* a eutrophic diatom, made up less than 10% of the diatom flora at 336 cm and disappeared in the next sample. The littoral cladocerans had a similarly brief population increase. This small change in the lake's biota relative to Little Round Lake and Sunfish Lake may be partially explained by size differences among the lakes. Although all three lakes are of a similar size today, at the time of the hemlock decline McKay Lake was approximately three times larger than it is now (Table 1). The littoral cladoceran species were the most affected group in McKay Lake which would be expected based on the much larger extent of the littoral zone.

The loss of hemlock from the lakes' catchment coincided with a change in the lakes' biota. Each lake reacted in a different way to this change in the terrestrial community. Little Round Lake underwent an acute eutrophic phase in which both the diatoms and littoral cladocerans were affected. In Sunfish Lake the planktonic cladocerans and bacterial populations were affected whereas the littoral cladocerans remained essentially unchanged. Both of these lakes recovered from the initial eutrophic phase quickly, but a prolonged mesotrophic state persisted during the hemlock minimum which lasted approximately 900 years. Conversely, McKay Lake appears hardly to have been affected by the loss of hemlock. The littoral cladocerans and diatoms were affected but their changes were small and short-lived. These differences among the lakes may be related to the lakes' size. McKay Lake was three times larger at the time of the hemlock decline then the other two lakes and probably experienced a greater dilution of incoming nutrients. Little Round Lake, the smallest of the three lakes, exhibited the most acute eutrophication.

Acknowledgements

This study was supported by the Natural Sciences and Engineering Research Council of Canada by grants to S.R.B. and J.P.S.

References

Adams, R. W. & H. C. Duthie, 1976. Relationships between sediment chemistry and postglacial production rates in a small Canadian lake. Int. Revue ges. Hydrobiol. 61: 21–36.

Boucherle, M. M. & H. Züllig, 1983. Cladocera remains as evidence of change in trophic state in three Swiss lakes. Hydrobiologia 103: 141–146.

Brown, S. R., 1956. A piston sampler for surface sediments of lake deposits. Ecology 37: 611–613.

Brown, S. R., R. J. Daley & R. N. McNeely, 1977. Composition and stratigraphy of the fossil phorbin derivatives of Little Round Lake, Ontario. Limnol. Oceanogr. 22: 336–348.

Brown, S. R., H. J. McIntosh & J. P. Smol, 1984. Recent paleolimnology of a meromictic lake: Fossil pigments of photosynthetic bacteria. Verh. int. Ver. Limnol. 22: 1357–1360.

Brugam, R. B., 1978. Human disturbance and the historical development of Linsley Pond. Ecology 59: 19–36.
Brugam, R. B., 1979. A re-evaluation of the Araphidineae/Centrales index as an indicator of lake trophic status. Freshwat. Biol. 9: 451–460.
Daley, R. J., S. R. Brown & R. N. McNeely, 1977. Chromotographic and SCDP measurement of fossil phorbins and the postgalcial history of Little Round Lake, Ontario. Limnol. Oceanogr. 22: 349–360.
Davis, M. B., 1965. Phytogeography and palynology of northeastern United States. In: H. E. Wright & D. G. Frey (eds.) The Quaternary of the United States, 377–401. Princeton Univ. Press.
Davis, M. B., 1981. Outbreaks of forest pathogens in Quaternary history. Proc. IV int. Conf., Lucknow, (1976–77), 3: 216–227.
Deevey, E. S., 1969. Cladoceran populations in Rogers Lake, Connecticut, during late- and post-glacial time. Mitt. int. Ver. Limnol. 17: 56–63.
Duthie, H. C. & J. S. H. Carter, 1970. The meromixis of Sunfish Lake, Southern Ontario. J. Fish Res. Bd Can. 27: 847–856.
Edmondson, W. T. & J. T. Lehman, 1981. The effect of changes in the nutrient income on the condition of Lake Washington. Limnol. Oceanogr. 26: 1–19.
Faegri, K. & J. Iversen, 1975. Textbook of pollen analysis. Blackwell Sci. Pub., Oxford 295 pp.
Frey, D. G., 1955. Langsee: a history of meromixis. Mem. Ist. ital. Idrobiol. suppl. 8: 141–164.
Likens, G. E. & M. B. Davis, 1975. Post-glacial history of Mirror Lake and its watershed in New Hampshire, U.S.A.: an initial report. Verh. int. Ver. Limnol. 19: 982–993.
Lotter, A. & M. M. Boucherle, 1984. A late-glacial and postglacial history of Amsoldingersee and vicinity, Switzerland. Schweiz. Z. Hydrol. 46: 192–209.
McIntosh, H. J., 1983. A paleolimnological investigation of the bacterial carotenoids of Sunfish Lake. M.Sc. Thesis, Queen's University, Kingston, Ontario, Canada.
Munch, C. S., 1980. Fossil diatoms and scales of chrysophytes in the recent history of Hall Lake, Washington. Freshwat. Biol. 10: 61–66.
Oliver, T. C., 1986. Postglacial history of McKay Lake based on algal microfossils. M.Sc. Thesis, Queen's University, Kingston, Ontario, Canada.
Smol, J. P., 1982. Postglacial changes in fossil algal assemblages from three Canadian lakes. Ph.D. Thesis, Queen's University, Kingston, Ontario, Canada.
Smol, J. P., 1983. Paleophycology of a high arctic lake near Cape Herschel, Ellesmere Island. Can. J. Bot. 61: 2195–2204.
Smol, J. P., S. R. Brown & R. N. McNeely, 1983. Cultural disturbances and trophic history of a small meromictic lake in Central Canada. Hydrobiologia 103: 125–130.
Smol, J. P. & M. M. Boucherle, 1985. Postglacial changes in algal and cladoceran assemblages in Little Round Lake, Ontario. Arch. Hydrobiol. 103: 25–49.
Sreenivasa, M. R. & H. C. Duthie, 1973. The postglacial diatom history of Sunfish Lake, southwestern Ontario. Can. J. Bot. 51: 1599–1609.
Tippett, R., 1964. An investigation into the nature of the layering of deep water sediments in two eastern Ontario lakes. Can. J. Bot. 42: 1693–1709.
Whitehead, D. R., H. Rochester, S. W. Rissing, C. B. Douglass & M. C. Sheehan, 1973. Late glacial and postglacial productivity changes in a New England Pond. Science 181: 744–747.
Wright, H. E., D. A. Livingstone & E. J. Cushing, 1965. Coring devices for lake sediments. In B. Kummel and D. Raup (eds), pp. 852–998. Handbook of Paleontological techniques. W. J. Freeman Co., San Francisco.
Züllig, H., 1982. Untersuchungen über die Stratigraphie von Carotinoiden in geschichteten Sediment von 10 Schweizer Seen zur Erkundung früherer Phytoplankton-Entfaltungen. Schweiz. Z. Hydrol. 44: 1–98.

Accepted 18 March 1986.

Hydrobiologia 143: 227–232, (1986)

Geology and sedimentary history of Lake Traunsee (Salzkammergut, Austria)

Jürgen Schneider[1], Jens Müller[2] & Michael Sturm[3]
[1]*Institut für Geologie und Dynamik der Lithosphäre, Goldschmidtstraße 3, D-3400 Göttingen, FRG*
[2]*Lehrstuhl für Geologie, Technische Universität München, Lichtenbergstraße 4, D-8046 Garching, FRG*
[3]*EAWAG-ETH Zürich, Überlandstraße 133, CH-8600 Dübendorf, Switzerland*

Keywords: lake sediments, clastic deposits, sediment echography, turbidities, sedimentary history

Abstract

Traunsee was formed by glacial overdeepening of a pre-existing fault system. Present-day morphology is characterized by a deep (189 m) narrow trough with steep slopes (>50°) in the southern part surrounded by the Northern Calcareous Alps. The northern part of the lake is bordered by flysch and glacial deposits with gentle slopes (<30°) and exhibits several ridges, basins and troughs.

During the late and postglacial period, more than 45 m of sediment has accumulated in the central basin. Sedimentation in the southern part of Traunsee is mainly controlled by the river Traun forming a prograding delta in the south and – within the past 50 years – by industrial tailings consisting mainly of calcite. Sediments are distributed by undercurrents and by turbidites. Cores from the central basin thus show an intercalation of Traun-derived dolomite-rich sediments with anthropogenic muds from the tailings deposited up to more than 6 km from its source.

Within the northern basin, land slides from the flysch region played an important role leading to drastic changes in the morphology of slopes and adjacent basins. These slumps have persisted until historic times. Sedimentation in the shallow sublittoral regions is dominated by benthic biogenic decalcification.

The frequency of turbidite sedimentation within the profundal basin decreased during the last 200 years probably due to man's activities in the drainage area such as regulation of rivers and torrents. Sedimentation rates during the past decades range from 2–3 cm/a in the southern basin to 0.4 cm/a in the northern part as shown by ^{137}Cs-dating.

Geologic setting, morphology and hydrography

The Traunsee is situated at the northern rim of the Alps. It is bordered to the south by the Northern Calcareous Alps and the northern part is surrounded by flysch and glacial deposits (Fig. 1, Baumgartner, 1984). The N-S extending basin follows a major transverse fault system which has been eroded and shaped during glaciation. Rock types surrounding the lake reflect the morphology of the present-day basin (see Fig. 2 and Fig. 3 a/b).

The steep slopes of the Northern Calcareous Alps continue below the lake level down to the profundal zone. Dives with the submarine GEO showed the existence of steep slopes (>50°) and even vertical walls with only local sediment cover. Accumulation of sediments from the major tributary – the river Traun – along the eastern border of the profundal zone causes a NW dipping of the basinal plain in the southern part. The narrow trough widens in the central part forming a 189 m deep central basin plain which is bordered in the north by ridges. Slope angles decrease distinctly in the northern part of the lake where moraines and flysch border the lake. Within this part of the lake several ridges reaching into shallow water depths are separated by smaller basins (see Fig. 2).

Traunsee has the second largest area (24.4 km^2) of all the Austrian Alpine lakes. Its volume totals 2228×10^6 m^3, the greatest depth is 189 m. Major

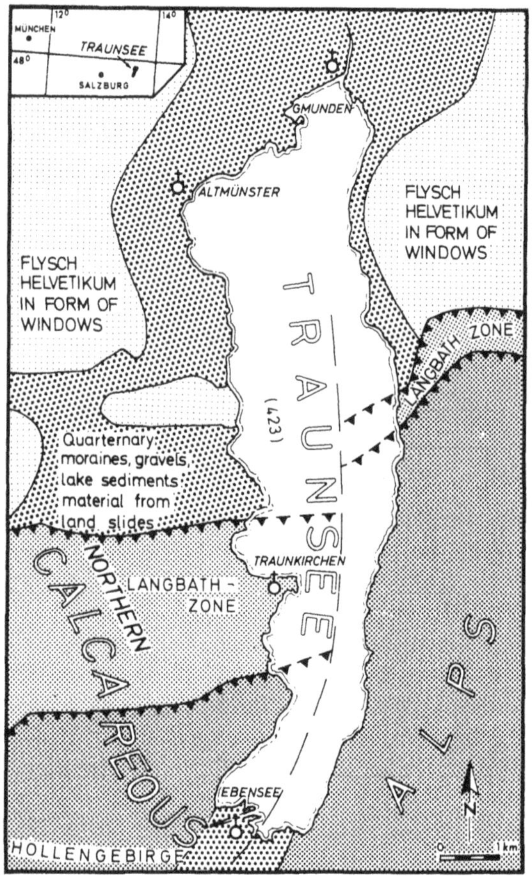

Fig. 1. Geological sketch map of the region around Traunsee (from Baumgartner, 1984).

inflow (80%) comes from the river Traun, draining an area of 1417 km^2. Mean discharge amounts to 65 m^3/sec transporting approx. 43000 t/a of suspended load into the lake. Bed load transport averages 36000 t/a (Baumgartner, 1984; Müller, Sossau & Zeh, 1983).

The flow of the river Traun into the lake causes a depression of the thermocline and a short residence time (0.9 a). The resulting exchange rate keeps Traunsee at a mesotrophic level in spite of substantial nutrient loading (Pechlaner & Sossau, 1982).

Sediment sources and sedimentation processes

Sediments are supplied by different sources and

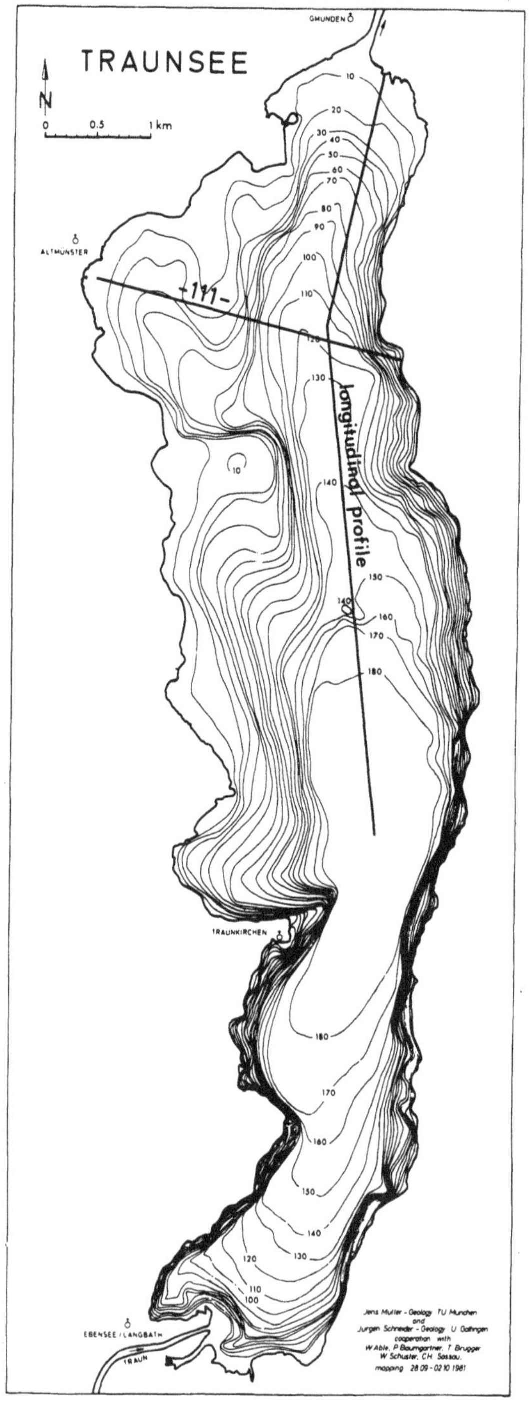

Fig. 2. Bathymetric map of Traunsee showing the position of two sediment echography profiles (Figs. 5 and 7) (from Müller & Schneider, 1984).

Fig. 3a. 3D computer graph of the Traunsee, viewed from North to South (from Müller & Schneider, 1984).

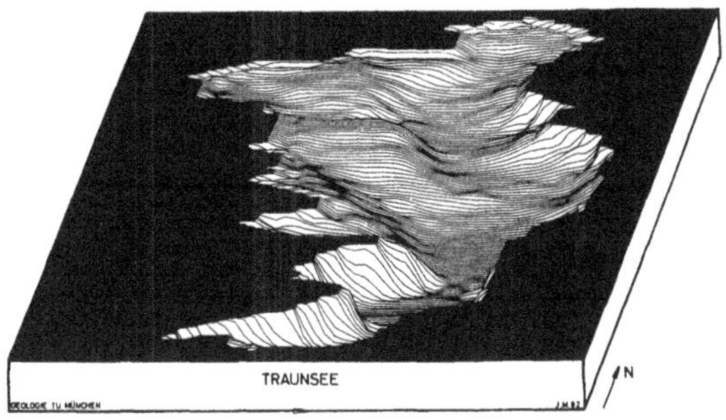

Fig. 3b. 3D computer graph of the Traunsee, viewed from South to North (from Müller & Schneider, 1984).

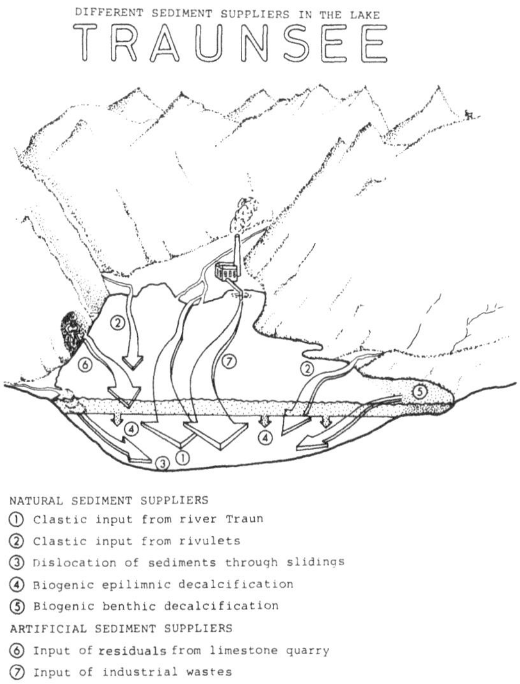

Fig. 4. Natural and artificial sediment suppliers of the Traunsee (modified after Schneider, Claes & Kersting, 1984).

mechanisms which are shown in Fig. 4 (see also Claes & Kersting, 1981). Allochthonous components are derived from natural sources such as the Traun river or other smaller rivulets and by sliding of unstable rock resp. sediment masses or by anthropogenic wastes. The latter come from a limestone quarry and from the combined tailings of soda works and salt works (Müller & Schneider, 1984).

Autochthonous sediment sources are biogenic epilimnetic decalcification and, in sublittoral regions, biogenic benthic decalcification.

1. Due to the petrography of the drainage area, dolomite is the most important clastic component of the river Traun. Deposition of Traun-derived sediments can thus be traced by the high dolomite content of basin sediments since no other substantial dolomite source exists. Most of the coarse suspended load and the bed load of the river Traun accumulates in the delta area forming foresets with 30° dip slopes. Episodic slumps in the foreset area continue into the profundal basin as turbidity currents often initiating additional slumps by erosion. Depending upon grain size, fine particles are transported by currents further north before they settle to the ground. Occasional turbidity underflows occur when river water is cold and extremely loaded with fines. In these instances, sediments are transported directly into the central basin.

The extension of turbidites and turbidity underflows is controlled by the basin morphology. Turbidites originating from the south are thus restricted to the northern end of the central basin due to the ridge separating the northern basin from the deeper southern basin (Sturm & Müller, 1984, Fig. 5).

2. Sediment input of rivulets is restricted both regionally and with respect to the amount deposited. Due to differences in the geology of the respective drainage area sediment distribution from the rivulets can be traced by their specific mineralogical association. Abundance in quartz characterizes catchment areas with flysch outcrops while those coming from moraine areas or from the Calcareous Alps are distinctly lower in quartz and dominated by dolomite and calcite (Fig. 6).

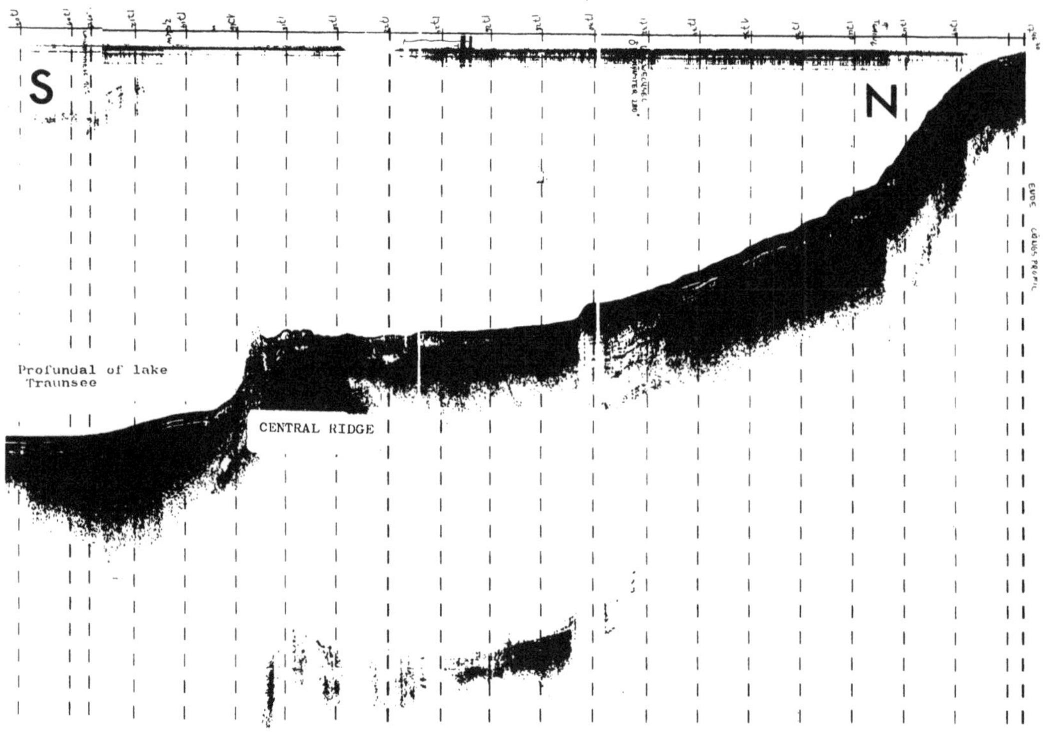

Fig. 5. Part of a longitudinal profile (see Fig. 2) from sediment echography (3.5 KHz ORE). The central ridge separates the profundal from the northern basin. Height of the ridge: about 30 m (from Müller & Schneider, 1984).

3. Landslides reaching into the lake are a common feature in the flysch zone at the northeastern side of the lake. These slumps have been active for centuries. The last catastrophic event occurred in 1910 when houses slid into the lake. Cores and profiling with 3.5 Khz echosounder showed numerous subaquatic slumps below a sequence of undisturbed lake sediments in the adjacent slope and basin documenting the frequency of slumps in the past (Fig. 7).

4. Biogenic epilimnetic decalcification is best developed during the summer. Increased phytoplankton production reduces CO_2 and HCO_3^- leading to an increase in pH and subsequent oversaturation causing the precipitation of calcite (e.g. Schröder, Windolph & Schneider, 1983). Sediment input by this mechanism is however considered to be rather low due to the dilution with allochthonous material in Traunsee.

5. Biogenic benthic decalcification is restricted to the photic zone of shallow sublittoral areas bordering the northwestern and northern part of Traunsee. Calcite precipitation is initiated by the activity of macro- and microphytes leading to the formation of lake marl and chalk in these areas (Schröder, 1982; Schneider, Schröder & Le Campion-Alsumard, 1983).

6. An additional source of calcite in the southern part of the lake comes from a limestone quarry (Karbach) operated by the soda works. Approx. 12500 t/a of calcite <1 mm is washed into the lake forming a distinct fan in the bottom sediments.

7. Calcite is also the major constituent of the industrial tailings accumulating in the bay of Ebensee in the southwestern corner of Traunsee. The tailings come from soda works and salt works located in Ebensee which have been pumping their solid and dissolved wastes into the lake for more than 50 years. These tailings are of special relevance for the lake ecosystem with respect to their input rate (max. 140000 t/a solid tailings, 244000 t/a dissolved wastes), their chemistry (highly alkaline pore waters, pH-values up to 12) and their unstable accumulation triggering turbidity currents and slumps reaching the central basin. Their alkaline environment thus influences benthic life in wide

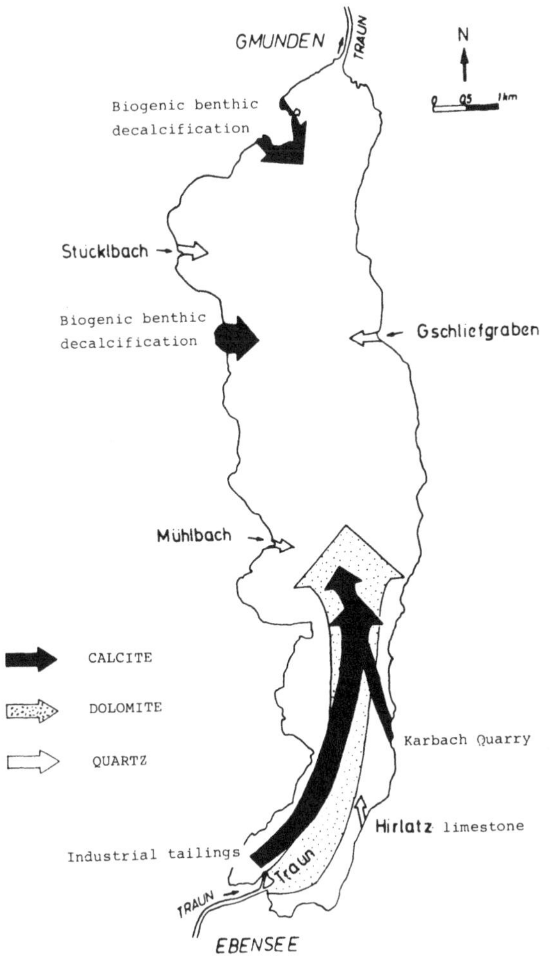

Fig. 6. The main suppliers of calcite, dolomite and quartz into the Traunsee (from Claes & Kersting, 1981).

portions (approx. 15% of lake floor) of the lake. (For further details see Müller, Schneider & Sturm, this volume.)

Sedimentary history and sedimentation rates

According to van Husen (1977) melting resp. retreat of the Traun glacier occurred rather fast and by 18000 years B.P. the lake was free of ice. A drill site situated approx. 1.5 km south of the present-day Traun-delta revealed the presence of delta foreset sediments down to a depth of 155 m overlaying bottomset sediments and glaciofluvial deposits (Baumgartner, 1984). Availability of abundant unconsolidated glacial deposits lacking soil and vegetation cover certainly favoured a high sedimentation rate at the beginning of the late glacial period. 3.5 KHz profiling showed within the present-day central basin sediment thickness exceeding 45 m. Glacial overdeepening thus reached down to 190 m above sea level; initial lake depth exceeded 235 m in this lake section.

Since these sediment thicknesses are beyond the

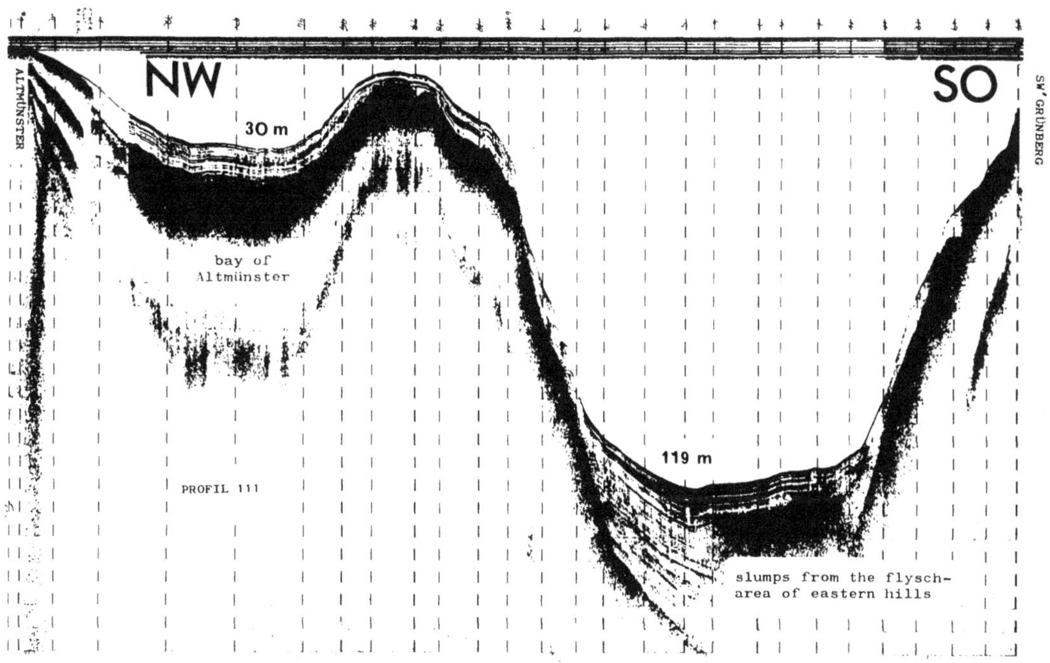

Fig. 7. Cross profile (see Fig. 2) from sediment echography (3.5 KHz ORE) through the northern part of Traunsee. Historical slumps from the flysch area are clearly documented (from Schneider, Claes & Kersting, 1984, ORE: Dr. J. Müller).

reach of conventional coring techniques information about the sedimentation during the early stages comes mainly from 3.5 kHz profiling (Müller & Schneider, 1984).

Records from the central basin reveal sequences of coherent parallel reflectors which document that basin fill occurred mainly by turbidity currents coming from the south. Profiles from the northern basin resp. trough indicate that subaqueous slumps originating from the flysch area played a major role in the past shaping slope and basin morphology (Fig. 7).

Piston coring down to a sediment depth of 6 m showed that turbidites and slumps prevailed in these sections as well. However, a decreasing frequency of thick (>5 cm) turbidites in the uppermost sediments of the central basin indicates that consolidation of erosion and weathering processes paralleled by the introduction of effective river regulation and avalanches breaks led to a decrease in these episodic sedimentation events during the past 200 years.

Turbidites originating from the industrial tailings led to a substantial increase in the sedimentation rate of the profundal basin within the past 50 years (Müller & Schneider, 1984).

Locally, more than 75% of the uppermost sediment depth consists of the alkaline mud. Sedimentation rates as calculated for this area range from 2–3 cm/a. ^{137}Cs-dating of cores from the northern basin where turbidites are less frequent showed that rates in this area amount to 0.4 cm/a (Sturm & Müller, 1984).

Acknowledgements

The investigations were supported by a grant of the 'Amt der Oberösterreichischen Landesregierung' (AZ: Bau 2-12106/18-1980/Wei/St.).

References

Baumgartner, P., 1984. Geologische Verhältnisse in der Umgebung des Traunsees. Limnologische Untersuchung Traunsee-Traun, Amt O. Ö. Landesregierung Bericht Nr. 12: 1–9.

Claes, M. & G. Kersting, 1981. Die Sedimente des Traunsees. (Österr. Kte. 1:25000, Bl. 66/2 Gmunden und Bl. 66/4 Ebensee Salzkammergut, Oberösterreich). Diplomarbeit Univ. Göttingen, 205 pp.

Husen, D. van, 1977. Zur Fazies und Stratigraphie der jungpleistozänen Ablagerungen im Trauntal. Jb. Geol. B.-A., Wien, 120/1: 1–130.

Müller, J., C. Sossau & U. Zeh, 1983. Die Schwebstoffe der Traun und des Traunsees. Limnologische Untersuchung Traunsee-Traun, Amt O. Ö. Landesregierung Bericht Nr. 11: 50 pp.

Müller, J. & J. Schneider, 1984. Die Industrieschlammablagerungen in der Bucht von Ebensee und im Profundal des Traunsees (Oberösterreich). Limnologische Untersuchung Traunsee-Traun, Amt O. Ö. Landesregierung Bericht Nr. 12a: 100 pp.

Pechlaner, R. & C. Sossau, 1982. Die Ergebnisse der fünfjährigen Studie 'Limnologische Untersuchung Traunsee-Traun'. Limnologische Untersuchung Traunsee-Traun, Amt O. Ö. Landesregierung Bericht Nr. 13: 234 pp.

Schneider, J., H. G. Schröder & T. Le Campion-Alsumard, 1983. Algal micro-reefs – coated grains from freshwater environments. In T. M. Peryt, (Ed.), Coated grains, Springer Berlin, Heidelberg: 284–298.

Schneider, J., M. Claes & G. Kersting, 1984. Die Sedimente des Traunsees. Limnologische Untersuchung Traunsee-Traun, Amt O. Ö. Landesregierung Bericht Nr. 12: 11–52.

Schröder, H. G., H. Windolph & J. Schneider, 1983. Bilanzierung der biogenen Karbonatproduktion eines oligotrophen Sees (Attersee, Salzkammergut – Österreich). Arch. Hydrobiol. 97: 356–372.

Schröder, H. G., 1982. Biogene benthische Entkalkung als Beitrag zur Genese limnischer Sedimente. Beispiel: Attersee (Salzkammergut, Oberösterreich). Dissertation Univ. Göttingen, 179 pp.

Sturm, M. & J. Müller, 1984. Die Untersuchung langer Sedimentprofile und die Verbreitung von Turbiditen im Traunsee. Limnologische Untersuchung Traunsee-Traun, Amt O. Ö. Landesregierung Bericht Nr. 12: 97–131.

Accepted 18 March 1986.

Hydrobiologia 143: 233–246, (1986)

Sediments and sedimentary history of Lake Attersee (Salzkammergut, Austria)

Ahmad-Reza Behbehani[1], Jens Müller[2], Roland Schmidt[3], Jürgen Schneider[1], Heinz-Gerd Schröder[4], Ines Strackenbrock[1] & Michael Sturm[5]
[1]*Institut für Geologie und Dynamik der Lithosphäre, Goldschmidtstraße 3, D-3400 Göttingen, FRG*
[2]*Lehrstuhl für Geologie, Technische Universität München, Lichtenbergstraße 4, D-8046 Garching, FRG*
[3]*Institut für Limnologie der Österreichischen Akademie der Wissenschaften, Gaisberg 116, A-5310 Mondsee, Austria*
[4]*Institut für Seenforschung und Fischereiwesen, Untere Seestraße 81, D-7994 Langenargen, FRG*
[5]*EAWAG-ETH Zürich, Überlandstraße 133, CH-8600 Dübendorf, Switzerland*

Keywords: lake sediments, late- and postglacial history, palynology, biogenic carbonate production, eutrophication, heavy metals

Abstract

Attersee represents a good example of a lake situated in the Northern forelands of the Northern Calcareous Alps and influenced by different sediment-supplying processes during the postglacial. Several compounds, of different origin, form the sediments of the basin. Clastics which are mainly composed of dolomites derive from the Northern Calcareous Alps. Clastic input of organic and inorganic particles is accomplished by rivers and landslides. They are responsible for the main input of siliciclasts like quartz, feldspar and mica. A high proportion of the sediment results from autochthonous biogenic carbonate precipitation. In the shallow sublittoral areas of the northern part of the lake benthic decalcification caused by encrusting macro- and micro-phytes is dominant, while in the southern and central parts of the lake epilimnetic decalcification caused by the blooming of phytoplancton is more important during summer. The total biogenic calcium carbonate production reaches about 11 000 to 12 000 metric tons a year.

Nutrients and residues of cyanophytes (*Oscillatoria rubescens*) deriving from the eutrophic lake Mondsee were washed into lake Attersee by the Mondseeache. High amount of phosphorus in the sediments of the southern basin depicts local eutrophication in the mouth area of the Mondseeache. The average sedimentation rate in lake Attersee can be determined by different dating methods. Sedimentation rates increased during the last 110 years from 1 mm a year to 1.8–2 mm a year as a result of human activities. Five main phases in the postglacial sedimentary history can be recognized: Würm moraines and finely banded varves (before 13 000 B.P.), the early Attersee stage (from 13 000 B.P. up to 1 200 B.P.), and the later Attersee stage after the Bavarian colonization (from 1 200 B.P. on). Using heavy metal and isotope analyses the sedimentary history can be reconstructed in more detail for the last 100 years.

Introduction

Sediments and their sites of generation and deposition are part of the exogenous system in the marine and terrestrial region. Because of their size and well defined morphology lakes are suitable areas for the investigation of recent sediments and their genesis. The genesis and amount of autochthonous carbonates, the input and distribution of clastic material, as well as human impact on the lake environment were investigated in lake Attersee.

2. Location, geography and geology

Lake Attersee is located in Austria about 40 km east of Salzburg in front of the Northern Calcareous Alps (altitude: 469.2 m above Adriatic Sea,

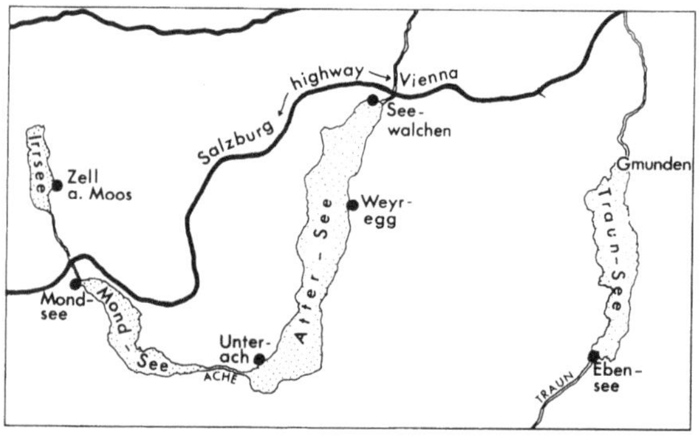

Fig. 1. Location map.

length: 20.1 km, mean width: 2.4 km, surface area: 45.9 km^2 length of coast: 53 km). The Attersee is the largest lake in the Salzkammergut-region. The Mondseeache draining from the eutrophic Mondsee into the oligotrophic Attersee, is the main Attersee inflow (more than 50% of the total incoming drainage, see Jagsch, 1971 and Moog, 1981). The only Attersee outlet is the river Ager, which flows towards the river Traun.

The Attersee basin is characterized by a threefold division (Fig. 2, compare Müller & Simony, 1985):

a. a southern basin (−170 m) which is separated by a rise from the central and northern part of the lake,

b. a morphologically heterogenous central part with deeper areas (−160 m) and a submarine rise to 49 m water depth (Laichberg),

c. a northern basin flattening to the North with a maximum depth of 130 m.

The topographic framework of lake Attersee area was formed during the last sequence of Alpine orogenesis in the Pliocene (Spengler & Pia, 1924; Schadler, 1959; Janoschek, 1964). During the period of Quaternary glaciation the Attersee region was influenced by three glacial advances of the Traun

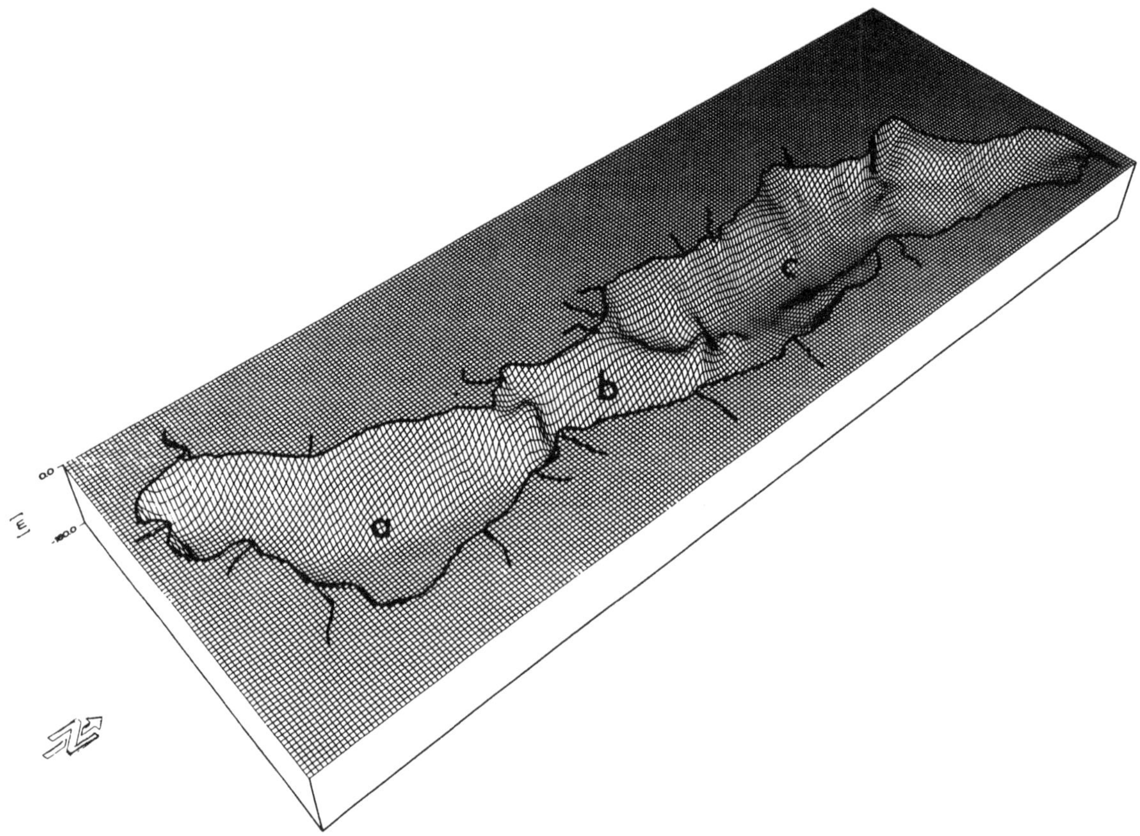

Fig. 2. Attersee basin, 3 dimensional computer graph (a–c: 3 main basins).

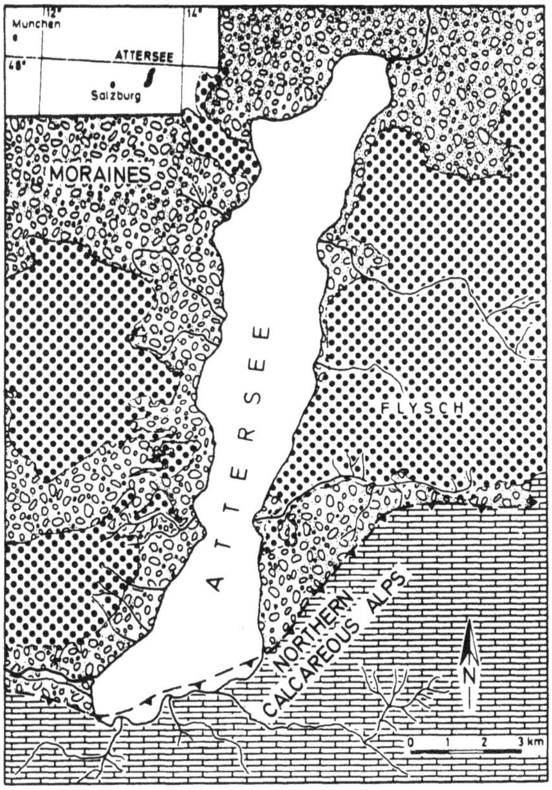

Fig. 3. Geological sketch map of the Attersee region (after Häusler & Schwarzenböck, 1970).

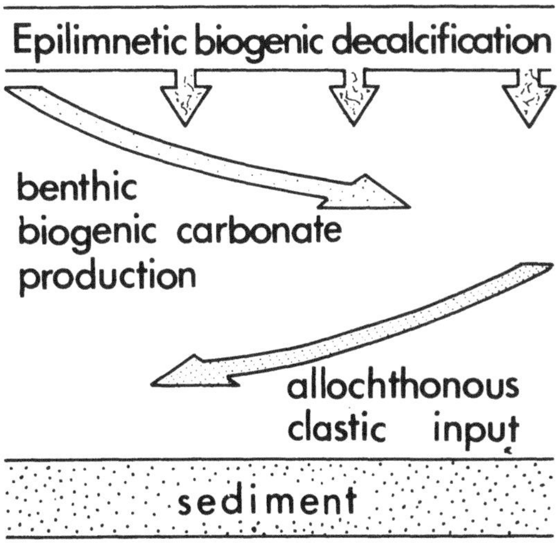

Fig. 4. Sediment supply processes in the Attersee.

glacier system. The motion of the glacier modified the basin morphology and formed the moraine ramparts of Mindel, Riss and Würm (Weinberger, 1955; Del Negro, 1969). The Attersee basin was finally shaped during the Würm glaciation. The overdeepened basin was filled with water and sediments during the last ice retreat.

The geological setting of the Attersee is divided into three main tectonic units (Fig. 3):

a. The Northern Calcareous Alps in the South
b. The flysch zone in the east
c. The moraine mounds in the North and West

The character of the lake Attersee shoreline is determined by its geological environment. Hence, there are steep banks in the southern and eastern areas of the Attersee whereas in the southern and western parts the slope angles of the shoreline are gentle and "shelf-like".

3. Sediment supplying processes and distribution of sediments

Generally three sediment supplying processes can be distinguished for lake Attersee (Fig. 4)

1) autochthonous carbonate production by biogenic epilimnetic decalcification
2) autochthonous carbonate production by biogenic benthic decalcification (see section 4)
3) allochthonous input to the lake from different source areas.

The input of allochthonous clastic material is accomplished principally by river influx. Besides the rivers, the allochthonous sediment particles are transported into the lake by landslides and mud flows (especially on the eastern bank of the Attersee), by bank erosion in the wave zone and by surface waters, groundwater, rain and wind.

4. Composition of the sediments

The mineralogical character of the suspended matter in the rivers draining into Attersee is determined by the different lithologies in each catchment. Dolomite is transported into the lake from the Calcareaous Alps, whereas quartz, feldspars (albite and microcline), clay minerals (kaolinite, illite, montmorillonite) and micas (chlorite and muscovite) derive from the flysch and moraine areas.

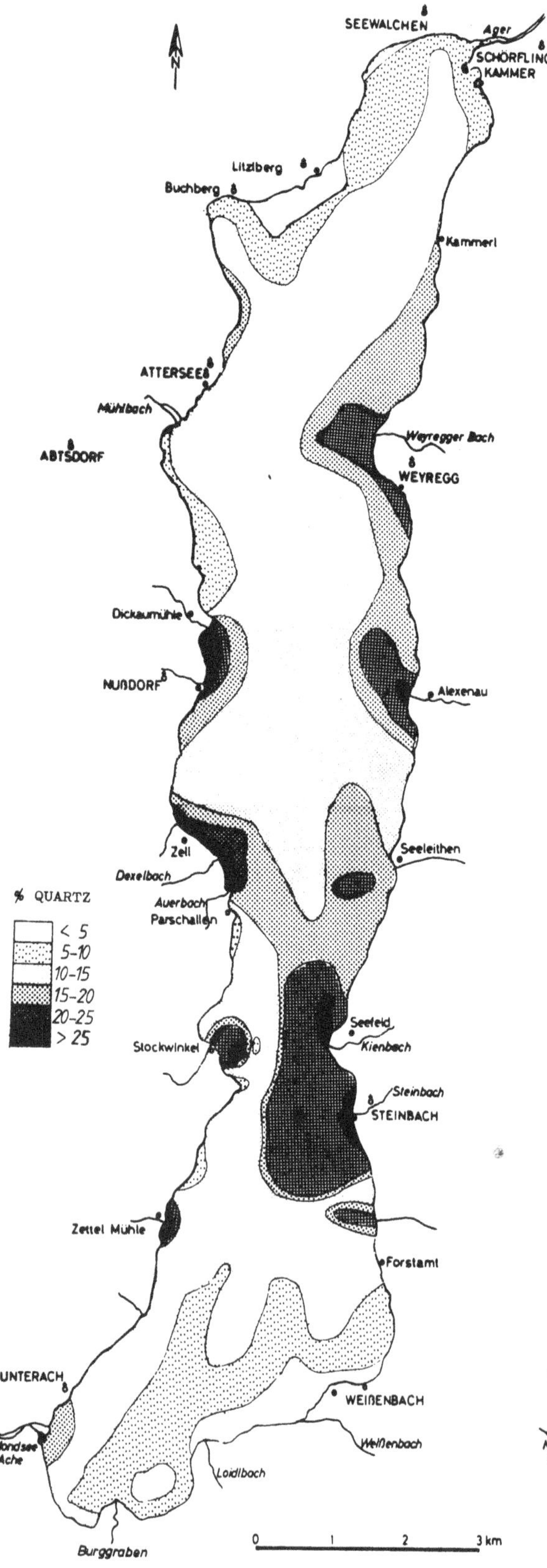
SEEWALCHEN
Ager
SCHÖRFLING
KAMMER
Litzlberg
Buchberg
Kammerl
ATTERSEE
Mühlbach
ABTSDORF
Weyregger Bach
WEYREGG
Dickaumühle
NUßDORF
Alexenau
Zell
Seeleithen
Dexelbach
Auerbach
Parschallen
% QUARTZ
< 5
5-10
10-15
15-20
20-25
> 25
Seefeld
Kienbach
Stockwinkel
Steinbach
STEINBACH
Zettel Mühle
Forstamt
UNTERACH
WEIßENBACH
Mondsee Ache
Weißenbach
Loidlbach
Burggraben
0
1
2
3 km

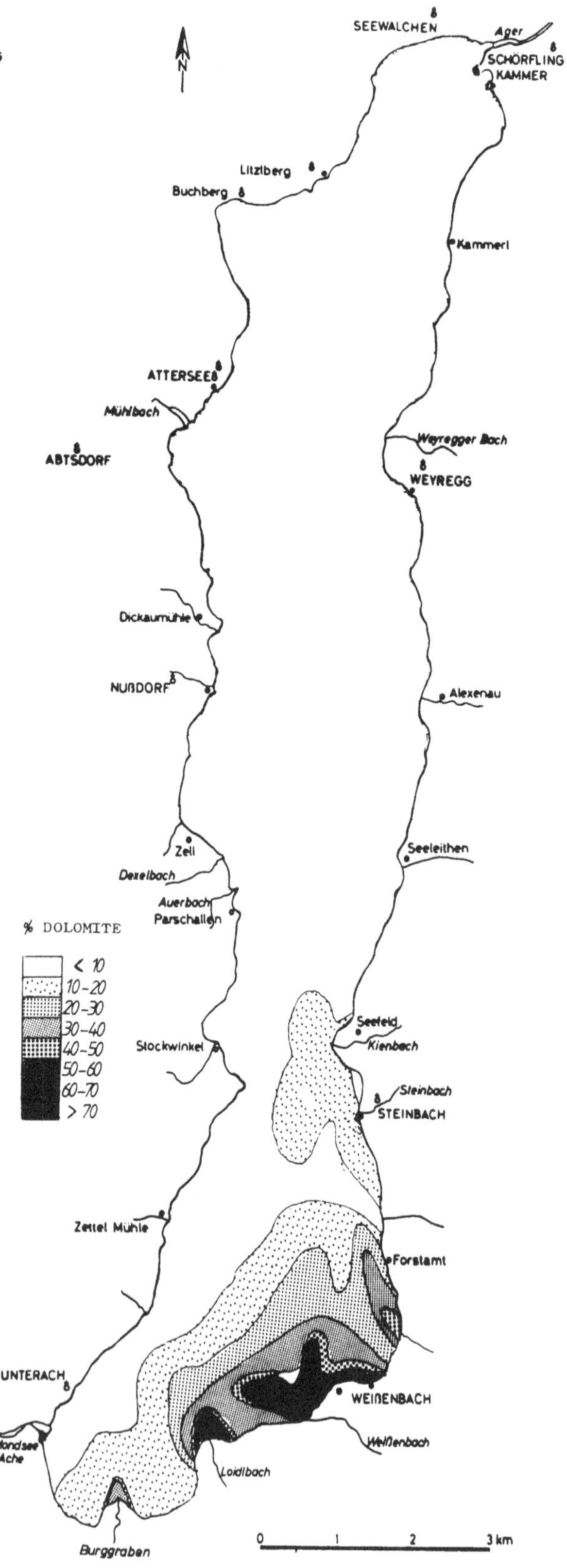
SEEWALCHEN
Ager
SCHÖRFLING
KAMMER
Litzlberg
Buchberg
Kammerl
ATTERSEE
Mühlbach
ABTSDORF
Weyregger Bach
WEYREGG
Dickaumühle
NUßDORF
Alexenau
Zell
Seeleithen
Dexelbach
Auerbach
Parschallen
% DOLOMITE
< 10
10-20
20-30
30-40
40-50
50-60
60-70
> 70
Seefeld
Kienbach
Stockwinkel
Steinbach
STEINBACH
Zettel Mühle
Forstamt
UNTERACH
WEIßENBACH
Mondsee Ache
Weißenbach
Loidlbach
Burggraben
0
1
2
3 km

The influence of the different source areas on the composition of the Attersee sediments is clearly shown by mapping the surface sediments (Fig. 5).

5. Sediment distribution mechanisms

The distribution mechanisms of the allochthonous clastics depend on the nature of the minerals (size, shape, density), the physical chemical parameters of the water-body and the lake basin morphology.

From May to October vertical differences in the water column of Attersee are caused by thermal stratification. This water layering has a considerable influence on the distribution mechanisms of the allochthonous sediments. A distinction was made between suspended sediment transported by over, inter and underflows (Fig. 6, Sturm & Matter, 1978; Lambert, 1976).

The occurrence of over and interflows depends on the temperature stratification of the waterbody whereas the occurrence of underflows and turbidity currents is not bound to temperature stratification. The sedimentary structures produced by turbidites and underflows (e.g. graded beds etc.) are very similar and cannot be traced back to their processes of origin.

6. Landslides on the eastern shore

The flysch mountains east of lake Attersee are particularly susceptible to landslides and mud flows because of their morphology and geology. They are composed of clayey to sandy sedimentary rocks and rise steeply from the shoreline of the lake.

The high frequency of mud flows into the Steinbach valley results from the presence of the overthrust zone of the Northern Calcareous Alps above the flysch zone (Janoschek, 1964). Sediment material deriving from these landslides can be found in the Attersee sediments by a well-placed core sampling (Chondrogianni & Schmidt, 1984; Strackenbrock, 1985).

7. Carbonate production and -sedimentation

The biogenic carbonates in lake Attersee are produced by two different processes: biogenic epilimnetic decalcification and biogenic benthic decalcification (Fig. 7).

7.1 Biogenic epilimnetic decalcification

During the summer months the increasing assimilation activities of blooming phytoplankton

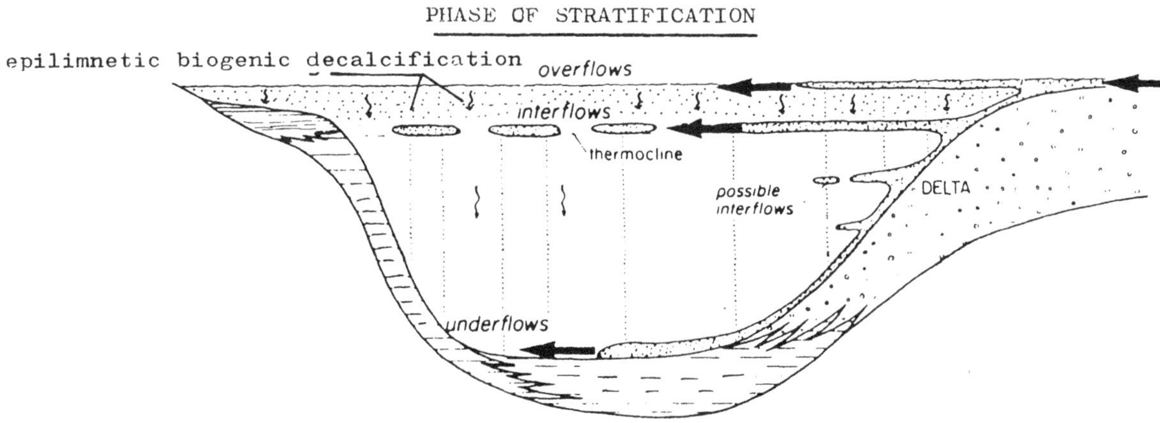

Fig. 6. Distribution mechanisms of allochthonous and autochthonous sediment particles in a lake under thermal stratification conditions (modified after Sturm & Matter, 1978).

Fig. 5. The distribution of index minerals according to the source areas is well documented within the uppermost 5 cm of the surface sediments (mapping after Niessen, 1981; Chondrogianni & Schmidt, 1984; Behbehani, 1984; Strackenbrock, 1985).

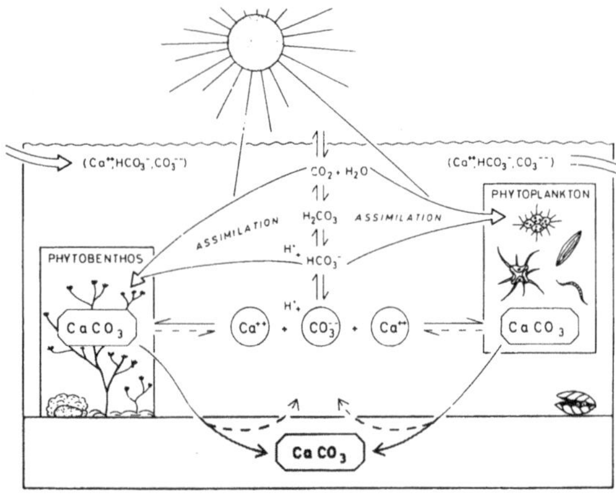

Fig. 7. Scheme of the hydrochemical/biological carbonate precipitation (from Schröder, Windolph & Schneider, 1983).

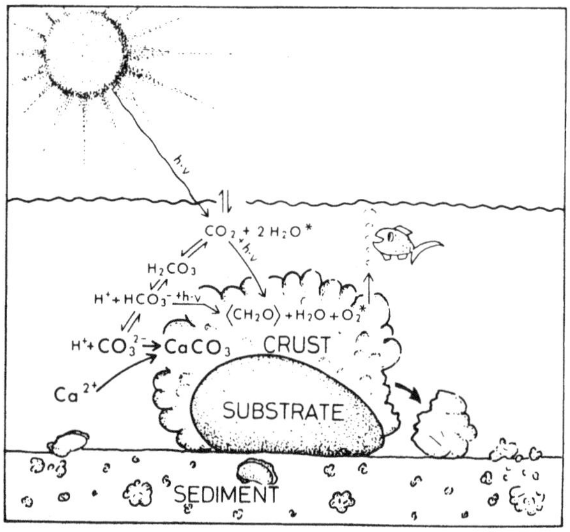

Fig. 8. Construction of lacustrine crusts and their contribution to the formation of sublittoral sediments (after Schneider & Schröder, 1980).

cause intense consumption of CO_2 and HCO_3^- in the epi- and metalimnion (0–20 m water depth). Due to these biological activities the water becomes highly oversaturated with respect to $CaCO_3$ and calcite is precipitated (Nipkow, 1920; Minder, 1922; Ruttner, 1947, 1962; Schöttle & Müller, 1968; Kelts & Hsü, 1978; Schröder & Schneider, 1980; Schröder, Windolph & Schneider, 1983). Within the Attersee area of 45.9 km^2 the epilimnetic biogenic decalcification produces about 6500 metric tons of $CaCO_3$ a year.

7.2. Biogenic benthic decalcification

The precipitation of $CaCO_3$ in the shallow sublittoral area (0 to about 12 m water depth) is due to the CO_2 and HCO_3^- assimilation of micro-, macro- and epiphytes. In the northern part of lake Attersee large parts of the sublittoral zone are covered by rounded and mostly flattened stones and gravels of morainic material. These gravels and stones (limestones, dolomites, sand and siltstones, mostly from glacial till) form the substrate for microphytes which produce calcareous crusts (Fig. 8). In between the stones, macrophytes (mostly charophytes) grow on muddy sediments (Fig. 7/8). These macrophytes form the living substrate for epiphytes which together with the macrophyte host plant produce calcareous overgrowths.

The crusts and overgrowths are destroyed by storms and waves and then contribute to the sedimentation of lake marl within the shallow sublittoral zones as well as to the sedimentation of marly muds in the deeper zones (Fig. 9, Schäfer, 1973; Schäfer & Stapf, 1978). On a sublittoral area of 1.3 km^2 benthic biogenic decalcification produces about 5000 metric tons of $CaCO_3$ a year. The whole biogenic calcium carbonate production is calculated as 11000 to 12000 metric tons a year or 250 metric tons per km^2 a year (Schröder, Windolph & Schneider, 1983).

8. Eutrophication

Most of the nutrients that reach lake Attersee are washed in by the river Mondseeache flowing from

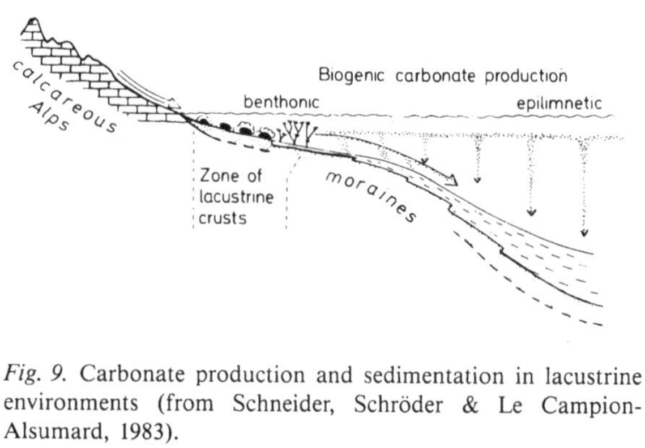

Fig. 9. Carbonate production and sedimentation in lacustrine environments (from Schneider, Schröder & Le Campion-Alsumard, 1983).

lake Mondsee into lake Attersee (more than 50% of the total phosphorus load, Müller, 1979b). Lake Mondsee was characterized by eutrophication and massive bloom of *Oscillatoria rubescens* until recently. These cyanophytes were regularly washed into the Attersee by the river Mondseeache and caused a slight eutrophication of the southern part of Attersee. Consequently an increasing development of phytoplankton occured in the southern part of lake Attersee (Müller, 1979).

The bay of Unterach in the southern basin has become slightly eutrophic (Behbehani, 1984) as indicated by:

1. lower transparency of water within the bay of Unterach,
2. massive grow of the alga *Cladophora* in front of the mouth of Mondseeache river.
3. increased growth of submersed plants within the bay.
4. poor colonization by calcifying macrophytes (Characeae),
5. decline in benthic biogenic decalcification
6. existence of Oligochaetes and other organisms within the sediments which can be used as eutrophication indicators
7. allochthonous *Oscillatoria* filaments within the sediments
8. relatively high amount of organic substances within the sediments
9. high amounts of total phosphorus within the sediments.

Some of the eutrophication indicators can be identified within the surface sediments as well as in deeper parts of sediment cores which represent the fossil equivalent. A comparison of oligotrophic and eutrophic hard water lakes shows that a remarkable development of calcareous crusts (micro- and macrophytes) does not occur in all hard water lakes (Kann, 1940). One reason could be the presence of nutrients like phosphates which can act as inhibitors to calcite nucleation as well as organic substances which can hinder the nucleation of $CaCO_3$ by complexing Ca-ions as known from the investigations of Suess (1970), Chave & Suess (1970), and Walter & Hanor (1979a, b).

Fig. 10 clearly shows the decreasing eutrophication/increasing oligotrophic trend in the Attersee from south to north. Investigations by Behbehani (1984) show that calcification of macrophytes is zero in front of the mouth of Mondsee and increases northwards. In the oligotrophic northern part of lake Attersee different types of Charophyta calcification are be found (Fig. 11).

In the southern part of the Attersee the more or less intensive calcification seems to be used as a qualitative indicator of eutrophication. The construction of a central sewage plant and a ring pipe for the Attersee and the Mondsee area should bring about an amelioration in the ecological situation. The increase of water transparency and the decrease of nutrients and of *Oscillatoria* blooms are the first indicators of success.

9. Sedimentation rates and sedimentation history

Several cores were analyzed in order to get information about sediment age and sedimentation rates. Three different dating methods were applied.

9.1. Pollen analysis

Because of the late and postglacial floral development it is possible to determine the age of sediments from three cores (Fig. 12). The following boundaries are important horizons:

a. late-glacial decrease of the non arboreal pollen, first occurence and expansion of *Juniperus* (juniper) and *Betula* (birch) 13000 B.P.

b. border between late and postglacial time 10000 B.P. (immigration of forest trees)

c. *Corylus* (hasel) peak of Boreal 8500 B.P.

d. immigration of *Fagus* (beech) and *Abies* (fir) during the Older Atlantic period 6500 B.P.

e. first (Neolithic) peak of forest clearance and immigration of *Carpinus* (hornbeam) 4700 B.P.

f. massive non arboreal pollen increase, colonization phase of the Bavarians in the 8th century A.D. and subsequently from about 1200 B.P.

Paleoclimatic indications from pollen and sedimentological evidence point to the same conclusions: About 6500 B.P. a more humid climate is indicated by the frequency of turbidites (see also Fig. 12 prequaternary pollen). After that time the frequency of pollen increases and simultaneously the frequency of turbidites with a simultaneous increase of pollen concentration occurs before the historical time of the non arboreal pollen peak (1200 B.P.)

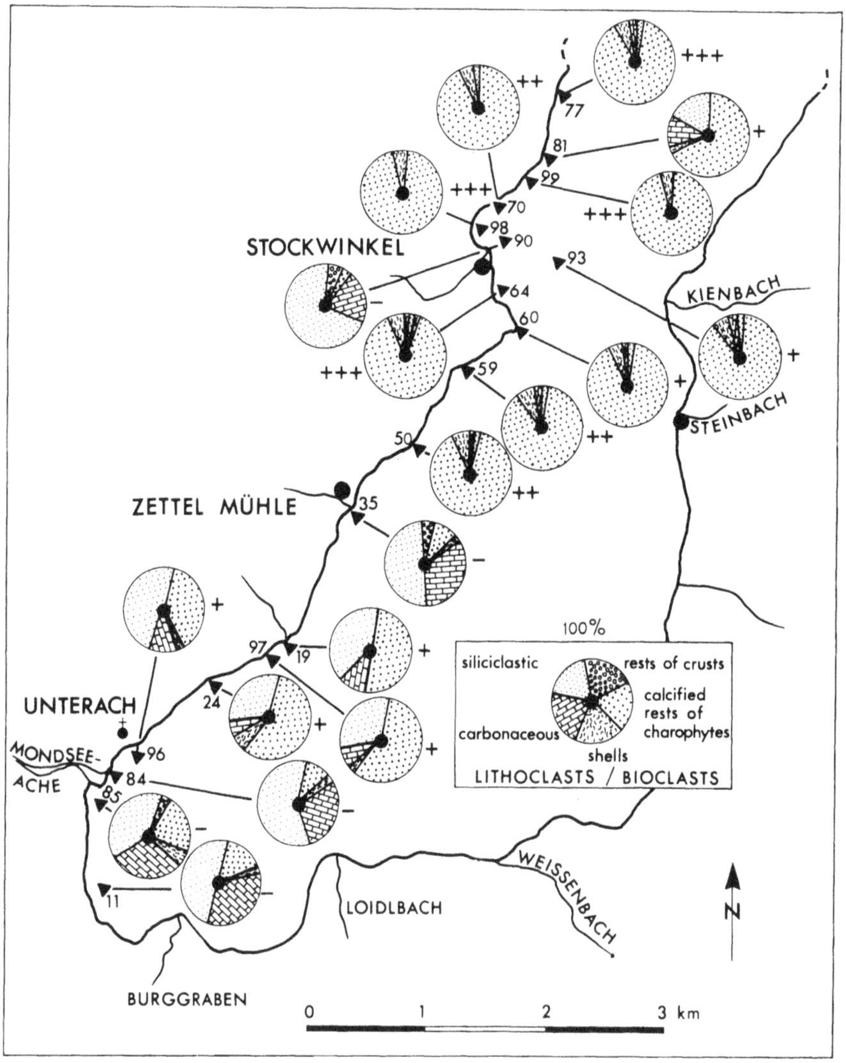

Fig. 10. Decreasing eutrophic/increasing oligotrophic trend in the Attersee from south to north indicated by northward increasing calcification of charophytes (evaluation of the sand-fraction from surface sediments, after Behbehani, 1984).

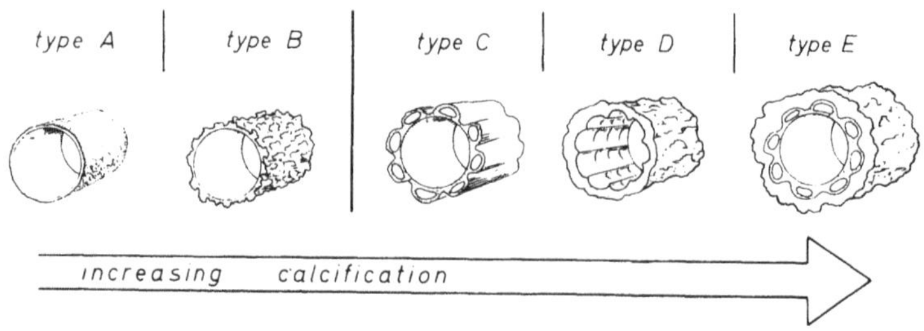

Fig. 11. Different types of macrophytes (Charophyta-) carbonate. Type A and B: unstructured stems, Type C–E: structured stems. A without epiphytes, B with epiphytes, C small epiphytic calcification, D strong epiphytic calcification with minor internal calcification, E strong epiphytic and internal calcification (from Schneider, Schröder & Le Campion-Alsumard, 1983).

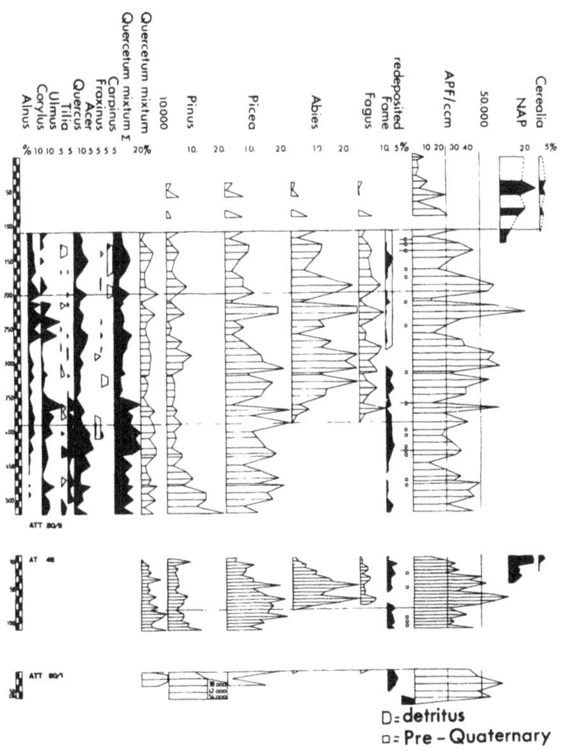

Fig. 12. Pollen analyses from 3 sediment cores (ATT 80/9, AT 48, ATT 80/1). NAP = non arboreal pollen, APF = absolute pollen frequency. For position of the cores see Fig. 17.

9.2 Heavy metal and isotope analyses

Since the industrial revolution heavy metals have become characteristic components in recent sediments. It is possible to correlate the enrichment of heavy metals in sediments with historically dated events.

Increased Zn (3.1 fold), Cd (5.6 fold) and Pb (1.9 fold) contents were observed in the upper 20 cm of the sediment column of lake Attersee (see Fig. 13). ^{210}Pb dating designates 1870–1880 as the initial stage of heavy metal pollution originating from anthropogenic sources (Fig. 13). ^{137}Cs dating designates 1960–1965 as the stage of maximum enrichment originating from the atmospheric atomic bomb tests (1963 test stop treaty; Fig. 13).

9.3 Slag enrichments

Slag particles in the lake sediments indicate coal and oil combustion by vessels on lake Attersee. A steamship line was first established in 1869 (Siakala, 1979). Between 1947 and 1958 diesel and electrically driven ships took the place of the steamboats. The slag particles occur above 20 cm sediment depth in the southern profundal zone as well as on the carbonate shelf in the North and reach their maximum concentration between 9 cm and 5 cm depth. The contents of slag particles decreases towards the sediment surface. Hence, the average sedimentation rate in the profundal zone of the Attersee amounts to about 2 mm/a within the last 100 years.

From these dates it is possible to reconstruct sedimentation rates in the Attersee:

1. The oldest documented stage in the north (13000–12000 B.P.) is characterized by small average sedimentation rates of 0.17 mm/a.
2. The average sedimentation rate is 0.3 mm/a from the beginning of the medieval phase of forest clearance (NAP rise 1200 B.P.).
3. During the last 100 years the sedimentation rate has increased to 2 mm/a.

The increasing sedimentation rate can be explained either by a changing allochthonous clastic input (e.g. increasing erosion through agricultural activity) or by a changing autochthonous carbonate production. It is also possible that both processes have accelerated.

10. Results: Sedimentary history of the last 15000 years

The postglacial history (after Würm glaciation) of the Attersee was investigated on the basis of sediment cores along a profile across the lake, sediment echography (ORE, 3.5 kHz) and by mapping the sediment types (see Figs. 14–17). Five main phases in the sedimentary history were clearly recognized by means of the sedimentological studies (see Fig. 17 and 18):

a. Würm moraines with glacially striated pebbles within glacial flour (see Fig. 17)

b. Finely banded varves formed during the alpine arctic lake stage before 13000 B.P. (Fig. 17)

c. From 13000 B.P. on: the Attersee stage with well developed epilimnetic and benthic carbonate production which is represented by higher carbonate contents (Fig. 17). Pollen Analyses (see Fig. 12) and sedimentology allow the recognition of finer subdivision into individual events. In the basin sediments and on the shelf different segments

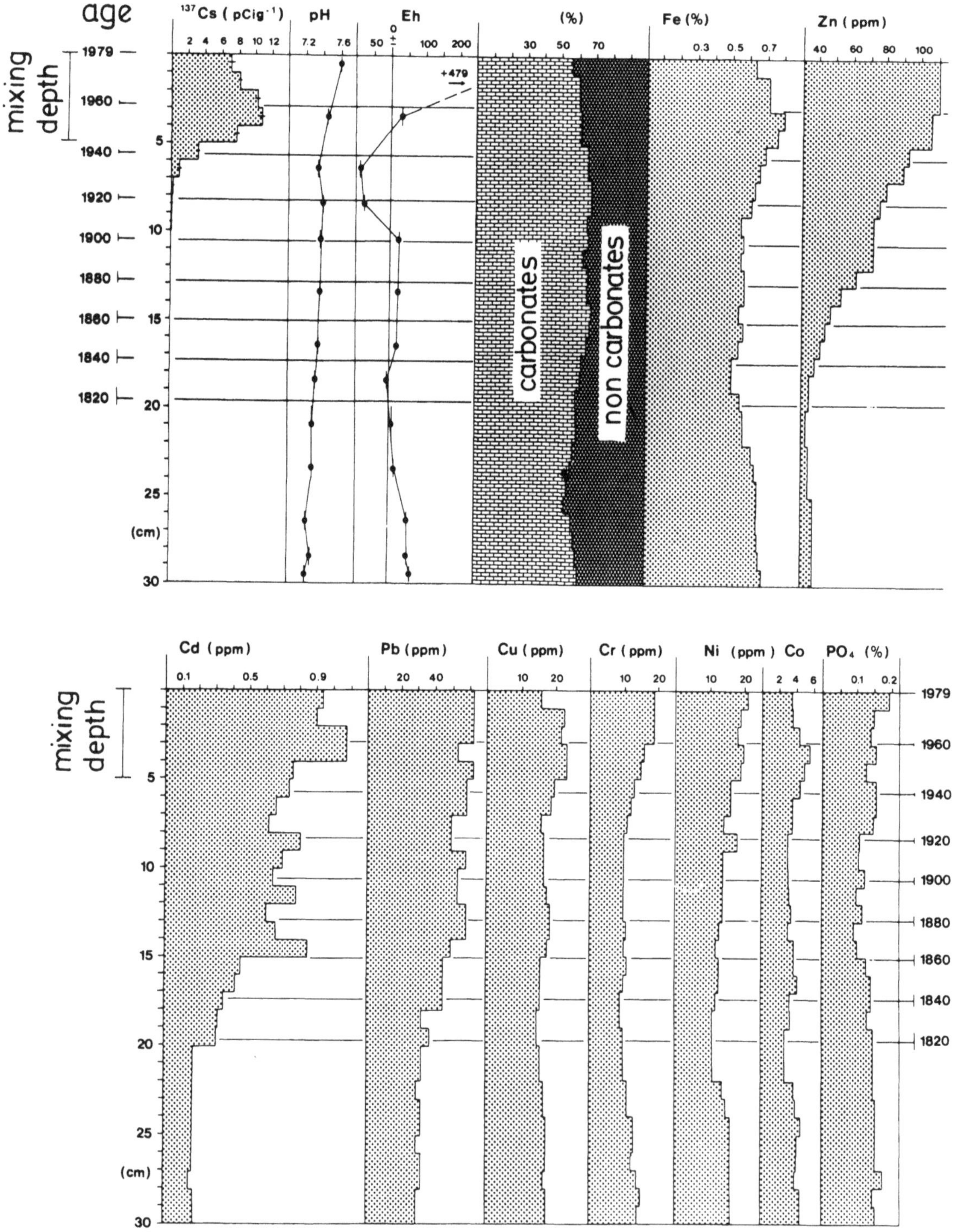

Fig. 13. Vertical distribution of carbonate content, heavy metals, ^{137}Cs and PO_4-concentrations within the core AT 1 (northwestern shelf of Attersee). pH and Eh were measured within the first hour after sampling. Time scale was calculated assuming a constant sedimentation rate (from Müller, Dominik, Schröder & Schneider, 1983).

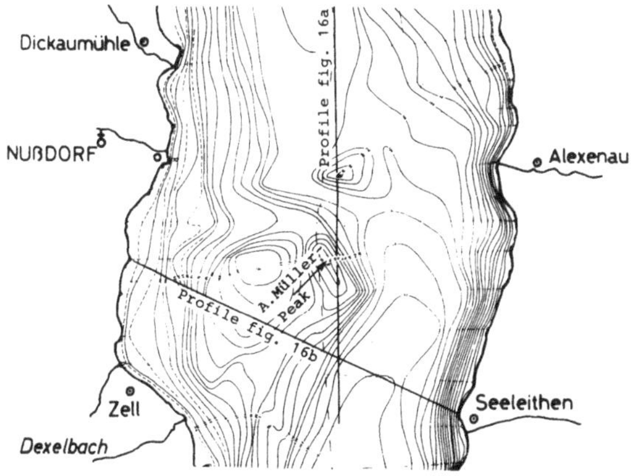

Fig. 14. Map of the central part of Attersee showing the position of the sediment echography profile of Fig. 15.

with increased or decreased turbidite activity can be identified by pollen analysis from 13000 to about 1200 B.P. (Fig. 17).

d. From 1200 B.P. on: the Attersee stage shows increased sedimentation rates and lower carbonate contents (Fig. 17) after forest clearance as a consequence of the Bavarian colonization.

e. During the last 100 years the sedimentation rate further increases and the sedimentary history can be reconstructed in more detail using heavy metal and isotope studies as well as slag particles from steamboats (see section 9.2, 9.3, Fig. 13).

The rough time scale of the first four phases is verified by pollen analyses, ^{14}C dating, sedimentology and sediment chemistry and is even macroscopically visible in sediment cores. The fifth phase of the last 100 years can be dated much more precisely by heavy metal and isotope studies (see section 6). Comparison of sediment cores from other parts of the Attersee and other lakes in the northern alpine foreland (see e.g. Müller & Schmidt, 1985) show that this rough chronology is also regionally applicable.

Acknowledgements

The investigations were supported by grants of the "Deutsche Forschungsgemeinschaft" (DFG AZ: Schn 16/8 to 12-1). We thank Prof. Dr. F. Oldfield (Liverpool) for critically reading the manuscript.

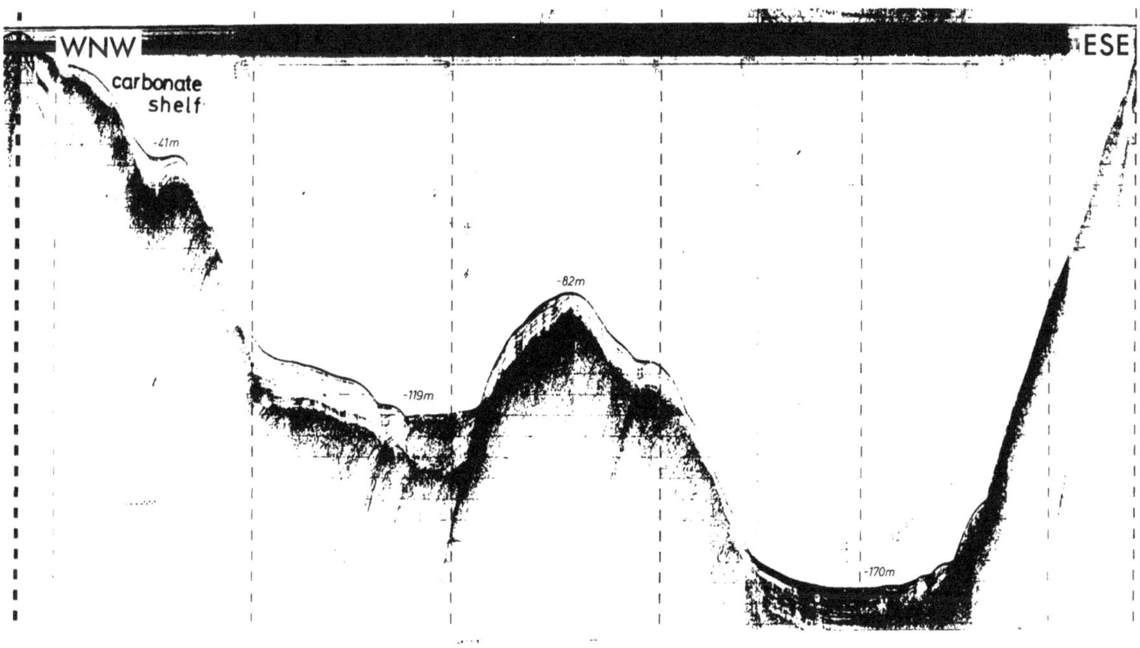

Fig. 15. Sediment echography (ORE, 3.5 KHz) showing the different sediment thickness in the basin respectively on the ridge and on the shelf region.

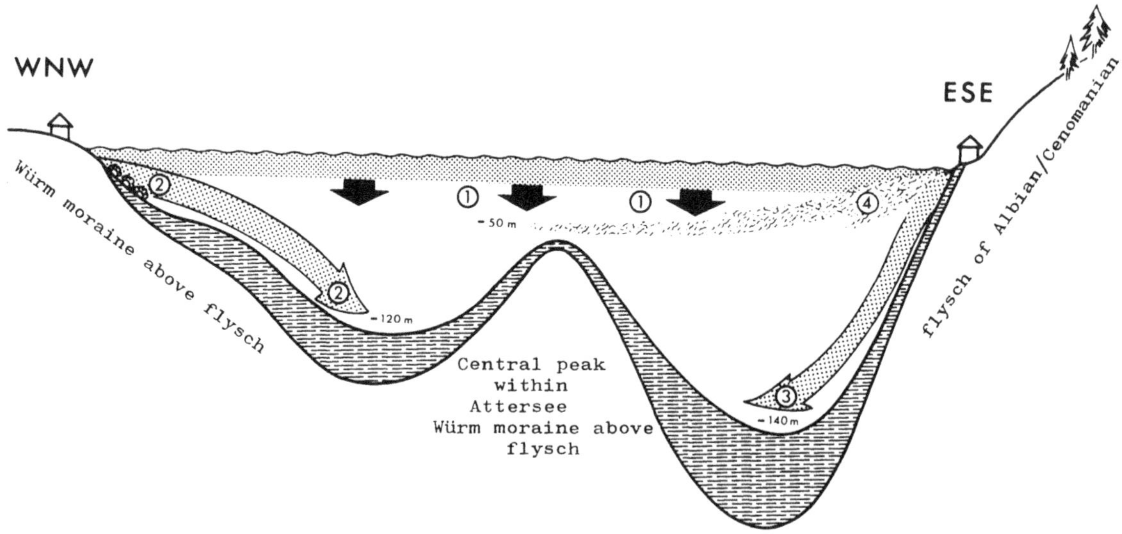

Fig. 16. Model of the different influences of the sediment contributing processes within the central part of lake Attersee.

1 $CaCO_3$ from the epilimnetic biogenic decalcification
2 $CaCO_3$ from the benthic biogenic decalcification
3 clastic input of silicates (mostly)
4 sporadic inflow of fine grained sediments during thunderstorms, heavy rain falls, snow break or through land slides

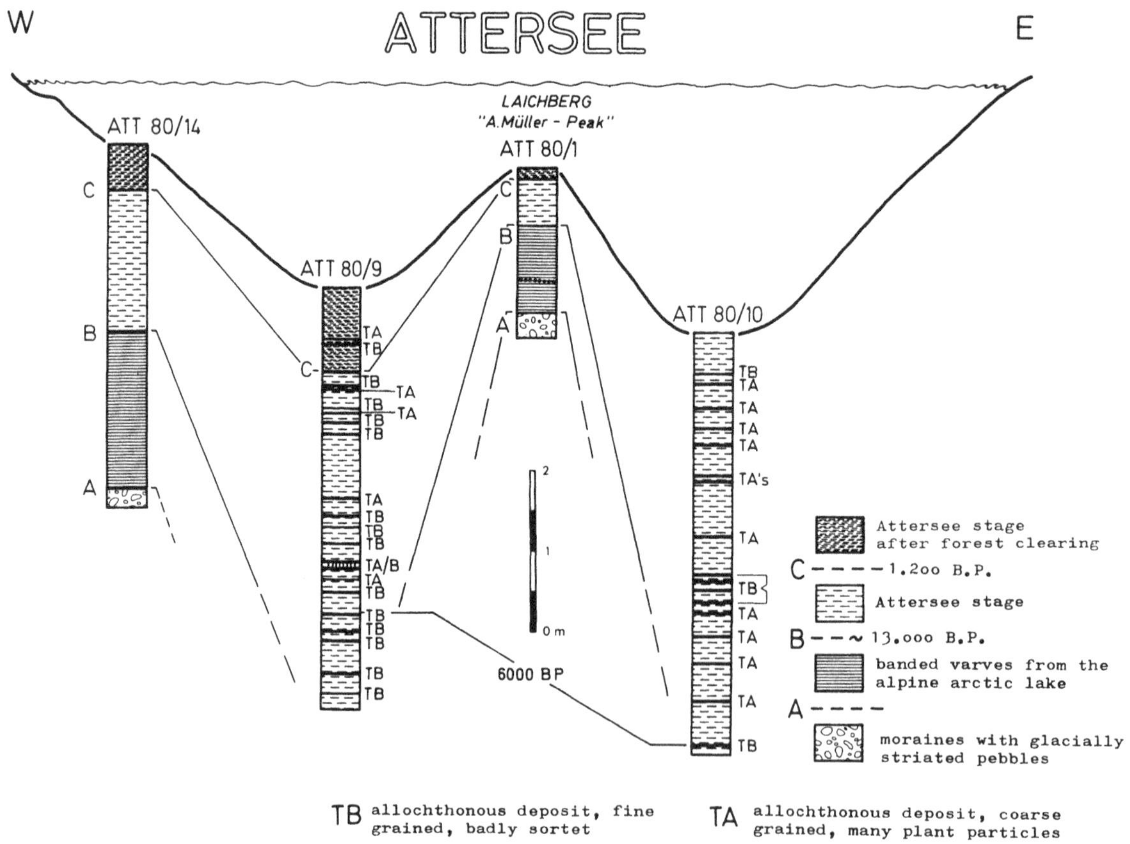

Fig. 17. 4 cores across the underwater ridge ("Armin Müller peak") from the central part of Attersee. The four main phases of sedimentary history are indicated.

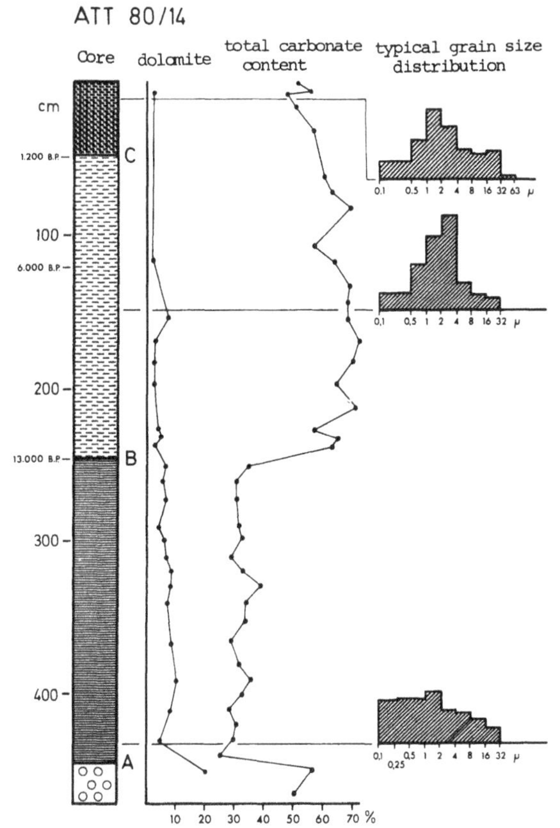

Fig. 18. Sedimentological analysis of a typical core from the shelf region (see Fig. 17). The four main phases of sedimentary history are well indicated by chemical, mineralogical and grain size distribution.

References

Behbehani, A. R., 1984. Sedimentologische Untersuchungen im südlichen Teil des Attersees (Österr. Kt. 1:25000, Bl. 64/4 Unterach, Salzkammergut, Oberösterreich). Diplome-thesis, Univ. Göttingen, 137 p.

Chave, K. E. & E. Suess, 1970. Calcium carbonate saturation in seawater: Effects of dissolved organic matter. Limnol. Oceanogr. 15: 633–637.

Chondrogianni, C. & H. Schmidt, 1984. Sedimentologische Untersuchungen im Attersee-Nordbecken (Österr. Kt. 1:25000, Bl. 65/2 Sankt Georgen im Attergau, Salzkammergut, Oberösterreich). Diplome-thesis, Univ. Göttingen, 208 p.

Del Negro, W., 1969. Bemerkungen zu den Kartierungen L. Weinberger's im Traungletschergebiet (Attersee- und Traunseebereich). Verh. Geol. B.-A., Wien, 1969: 12–15.

Golubic, S. & J. Schneider, 1979. Carbonate dissolution. In: P. A. Trudinger & D. J. Swanine, (Eds), Biogeochemical cycling of mineral-forming elements. Studies in Environment Science 3, Elsevier, Amsterdam-Oxford-New York: 107–129.

Häusler, H. & H. Schwarzenböck, 1970. Geologische Karte 1:50000, Karte 12. Amt O.Ö. Landesregierung, Abt. Wasserbau, Hydrogr. Dienst, Ing. Konsulent Dr. techn. H. Flögl, Linz, Dez. 1969.

Jagsch, A., 1977. Chemische Untersuchungen der wichtigsten Zuflüsse und des Ausrinns im Jahre 1976 und Nährstoff-Frachtberechnungen, Attersee. Vorläufige Ergebnisse des OECD-Seeneutrophierungs- und des Maß-Programms, 1978: 33–40.

Janoschek, W., 1964. Geologie der Flyschzone und der helvetischen Zone zwischen Attersee und Traunsee. Jb. Geol. B.A., Wien, 107: 161–214.

Kann, E., 1940. Ökologische Untersuchungen an Litoralalgen ostholsteinischer Seen. Arch. Hydrobiol. 37: 197–269.

Kelts, K. & K. J. Hsü, 1978. Freshwater carbonate sedimentation. In: A. Lerman (Ed.): Lakes, Chemistry, Geology, Physics. Springer, New York, Heidelberg, Berlin, 295–329.

Lambert, A., K. Kelts & N. F. Marshall, 1976. Measurements of density underflows from Walensee, Switzerland. Sedimentology 23: 87–105.

Minder, L., 1922. Über biogene Entkalkung im Zürichsee. Verh. Int. Ver. Limnol. 1: 20–32.

Moog, O., 1981. Wasserbilanzierung des Ager-Seenketten-Systems: Stand 1980. Arb. Lab. Weyregg, 5: 43–50.

Müller, G., J. Dominik, H.-G. Schröder & J. Schneider, 1983. Zur Chronologie anthropogener Schwermetalle in den Sedimenten des Attersees (Österreich). Tschermaks Min. Petr. Mitt., 31: 193–206.

Müller, G., 1979a. Grundlagendaten für Fuschlsee, Mondsee und Attersee sowie das gesamte Einzugsgebiet. Arb. Lab. Weyregg, 3: 10–14.

Müller, J. & R. Schmidt, 1985. Evolution of postglacial sedimentation in an alpine lake: Funtensee, Northern Calcareous Alps. Z. Geltscherkol. Glazialgeol., 21: 51-57.

Müllner, J. & F. Simony, 1895. Die Seen des Salzkammergutes. In: A. Penck & E. Richter, (Eds), Atlas der österreichischen Alpenseen. Hölzel, Wien.

Niessen, F., 1981. Sedimentologische Untersuchungen vor dem Weißenbach-Delta im Südbecken des Attersees. Diplome-thesis, Univ. Göttingen, 124 p.

Nipkow, F., 1920. Vorläufige Mitteilungen über Untersuchungen des Schlammabsatzes im Zürichsee. Schweiz. Z. Hydrol. 1: 100–122.

Ruttner, F., 1947. Zur Frage der Karbonatassimilation der Wasserpflanzen, 1. Teil: Die beiden Haupttypen der Kohlenstoffaufnahme. Österr. Bot. Z. 94: 265–294.

Ruttner, F., 1962. Grundriß der Limnologie. W. De Gruyter, Berlin, 332 pp.

Schadler, J., 1959. Zur Geologie der Salzkammergutseen. In: Flüsse und Bäche im Salzkammergut. Schr. Österr. Fischereiverb. Salzburg, 2: 36–54.

Schäfer, A., 1973. Zur Entstehung von Seekreide Untersuchungen am Untersee (Bodensee). N. Jb. Geol. Pal./Mh. 4: 216–230.

Schäfer, A. & K. R. G. Stapf, 1978. Permian Saar-Nahe Basin and recent Lake Constance (Germany): two environments of lacustrine algal carbonates. Spec. Publ. int. Ass. Sediment, 2: 83–107.

Schneider, J., H. G. Schröder & T. Le Campion-Alsumard, 1983. Algal micro-reefs coated grains from freshwater environments. In: T. M. Peryt, (Ed.), Coated Grains. Springer Verlag Berlin, Heidelberg: 284–298.

Schneider, J. & H. G. Schröder, 1980. Calcification des cyanophycées et leur contribution à la genèse des sédiments calcaires. In: L. Humbert, (Ed.), Cristallisation-Déformation-Dissolution des carbonates, 17./18. Nov. 1980, Bordeaux: 421–428.

Schöttle, M. & G. Müller, 1968. Recent carbonate sedimentation in the Gnadensee (Lake Constance) Germany. In: G. Müller & Friedman (Eds), Recent developments in carbonate sedimentology in Central Europe. Springer, Berlin, 148–156.

Schröder, H. G., 1982. Biogene benthische Entkalkung als Beitrag zur Genese limnischer Sedimente. Beispiel: Attersee (Salzkammergut, Oberösterreich). Diss. Univ. Göttingen, 178 p.

Schröder, H. G., H. Windolph & J. Schneider, 1983. Bilanzierung der biogenen Karbonatproduktion eines oligotrophen Sees (Attersee, Salzkammergut – Österreich). Arch. Hydrobiol., 97: 3–17.

Schröder, H. G. & J. Schneider, 1980. Hydrochemische Untersuchungen zum Karbonatkreislauf und zur Sedimentationsgeschichte des Attersees. Arb. Lab. Weyregg, 4: 235–257.

Siakala, H., 1979. Wirschaftsgeographie der Randgemeinden des Attersee. Diplome-thesis Geogr. Inst. Wirtsch. Univ. Wien.

Spengler, E. & J. Pia, 1924. Geologischer Führer durch die Salzburger Alpen und das Salzkammergut. Samml. geol. Führer, 26, Borntraeger, Berlin.

Strackenbrock, I., 1985. Sedimentologische Untersuchungen im mittleren Teil des Attersees. (Österr. Kt. 1:25000, Bl. 64/4 Unterach, Salzkammergut, Oberösterreich). Diplome-thesis Univ. Göttingen, 89 p.

Sturm, M. & A. Matter, 1978. Turbidites and varves in Lake Brienz (Switzerland): deposition of clastic detritus by density currents. In: A. Matter & M. E. Tucker, (Eds), Modern and Ancient Lake Sediments. Spec. Publ. int. Ass. Sediment., 2: 147–168.

Suess, E., 1970. Interaction of organic compounds with calcium carbonate. 1. Association phenomena and geochemical implications. Geochim. Cosmochim. Acta, 34: 157–168.

Walter, L. M. & J. S. Hanor, 1979a. Orthophosphate: Effect on the relative stability of aragonite and magnesian calcite during early diagenesis. J. Sed. Petrol. 49: 937–944.

Walter, L. M. & J. S. Hanor, 1979b. Effect of orthophosphate on the dissolution kinetics of biogenic magnesian calcite. Geochim. Cosmochim. Acta, 43: 1377–1385.

Weinberger, L., 1955. Beiträge zur Pleistozänforschung in Österreich, Exkursionen durch das österreichische Salzachgletschergebiet und die Moränen des Irr- und Atterseezweiges des Traungletschers. Verh. Geol. B.-A., Wien, Sonderheft D: 7–40.

Accepted 18 March 1986.

Hydrobiologia 143: 247–254, (1986)

Diatom distribution and diatom inferred pH in the sediment of four alpine lakes

Klaus Arzet[1], Christian Steinberg[1], Roland Psenner[2] & Norbert Schulz[3]
[1]*Bayerisches Landesamt f. Wasserwirtschaft, Lazarettstr. 67, 8000 München 19, FRG*
[2]*University Innsbruck, Dept. Limnology, Technikerstr. 13, 6020 Innsbruck, Austria*
[3]*Kärntner Institut f. Seenforschung, Flatschacherstr. 70, 9010 Klagenfurt, Austria*

Keywords: diatoms, sediment, lake acidification, high altitude lakes, *Melosira* species

Abstract

The diatom distribution in the sediment of four alpine lakes in the central alpine area (Ötztaler Alpen, Tyrolia, and Mölltal, Carinthia) has been analysed in order to assess possible influences of atmospheric pollutants on remote lakes. The lakes are located more than 2000 m above sea level on poorly buffered bedrock (granite, gneiss and mica schist) and are therefore susceptible to acidification by acid deposition. The pH history of the lakes has been calculated by using Index B from Renberg & Hellberg (1982), which has been adjusted to central European conditions and applied in this region for the first time. The results indicate that there have been no pH decreases in three of the lakes, while in one lake diatom composition and distribution change significantly towards the sediment surface. But taxonomical problems and difficulties in the pH classification of the rarely found diatoms complicate the interpretation of the results. Important taxa will be discussed in view of their ecological significance.

Introduction

Acidification of lakes by atmospheric pollutants has become a widespread problem in Europe as well as in North America (Dickson, 1981; Arzet *et al.*, 1986; Davis *et al.*, 1983). Little is known so far about the situation in poorly buffered, high altitude lakes in the central alpine area of Europe. Some studies have reported a high sensitivity of those lakes located on bedrock with poor buffering capacities towards acidic precipitation (Stumm & Righetti, 1982; Mosello, 1984). However these studies only gave a first insight on the water chemistry of high alpine mountain lakes on a regional basis. Most lakes have never been sampled before and available data for some lakes is difficult to compare with the present data, because of the use of different methods. Since information about these lakes is lacking, we have studied the sediment cores of four lakes with respect to their diatom composition in order to assess possible, environmental impacts. Diatoms are well known for their sensitivity towards water quality changes. In recent years paleolimnologists have successfully used the fossil remains of diatoms in lake sediments as indicators of recent pH changes (Renberg & Hellberg, 1982). This study was conducted to examine, if diatoms in high altitude lakes refer to the same environmental conditions as at lower elevations and if there is evidence for recent acidification of these selected lakes.

Study sites

We have sampled four lakes in the central alpine area (Fig. 1): Oberer Plenderlesee and Schwarzsee ob Sölden in the 'Ötztaler Alpen', Tyrolia, and Gippersee and Goaßelesee in Mölltal, Carinthia. The water chemistry data of the lakes are presented in Table 1. All lakes are located more than 2000 m above sea level. Since soils and vegetation at this elevation are poorly developed in comparison to lower sites, the chemical composition of the

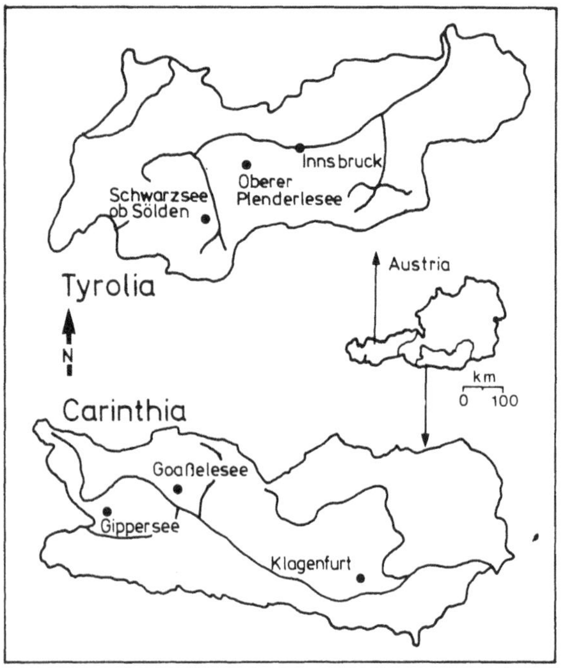

Fig. 1. Map of the study sites.

Table 1. Present water chemistry data for the lakes studied.

	SOS	OPL	Go	Gi
conductivity (μs_{20} cm^{-1})	11.2	24.3	–	–
pH	5.2–5.38	6.98	6.28	6.2
alkalinity (μeq l^{-1})	13	114	90	150
Ca	50.9	133.7	120	170
Mg	15.6	36.2	16.7	33.0
Na	10.0	19.6	23.9	20.4
K	3.3	27.0	19.0	4.9
NH_4	2.9	1.4	1.4	1.9
H^+	4.2	0.1	–	–
Al	2	0	–	–
SO_4	62.4	101.9	74.2	–
NO_3	11.1	27.5	3.7	2.7
Cl	10.7	4.2	4.2	–
HCO_3	17	114	90	150
O_2 (mg l^{-1})	10.0		8.98	8.95

SOS = Schwarzsee ob Sölder (2792 m), OPL = Oberer Plenderlesee (2344 m), Go = Goaselesee (2145 m), Gi = Gippersee (2541 m).

lakewater is probably mainly influenced by the geology of the catchments. The geological underground in the different lake areas is composed of granite, gneiss and mica schist.

Material and methods

The sediment cores were taken with a Kajak sampler at the deepest parts of the lakes and cutted into 1 cm sections by upward extrusion with a specially prepared sediment cutter immediately after sampling. Subsamples of these sediment sections were prepared for diatom analyses by boiling in 30% H_2O_2 to remove organic matter. The cleaned diatom valves were pipetted on cover slips and mounted in Naphrax. 400 to 600 valves were counted per slide. Diatom taxa were classified into pH groups (acidobiontic, acidophilous, circumneutral and alkalipilous) according to their ecological preference (Hustedt, 1939). The ecological classification of the different species was taken from the commonly used literature. The diatom inferred pH values in the sediment cores were obtained by using a linear regression equation based on 33 sediment surface diatom assemblages and surface water pH values (see Steinberg *et al.*, 1984). The pH of the lake water then can be expressed as a function of the diatom assemblages (Renberg & Hellberg, 1982):

$$pH = 6.25 - 0.76 \times \log \text{Ind B}$$
$$r = 0.95 \quad S_e = 0.26$$

Sediment dating was carried out by using ^{137}Cs fall-out as a stratigraphic marker.

Results

The distribution of ^{137}Cs in the sediment of the lakes does not yield very much information on the dating of the cores. Unfortunately all lakes except for Goaßelesee show a surface maximum in ^{137}Cs concentration (Fig. 2). The ^{137}Cs peak is generally referred to the 1963 maximum fall-out. Goaßelesee has a peak in 2 cm and seems to have a little higher rate of sediment accumulation. More information

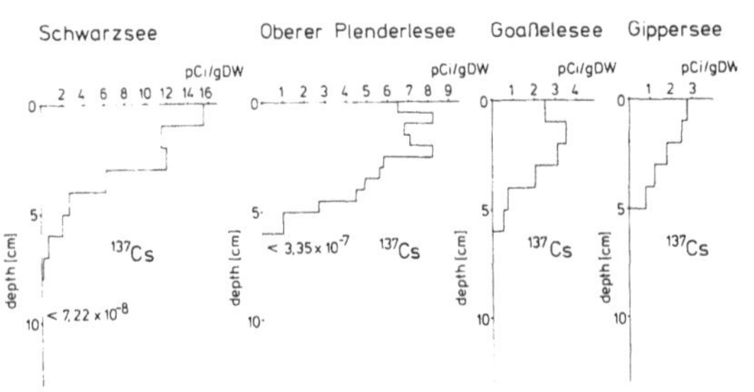

Fig. 2. ^{137}Cs profiles in the sediment of the lakes.

can be obtained by assigning a 1954 date to the first significant increase in ^{137}Cs content. Surface peak maxima are possibly due to very low rates of sediment accumulation. The presence of ^{137}Cs in lower sediment layers indicates a downward diffusion into the sediment. Similar ^{137}Cs profiles have been reported by Davis *et al.* (1984) from lakes in Norway and New England (U.S.). It is planned to perform also ^{210}Pb dating, which is presently the most favoured method for dating sediment cores.

The results of the diatom stratigraphy are presented in Figs. 3 to 6. Oberer Plenderlesee has the highest measured pH of the four sampled lakes (Table 1), which is also reflected by the diatom

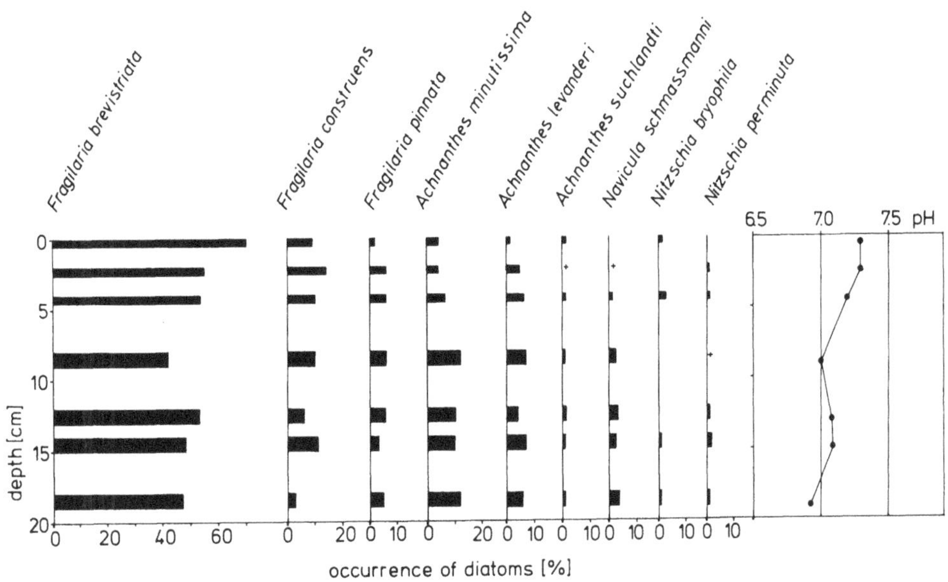

Fig. 3. Relative frequency distribution of diatoms in the sediment of Oberer Plenderlesee.

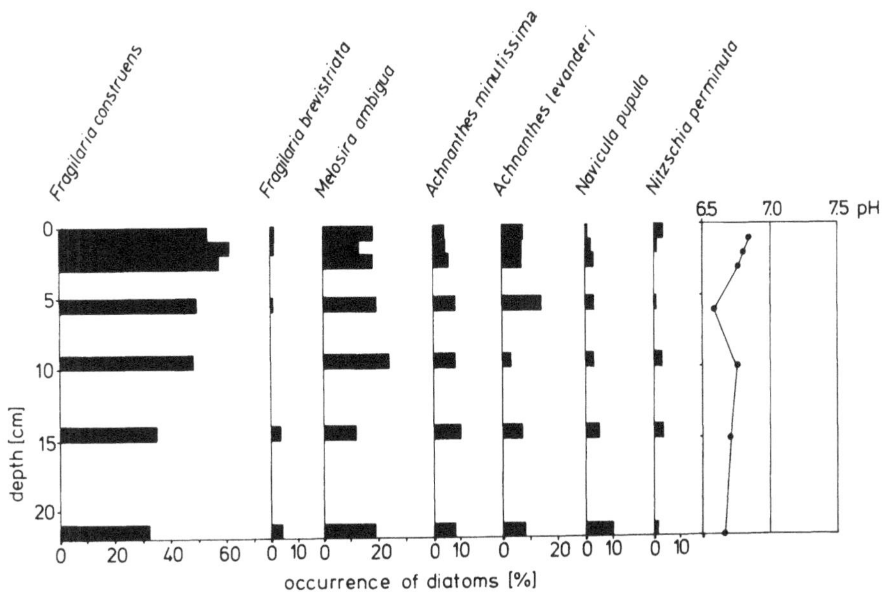

Fig. 4. Relative frequency distribution of diatoms in the sediment of Gippersee.

composition (Fig. 3). The most important diatoms throughout the entire core are the alkaliphilous species *Fragilaria brevistriata* Grun., *Fragilaria pinnata* Ehr. and *Fragilaria construens* (Ehr.) Grun. Other less important taxa include *Achnanthes minutissima* Kütz., *Achnanthes levanderi* Hust., *Achnanthes suchlandti* Hust. and *Navicula schmassmanni* Hust., which seem to prefer circumneutral to slightly acidic waters. Hustedt (1943) has observed similar diatom compositions in poorly buffered, high altitude lakes of the 'Davoser Alpen' in Switzerland. In bottom samples the diatom inferred pH varies between 6.9 and 7.1, while towards the surface it increases up to 7.3, which is due to the

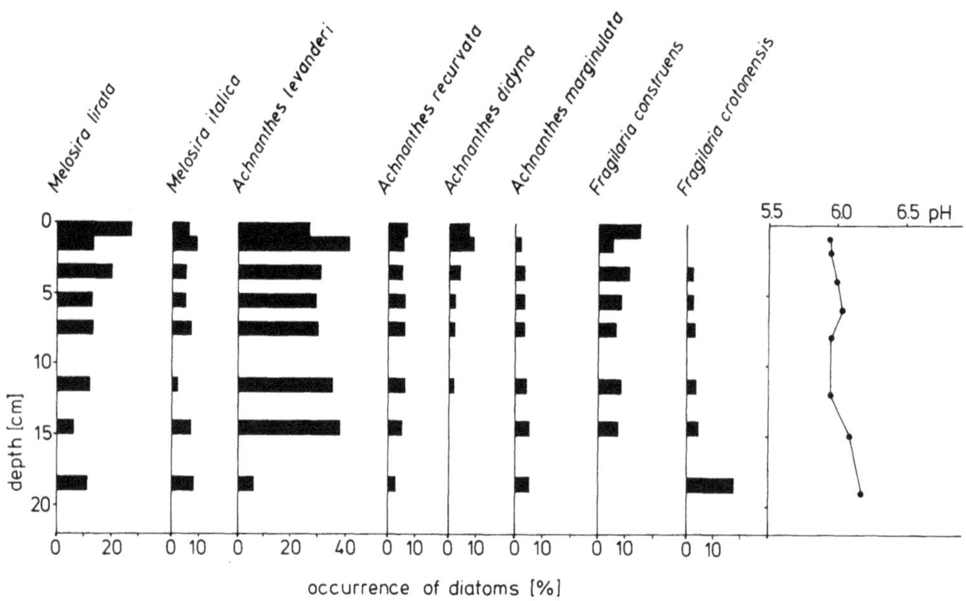

Fig. 5. Relative frequency distribution of diatoms in the sediment of Goaßelesee.

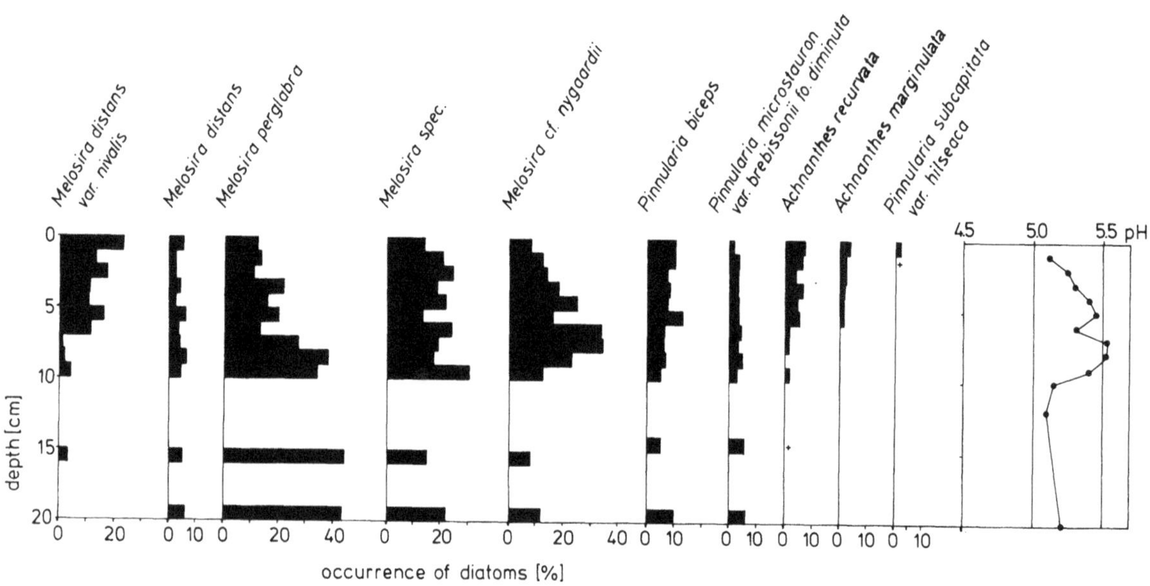

Fig. 6. Relative frequency distribution of diatoms in the sediment of Schwarzsee ob Sölden.

percentage increase of *F. brevistriata*. The calculated pH values based on Index B are in good agreement with the present measured pH values of the lake. The pH increase near the sediment surface is within the noise of the curve, but might be related to slight eutrophication by increased atmospheric nutrient loading (see sediment chemistry by Steinberg *et al.*, 1986). The results indicate pretty stable conditions as regards the diatom population, and influences of lake acidification have not been detected.

The lakewater of Gippersee has a similar alkalinity to Oberer Plenderlesee, while the pH of 6.2 was comparably low though measured only once on the occasion of the sediment coring. The diatom flora indicates 0.5 higher pH values, which corresponds well with the present alkalinity. More regular pH measurements throughout the year would probably result in a higher average pH. The diatom assemblages also resemble to the above discussed lake (Fig. 4). In Gippersee the dominant diatom species is *F. construens*, besides *Melosira ambigua* O. Müller, which is circumneutral. *A. minutissima* and *A. levanderi* are also moderately abundant. Moreover *Navicula pupula* Kütz. and *Nitzschia perminuta* Grun. are worth mentioning. The calculated pH values fluctuate between 6.6 and 6.8 throughout the core and deviate 0.4 to 0.6 units from the actual measured pH. Again the diatom inferred pH indicates a slight pH increase towards the sediment surface. No indications have been found for decreasing pH values.

In Goaßelesee the diatom composition is quite different from that in the above mentioned lakes (Fig. 5). The present measured pH is 6.2 and alkalinity is about 90 μeq/l. The most important diatoms in the sediment are *A. levanderi, M. lirata* (Ehr.) Kütz. and *Melosira italica* (Ehr.) Ralfs. While *M. lirata* is an acidophilous species, the other two are considered to be circumneutral. In addition *Achnanthes recurvata* Hust., *Achnanthes didyma* Hust. and *Achnanthes marginulata* Grun., all known as acidophilous taxa, are moderately abundant. *F. construens* is only present in low percentages. In the bottom sample *Fragilaria crotonensis* Kitton is the dominant species, but reaches low percentages in all upper samples. The diatom inferred pH values vary between 6.3 at the bottom and 5.9 at the surface. This pH decline is certainly within the standard deviation of the method and a lake acidification trend cannot be postulated, however the pH trend in this lake is negative. Moreover this possible pH decrease would have taken place over a longer time period and therefore would be probably due to natural processes in the catchment (i.e. leaching of soils, low weathering of bedrock). The diatom composition indicates slightly more acidic conditions (lower alkalinity), in comparison to the former lakes, which is also documented by the water chemistry analyses (Table 1).

Schwarzsee is the most acidic lake of all the studied lakes with a pH of 5.2 to 5.4 and very low alkalinity (Table 1). This is also reflected by the diatom composition, which is totally different to the other lakes (Fig. 6). The most dominant diatom genus is *Melosira* with a large variety of seldom reported *Melosira* species. The taxonomy is based on studies conducted by Florin (1981) and Camburn & Kingston (1986). In the bottom samples *Melosira perglabra* Østrup is the most abundant species (Plate 1, Figs. 4 to 6), besides *Melosira nygaardii* Camburn (Plate 1, Fig. 7), *Melosira distans* (Ehr.) Kütz. and *Melosira* spec. (Plate 1, Figs. 8, 9). Moreover *Pinnularia biceps* Greg. and *Pinnularia microstauron* var. *brebissonii* f. *diminuta* Grun. are represented. Towards the sediment surface *M. perglabra* and *M. nygaardii* decrease, while *M. distans* var. *nivalis* (W. Smith) Kirchner becomes the most frequent species (Plate 1, Figs. 1 to 4). Other taxa that increase towards the surface include *A. recurvata, A. marginulata* and *Pinnularia subcapitata* var. *hilseana* (Janisch) O. Müller. This shift in the diatom succession of the core is very difficult to interpret, since most of the important species are ecologically insufficiently known. Most ecological knowledge about certain species derives from studies conducted by Jørgensen (1948), Nygaard (1956) and more recently by Camburn & Kingston (1986). *M. perglabra* and *M. distans* var. *nivalis* are both considered to be either acidophilic or acidobiontic. Jørgensen reported (1948) the occurrence of *M. perglabra* from lakes in Denmark with a pH between 5.0 to 6.3. Nygaard (1956) has found it in the sediment and in an plankton sample of Lake Gribso, which had a pH of 3.9 to 6.5. Recently Camburn & Kingston (1986) have discovered it in six lakes with a pH between 4.4 to 6.1, where it commonly occurred. The ecological knowledge about *M. distans* var. *nivalis* is even less complete. Camburn & Kingston (1986) have found it only in

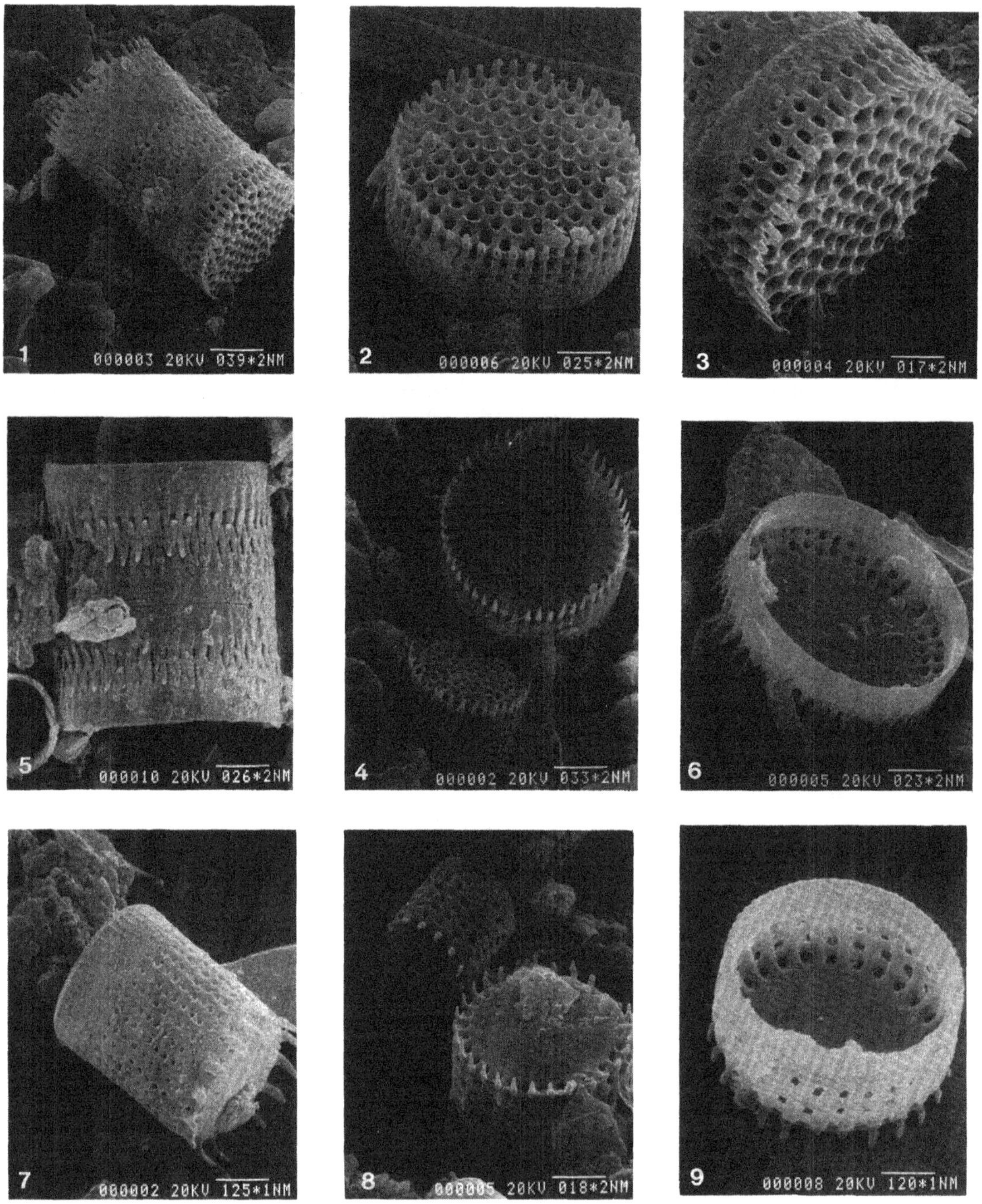

Plate 1. Scanning electronic micrographs. *Fig. 1. Melosira distans* var. *nivalis* (W. Smith) Kirchner, showing girdle view with two elongate punctae in one striae, ×3800. *Fig. 2. M. distans* var. *nivalis,* showing valve face with irregularly shaped areolae, ×6800. *Fig. 3. M. distans* var. *nivalis,* showing girdle view and valve face, ×8400. *Fig. 4. M. perglabra* Østrup, showing valve face with narrow band of areolae at the periphery of the valve face and small *M. distands* var. *nivalis* ×4500. *Fig. 5. M. perglabra,* girdle view, with narrow band of areolae extending onto valve mantle, ×5600. *Fig. 6. M. perglabra*, showing inside of valve, ×6400. *Fig. 7. M. nygaardii* Camburn, showing valve mantle with striae parallel to the pervalvar axis and row of spines around the circumference of the valve face, ×12000. *Fig. 8. M.* spec., showing valve mantle with fairly large areolae (2 to 6 in one striae) and a row of spines around the circumference of the valve face, ×7900. *Fig. 9. M.* spec., showing inside of valve, ×12000.

McNearney Lake (pH 4.37, alkalinity: −41 μeq/l) during their study, which was conducted in conjunction with a broadly interdisciplinary paleolimnological investigation of recent lake acidification (PIRLA) in five acid sensitive regions of North America. There it was associated with *M. perglabra*. In earlier publications this species was considered as a variation of *M. distans,* although the varietal epithet has not been widely used in recent literature. *M. nygaardii* is a new species described by Camburn & Kingston, which apparently is widespread in the U.S. as well than in Europe. Water chemistry data from the type locality, Sand Lake, Michigan, indicate a pH of 6.37 and total alkalinity of 37 μeq/l. This species probably has been misidentified or overlooked in the past. Based on the literature and our own experience we have classified *M. perglabra* and *M. distans* var. *nivalis* as acidophilous and *M. nygaardii* as circumneutral species in our calculations. *M. distans* var. *nivalis* possibly is more tolerant towards acidic conditions. Chemical sediment stratigraphy of this lake gives evidence for the input of atmospheric pollutants and increase of several heavy metals towards the sediment surface (from 5 to 8 cm upwards) occur at the same time as the observed diatom changes (see Steinberg *et al.*, 1986). However calculated pH values on assumed pH categories failed to give a real pH trend (Fig. 6).

Conclusions

Our results substantiate the application of the sediment inferred pH on the basis of subfossil diatom assemblages in high altitude, alpine lakes. Hustedt's concept (1943) for the pH dependancy of diatoms has been confirmed, while high elevation and low temperature of the study sites did not seem to affect the diatom composition and distribution primarily. Our data agree very well with the findings of Hustedt in high, alpine lakes. Almost no planktonic forms were present. No significant pH changes have been found in all the lakes, while in Schwarzsee changes in the diatom distribution might refer to decreasing pH. But the interpretation of the diatom data of this lake is complicated by the poor knowledge of the taxonomy and ecology of the most abundant diatoms. Recent pH-measurements of the precipitation in high altitude, alpine regions in Tyrolia, Austria, indicate that the rain water at high altitudes is less acidic than at lower elevations (pH 5.2, R. Psenner, personal communication). This fact possibly could explain, why high alpine lakes in this region so far do not show any significant signs of acidification.

Acknowledgements

This study was supported by the Umweltbundesamt, Berlin, F.R.G., UFO-Plan nr. Wasser 10204333. We would like to thank Brigitte Heindel for technical assistance. Special thanks are due to Rolf Klee for taking the Scanning Electronic Micrographs.

References

Arzet, K., D. Krause-Dellin & C. Steinberg, 1986. Acidification of four lakes in the Federal Republic of Germany as reflected by diatom assemblages, cladoceran remains and sediment chemistry. In: J. Smol, R. Battarbee, R. B. Davis & J. Merilainen (eds), Diatoms and Lake Acidity. W. Junk, The Hague, The Netherlands: 227–250.

Camburn, K. E. & J. E. Kingston, 1986. The genus *Melosira* from soft-water lakes with special reference to Northern Michigan, Wisconsin, and Minnesota. In: J. Smol, R. Battarbee, R. B. Davis & J. Merilainen (eds), Diatoms and Lake Acidity. W. Junk, The Hague, The Netherlands: 17–34.

Davis, R. B., S. A. Norton, C. T. Hess & D. F. Brakke, 1983. Paleolimnological reconstruction of the effects of atmospheric deposition of acids and heavy metals on the chemistry and biology of lakes in New England and Norway. Hydrobiologia 103: 113–124.

Davis, R. B., C. T. Hess, D. Norton, W. Hanson, K. D. Hoagland & D. S. Anderson, 1984. ^{137}Cs and ^{210}Pb dating of sediments from soft-water lakes in New England and Scandinavia. A failure of ^{137}Cs dating. Chemical Geology 44: 152–185.

Dickson, W., 1981. Freshwater acidification in Scandinavia and Europe: an overview. In: R. H. Fazzolare & C. B. Smith (eds), Beyond the energy crisis – Opportunity and Challenge, New York.

Florin, M., 1981. The taxonomy of some *Melosira* species. A comparative morphological study. II. In: R. Ross (ed.), Proc. 6th Symposium on Recent and Fossil Diatoms, Budapest, September 1–5, 1980, pp. 43–74.

Hustedt, F., 1939. Systematische und ökologische Untersuchungen über die Diatomeenflora von Java, Bali und Sumatra. Arch. Hydrobiol. Suppl. 15: 638–790, 274–394.

Hustedt, F., 1943. Die Diatomeenflora einiger Hochgebirgsseen der Landschaft Davos in den Schweizer Alpen. Int. Revue Ges. Hydrobiol. 43: 124–197, 235–280.

Jørgensen, E. G., 1948. Diatom communities in some Danish lakes and ponds. Danske Videnskabernes Selskab Biologiske Skrifter, Bind V, Nr. 2: 1–140.

Mosello, R., 1984. Hydrochemistry of high altitude alpine lakes. Schweiz. Z. Hydrol. 46/1: 86–99.

Nygaard, G., 1956. Ancient and recent flora of diatoms and chrysophyceae in Lake Gribsø. Folia limnol. scand. 8: 32–99.

Renberg, I. & T. Hellberg, 1982. The pH history of lakes in southwestern Sweden, as calculated from the subfossil diatom flora of the sediments. Ambio 11: 30–33.

Steinberg, C., K. Arzet & D. Krause-Dellin, 1984. Gewässerversauerung in der Bundesrepublik Deutschland im Lichte paläolimnologischer Studien. Naturwissenschaften 71: 631–633.

Steinberg, C., W. Beckstette, R. Psenner & N. Schulz, 1986. Chemical sediment stratigraphy of high alpine lakes. Hydrobiologia (in press).

Stumm, W. & G. Righetti, 1982. Tessiner Bergseen: Saurer Regen, saure Traufe. Neue Zürcher Zeitung, Forschung und Technik, Fernausgabe Nr. 231.

Accepted 18 March 1986.

Hydrobiologia 143: 255–265, (1986)

Late Quaternary diatoms in the sediments of Przedni Staw Lake (Polish Tatra Mountains)

Barbara Marciniak
Institute of Geological Sciences, Polish Academy of Sciences, 02-089 Warszawa, Al. Żwirki i Wigury 93, Poland

Keywords: high mountain lake, Late Glacial, Holocene, diatom succession, biostratigraphy

Abstract

The paper presents a diatom succession in the Late Glacial and Holocene sediments of the Przedni Staw lake in the Pieć Stawów Polskich Valley High Tatra Mts. Ten diatom phases are distinguished and presented against a pollen diagram from the analyzed core showing continuous sedimentation from the Oldest Dryas to the Subatlantic period. The diatom flora in the Late Glacial sediments is predominantly littoral – *Fragilaria, Navicula, Nitzschia, Diploneis, Pinnularia* and *Amphora* species. The Holocene sediments enclose more abundant planktonic species such as *Cyclotella quadriiuncta, Asterionella formosa* and *Melosira distans.*

Introduction

Postglacial lakes in the Pieć Stawów Polskich Valley occur above the present timberline in a pasture and dwarf mountain pine zone at altitudes of over 1600 m (Fig. 1). Among the Tatra lakes, the Przedni Staw Lake (1668 m a. s. l.) is the sixth largest one in terms of surface area, volume, and shoreline length as well as mean and maximum (34.6 m) depth (Szaflarski, 1936).

Diatom analyses were based on samples from core no. 2, taken in 29.1 m of water. The core was 3.08 m long and contained organic gyttja down to 2.01 m, and clays below.

The sediments of the Przedni Staw Lake were subjected to sedimentological, geochemical, mineralogical, pollen, *Cladocera* and preliminary diatom analysis (Wicik, 1979, 1984; Marciniak, 1982; Marciniak & Cieśla, 1983; Krupiński, 1984; Szeroczyńska, 1984; Wieckowski, 1984). The contact of organic and minerogenic sediments (at about 2 m depth) was radiocarbon dated to 9900 ± 120 BP (Wicik, 1979, 1984; Krupiński, 1984). For the diatom analyses 49 samples were collected at 5 cm intervals (Fig. 2).

The following diatoms of the Przedni Staw Lake sediments are described as new species and varieties: *Fragilaria pseudoconstruens, F. pseudoconstruens* var. *bigibba, F. pseudoconstruens* var. *rhombica* and *F. microstriata.* Besides, difficulties in the determination of *Fragilaria elliptica* and *F. construens* var. *venter* as well as of *Cyclotella quadriiuncta* were noted (Marciniak, 1982).

Diatom succession

Ten diatom phases were distinguished in the sediments – numbered PS1 to PS10 on the basis of relative frequency counts of dominant or characteristic diatom taxa and their distribution in the core. They were correlated with pollen and geochemical zones (Fig. 2). Taxa included in the diagram can be arranged according to ecological groupings, pH preferences or life-forms (Battarbee, 1979). In the Przedni Staw Lake sediments the ecological groupings and a use of the indices X or B (Renberg & Hellberg, 1982; Battarbee, 1984) are very difficult because most dominant diatoms have the 'unknown' taxonomic position and indetermined ecology.

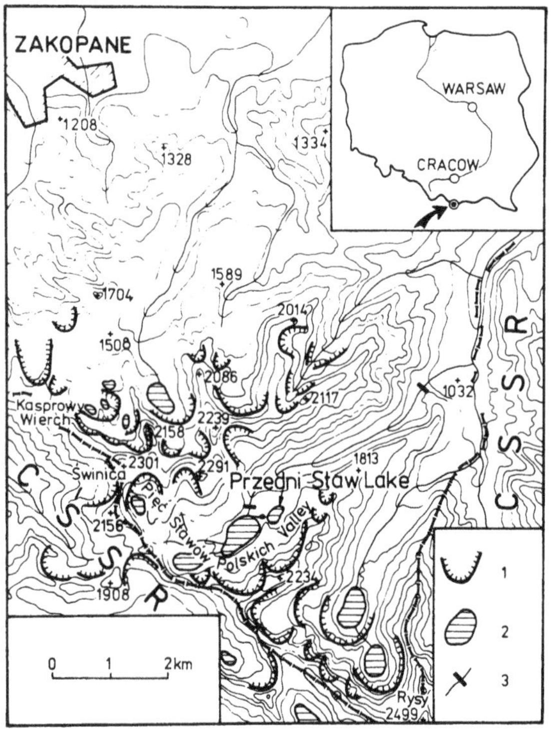

Fig. 1: Location of the Przedni Staw Lake at the background of the major elements of glacial relief in the High Tatra Mts., after Lencewicz (fide Lencewicz et Kondracki, 1964), 1 – glacial cirques, 2 – lakes, 3 – streams and waterfalls.

Diatom phase PS1

The diatom phase PS1 is represented by only 24 diatom taxa in sample 61 (not presented in the diagram) with *Fragilaria, Denticula, Stauroneis, Meridion, Cymbella* the main genera. This phase is thought to immediately succeed a glacial retreat from the area. These sediments were found by Krupiński (1984) to have been probably deposited during the Oldest Dryas.

Diatom phase PS2

Diatom phase PS2 is tripartite (Fig. 2). During the first subphase, PS2a, diatom content (especially of *Fragilaria, Navicula* and *Cyclotella*) increases, but during subphase PS2b diatom frequencies decrease. A renewed rise in diatom content during subphase PS2c is marked by *Cyclotella* spp. which, together with *Amphora libyca* (Plate V, Fig. 1) and *Amphora parallelistriata* (Plate V, Figs. 4-6) are most abundant in this part of the sequence. These species are accompanied by *Denticula tenuis* var. *crassula,* which is common in lakes of the Alps (Krasske, 1932; Hustedt, 1943) and in small deep lakes in Alaska (Foged, 1971). Foged (1953, 1955) also reports *Denticula tenuis* in shallow, alkaline

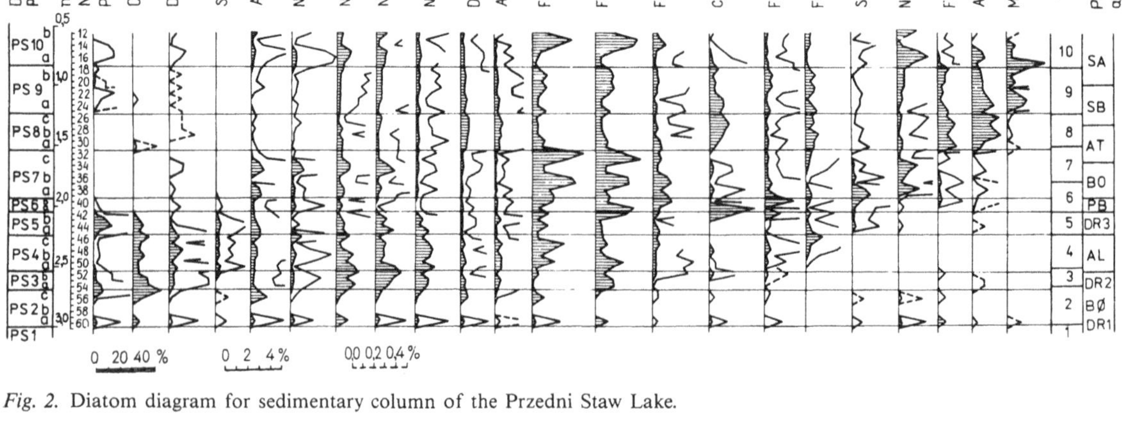

Fig. 2. Diatom diagram for sedimentary column of the Przedni Staw Lake.

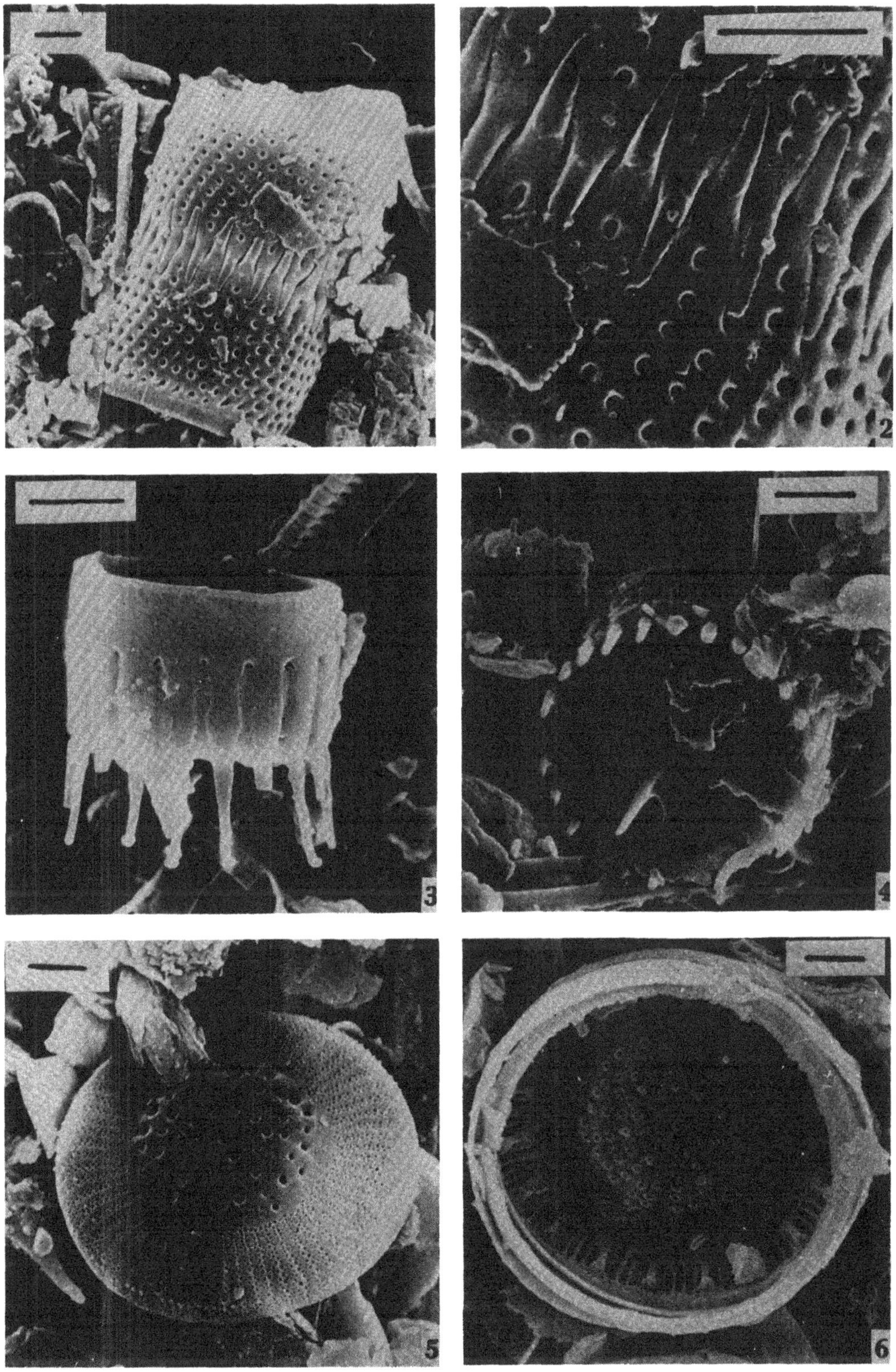

Plate I. Figs. 1–2. *Melosira distans* (Ehr.) Kütz. Figs. 3–4. *Melosira distans* var. *humilis* A. Cleve (*Melosira distans* var. *blelhamensis* Evans). Figs. 5–6 *Cyclotella quadriiuncta*? (Schröter) Hust. – all scales = 2 μm.

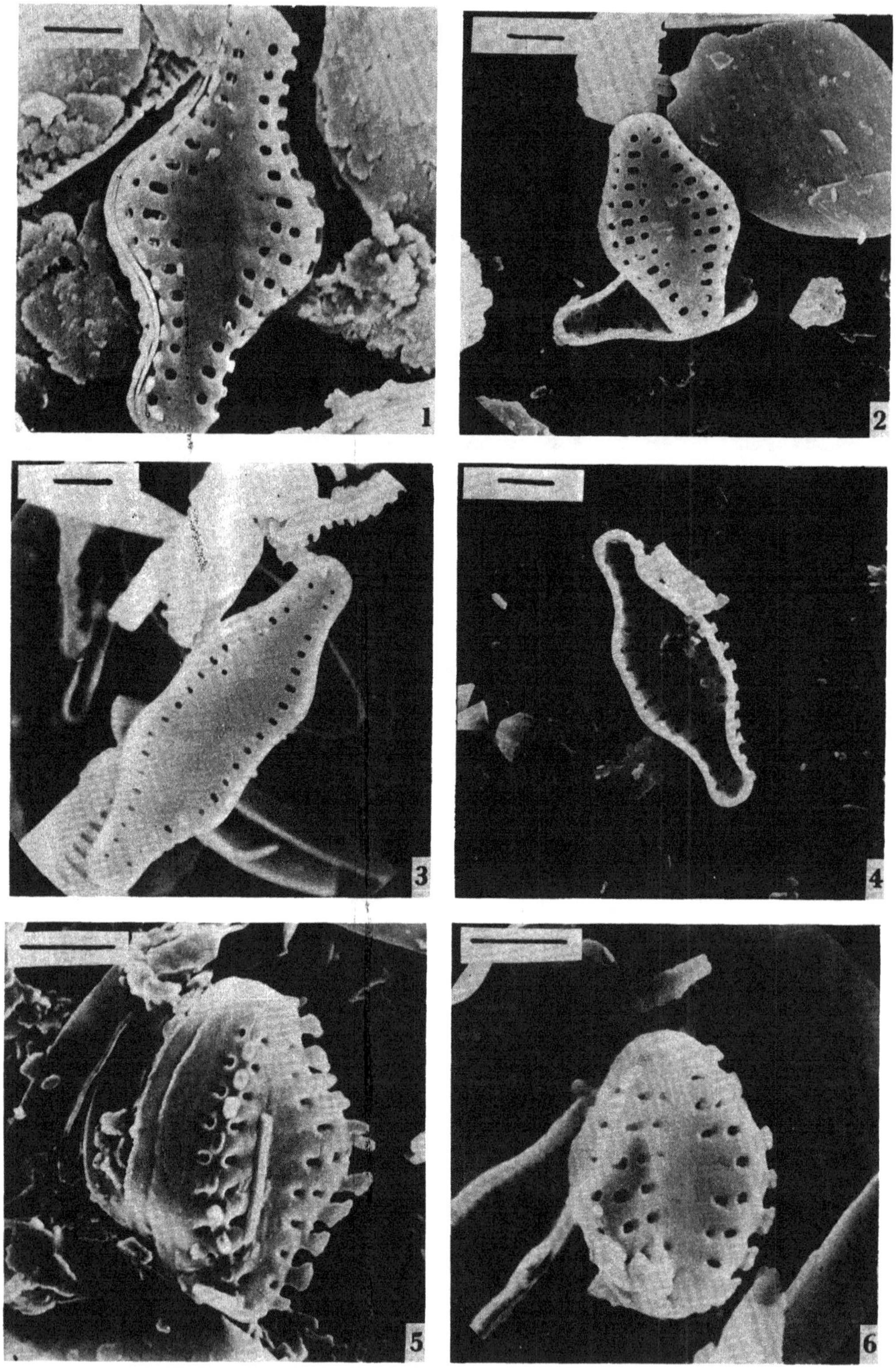

Plate II. Figs. 1–2. *Fragilaria pseudoconstruens* Marciniak (*Fragilaria construens* (Ehr.) Grun. sensu Foged, 1958). Fig. 3. *Fragilaria pseudoconstruens* var. *bigibba* Marciniak (*Fragilaria robusta* Manguin, after Manguin, 1954). Fig. 4. *Fragilaria pseudoconstruens* var. *bigibba* f.? (*Fragilaria construens* var. *binodis* f. *borealis*, after Foged, 1974). Fig. 5. *Fragilaria pseudoconstruens* var. *rhombica* Marciniak. Fig. 6. *Fragilaria pseudoconstruens* var. *rhombica* f.?

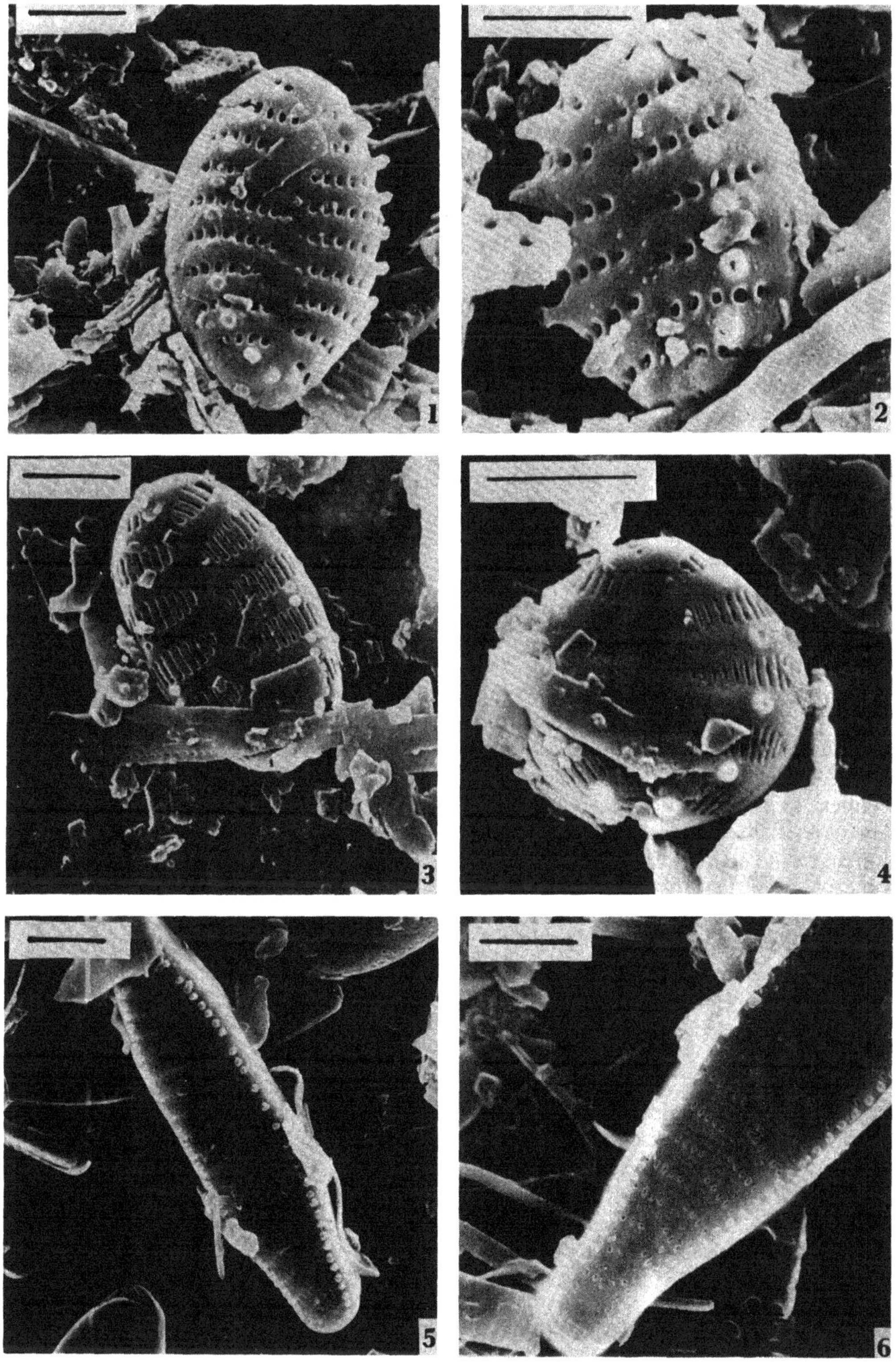

Plate III. Figs. 1–2. *Fragilaria elliptica* (Schumann) Haworth. Fig. 3. *Fragilaria pinnata* Ehr. Fig. 4. *Fragilaria pinnata* f. *subrotunda* Mayer (or *Fragilaria leptostauron* f. *minuta* A. Cleve, cf. Okuno, 1964). Figs. 5–6. *Fragilaria alpestris*? Krasske.

ponds in Greenland (with pH 7.6–9). Bradbury & Whiteside (1980) defined the *Denticula tenuis* zone in the Triangle Lake sediments to coincide with open vegetation (Birks, 1980) on moraines of the Klutlan Glacier. A low organic content in this zone suggests a presence of shallow and probably ephemeral aquatic habitats. In all likelihood, the planktonic diatom taxa present (e.g. *Cyclotella* spp.) are connected with an amelioration in light and temperature conditions which resulted in a somewhat longer vegetative period at the end of this phase (subphase PS2c). The floristic assemblages in these sediments were referred to the Older Dryas but a possible identification with the Bølling is also suggested (Krupiński, 1984).

Diatom phase PS3

This phase is typified by an increase in the relative frequency of *Navicula* spp., *Fragilaria pseudoconstruens* (Plate II, Figs. 1–2), *Nitzschia* spp., *Pinnularia pulchra* (or *P. nodosa*) and *Fragilaria alpestris*? (Plate III, Figs. 5–6) with a synchronous decrease of *Cyclotella* sp. *Fragilaria pseudoconstruens* is probably identical with *Fragilaria construens* from Greenland (Foged, 1958). Increasing relative frequencies of periphytic as well as of aerophilous or terrestrial species suggest more intensive erosion and wind action during this phase. In the Tatra vegetation of that time (Older Dryas) the predominating role was gained by *Gramineae* (that formed grasslands) with abundant *Artemisia,* (typical heliophytes become more significant) and *Selaginella selaginoides* (Krupiński, 1984).

Diatom phase PS4

Diatom phase PS4 still contains many of the taxa predominant in the previous phase (*Navicula* spp., *Nitzschia* spp. and *Fragilaria pseudoconstruens*). Later on, *Fragilaria elliptica* (Plate III, Figs. 1–2). and *F. microstriata, Diploneis* sp. (Plate IV, Figs. 4–6), frequencies increase. *Fragilaria elliptica* is also noted by Haworth (1975, 1976) in Late Glacial sediments from Northern Scotland. *Diploneis* sp. is very similar to *Diploneis elliptica* from Iceland (Foged, 1974) and to *Diploneis boldtiana* from Alaska (Foged, 1971). There are also numerous *Cyclotella, Navicula, Amphora, Stauroneis, Denticula, Achnanthes* frustules as well as many other taxa suggesting an increase in diatom diversity during this zone. It seems to results from physical and chemical changes in the water column due to a climatic amelioration. Diatom phase PS4 is referred to the Allerød, marked in the pollen diagram by the increase of *Betula,* then a rapid increase of *Pinus* and very distinct decrease of Gramineae and *Artemisia* (Krupiński, 1984). Such a change is also suggested by the development of a rather impoverished Cladocera fauna at this stage (Szeroczyńska, 1984).

Diatom phase PS5

The diatom assemblage in this zone is somewhat similar to that in diatom phase PS3. An increase in the relative frequency of the small *Navicula* spp. and *Pinnularia pulchra* (or *P. nodosa*) is recorded, as well as a decrease of the diatom taxa predominant during the previous phase (*Fragilaria pseudoconstruens, F. microstriata, F. elliptica* and *Diploneis* sp.). There are also changes in the frequency of the rare species, for example an increase in *Achnanthes gracillima* (Plate IV, Figs. 1–3). This taxon is noted in Japan, Alaska, North Europe (Hustedt, 1927; Foged, 1968, 1972, 1974; Haworth, 1976). This species is probably circumpolar, it occurs especially in oligotrophic, circumneutral to slightly acid localities (Foged, 1968). Phase PS5 is characterized by its poorer littoral diatom flora and by assemblage suggesting less favourable conditions for planktonic species. This may have resulted from longer winter ice cover and lower temperatures (see also Pechlaner, 1971). The pollen spectrum indicates a climatic cooling and drying marked by the increase of scrubby and herbaceous vegetation (NAP), mainly *Artemisia* (over 20%), Gramineae (12%) as well as an increase of *Betula* and *Salix,* and decrease of *Pinus* in Younger Dryas (Krupiński, 1984).

Diatom phase PS6

Phase PS6 is characterized by a dramatic increase in *Cyclotella quadriiuncta?* (Plate I; Figs. 5–6) and *Fragilaria pseudoconstruens* (Subphase PS6a) and then an increase in *F. pinnata* (Plate III, Figs. 3–4), *F. elliptica* (Plate III, Figs. 1–2), *F. pseudoconstruens* var. *bigibba* (Plate III, Fig. 3) and other taxa (subphase PS6b).

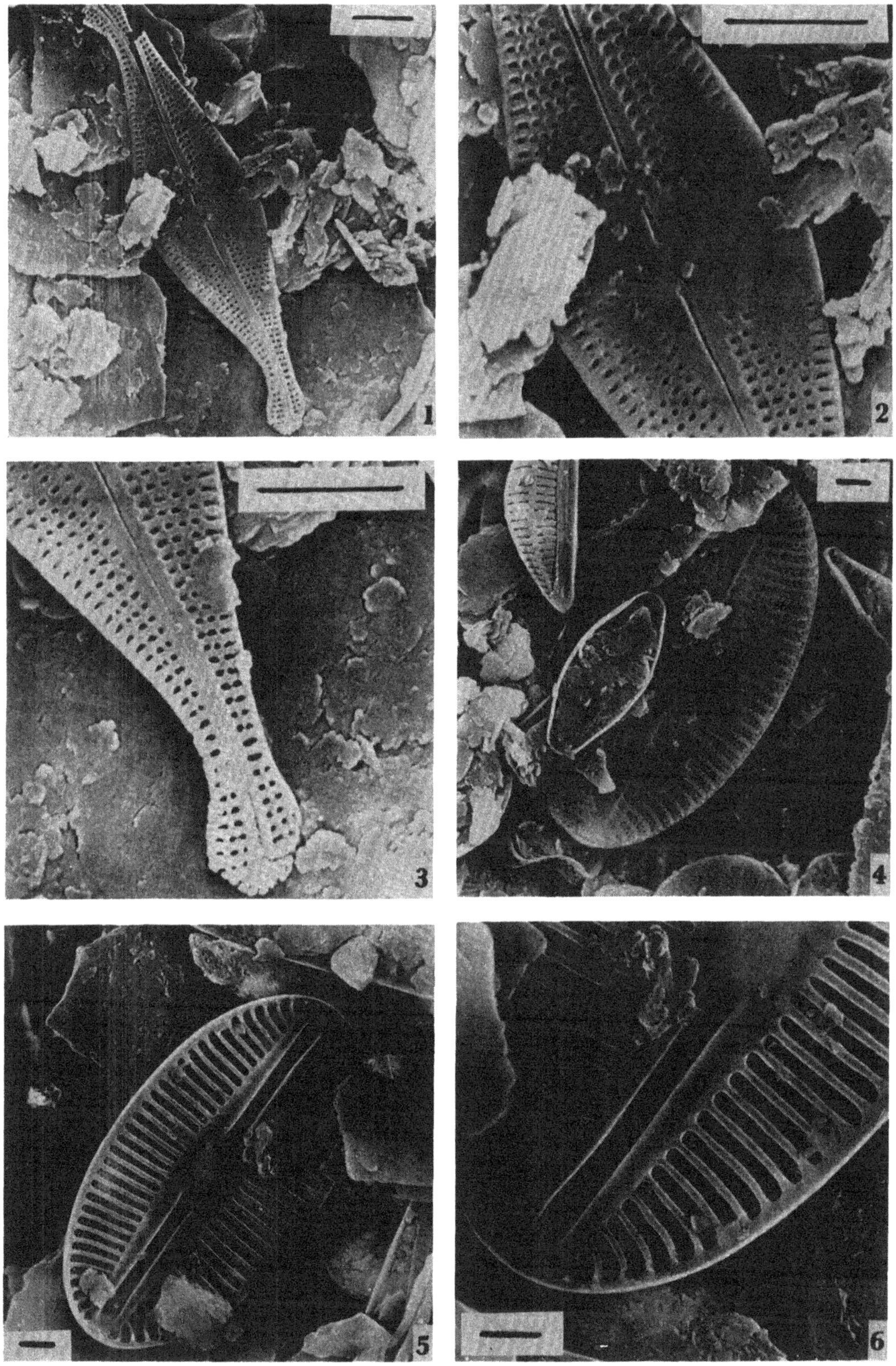

Plate IV. Figs. 1–3. *Achnanthes gracillima* Hust. Figs. 4–6. *Diploneis* sp.

Fragilaria pseudoconstruens var. *bigibba* is very similar to *Fragilaria construens* var. *binodis* from Tatra (Starmach, 1973) and from Iceland (Foged, 1974), and to *Fragilaria robusta* (Manguin, 1954). These changes mark the beginning of the Holocene (Preboreal Period). An enormous expansion of Musci is typical of the turn of the Younger Dryas and the Preboreal Period (Krupiński, 1984), Brief peak representation of probably planktonic species (e.g. *Cyclotella quadriiuncta*?) may have been caused by a rapid warming and a rise of the water level (subphase PS6a). At the same time a sudden increase in the Cladocera fauna occurred in the lake (Szeroczyńska, 1984). Moreover there was also a significant increase in the organic content of the sediment (Wicik, 1979, 1984; Marciniak & Cieśla, 1983).

Diatom phase PS7

This phase is highly changeable with a diatom flora dominated by *Fragilaria* species (mainly *F. elliptica,* F. *pseudoconstruens, F. pseudoconstruens* var. *bigibba, F. pseudoconstruens* var. *bigibba* f.? (Plate II, Fig. 4). There is also an increase in the representation of *Amphora parallelistriata, Navicula* spp. and *Stauroneis anceps* (Fig. 2). *Fragilaria pseudoconstruens* var. *bigibba* f.? is similar to *Fragilaria brevistriata* var. *subcapitata* (Mölder & Tynni, 1970), *Fragilaria inflata* (Starmach, 1973) and *Fragilaria construens* var. *binodis* f. *borealis* (Foged, 1974). There are slight similarities between the diatom flora in the zone (Boreal Period) and that in the Late Glacial phases, particularly in the cases of the *Fragilaria, Navicula* and *Amphora* genera. On the other hand, changes in a vegetation cover during the Boreal Period (dynamic development of *Ulmus,* appearance of other elements of the oak mixed forest and, abundance of *Corylus*) indicate the warmer climate than at present (Krupiński, 1984).

Diatom phase PS8

Phase PS8 can be divided into three subphases, represented by planktonic species *Asterionella formosa* and probably *Cyclotella quadriiuncta* (Fig. 2). *Fragilaria brevistriata* and *F. microstriata* are also abundant. An increase in the relative frequency of planktonic diatoms and decrease of *Fragilaria elliptica, F. pseudoconstruens* and *Navicula* spp. points to more favourable climatic and edaphic conditions during the Holocene optimum (particularly at the transition between the Atlantic and Subboreal periods) and probably a high water level. According to Krupiński (1984) a mean temperature of the warmest month during the Atlantic Period was definitely higher (about 2 – 3 °C), than in nowadays, winters were mild and vegetative season was longer.

Diatom phase PS9

This phase consists of two subphases. In subphase PS9a the plankton flora is composed of *Asterionella formosa, Cyclotella quadriiuncta?* and *Melosira distans* (Plate I, Figs. 1 – 2). By contrast in the subphase PS9b, there is a small increase in *Fragilaria pseudoconstruens* and *F. elliptica* representation. These changes may suggest a nutrient-poor (oligotrophic) environment and possibly a slight drop in the lake water level at the end of this phase. During the Subboreal Period the climate in the Tatra Mts. became more severe and continental, as proved also by changes in the forest communities in which *Corylus* was completely absent, being dislodged by *Picea* that needs frosty winters (Krupiński, 1984).

Diatom phase PS10

In this phase, a more clearly marked deterioration of climatic and edaphic conditions seems to be recorded. There is an increase in the frequency of *Melosira distans, M. distans* var. *humilis* (Plate I, Figs. 3 – 4, Florin, 1981), *Navicula* spp., a decrease of *Fragilaria brevistriata* and *Cyclotella quadriiuncta*? In subphase PS10b, these taxa are replaced by the *Fragilaria pseudoconstruens, F. pseudoconstruens* et f.?, *F. pseudoconstruens* var. *rhombica* (Plate II, Figs. 5), *F. pseudoconstruens* var. *rhombica* f.? (Plate II, Fig. 6) and *Amphora inariensis* (Plate V, Figs. 2 – 3). This taxon is very similar to *Amphora pediculus* (Krammer, 1980). Subphase PS10b is roughly comparable to the Early Holocene (diatom phase PS7) and the Late Glacial where similar diatom assemblages, with few planktonic diatoms are observed, a condition resembling some present-day alpine lakes (Pechlaner, 1971) and Late Glacial alpine lake (Marciniak, 1985).

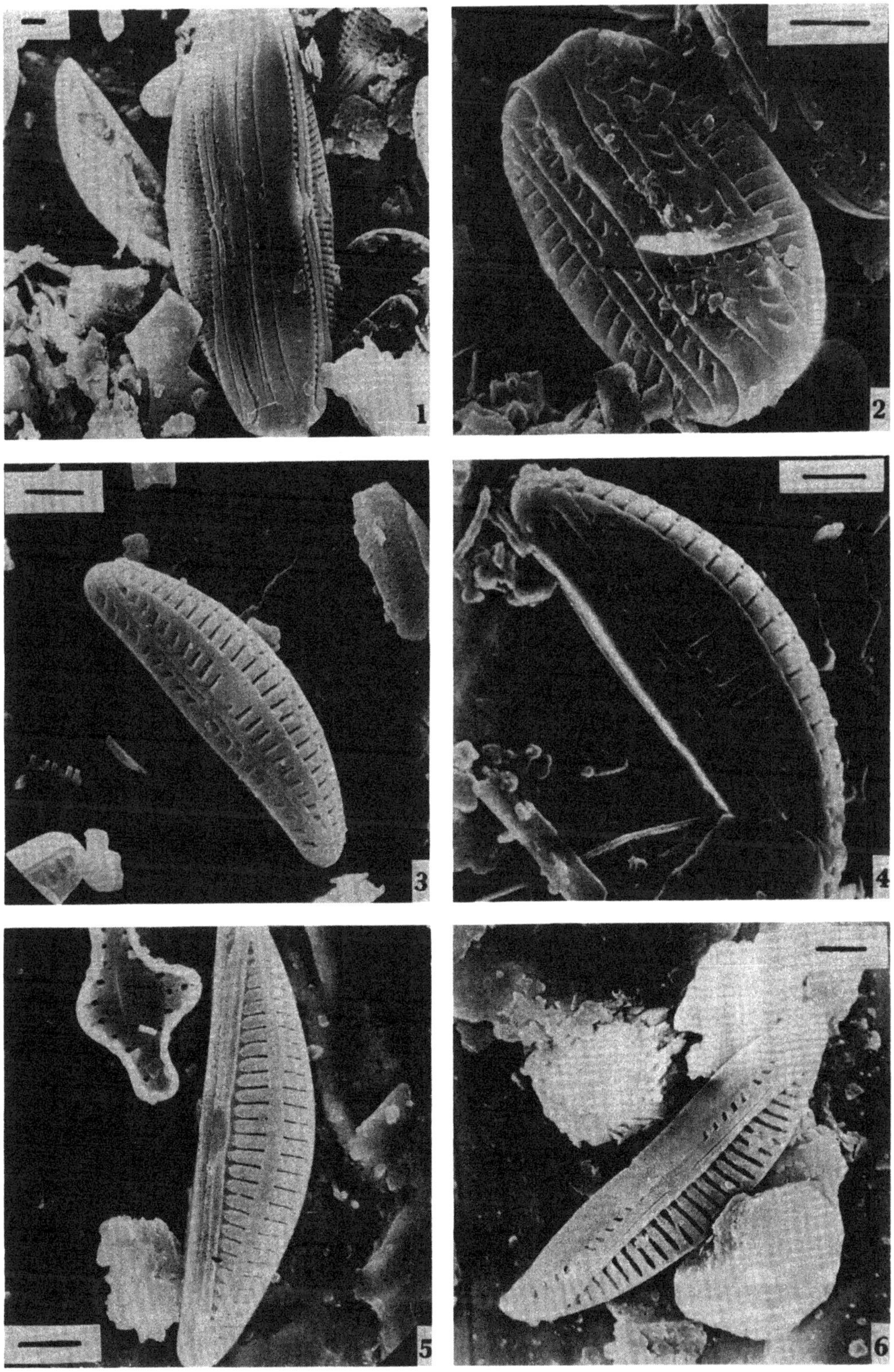

Plate V. Fig. 1. *Amphora libyca* Ehr. Figs. 2–3. *Amphora inariensis* Krammer. Figs. 4–6. *Amphora parallelistriata* Manguin.

Summary

Diatom analysis of a 3.08 m long core of sediments from Przedni Staw Lake exhibits five Late Glacial (PS1 – PS5) and five Holocene phases (PS6 – PS10).

The diatom flora of the Late Glacial sediments (phases PS1 and PS3 – PS5) was dominated by littoral taxa. Species of the genera *Fragilaria, Navicula, Nitzschia, Diploneis, Pinnularia* and *Amphora* were the most abundant. Only the diatom subphase PS2c contained numerous *Cyclotella* spp., probably planktonic ones. A diatom assemblage with a similar *Fragilaria* flora was also noted by Marciniak (1985) in the Late Glacial sediments of the alpine lake Hobschensee.

At the beginning of the Holocene (diatom phase PS6) in the Przedni Staw Lake sediments, a rapid increase in *Cyclotella quadriiuncta*? frequencies is observed. However, phase PS7 includes abundant species of *Fragilaria, Navicula, Amphora,* suggesting comparisons with the Late Glacial and Subatlantic diatom succession. The optimal phase of diatom development (phase PS8) shows a peak of *Asterionella formosa* frequencies noted at the transition from the Atlantic to the Subboreal period. Abundant frustules of *Melosira distans* are recorded during phase PS9, and its increase along with the *Fragilaria* and *Navicula* taxa in phase PS10, points to progressive cooling and acidification of lake waters during the Late Holocene.

Unfortunately, there are insufficient data to explain the distribution of diatoms in high mountain lake sediments. In order to understand the described succession further studies of taxonomy and ecology of the high mountain diatoms are needed.

Acknowledgements

The authoress is much indebted to Professor Frank Oldfield and Dr. Richard W. Battarbee for valuable comments and critical readings of the manuscript.

References

Battarbee, R. W., 1979. Diatoms in lake sediments. In: Paleohydrological changes in the temperate zone in the last 15 000 years (ed. B. E. Berglund), 177 – 225, Lund University, Sweden.

Battarbee, R. W., 1984. Diatom analysis and the acidification of lakes. Phil. Trans. R. Soc. Lond. B 305: 451 – 477.

Birks, H. J. B., 1980. Modern Pollen Assemblages and Vegetational History of the Moraines of the Klutlan Glacier and its Surroundings, Yukon Territory, Canada, Quaternary Research 14: 101 – 129.

Bradbury, J. Platt & M. C. Whiteside, 1980. Paleolimnology of Two Lakes in the Klutlan Glacier Region, Yukon Territory, Canada, Quaternary Research 14: 149 – 168.

Florin, M.-B., 1981. Taxonomy of some *Melosira* species. A comparative morphological study. In: R. Ross (ed.) Proc. Sixth Symp. Recent Fossil Diatoms, Budapest 1980, Koeltz (Publ.), 43 – 67.

Foged, N., 1953. Diatoms from West Greenland. Meddr Grønland 148: 1 – 86.

Foged, N., 1955. Diatoms from Peary Land, North Greenland. Meddr Grønland 128: 1 – 90.

Foged, N., 1958. The Diatoms in the Basalt Area and Adjoining Areas of Archean Rock in West Greenland. Meddr Grønland 156: 1 – 146.

Foged, N., 1968. The Freshwater Diatom Flora of the Varanger Peninsula, North Norway. Acta Borealia A. Scientia 25: 1 – 64.

Foged, N., 1971. Diatoms found in a bottom sediment sample from a small deep lake on the Northern Slope, Alaska. Nova Hedwigia 21: 923 – 1034.

Foged, N., 1973. Diatoms from Southwest Greenland. Meddr. Grønland 194: 1 – 84.

Foged, N., 1974. Freshwater diatoms in Iceland. Bibliotheca Phycologica 15: 1 – 118.

Foged, N., 1977. Freshwater diatoms in Ireland. Bibliotheca Phycologica 34: 1 – 123.

Haworth, E. Y., 1975. A scanning electron microscope study of some different frustule forms of the genus *Fragilaria* found in Scottish late-glacial sediments. Br. phycol., J., 10: 73 – 80.

Haworth, E. Y., 1976. Two late-glacial (Late Devensian) diatom assemblage profiles from Northern Scotland. New Phytol., 77: 227 – 256.

Hustedt, F., 1927. Bacillariales aus dem Aokikosee in Japan. Arch. Hydrobiol., 18: 155 – 172.

Hustedt, F., 1943. Die Diatomeenflora einiger Hochgebirgseen der Landschaft Davos in den Schweizer Alpen. Arch. Hydrobiol., 43: 124 – 197.

Krammer, K., 1980. Morphologic and Taxonomic Investigations of Some Freshwater Species of the Diatom Genus *Amphora* Ehr., Bacillaria, 3: 197 – 225.

Krasske, G., 1932. Beiträge zur Kentnnis der Diatomeenflora der Alpen. Hedwigia, 72: 92 – 134.

Krupiński, K. M., 1984. Evolution of Lake Glacial and Holocene Vegetation in the Polish Tatra Mts., based on Pollen Analysis of sediments of the Przedni Staw Lake. Bull. Pol. Acad. Sci., Earth Sci. 31: 37 – 48.

Lencewicz, S. & J. Kondracki, 1964. Geografia Fizyczna Polski. PWN, Warszawa, 486 pp.

Manguin, E., 1954. Contribution a la connaissance biologigue des boues lacustres, lac Pavin (Puy-de-Dôme). Ann. Ec. Nat. Eaux et Forêts. 14: 5–19.

Marciniak, B., 1982. Late Glacial and Holocene new diatoms from a glacial lake Przedni Staw in the Pieć Stawów Polskich Valley, Polish Tatra Mts., Acta Geol. Acad. Sci. Hungar. 25: 161–171.

Marciniak, B., 1985. Diatoms in bottom sediments of Lake Hobschen (Preliminary report). INQUA Eurosiberian Subcom. Stud. Holocene and IGCP Project 158, Excursion Guide 'Symposium in Switzerland' 85', Bern, 92–93.

Marciniak, B & A. Cieśla, 1983. Diatomological and geochemical studies on Late Glacial and Holocene sediments from the Przedni Staw Lake in the Dolina Pieciu Stawów Polskich Valley (Tatra Mts). Kwart. Geol. 27: 123–150.

Mölder, K. & R. Tynni, 1970. Über Finnlands rezente und subfossile Diatomeen IV. Bull. Geol. Soc. Finland 42: 129–144.

Okuno, H., 1964. Fossil diatoms. In. J.-G. Helmcke, W. Krieger, Diatomeenschalen im elektronenmikroskopischen Bild. 5; pl, 414–513, Weinheim.

Pechlaner, R., 1971. Factors that control the production rate and biomass of phytoplankton in high-mountain lakes. Mitt. Internat. Verein. Limnol. 19: 125–145.

Renberg, I. & T. Hellberg, 1982. The pH history of lakes in south-western Sweden, as calculated from the subfossil diatom flora of the sediments. Ambio 11: 30–33.

Starmach, K., 1973. Benthic Algae of the Great Lake in the Valley of the Five Polish Lakes in the Tatra Mountains. Fragm. Flor. Geobot. 19: 481–511.

Szaflarski, J., 1936. Morphometrie des lacs de la Tatra Polonaise. Wiad. Słuzby Geogr. 1: 5–37.

Szeroczyńska, K., 1984. Results of examination of *Cladocera* remnants in lacustrine sediments of Dolina Pieciu Stawów Polskich. Prace i stud. Geogr. Uniw. Warsz. 5: 93–102.

Wicik, B., 1979. Postglacial accumulation in the High Tatra lakes. Przeg. Geol. 7, (315): 403–404.

Wicik, B., 1984. The Tatra lake deposits and their accumulation stages. Prace i stud. Geogr. Uniw. Warsz. 5: 55–69.

Wieckowski, K., 1984. Macroscopic characteristics of the lacustrine deposits of the Tatra lakes. Prace i Stud. Geogr. Uniw. Warsz. 5: 39–54.

Accepted 13 May 1986.

Hydrobiologia 143: 267–276, (1986)

Scandinavian limnochronology of sediments and heavy metals

Farid El-Daoushy
Institute of Physics, Box 530, S-751 21 Uppsala, Sweden

Keywords: Pb-210 fluxes, heavy metals, accumulation rates, lakes, peatlands, marine coasts

Abstract

Sediment and peat accumulation rates as well as heavy metal fluxes are examined in Scandinavian lakes, marine coasts and peatlands. Modern accumulation rates are determined using Pb-210 and other chronological tools. Mean deposition rates in the past 150 ± 20 yrs are 3–95, 40–190 and 1–5 $mg \cdot cm^{-2} \cdot yr^{-1}$ for lakes, marine coasts and peatlands respectively. The Pb-210 fluxes at the investigated sites are quite variable. The lowest values are found for peatlands 0.04–0.20 $pCi \cdot cm^{-2} \cdot yr^{-1}$ while coastal marine environments showed 0.35–0.70 $pCi \cdot cm^{-2} \cdot yr^{-1}$. A wide range of variation of Pb-210 fluxes is obtained for the studied lake sediments 0.01–0.51 $pCi \cdot cm^{-2} \cdot yr^{-1}$.

The chronological records of Pb, Zn, Cu, Cd and Hg show that heavy metal fluxes are much higher in coastal marine areas. Inland lakes and ombrotrophic peat-bogs indicate that Pb is introduced exclusively through the atmosphere and decrease exponentially from South to North Scandinavia. However, the other heavy metals are transported to freshwater systems by different hydrologic and atmospheric processes. Zn and Fe exhibit some correlation with lake-drainage area.

Introduction

The increasing industrial, agricultural and other human consumption of natural resources created urgent needs for handling waste problems and assessing their effects on various aquatic systems. Palaeoecological and palaeolimnological records of selected sediments and peats show that anthropogenic pollution has been going on for centuries and became more remarkable during the past decades. The global nature of such environmental problems calls for further modelling of multisystem interactions and quantitative estimation of pollution in relation to space and time. Since late 1960s and 1970s major chronological investigations have been carried out on flux densities of various organic and inorganic remains in sediments and peats. The available chronological data on heavy metals are relatively scarce for peatland and coastal-marine environments compared to freshwater lakes. More research is still required to enhance the accuracies of the existing chronological tools and widen their applicabilities (Alderton, 1985) for better modelling of atmospheric and aquatic flux densities.

The Pb-210 method (Goldberg, 1963; Robbins, 1978; Krishnaswami & Lal, 1978; Oldfield & Appleby, 1984) has been used for dating lake and coastal marine sediments as well as peatland deposits. Tracing internal and external processes influencing various aquatic environments is another possible application of the Pb-210 technique (Robbins, *loc. cit.*; El-Daoushy & Tolonen, 1984; McCaffrey & Thomson, 1980; Dearing *et al.*, to be published; Lewis, 1977; Chanton *et al.*, 1983; Davis *et al.*, 1982; Evans & Rigler, 1983). However, additional chronological tools and tracers are necessary for critical evaluation of the Pb-210 method and credible construction of time-scales in calendar years.

The influence of atmospheric and aquatic

processes, as well as the composition and structure of depositional sequences, on local and global Pb-210 fluxes is not fully examined. Knowledge on such details would certainly increase our understanding of the Pb-210 cycle in nature.

Description of sites and materials

Fig. 1 shows the sampling sites of 41 Scandinavian cores which were collected from twenty lakes, six brackish and marine water-systems, and two peatlands. The aquatic systems 1–4 and 13–21 are typical Swedish woodland lakes with average depths of 1–5 m and surface areas of 10–150 hectares. Lakes Väster–Täckelsjön, Ellisjaur and Ala Makkarijärvi have the largest drainage areas while lake Nattajärvi has the lowest. These woodlands lakes are generally well protected against direct discharge of waste water and far from local sources of atmospheric pollution. Moreover, the bedrock of these lakes has a high resistance against weathering and is covered with thin layers of moraine. These lakes have low accumulation rates (Table 1) and

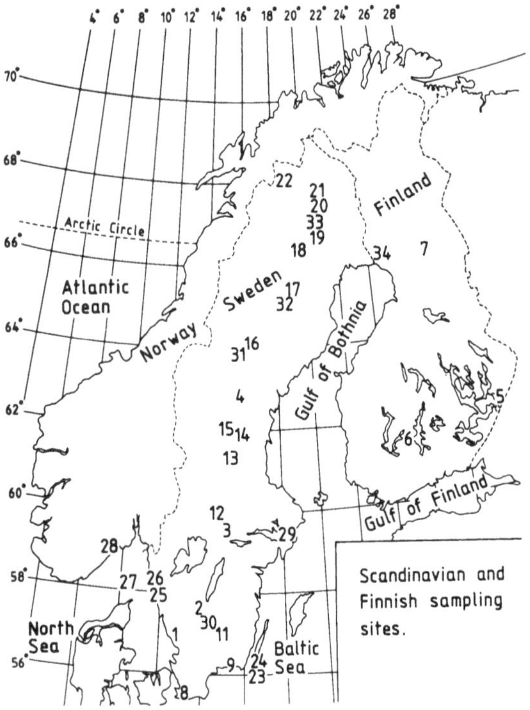

Fig. 1. Scandinavian and Finnish sampling sites.

their primary production is very limited because of lack of nutrients. The drainage areas of such lakes have not been affected by modern human activities such as land-use and tourism. For these and other reasons they were considered to be suitable for atmospheric pollution studies (El-Daoushy *et al., op. cit.*). Nevertheless, about 20% of the drainage area of Lake Ramsjön is cultivated and populated land (Edberg, 1980). More detailed description of such lakes is given elsewhere (El-Daoushy & Johansson, 1983; El-Daoushy *et al.*, in preparation). Lakes 7–12 have been influenced by modern human activities such as mining, farming and other forms of land management (Battarbee & Digerfeldt, 1976; Olsson *et al.*, 1983; Dearing *et al., loc. cit.*; El-Daoushy *et al.*, in preparation). However, they are about the same size as previously mentioned lakes. Lake Vasajaure is a highland lake and primarily used for deglaciation studies (Karlén, private communication). The coastal marine sites, as well as the investigated lake cores, are situated within accumulation bottoms. The deepest core was collected from Skagerrak at about 250 m water depth (Swinder, private communication). However, sites 23 and 25 are transportation bottoms (Notter & Göthberg, private communication), save 24 which is located on Öland's shore where highly organic matter are deposited (Königsson, private communication). Edsviken Bay is a brackish water body with annually varved sediments (El-Daoushy & Axelsson, in preparation).

In most cases the sediments were collected from flat bottoms with maximum depths and more or less stable sedimentary conditions. The sediment cores were carefully sampled using gravity coring units with tube carriers of plexiglass. The water and organic contents of the lake and peat deposits are generally high except lake Saxen, Lake Vasajaure and the upper layers of Lake Säynäjälampi (Table 1). Skagerrak has the lowest organic content. Both the organic and water content profiles of the Frierfjord Fjord and Edsviken Bay were found to be highly variable (Jacobsen & Axelsson, private communication).

The peats were sampled from ombrotrophic (rain-fed) peat-bogs using very thin-walled steel corers. The majority of the peat-mosses have grown under ombrotrophic conditions and derived their water supply exclusively from atmospheric precipitation. Other criteria for selecting peatlands and lakes for atmospheric-pollution studies are given elsewhere (El-Daoushy & Tolonen, *loc. cit.*). The given localities have peat hummocks with water table below the peat surface except site 30 which comprises a peat hollow as well. Details on the botanical composition, growing conditions and degree of humification are published by El-Daoushy & Tolonen, *op. cit.* and Franzén, 1985.

Laboratory procedures and methods

The cores were sliced into suitable intervals, hot or freeze dried and then kept in acid-washed plastic boxes. Organic contents (Table 1) are rough estimates since they were analysed using loss on ignition. Both water content and organic content are of special value in surface-layer studies of aquatic deposits since information on lake-bottom type, bulk-density and porosity could be obtained. Analyses of grain-size distribution of Frierfjorden Fjord and benthic fauna of Gullmarsfjorden Fjord are in progress. Such data are of importance to explain Pb-210 and heavy metal anomalies. Absolute activities of total Pb-210 were determined using isotope dilution (Po-210/Po-208) and alpha-spectrometry. Quantitative estimates of Ra-226 are done by radon emanation and gross-alpha counting. Si-surface barrier detectors and ionization chambers are utilized for such measurements. The longterm stability as well as the background of the counting systems were carefully examined (El-Daoushy, 1978, 1981a and 1981b). This allowed the calculation of total uncertainties for unsupported Pb-210 activities including all possible errors (El-Daoushy & Garcia-Tenorio, in preparation).

Other chronological tools have been used to evaluate and/or extend the Pb-210 time-scales. Cs-137, magnetic measurements, C-14, x-rays, historical and environmental events, pollen horizons and moss-increment dating were also used in this study (El-Daoushy, in press). In most cases, the low accumulation rates limited the applicability of other alternative methods. However, heavy metal profiles gave some useful information on various sources and processes controlling the depositional conditions of the aquatic systems.

The investigated heavy metals were analysed by conventional atomic absorption spectrometries (Johansson, 1980).

Results and discussion

About 500 Pb-210 and Ra-226 measurements have been carried out in this study. A summary of the obtained results is given in Table 1. The evaluation of supported Pb-210 in the studied deposits was based on Ra-226 concentrations in the cores and/or the total Pb-210 in deposits older than 150–200 yrs. Only in ideal case of closed systems is supported Pb-210 in secular equilibrium with the Ra-226. However, Ra-226 could be a good guide to select region of secular radioactive equilibrium between Pb-210 and Ra-226. Supported Pb-210, as determined from the total Pb-210 in deposits older than 150–200 yrs, and Ra-226 were very low for ombrotrophic peats. This indicated that the analysed peat bogs were totally dependent on atmospheric precipitation. In 25% of the examined sediments, Ra-226 were found to be either variable or in disagreement with the supported Pb-210. The Ra-226 and supported Pb-210 activities of brackish and marine sediments displayed major anomalies. In such deposits, the estimation of unsupported Pb-210 activities was critical because the total Pb-210 concentrations were not high. Studies on Gullmarsfjorden Fjord, Edsviken Bay and the Baltic Sea illustrated the necessity of careful examination of supported Pb-210 in such environments. The supported Pb-210 was higher than Ra-226 in Gullmarsfjorden Fjord and Edsviken Bay. Both supported Pb-210 and Ra-226 of site 23 were found variable. In woodland lakes, the supported Pb-210 was generally low and proved secular equilibrium with Ra-226. Lakes Skärvsjön, Björken and Tunturijärvi showed some disequilibrium between Ra-226 and supported Pb-210, however, this was not a serious problem in estimating unsupported Pb-210. Lakes with variable sediment input (organic/inorganic) revealed unstable delivery of supported Pb-210 to their bottoms. This was observed for lakes Växjösjön, Saxen and Havgårdssjön and caused difficulties in calculating the unsupported Pb-210. The data obtained from lakes Björken, Havgårdssjön, Växjösjön and Tunturijärvi showed that the supported Pb-210 have not been constant. Both total Pb-210 and Ra-226 are continuously decreasing although secular equilibrium could be recognized in some or several layers of the deeper parts of the examined cores. In such cases independent evaluations of supported Pb-210 using

Table 1. Pb-210 parameters for various aquatic sites in Scandinavia and Finland (Fig. 1). Mean deposition rates, in past 150 ± 20 yrs, are calculated using the C.I.C. and C.R.S. models. Actual initial unsupported Pb-210 were determined for layers just below the surface ones (in most cases). Actual values are, therefore, slightly different.

Coring site, number on Fig. 1, core code	Calculated unsup. Pb-210		Actual initial unsupported Pb-210, $pCi \cdot g^{-1}$	Supported Pb-210*		Mean deposition rate, $mg \cdot cm^{-2} \cdot yr^{-1}$		Average organic contents, %
	Mean flux, $pCi \cdot cm^{-2} \cdot yr^{-1}$	Initial activity, $pCi \cdot g^{-1}$		Estimated from total Pb-210, $pCi \cdot g^{-1}$	Ra-226 activity, $pCi \cdot g^{-1}$	C.I.C.	C.R.S.	
Lake Tussjön, 1	≅0.01	≤2.7 ± 0.5	>1.0 ± 0.1	0.36 ± 0.06	-	≅5	4.6 ± 0.6	46
Lake Skärvsjön, 2, S1	0.26 ± 0.03	66.6 ± 7.3	>28.9 ± 2.4	0.95 ± 0.10	2.0 ± 0.1	5.0 ± 0.4	4.1 ± 0.6	34
Lake Skärvsjön, 2, S2	0.09 ± 0.01	35.9 ± 5.8	>13.3 ± 1.9	0.95 ± 0.10	-	3.0 ± 0.4	3.5 ± 0.5	36
Lake Björken, 3	0.16 ± 0.02	33.4 ± 5.4	>18.4 ± 1.7	(3.9 ± 0.6 – 6.2 ± 0.5)	(4.9 ± 0.2 – 6.7 ± 0.3)	6 ± 2	6.0 ± 0.8	22
Lake Väster-Täckelsjön, 4	0.17 ± 0.02	22.3 ± 2.7	13.4 ± 1.0	1.6 ± 0.15	2.0 ± 0.1	11 ± 2	8.2 ± 1.1	31
Lake Säynäjälampi, 7	Short core	-	1.8 ± 0.3	-	(0.08 ± 0.04 – 1.9 ± 0.1)	(sedimentation rate 1 mm · yr⁻¹)		5 – 50
Lake Havgårdssjön, 8, BPM	0.51 ± 0.06	11 ± 3	≥4.1 ± 0.1	(0.7 ± 0.1 – 1.2 ± 0.1)	(0.81 ± 0.05 – 1.6 ± 0.07)	55 ± 11	49.7 ± 6.6	20
Lake Havgårdssjön, 8, C3	-	-	≈3.9 ± 0.1	decreasing	decreasing	17 ± 3[$]	15.3 ± 2.0[$]	
Lake Hallsjön, 9	0.24 ± 0.02	31.5 ± 2.7	≤14.4 ± 0.9	1.3 ± 0.1	1.3 ± 0.3	10 ± 3	9.5 ± 1.3	46
Lake Sännen, 9	0.33 ± 0.04	25.4 ± 4.6	≥8.4 ± 0.7	≤1.55 ± 0.11	1.55 ± 0.04	9 ± 3	10.7 ± 1.4	27
Lake Växjösjön, 11	-	12.7 ± 1.8	7.3 ± 0.5	(1.4 ± 0.1 – 3.6 ± 0.4)	(2.0 ± 0.1 – 3.5 ± 0.2)	(sedimentation rate 5.5 mm · yr⁻¹)		-
Lake Saxen, 12[S]	≅0.06	-	Variable	-	(1.9 ± 0.1 – 3.1 ± 0.1)	20 ± 10[$]	10.0 ± 1.3[$]	8
Lake Uggsjön, 13	0.37 ± 0.04	143 ± 19	>73.3 ± 9.5	1.1 ± 0.1	1.2 ± 0.3	3.5 ± 0.5	4.4 ± 0.6	33
Lake Ramsjön, 14	0.21 ± 0.04	1.9 ± 0.3	1.4 ± 0.2	1.7 ± 0.1	-	94 ± 15	67.0 ± 8.9	22
Lake Karsvattnet, 15	0.09 ± 0.01	25.9 ± 2.1	>14.9 ± 1.1	0.64 ± 0.02	0.4 ± 0.6**	3.3 ± 0.6	2.9 ± 0.4	50
Lake Rensjön, 16	0.10 ± 0.01	23.5 ± 3.1	>11.8 ± 1.0	1.2 ± 0.1	1.1 ± 0.2	4.5 ± 0.7	5.9 ± 0.8	38
Lake Ellisjaur, 17	0.26 ± 0.03	56.1 ± 5.0	≤18.4 ± 1.6	2.6 ± 0.2	2.2 ± 0.1**	7 ± 1	9.4 ± 1.3	37
Lake Sårvatj, 18	0.16 ± 0.02	32.3 ± 3.0	>21.7 ± 1.6	0.8 ± 0.1	0.6 ± 0.2	6 ± 0.7	5.1 ± 0.7	45
Lake Nattajärvi, 19	0.12 ± 0.01	23.6 ± 2.2	≥12.3 ± 2.0	0.9 ± 0.1	0.70 ± 0.05**	7 ± 3	10.8 ± 1.4	41
Lake Ala Makkarijärvi, 20	0.20 ± 0.02	34.3 ± 2.1	≥16.8 ± 1.3	0.52 ± 0.04	0.30 ± 0.04	11 ± 2	13.3 ± 1.8	35
Lake Tunturijärvi, 21	0.08 ± 0.01	6.9 ± 2.0	>5.0 ± 0.6	≥1.55 ± 0.20	(1.5 ± 0.1 – 3.6 ± 0.2)	13 ± 3	10.9 ± 1.5	23
Lake Vasajaure, 22	Deficient Pb-210 in the surface layers			(1.4 ± 0.2 – 2.6 ± 0.2)	(1.1 ± 0.1 – 2.7 ± 0.2)	(sedimentation rate 0.1 mm · yr⁻¹)		10
The Baltic Sea, 23	≅0	≅0	≅0	(0.45 ± 0.04 – 2.83 ± 0.12)	(0.40 ± 0.04 – 2.1 ± 0.1)	≅0	≅0	-

Table 1. Continued.

Ölands coast, The Baltic Sea, 24	Short core	-	2.9±0.1	-	0.2 ±0.1	-	-	~95
Stenungsund Sound, 25, H3	≅0.16	2.5± 0.1	>2.1±0.1	-	0.95±0.05	52 ± 6	-	7
Stenungsund Sound, 25, Hg I	≅0.39	3.2± 0.5	>2.6±0.2	-	0.55±0.2	40 ±10	-	15
Stenungsund Sound, 25, Hg II	≅0.26	3.2± 0.5	>2.3±0.2	-	0.75±0.1	38 ± 9	-	8
Gullmarsfjorden, 26, C 3	0.60±0.04	5.3± 1.0	2.5±0.2	0.50±0.05	0.35±0.2	115 ± 7	106 ±14	11
Gullmarsfjorden, 26, C 1	0.61±0.03	4.8± 1.0	2.3±0.2	0.65±0.10	0.50±0.15	166 ±10	118 ±16	9
Skagerrak, 27	0.55±0.04	7.7± 0.6	>6.8±0.4	0.55±0.04	0.55±0.03	111 ±16	89.0±11.9	2.4
Frierfjorden Fjord, 28, F 1	0.50±0.04	6 ± 2	4.0±0.3	0.6 ±0.1	0.4 ±0.2	110 ±22	113 ±15	6.3
Frierfjorden Fjord, 28, F 6	0.70±0.06	8 ± 2	≤4.2±0.2	0.6 ±0.1	-	187 ±20	187 ±25	4.5
Edsviken Bay, The Baltic Sea, 29	0.35±0.03	8.8± 4.2	≶3.3±0.3	1.0 ±0.1	(0.45±0.05 – 1.4 ±0.1)	60 ± 8	61.3± 8.2	11.3
Kunonniemensuo bog, 5, core F9	0.15±0.01^{+}	>6^{+}	5.6±0.2	0.08±0.01	0.12±0.04	39.5± 7.0^{+}	21.0± 2.8^{+}	98
Kärpänsuo bog, 6, core 1	0.12±0.02^{+}	>6^{+}	5.3±0.2	0.20±0.03	0.12±0.04	21.0± 5.0^{+}	9.3± 1.2^{+}	98
Store mosse bog, 30, Hummock	0.10±0.04	≥15	≶16.2±2.2	- -	0.1 ±0.1^{++}	≅2.6$^{\$}$	2.3± 0.3$^{\$}$	~95
Store mosse bog, 30, Hollow	0.20±0.07	≥10	≤6.5±0.5	- -	0.09±0.05^{++}	≅4.4$^{\$}$	3.9± 0.5$^{\$}$	~95
Fen I, Västerbotten, 31	0.12±0.04	≥7	≶6.0±0.5	- -	0.3 ±0.2^{++}	≅3.8$^{\$}$	3.0± 0.4$^{\$}$	~95
Fen II, Norrbotten, 32	0.07±0.03	≅8	≥6.8±0.5	- -	0.2 ±0.2^{++}	≅3.1$^{\$}$	2.5± 0.3$^{\$}$	~95
Fen III, Norrbotten (Saltmyran), 33	0.04±0.02	≅4	>2.6±0.2	- -	0.2 ±0.1^{++}	≅1.3$^{\$}$	1.0± 0.1$^{\$}$	~95

* Single values are averages, otherwise supported Pb-210 is variable.

$^{+}$ Average of two laboratories.

$ Sediment and peat densities are assumed.

** One measurement.

$^{++}$ Supported Pb-210 is very low and is taken equal to Ra-226.

S Average of two cores, however both cores are short and very disturbed.

either total Pb-210 or Ra-226 might yield erroneous supported Pb-210 values. Variable sediment composition has been found to influence Pb-210 and Ra-226 profiles of marine (Frierfjorden Fjord and The Baltic Sea sites) and lake (Växjösjön, Säynäjälampi, Saxen and Havgårdssjön) cores, evaluation of data is to be published (El-Daoushy; El-Daoushy *et al.*). A detailed treatment of the equilibrium conditions between Ra-226 and Pb-210 in aquatic deposits is given elsewhere (El-Daoushy, in preparation). In some cases the collected cores were short and supported Pb-210 is only based on Ra-226 measurements. Accordingly unsupported Pb-210 fluxes are given as rough estimates without uncertainties (Säynäjälampi, Saxen, Stenungsund and Ölands coast).

Semi-logarithmic relations between unsupported Pb-210 activities, $pCi \cdot g^{-1}$, and compaction-corrected depths, $mg \cdot cm^{-2}$, showed 'disturbed' zones near the sediment-water interface. The C.I.C. model and the method of the least-squares were used to calculate mean deposition rates except for two cores (lake Säynäyälampi and lake Växjösjön) where only mean sedimentation rates are given. However, the mentioned concentration anomalies (due to mixing, dilution, dissolution, etc.) were rejected. Previous studies indicated that oxidation/reduction transitions near water-sediment interfaces have very limited effect on Pb-210 profiles (Robbins, *op. cit.*; El-Daoushy & Tolonen, *op cit.*). The observed high uncertainties of mean deposition rates (Table 1) were slightly affected by the individual errors of the unsupported Pb-210 activities. These uncertainties provide the actual limits of variation in the mean accumulation rates during the examined time intervals. The initial concentrations of the unsupported Pb-210 were also determined using the preceding least-squares fitting. The obtained least-squares activities were corrected for obvious bioturbation in their profiles (lake Sännen, Lake Ramsjön and Gullmarsfjorden Fjord). However, the obtained averages were higher than the observed activities of surface layers. Moreover, the least-squares initial concentrations exposed high uncertainties. The calculations suggest that typical initial Pb-210 activities may not be available because of variable sedimentation rates or existing surface anomalies. Table 1 gives a summary on the extrapolated averages of unsupported Pb-210 concentrations as well as the measured initial activities of the examined cores. A wide variation in Pb-210 concentrations could be found for lake sediments, nevertheless ombrotrophic peats and marine sediments have generally lower initial Pb-210 activities.

The C.R.S. model was applied to get unsupported Pb-210 fluxes (Table 1). There are poor correlation between the total unsupported Pb-210 and the total organic matter (Fig. 2) in the investigated cores. The Pb-210 fluxes have no correlation with the total inorganic matter of lake sediments (Fig. 3). Marine sediments, on the other hand, may show higher unsupported Pb-210 fluxes ($0.35-0.70$ $pCi \cdot cm^{-2} \cdot yr^{-1}$) than lake sediments ($0.01-0.51$ $pCi \cdot cm^{-2} \cdot yr^{-1}$). However, the data given here show that unsupported Pb-210 fluxes, which could be estimated with good accuracies, are quite variable for lake sediments. Lake sediment accumulation-rates and heavy metal fluxes did not indicate any focusing for sites having high Pb-210 fluxes (El-Daoushy *et al., loc. cit.*). The Pb-210 fluxes of ombrotrophic peats give evidence that

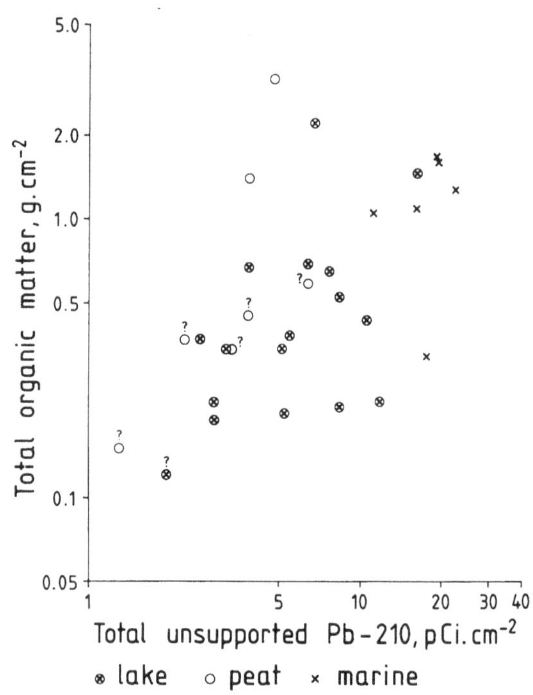

Fig. 2. The total unsupported Pb-210, $pCi \cdot cm^{-2}$, against total organic matter, $g \cdot cm^{-2}$, in some investigated cores from Scandinavian aquatic systems. Bulk densities of the cores with question mark are assumed.

lake and marine sediments may not reflect direct atmospheric fluxes of Pb-210. Recent laboratory experiments showed that humic and fulvic acids are efficient carriers of unsupported Pb-210. However some minor amount of unsupported Pb-210 could be found in humate deposits which are strongly bound to inorganic materials. A more detailed analysis of Pb-210 speciation in aquatic sediments is to be found elsewhere (El-Daoushy & Garcia-Tenorio). The accumulation of such humus material under different sedimentological and geochemical conditions plays a major role in controlling Pb-210 fluxes in aquatic systems. Nevertheless more data is required in order to assess the influence of different limnological and sedimentological conditions on the Pb-210 cycle in nature.

The C.R.S. mean deposition rates (Table 1) were determined in a similar way as the Pb-210 fluxes. The total accumulated material containing unsupported Pb-210 were calculated and divided by the corresponding time-interval which according to the C.R.S. model was on the average 150 ± 20 yrs. The mean deposition-rates, as determined using the C.I.C. and the C.R.S. models, are in general agreement. However, while comparing the mean deposition rates the uncertainties in their values were considered. Although the mentioned agreement in mean deposition rates, major discrepancies have been obtained for the C.I.C. and the C.R.S. chronologies (Dearing *et al., op cit.*; El-Daoushy *et al.*, to be published).

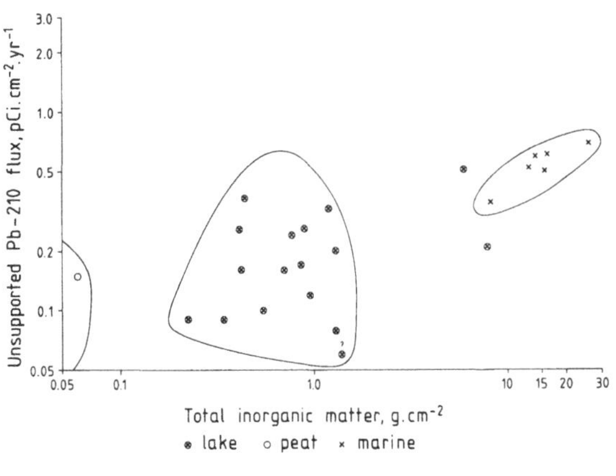

Fig. 3. The unsupported Pb-210 flux, pCi·cm^{-2}·yr^{-1}, and total inorganic matter, g·cm^{-2} of the studied Scandinavian sites. The question mark indicates that bulk densities for the given core are assumed.

Elements such as lead, zinc, copper and probably cadmium seem to behave like Pb-210. The analogy in their geochemistry makes it possible to use Pb-210 to trace these other metals. As the sources of Pb-210, lead, zinc and copper are quite different so they must be chemically homogenized before deposition. Earlier studies showed that these heavy metals, including Pb-210, are completely mixed through atmospheric and aquatic processes and Pb-210 could be an effective tracer (Lewis, *loc. cit.*; McCaffrey & Thomson, *loc. cit.*; El-Daoushy & Tolonen, *op. cit.*). Mobilization of metals and organic compounds in aquatic deposits is assumed to be of minor importance. However, Fe and Mn are easily mobilized from reduced lower zones to the upper oxidized zones and they could be used to give some indications on aquatic oxidation-reduction conditions (Alderton, *loc. cit.*). Mn showed to be redistributed in the studied peats (Table 2). For the marine sites, sediments were collected from regions deeper than, 60 m or so and are expected to be slightly influenced by red./ox. conditions. This could also be seen from the Mn-distribution (Table 2). However, the red./ox. conditions in the studied lakes were not fully examined.

Table 2 gives a summary of the total accumulated amounts, and background values in the past 150 years and per m^2, of heavy metals in seventeen sites from Scandinavian and Finnish aquatic systems (El-Daoushy & Tolonen, *op. cit.*; El-Daoushy *et al., op. cit.*). The chronological pattern of lead, zinc, copper, cadmium and mercury (El-Daoushy & Johansson, *loc. cit.*; El-Daoushy *et al., op. cit.*) in the studied lakes demonstrated the increasing contents of these heavy metals due to atmospheric inputs and drainage-area related weathering effects. Lower background values may be obtained in older deposits as earlier atmospheric pollution, due to coal and wood burning, could be expected. The total accumulated amounts of lead are high for southern Swedish inland lakes (Lake Skärvsjön) and exponentially decreasing towards the northern regions of Sweden with minimum values for Lake Tunturijärvi. The estimated background in the same time interval are considerably lower indicating that lead is mainly introduced to inland lakes via atmospheric processes. Lakes 19, 20 and 21 showed no atmospheric inputs in the past 150 years. The northern lakes are far from heavily populated areas which are mostly concentrated in

Table 2. Summary of total accumulated amounts of heavy metals (the left side of each column) per m² in the past 150 yrs. Thirteen sites from twelve woodland lakes 1 – 4, 13 and 15 – 21, two peatbogs 5 and 6, and two marine sites 26 and 27 are included. The background values (right side of each column) are estimated, for the same period of time, using lake sediments older than 300 yrs, marine sediments of > 200 yrs and peat deposits of 150 – 200 yrs.

Coring site, number on figure 1	Total/Background amounts of heavy metals in the past 150 yrs													
	Pb, $mg \cdot m^{-2}$		Zn, $mg \cdot m^{-2}$		Cu, $mg \cdot m^{-2}$		Cd, $mg \cdot m^{-2}$		Hg, $\mu g \cdot m^{-2}$		Fe, $g \cdot m^{-2}$		Mn, $g \cdot m^{-2}$	
Tussjön, 1	<860	285	1295	390	191	110	15.3	3.1	<1349	370	51	90	-	-
Skärvsjön, 2, core 1	1551	345	1809	1025	156	105	<12.7	3.5	1277	290	581	460	-	-
Skärvsjön, 2, core 2	1010	330	1908	960	145	90	12.3	3.3	609	490	498	410	-	-
Björken, 3	960	315	1739	1260	220	180	6.7	3.6	965	575	429	500	-	-
Väster-Täckelsjön, 4	281	90	3284	3320	184	195	3.8	4.5	3090	2600	1754	1475	-	-
Uggsjön, 13	603	45	1682	1050	198	225	6.6	3	942	725	483	640	-	-
Karsvattnet, 15	260	30	870	820	67	75	2.8	1.5	550	350	125	120	-	-
Rensjön, 16	507	180	2033	1850	190	190	10.1	6	1944	1420	415	400	-	-
Ellisjaur, 17	281	60	3711	3950	162	155	11.1	10.6	982	500	726	775	-	-
Sårvatj, 18	272	46	886	915	136	145	6.0	4.5	765	280	237	260	-	-
Nattajärvi, 19	<114	113	1098	800	104	85	<7.1	7.2	768	460	3381	3564	-	-
Ala Makkarijärvi, 20	<121	120	1099	990	1025	1055	<9.0	8.9	1244	980	410	339	-	-
Tunturijärvi, 21	<81	82	1820	1500	619	735	<5.2	5	1178	1020	1114	800	-	-
Skagerrak, 27	5210	2880	3380	1470	-	-	-	-	-	-	-	-	-	-
Gullmarsfjorden, Hällebäck, 26	6200	3170	20840	9810	3580	2370	56	45	19130	9150	4780	3960	139	60
Kärpänsuo, core 1, 6	220	76	700	72	50	≤23	-	-	-	-	7.0	1.7	0.8	0.05
Kunonniemensuo, core F 9, 5	220	<79	650	~196	40	~15	-	-	-	-	7.3	~2.4	3.4	0.40

southern and coastal regions. Petrol-deriven vehicles, using tetraethyllead as antiknocking agent, are more dominant in southern parts of Sweden. The investigated ombrotrophic peats gave similar results for Pb compared to inland lakes in the same latitudes (Fig. 1). The heavy metal results of ombrotrophic peats emphasized the importance of anthropogenic atmospheric-pollution in perturbing natural aquatic systems. Power stations and industrial exhausts are the main sources of modern atmospheric pollution. However, automobile oil-fuel burning represents the major source of atmospheric lead (Jacobson, 1981). Petit, 1974 showed that stable lead contents in a small Belgian pond increased from the older to more recent sediments. According to him the isotopic composition of the sediments near the water/sediment interface was identical to that of present atmospheric lead. His background stable-lead showed very little contributions of atmospheric lead. However, the marine sites showed very high accumulation of Pb. The atmospheric lead, of direct or indirect origin, in marine sediments is higher by a factor of two and non-atmospheric lead seems to be the same. Atmospheric Zn could only been found in southern Swedish inland-lakes, marine coasts and ombrotrophic peats. Drainage areas of inland lakes may have influenced the accumulation of Zn in their sediments. The same correlation with drainage area was also found for Fe. The low amounts of Fe in ombrotrophic peats indicate that Fe is introduced to lakes and marine coasts through aquatic processes and atmospheric Fe is extremely low. Some atmospheric Cu could be estimated for southern inland lakes, marine coasts and peat lands. However, non-atmospheric sources seemed to be of major importance. This could be seen in Lakes Ala Makkarijärvi and Tunturijärvi. Atmospheric Cd is documented in the analysed aquatic sediments. Other inputs of Cd are likely to be important as well. Atmospheric Hg could be found in all examined inland lakes and show somewhat complicated areal distribution. Marine sediments of site 26 show very high amounts of Hg compared to inland lakes. The amounts accumulated during the past 150 years make 50% of the total Hg at site 26. It seems likely that the investigated marine sites have been influenced by indirect atmospheric inputs through rivers and other incoming water especially with increasing acidification. Direct discharge may be also possible. However the results of the marine environments are very limited and may not represent the general situation in Skagerrak and related coastal areas.

Conclusions

Lake and coastal-marine sediments as well as ombrotrophic peats from Scandinavia and Finland show that typical initial unsupported Pb-210 activities may not be available for vertical depositional successions. Surface anomalies and actual variations in accumulation rates induce significant uncertainties in unsupported Pb-210 initial concentrations. This provide limitations in Pb-210 dating using the C.I.C. model.

A wide range of variations in Pb-210 fluxes are found in Scandinavian and Finnish aquatic systems. However, more data are still required in order to assess various limnological and sedimentological processes influencing the Pb-210 cycle in aquatic systems.

Ombrotrophic peats, inland lakes and some coastal marine sites could give valuable quantitative information on various sources of heavy metal pollution. The atmosphere is an effective carrier of Pb, however Zn, Hg, Cu and Cd may be transported through it. Lake drainage areas could influence the accumulation of Fe and Zn. Further quantitative evaluations based on integrated lake studies rather than single cores would open further possibilities for multi-system interactions modelling.

Acknowledgements

The main part of these investigations was supported by the Swedish Natural Science Research Council and done at the Institute of Physics, Uppsala University. Many valuable contributions, concerning core sampling and heavy metal analyses, have been carried out by other research institutes and the National Swedish Environment Protection Board.

References

Alderton, D. H. M., 1985. Sediments. In MARC report 31, Historical Monitoring. University of London; Lond.: 1–95.

Battarbee, R. C. & G. Digerfeldt, 1976. Palaeoecological studies of the recent development of Lake Växjösjön. Arch. Hydrobiol. 77: 330–346.

Chanton, J. P., G. S. Martens & G. W. Kipput, 1983. Lead-210 sediment geochronology in a changing coastal environment. Geochim. Cosmochim. Acta 47: 1791–1804.

Davis, A. O., J. N. Galloway & D. K. Nordstrom, 1982. Lake acidification: Its effect on lead in the sediment of two Adirondack lakes. Limnol. Oceanogr. 27: 163–167.

Edberg, N., 1980. Ålderbestämning av sediment med ^{137}Cs. Tillämpningar och problem. In F. Ø. Andersen, L. H. Kofoed & E. Lastein (eds), The 8th Nordic Symposium on Sediments. Biological Institute, Odense University, Odense: 20–30.

El-Daoushy, M. F. A. F., 1978. The determination of Pb-210 and Ra-226 in lake sediments and dating applications. Uppsala University Institute of Physics Reports, Uppsala: UUIP-979, 45 pp.

El-Daoushy, M. F. A. F., 1981a. Stability of two silicon detectors in low-level alpha counting. In: Methods of Low-level Counting and Spectrometry, I.A.E.A., Vienne: 151–160.

El-Daoushy, M. F. A. F., 1981b. An ionization chamber and a Si-detector for lead-210 chronology. Nucl. Instr. Methods 188: 647–656.

El-Daoushy, F., 1986. The value of ^{210}Pb in dating Scandinavian aquatic and peat deposits. Radiocar. 28 (in press).

El-Daoushy, F. & K. Johansson, 1983. Radioactive lead-210 and heavy metal analyses in four Swedish lakes. Ecol. Bull. (Stockholm) 35: 555–570.

El-Daoushy, F. & K. Tolonen, 1984. Lead-210 and heavy metal contents in dated ombrotrophic peat hummocks from Finland. Nucl. Instr. Meth. Res. 223: 392–399.

Evans, R. D. & F. H. Rigler, 1983. A test of ^{210}Pb dating for the measurement of whole lake soft sediment accumulation. Can. J. Fish. aquat. Sci. 40: 506–515.

Franzén, L., 1985. Peat in Sweden – a method to calculate the resources (Thesis). Department of Physical Geography, University of Göteborg, Göteborg: 241 pp.

Goldberg, E. D., 1963. Geochronology with lead-210. In: Radioactive Dating, I.A.E.A., Vienne: 121–131.

Jacobsen, O. S., 1981. The historical development of air pollution as reflected in ombrotrophic peat bogs in Denmark. In: Proc. Inter. UNEP-SCOPE Meet., Dynamics of Wetlands and Shallow Water Bodies, USSR, 1981.

Johansson, K., 1980. Tungmetaller i sura skogssjöar. The National Swedish Environment Protection Board, SNV-PM 1359, Solna: 70 pp. [In Swedish, English Summary].

Krishnaswami, S. & D. Lal, 1978. Radionuclide limnochronology. In A. Lerman (ed.), Lakes: Chemistry, Geology, Physics. Springer-Verlag, N.Y.: 153–177.

Lewis, D. M., 1977. The use of ^{210}Pb as a heavy metal tracer in the Susquehanna River system. Geochim. Cosmochim. Acta 41: 1557–1564.

McCaffrey, R. J. & J. Thomson, 1980. A record of the accumulation of sediment and trace metals in a connecticut salt marsh. Adv. Geophys. 22: 165–236.

Oldfield, F. & P. G. Appleby, 1984. Empirical testing of ^{210}Pb-dating models for lake sediments. In E. Y. Haworth & J. W. G. Lund (eds), Lake Sediments and Environmental History. Leicester University Press, Leicester: 93–124.
Olsson, I. U., F. El-Daoushy & Y. Vasari, 1983. Säynäjälampi and the difficulties inherent in dating of sediments in a hard-water lake. Hydrobiologia 103: 5–14.
Petit, D., 1974. ^{210}Pb et isotopes stables du plomb dans des sediments lacustres. Earth Planet. Sci. Lett. 23: 199–205. [In French, English summary].
Robbings, J. A., 1978. Geochemical and geophysical applications of radioactive lead. In J. O. Nriagu (ed.), The Biogeochemistry of Lead in the Environment, Part A. Elsevier, Amsterdam: 285–393.

Accepted 7 April 1986.

Hydrobiologia 143: 277–286, (1986)

A geographical survey of littoral crustacea in Norway and their use in paleolimnology

Steinar Sandøy & Jens Petter Nilssen
Department of Biology, Division of Zoology, University of Oslo, P.O. Box 1050, Blindern, N-0316 Oslo 3, Norway

Keywords: littoral Crustacea, pH, conductivity, altitude, Norway

Abstract

Littoral Crustacea from 400 sites all over Norway were related to lake pH, conductivity, altitude and major biogeography. Being tolerant animals, littoral Crustacea do not fall into clear pH classes as do many Bacillariophyceae or Chrysophyceae. Littoral Crustacea were subdivided into 4 main pH classes: acidobiontic, acidophilous, circumneutral and indifferent, the last group containing most species. Species typical for acidic lakes in the region were: *A. excisa, S. serricaudatus, A. curvirostris, A. nana, S. mucronata* and *D. brachyurum*; whereas *D. longispina, S. vetulus, A. intermedia* and *A. rectangula* were more common in circumneutral waters. Many littoral Crustacea were shown to be related to altitude, which should be considered in studies on lake histories. Some species showed in addition a restricted geographical distribution. Conductivity, with values recorded in this study, played a minor role in distribution of taxa. More ecological studies are necessary to fully utilize the potential of cladoceran remains in understanding lake histories.

Introduction

Remnants of plants and animals recorded in sediments are useful in interpreting lake histories (Frey, 1964). In research on lake acidification diatom shells are frequently used (Davis & Berge, 1980; Renberg & Hellberg, 1982; Battarbee, 1984), whereas animal remains seldom are included in the analysis.

Shells, head shields, postabdomens and claws of Cladocera are the most frequent animal remains found in lake sediments, both within pelagic species like *Bosmina* spp., *Daphnia* spp. and *Diaphanosoma brachyurum* and littoral ones, mainly Chydoridae (Goulden, 1969). One of the major problems in animal analysis has been to relate core findings to animal ecology. The best known ecology is that of the genera *Bosmina* and *Daphnia,* where the relationship to abiotic as well as biotic factors are known for many species. The ecological demands of the Chydoridae, the group most commonly used in paleolimnological analysis, are not well known. Some good studies exist (Quade, 1968; Whiteside, 1970; Goulden, 1971), but very little is known about the geographical and ecological distribution of chydorid species in the areas most affected by acidic precipitation and with strongly acidified lakes, as in the southern part of Scandinavia.

In order to relate Cladoceran community composition of sediment cores to acidification history we need knowledge on the relationship of different species or community types to pH. One way to obtain this would be to make a survey of species composition of Cladocera in lakes with various pH.

Because of meteorological conditions acid rain is restricted to definite geographical areas, and mountain regions are generally more affected than lowland regions. The effect of acidic fallout on lakes and rivers is largely determined by the general geology of the catchment area, which also make mountain regions more vulnerable. Specific conductivity, partly related to geology as well as to human activities, indicate the neutralizing capacity of the water. A lake with high $CaCO_3$ content is not in danger of being acidified, even if the area receives strongly acidic fallouts. The different Cladocerans are also related to geographical distribution, altitude above sea level and conductivity. The altitudinal distribution of species may also be useful in interpreting climatic and habitat changes.

Material and methods

We have here used published data and two unpublished thesis dealing with more than 400 lakes located in South- and Eastern Norway, Mid-Norway and southern part of North-Norway (Fig. 1). Animals were sampled with a plankton net, mesh size 90 μm, both in the pelagial and the littoral zone. We have noted the occurrence of the different species in each lake, quantitative data were only available from a few of the localities.

pH was measured with both potentiometrical and colourimetrical methods. All pH-measurements in the range 4.0–4.5 were made with a colour comparator, suggesting that pH in at least some of these localities was underestimated. Blakar & Digernes (1984) have shown that the addition of indicators may have a considerable effect on pH in weakly buffered solutions like most of the lakes here. The pH of a lake normally undergoes considerable variations throughout the year. If a locality is being surveyed once only, the measured pH will depend to a great extent on the season and on the weather conditions of that particular day. Many of the localities in this study were visited once only, but we assume the uncertainties mentioned above will partly be masked by the number of localities included. Conductivity measurements were corrected for contribution from H^+ ions.

Fig. 1. The investigated regions. Symbols as in Fig. 4.

The *Alona*-species, *A. costata* and *A. rustica* were treated together. The species are morphologically closely related (Frey, 1965) and in many of the lakes they were not distinguished by the authors.

Hustedt (1939) subdivided diatoms in five categories according to pH preference/tolerance:

Acidobiontic (abc) – species which occur at pH below 7, with optimal occurrence at pH 5.5 and below.

Acidophilous (acf) – species which occur at about pH 7, but mainly at pH below 7.

Circumneutral (circ) – species which occur at about pH 7.

Alkaliphilous (alkf) – species which occur at about pH 7, but mainly at pH above 7.

Alkalibiontic (alkb) – species which occur at pH values above 7.

In this study we will subdivide the Cladocera, using the above pH-categories. In addition we group species which are evenly distributed, irrespective of pH, as indifferent (ind). Only a few of the localities had pH above 7.0, none above 7.5, which made it difficult to decide preferences for pH above neutrality. Therefore, some of the species denoted circumneutral possibly are alkaliphilous.

Results and discussion

The pelagic/littoral species *Bosmina longispina* was found in all types of lakes and was common also in the most acidic lakes. *Daphnia longispina* preferred lakes with circumneutral pH and was rare in lakes with pH below 5.5. Distribution of both species in relation to pH agrees well with other investigations (Hörnström, 1973; Nilssen, 1976, 1980;

Hobæk & Raddum, 1980). Distribution of *D. longispina* showed preference both for low and high altitudes. This may partly be due to the fact that many of our high-mountain lakes were situated in less acidified areas and that lowland lakes are normally less vulnerable to acidification.

Diaphanosoma brachyurum seemed to be most common in the acidic lakes and relatively rare in lakes with pH above 6.0. Being a warm water species (Herzig, 1984), *D. brachyurum* was not found in mountain areas, was rare above 600–800 m and absent in North-Norway (cf. Nilssen, 1976). It showed preference for southern and eastern parts of Norway (Fig. 4), the areas with the highest numbers of acidified lakes. Therefore its low-pH preference may largely be an effect of its geographical distribution. This shows, however, that *D. brachyurum* may tolerate acidic conditions as stated by Flössner (1972) and Eie (1974).

Among the littoral crustaceans *Alonella excisa, Alona costata/rustica, Streblocerus serricaudatus, Acantholoberis curvirostris* and *Scapholeberis mucronata* were by far most common in the acidic lakes. *A. excisa* was also found in some of the circumneutral lakes, in about 20% of the lakes with pH between 6.0 and 7.5, but it occurred in more than 90% of the lakes with pH less than 5.0 and thus seems to be most characteristic for acidic lakes. We group all these species as acidobiontic.

Alonella nana also was a common species in the most acidic lakes, but its preference for acidic conditions was not as clear as that of *A. excisa. Alonella exigua, Graptoleberis testudinaria* and *Peracantha truncata* were relatively rare in all types of lakes, but seemed to occur more often in acidic than in circumneutral lakes. *A. exigua* was found in about 50% of lakes with pH below 4.5, *P. truncata* occurred in more than 50% of lakes with pH between 5.5 and 6.0. We group the last four species as acidophilous. Most of the species grouped as acidobiontic or acidophilous were found in all the investigated regions (Fig. 4), indicating that the pH relationship was not an artificial effect of biogeography.

Littoral cladocerans with preference for lakes with circumneutral pH were *Alona rectangula, Alona intermedia, Simocephalus vetulus* and partly *Ophryoxus gracilis*. The above species were relatively rare in all type of lakes, but were especially rare below pH 5.5, and none were found in lakes with pH 4.0–4.5. *A. rectangula* and *A. intermedia* were most common in the geographic areas most affected by acidification (Fig. 4). This shows that their more frequent occurrence in the circumneutral lakes is an effect of avoidance of acidic waters. Distribution of *A. rectangula* in relation to pH indicate that it possibly is an alkaliphilous species. Biogeography and the distribution of acidic lakes in Norway may explain the generally rare occurrence of these species.

Eurycercus lamellatus was evenly distributed at pH above 4.5 and almost absent between pH 4.0 and 4.5. As an attractive prey for fish, one would expect this species to be more common in acidic lakes, usually devoid of fish. We may therefore suggest that *E. lamellatus* prefers less acidic lakes, even if Fig. 2 may indicate its indifferent relationship to pH, at least above pH 4.5. Figure 3 shows that *E. lamellatus* was most common in lakes with relatively high conductivity, while it seems to be indifferent to altitude (Fig. 4).

The other investigated species seem to be indifferent to pH. *Acroperus elongatus* is, together with *B. longispina,* the most widely distributed cladoceran species in Norway and was found in about 85% of the investigated lakes, regardless of lake type. *Acroperus harpae* and *Chydorus sphaericus* also were very common, obviously with no preference to pH, conductivity, altitude or geographic regions.

None of the species in the acid groups were totally absent in lakes with higher pH. Some of the circumneutral species were, however, not represented in the most acidic lakes. Littoral crustaceans thus seem to be fairly tolerant animals and more difficult to put into clear pH-classes than the Bacillariophyceae or Chrysophyceae.

Few studies have been carried out where distribution of chydorids are related to pH. Krause-Dellin & Steinberg (1984) sampled 8 lakes in Germany and, with few exceptions, their results are in accordance with the present paper. *A. excisa,* the species showing the most clear preference for acidic lakes in this study, was the only species grouped as acidobiontic in the German study. Brakke, Davis & Kenlan (1984) studied cladoceran remains in New England lakes. Their species-pH list contains many question marks, and is not in agreement with our data. Geographic varieties or obscure taxonomy may possibly explain the discrepancies in the data

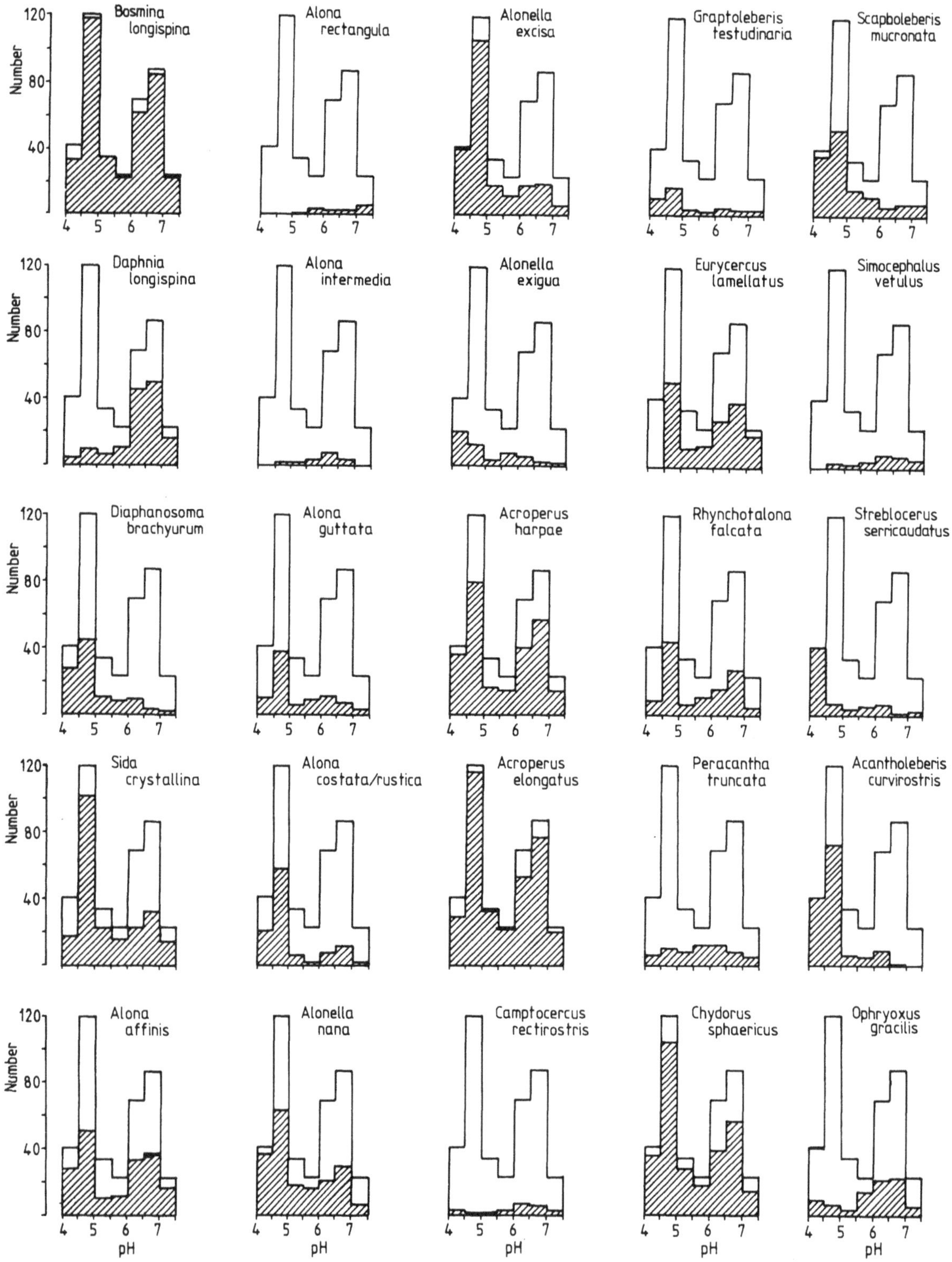

Fig. 2. The frequency of occurrence of cladoceran species in relation to pH. The total area of the histograms represents all investigated localities, striped areas show the number of localities were the species occurred.

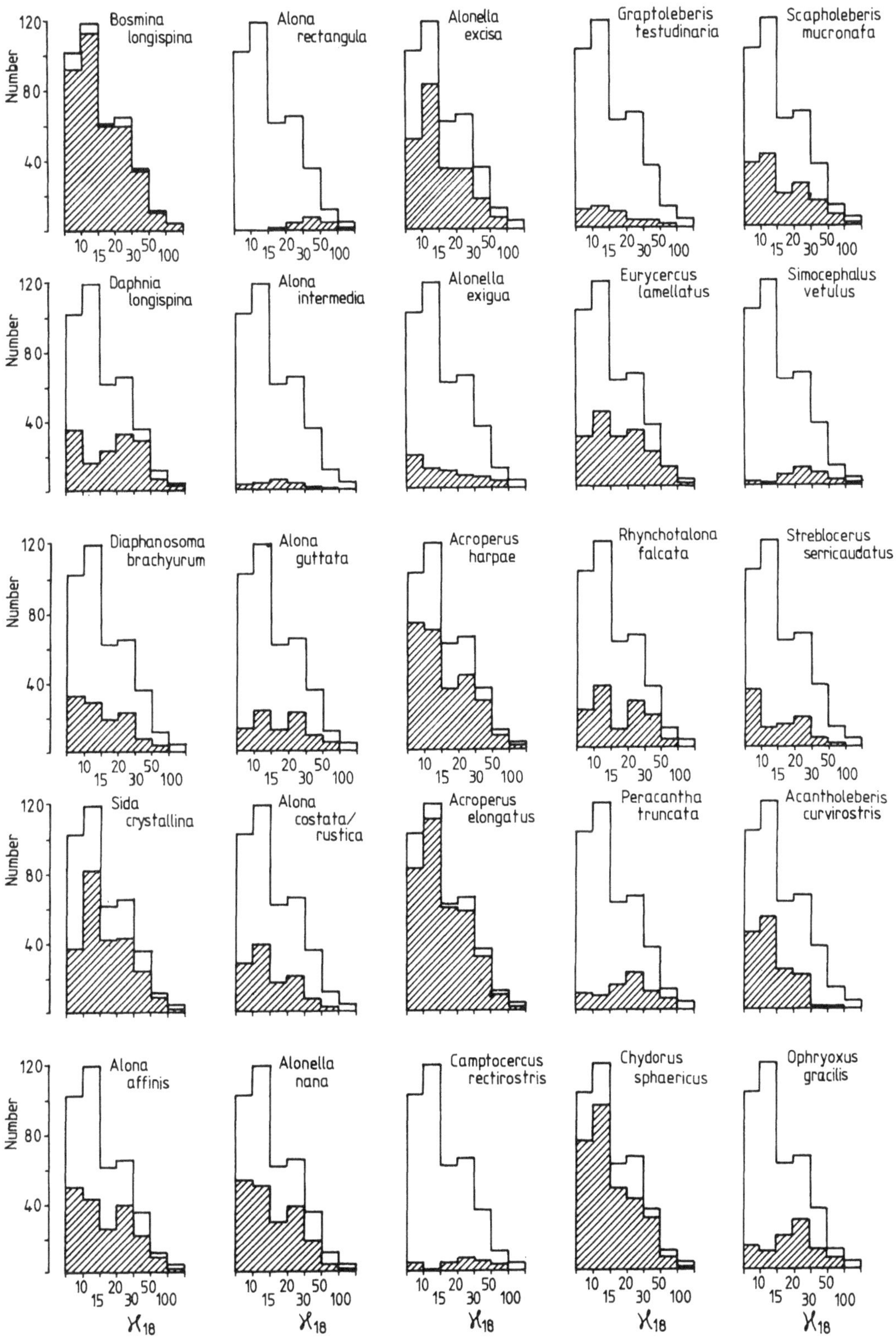

Fig. 3. The frequency of occurrence of cladoceran species in relation to conductivity (H_{18}). Legend as in Fig. 2.

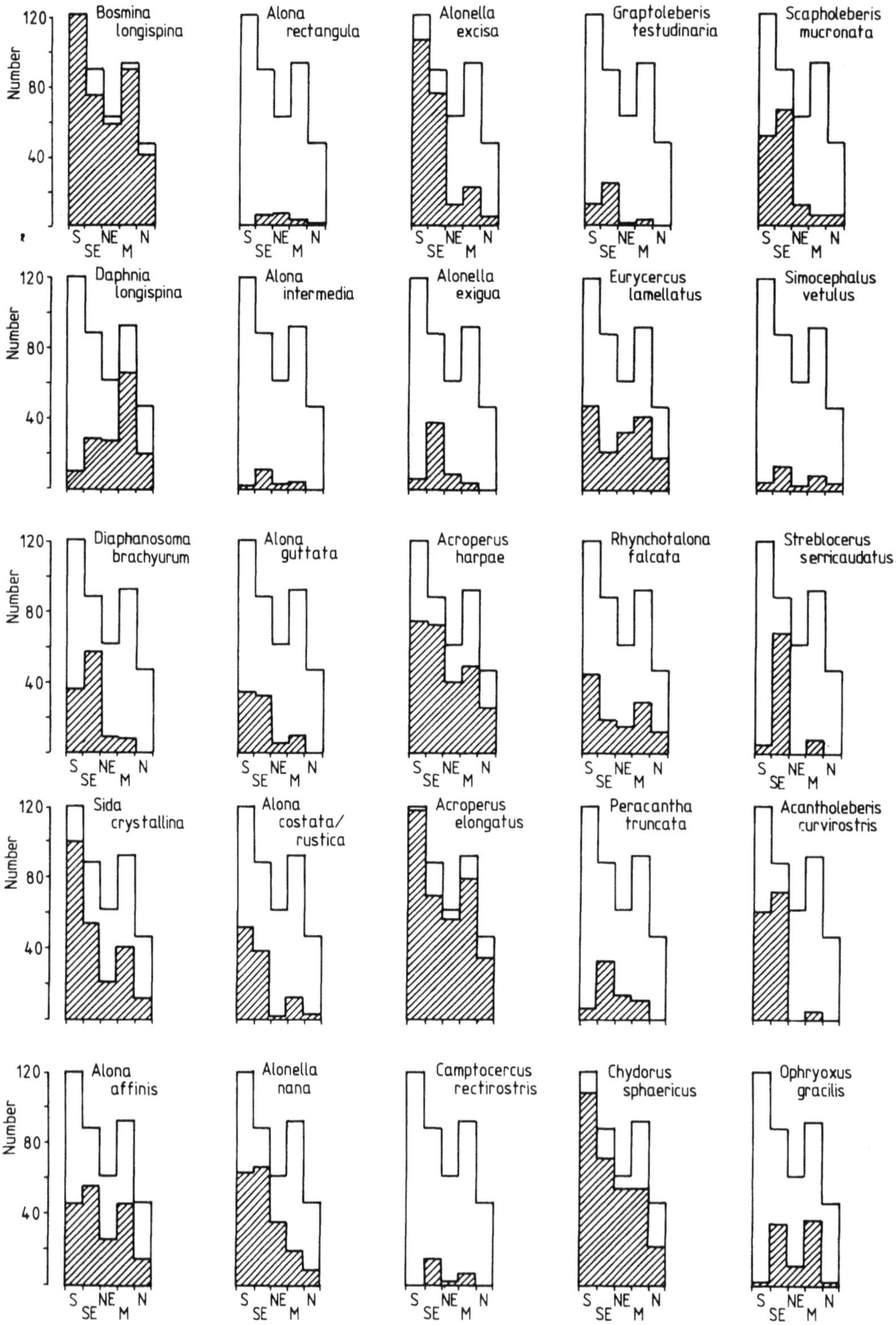

Fig. 4. The frequency of occurrence of cladoceran species in relation to geographical areas. Legend as in Fig. 2. S – South-Norway, SE – South-East Norway, NE – North-East Norway, M – Mid-Norway, N – North-Norway. See also Fig. 1.

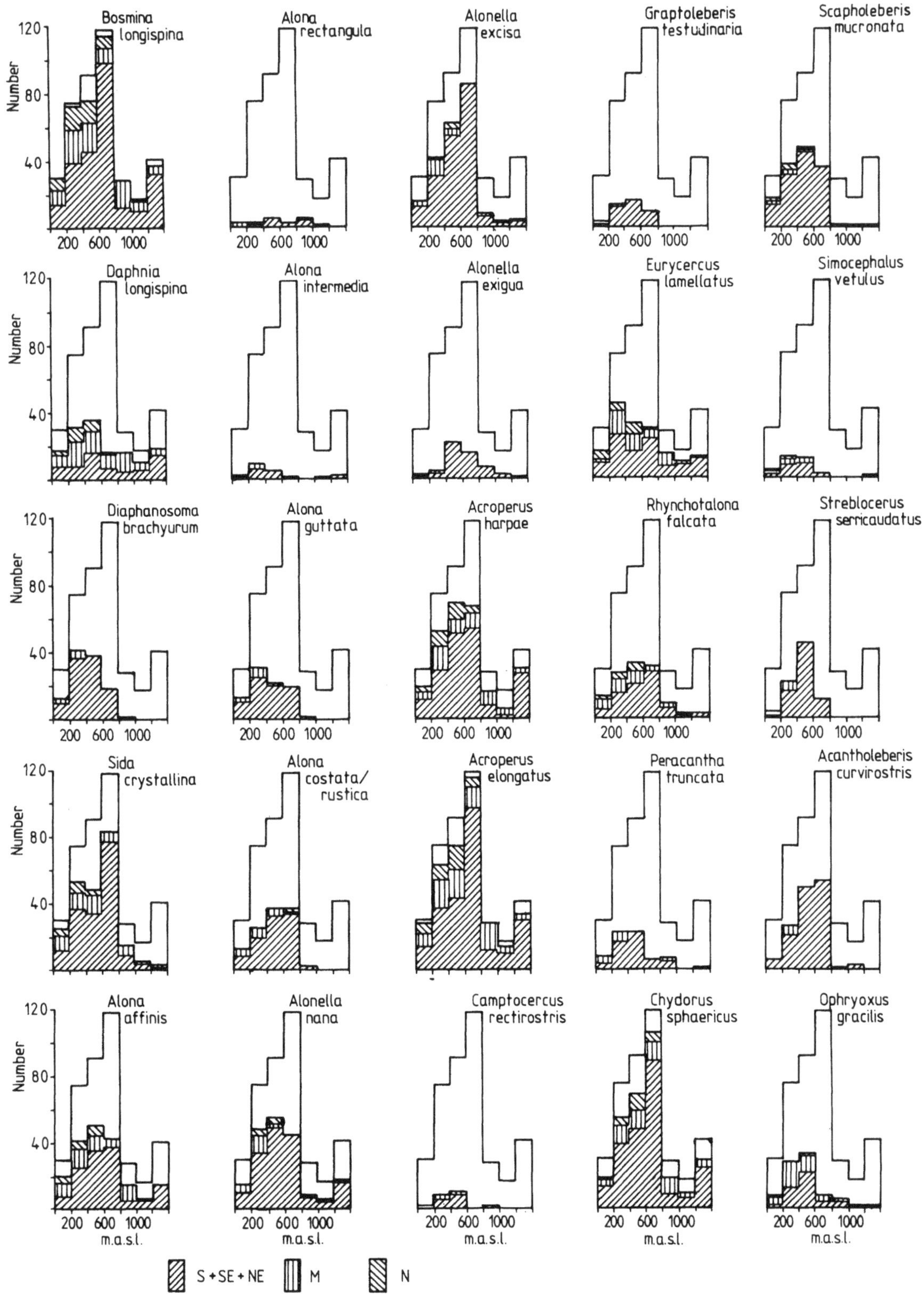

Fig. 5. The frequency of occurrence of cladoceran species in relation to height a.s.l. Legend as in Fig. 2. Symbols as in Fig. 4.

(cf. Frey, 1982). One should be especially careful when comparing the 'same species' on the European and North-American continents. *A. nana* is more common in acidic regions in Scandinavia, but decreases with acidity in some North-American lakes (Paterson, 1984).

S. vetulus, O. gracilis, A. rectangula and to some extent *Rhynchotalona falcata,* all of which were either indifferent to or preferred circumneutral pH, showed some preference to conductivities above 20 μS/cm. However, most of the species were also found at lower conductivities and this parameter seemed to play a minor role in distribution of taxa.

Species restricted to lower altitudes were: *D. brachyurum, Alona guttata, A. costata/rustica, S. serricaudatus, A. curvirostris* and *G. testudinaria.* Except for *A costata/rustica,* none of them were found in North-Norway. In South-Norway they were restricted to altitudes below 800 m, and below 600 m in Mid-Norway.

To be of any use in paleolimnology, a species must leave recognizable remains in the sediments. Some species mentioned in this study, mainly the species of the family Macrothricidae, i.e. *S. serricaudatus, A. curvirostris* and *O. gracilis,* are represented by occasional remains only (cf. Frey, 1964) and can therefore be of limited value in core analysis. Paterson (1984) found some of these species of value in analysing the core in acidified lakes.

There is a strong need for further knowledge on the ecology of the cladoceran species leaving remains in the sediments of lakes. The reason for the scarcity of such information is that quantitative sampling of littoral cladocerans is much more difficult than that of pelagic species (cf. Quade, 1968). Table 1 shows the relationship to pH and altitude of the present material. This is to be considered as a first approximation, and will be modified when more data become available.

The different species of Cladocera do not leave remains in the sediments to the same extent. Therefore no straightforward relationship exists between samples of living animals and remains in sediments. This problem can be elucidated with a thorough study on samples from surface sediments. We are carrying out such a study by comparing composition of cladoceran remains of surface sediments in a number of lakes at all altitudes in different parts of Norway, from which hydrochemical data are available.

Table 1. Cladocera from Norwegian lakes.

Species	pH-group	Preferred altitude
Bosmina longispina	ind	
Daphnia longispina	circ	High-Low
Diaphanosoma brachyurum	acf	Low
Sida crystallina	ind	
Alona affinis	ind	
A. rectangula	circ	
A. intermedia	circ	
A. guttata	ind	Low
A. costata/rustica	acb	Low
Alonella nana	acf	
A. excisa	acb	
A. exigua	acf	
Acroperus harpae	ind	
A. elongatus	ind	
Camptocercus rectirostris	ind	
Graptoleberis testudinaria	acf	Low
Eurycercus lamellatus	circ	
Rhynchotalona falcata	ind	
Peracantha truncata	acf	
Chydorus sphaericus	ind	
Scapholeberis mucronata	acb	
Simocephalus vetulus	circ	
Streblocercus serricaudatus	acf	Low
Acantholeberis curvirostris	acf	Low
Ophryoxus gracilis	circ	

This paper showed that some species have preferences for distinct pH-intervals and with more research we hope to relate species communities to distinct pH-intervals. We may then also detect eventually ecological differences between geographical varieties of the same species. Finally we hope to be able to construct a coefficient to predict earlier lake pH based on littoral Crustacea community structure, as is done for diatoms (Renberg & Hellberg, 1982; Battarbee, 1984).

Acknowledgements

We are greatful to Dr. Elizabeth Haworth for improving the English. The study was partly supported by the SWAP Palaeolimnology Programme.

References

(The reports and thesis used in collecting the data are marked with an asterix.)

Battarbee, R. W., 1984. Diatom analysis and the acidification of lakes. Phil. Trans. r. Soc. Lond. B 305, 451–477.

*Bjerke, G. & G. Halvorsen, 1982. Hydrografi og evertebrater i innsjøer og elver i Hemsedal 1979. Kontaktutv. vassdragsreg. Univ. Oslo, Rapp. 49, 50 pp. (In Norw.)

Blakar, I. A. & I. Digernes, 1984. Evaluation of acidification based on former colorimetric determination of pH: The effect of indicators on pH in poorly buffered water. Verh. int. Verein Limnol. 22, 679–685.

Brakke, D. F., R. B. Davis & K. H. Kenlan, 1984. Acidification and changes over time in the chydorid Cladocera assemblage of New England lakes. In G. Hendry, (ed.), Aquatic effects of acid rain. Michigan Univ. Press, Ann Arbor, pp. 85–104.

Davis, R. B. & F. Berge, 1980. Atmospheric deposition in Norway during the last 300 years as recorded in SNSF Lake Sediments, II. Diatom stratigraphy and inferred pH. In D. Drabløs & A. Tollan (eds), Ecological impact of acid precipitation. Proc., Int. Conf. Ecol. Impact Acid Precip., SNSF Project, Oslo, Norway, 270–271.

*Eie, J. A., 1974. A comparative study of the crustacean communities in forest and mountain localities in the Vassfaret area (southern Norway). Norw. J. Zool. 22, 177–205.

*Eie, J. A., 1982. Hydrografi og evertebrater i elver og vann i Grimsavassdraget, Oppland og Hedmark, 1980. Kontaktutv. vassdragsreg. Univ. Oslo, Rapp. 49, 51 pp. (In Norw.)

Flössner, D., 1972. Krebstiere, Crustacea. Kiemen- und Blattfüsser, Branchiopoda. Fischläuse, Branchiura. Die Tierwelt Deutschlands 60 Teil. VEB Gustav Fischer Verlag, Jena. 501 pp.

Frey, D. G., 1964. Remains of animals in quarternary lake and bog sediments and their interpretation. Arch. Hydrobiol. Beih. Ergebn. Limnol. 2: 1–114.

Frey, D. G., 1965. Differentiation of *Alona costata* SARS from two related species (Cladocera, Chydoridae). Crustaceana 8: 159–173.

Frey, D. G., 1982. Questions concerning cosmopolitanism in Cladocera. Arch. Hydrobiol. 93: 484–502.

Goulden, C. E. 1971. Environmental control of the abundance and distribution of the chydorid Cladocera. Limnol. Oceanogr. 16: 320–331.

*Halvorsen, G., 1980. Planktoniske og littorale krepsdyr innenfor vassdragene Etna og Dokka. Kontaktutv. vassdragsreg. Univ. Oslo, Rapp. 11, 95 pp. (In Norw.).

*Halvorsen, G., 1981. Hydrografi og evertebrater i Lyngdalsvassdraget i 1979 og 1980. Kontaktutv. vassdragsreg. Univ. Oslo, Rapp. 26, 89 pp. (In Norw.).

*Halvorsen, G., 1982. Ferskvannsbiologiske undersøkelser i Joravassdraget, Oppland, 1980. Kontaktutv. vassdragsreg. Univ. Oslo, Rapp. 38 I, 59 pp. (In Norw.).

*Halvorsen, G., 1983. Hydrografi og evertebrater i Kosånavassdraget 1981. Kontaktutv. vassdragsreg. Univ. Oslo. Rapp. 62, 62 pp. (In Norw.).

*Halvorsen, G., 1983. Ferskvannsbiologiske undersøkelser i Råkåvatnområdet, Lom og Skjåk, Oppland. Kontaktutv. vassdragsreg. Univ. Olso, Rapp. 64, 43 pp. (In Norw.).

Herzig, A., 1984. Temperature and life cycle strategies of *Diaphanosoma brachyurum:* An experimental study on development, growth, and survival. Arch. Hydrobiol. 101: 143–178.

Hobæk, A. & G. Raddum, 1980. Zooplankton communities in acidified lakes in South Norway. SNSF-project IR 75/80, 132 pp.

Hustedt, F., 1939. Systematische und ökologische Untersuchungen über die Diatomen-Flora von Java, Bali, und Sumatra. III Die ökologische Faktoren und ihren Einfluss auf die Diatomen-Flora. Arch. Hydrobiol. Suppl. 16: 274–394.

Hörnstrom, E., C. Ekström, U. Miller & W. Dickson, 1973. Försuringens inverkan på västkustjöar. Fytoplankton, Zooplankton, Diatomeer, Kemi. Inform. Inst. Freshw. Res. Drottningholm 4, 97 pp. (In Swedish, abstract in English).

*Jørgensen, I., 1972. Forandringer i strukturen til planktoniske og littorale Crustaceasamfunn under gjengroing av humusvann i området Nordmarka og Krokskogen ved Oslo, korrelert med hydrografiske data. Unpupl. Thesis Univ. Oslo. 82 pp. (In Norw.).

*Koksvik, J. I., 1975. Årstidsvariasjoner og døgnrytmikk hos littorale Cladocera (Crustacea) i Målsjøen, Sør Trøndelag. Unpupl. Thesis Univ. Trondheim. 130 pp. (In Norw.).

*Koksvik, J. I., 1976. Hydrografi og evertebratfauna i Vefsnavassdraget 1974. Royal Norw. Society of Science and Letters, Rapp. Zool. ser. 1976–4, 96 pp. (In Norw.).

*Koksvik, J. I., 1977. Ferskvannsbiologiske og hydrografiske undersøkelser i Saltfjell-/Svartisområdet. I. Stormdalen, Tespdalen og Bjøllådalen. Royal Norw. Society of Science and Letters, Rapp. Zool. ser. 1977–2, 60 pp. (In Norw.).

*Koksvik, J. I., 1977. Ferskvannsbiologiske og hydrografiske undersøkelser i Saltfjell-/Svartisområdet. II. Saltdalsvassdraget. Royal Norw. Society of Science and Letters, Rapp. Zool ser. 1977–16, 62 pp. (In Norw.).

*Koksvik, J. I., 1978. Ferskvannsbiologiske og hydrografiske undersøkelser i Saltfjell-/Svartisområdet. III. Vassdrag ved Svartisen. Royal Norw. Society of Science and Letters, Rapp. Zool. ser. 1978–5, 57 pp. (In Norw.).

*Koksvik, J. I., 1978. Ferskvannsbiologiske og hydrografiske undersøkelser i Saltfjell-/Svartisorådet. IV. Beiarvassdraget. Royal Norw. Society of Science and Letters, Rapp. Zool. ser. 1978–9, 66 pp. (In Norw.).

*Koksvik, J. I., 1978. Ferskvannsbiologiske og hydrografiske undersøkelser i Saltfjell-/Svartisområdet. V. Misværvassdraget. Royal Norw. Society of Science and Letters, Rapp. Zool. ser. 1978–12, 43 pp. (In Norw.).

*Koksvik, J. I., 1979. Hydrografi og ferskvannsbiologi i Eitråga, Grane og Vefsn kommuner. Royal Norw. Society of Science and Letters, Rapp. Zool. ser. 1979–9, 34 pp. (In Norw.).

*Koksvik, J. I. & T. Dalen, 1979. Hydrografi og ferskvannsbiologi i Krutvatn og Krutåga, Hattfjelldal kommune. Royal Norw. Society of Science and Letters, Rapp. Zool. ser. 1979–10, 45 pp. (In Norw.).

Krause-Dellin, D. & C. Steinberg, 1984. Evidence of lake acidification by a novel biological pH-meter. Envir. Tech. Letts. 5: 403–406.

Nilssen, J. P., 1976. Community analysis and altitudinal distribution of limnetic Entomostraca from different areas in southern Norway. Pol. Arch. Hydrobiol. 23: 105–122.

Nilssen, J. P., 1980. Acidification of a small watershed in southern Norway and some characteristics of acidic aquatic environments. Int. Revue ges. Hydrobiol. 65: 177–207.

*Nøst, T., 1981a. Ferskvannsbiologiske og hydrografiske undersøkelser i Drivavassdraget 1979–80. Royal Norw. Society of Science and Letters, Rapp. Zool. ser. 1981-10, 77 pp. (In Norw.).

*Nøst, T., 1981b. Ferskvannsbiologiske og hydrografiske undersøkelser i Ognavassdraget 1980. Royal Norw. Society of Science and Letters, Rapp. Zool. ser. 1981-25, 53 pp. (In Norw.).

*Nøst, T., 1982. Hydrografi og ferskvannsevertebrater i Raumavassdraget 1982. Royal Norw. Society of Science and Letters, Rapp. Zool. ser. 1982-2, 74 pp. (In Norw.).

Paterson, M. J., 1984. Paleolimnological analysis of recently acidified lakes in the Adirondacs mountains. Unpubl. Ph. D. thesis, Univ. Indiana, Bloomington.

Quade, H. W., 1968. Cladoceran faunas associated with aquatic macrophytes in some lakes in Northwestern Minnesota. Ecology 50: 170–179.

Renberg, I. & T. Hellberg, 1982. The pH history of lakes in southwestern Sweden, as calculated from the subfossil diatom flora of the sediments. Ambio 11: 30–33.

*Sandlund, O. T. & G. Halvorsen, 1980. Hydrografi og evertebrater i elver og vann i Kynnavassdraget, Hedmark, 1978. Kontaktutv. vassdragsreg. Univ. Oslo, Rapp. 14, 80 pp. (In Norw.).

*Spikkeland, I., 1977. Acidotrofe vann og dammer i Bygland, Aust-Agder. En undersøkelse av hydrografi og limnetiske og littorale crustacesamfunn. Unpubl. Thesis Univ. Oslo. 119 pp. (In Norw.).

*Spikkeland, I., 1979. Hydrografi og evertebratfauna i innsjøer i Tovdalsvassdraget 1978. Kontaktutv. vassdragsreg. Univ. Oslo, Rapp. 8, 93 pp. (In Norw.).

*Spikkeland, I., 1980. Hydrografi og evertebratfauna i vassdragene i Sjåvatnområdet, Telemark 1979. Kontaktutv. vassdragsreg. Univ. Oslo, Rapp. 18, 49 pp. (In Norw.).

*Spikkeland, I., 1980. Hydrografi og evertebratfauna i vassdragene på Lifjell, Telemark 1979. Kontaktutv. vassdragsreg. Univ. Oslo. Rapp, 19, 55 pp. (In Norw.).

*Spikkeland, I., 1983. Hydrografi og evertebratfauna i Sokndalsvassdraget 1982. Kontaktutv. vassdragsreg. Univ. Oslo, Rapp. 66, 79 pp. (In Norw.).

Whiteside, M. C., 1970. Danish chydorid Cladocera: Modern ecology and core studies. Ecol. Monogr. 40: 79–118.

Accepted 18 March 1986.

Hydrobiologia 143: 287–291, (1986)

Preliminary study of the development of the subfossil chironomid fauna (Diptera) of Lake Taylor, South Island, New Zealand, during the younger Holocene

Barbara Schakau
Department of Zoology, University of Canterbury, Christchurch, New Zealand. Present address: Department of Geography, Monash University, Clayton Victoria 3168, Australia

Keywords: paleolimnology, chironomidae, subfossil remains, Aranuian, lake Taylor

Abstract

The composition and succession of the subfossil chironomid fauna has been studied in a sediment core from Lake Taylor. The sediments of the core are all within the *Nothofagus* phase of the last 3000 years within the Aranuian. In the older sediments of the core the chironomid fauna was dominated by Tanytarsini and Orthocladiinae whereas Chironomini dominated in the more recent sediments. The species diversity is very high compared with the results of other studies of the recent chironomid fauna of lakes in the South Island. The most abundant head capsules belong to the taxa *Chironomus zealandicus, Tanytarsus vespertinus, Corynocera* sp., and *Eukiefferiella* sp. Several additional taxa which have not yet been described also occurred in large numbers. It is not possible to correlate any definite changes in the composition of the subfossil chironomid fauna with a change in climatic conditions or the trophic status of the lake.

Introduction

In North America and Europe, analyses of subfossil chironomid remains from lake sediments have proved to be a useful paleolimnological technique for indicating the changes of limnological conditions during the developmental history of lakes. In New Zealand, there exist only two paleolimnological studies. Deevey (1955) analysed a short core from the Pyramid valley upper swamp deposit in the South Island without a detailed chironomid analysis. Boubee (1983) studied the developmental history of Lake Maratoto, North Island, using chironomid remains. The results of his study showed a succession of the subfossil chironomid fauna as a response to changes in the typological development of the lake.

The intention of my study has been to gain information about the distribution and succession of chironomid taxa in the past of South Island lakes.

Study area

Lake Taylor (Fig. 1) is one of the high country lakes of the upper Hurunui valley, North Canterbury. The lake is at latitude of 42° 46′ South, longitude 172° 14′ East. The area of the upper Hurunui valley was greatly influenced by the last glaciation (the Otira) and L. Taylor has existed since the end of the Poulter advance – the last glacial advance of Otira – which came to an end about 14000 years B.P. (Suggate 1965). The lake has an area of approx. 1.85 km^2 and a maximum depth of 40 m (Irwin, 1975). The chemical quality of the lake is typical of Canterbury high country lakes: low in major ions and low in nitrogen and phosphorus. As stated by previous studies, the trophic status of the lake ranges between oligotrophic and mesotrophic (Bowden, 1977). Thermal stratification did not occur during previous sampling programmes.

Additionally, short sediment cores from Lake Monowai, one of the deep, cold, and oligotrophic lakes in the south of the South Island, and from Lake Grasmere in the Waimakariri River catch-

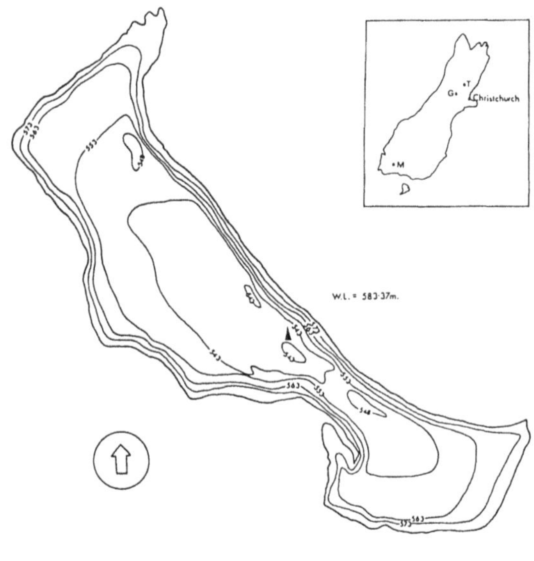

Fig. 1. Lake Taylor: Bed contours, (m.s.l.), scale 1:150 000. The black triangle indicates the sampling site of the sediment core. The inlet shows the South Island of New Zealand with the location of the 3 studied lakes (M = Lake Monowai, G = Lake Grasmere, T = Lake Taylor).

ment, have been studied to provide material for a comparison of different subfossil chironomid communities (Fig. 1).

Material and methods

The sediment core of Lake Taylor was sampled by the Oceanographic Institute, Wellington, using a Hunt piston corer. The sampling site was in mid-lake at a water depth of 37 m. The 1.85 m long core was subsampled at 10 cm intervals, each subsample consisting of 1 ml dry sediment. For the determination of chironomid remains, the subsamples were treated as described by Boubee (1983). Both New Zealand identification keys (Forsyth, 1971, Stark, 1981b) and overseas keys (Hofmann, 1971, Oliver *et al.,* 1978, Wiederholm, 1983) were used to identify the chironomid remains. The interpretation of the results was based on the relative abundance of the chironomid taxa in the subsamples.

The composition of the subfossil fauna in the subsamples was compared by using Sørenson's (1948) Quotient of Similarity. Afterwards, a cluster analysis was carried out (Southwood, 1966).

According to the radiocarbon dating, the oldest sediments of the core were deposited 3000 years B.P. (Institute of Nuclear Science, Wellington, File CA/4/1). A pollen analysis showed that all studied samples were dominated by *Nothofagus fusca* type pollen and all sediments of the core came within the *Nothofagus* phase of the Aranuian.

The Aranuian (Suggate & West, 1967) began with the first warming after the last glacial advance and corresponds with the late- and postglacial period in Central Europe. During the Aranuian, the succession of vegetation in different South Island districts was through grassland-shrubland to shrubland and then to forest with a succession from *Podocarpus* to *Nothofagus* forest (Burrows, 1979). There exists no pollen diagram of the Hurunui catchment but present evidence indicates that *Nothofagus* spread in the Waimakariri valley approx. 6000 years B.P. (Moar 1971) and the same timing may apply in the Hurunui valley (Moar, pers. comm.).

Results

A total of 1831 chironomid head capsules representing 40 taxa was identified from the 20 subsamples of the Lake Taylor core. The total number in the subsamples varied between 21 to 196 head capsules with an average of 91 head capsules/ml dry sediment. The Tanytarsini were the most important subgroup with 30.5% relative abundance, followed by the Chironomini with 27.5% and Orthocladiinae with 25.7%. Tanypodinae occurred with a relative abundance of 10.5% and Podonominae with only 0.2%.

The abundance of chironomids showed strong fluctuations (Fig. 2) but in the older sediments of the core (from 185 cm to 110 cm), the number of head capsules was generally larger than in the more recent sediments, and the chironomid fauna was dominated during these times by Tanytarsini and Orthocladiinae whereas a slight domination of Chironomini took place from 70 cm up to the top of the core.

Of the 40 taxa distinguished, 18 taxa belong to yet undescribed species. The head capsules of these taxa make up 27% of the total number of chironomid remains. Only 5 taxa of the 18 distinguishable Orthocladiinae taxa could be identified to generic level. Thus, 15% of the Orthocladiinae head capsules were not determinable. In the sub-

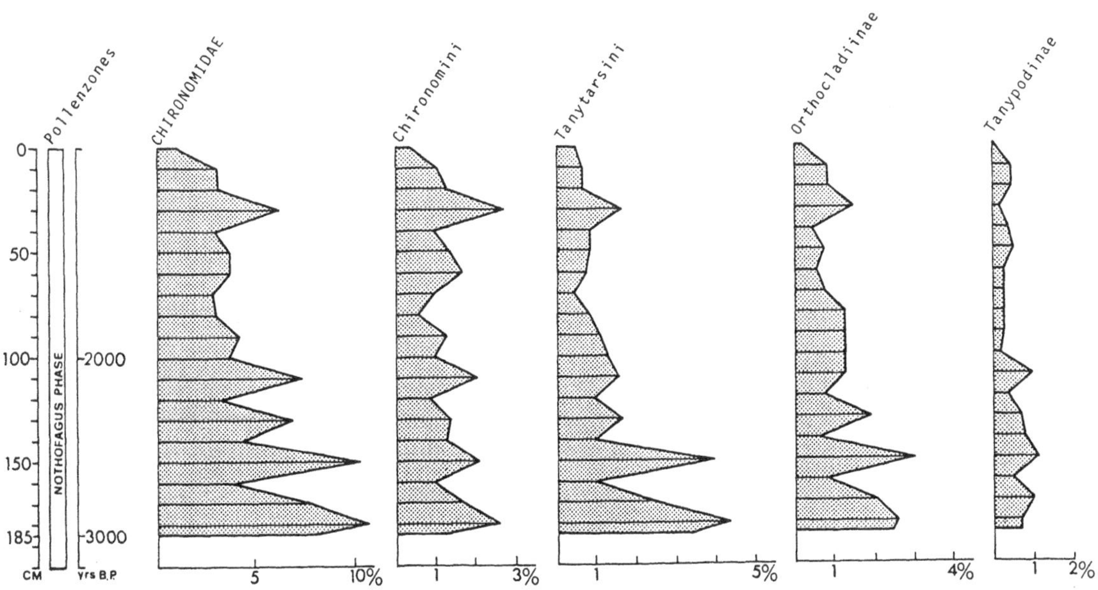

Fig. 2. The relative abundance (% of the total number) of total chironomids and chironomid subgroups in the sediment core of Lake Taylor. Radiocarbon dating according to the Institute of Nuclear Science, Wellington.

group Tanytarsini. 2 of the 8 taxa were undescribed, 7.7% of the Tanytarsini remains belonged to these 2 taxa. 4 undescribed Chironomini taxa were found in the relative abundance of 4.6%. A total number of 11 Chironomini taxa was distinguishable. Tanypodinae presented the greatest taxonomic difficulties; most of the head capsules were too much fragmented to make an identification possible, except for some head capsules which belong to *Macropelopia umbrosa, Gressittius antarcticus* and *Psectrotanypus* sp.

The most abundant chironomid genera (Fig. 3) were *Chironomus zealandicus* (14.7%), *Tanytarsus vespertinus* (14.0%), *Corynocera* (7.5%), *Eukiefferiella* (4.0%), *Cricotopus* (3.7%), and *Paucispinigera* spec. a (3.3%). Some of the undescribed taxa occurred in quite large numbers, e.g. Chironomini A (4.0%), Tanytarsini A (7.5%), and Orthocladiinae A (3.5%). The relative abundance of 26 taxa was smaller than 1% of the total remains. Some of the rare taxa were *Cladopelma, Calopsetra funebris, Stempellina, Synorthocladius* and *Corynoneura. Stempellina* and *Synorthocladius* are not yet recorded in New Zealand. I identified these taxa with the help of overseas keys.

The subfossil chironomid fauna of Lake Grasmere consisted of 24 taxa (incl. 6 undescribed taxa). Tanytarsini were the main component of the fauna with a relative abundance of 35%. The dominant taxa were *Corynocera* sp. (29%) and *Chironomus zealandicus* (19%). The composition of the subfossil chironomid fauna of Lake Monowai differed from those of both North Canterbury lakes. A total number of 30 chironomid taxa was determinable (incl. 7 undescribed taxa). Orthocladiinae were the dominant component of the fauna (52% relative abundance). Tanytarsini occurred only with a relative abundance of 5%. The so far highest percentage of Podonominae (8.8%) was found and 2% Diamesinae head capsules. The composition of the subfossil chironomid fauna of Lake Monowai with the higher percentage of Orthocladiinae and Podonominae, which are more cold adapted taxa, resembles the composition of the Chironomidae fauna of New Zealand's subantarctic islands (Sublette & Wirth, 1980).

Discussion

Most of the dominant taxa of the subfossil chironomid fauna of Lake Taylor were distributed relatively regularly in the core. *Chironomus zealandicus* appeared in nearly every subsample in high

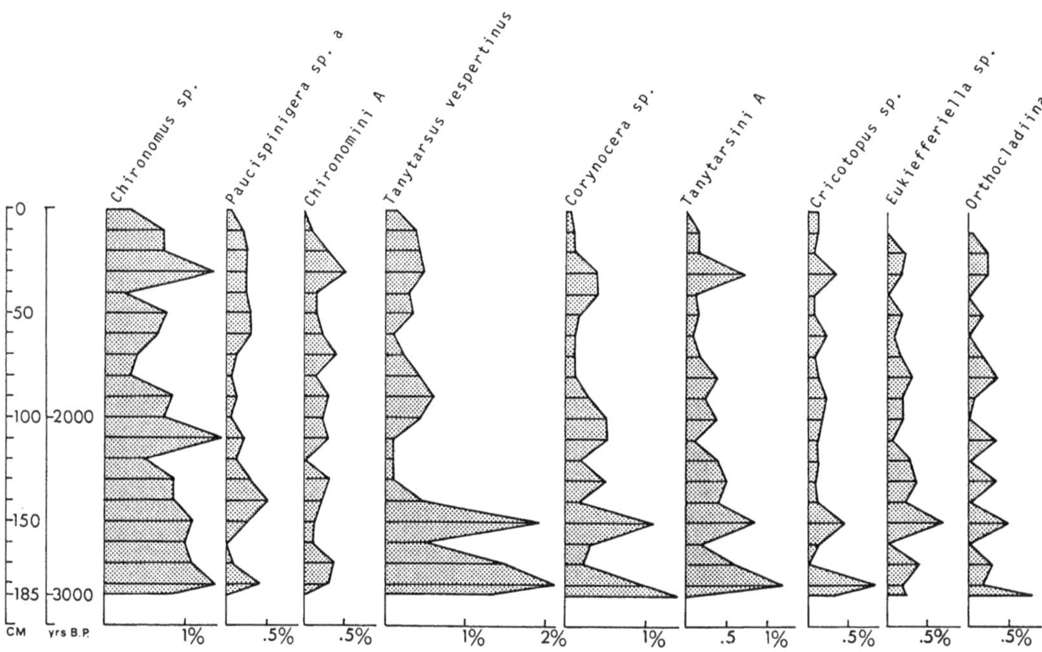

Fig. 3. The relative abundance of dominant chironomid taxa in the sediment core of Lake Taylor.

abundance whereas all 3 dominant Tanytarsini taxa had their maximum abundance in the older sediments and then decreased in numbers. Except for a slight shift in dominance from Tanytarsini to Chironomini in the last 2500 years, the composition of the chironomid fauna remained stable throughout most of the last 3000 years of the developmental history of Lake Taylor. The cluster analysis dendrogram of the 20 subsamples (Fig. 4) based on Sørenson's Quotient of Similarity sug-

Fig. 4. Dendrogram showing the result of the Cluster analysis of the 20 subsamples on the abundances of the subfossil chironomid taxa. The vertical axis gives the percentage values of Sørenson's Quotient of Similarity. The horizontal numbers indicate the numbers of subsamples (1 = subsample from the top of the core, 20 = subsample from the bottom).

gests that there is no faunal zonation. Therefore, it is not possible to correlate any definite changes in the composition of the subfossil chironomid fauna with a change in climatic conditions of trophic status of the lake.

It can be suggested that in comparison with the composition of the subfossil chironomid fauna of oligotrophic Lake Monowai, the relatively high percentages of Chironomini and Tanytarsini of Lake Taylor, especially *Chironomus zealandicus,* indicate a mildly-mesotrophic to mesotrophic status of the lake throughout the last 3000 years of the latter's history. The subfossil chironomid fauna of mesotrophic Lake Grasmere was also dominated by Chironomini and Tanytarsini, with *Chironomus zealandicus* as the dominant component of the fauna in the more recent sediments. *Chironomus zealandicus* can sustain quite low oxygen concentrations and has high abundances in highly productive lakes (Graham, 1976).

It can be regarded as a main result of my study that the species richness of the chironomid fauna of the South Island appears to be much higher than it had been stated before (Timms, 1982, compare also with Stark, 1981a).

Although the high number of undescribed species restricted the analysis of subfossil chironomids, this type of analysis could encourage further searches for living larvae of these species and, therefore, lead to an increase of the taxonomic and ecological knowledge of New Zealand's Chironomidae.

Acknowledgements

I thank Deutscher Akademischer Austauschdienst for the scholarship which made this study possible, Dr. Vida Stout, University of Canterbury, for valuable advice and support, and the Department of Zoology, University of Canterbury, for help received during my stay. Mr. J. Irwin, D.S.I.R., Oceanographic Institute, Wellington, placed the sediment samples at my disposal; Dr. N.T. Moar, Lincoln College, Lincoln, gave me the results of the pollen analysis; the Institute of Nuclear Science, Wellington, carried out the radiocarbon dating. Prof. Weigmann, Freie Universität, Berlin, gave me the possibility to conclude my study in his institute. Gertrud Schlag, Freie Universität, Berlin, read the manuscript critically and made valuable suggestions.

References

Boubee, J. A. T., 1983. Past and Present Fauna of Lake Maratoto with Special Reference to the Chironomidae. Ph. D. thesis, University of Waikato, New Zealand.

Bowden, M. J., 1977. The water resources of the Hurunui catchment. North Canterbury Catchment Board and Regional Water Board, Christchurch, New Zealand.

Burrows, C. J., 1979. A Chronology For The Cool-Climate Episodes In The Southern Hemisphere 12000–1000 Yrs. B.P. Palaeogeography, Palaeoclimatology, Palaeoecology, 27: 287–347.

Deevey, E. S., 1955. Paleolimnology of the upper swamp deposit, Pyramid Valley. Rec. Canterbury Mus. (N.Z.) 6: 291–344.

Forsyth, D. J., 1971. Some New Zealand Chironomidae (Diptera). J. r. Soc. N. Z. 1: 113–144.

Graham, A. A., 1976. Ecology and Production of *Chironomus zealandicus* in Lake Hayes. M. C. thesis, Dep. of Zoology, Otago University, New Zealand, 96 p.

Hofmann, W., 1971. Zur Taxonomie und Palökologie subfossiler Chironomiden (Dipt.) in Seesedimenten. Arch. Hydrobiol. Beih. Ergebn. Limnol. 6: 1–50.

Irwin, J., 1975. Checklist of New Zealand Lakes. N. Z. Oceanogr. Inst. Mem. 74.

Moar, N. T., 1971. Contributions to the Quaternary History of the New Zealand Flora. 6. Aranuian Pollen diagrams from Canterbury, Nelson and North Westland, South Island. N. Z. J. Bot. 9: 80–145.

Oliver, D. R., D. McClymont and M. E. Roussel, 1978. A key to some larvae of Chironomidae (Diptera) from the Mackenzie and Porcupine River watersheds. Fish. mar. Serv. techn. Rept 791.

Sørenson, T., 1948. A method of establishing groups of equal amplitude in plant sociology based on similarity of species content and its application to analyses of the vegetation on Danish commons. Biol. Skr. K. dansk vidensk. Selsk. N. S. 5: 1–34.

Southwood, T. R. E., 1966. Ecological methods with particular reference to the study of insect populations. Methuen and Co. Ltd., London, 391 p.

Stark, J. D., 1981a. Trophic Interrelationships, Life-Histories and Taxonomy of some Invertebrates associated with Aquatic Macrophytes in Lake Grasmere. Ph. D. thesis, University of Canterbury, Christchurch, N. Z., 256 p.

Stark, J. D., 1981b. Chironomidae (nonbiting midges). In: M. J. Winterbourn & L. D. Gregson: A Guide to the Aquatic Insects of New Zealand. Bull. ent. Soc. N. Z. 5: 60–67.

Sublette, J. E. & W. V. Wirth, 1980. The Chironomidae and Ceratopogonidae (Diptera) of New Zealand's subantarctic islands. N. Z. J. Zool. 7: 299–378.

Suggate, R. P., 1965. Late Pleistocene Geology of the Northern Part of the South Island, N. Z. Bull. N. Z. Geol. Surv. 77, 91 p.

Suggate, R. P. & R. G. West, 1967. The substitution of local stage names for Holocene and Post-glacial. Quaternaria 9: 245–246.

Timms, B. V., 1982. A study of the benthic communities of 20 lakes in the South Island, N. Z. Freshwat. Biol. 12: 123–138.

Wiederholm, T. (Ed.), 1983. Chironomidae of the Holarctic region. Keys and diagnoses. Part. 1. Larvae. Ent. Scand. Suppl. 19, 457 p.

Accepted 18 March 1986.

Hydrobiologia 143: 293–300, (1986)

Biogenic silica record in the sediments of Little Round Lake, Ontario

Teresa L. Newberry & Claire L. Schelske
Great Lakes Research Division, The University of Michigan, Ann Arbor, MI 48109, USA

Keywords: trophic history, nutrient enrichment, diatom, chrysophyte, paleolimnology

Abstract

The biogenic silica (BSi) record has been determined in the sediments of Little Round Lake, Ontario in order to review its postglacial development and study the relationship between BSi and diatoms. BSi concentrations in the sediment stratigraphy were found to correspond for the most part to the trophic history of Little Round Lake. Calculation of accumulation rates for BSi improved the correspondence of the BSi profile to the trophic history. Thus, BSi is a valuable paleoindicator when concentration and flux profiles are considered concurrently. Regression analysis of BSi and diatoms revealed that the concentration of BSi in the sediments was not a simple function of diatom numbers or biovolume, but that factors such as a correction for the rate of dissolution and abundance of chrysophycean scales and cysts were also important.

Introduction

The biogenic silica (BSi) record in the sediments represents the difference between diatom productivity and the amount of BSi either dissolved in the water column or in the sediment (Engstrom & Wright, 1984). Most paleolimnological studies have found good correspondence between the abundance of diatom valves and BSi concentrations despite the potential for the dissolution of diatoms (Bradbury & Winter, 1976; Renberg, 1976; Flower, 1980; Glover, 1982; Schelske *et al.,* 1983; Engstrom *et al.,* 1985). The generally good agreement between BSi and diatoms supports the premise that the BSi measured chemically has the potential of being a good indicator of diatom productivity of the past.

Relating BSi profiles to the productivity of the whole lake presents a more complex problem than simply relating it to the number of diatom valves, and, therefore, has not been as successful. Digerfeldt (1972) found an increase in the accumulation rate of BSi coincident with the cultural eutrophication in the sediments of Lake Trummen, Sweden, but in the early postglacial sediments, low levels of BSi do not agree with other evidence for relatively high productivity. Renberg (1976) also reports unexpectedly low BSi accumulation and concentrations during the early phases of Lake Prastsjön's isolation from the Baltic (~3200 BP). During these periods, poor preservation is a likely cause for the poor agreement between BSi and other measures of lake productivity (Engstrom & Wright, 1984). Schelske *et al.* (1983) believe that silica-limited diatom growth was the cause for reduced BSi storage in Lake Erie, Lake Michigan, and Lake Ontario during periods when these lakes were experiencing increased productivity.

The objective of this study was to determine the BSi record in the sediments of Little Round Lake in order to (1) review its postglacial development and (2) study the relationship between BSi and diatoms. Little Round Lake represents an excellent opportunity to better understand the role of BSi as a paleoindicator because so much is already known about the postglacial history of the lake (McNeely, 1973; Daley *et al.,* 1977; Munro, 1978; Smol *et al.,* 1983; Smol & Boucherle, 1985).

Description and history of Little Round Lake

Little Round Lake (44°47′30″N, 7°41′30″W) is a small, deep, meromictic lake located in southeastern Ontario. It is about 250 m wide and 450 m long with a maximum depth of about 17 m. The lake is presently oligo-mesotrophic with a distinct chemocline at a depth of 11 m.

Smol & Boucherle (1985) described the lake's postglacial history in terms of four 'zones' depicted by changes in pollen composition (Table 1). During the early postglacial period (Zone 1), the lake was oligotrophic. The lake became increasingly enriched throughout Zones 2 and 3a reflecting the higher concentration in deciduous litter and a more favorable climate for nutrient mineralization and export. Smol & Boucherle (1985) presented evidence which supports the development of meromixis during Zone 2 as postulated by Daley *et al.* (1977). Their data, however, show evidence that a breakdown in meromixis also occurred between 200 and 202 cm. The hemlock decline, Zone 3b, signifies a period of accelerated eutrophication. The algal and cladoceran microfossils did not record a return to more oligotrophic conditions with the return of hemlock in Zone 3c, but instead recorded large fluctuations in trophic status. These oscillations have been attributed to recurrent breakdowns in meromixis possibly due to an intermittent outlet. Entrainment of monolimnetic waters would increase phosphorus concentrations three-fold and nitrogen concentrations ten-fold according to calculations performed by McNeely (1973). Therefore, these shifts to holomixis would have a substantial effect on the trophic status of the lake. Zone 4 marks the beginning of human settlement and initial enrichment as a consequence of logging. Smol *et al.* (1983) attribute the strengthening of the chemocline to seepage of salt applied to the roads near the lake. Due to the increased stability of meromixis, Little Round Lake has returned to oligotrophic conditions instead of becoming increasingly eutrophic as is characteristic of lakes in its region (Smol *et al.,* 1983).

Methods and materials

Subsamples from a 4.35-m sediment core taken by Smol & Boucherle (1985) were used for the BSi analysis in this study. The first meter was collected with a Brown piston corer and the remaining sediment was taken with a Livingstone corer. The cores were extruded in the laboratory, split longitudinally and sectioned at 1-cm intervals. Samples 0–20 cm were taken from the 'Frigid Finger' short core obtained by Smol & Boucherle (1985).

BSi was extracted in a 1.0% sodium carbonate solution (Schelske *et al.,* 1985) using a time series technique to correct for mineral interference (Krausse *et al.,* 1983). BSi determination was done at selected intervals on these cores down to the depth which marked the beginning of organic deposition at 270 cm. The analysis of diatom and cladoceran remains by Smol & Boucherle (1985) at 5-cm and some 1-cm intervals depleted sample material leaving these intervals unavailable for BSi analysis. Therefore, samples for silica analysis were chosen as close as possible to these 5-cm intervals plus 1 to 2 additional samples in between. In total, 106 samples were analyzed at 2- to 3-cm intervals throughout the core, except from 0–8 cm and 158–172 cm where no sample material was available. Fifteen samples were run in triplicate to determine reproducibility. This gave a coefficient of variation of ±7.3%. BSi in this study is expressed as SiO_2.

Dates which had been obtained by matching pollen stratigraphies from nearby lakes (Table 1) were used to calculate average sedimentation rates for

Table 1. Description of paleovegetation and stratification of Little Round Lake (after Smol & Boucherle 1985).

Date (B.P.)	Zone	Paleovegetation	Description of stratification
125	4	Culturally disturbed	Stable meromixis
2500	3c	Mixed deciduous coniferous forest	Fluctuations in meromixis
4500	3b		Stable meromixis
7500	3a		Brief breakdown of meromixis between Zones 2 and 3a
10600	2	Coniferous forest	Biogenic development of meromixis
11180	1	Dwarf-shrub tundra with spruce	Holomixis

each zone as well as BSi flux to the sediment.

In order to assess the relative contribution from each diatom species to the total BSi, three different approaches were used:

(1) Biovolume was calculated using volume estimates for individual species based on the best regular geometric approximation of cell shape for each taxon (E. F. Stoermer, unpublished data).[1] This calculation was made on a sample size of 36.

(2) Estimates of frustule volume were obtained from Sicko-Goad *et al.* (1984) and E. Theriot (personal communication).[2] The sample size used for this calculation was 36.

(3) Diatom surface area was calculated using either the length and width or the diameter of each species from Patrick & Reimer (1966, 1975) and applying these parameters to a best regular geometric approximation of cell shape for each species. Area estimates for chrysophycean cysts were obtained by measuring the average dimensions of 10 individuals taken from 5 different depths and for chrysophycean scales by measuring 5 individuals taken from 5 different depths. Cysts were assumed to approximate spheres and scales were assumed to approximate discs. The sample size used for these calculations was 18.

Statistical analyses were done using MIDAS which is a statistical package available through the University of Michigan's Michigan Terminal System. The dataset was found to violate the assumptions of homogeneity but to be close to normally distributed. Given the robustness of the regression model and the significance of the decisions made, the effects of the violated assumptions were assumed to be negligible. Since the biological analysis was done at 5-cm intervals whereas the chemical analysis was done at 2- or 3-cm intervals, only those sample depths where analyses differed by less than 1 cm were used for regression analysis. This reduced the total sample size to 59 unless otherwise noted. Of these, both chemical and biological analyses were performed at identical depths in only 23 samples.

[1]Great Lakes Research Division, University of Michigan, Ann Arbor, MI 48109

[2]Great Lakes Research Division, University of Michigan, Ann Arbor MI 48109

Results

Biogenic Silica Profiles

In general, the BSi profile (Fig. 1) appears to correspond well to the trophic development as proposed by Smol & Boucherle (1985). The average weight percent BSi (wt% BSi) was the lowest during the late glacial/early postglacial periods (Zone 1) when the lake was oligotrophic. It increased throughout Zones 2 and 3a as the lake became increasingly enriched. A sharp decline in hemlock demarks the beginning of Zone 3b and signifies a period of increased productivity. This is manifest by maximum concentrations of BSi in Zone 3b. In Zone 3c, the hemlock is re-established as the dominant conifer and wt% BSi shows a corresponding decrease as nutrient levels decrease.

Wt% BSi fluctuates over most of Zones 2 and 3. This oscillating pattern has an increased amplitude and frequency during Zone 3c. These shifts in wt%

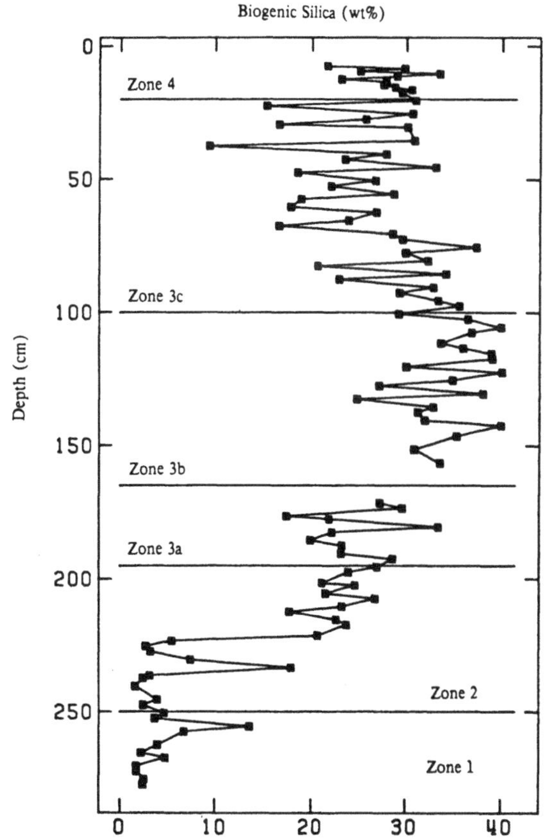

Fig. 1. Weight percent biogenic silica *vs.* depth.

BSi are consistent with evidence found by Smol & Boucherle (1985) on the basis of shifts in cladoceran community structure that suggests alternating states of meromixis and holomixis for Little Round Lake during Zone 2 with these shifts becoming more rapid in Zone 3c. Although the frequency of shifts could not be determined on the basis of biological evidence, better resolution was possible in the BSi analysis because more intervals were sampled.

After cultural disturbance at the beginning of Zone 4, productivity is believed to have increased, but this is not reflected by an increase in BSi concentrations. However, the largest flux of BSi to the sediments is found in this zone (Fig. 2). This increased flux in Zone 4 is explained by the increased sedimentation rate resulting from deforestation. With the exceptions of an unexpected decrease in BSi input in Zone 3a and two pulses at 234 and 256 cm, the flux profile is consistent with the

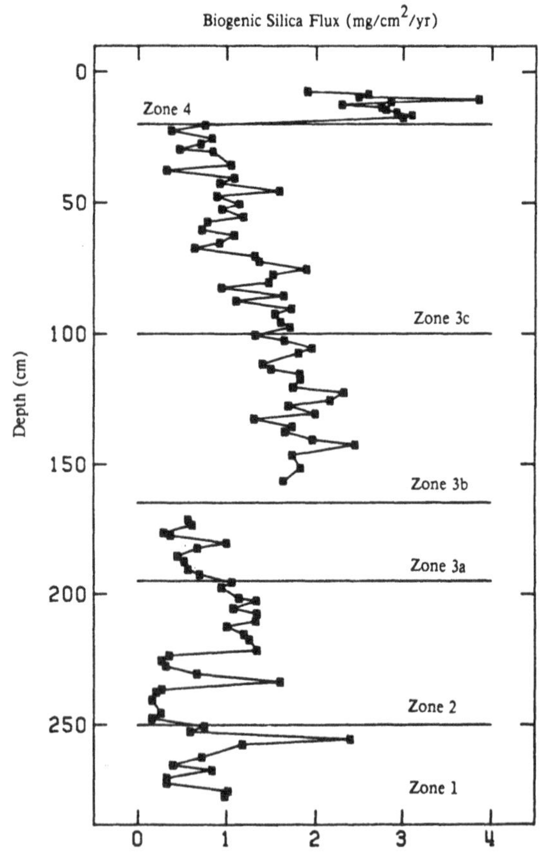

Fig. 2. Biogenic silica flux *vs.* depth.

trophic development as well as with the fluctuations in meromictic status. The reason for the unexpected decrease in BSi flux during Zone 3a could be that the assumption of an 'average' rate of sedimentation is not valid in this zone.

Several models for predicting BSi storage in the sediments have been considered. Of the models tested to determine whether there was a significant relationship between BSi and different measures of cell abundance, significant linear relationships were obtained only with number of diatoms (Table 2). It is somewhat surprising that regressions with adjustments for biovolume and frustule volume were not significant because it is known that the amount of BSi per cell varies more than an order of magnitude (Einsele & Grim, 1938) and that the frustule volume varies from 5 to 50% of the total cell volume (Sicko-Goad *et al.,* 1984). A possible explanation for the nonsignificant linear relationship may be the use of smaller sample size. For example, the regression equation BSi = f (number of diatoms) where $n = 59$ had a much greater significance ($p = 0.0002$) than the same regression equation where $n = 18$ ($p = 0.0759$). Because the sediments contain high numbers of chrysophycean cysts and spores, it is also surprising that their inclusion in the model was not significant.

It can be seen in Fig. 3a that the number of valves per gram of BSi (valves/g BSi) is relatively constant with depth except for a narrow spike in each of Zones 1, 2, 3b, and 4. Further examination of these spikes was considered to be important to finding explanations for the poor correlations in Table 2. Comparison of this plot with a plot of the ratio of diatoms:chrysophytes as a function of depth (Fig. 3b), shows that the peaks in Zones 4 and 3b coincide closely with those depths where diatoms comprised almost 100% of the total sample. An increase in the number of valves/g BSi would be expected at these depths since there would be a decrease in the BSi contribution from the chrysophycean scales and cysts. Likewise, the valves/g BSi profile is similar to the profile of the ratio of diatoms:chrysophytes between the top of the core and the middle of Zone 2. It would seem, therefore, that chrysophytes are contributing significantly to the amount of BSi which was detected in the chemical analysis in this section of the core.

Below 225 cm, before the development of meromixis, increased dissolution of diatoms due to

Table 2. Results of regression analysis.

BSi as a function of (X):	Sample size	Regression coefficients†		Correlation parameters	
		a_1	a_2	r-Value	p-Value
Number of diatoms (offset by 1 cm or less)	59	−0.07 ±3.97	0.58 ±0.14	0.469	0.0002
Number of diatoms (same intervals)	23	18.44 ±14.41	1.39 ±0.51	0.516	0.0117
Number of diatoms	36	−1.11 ±7.33	0.67 ±71.44	0.397	0.0165
Diatom biovolume	36	1393.00 ±1973.90	175.22 ±71.44	0.388	0.0195
Diatom frustule volume	36	1157.80 ±303.27	−7.46 ±10.98	0.116	0.5012
Number of diatoms	18	3.49 ±4.44	0.312 ±0.16	0.429	0.0759
Diatom surface area	18	155.90 ±1011.00	50.68 37.39	0.321	0.1942
Surface area of diatoms plus chrysophytes	18	2220.90 ±1159.10	69.72 42.87	0.377	0.1234
Number of diatoms plus chrysophytes	18	8.00 ±6.08	0.37 ±0.22	0.383	0.1171

† Regression equation: $Y = a_1 X \pm a_2$.

fragmentation most likely occurred (Hurd, 1972). Dissolution of diatoms would decrease the amount of BSi/valve and would thereby increase the number of valves/g BSi which would be observed in Zones 1 and 2. A large peak in the valves/g BSi occurring in Zone 2 corresponds to an interval of high carbonate deposition (Smol & Boucherle, 1985). This leads to speculation that the high pH conditions necessary for carbonate precipitation may have promoted accelerated dissolution of diatoms during these same intervals.

In summary, it appears that before the development of meromixis, the dissolution of diatoms may have affected the amount of BSi in the sediments and that after the development of meromixis, the relatively high abundance of chrysophycean scales and cysts may have contributed to the BSi concentrations in the sediments. Based on the above findings, the correct model for predicting the amount of BSi in the sediments is most likely BSi = f (number of diatoms, chrysophytes, dissolution correction).

Discussion

Since silica depletion can occur in eutrophic conditions when phosphorus enrichment of P-limited waters causes an increased silica demand for diatom production (Schelske *et al.,* 1983), it is possible that BSi profiles may not always reflect the trophic history of a lake. However, in Little Round Lake, the BSi profiles correspond fairly well to the trophic history proposed by Smol & Boucherle (1985). Furthermore, McNeely (1973) reports that the current silica concentrations in the water column are as high as 6 mg/L in early spring and only decline to 3 mg/L in the summer. Little Round Lake, therefore, has most likely never been silica-limited for extended periods of time. This would be expected since periods of eutrophication have been brief and sporadic and silica supplies are most likely renewed rapidly by external inputs.

Presentation of BSi in terms of flux instead of concentration improved the correspondence of the BSi profile to the trophic history of the lake. For example, the correction for changes in sedimentation rates is necessary to obtain the effect of cultural disturbance in Zone 4. These results are consistent with those of Schelske *et al.* (1985) and Engstrom *et al.* (1985) in that accumulation rates of BSi agreed more closely to trophic histories than BSi concentrations. The only exception occurs in Zone 3a. Here the assumption of a constant rate of

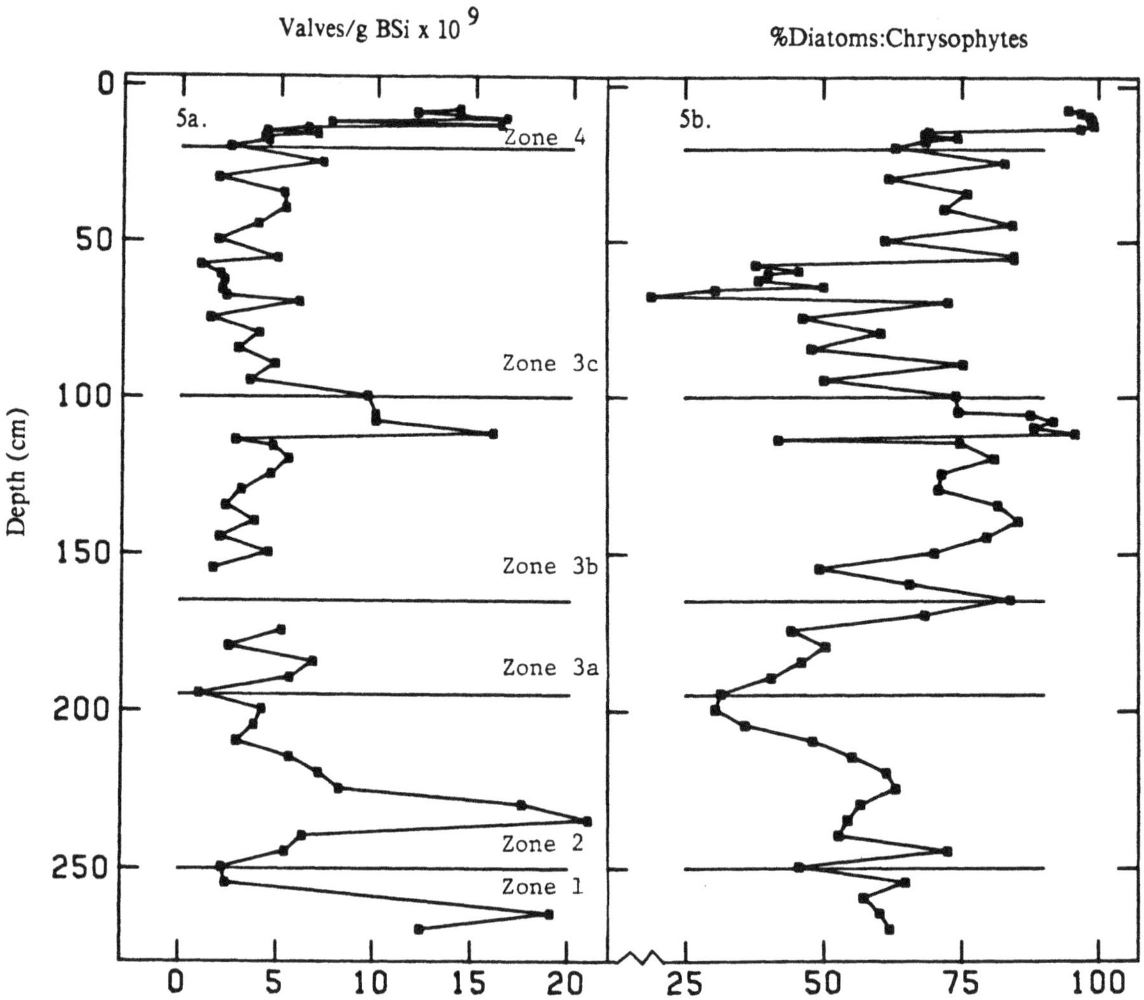

Fig. 3a. Number of valves per gram biogenic silica *vs.* depth.

Fig. 3b. Percent diatoms:chrysophycean scales and spores *vs.* depth.

sedimentation within each zone is most likely invalid.

Since the dissolution of ferric hydroxide coatings on frustules under reducing conditions has been reported to cause an increase in the rate of dissolution of frustules (Lewin, 1961), increased dissolution of diatoms in anoxic monimolimnia might be expected. However, studies by Kamp-Nielson (1974), Graneli (1978) and Yamada & D'Elia (1984) showed no difference in the flux of silica from the sediment under both anaerobic and aerobic conditions. Also, Merilainen (1971) observed corrosion of diatom frustules beneath the monimolimnion in four meromictic lakes, but noted that no serious dissolution has been found in meromictic Lake Mary in Wisconsin, U.S.A. and Längsee in Carinthia, Austria. In Little Round Lake, diatoms in the sediments were better preserved after the development of meromixis (Smol & Boucherle, 1985 and personal observation). Since iron concentrations in the monimolimnion of Little Round Lake are very low, the good preservation would be consistent with the speculation by Merilainen (1971) on the existence of a correlation between high iron concentrations and dissolution of diatom frustules. In contrast to Lake Mary and Längsee, sulfide concentrations were found to be low in Little Round Lake. McNeely (1973) believed that precipitation of iron sulfide was occurring in the bottom of Little Round Lake as reported by Vallentyne (1963) since iron concentrations were close to the thermodynamic equilibrium concentrations.

Previously reported values of the number of diatom valves equivalent to one gram of SiO_2 range from $1.5 \sim 4.4 \times 10^9$ (Renberg, 1976; Parker & Edgington, 1976; Flower, 1980; Glover, 1982; Engstrom *et al.,* 1985). The value found in this study is 6.2×10^9 valves/g BSi. Engstrom & Wright (1984) attributed the possible causes of the differences found in various studies to the partial dissolution of the diatoms, differences in the taxonomic composition of the flora, or analytical error in silica determinations or diatom counts. Figure 4a shows that most of the individual values for Little Round Lake fall within the range reported by other workers. The exceptions coincide with those intervals where diatoms comprise almost 100% of the sample or occur in Zones 1 and 2 where there was an increase in the dissolution rate of diatoms.

A comparison of the profiles of diatom:chrysophytes and valves/g BSi indicates that after the development of meromixis, chrysophytes have an effect on the number of valves/g BSi. This is concluded because the trends in valves/g BSi and percent diatoms/chrysophytes in this section of the profiles are similar (Fig. 3a and 3b). As a consequence of their interrelationship, one would have expected that the correlations of BSi as a function of diatoms plus chrysophytes would have been significant. One reason for this inconsistency may be that the BSi content of chrysophytes was not estimated accurately or another that the method for the extraction of BSi did not completely extract BSi from chrysophytes, especially the cysts. Cysts, for example, are globular and most likely contain much more BSi per individual than do diatoms.

Conclusions

The BSi data reported here correspond fairly well with the trophic history of Little Round Lake. Chemical analyses of BSi provided definition of the fluctuations in meromixis hypothesized by Smol & Boucherle (1985) on the basis of shifts in cladoceran populations. Using accumulation rates for BSi instead of concentrations of BSi improved the correspondence to the trophic history in the section representing cultural disturbance (~ 200 years B.P.) when logging activities increased the sedimentation rate. The best model for predicting the amount of BSi in the sediments of Little Round Lake most likely is BSi = f (number of diatoms, chrysophytes, dissolution correction). Further studies are necessary to determine the BSi content of chrysophytes as well as to calculate a correction for the rate of dissolution of diatoms and chrysophytes under varying physical conditions in order to quantify the above relationship.

Acknowledgements

The authors gratefully acknowledge Mary Boucherle and John Smol for providing splits of their core sample. This is Contribution No. 456, Great Lakes Research Division, University of Michigan.

References

Bradbury, J. P. & T. C. Winter, 1976. Areal distribution and stratigraphy of diatoms in the sediments of Lake Sallie, Minnesota. Ecology 57: 1005–1014.

Daley, R. J., S. R. Brown & R. N. McNeely, 1977. Chromatographic and SCDP measurement of fossil phorbins and the postglacial history of Little Round Lake, Ontario. Limnol. Oceanogr. 22: 349–360.

Digerfeldt, G., 1972. The postglacial development of Lake Trummen. Folia limnol. scand. 16: 1–104.

Einsele, W. & J. Grim, 1938. Über den Kieselsäuregehalt planktischer Diatomeen und dessen Bedeutung für einige Fragen ihrer Ökologie. Z. Bot. 32: 545-590.

Engstrom, D. R. & H. E. Wright Jr., 1984. Chemical stratigraphy of lake sediments as a record of environmental change. Lake Sediments and Environmental History. Studies in Palaeolimnology and Palaeoecology in Honour of Winifred Tutin (ed. Lund and Haworth) pp. 11–67. Leicester University Press.

Engstrom, D. R., E. B. Swain & J. C. Kingston, 1985. A paleolimnological record of human disturbance from Harvey's Lake, Vermont: geochemistry, pigments and diatoms. Freshwat. Biol. 15: 261–288.

Flower, R. J., 1980. A study of sediment formation, transport and deposition in Lough Neagh, Northern Ireland, with special reference to diatoms. Ph.D. Dissertation, New University of Ulster.

Glover, R. M., 1982. Diatom fragmentation in Grand Traverse Bay, Lake Michigan and its implications for silica cycling. Ph.D. thesis, University of Michigan.

Graneli, W., 1979. The influence of *Chironomus plumosus* larvae on the exchange of dissolved substances between sediment and water. Hydrobiologia 66: 149–159.

Hurd, D. C., 1972. Factors affecting solution rate of biogenic opal in seawater. Earth and Planetary Science Letters 15: 411–417.

Kamp-Nielson, L., 1974. Mud-water exchange of phosphate and

other ions in undisturbed sediment cores and factors affecting the exchange rates. Arch. Hydrobiol. 73: 218–237.
Krausse, G. L., C. L. Schelske & C. O. Davis, 1983. Comparison of three wet-alkaline methods of digestion of biogenic silica in water. Freshwat. Biol. 13: 73–81.
Lewin, J. C., 1961. The dissolution of silica from diatom walls. Geochim. Cosmochim. Acta 21: 182–198.
McNeely, R. N., 1973. Limnological investigations of a small, meromictic lake, Little Round Lake, Ontario. Ph.D. thesis. Queen's University.
Merilainen, J., 1971. The recent sedimentation of diatom frustules in four meromictic lakes. Ann. bot. fenn. 8: 160–176.
Munro, K. A., 1978. Phytoplankton dynamics in meromictic Little Round Lake, southeastern Ontario, with emphasis on the growth of *Cryptomonas rostratiformis* in nature and in culture. M.Sc. Thesis. Queen's University.
Parker, J. I. & D. N. Edgington, 1976. Concentration of diatom frustules in Lake Michigan sediment cores. Limnol. Oceanogr. 21: 887–893.
Patrick, R. & C. W. Reimer, 1966. Diatoms of the United States. Vol. I. Acad. Nat. Sci. Phil., Monograph No. 13. 688 pp.
Patrick R. & C. W. Reimer, 1975. Diatoms of the United States. Vol. II. Acad. Nat. Sci. Phil., Monograph No. 13. 213 pp.
Renberg, I., 1976. Paleolimnological investigations in Lake Prastsjön. Early Norrland 9: 113–159.
Schelske, C. L., E. F. Stoermer, D. J. Conley, J. A. Robbins & R. M. Glover, 1983. Early eutrophication in the Lower Great Lakes: New evidence from biogenic silica in sediments. Science 222: 320–322.
Schelske, C. L., D. J. Conley & W. F. Warwick, 1985. Historical relationships between phosphorus loading and biogenic silica accumulation in Bay of Quinte sediments. Can. J. Fish. aquat. Sci. 42: 1401–1409.
Sicko-Goad, L. M., C. L. Schelske & E. F. Stoermer, 1984. Estimation of intracellular carbon and silica content of diatoms from natural assemblages using morphometric techniques. Limnol. Oceanogr. 29: 1170–1178.
Smol, J. P., S. R. Brown & R. N. McNeely, 1983. Cultural disturbances and trophic history of a small meromictic lake from Central Canada. Hydrobiologia 103: 125–130.
Smol, J. P. & M. M. Boucherle, 1985. Postglacial changes in algal and cladoceran assemblages in Little Round Lake, Ontario. Hydrobiologia 103: 25–49.
Vallentyne, J. R., 1963. Isolation of pyrite spherules from recent sediments. Limnol. Oceanogr. 8: 16–30.
Yamada, S. S. & C. F. D'Elia, 1984. Silicic acid regeneration from estuarine sediment cores. Mar. Ecol. Prog. Ser. 18: 113–118.

Accepted 18 March 1986.

Hydrobiologia 143: 301–307, (1986)

Litho- and palynostratigraphy at Lobsigensee: Evidences for trophic changes during the Holocene

Studies in the Late-Quaternary of Lobsigensee No 13

B. Ammann
Systematisch-Geobotanisches Institut Universität Bern, Switzerland

Keywords: lake history, lithostratigraphy, palynostratigraphy, laminations, eutrophication, human impact

Abstract

In the pond of Lobsigensee (2 ha, on the Swiss Plateau) lithostratigraphic and palynostratigraphic units for the Late-glacial and the Holocene have been described. During the Late-Glacial the sediment types are mainly controlled by climate (the end of clay deposition coincides with the end of the Oldest Dryas, i.e. with reforestation at the beginning of Bölling) or by facies (lake marl in the littoral, gyttja in the profundal during Bölling and Alleröd). During the Holocene the facies and the changes in the trophic status, partly combined with water level changes, seem to determine the types of sediment. Two phases of natural eutrophication are deduced from laminated sections during the Boreal and the early Older Atlantic. Towards the end of the Younger Atlantic an early example of a probably anthropogenic eutrophication is found in the layers contemporaneous with the neolithic settlement of the Cortaillod culture. Since the Roman colonization the amount of erosional input has markedly increased.

Introduction

At Lobsigensee – a small closed basin on the tertiary Molasse covered by till on the Swiss Plateau – multidisciplinary studies are in progress to elucidate the paleoecology of the lake and its surroundings since the last glaciation. Among other biostratigraphies, pollen stratigraphies have been developed from 12 cores along two transects (see Fig. 1). They reveal 10 local pollen assemblage zones (= local paz) for the Late-Glacial and 20 local paz for the Holocene (Ammann & Tobolski, 1983; Ammann, 1984, 1985). Conclusions can be drawn in terms of regional and local vegetation, climate, lake level changes, trophic status and patterns of pollen deposition within the basin. The present communication shall concentrate on the evidences for changes in the lake level and in the trophic status during the Holocene. Fig. 1a presents the situation of Lobsigensee, covered by Rhone ice during the Last Glaciation, and Fig. 1b the coring sites along the two cross sections LQ and LL.

Description of cross section LQ

The main cross section LQ through the sediments of Lobsigensee is represented in Fig. 2. From the longitudinal cross section LL the most important core LL-160 (see Löffler, 1986; Züllig, 1986, this volume) is projected into LQ. The original sediment description was worked out according to the simplified Troels-Smith-system proposed by Aaby 1979. But for the present synopsis a further reduction of elements was necessary, especially within the symbols for peat and for gyttjas (see caption in Fig. 2); a somewhat more detailed version of the lithostratigraphy is presented in Ammann 1985.

The topmost sediment was often disturbed, either by tillage (cores on land, LQ-20 to LQ-150) or during the coring process under water. All cores were taken by means of two modified Livingstone corers (Merkt & Streif, 1970). In Fig. 2 the columns to the right of the lithostratigraphy show the local paz in Arabic numerals (L1 to L30, see Ammann, 1985) and/or the regional pollen zones according to

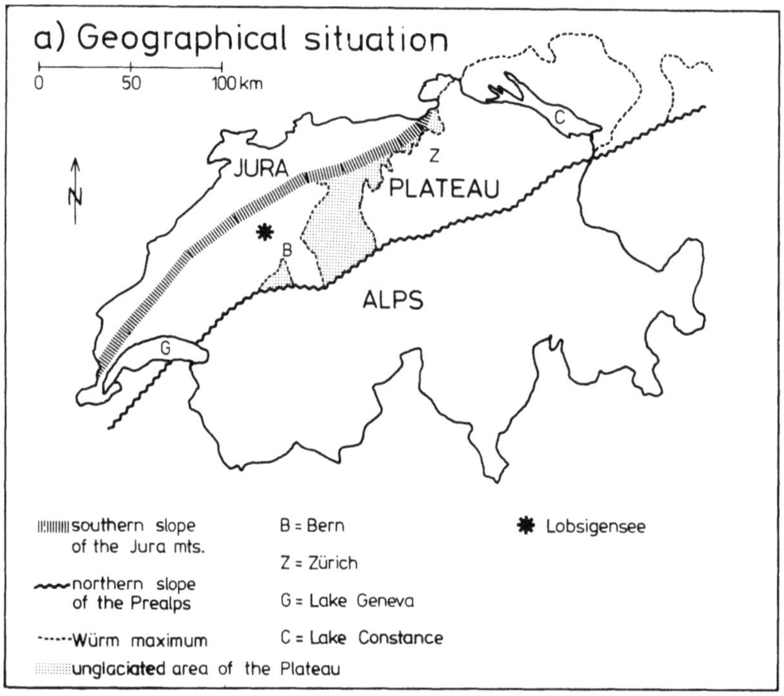

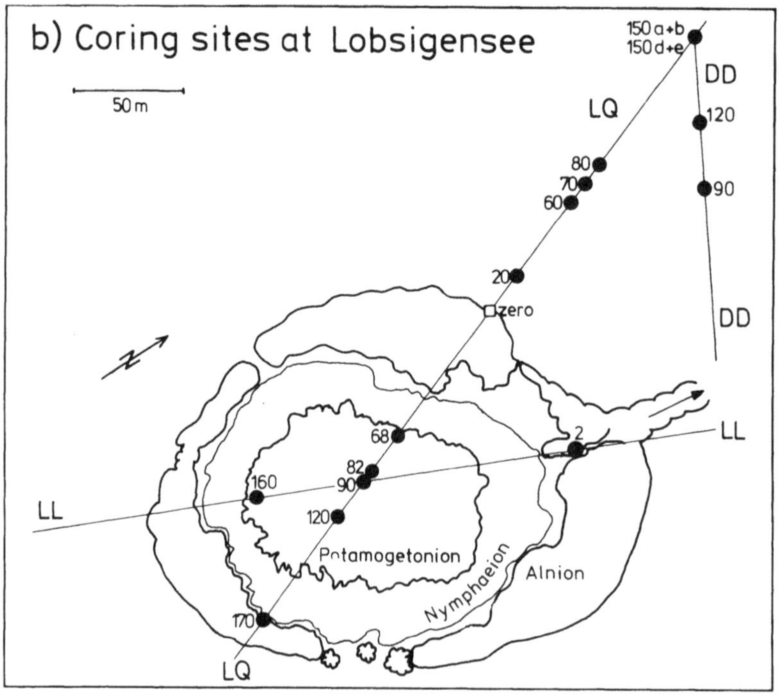

Fig. 1. The site of Lobsigensee, a) Geographical location on the Swiss Plateau. b) Cross sections LQ and LL with sites of cores taken by modified Livingstone samplers. DD = ditch for drainage pipes.

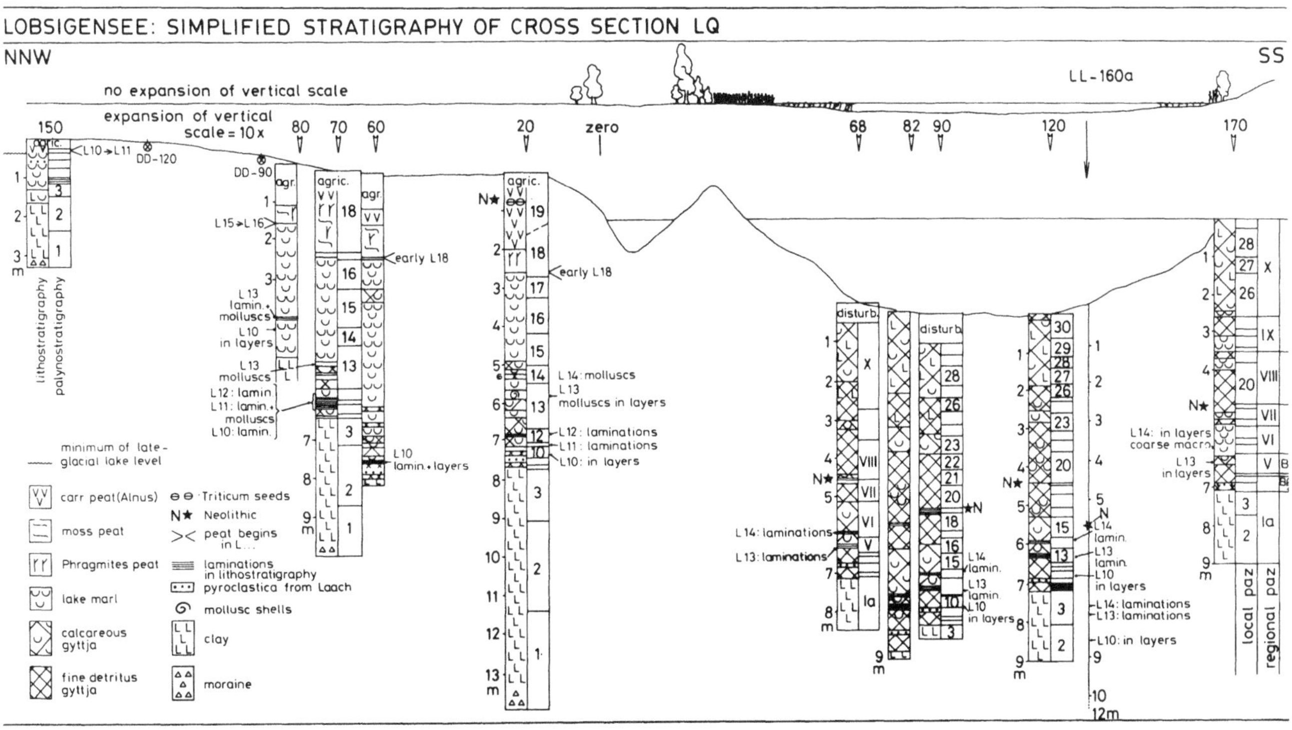

Fig. 2. Simplified litho- and palynostratigraphy of the cross section LQ at Lobsigensee.

Firbas 1949, 1954 in Roman numerals.

As main features we can observe:

– The lake level during the Late-Glacial (local paz L1 to L10) was at least 2 m higher than today. During the Late-Glacial the surface was about 10 ha (today 2 ha), the maximum depth in early Late-Glacial was 17 m (today 2.7 m).

– The SSE-shore is relatively steep, the NNW-shore exhibits a former littoral terrace (LQ-20 to LQ-150); this reflects the general topography and the geology of the gently folded Molasse (Kellerhals & Tröhler, 1981).

– In all cores the clay of the Oldest Dryas was reached (local paz L1 to L3), in some cores even the moraine of the Last Glaciation (Würm), see cores LQ-150, LQ-70 and LQ-20.

– The upper limit of the clay always corresponds to the upper limit of the Oldest Dryas, i.e. the transition to the Bölling. At that time reforestation by juniper and tree-birches took place, soils became stabilized and the inwash of clay stopped.

– The sediment of the Bölling (local paz L4 to L7) and Alleröd (L8 and L9) consists of lake marl in the most littoral cores (LQ-80 and LQ-150) and of gyttja in all the other cores.

– On the littoral terrace hiatuses and 'condensation' of local paz are registered (L4 to L9); these features will be discussed in another paper.

– In general calcareous gyttjas or even lake marl dominate the littoral cores and fine detritus gyttjas dominate the profundal cores.

– During the Younger Dryas (L10) laminations and some alternations of coarser layers are found for the first time. Whereas earlier papers dealt mainly with results from the Late-Glacial and early Holocene sediments (Ammann *et al.*, 1983, 1985) we here want to focus on a prominent feature of the holocene sediment, namely the laminations:

– During the Preboreal (L11) laminations are visible in LQ-20; they are also present over a bed of densely packed mollusc shells in LQ-70.

– During the first part of the Boreal (L12) only two cores show laminated sediments, LQ-70 and LQ-20.

– Nearly all cores have laminated sediments during the second part of the Boreal (L13); in

LQ-20, LQ-70 and LQ-80 they are combined with mollusc beds.

– The next event with laminations occurs during the earliest part of the Older Atlantic (L14), but is registered only in the profundal cores LQ-68 to LQ-170. In the cores of the littoral terrace however the deposit is more or less pure lake marl. The core LQ-20 may be considered as a transition: some bands with more organic material and two layers of molluscs are found in the lake marl. The core LQ-170 at the SSE-end of the transect exhibits instead of laminations some coarse layers of plant macrofossils in the lake marl during L14.

– A narrow but very marked section with laminations is coincident with the local paz L19, marked with an * and N in Fig. 2. It is the sediment deposited during the neolithic settlement of the Cortaillod culture at Lobsigensee. The village of lake dwellers has not yet been excavated but is known from sporadic findings of ceramics and stone axes and from charred grains of wheat as found in LQ-20.

– From archaeological findings (Stöckli, pers. comm.) and two radiocarbon datings (Oeschger *et al.*, 1985) we know that the settlement has an age of about 5000 conventional radiocarbon years.

– The beginning of peat formation is naturally a metachronous process in the basin: in the most littoral core LQ-150 it is synchronous with the transition from Late-Glacial to Holocene i.e. from Younger Dryas to Preboreal (L10 to L11). Towards the lake it becomes younger: in core LQ-80 during the Older Atlantic (L15 to L16), in cores LQ-70, LQ-60 and LQ-20 in the middle of the Younger Atlantic (early L18).

The nature of laminations and their conditions of formation

The laminations found in the above described stratigraphic positions are conspicuous to the eye by their colors: grey-whitish lake marl alternates with dark brown fine detritus gyttja. But the limits of the layers are unsharp and often uneven or slightly undulating; therefore some layers are lenticular rather than horizontally continuous. Counting the layers would be hazardous. Thin sections in araldit (Merkt, 1971) confirm these observations. The thickness of a twin-layer (one organic + one carbonatic layer) varies between 0.3 and 0.9 mm. Most probably the organic layer was deposited during autumn and winter, the carbonatic one during spring and summer (Kelts & Hsü, 1978; Sturm, 1985, 1986); no distinct diatom layers are visible.

Laminations and their formation have been studied in a vast series of lake types (e.g. by Welten, 1944; Tippett, 1964; Simola, 1977; Saarnisto, 1979; Renberg, 1976, 1981; Renberg & Segerström, 1981; Swain, 1978; Tolonen, 1978; Sturm, 1979, 1985, and others). One condition for preservation of laminae of any kind is valid for all lake types, namely the absence of bioturbation. This is usually connected with the lack of oxygen at least during the summer at the bottom of the lake. Anoxis may be due to meromictic or eutrophic conditions. Considering Fig. 2 we can also ask: why does carbonate, which is usually precipitated in the littoral, reach the center of the lake during periods of laminations (e.g. during L13 and L14 in cores LQ-68 to core LQ-120 and LL-160)?

Either or both of the following reasons may apply:

– The littoral influence became marked in the profundal zone because the littoral zone got closer, i.e. the lake level fell.

– The bicarbonate equilibrium is not only controlled by factors such as temperature and pH but also by the phosphorus concentration (Kunz, 1983; Kunz & Stumm, 1984; Sturm, 1985). A pulse of phosphorus input would thus first inhibit the calcite precipitation; after consumption of the phosphates by the phytoplankton in summer the precipitation of large calcite crystals out of the oversaturated solution would follow.

Discussion of the processes at Lobsigensee

The beginning of peat formation does not necessarily point to a falling lake level but can be the result of infilling processes and overgrowth under more or less constant lake level. But at our site the bottom layer of peat gets both younger and deeper towards the lake. Unfortunately the estimation of the water level from the types of peat is not very exact and seasonal water level fluctuations may be important. But generally we can assume that the mean water table was near the transition from Phragmites peat to carr peat (here mainly alder).

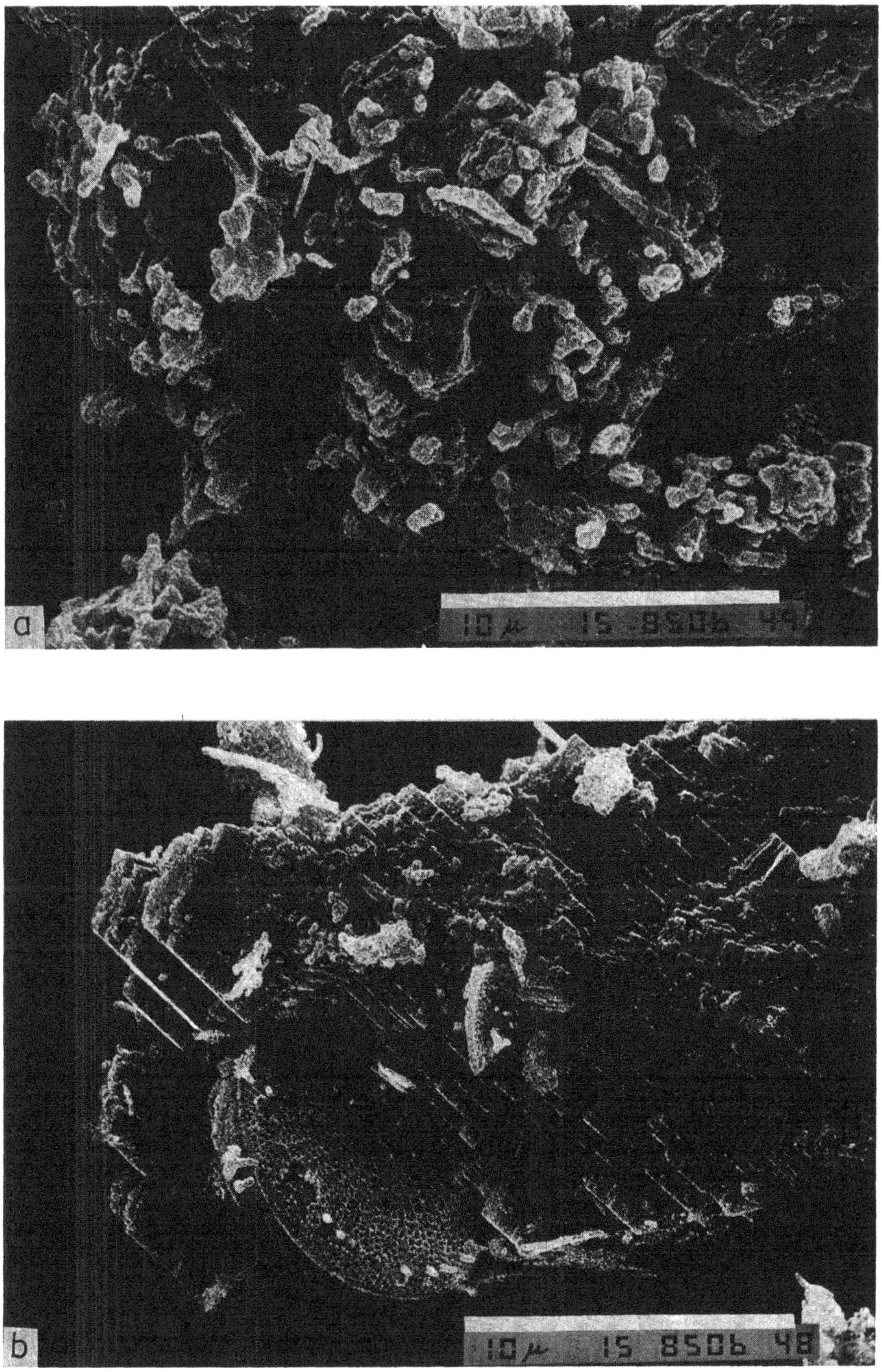

Fig. 3. a) Micritic $CaCO_3$-crystals from the littoral lake marl in core LQ-150a (Lobsigen, local paz L7, Bølling). b) Large, blocky calcite crystals from the profundal core LQ-90 (Lobsigen, local paz L19, Younger Atlantic, neolithic settlement of the Cortaillod culture). A frustule of Stephanodiscus cf. hantzschii functioned as a nucleus. (REM-micrographs taken on Jeol JSM-T 300 at the Systematisch-Geobotanisches Institut Bern).

During April 1981 an artificial ditch for drainage pipes was dug (in Fig. 1b indicated with DD); thus we could sample the transition from lake marl to carr peat in the upper part of the littoral terrace: at point LQ-150 first peat formation at transition L10/L11, at point DD-120 first peat formation during L11 (Preboreal), at point DD-90 first peat formation at transition L11/L12 (PB/BO), at point LQ-80 first peat formation at transition L15/L16 (Older Atlantic), from point LQ-70 to LQ-20 formation in early L18 (Younger Atlantic). This shows that in the early Holocene (Preboreal = L11 and Boreal = L12 and L13) the upper section of the littoral platform became overgrown and thus the volume of the lake became smaller. Remarkable also are the findings of densely packed mollusc layers during L11 (in core LQ-70) and during L13 (in cores LQ-80, LQ-70 and LQ-20) which must be littoral formations. They are possibly remnants of short term fluctuations of unusually low water levels.

As a working hypothesis we therefore suggest: during the climatic warming of the Preboreal and Boreal and with a falling lake level and overgrowth the water body of Lobsigensee became – at least for some periods – small enough to become naturally eutrophic. Eutrophication (or meromixis) is confirmed by Züllig's (1985, 1986 this volume) findings of fossil pigments from *Oscillatoria rubescens* in core LL-160 (laminated part of Boreal) and in core LQ-68 (laminations from early Older Atlantic).

The other marked laminations coincide with the local paz L19. Six holocene pollen diagrams prove, that during L18 the first neolithic settlers were present in the area and that during L19 they lived at the shores of Lobsigensee. As especially noteworthy plant macrofossils Tobolski 1985 recorded the seeds of *Fragaria vesca* and *Papaver somniferum* in L18 of the littoral core LL-2. In core LQ-20 the charred seeds of *Triticum aestivum* have been found in L19. From Fig. 2 we can see that during the zone L18 peat formation started on the lower part of the littoral terrace. According to the stratigraphy of core LQ-20 the lake level was then even lower than today (by about 80 cm).

The size and aspect of the calcite crystals in these neolithic laminations differ distinctly from the ones in ordinary littoral lake marl as Fig. 3 illustrates. Remembering the results presented by Kunz 1983 and Sturm 1985, we propose our second working hypothesis: these large blocky calcite crystals can be a result of a fossil P-input by the neolithic lake dwellers into a lake of a smaller volume and would thus be an example of an early human impact on a lacustrine environment. A reduction in the volume of the water body obviously played a role in the formation of all laminated sections (L11 to L19) as they were preceded by phases of a falling lake level and steps of peat formation on the littoral terrace (during L11 to L18). In addition, the two main steps of eutrophication were caused one by the climatic warming in the early Holocene and the other by an antropogenic phosphorus input during a neolithic phase of settlement. This means that we have an example of a natural and of a more man-made eutrophication respectively.

In all holocene pollen diagrams from Lobsigensee (e.g. Ammann, 1985) two major levels of changes among herbaceous plants are recorded: one at the level of the neolithic settlement (L19) and another at the level of the Roman colonization (L23). Oeschger *et al.* 1985 could show by means of a series of 10 radiocarbon dates through the Holocene that this latter signal is also visible in the age-depth relation: shortly before 2000 B.P. (conventional) the sedimentation rate increases by a factor of 3.7. The increasing portion of inwashed clay in the Roman and post-Roman sediment pinpoint to the same agent – intensified deforestation by man.

Acknowledgements

My cordial thanks to all who helped with this project: G. Lang, H. Löffler, A. Lotter, M. Moser, F. Oldfield, C. Scherrer, M. Sturm, E. Styner, E. Venanzoni, L. Wick, H. Züllig.

References

Aaby, B., 1979. Characterization of peat and lake deposits. In: B. E. Berglund (ed.), Palaeohydrological changes in the temperate zone in the last 15000 years. Subproject B. Lake and mire environments. Project Guide. Dept. of Quaternary Geology, Lund: 77–98.

Ammann, B., 1984. Prozente, Konzentrationen und Durchschnitts-Influx der Pollenzonen im Spätglazial vom Lobsigensee. (Studien zum Spät-Quartär des Lobsigensees 10). In: G. Lang (ed.), Festschrift Max Welten, Diss. Bot. 72: 11–44.

Ammann, B., 1985. Introduction and palynology: vegetational history and core correlation at Lobsigensee (Swiss Plateau). In: G. Lang (ed.), Swiss Lake and Mire Environments during the last 15000 years. Diss. Bot. 87: 127–134.

Ammann, B. & K. Tobolski, 1983. Vegetational development during the Late-Würm at Lobsigensee (Swiss Plateau). Studies in the Late-Quaternary of Lobsigensee 1. Revue Paléobiol. 2: 163–180.

Ammann, B., L. Chaix, U. Eicher, S. A. Elias, M.-J. Gaillard, W. Hofmann, U. Siegenthaler, K. Tobolski & B. Wilkinson, 1983. Vegetation, insects, molluscs and stable isotopes from Late-Würm deposits at Lobsigensee (Swiss Plateau). Revue Paléobiol. 2: 221–227.

Ammann, B., M. Andrée, L. Chaix, U. Eicher, S. A. Elias, W. Hofmann, H. Oeschger, U. Siegenthaler, K. Tobolski, B. Wilkinson & H. Züllig, 1985. An attempt at a palaeoecological synthesis. In: G. Lang (ed.), Swiss Lake and Mire Environments during the last 15000 years. Diss. Bot. 87: 165–170.

Firbas, F., 1949. Spät- und nacheiszeitliche Waldgeschichte Mitteleuropas nördlich der Alpen. 1. Band: Allgemeine Waldgeschichte. Fischer, Jena 480 pp.

Firbas, F., 1954. Die Synchronisierung der mitteleuropäischen Pollendiagramme. Danmarks Geol. Undersögelse II Raakke. 80: 12–21.

Kelts, K. & K. J. Hsü, 1978. Freshwater Carbonate Sedimentation. In: A. Lerman (ed.), Lakes Chemistry, Geology, Physics. Springer New York Berlin: 295–323.

Kellerhals, P. & B. Tröhler, 1981. Geologischer Atlas der Schweiz 1:25000, Blatt Lyss. Schweiz. Geol. Kommission.

Kunz, B., 1983. Heterogene Nukleierung und Kristallwachstum von $CaCO_3$ (Calcit) in natürlichen Gewässern. Diss. ETH Zürich Nr. 7355, Zürich, 159 pp.

Kunz, B. & W. Stumm, 1984. Kinetik der Bildung und des Wachstums von Calciumcarbonat. Vom Wasser 62: 279–293.

Löffler, H., 1986. An early meromictic stage in Lobsigensee (Switzerland) as evidenced by ostracods and *Chaoborus* (this volume).

Merkt, J., 1971. Zuverlässige Auszählung von Jahresschichten in Seesedimenten mit Hilfe von Gross-Dünnschliffen. Arch. Hydrobiol. 69: 145–154.

Merkt, J. & H. Streif, 1970. Stechrohr-Bohrgeräte für limnische und marine Lockersedimente. Geol. Jb. 88: 137–148.

Oeschger, H., M. Andrée, M. Moell, T. Riesen, U. Siegenthaler, B. Ammann, K. Tobolski, B. Bonani, H. J. Hofmann, E. Morenzoni, M. Nessi, M. Suter & W. Wölfli, 1985. Radiocarbon chronology at Lobsigensee. Comparison of materials and methods. In: G. Lang (ed.), Swiss Lake and Mire Environments during the last 15000 years. Diss. Bot. 87: 135–139.

Renberg, I., 1976. Annually laminated sediments in Lake Rudetjärn, Medelpad province, northern Sweden. Geol. Fören. Stockholm Förhandl. 98: 355–360.

Renberg, I., 1981. Formation, structure and visual appearance of iron-rich, varved lake sediments. Ver. int. Ver. Limnol. 21: 94–101.

Renberg, I. & U. Segerström, 1981. Applications of varved lake sediments in palaeoenvironmental studies. Wahlenbergia 7: 125–133.

Saarnisto, M., 1979. Studies of annually laminated Lake sediments. In: B. E. Berglund (ed.), Palaeohydrological changes in the temperate zone in the last 15000 years. Subproject B. Lake and Mire Environments. Project Guide 2: 61–80.

Simola, H., 1977. Diatom succession in the formation of annually laminated sediment in Lovojärvi, a small eutrophicated lake. Ann. Bot. Fennici 14: 143–148.

Sturm, M., 1979. Origin and composition of clastic varves. In: Ch. Schlüchter (ed.), Moraines and Varves. Balkema, Rotterdam, 281–285.

Sturm, M., 1985. Schwebstoffe in Seen. Mitteilungen der EAWAG 19: 9–15.

Swain, A. M., 1978. Environmental changes during the past 2000 years in North-Central Wisconsin: Analysis of Pollen, Charcoal and Seeds from Varved Lake Sediments. Quat. Res. 10: 55–68.

Tippett, R., 1964. An investigation into the nature of the layering of deepwater sediments in two Eastern Ontarion Lakes. Can. J. Bot. 42: 1693–1704.

Tobolski, K., 1985. Plant macrofossils from Lobsigensee. In: G. Lang (ed.), Swiss Lake and Mire Environments during the last 15000 years. Diss. Bot. 87: 140–143.

Tolonen, M., 1978. Palaeoecology of annually laminated sediments in Lake Ahvenainen, S. Finland. I. Pollen and charcoal analysis and their relation to human impact. Ann. bot. fenn. 15: 177–208.

Troels-Smith, J., 1955. Characterization of unconsolidated sediments. Danmarks Geol. Undersögelse, LV. Raekke 3 (nr. 10, 78 pp + 13 tables).

Welten, M., 1944. Pollenanalytische, stratigraphische und geochronologische Untersuchungen aus dem Faulenseemoos bei Spiez. Veröff. geobot. Inst. Rübel Zürich 21, 201 pp.

Züllig, H., 1985. Carotenoids from plankton and phototrophic bacteria in sediments as indicators of trophic changes: evidence from the Late-glacial and early Holocene of Lobsigensee. In: G. Lang (ed.), Swiss Lake and Mire Environments during the last 15000 years. Diss Bot 87: 143–147.

Züllig, H., 1986. Carotenoids from plankton and phototrophic bacteria in sediments as indicators of trophic changes in Lake Lobsigen during the last 14000 years (this volume).

Accepted 18 March 1986.

Hydrobiologia 143: 309–314, (1986)

An early meromictic stage in Lobsigensee (Switzerland) as evidenced by ostracods and *Chaoborus*

Studies of the late quaternary of Lobsigensee No. 12

Heinz Löffler
Department of Limnology, Institute of Zoology, University of Vienna, Althanstr. 14, A-1090 Wien, Austria

Keywords: ostracods, *Chaoborus*, meromixis

Abstract

Lobsigensee is a small kettle on the western Swiss Plateau in a tectonic depression of the tertiary Molasse. At present its maximum depth is 2.7 m. Analysis of ostracods and other organisms (*Chaoborus*) from a 13 m core indicates that the lake became meromictic during the Bølling and returned again to a holomictic stage during the Boreal when the lake was approximately 10 m deep. The meromictic period is characterized by the disappearance of ostracods – especially of *Cytherissa lacustris* and *Limnocythere sanctipatricii* – and the appearance of *Chaoborus*, whereas the onset of the second holomictic stage is marked by the recolonization of the profundal zone by ostracods not present before in the lake (*Metacypris cordata, Darwinula stevensoni*). A very similar case of 'fossil' meromixis has been found in Kleinsee in Carinthia.

Introduction

Lobsigensee on the western Swiss Plateau (Fig. 1) has recently become an important site for paleolimnological studies (Ammann & Tobolski, 1983; Ammann *et al.*, 1983, 1984, 1985; Hofmann, 1985a; Hofmann, 1985b; Züllig, 1985, etc.). As a result of these investigations it was possible to draw several conclusions concerning the climatic history and its influence on the trophic state of the small lake. It was shown that major steps of warming at the beginning of the Bølling and at the onset of the Holocene resulted in eutrophication and, as indicated by fossil pigments (Züllig, 1985), a meromictic period was therefore postulated for the Boreal. Since ostracods are extremely useful for the recognition of meromixis and eutrophication processes in lakes (Löffler, 1969, 1975, 1977), it was of interest to analyze the development of the ostracod fauna of Lobsigensee and to compare the results with the information from this lake obtained so far.

Material and methods

Lobsigensee is a small (2 ha) kettle with a maximum depth of 2.7 m (during the early Late Glacial it was at least 17 m deep) with a rich macrophytic vegetation (Potamogetonion, Nymphaeion, Phragmition and Alnion glutinosae). From several cores obtained by a modified Livingstone sampler, core 160 a (western margin of the open lake) was used for a.o. diatom, pigment and ostracod analysis. For the inspection of ostracods, 50 ml (at 893 only 10 ml) from sections between 1298 and 90 cm depth were selected and washed through a 50 μm filter. Ostracods and remains of other organisms were counted and the results are presented in Fig. 2a, b, c. It is obvious that due to the procedure used, the figures for *Spongilla*, small diatoms and *Pediastrum* are unrealistic.

Results (Fig. 2a, b, c)

A total of 17 (18?) subfossil species was found. Of these, several are strictly confined to the Dryas 1

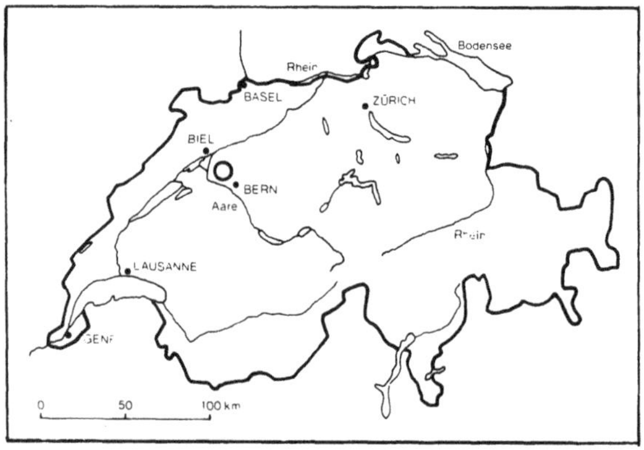

Fig. 1. Geographic location of Lobsigensee (circle).

(*Cytherissa lacustris, Limnocythere sanctiparicii, Leucocythere mirabilis, Ilyocypris sp., Eucypris* cf. *pigra* and *Potamocypris sp.*) and start to disappear and to become extinct before the Bølling. From that time until the middle of the Boreal all ostracods with the exception of species able to swim (*Cypria ophthalmica, Cyclocypris sp.*) are absent. Only during the late Boreal does the colonization of the profundal with ostracods start again. Among the species involved in this recolonization are some candonids (*Candona candida, C. neglecta, C. caudata*) which were present during the Dryas 1 whereas the other species (*Limnocythere inopinata, Metacypris cordata* and *Darwinula stevensoni*) obviously are new immigrants to the lake. Rare species like *Candona fragilis* (at a depth of 90 cm), *Cyprinae sp.* and *Cypridopsis vidua* may also belong to this group. The latter species, however, is likely to be confined to the macrophytic zone, and its shells found for the first time in the profundal and at the end of the Boreal are almost certainly specimens which were moved there by currents.

Other organisms

Studies so far carried out included the investigation of bacteria and algae (fossil pigments, diatoms), *Bosmina,* chydorids, Trichoptera, Coleoptera, chironomids, *Chaoborus* and molluscs. their distribution in the sediment has been partly explained by the climatic history (e.g. chironomids) and partly also by the trophic conditions (e.g. *Bosmina,* chydorids). Among the organisms presented in Fig. 2a, b, c the Oribatei occurred mainly during the time when most of the ostracods were absent. Their presence may be explained by the development of mosses though any information about Musci is still lacking. Of even more interest is the first appearance of fish scales (most likely *Perca fluviatilis*) during the Boreal when the recolonization of the profundal with ostracods took place. *Campylodiscus sp.* and *Spongila* sp. appeared at the same time. In addition, the sudden occurrence of *Chaoborus flavicans* during the Bølling (according to Hofmann, 1985b during the Dryas 3!) soon after most of the ostracods had become extinct is of considerable significance. *Chaoborus flavicans* is well known for its preference for lakes with anoxic deep water or anoxic loose sediment to which the animals migrate during the day. As in Lobsigensee *Chaoborus flavicans* was an early immigrant to small meromictic lakes in Carinthia at the end of the Pleistocene.

Discussion

According to the distribution of ostracods it seems most likely that meromixis in Lobsigensee started during the Bølling or even slightly before. The almost complete extinction of the early ostracod fauna during the late Dryas 1 is paralleled by the disappearance of several cold stenothermic chironomid taxa before and during the Bølling (Hofmann, 1985b). At that time the lake was still about 12 m deep and therefore must have had a cold hypolimnic body of water which enabled the survival of cold stenothermic ostracods and chironomids. The sudden occurrence of *Chaoborus flavicans* during the Bølling suggests a deterioration of the oxygen condition of the profundal zone.

The end of the meromictic stage is marked by the recolonization of the profundal zone by ostracods during the Boreal, when the lake was only about 10.5 – 11 m deep. Based on fossil pigments a.o. from *Oscillatoria rubescens* Züllig (1985) has given support to this last period of the meromictic stage but he also mentions an earlier step of eutrophication at the beginning of the Bølling. This step, in combination with a change in the thermal conditions from cold monomictic towards dimictic behaviour of Lobsigensee, would account for the onset of meromixis. A further strong argument for the meromixis of Lobsigensee during the latest Pleistocene and Preboreal is the oligotrophic condition in

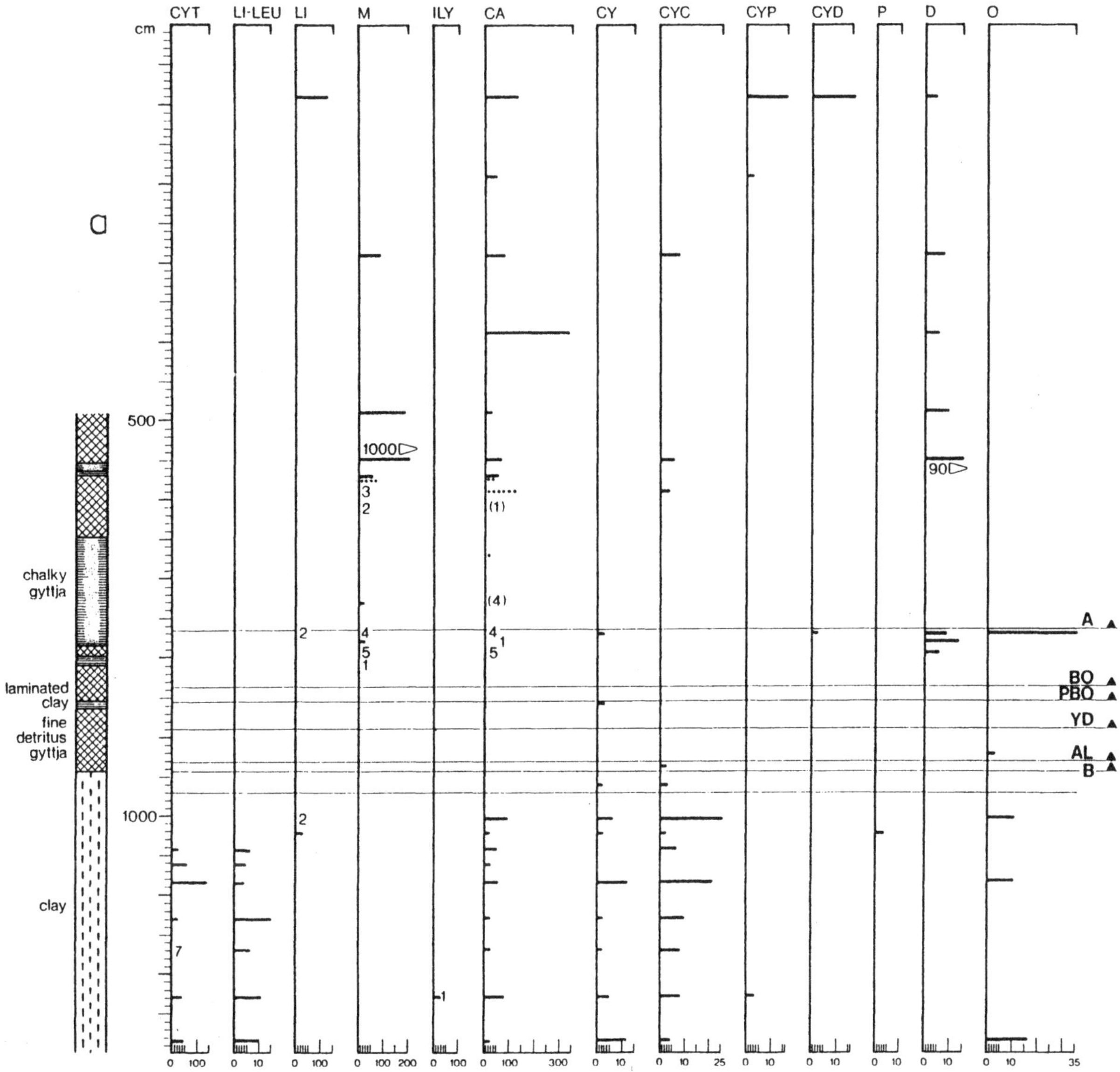

Fig. 2a, b, c. Distribution of the remains of some organisms in core 160 a. To the left the type of sediment is indicated (information available only from 500 cm downward).

B: Bølling, AL: Allerød, YD: Younger Dryas, PBO: Preboreal, BO: Boreal, A: Atlanticum.

(a) Cyt: *Cytherissa lacustris*, Li-Leu: *Limnocythere sanctipatricii* – *Leucocythere mirabilis*, Li: *Limnocythere* juv. (most likely *L. inopinata* at 90 cm), M: *Metacypris cordata, Ily: Ilyocypris sp.* Ca: *Candona spp.,* Cy: *Cypria ophthalmica*, Cyc: *(C) Cyclocypris sp.,* Cyp: Cypriniae sp., Cyd: *(C) Cypridopsis vidua,* P: *Potamocypris sp.,* D: *Darwinula stevensoni,* O: Fragments of ostracod shells.

(b) S: *Sida crystallina*, Dae: Ephippia of *Daphnia sp.,* Sim: Ephippia of *Simocephalus sp.,* Cer: Ephippia of *Ceriodaphnia,* B: *Bosmina sp.*, Chyd: Chydoridae, Cha: *Chaoborus flavicans*, Chi: Chironomids, head-capsulae, Or: Oribatei, Mo: Fragments of molluscs (dotted lines: *Pisidium sp.*, otherwise gastropods), Br: Bryozoans (*Plumatella*).

(c) Tu: egg cases of Turbellaria, Sp: Sponges, T: Testacea, Pi: Scales of *Perca fluviatilis*, Camp: *Campylodiscus sp.*, Diat: other diatom species, Ped: *Pediastrum*, Ch: Charales, F: Ferns.

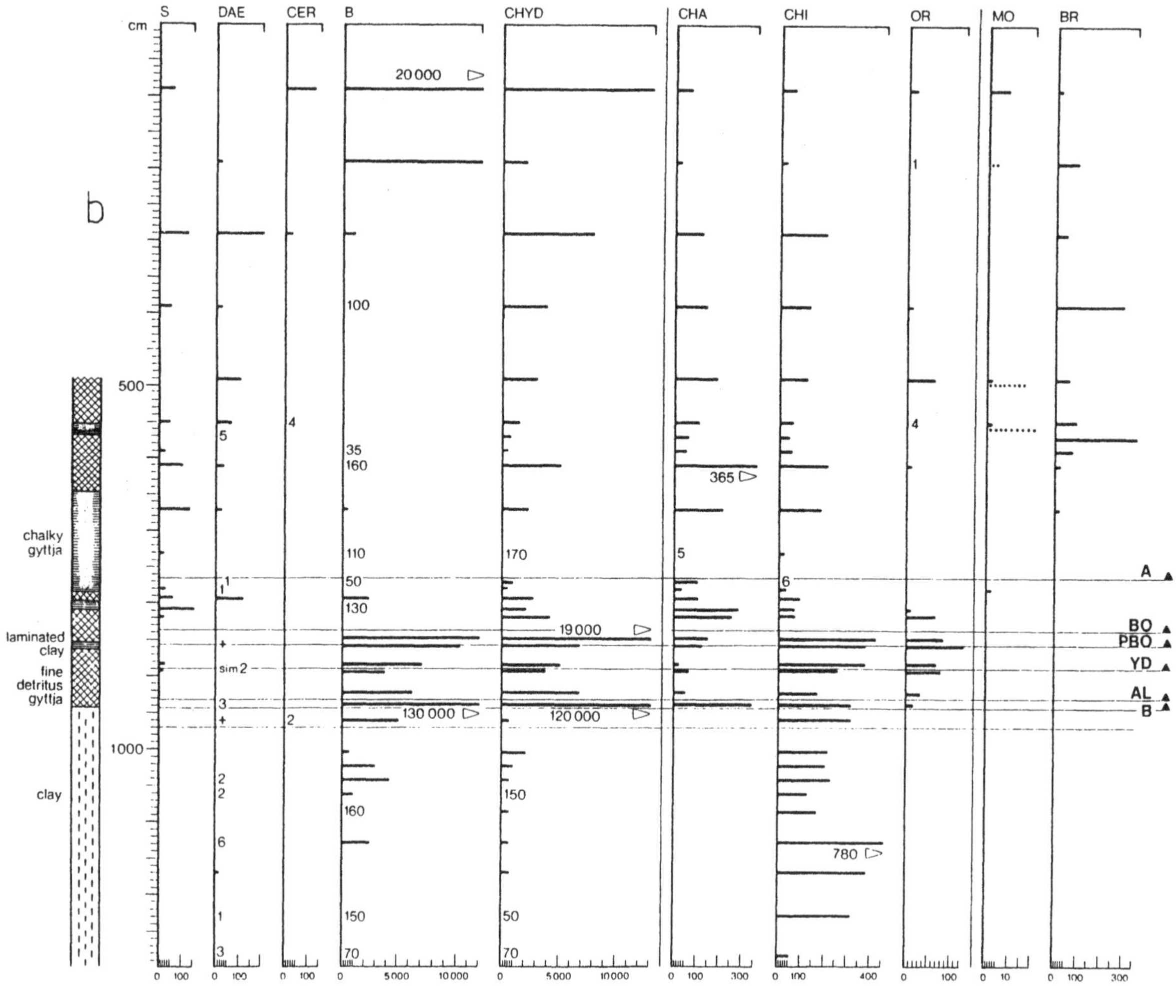

the Younger Dryas proved by pigments (Züllig, 1985).

A striking similar development of circulation patterns was found in Kleinsee (Carinthia), a kettle which at present has a maximum depth of 9.3 m (during the early late Glacial it was at least 16 m deep) and an area of 1.2 ha (Löffler, 1977). As in Lobsigensee a group of ostracod species (*Cytherissa lacustris, Ilyocypris sp.* and *Eucypris sp.*) disappears during the early Bølling, and *Chaoborus* enters the lake – somewhat later namely during the Allerød – when the maximum depth was about 13 m. Recolonization of the profundal zone by ostracods, however, took place only during the Subatlanticum but at a time when Kleinsee, like Lobsigensee, had a depth of approximately 11 m. It is also of interest that species involved in this recolonization are the same in Lobigensee: *Limnocythere inopinata, Metacypris cordata* and *Darwinula stevensoni.*

Finally it may be mentioned that in the lakes of Carinthia which are still meromictic (Goggausee, Klopeiner See and Längsee) very similar events were observed. In all of these cases the disappearance of an early ostracod fauna is followed by the immigration of *Chaoborus.* In small lakes this happened between Bølling and Allerød whereas in lakes with a size class of 1–2 km^2 it took place between Allerød and Preboreal.

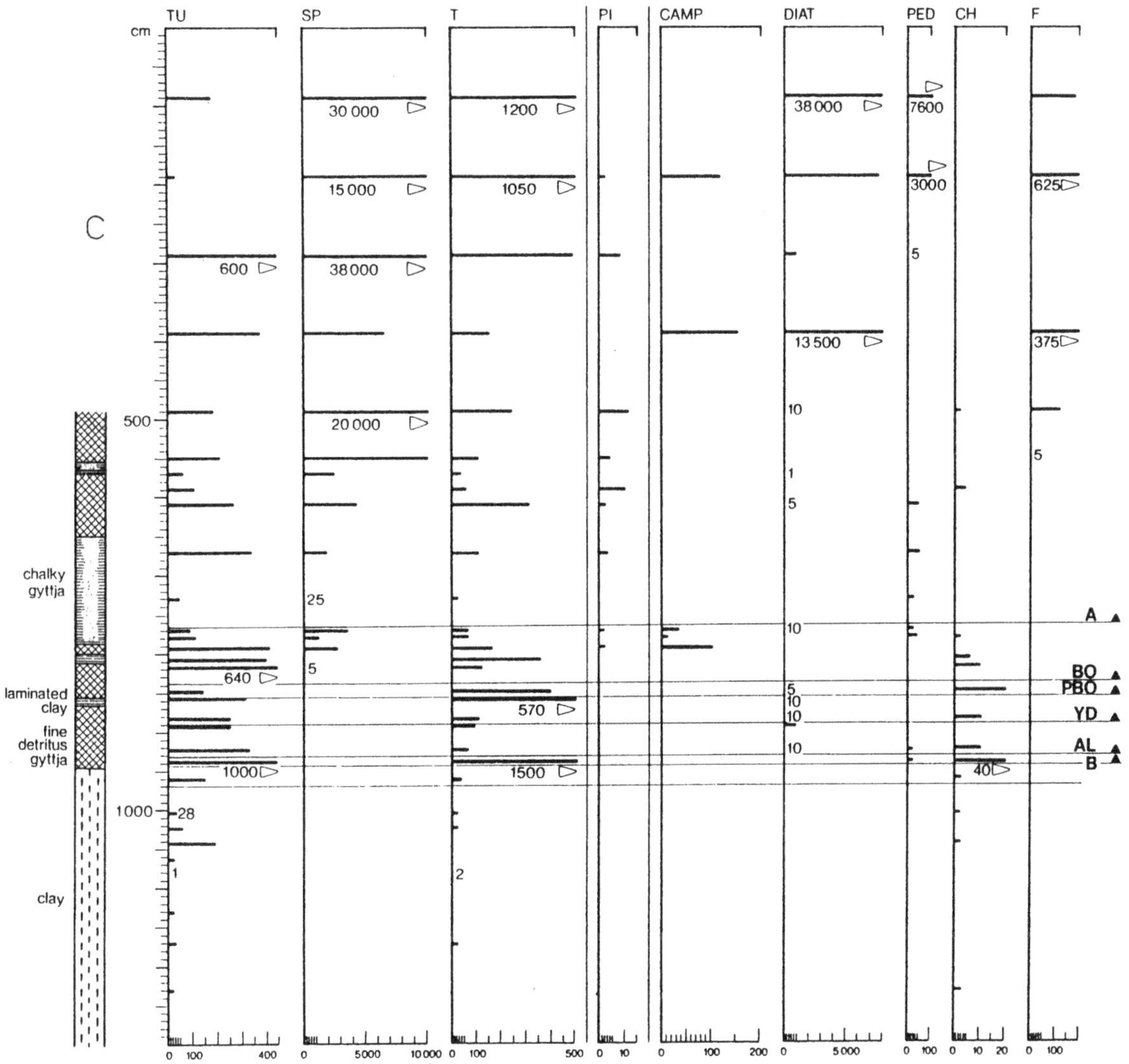

Acknowledgement

The author is most grateful to Dr. B. Ammann for the collection of core samples and most recent information about Lobsigensee.

References

Ammann, B., 1984. Prozente, Konzentrationen und Durchschnittsinflux der Pollenzonen im Spätglazial von Lobsigensee. Diss. Bot. 72 (Festschrift Welten): 11–44.

Ammann, B. & K. Tobolski, 1983. Vegetational development during the Late-Würm at Lobsigensee (Swiss Plateau). Studies in the Late Quaternary of Lobsigensee 1. Rev. Paléobiol. 2: 163–180.

Ammann, B., L. Chaix, U. Eicher, S. A. Elias, M.-J. Gaillard, W. Hofmann, U. Siegenthaler, K. Tobolski & B. Wilkinson, 1983. Vegetation, insects, molluscs and stable isotopes from Late-Würm deposits at Lobsigensee (Swiss Plateau). Studies in the Late Quaternary of Lobsigensee 7. Rev. Paléobiol. 2: 221–227.

Ammann, B., L. Chaix, U. Eicher, S. A. Elias, M.-J. Gaillard, W. Hofmann, U. Siegenthaler, K. Tobolski & B. Wilkinson, 1984. Flora, fauna und stable isotopes in Late-Würm deposits at Lobsigensee (Swiss Plateau). In: N.-A. Mörner & W. Karlen (eds), Climatic Changes on a Yearly to Millenial Basis: 69–73.

Ammann, B., M. Andrée, L. Chaix, U. Eicher, S. A. Elias, W. Hofmann, H. Oeschger, U. Siegenthaler, K. Tobolski, B. Wilkinson & H. Züllig, 1985. Lobsigensee – Late-Glacial and Holocene Environments of a Lake on the Central Swiss Plateau. Studies in the Late Quaternary of Lobsigensee 11. Diss. Bot. 87: 127–170.

Hofmann, W., 1985a. Developmental history of Lobsigensee: subfossil Cladocera (Crustacea). In: Ammann, B. *et al.*, Lobsigensee – Late-Glacial and Holocene Environments of a Lake on the Central Swiss Plateau. Studies in the Late Quaternary of Lobsigensee 11. Diss. Bot. 87: 150–153.
Hofmann, W., 1985b. Developmental history of Lobsigensee: subfossil Chironomidae (Diptera). In: Ammann, B. *et al.*, Lobsigensee – Late-Glacial and Holocene Environments of a Lake on the Centra Swiss Plateau. Studies in the Late Quaternary of Lobsigensee 11. Diss. Bot. 87: 154–156.
Löffler, H., 1969. Recent and subfossil distribution of *Cytherissa lacustris* (Ostracoda) in Lake Constance. Mitt. Int. Ver. Limnol. 17: 240–251.
Löffler, H., 1975. The onset of meromictic conditions on Goggausee, Carinthia. Verh. Int. Ver. Limnol. 19: 2284–2289.
Löffler, H., 1977. 'Fossil' meromixis in Kleinsee (Carinthia) indicated by Ostracodes. In: H. Löffler & D. Danielopol (eds), Aspects of Ecology and Zoogeography of Recent and Fossil Ostracoda – Proc. 6th Int. Symp. on Ostracods (1976). Dr. W. Junk b.v. Publ., The Hague: 321–325.
Züllig, H., 1985. Carotenoids from plankton and phototrophic bacteria in sediments as indicators of trophic changes: evidence from the Late-glacial and the early Holocene of Lobsigensee. In: Ammann, B. *et al.*, Lobsigensee – Late-Glacial and Holocene Environments of a Lake on the Central Swiss Plateau. Studies in the Late Quaternary of Lobsigensee 11. Diss. Bot. 87: 143–147.

Accepted 13 May 1986.

Hydrobiologia 143: 315–319, (1986)

Carotenoids from plankton and photosynthetic bacteria in sediments as indicators of trophic changes in Lake Lobsigen during the last 14000 years

H. Züllig
Rheineck, Switzerland

Keywords: lake history, plankton carotenoids, photosynthetic carotenoids, carotenoid stratigraphy, natural and anthropogenic eutrophication

Abstract

Previous examination of drilling cores showed that the majority of the carotenes and carotenoids originally present in the various planktonic organisms and phototrophic bacteria are preserved in sediments. The indicator pigments for algae phyla are: Lutein for Chlorophyta, Myxoxanthophylls and their derivatives for Cyanophyta, Fucoxanthin for Chrysophyta, and Alloxanthin for Cryptophyta.

The pigments in the deepest sediment sample (late glacial time, clay, 12.83 m) consists primarily of Alloxanthin, secondly of Lutein and β-carotene and thirdly of traces of Okenone, Speroidenone and other such bacterial carotenoids. The first plankton organisms were thus Cryptophyta and some Chlorophyta. The presence of the phototrophic bacteria pigments indicates that at the time of sediment formation, anaerobic conditions prevailed at the lake bottom.

The Holocene era commences at a depth of 8.55 m and is characterized by the first occurrence of Myxoxanthophyll and Echinenone from Cyanophyta, as well as by a rapid increase of β-carotene, Lutein and Alloxanthin. The pronounced occurrence of *Oscillatoria rubescens* ('blood of the Burgundies'), characterized by Oscillaxanthin at 8.21 m must be considered for Swiss lakes as a very surprising discovery.

The intensive plankton production again lead to stringent anaerobic conditions. Predominant among phototrophic bacteria pigments were Okenone, Spheroidene and Rhodopin.

Introduction

Stratigraphic pigment analysis of recent sediments from 10 Swiss lakes have shown that the majority of carotenes and carotenoids of former plankton populations are partly sedimented with the dead biomass and are mostly preserved in the sediment (Züllig, 1982).

Liaaen-Jensen & Andrewes (1972) wrote: 'Although some carotenoids are specific to higher plants and animals, the majority of naturally occurring carotenoids are encountered in microorganisms, which no doubt have demonstrated the highest ingenuity as to structural modifications of the carotenoid molecule'.

Based on the spectra of plankton-carotenoids in sediments the former occurrence of organisms can be reconstructed with the level of classification being sometimes to the family and occasionally even to the genera and species.

Leading pigments for the division of algae are:

1) Lutein for Chlorophyta.
2) Myxoxanthophyll and its derivatives for Cyanophyta.
3) Fucoxanthin for Chrysophyta.
4) Alloxanthin for Cryptophyta.
5) Peridinin for Pyrrhophyta.

Based on these observations it is possible to reconstruct the history of primary production in lakes. A critical discussion is given by Swain (1985).

Likewise, carotenoids of purple non-sulfur and purple sulfur bacteria have also been found in sediments of shallow lakes (Brown, 1968; Brown *et al.*, 1984; McIntosh, 1983; Züllig, 1984). Schmidt *et al.* (1965) have shown that, to a limited degree, the previous presence of photosynthetic bacteria can be traced and classified to the family and species by means of the pigments.

The following is a selection of bacterial pigments and their *main* occurrence in some photosynthetic bacteria (Liaaen-Jensen, 1965; Pfennig, 1978; Schmidt & Schiburr, 1970):

1) Okenone: *Chromatium okenii, C. weissei, Thiopedia* spp.
2) Spheroidene, Spheroidenone, OH-Spheroidene, OH-Spheroidenone: *Rhodopseudomonas spheroides.*
3) Rhodopin, Rhodopinol, Rhodopinal: *Chromatium warmingii.*
4) Spirilloxanthin: *Rhodospirillum rubrum, Rhodopseudomonas palustris, Thiocapsa roseopersicina, Amoebacter roseus, A. pendens, Thiodictyon elegans,* etc.
5) Rhodovibrin: *Rhodospirillum photometricum, R. rubrum, Rhodopseudomonas palustris.*
6) Lycopen, Lycopenal: *Lamprocystis roseopersicina.*
7) Isorenieratene: *Phaeobium* species.
8) Chlorobactene: *Chlorobium* species.

[newly classified by Pfennig & Trüper (1983)]

The occurrence of some bacterial pigments provides evidence for permanent or transitory anoxic conditions at the sediment surface or at the sediment-water interface.

The main carotenoids Spheroidene and Spheroidenone of *Rhodopseudomonas spheroides* may be considered as having special significance. Under strong anoxic conditions the bacteria in pure culture produce only yellow Spheroidene. Red Spheroidenone is produced as soon as the smallest amount of oxygen is present. If only Spheroidene is present and Spheroidenone absent it is assumed that this is an indicator of a meromictic state. It is interesting, that Spirilloxanthin could not be found in the samples.

Methods

The analysis of pigments was carried out by extraction of the wet sediment with acetone and alcohol at a ratio of 1:1, followed by separation with thin-layer-chromatography (Kieselgel-Plate 60, 20×20, Marck) in a N_2-atmosphere. Planktonic algal pigments were developed by several steps (Züllig, 1982):

1) Plates were developed twice with hexane/acetone/isopropanol (80:20:5) to a height of 16 cm with drying in between.
2) Plates were developed once with hexane/acetone/isopropanol (70:40:15) to a height of 16 cm and followed by drying.
3) Finally, plates were developed to a height of 2 cm with acetone/alcohol (70:40).

Bacterial carotenoids were developed twice with benzine (Cp 110/140 °C)/acetone (80:20) to a height of 16 cm with drying in between (Züllig, 1985).

The summary determination of carotene and carotenoids denoted here as 'crude carotenoids' was made by absorption measurements of an acid extract, at 450 and 600 nm.

The following formula was used to calculate crude carotenoids and it gives an approximate scale of bioproduction, proposed by Züllig (1985):

$$\dot{C}C \text{ mg} = \frac{(E_{450} - E_{600}) \times V_{ml} \times 10}{E_{1\,cm}^{1\%} = 2250}$$

CC = crude carotenoids
V_{ml} = volume of extract
$E_{1\,cm}^{1\%}$ = 2250 = average extinction coefficient of 1% extract of sediment in a 1 cm layer (Züllig, 1981).

Results and discussion from Lobsigensee

The Swiss lake or pond Lobsigen, near Bern has a present-day surface area of 2 ha and a maximum depth of 2.7 m (during the early Late-glacial these were respectively about 10 ha and at least 17 m), (Ammann, 1985). We obtained samples from 2 cores for pigment stratigraphic analysis from Dr. Brigitta Ammann, Geobotanical Institute, University Bern (supported by the Swiss National Science Foundation Proj. No. 3.336-78). The results of the Late-glacial and *early Holocene* have already been presented at the INQUA Symposium (1985) in Switzerland (Ammann, 1985; Züllig, 1985).

The values of 'crude carotenoids' are presented on the left hand side of Fig. 1 and the single carotenoid values are next to them.

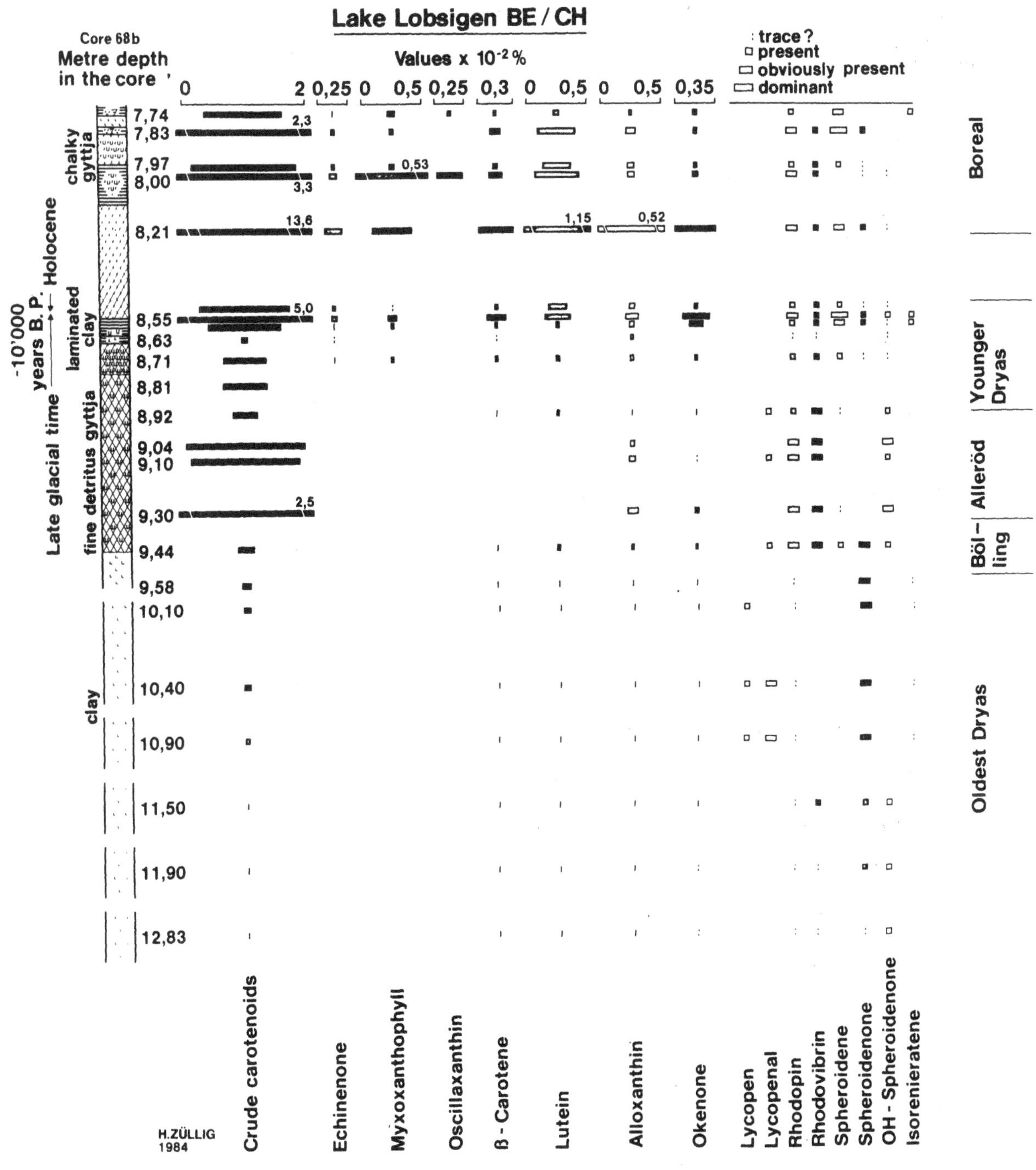
Lake Lobsigen BE / CH
Core 68b
Metre depth in the core
Values x 10^-2 %
: trace ?
present
obviously present
dominant
0 2 0,25 0 0,5 0,25 0,3 0 0,5 0 0,5 0,35
7,74
7,83
7,97
8,00
8,21
8,55
8,63
8,71
8,81
8,92
9,04
9,10
9,30
9,44
9,58
10,10
10,40
10,90
11,50
11,90
12,83
2,3
3,3
0,53
13,6
1,15
0,52
5,0
2,5
chalky gyttja
laminated clay
fine detritus gyttja
clay
-10'000 years B. P.
Holocene
Late glacial time
Boreal
Younger Dryas
Alleröd
Böl-ling
Oldest Dryas
Crude carotenoids
Echinenone
Myxoxanthophyll
Oscillaxanthin
ß - Carotene
Lutein
Alloxanthin
Okenone
Lycopen
Lycopenal
Rhodopin
Rhodovibrin
Spheroidene
Spheroidenone
OH - Spheroidenone
Isorenieratene
H.ZÜLLIG 1984

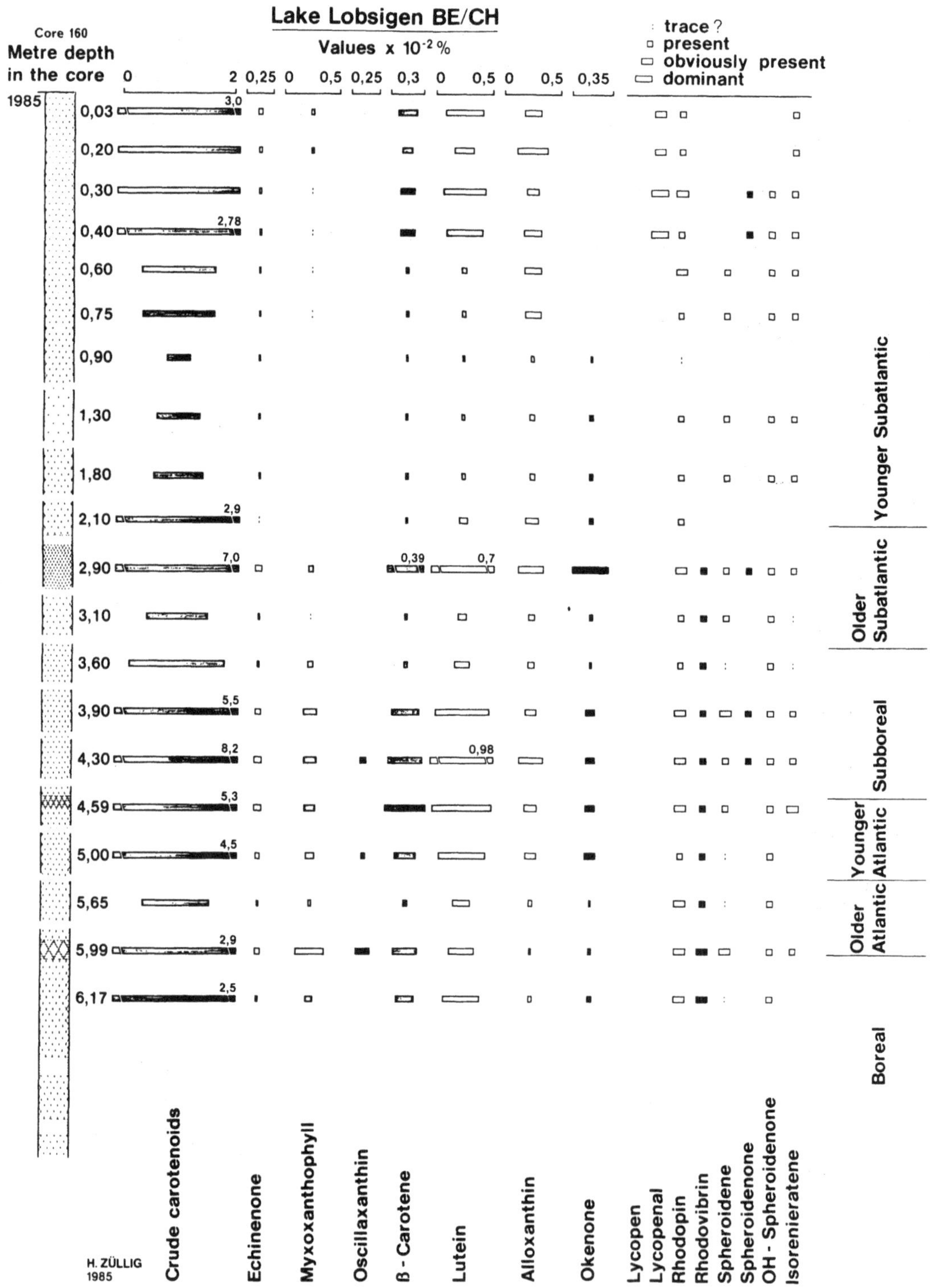
Lake Lobsigen BE/CH
Values x 10⁻² %
Core 160
Metre depth
in the core
1985
trace ?
present
obviously present
dominant
0,03
0,20
0,30
0,40
0,60
0,75
0,90
1,30
1,80
2,10
2,90
3,10
3,60
3,90
4,30
4,59
5,00
5,65
5,99
6,17
Crude carotenoids
Echinenone
Myxoxanthophyll
Oscillaxanthin
ß - Carotene
Lutein
Alloxanthin
Okenone
Lycopen
Lycopenal
Rhodopin
Rhodovibrin
Spheroidene
Spheroidenone
OH - Spheroidenone
Isorenieratene
Younger Subatlantic
Older Subatlantic
Subboreal
Younger Atlantic
Older Atlantic
Boreal
H. ZÜLLIG
1985

During the OLDEST DRYAS the lake had low production of Cryptophyta and Chlorophyta plankton (characterized respectively by Alloxanthin and Lutein). The deepest sample already contained traces of photosynthetic bacterial pigments, Okenone, Spheroidenone but not Spheroidene. *Lamprocystis roseopersicina,* characterized by Lycopene and Lycopenal, appears briefly. Possibly, anoxic conditions occurred temporarily.

In the BÖLLING and early ALLERÖD there were high values of crude carotenoids and Alloxanthin. Lutein and β-Carotene were almost completely absent. In the lower part of this zone a high concentration of Okenone appears for the first time, then bacterial pigments decreased in general, only Rhodopin and Rhodovibrin being clearly present. A variety of photosynthetic bacteria was possibly involved in the decomposition of fine detritus gyttja. The absence of β-Carotene and Lutein suggest that this was a cool period, perhaps with a rather long lasting ice cover.

In the early YOUNGER DRYAS a reappearance of Chlorophyta is indicated by Lutein. A new phase starts in the late YOUNGER DRYAS through the appearance of the Cyanophyta, indicated by Myxoxanthophyll and Echinenone, as well as the rapid increase of β-Carotene, Lutein and Alloxanthin. In a laminated zone high values of Okenone and Spheroidene are found, indicating strictly anaerobic conditions.

This trophic development led to a peak in the BOREAL with the highest value of crude carotenoids. A surprising result was the discovery of fluctuating quantities of Oscillaxanthin in laminated zones, indicating blooms of *Oscillatoria rubescens* or *Oscillatoria aghardii.* In Swiss lakes only *Oscillatoria rubescens* has been observed. This blue-green coloured Swiss lakes red/brown during this century and removed oxygen from the deeper layers of the lakes. It could be described as a veritable epidemic! We think that increased production of *Oscillatoria rubescens* was partially responsible for anaerobic conditions and probably meromixis at which stage Spheroidene was observed and Spheroidenone was absent.

It seems that this natural eutrophication was induced by an increase in temperature during the PREBOREAL and BOREAL and also by a decrease of water levels (see Ammann, 1986, this volume) in these periods.

During the OLDER ATLANTIC (Fig. 2) there was a decrease in production followed by an increase during the YOUNGER ATLANTIC. In the middle of the SUBBOREAL, trophic production reached a new peak, comparable with the beginning of the Boreal, but *Oscillatoria rubescens* was seen for the last time. Chlorophyta and Cryptophyta dominated. A great diversity of bacteria pigments was observed. The remarkable presence of Spheroidenone indicates that possibly temporary oxic conditions were found on the lake bottom. This eutrophication was probably produced as a result of a human settlement in the early Stone and Bronze age (first anthropogenic eutrophication). Then the trophic status decreased as shown by a minimum of pigments in the middle of the OLDER SUBATLANTIC, followed rapidly by a peak in plankton production and bacteria pigments.

In the beginning of the YOUNGER SUBATLANTIC a rapid impoverishment of plankton was observed. The blue-green algae disappeared as did most of the photosynthetic bacteria, including Okenone. This phenomenon can be shown in other lakes. In the last 200–300 years we found a new increase of pigments, probably as a result of new human agricultural activity. Main pigments are β-Carotene, Lutein and Alloxanthin. Echinenone and Myxoxanthophyll were found in trace amounts only. Okenone remained absent. A new combination of bacteria pigments was formed by Lycopenal, Rhodopin and Isorenieratene and indicates probably a quite new development in the lake. This new biological development was probably induced by a decrease of maximum water depth from 4.8 to 2.7 m and thus a more efficient penetration of oxygen to the lake bottom during approximately the last 1000 years.

The analytic values are indicated in % of dry weight. Using these values, true comparison is possible only if sedimentation rates are comparable. Oeschger *et al.* (1985) (see Ammann, 1986, this volume) discovered that shortly before 2000 B.P. the sedimentation rate increased by a factor of 3.7 (due to erosional inwash since the Roman colonization, intensified deforestation by man). Accordingly the values of carotenoids since then would have to be multiplied by an average factor of 3.7, which would indicate a new era of eutrophic development during the last centuries.

The stratigraphic analysis of carotenoids of

plankton and photosynthetic bacteria in sediments formed during the last 14000 years of the Lake Lobsigen, show the importance of these pigments for paleolimnological research. They provide a new way of investigating trophic history over long periods.

References

Ammann, B., 1985. Lobsigensee. Introduction and Palynology: Vegetational history and core correlation at Lobsigensee (Swiss Plateau). In G. Lang (ed.), Swiss lake and mire environments during the last 15000 years. Diss. Bot. 87: 127–134. J. Cramer, Vaduz.

Brown, S. R., 1968. Bacterial carotenoids from freshwater sediments. Limnol. Oceanogr. 13: 233–241.

Brown, S. R., H. J. Mc Intosh & J. P. Smol, 1984. Recent paleolimnology of a meromictic lake. Fossil pigments of photosynthetic bacteria. Verh. Internat. Verein. Limnol. 22: 1357–1360.

Liaaen-Jensen, S., 1965. Bacterial carotenoids XVIII. Arylcarotenes from Phaeobium. Acta chem. Scand. 19: 1025–1030.

Liaaen-Jensen, S. & A. G. Andrewes, 1972. Microbial Carotenoids. Ann. Rev. Microbiol. 26: 226.

Mc Intosh, H. H., 1983. A paleolimnological investigation of the bacterial carotenoids of Sunfish Lake. M.Sc. Thesis, Queen's University, Kingston, Ontario, 129 pp.

Pfennig, N. & H. G. Trüper, 1983. In 'Die Binnengewässer' 28: 114.

Pfennig, N., 1978. General physiology and ecology of photosynthetic bacteria, pp. 3–18. In R. K. Clayton & W. R. Sistrom, 1978. The photosynthetic bacteria, VII: Biosynthesis 740–747. Plenum Press, New York.

Schmidt, K., N. Pfennig & S. Liaeen-Jensen, 1963. Carotenoids of Thiorhodaceae. Arch. Mikrobiol. 46, 138: 132–136.

Hydrobiologia 143: 321–325, (1986)

Stratigraphy of specific algal and bacterial carotenoids in the sediments of Lake Varese (N. Italy)

Piero Guilizzoni, Andrea Lami, Delio Ruggiu & Giuliano Bonomi
C.N.R. Istituto Italiano di Idrobiologia, Largo V. Tonolli 50/52 – 28048 – Pallanza, Italy

Keywords: paleolimnology, sedimentary carotenoids, diatoms

Abstract

A study on the succession of some phytoplankton populations in the very eutrophic Lake Varese over the past 80–100 years, was based on thin-layer chromatography of four group-specific carotenoids (echinenone, lutein, myxoxanthophyll and oscillaxanthin). The development of diatoms was traced by microscope counts of their frustules. In addition, the carotenoid okenone, peculiar to some *Chromatium* species, was identified and related to lake trophic conditions (the *Chromatium* species are obligate anaerobic bacteria, and use H_2S).

Introduction

During the past 2–3 decades, a great number of paleolimnological studies have dealt with sedimentary chlorophylls and total carotenoids (Guilizzoni *et al.*, 1982; Swain, 1985). There are, however, relatively few studies on the occurrence and distribution of specific algal carotenoids in lake sediments (Züllig, 1982). Among these pigments, the carotenoids of various blue-green algae have been studied extensively, this because Cyanophyta occupy a central position in the eutrophication problem of inland waters. For example, it is known that the pigment oscillaxanthin is largely restricted to some species of the blue-green alga *Oscillatoria* and is particularly abundant in *O. rubescens*. Thus, the concentration of this xanthophyll in the core profile can be useful in detecting the past development of this taxonomic group (Griffiths, 1978; Engstrom *et al.*, 1985).

In addition to plant pigments, those of photosynthetic bacteria have been successfully separated from their original population and used as valuable paleoecological indicators (Brown *et al.*, 1984; Züllig, 1984 and 1985).

Although several problems still remain partially unresolved (e.g. diagenesis, preservation/degradation), no doubt exists about the usefulness of this approach, particularly when the pigment data are related to concurrent investigations that include other chemical and biological parameters.

In the study presented here, along the lines followed by Züllig (1981, 1982) and Brown *et al.* (1984), we have attempted to trace the development over the last century of some phytoplanktonic groups, and of photosynthetic purple sulfur bacteria (*Chromatium* spp.) in the subalpine hypertrophic Lake Varese (N. Italy). This was accomplished by analysis of group specific carotenoids in a sediment core. The study was supplemented by counting frustules of diatoms from the Centricae and Pennatae. Since all these organisms are strongly affected by the process of eutrophication, their changes with time will reflect the trophic status of the waters.

Study area and methods

Lake Varese is a very eutrophic, relatively small ($15\ km^2$) well buffered (total alkalinity: $2-3\ meq \cdot l^{-1}$) water body, with maximum depth of 26 m ($\bar{z}$ = 11 m). The lake is generally stratified from May to November, during wich time the hypolimnion undergoes a strong deoxygenation. Spectacular *Microcystis aeruginosa* blooms commonly take place in the superficial waters.

A 60-cm sediment core was collected on September 1983 from the deepest area of the lake.

Thin layer chromatography of carotenoids was performed on 1.5 cm sediment sections in the first 9 cm of the core, every 3 cm from 9 to 15 cm, every 5 cm from 15 to 30 cm and every 10 cm in the remaining part. The pigments were extracted and their amount (mg) calculated according to Züllig (1982); the concentration was expressed as percent of carotenoids per g organic weight. Carotenoids from pure strains of algae (NIVA's culture collection, Blindern, Norway) were used as reference standards.

Sample preparation for microscope counts of diatom frustules was based on Battarbee's method (1973) with minor modifications (Candido *et al.,* in press).

Results and discussion

Echinenone, myxoxanthophyll and oscillaxanthin were studied as characteristic carotenoids of blue-green algae; lutein and okenone as distinctive pigments of green plants and of some purple sulfur bacteria of the genus *Chromatium*, (Fig. 1). The more ubiquitous β-carotene and the pool of total carotenoids were ascertained in order to obtain a general picture of the evolution of the whole phytoplankton community.

The sediments of Lake Varese show clear laminae only for the first 15–20 cm, were the percent of organic matter varies from the basic value of 11% to 21% dry wt. (Lami, 1985). Calcium carbonate concentration is high throughout (ca. 20% dry wt. in the lowest part of the core) but increases markedly during the last 15–20 years (ca. 45% in the surface: Lami, 1985). Such changes are interpreted as a product of biogenic origin (Adams *et al.,* 1978).

In spite of the available literature on sedimentation rates in Lake Varese, calculated as 1.3 cm·yr^{-1}

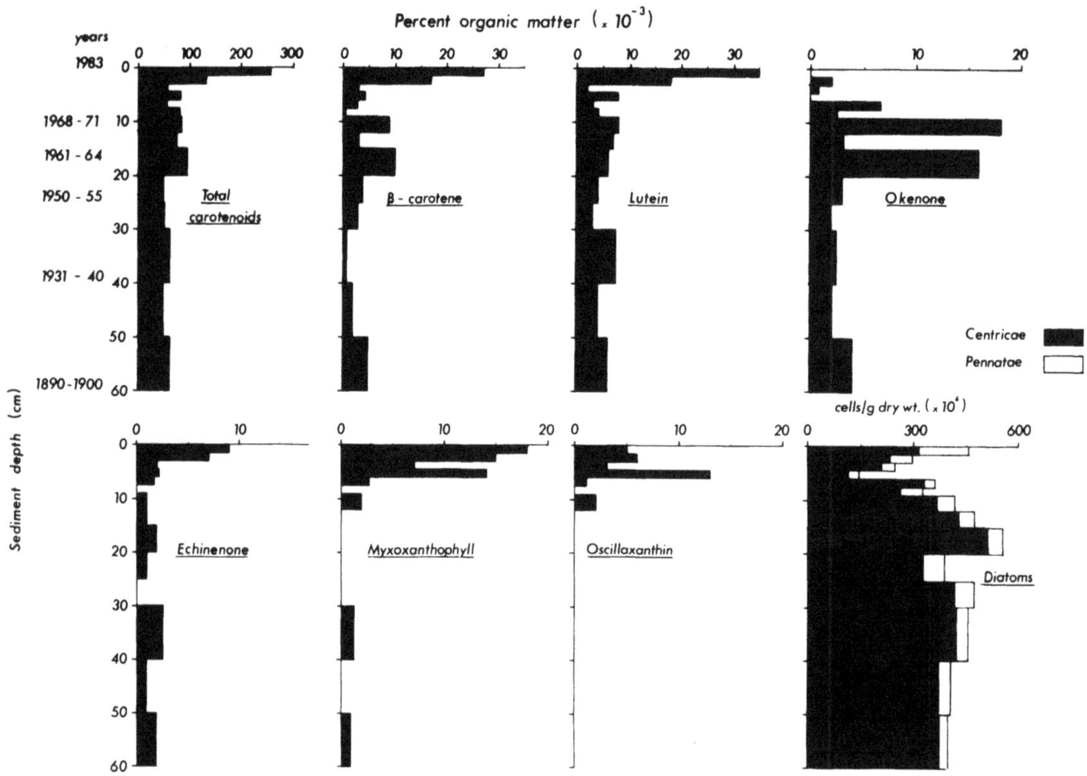

Fig. 1. Carotenoid stratigraphy of the sediments of Lake Varese. Depth distribution of diatoms is also shown.

by the ^{137}Cs method (Ravera & Premazzi, 1971), the pigment and diatom data cannot be expressed as mg or number of cells sedimented per year, but only as percentages on an organic and dry weight basis respectively (Fig. 1). This was done because we do not know a mean sedimentation rate (although surely lower than the present) for the period prior to the 1960's. Approximate core dating (Fig. 1) was estimated by considering the effect of vertical compaction (Håkanson & Källstrom, 1978) on about $2\ cm \cdot yr^{-1}$ of wet material recently deposited (Lami, 1985).

The total carotenoids, β-carotene and lutein profiles are very similar with maximum values in the superficial layers (Fig. 1). Oscillaxanthin was detected in appreciable quantities from 9–12 cm upwards and its values parallel those of myxoxanthophyll. The appearance of oscillaxanthin is indicative of the onset of blooms of *Oscillatoria* in the lake that began in the late 1960's. The presence of *O. rubescens,* however, was noted by Tonolli & Bonomi (1965) as early as 1953. The apparent discrepancy with the core analysis is probably due to (i) the low density of the species in the 1950's (ca. 30×10^6 cells $\cdot m^{-3}$), which in turn results in quantities of pigment too small to be easily measured chromatographically, and (ii), to the patchy distribution of the carotenoids in different areas of the lake facilitated by the filamentous nature of *Oscillatoria* which might cause it to be easily moved about by currents. Judging from the oscillaxanthin content, this blue-green alga reached its highest population density about 1975 (4.5–6 cm layer) and then decreased in more recent times. The reduction probably corresponds to an increase in different species of Cyanophytes evidenced by increases in myxoxanthophyll and echinenone (Fig. 1). Shifts in the ratio oscillaxanthin/myxoxanthophyll provide further evidence for increased species variability in the last 15 years or so. These data are consistent with the field phytoplankton results (Ruggiu *et al.,* 1981) and with the current annual inspections that reveal continuous and heavy blooms of *Microcystis aeruginosa* from the beginning of summer to December. It is important to note that echinenone and myxoxanthophyll distributions in the sediment core indicate the quite abundant concentrations of these organisms during the last century.

Lutein, the main pigment of Clorophyceae has tripled its basic concentration of ca. $5 \times 10^{-3}\%$ organic matter in recent years, reaching with β-carotene the highest concentration (ca. $25-30 \times 10^{-3}\%$ o.m.). The most important species in the class is *Sphaerocystis schröteri.* In general, green algae (together with diatoms) are the most important group in spring and at the end of winter.

During this century the diatom populations also show a very high cell density per g dry wt. (Fig. 1): these values are 3–4 times higher than in Lake Mergozzo, a small, moderately productive lake of low alkalinity situated in the same geographic area as Lake Varese (Candido *et al.,* in press). A sharp reduction of the cell density of the Centricae, the dominant group in Lake Varese, occurs between 7 and 15 cm (about 1965–1975: Fig. 1). It seems likely that the massive development of blue-green algae displaced the diatoms in that period.

Stephanodiscus hantzschii, and to a lesser extent *Melosira granulata,* both well-known species of eutrophic lakes (Margalef, 1983), occur with the unexpected *Cyclotella* spp. the dominant form in Lake Varese (Lami, 1985). This may reflect frustule dissolution in hard water (Brugam, 1983). The A/C (araphidineae/centrales) ratio is always below 1 and therefore not in agreement with the classification proposed by Stockner (1971), and the Centricae species are always dominant over those belonging to the Pennatae group. These features are discussed in a forthcomming paper (Lami *et al.,* in press). The phytoplankton community, as a whole, increased dramatically from about the mid-1970's. The late 1950's was the time when the lake became very eutrophic. Bonomi (1962a), for instance, comparing the profundal benthos described for the period 1956–1957 with that studied again in 1962 (Bonomi, 1962b), demonstrated that *Chironomus anthracinus* had disappeared from the profundal zone. He attributed this catastrophic change to the extreme oxygen deficits that were developing during late winter, the pupation-period of *C. anthracinus* in Lake Varese (Bonomi, 1964 and 1969). Recent paleolimnological data (Bonomi, unpublished) confirm that *Chironomus* head capsules are present up to the 12 cm depth in the core but are completely missing in the more recent sediments. The development of greater anaerobiosis is also supported by some additional chemical and biological investigations (Lami, 1985) and by the okenone stratigraphy (Fig. 1). The high concentration of this

Chromatium carotenoid in all deep sediment core sections indicates an anaerobic hypolimnetic condition for the entire period represented by the core (Züllig, 1984). From 15–20 cm level (1960's) up to the superficial layer, the okenone concentration showed wide variation with near zero values in some sections and very high concentrations in others. The highest concentrations exceeded four times the pre-1960 values. Parking & Brock (1980) and Brown *et al.* (1984) have both noted the disappearance of photosynthetic bacteria when light is limiting. The heavy algal blooms of L. Varese in recent years may have reduced the light penetration so completely as to cause population crashes of *Chromatium.*

Both total and individual carotenoid concentrations measured in Lake Varese are in general higher than those reported by Züllig (1982) for some eutrophic Swiss lakes. However, Lake Varese is in one of the most populated and industrialized areas in Italy.

On the other hand, from the present study it appears that Lake Varese is a lake 'naturally productive' as suggested by its morphometric characteristics (low mean depth, large shallow area, etc.) and by the geology of its drainage basin.

In this context, the phytoplankton production of 160 g $C \cdot m^{-2} \cdot yr^{-2}$, estimated by Guilizzoni *et al.* (1983) for a period prior to 50–70 years ago, is well in accordance with the conclusions of this work. The further increase in production occurring in more recent years must be viewed as the result of intensified human activities acting upon a naturally productive lake.

Acknowledgements

We want to express our gratitude to Dr. Hans Züllig (Rheineck-Switzerland) for his assistance in the TLC carotenoids analysis. The critical reading of the text by Dr. J. E. Sanger (Dept. of Botany-Bacteriology, Ohio Wesleyan University, Delaware, U.S.A.), Prof. M. S. Adams (Dept. of Botany, University of Wisconsin, Madison, U.S.A.) and Dr. H. Züllig is fully acknowledged.

References

Adams, M. S., P. Guilizzoni & S. Adams, 1978. Sedimentary pigments and recent primary productivity in Northern Italian lakes. Mem. Ist. ital. Idrobiol. 36: 267–285.

Battarbee, R. W., 1973. A new method for the estimation of absolute microfossil numbers, with reference especially to diatoms. Limnol. Oceanogr. 18: 647–653.

Bonomi, G., 1962a. La dinamica produttiva delle principali popolazioni macrobentoniche del Lago di Varese. Mem. Ist. ital. Idrobiol. 15: 207–254.

Bonomi, G., 1962b. Le grandi modificazioni nella fauna macrobentonica del Lago di Varese intervenute nel periodo 1957–1962. Acqua Industriale. 4: 1–4.

Bonomi, G., 1964. Un nuovo aspetto dell'evoluzione del Lago di Varese: la comparsa di situazioni meromittiche. Mem. Ist. ital. Idrobiol. 17: 231–246.

Bonomi, G., 1969. Una ulteriore fase di declino del Lago di Varese. Naturae Montagna 9: 17–22.

Brown, S. R., H. J. McIntosh & J. P. Smal, 1984. Recent paleolimnology of a meromictic lake: fossil pigments of photosynthetic bacteria. Ver. int. Ver. Limnol. 22: 1357–1360.

Brugam, R. B., 1983. The relationship between fossil diatom assemblages and limnological conditions. Hydrobiologia 98: 223–235.

Candido, A., P. Panzani & D. Ruggiu, (in press). Diatom remains in sediment core 1 m long from the subalpine Lake Mergozzo (Italy). Mem. Ist. ital. Idrobiol. 41.

Engstrom, D. R., E. B. Swain & J. C. Kingston, 1985. A paleolimnological record of human disturbance from Harvey's Lake, Vermont: geochemistry, pigments and diatoms. Freshwat. Biol. 15: 261–288.

Griffiths, M., 1978. Specific blue-green algal carotenoids in sediments of Easthwaite water. Limnol. Oceanogr. 23: 777–784.

Guilizzoni, P., G. Bonomi, G. Galanti & D. Ruggiu, 1982. Basic trophic status and recent development of some Italian lakes as revealed by plant pigments and other chemical components in sediment cores. Mem. Ist. ital. Idrobiol. 40: 79–98.

Guilizzoni, P., G. Bonomi, G. Galanti & D. Ruggiu, 1983. Relationship between sedimentary pigments and primary production: evidence from core analyses of twelve Italian lakes. Hydrobiologia 103: 103–106.

Håkanson, L & A. Källström, 1978. An equation of state for biologically active lake sediments and its implications for interpretations of sediment data. Sedimentology 25: 205–226.

Lami, A., 1985. Il Lago di Varese: ricostruzione della sua recente evoluzione trofica attraverso uno studio paleolimnologico. Tesi di laurea in Scienze Biologiche, Univ. di Milano: 192 pp.

Lami, A., D. Ruggiu, P. Guilizzoni & A. Panzani, (in press). Subfossil diatom remains in a short sediment core of the eutrophic Lake Varese, Italy. Mem. Ist. ital. Idrobiol., 42.

Parkin, T. B. & T. O. Brock, 1980. Photosynthetic bacterial production in lakes: the effects of light intensity. Limnol. Oceanogr. 25: 711–718.

Ravera, O. & G. Premazzi, 1971. A new method to study the history of any persistent pollution in a lake by the concentration of ^{137}Cs from fall-out. Proceedings of the Intern. Symp. on 'Radioecology applied to the protection of man and his environment', Rome 7–10 September 1971 Commission of the European Communities 'u' 697 D.G.3: 703–719.

Ruggiu, D., C. Saraceni & R. Mosello, 1981. Fitoplancton, produzione primaria e caratteristiche chimiche di un lago fortemente eutrofizzato: il Lago di Varese. Mem. Ist. ital. Idrobiol., 39: 47–64.
Stockner, J. G., 1971. Preliminary characterization of lakes in the experimental lakes area, North-western Ontario, using diatom occurrence in sediments. J. Fish. Res. Bd Can. 28: 265–275.
Tonolli, V. & G. Bonomi, 1965. Stato biologico del Lago di Varese nelle situazioni passata e presente. In Amm. Prov. Varese (ed.), L'inquinamento del lago di Varese, indagini-rilevamenti-proposte: 89–108.
Züllig, H., 1981. On the use of carotenoid stratigraphy in lake sediments for detecting past developments of phytoplankton. Limnol. Oceanogr. 26: 970–976.
Züllig, H., 1982. Utersuchungen über die Stratigraphie von Carotinoiden im geschichteten Sediment von 10 Schweizer Seen zur Erkundung früherer Phytoplankton-Entfaltungen. Schweiz. Z. Hydrol. 44: 1–98.
Züllig, H., 1984. Vorläufige Mitteilungen über das Vorkommen des aus Purpurbakterien stammenden Pigmentes Okenon in Sedimenten. Schweiz. Z. Hydrol. 46: 297–300.
Züllig, H., 1985. Carotenoids from plankton and photosynthetic bacteria in sediments as indicators of trophic changes: evidence from the late glacial and the early Holocene of Lobsigensee. In G. Lang (Ed.), Swiss lake and mire environments during the last 15000 years. J. Cramer, Vaduz: 143–147.

Accepted 18 March 1986.

Hydrobiologia 143: 327–329, (1986)

The content of chlorophyll in the sediment of the volcanic maar lakes in the Eifel region (Germany) as an indicator for eutrophication

Wolfgang A. A. Möller[1] & Burkhard W. Scharf[2]
[1]*Rodensteinstraße 8, 6101 Brensbach, FRG*
[2]*Fachhochschule, Rüdesheimerstraße 64–68, 6550 Bad Kreuznach, FRG*

Keywords: sediment, chlorophyll, primary production, trophic state, volcanic lake, glacial periode, post-glacial period

Abstract

The content of chlorophyll a and its degradation products has been measured in cores of nine volcanic lakes in the Eifel region. The primary production increased at the beginning of Post Glacial Period but degreased after Atlantic Period. During the last three decades the production of biomass increased again due to anthropogenic eutrophication.

Introduction

There are very old limnetic investigations of the volcanic maar lakes of the Eifel region (Thienemann 1914/15), but only a few measurements of the primary production of these lakes. We have therefore investigated the content of chlorophyll a and its degradation products in the sediment as an indicator of eutrophication. For hydrology, morphometry and general limnetic characteristics of these maar lakes see Scharf (1983) and Lawa (1985).

Methods

We took short cores with a 'Kajak-Sampler' in all nine maar lakes and we investigated three up to 28 m long cores from the Meerfelder Maar, Schalkenmehrener Maar and Holzmaar (Irion & Negendank, 1984). The content of chlorophyll a and phaeopigments was analysed according to a method of Bayerisches Landesamt Für Wasserwirtschaft (unpublished). The samples were dried, pulverized and dissolved in 20 ml of an ethanol (90%) – acetone (10%) mixture. After heating in hot water (60 °C; 1,5 h) the extract was separated in a centrifuge and measured with a spectrophotometer. Further details see Möller (1985).

Time horizons in the Weinfelder Maar and the Gemündener Maar sediments were provided by Steinberg (unpublished) by means of ^{137}Cs-measurements. He found the 1963-horizon in the sediment of the Weinfelder Maar at a depth of 3 cm and in the Gemündener Maar at 4 cm.

In some maar lakes we could discover a connection of a well known historical event with a change in their trophic state. For example a sinking of the water level (Laacher See, Meerfelder Maar) or a sudden increase of the catchment area (Ulmener Maar) gave evidence of the moment of the change in trophic state (Möller, 1985).

As a time horizon in the long cores we could use the 'Laacher Bimstuff' wich is formed in the Alleröd Period about 12000 years before present (Negendank, 1984) and pollen chronology (Usinger, 1984).

Results and discussion

We have re-examined the method of sample preparation. The investigation showed that the concentration of chlorophyll and phaeopigments depends on the temperature and the time during the drying process (Fig. 1). Therefore it is very neces-

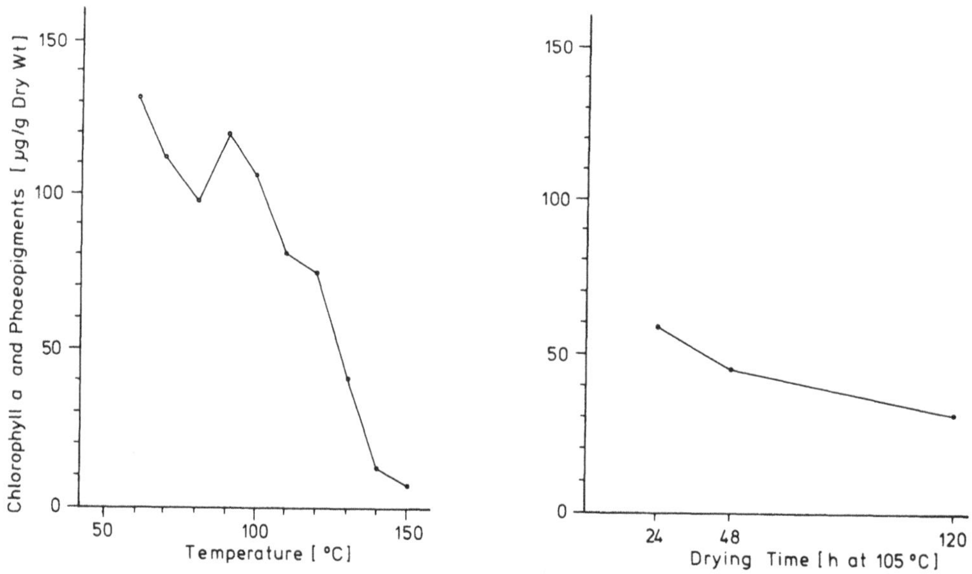

Fig. 1. Dependence of the content of chlorophyll a and phaeopigments in a homogeneous sample on the temperature and the time during the drying process.

sary to control the conditions of preparation in order to obtain comparable results and to standardize the method.

All the sediment cores showed an identical result referring to their tendency. The content of chlorophyll a and phaeopigments generally increases in the youngest part of the cores of all maar lakes, as in core from the Weinfelder Maar (Fig. 2). The increase of the primary production can be explained by anthropogenic eutrophication.

An index for the trophic state of a lake can be given by means of content of chlorophyll a and phaeopigments as the result of comparing all the short cores of the different maar lakes (Tab. 1).

The content of chlorophyll a and paeopigments in the long cores of the Meerfelder Maar, Schalkenmehrener Maar and Holzmaar reflects the change of climate during Glacial and Postglacial Period. The investigation of the Holzmaar is figured as an example (Fig. 3). The primary production was low during the Glacial Period, increased in the Postglacial Period and decreased after the Atlantic Period.

The result of the short cores dried in heat are not directly comparable with those of the long cores dried by freezing. The concentrations of chlorophyll and phaeopigments of the freeze dried samples are essentially higher than those of the samples dried by heating.

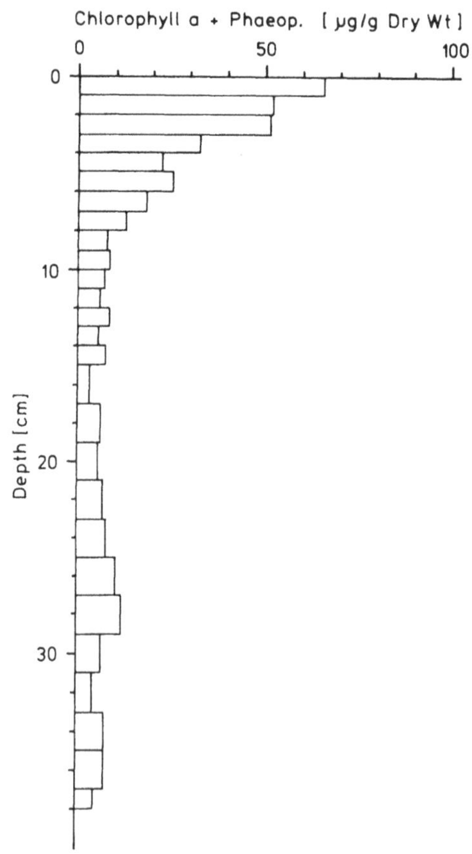

Fig. 2. Content of chlorophyll a and phaeopigments in the sediment of the Weinfelder Maar (from Möller 1985).

Table 1. Relation of the content of chlorophyll a and phaeopigments in the sediment of some volcanic maar lakes to the trophic state.

trophic state	chlorophyll a + phaeopigments in $\mu g/g$ dry matter
oligotrophic	13
mesotrophic	13 – 60
eutrophic	60 – 120
hypertrophic	120

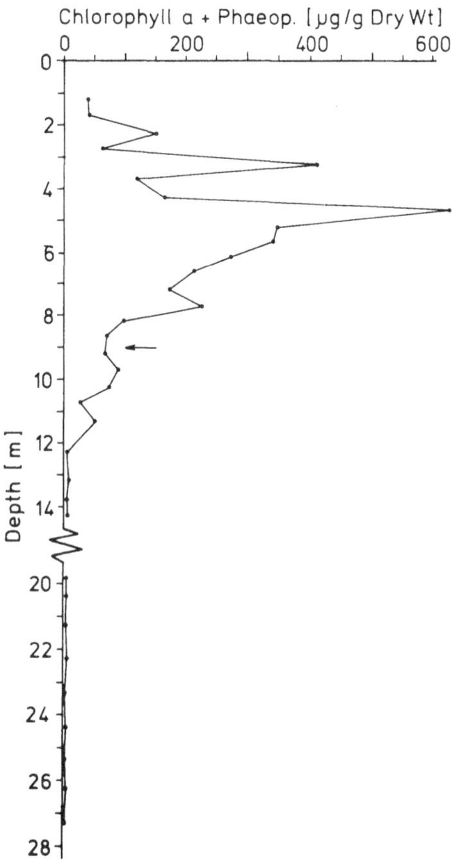

Fig. 3. Content of chlorophyll a and phaeopigments in the sediment of the Holzmaar. The arrow indicates the 'Laacher Bimstuff' (from Möller 1985).

Acknowledgments

We thank Prof. Dr. J. Negendank and Dr. G. Irion for their samples of the Meerfelder Maar, Schalkenmehrener Maar and Holzmaar, Dr. B. Bahrig for his samples of the Laacher See, and Dr. H. Kalweit for the possibility to do the investigations in the Landesamt für Wasserwirtschaft Rheinland-Pfalz.

References

Irion, G. & J. F. W. Negendank, (eds), 1984. Das Meerfelder Maar-Untersuchungen zur Entwicklungsgeschichte eines Eifelmaares. Cour. Forsch. Inst. Senckenberg 65: 1 – 101. Frankfurt.

Lawa (Länderarbeitsgemeinschaft Wasser), 1985. Seen in der Bundesrepublik Deutschland. 190 pp., Essen, Woeste Druck V.

Möller, W. A. A., 1985. Der Chlorophyllgehalt im Sediment verschiedener Eifelmaare als Eutrophierungsanzeiger. Dipl. Arbeit, FH Bad Kreuznach.

Negendank, J. F. W., 1984. Die Untersuchung der Schwerminerale der Seesedimente des Meerfelder Maares und des 'Laacher Bimstuffes' in den Sedimenten des Meerfelder Maares, des Hinkelsmaares und der Hitsche. Cour. Forsch. Inst. Senckenberg 65: 41 – 47. Frankfurt.

Scharf, B. W., 1983. Hydrographie und Morphometrie einiger Eifelmaare. Beiträge z. Landespfl. Rheinl.-Pfalz 9: 54 – 65 Oppenh.

Thienemann, A., 1914/15. Physikalische und chemische Untersuchungen in den Maaren der Eifel. Verh. naturhist. Ver. preuss. Rheinl. Westf. 70: 249 – 302 und 71: 273 – 389. Bonn.

Usinger, H., 1984. Pollenanalytische Untersuchungen zum Alter des Meerfelder Maares und zur Vegetationsentwicklung in der Westeifel während der ausklingenden Eiszeit. Cour. Forsch. Inst. Senckenberg 65: 49 – 66. Frankfurt.

Accepted 18 March 1986.

Hydrobiologia 143: 331–336, (1986)

Phosphorus budgets and forms of phosphorus in the Sau reservoir sediment: An interpretation of the limnological record

J. Armengol[1], M. Crespo[1], J. A. Morgui[1] & A. Vidal[2]
[1]*Dept. Ecologia, Fac. Biologia, Universidad de Barcelona, Avda. Diagonal, 645, 08028 Barcelona, Spain*
[2]*Abastecimiento de agua del Ter. Carretera Dosrius, 168, Cardedeu, Barcelona, Spain*

Keywords: paleolimnology, sedimentation rates, phosphorus inputs, apatite-phosphorus, organic phosphorus, inorganic non apatite phosphorus, eutrophication

Abstract

The phosphorus compounds in the sediment of Sau reservoir have been analysed and related to the annual phosphorus budgets from 1965 to 1969. With the increase of eutrophy, the phosphorus inputs leave the water in two ways: to the sediment and to the water outflow. The second way becomes more important throughout the eutrophication process.

Of the three phosphorus forms studied, apatite-P is related to annual water flow, which is also directly related to sedimentation rate. By contrast, the other two phosphorus forms (organic-P and inorganic non apatite-P) are linked to phosphorus outputs. These relationships enable us to propose an historical sequence for the sediment. Finally, the dating of sediment allows us to develop an equation relating the total phosphorus in the sediment to the annual phosphorus inputs and water flow into the reservoir (r = 0.94).

Introduction

Since autumn 1963, when Sau reservoir was filled for the first time, the evolution of its physical, chemical and biological characteristics has been studied on a monthly sampling basis. The data obtained and only partially published (Vidal, 1969, 1970, 1972, 1973, 1976), have been used here for the interpretation of the historical record in the sediments. The extremely high rates of sedimentation in Sau make it a suitable ecosystem for detailed paleolimnological reconstruction.

In a previous paper (Armengol *et al.*, 1984), we analysed the relationships between phosphorus compounds, organic carbon and total nitrogen for the same data set we use here. In the earlier paper, a tentative method was proposed for estimating the annual sedimentation rates based on the water flowing into the reservoir. Now we try to answer two more questions related to the sediment composition: 1) What is the role of the sediments in the biogeochemical phosphorus cycle in an ecosystem, such as Sau, experiencing increasing eutrophy? and 2) Is it possible to reconstruct the recent limnological history of Sau from the phosphorus compound profiles? In response to the first question, an input-output model is proposed for an indirect estimation of the importance of the sediments in the annual phosphorus budget. In the second case, an accurate sediment dating is necessary before we can explore the relationships between the phosphorus budgets and the sediment composition.

Material and methods

Using the methodology of Sommers *et al.* (1972) and Williams *et al.* (1971), the concentration of three different forms of phosphorus in a 46.5 cm long core was analysed. In the 31 1.5 cm thick samples into which the core was divided, the concentrations of organic phosphorus, inorganic non apatite-phosphorus (INA-P) and apatite-phosphorus (Ap-P) were determined (see Fig. 1). Organic carbon and total nitrogen were also measured by the Walkley & Black, and Kjeldhahl

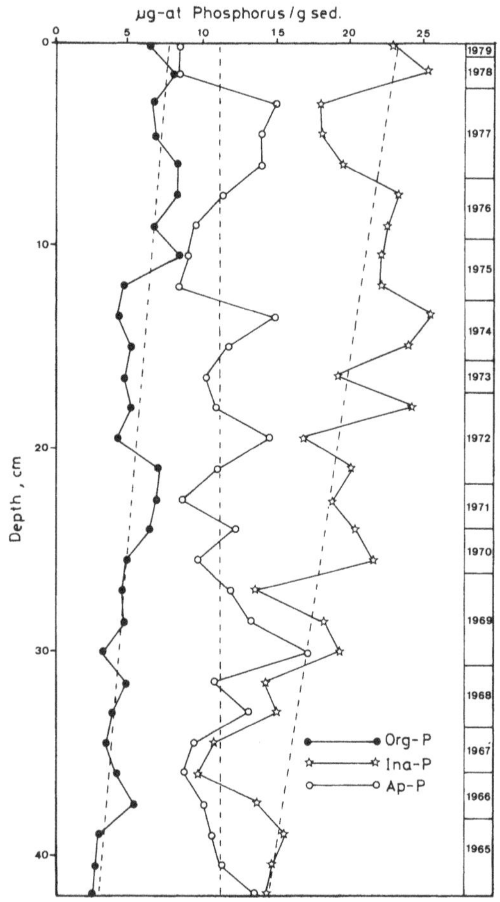

Fig. 1. Vertical distribution of phosphorus compounds in the sediment. Dotted lines show the vertical trend of each fraction. In the right hand margin the proposed dating is indicated.

methods (Jackson, 1970) respectively.

Annual soluble reactive phosphorus (SRP) inputs and outputs were calculated by means of monthly water analyses multiplied respectively by the water flow into and out of the reservoir. From these data, Fig. 2 shows the annual SRP inputs and water inflow into the reservoir for the period 1965–1979.

Results

The role of sediments in the annual phosphorus budgets

Since 1965 the annual SRP inputs in Sau have increased from values near 30 Tm yr^{-1} during the period 1965 to 1967 to more than 140 Tm yr^{-1} from 1974 to 1979 (see Fig. 2). This has been a consequence of increased human impact in the upstream river basin. In response to this change the phytoplankton standing crop also has become larger during the same period (Vidal, 1976). But not all the SRP inputs were transformed into biomass. As we can see in Fig. 3, the results of a regression analysis show strong correlation coefficents between SRP inputs and outputs ($r = 0.94$) and between SRP inputs and phosphorus budgets ($r = 0.88$). In both cases, the results are an expression of the same process, that is the increase of the amount of phosphorus taken out of water in two ways: sedimentation and water outflow. Nevertheless, the importance of these outputs differs when the SRP inputs increase. During the 1965–1967 period 81% of the inputs were accumulated in the sediment; these percentages decrease during the 1968–1973 period to 59% and reach 46% between 1974 and 1979. A similar, but inverse trend is shown by the SRP outputs.

Dating the sediments

In order to date the sediments, we use the hypothesis of Armengol *et al.* (1984) which propose a relationship between the annual water flow into the reservoir and apatite-phosphorus concentration. According to this hypothesis, the sedimentation rates for Sau fell between 1.5 cm in dry years and 3–4.5 cm in those of heavy rainfall. Accepting these values as a first approximation, we try to find through analysis the best relationship between: 1) water flow and rates of sedimentation, and 2) water flow and apatite-P concentrations that, at the same time, maximizes the correlation between SRP inputs and INA-P, and SRP with organic-P. In other words, we used a set of variables (water flow, sedimentation rates and apatite-P) to infer the history of sedimentation, and a second set (SRP inputs, INA-P and organic-P) for another estimate that can be used as a test. It is important to note that the variables of the first set have no significant correlation coefficients with those of the second set. For this reason we consider that the sets are independent and use this to test the results. Fig. 4 shows the relationships between the first set of variables, and Fig. 5 the second, also including organic-C and total nitrogen. In both cases correla-

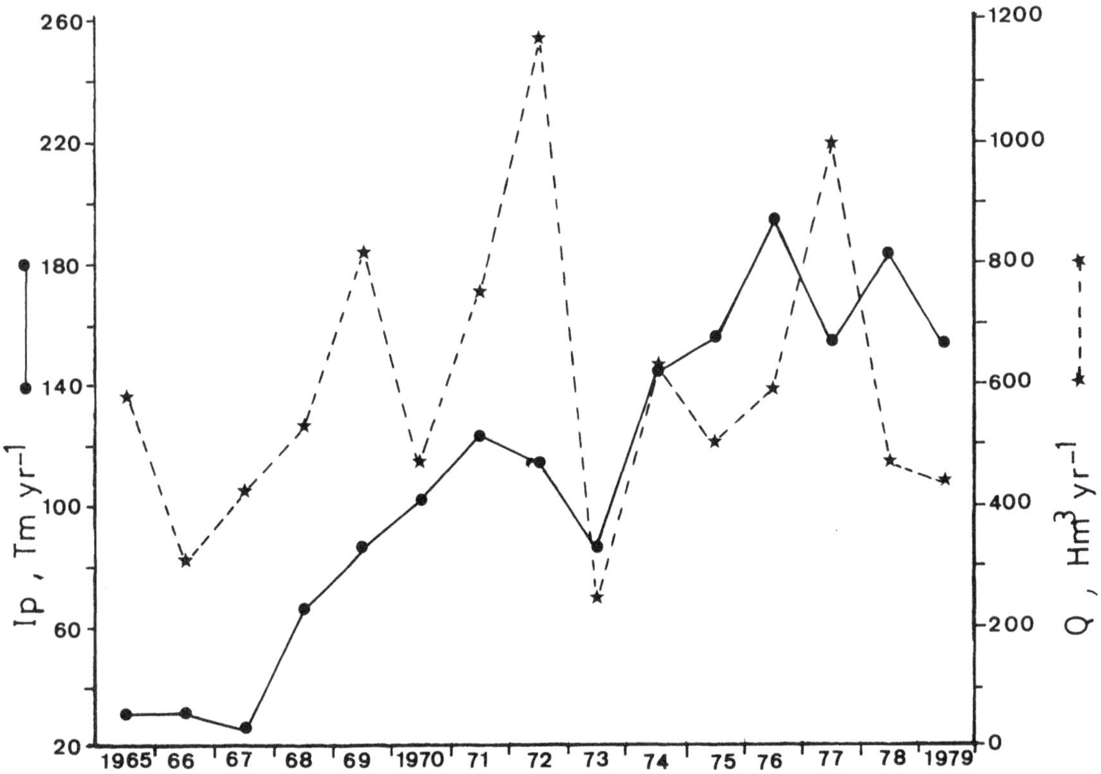

Fig. 2. Annual SRP inputs (I_p) and water flow into the reservoir (Q) for the period included in the sediment core.

tions are more significant for the second set than for the first one.

The interpretation of the limnological record

Although each phosphorus compound can be related to a specific input into the reservoir, the total phosphorus in the sediment is the sum of all the fractions. In fact the total phosphorus sedimented each year can be estimated from the total inputs considered before. The multiple power regression equation that relates the total phosphorus in the sediment with the reservoir inputs is as follows:

$$P_T = 8.20\ I_p^{0.25}\ Q^{0.05};\ r = 0.95^{**}\ p < 0.01$$

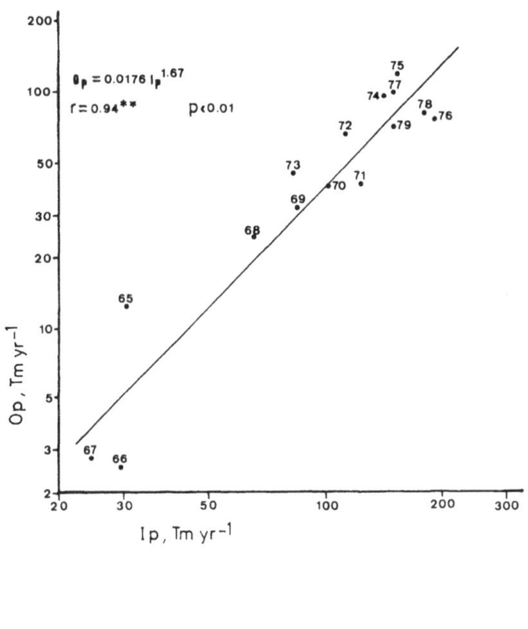

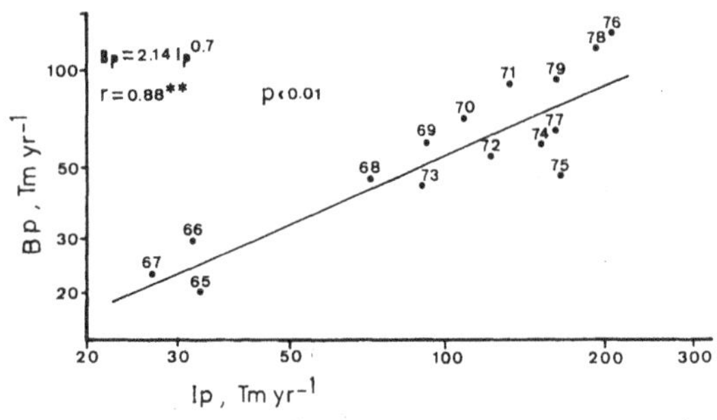

Fig. 3. Relationships of SRP inputs (I_p) with SRP outputs (O_p) and with the phosphorus budgets (B_p).

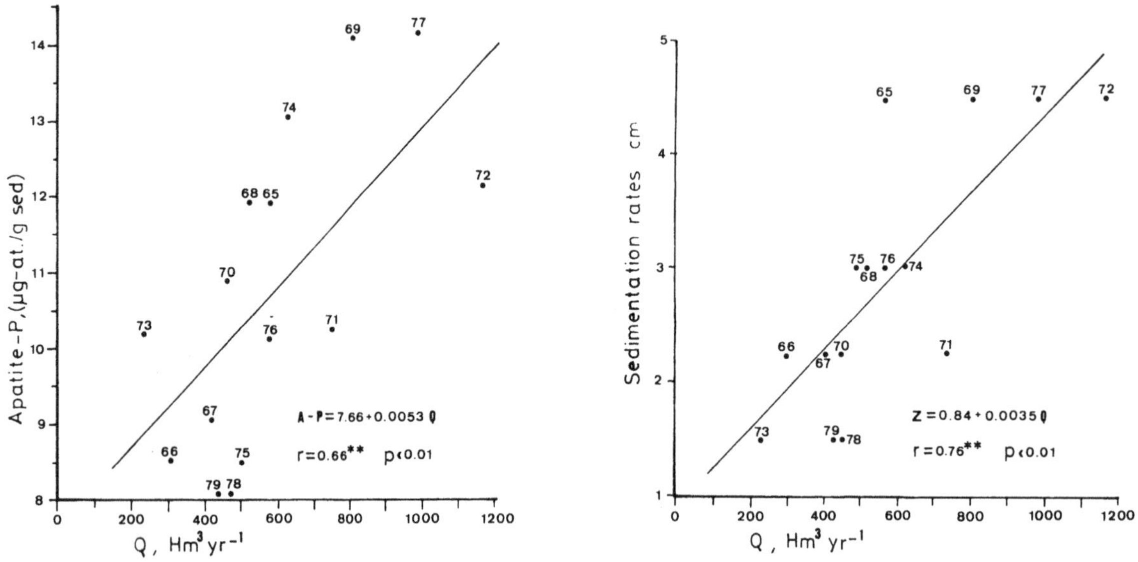

Fig. 4. Regression analysis of the water flow into the reservoir (Q) with the apatite-P and with the sedimentation rates.

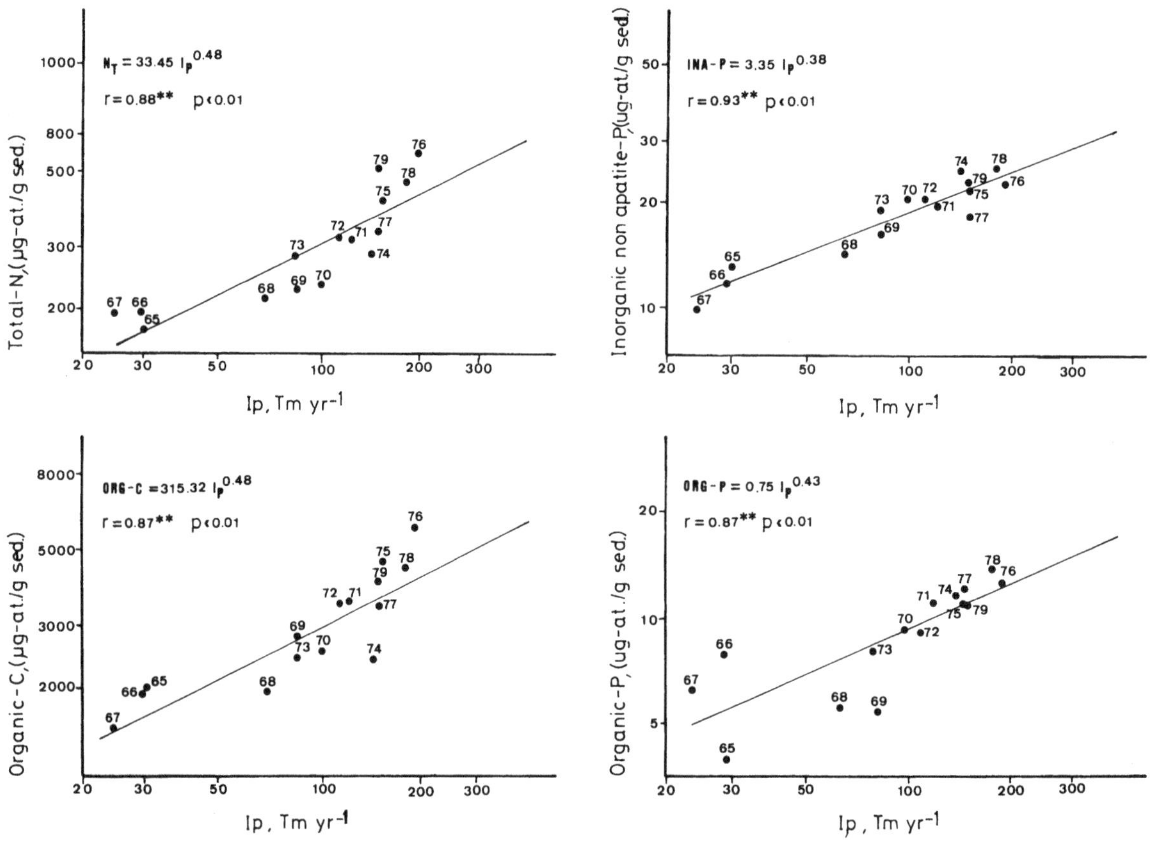

Fig. 5. Regression analysis of the SRP inputs into the reservoir (I_p) with the organic-P, inorganic non apatite-P, organic carbon and total nitrogen.

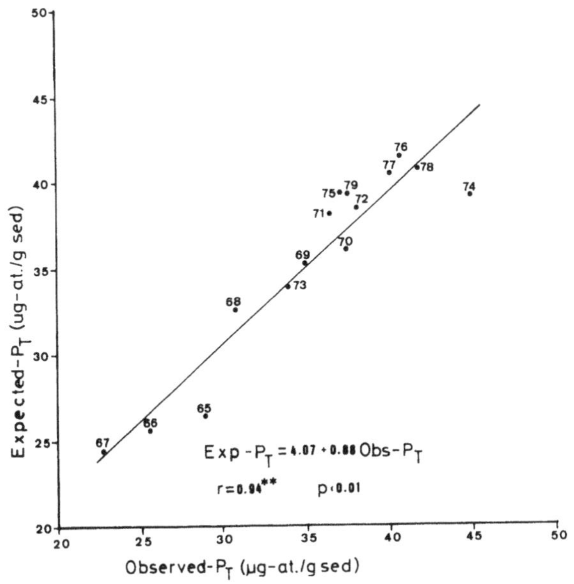

Fig. 6. Relationships between observed and expected annual total phosphorus accumulated in the sediment according to the equation proposed in the text.

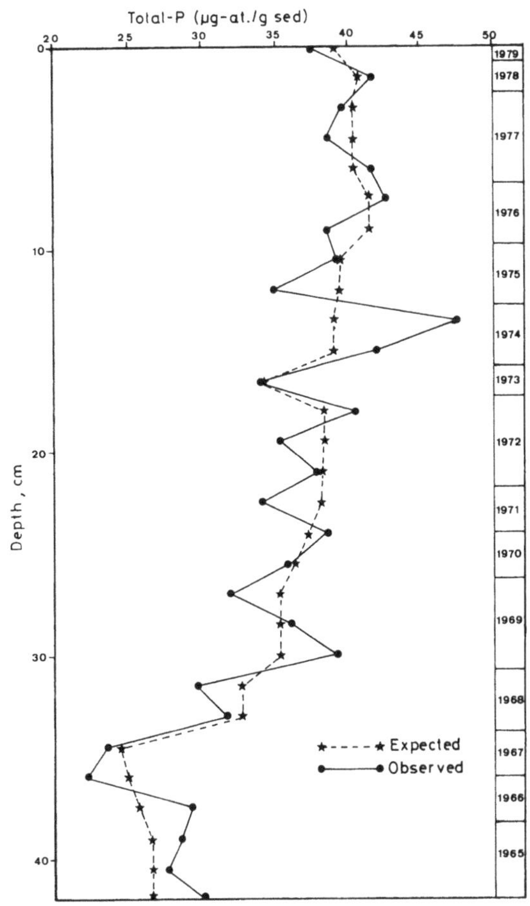

Fig. 7. Total observed phosphorus concentrations compared with those calculated using the equation proposed in the text, and the dating shown in the right hand margin.

where P_T is the total phosphorus in µg-at g^{-1} dry weight sediment, I_p is the annual SRP inputs in Tm yr^{-1} and Q the water flow in Hm^3. The Fig. 6 shows the fit between the expected and observed concentrations of total phosphorus. The next step has been to reconstruct the total phosphorus distribution in the sediment using the sedimentation rates and the expected total phosphorus according to the above equation, as can seen in the Fig. 7.

These relations can also be extended forward for the purpose of prediction. Table 1 gives the SRP input estimation since 1964 knowing the sedimentation rate, the water flow and the total phosphorus in the sediment. But at the same time it shows the expected sedimentation rates and sediment phosphorus composition for the period 1980–1984, after the sample core was taken.

Table 1. Expected values (*) of total phosphorus in the rates of sedimentation and SRP inputs for periods before and after the coring date calculated with the equation proposed in the text. Water flow (Q) is known for all the years.

Year	Q	P_T	Sedimentation rates	I_P
1964	631.6	30.51	3.00	2.69*
1980	370.7	39.03*	2.14*	157.2
1981	376.2	41.23*	2.16*	195.3
1982	1040.8	47.71*	4.48*	285.8
1983	350.8	38.85*	2.07*	156.1
1984	549.3	43.08*	2.76*	215.7

Discussion

The role of the sediments as a phosphorus trap, and the resulting decrease in water phosphorus concentration, is very important. In Sau this phosphorus loss shows a similar pattern to that due to the adsorption process found in laboratory experiments by Lijklema (1977) and Golterman (1984). Nevertheless, natural systems, like Sau, have a higher efficiency due to other processes like precipitation of apatite-P or organic matter.

The increasing importance of the SRP outputs during the eutrophication process can be a consequence of the limited capacity to transform phosphorus and deposit it in the sediment. In reservoirs this effect can be more important than in lakes due to the profundal outflow. The phosphorus concentrations in the sediment also suggest the minor importance of phosphorus release. The high sedimentation rates, the use of annual study periods which may not reveal shorter term processes, and the possible addition of this type of phosphorus to the SRP outputs may all help to explain the apparent minimization of the effect of phosphorous release.

Diagenetic transformations among phosphorus compounds seems to be another unimportant process. Transformation of the other forms of phosphorus into apatite-P takes too long relative to the age of the reservoir. The high correlation coefficients between SRP inputs and both INA-P and organic-P could not be expected if important diagenetic changes had taken place. The study of the organic matter in the sediment confirms these results because a great part of the organic molecules are primarily derived compounds of the phytoplankton with little structural change (Aceves, 1985).

The form in which phosphorus is accumulated into the sediment is another of the problems that arise from this study. Two forms of phosphorus compounds, INA-P and organic-P are directly related to biological processes that take place in the water. INA-P accumulates three times faster than organic-P, but the latter shows a somewhat faster response in relation to eutrophication (see Fig. 5). This is considered to be a direct effect of the growing phosphorus input (Schindler *et al.*, 1978). However, the primary production is an asymptotic process, and if we accept the organic matter of the sediment (C, N and P) as an expression of it, its increase also has to have the same limit. When the maximum primary production (near 400 mg of active chlorophyll m^{-2} or 600 g C m^2 according to Margalef, 1983) is reached, the model will depart from linearity and will become less usefull. Perhaps this limit has already been reached if we consider that the mean SRP inputs for the period 1980–1984 (see Table 1) have been of 202 Tm yr^{-1}. Further studies of the relationship between phosphorus inputs and water chlorophyll concentrations, using data available for the period 1965–1972 (Vidal, 1976) may help to define the limits of application of the model.

Acknowledgements

Financial support for the study of the river Ter and its reservoirs has been provided by the Comisión Asesora de Investigación Científica y Técnica, project no. 478/81. Our thanks are due to Dr. R. Margalef for critical improvements to the manuscript.

References

Aceves, M., 1985. Caracterització d'ambients lacustres per mitjà d'indicadors biogeoquímics. Tesis de licenciatura. Facultad de Química, Universidad de Barcelona.

Armengol, J., M. Crespo & J. A. Morgui, 1984. Phosphorus compounds in the sediment of the Sau reservoir (Barcelona, NE Spain) throughout its twenty-year existence. Verh. int. Ver. Limnol. 22: 1536–1540.

Golterman, H. L., 1984. Sediments, modiphying and equilibrating factors in the chemistry of freshwater. Verh. int. Ver. Limnol. 22: 23–59.

Jackson, M. L., 1970. Análisis químico del suelo. (2nd Ed.). Ed. Omega, Barcelona, 662 pp.

Lijklema, L., 1977. The role of iron in the exchange of phosphate between water and sediments. In H. L. Golterman (ed.), Interactions between sediments and freshwater. Junk Publ., The Hague: 313–317.

Margalef, R., 1983. Limnología. Ed. Omega. Barcelona, 1010 pp.

Schindler, D. W., E. J. Fee & T. Ruszczynski, 1978. Phosphorus input and its consequences for phytoplankton standing crop and production in the Experimental Lakes Area and similar lakes. J. Fish. Res. Board. Can. 35(2): 190–196.

Sommers, L. E., R. F. Harris, J. D. Williams, D. E. Armstrong & J. R. Syers, 1972. Fractionation of organic phosphorus in lake sediments. Soil Sci. Soc. Amer. Proc. 36: 51–54.

Vidal, A., 1969. Evolution d'un lac de barrage dans le NE de l'Espagne pendant les quatre premières années de service. Verh. int. Ver. Limnol. 17: 191–200.

Vidal, A., 1970. Fitoplancton del embalse de Sau (Barcelona) durante 1969. Pirineos 96: 51–76.

Vidal, A., 1972. Dinámica del hipolimnion anóxico en el embalse de Sau (Barcelona). Pirineos 105: 129–169.

Vidal, A., 1973. Developpement et évolution du phytoplankton dans le réservoir de Sau. XI Congress Internat. Comun. Large Dams, Madrid, 23 pp.

Vidal, A., 1976. Eutrofización del embalse de Sau en el transcurso de sus primeros años (1963–1972). Confed. Hidrogr. del Pirineo Oriental. Ministerio de Obras Públicas. Cardedeu. Barcelona. 43 pp.

Williams, J. D. H., J. K. Syers, R. F. Harris & D. E. Armstrong, 1971. Fractionation of inorganic phosphate in calcareous lake sediments. Soil Sci. Soc. Am. Proc. 35: 250–255.

Accepted 18 March 1986.

Hydrobiologia 143: 337–342, (1986)

Sediment accumulation in Lakes Lilla Ullfjärden and Stora Ullfjärden, Sweden

Laila Bodbacka
Department of Physical Geography, University of Uppsala, Box 554, S-751 22 Uppsala, Sweden

Keywords: annually laminated sediments, X-ray radiography, sediment accumulation, eutrophication

Abstract

X-ray radiography is a superior method of showing structures in recently deposited sediments. Annually laminated sediments have been proved to exist in Lakes Lilla Ullfjärden and Stora Ullfjärden using this method.

From X-ray radiographs of laminated sediment cores sampled nearly every year since 1977 in Lake Lilla Ullfjärden and since 1978 in Lake Stora Ullfjärden, it is possible to distinguish the seasonal variation in sediment deposition in the lakes. The variation in time and space in sediment accumulation is possible to estimate. The rate of sediment accumulation has been calculated to be 320 g solids/$m^2 \cdot y^{-1}$ for the deepest part of Lake Lilla Ullfjärden and 920 g solids/$m^2 \cdot y^{-1}$ for the central part of Lake Stora Ullfjärden for the period 1978 to 1983. It is also possible to trace the effect of eutrophication in time and space. An estimation of the appearance of the oxygen deficit in Lilla Ullfjärden gives a time around 1885.

Introduction

Annually laminated (varved) sediments are found in both lacustrine and marine environments, due to the seasonal variations in sediment influx and biological production. The laminated deposits are mainly found in the deeper parts of the sedimentation basins, where the laminated structures may be preserved due to oxygen deficit, which reduces bioturbation. Fossil examples are documented by many authors. Recently deposited, annually laminated sediments, as in the lakes studied here, are less well known. This circumstance is mainly due to methodical difficulties in studying the structures in the upper unconsolidated part of the sediment cover, and due to difficulties in distinguishing between annual and non-annual periodicity. It is often difficult to prove the existence of varves which represent a true annual periodicity in sedimentation.

Through comparative X-ray radiographic analyses of sediment cores, from separate times and locations in the same basin, it is possible to determine the continuing accumulation pattern and to separate annual and non-annual periodicity. Through stereographic X-ray analyses of rectangular sediment cores it is possible to obtain a three-dimensional picture of primary and secondary sediment structures even in sediments with high water content and in-gas rich sediments, before the sediment core is extruded. These methods are used to prove the existence of annual lamination in the studied lakes.

Lake descriptions

The studied lakes, Stora Ullfjärden and Lilla Ullfjärden (Fig. 1) are in reality two parts of the Lake Mälaren water system, but can be treated as separate lakes due to the narrow sounds separating them from each other and from the rest of Lake Mälaren.

The lakes are situated along a fault oriented in

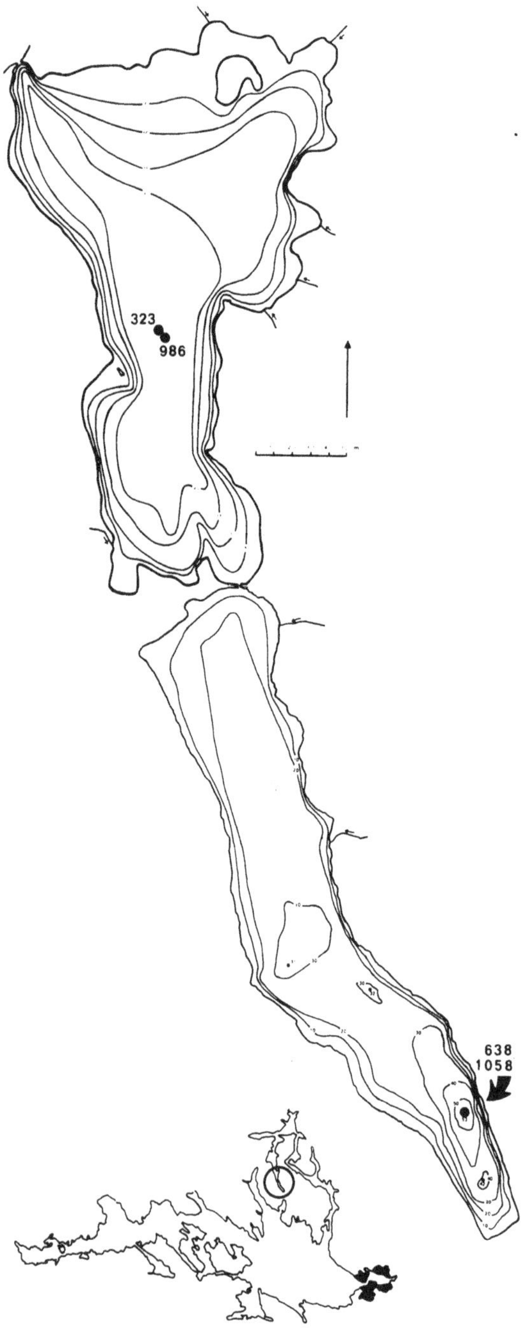

Fig. 1. Lake Mälaren (inset) with Lake Stora Ullfjärden (top) preliminary bathymetric map (5 m contour interval), after Bodbacka, 1985, and Lake Lilla Ullfjärden (below) bathymetric map (10 m contour interval), after a map drawn by Arnemo, Hammar & Kjällman, Lake Mälaren Research, 1966. Locations of sediment cores are indicated.

Table 1. Lake parameters.

Lake	Stora Ullfjärden	Lilla Ullfjärden
Volume	41.1×10^6 m^3	42.2×10^6 m^3
Area (lake)	2.7 km^2	1.9 km^2
Mean depth	15.2 m	22.2 m
Maximum depth	24 m	53 m
Length	2.9 km	3.8 km
Mean width	0.93 km	0.50 km
Drainage area	34.4 km^2	8.5 km^2

the NNW–SSE direction. An esker forms the western shores of the lakes. The eastern shores are mainly very steep and consist of bare rock and moraine. The bottom topography in Lake Lilla Ullfjärden is characterized by steep slopes and separate deep basins. In Lake Stora Ullfjärden the deepest part is more even, and the lake slowly becomes more shallow in the northern part. Table 1 shows some morphological parameters of the lakes.

Both lakes are dimictic and develop an oxygen deficit in the hypolimnion during both winter and summer stratification. The environmental influence is different for each lake. Lake Lilla Ullfjärden is free from waste water inflow. The main water supply is ground water from the esker, as indicated in the sediment by the accumulation of iron and manganese concretions at about 20 m depth along the esker. There is a net loss of water to Lake Stora Ullfjärden. The drainage area is dominated by woodland.

About 60% of the drainage area of Lake Stora Ullfjärden is agricultural land. The supply of surface water is larger in this lake, and occasionally there are inflows of contaminated water from the part of Lake Mälaren north of Lake Stora Ullfjärden.

Methods

Sediment sampling

The sediment sampler used is a gravity corer developed by Axelsson & Håkanson (1978). It consists of rectangular exchangeable coring tubes made of acrylic glass screwed into a corer head with a valve system. The corer allows free passage of water during lowering and an undisturbed penetration into the sediment.

X-ray radiography

The rectangular coring tubes are specially designed for scanning the unextruded sediment cores by X-ray radiography (Axelsson, 1979). The X-ray radiographic method, described by Axelsson (1983), is especially valuable when studying the upper part of a deposit with a high water content. The radiographs are based on the following data: Philips K 140 Be X-ray apparatus set at 50 kV and 2 mA, focal spot size 0.6×0.6 mm, focal distance 100 cm, and Agfa-Gevaert X-ray film Structurix D7Pb. The film density curves are digitized and used for digital image processing and for computer controlled plotting of the vertical variation in density correlated sedimentological parameters. Stereo radiographs are made to get a three-dimensional image of the laminae and the iron concretions in the sediment.

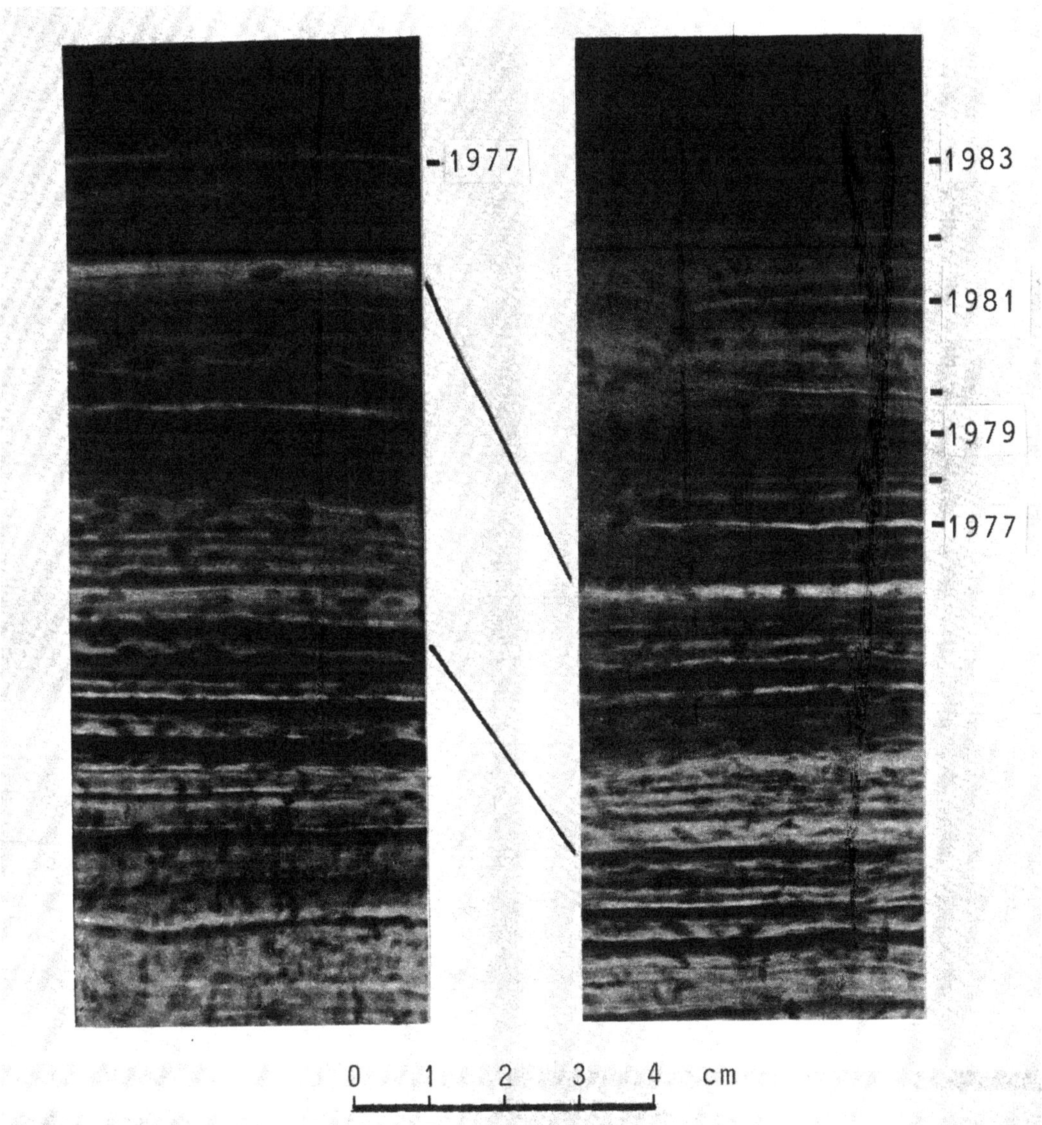

Fig. 2. Radiographs of annually laminated sediments in Lake Stora Ullfjärden. Upper parts of (from left) sediment cores 323 (March 1978) and 986 (March 1984) with adjoining bottom water. The dates show the spring laminae.

Preliminary results

Lake Stora Ullfjärden

The sediment cores below a water depth of about 13 m in this lake are laminated on top. The thickness of the laminated part increases with water depth. Below the laminated top layer the sediment is bioturbated. Fig. 2 shows radiographs of the upper part of the laminated sediments in cores 323 (March 1978) and 986 (March 1984). The sampling points shown in Fig. 1 are at about the same depths (22 m).

The radiographs show alternating dark, softer and light, harder laminae. The rounded spots are gas bubbles. One set of laminae (varve) representing one year's deposit can include, from the bottom, grey-white, dark grey, clay-grey and green-brown laminae. Other combinations of laminae exist, but the grey-white lamina is always present in every set of laminae. The grey-white laminae consist

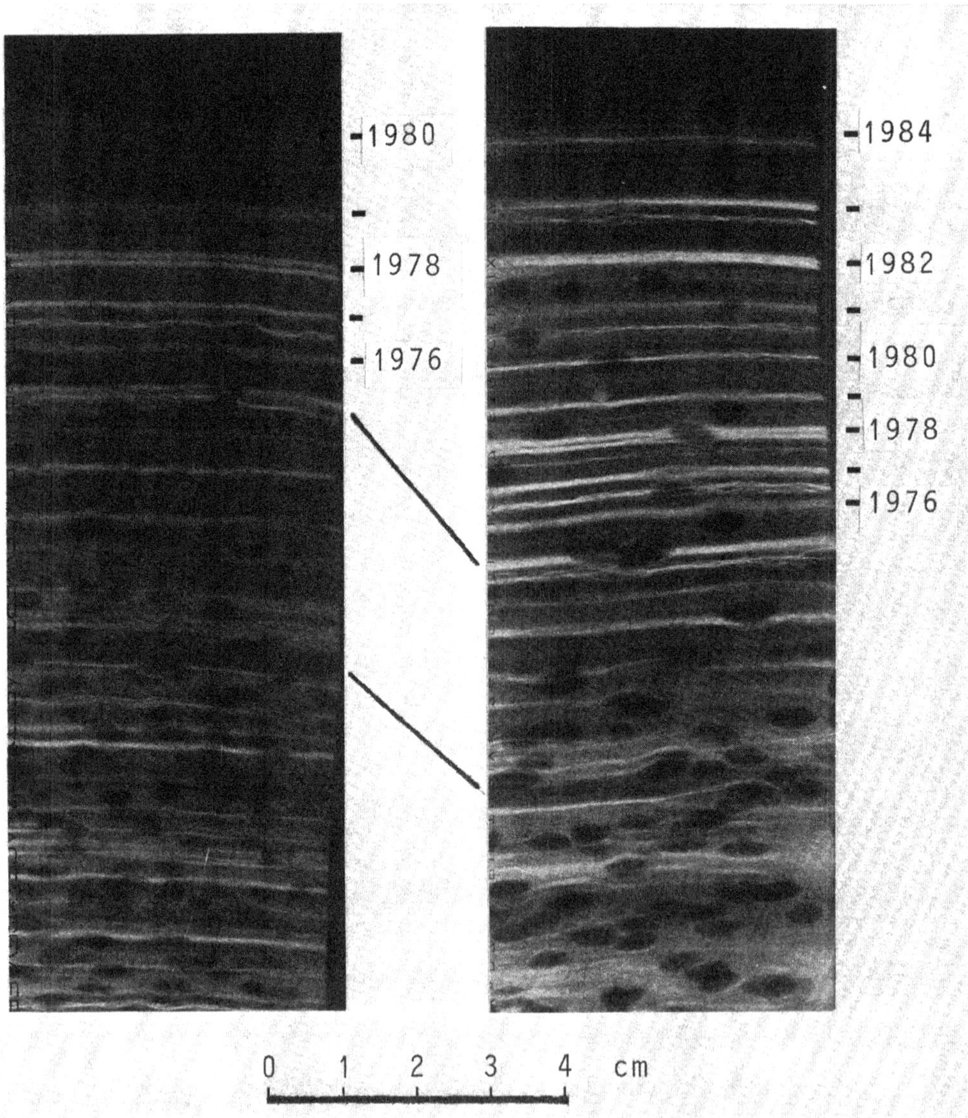

Fig. 3. Radiographs of annually laminated sediments in Lake Lilla Ullfjärden. Upper parts of (from left) sediment cores 638 (September 1980) and 1058 (October 1984) with the adjoining bottom water. The dates show the spring laminae.

mainly of diatoms and are formed during the spring. The darker laminae consist of other algae and plankton. The clay-grey laminae are very fine-grained and consist of clay from the surrounding arable land (Bodbacka, 1985).

The even floor of the deeper part of the lake makes it easy to correlate individual laminae over the whole lake. For the central part of the lake (Fig. 2) the rate of sediment accumulation is about 920 g solids/m$^2 \cdot$ y^{-1} or 9.2 mm y^{-1} for the period 1978–83 and 400 g solids/m$^2 \cdot$ y^{-1} or 10 mm y^{-1} for the period 1983–85. The same decreasing tendency is found in other cores both from the northern part of the lake, where the rate of sediment accumulation is about 900 g solids/m$^2 \cdot$ y^{-1} or 10 mm y^{-1} (1978–83) versus 400 g solids/m$^2 \cdot$ y^{-1} or 10 mm y^{-1} (1983–85), and from the southern part, where the rate of sediment accumulation is about 900 g solids/m$^2 \cdot$ y^{-1} or 9.1 mm y^{-1} (1978–83) versus 450 g solids/m$^2 \cdot$ y^{-1} or 10 mm y^{-1} (1983–85).

Lake Lilla Ullfjärden

In this lake the sediment cores from a water depth below about 23 m show a very similar stratification. On top is a laminated part, and below that, sediment with a chaotic, bioturbated structure. The thickness of the laminated part increases with water depth. The radiographs 638 (September 1980) and 1058 (October 1984) in Fig. 3 are from the deepest part of the lake (water depth 53 m). The lamination is very distinct and it is easy to distinguish each set of laminae representing one year of deposition. One set of laminae is built up, from below, of white, dark-grey, black and greybrown laminae. Sometimes there is another white lamina within the black lamina. The white laminae consist mainly of diatoms, deposited during spring or after a second bloom during the autumn. The darker laminae are formed by different species of algae and plankton (Bodbacka, 1985). It is possible to correlate individual laminae in this lake too.

The rate of sediment accumulation varies in the different parts of the lake. In the central part of the lake (30 m depth), the rate of sediment accumulation, for the period 1978–83 is about 200 g solids/m$^2 \cdot$ y^{-1} (8.1 mm y^{-1}) and in the southern part of the lake (35 m depth) the corresponding value is about 430 g solids/m$^2 \cdot$ y^{-1} (6.2 mm y^{-1}). The rate of sediment accumulation in the deepest part of the lake (53 m) is about 320 g solids/m$^2 \cdot$ y^{-1} (7.4 mm y^{-1}) for the same period.

For the period 1940–80 the rate of sediment accumulation is 280 g solids/m$^2 \cdot$ y^{-1}. This gives an approximate date of 1885 for the appearance of the oxygen deficit in the deepest part of the lake (Bodbacka, 1985).

Discussion

X-ray radiography is a superior method of showing structures in recently deposited sediments. The X-ray radiographic technique is fast and accurate for the calculation and recording of density-correlated sedimentary properties and it is particularly applicable to sediments with a low bulk density (Axelsson, 1983). According to Axelsson, the accuracy of the calculations increases with decreasing density of the deposits. The technique is, therefore, especially valuable when studying the uppermost, often very loose part of a sediment deposit.

The sediments of the studied lakes (Lilla Ullfjärden and Stora Ullfjärden) are very loose, the water content at the surface being over 95%. The accuracy of the X-ray radiographic technique is perhaps shown by an example from Lake Lilla Ullfjärden. For the period 1940–1980 the value of the rate of sedimentation is 280 g solids/m$^2 \cdot$ y^{-1}. This corresponds well to Tolstoy's (1972) estimate of primary production in 1968 of 115 g C m^{-2} (lake area), which is about 290 g solids/m$^2 \cdot$ y^{-1}.

The much higher rate of sediment accumulation in Lake Stora Ullfjärden is probably due to the larger catchment area and the fact that the drainage area is 60% agricultural land.

Willén & Tirén (1984), in their compilation of results from previous studies in Lake Lilla Ullfjärden, found the considerably increasing sulphate and nutrient levels since the 1930's, and the change in composition of zooplankton, to be typical of eutrophication. The increased eutrophication in both lakes is mainly caused by increased use of fertilizers in farming, changed farming methods and increasing atmospheric deposition of nutrients, as discussed by Willén & Tirén (1984) for Lake Lilla Ullfjärden. It is not surprising that Lake Stora Ullfjärden with its larger, more agriculturally centered catchment area, gives even more evidence of eutrophication, although during the last years (1983–1985) the rate of eutrophication has decreased.

References

Axelsson, V., 1979. Sjöars sediment studeras med röntgenteknik. Ymer 99: 133–144.

Axelsson, V., 1983. The use of X-ray radiographic methods in studying sedimentary properties and rates of sediment accumulation. Hydrobiologia 103: 65–69.

Axelsson, V. & L. Håkanson, 1978. A gravity corer with a simple valve system. J. Sed. Petrol. 48: 630–633.

Bodbacka, L., 1985. Annually laminated sediments in two basins of Lake Mälaren (Lilla Ullfjärden and Stora Ullfjärden) studied by X-ray radiography. Geogr. Ann. 67A: 145–150.

Tolstoy, A., 1972. Chlorophyll-a and primary production of phytoplankton in two basins of Lake Mälaren (Lilla Ullevifjärden and Ekoln). Naturvårdsvrekets limnologiska undersökning. Rapport 52. 55 pp.

Willén, T. & T. Tirén, 1984. Lilla Ullfjärden- en sjöbeskrivning. Statens naturvårdsverk PM 1769. 121 pp.

Accepted 18 March 1986.

Hydrobiologia 143: 343–348, (1986)

Chemical sediment stratigraphy of four high alpine lakes in Austria

Christian Steinberg[1], Wolfgang Beckstette[1], Roland Psenner[2] & Norbert Schulz[3]
[1]*Bayerisches Landesamt für Wasserwirtschaft, Lazarettstr. 67, D-8000 München 19, FRG*
[2]*University Innsbruck, Department of Limnology, Technikerstr. 13, A-6020 Innsbruck, Austria*
[3]*Institut für Seenforschung, Flatschacherstr. 70, A-9010 Klagenfurt, Austria*

Keywords: sediment, chemical stratigraphy, high altitude lakes, eutrophication, acidification

Abstract

Chemical sediment stratigraphy of four high alpine lakes is discussed with respect to atmogenic eutrophication as well as acidification. All lakes show more or less clearly paleoindications of atmospheric, anthropogenic impacts, mainly as slight eutrophication (Schwarzsee and Upper Plenderlesee, Tyrolia, and Gippersee, Carinthia). In Schwarzsee and Goaßelesee (Carinthia), there are weak but not yet significant indications of slight lake acidification, additionally.

Introduction

Chemistry of wet and dry precipitation strongly effects water chemistry of high alpine lakes on poorly buffered bedrocks (e.g. Stumm & Righetti, 1982; Stumm *et al.*, 1983). As development of soil and higher vegetation in high altitude regions probably can be almost completely disregarded, effects of precipitation on these lakes are even more pronounced than those on lakes of lower altitudes, for instance in secondary chains of mountains. As a consequence, atmospheric pollution may result in eutrophication as well as acidification in the high altitude lakes.

Since many of the Austrian high alpine lakes have never been sampled with regard to sediments before, we studied the chemical sediment stratigraphy of four high alpine lakes in Austria in order to assess the effects of atmospheric pollutants on these ecosystems.

Study sites and methods

Some features of the samples lakes are listed in Table 1. Further informations, especially on geographic locations and lake water chemistry is given by Arzet *et al.* (this volume).

Sediment cores were taken by a gravity (Kajak) corer and sliced into 1-cm sections (Upper Plenderlesee: 0.5-cm sections) immediately after coring. Dried subsamples were extracted for chlorophyll *a* and derivatives with ethanol (90%). Extracts were measured at 665 nm. Dry matter (DM) was digested for metal analyses by HNO_3. The elements Na, K, Mg, Ca, Ba, Al, Fe, Zn, Cu, Pb, Cd, Cr, Ni, Hg, and As were determined by applying different kinds of atomic absorption and (ICP)-atomic emission spectroscopy (Perkin Elmer). Residue on ignition (RI) and total phosphorus-content were determined according to German Standard Methods. All data refer to dry matter. Several graphs haven been expanded for better comparability. Factors of expansion are indicated in the figures.

Results and discussion

Results of the chemical sediment stratigraphy are given in Table 2 and in Figs. 1 to 4. Al is quantitively the most important metal in the sediments, except in Upper Plenderlesee, followed by Fe, Mg,

Table 1. Geographical data of the sampled lakes.

Lake	Altitude (m)	Watershed geology	District in Austria
Schwarsee ob Sölden	2792	gneiss, granite	Tyrolia
Upper Plenderlesee	2344	gneiss, granite	Tyrolia
Gippersee	2541	granite, gneiss mica schists	Carinthia
Goaßelesee	2145	granite, gneiss	Carinthia

Table 2. Range of mean sediment components in μequ kg^{-1} DM.

Lake	Phosphorus	Aluminum	Iron	Magnesium	Calcium	Potassium	Sodium
Schwarsee	65 – 81	925 – 1300	630	400 – 580	50 – 100	130 – 200	ca. 10
Up. Plenderlesee	ca. 32	ca. 220	890 – 1250	ca. 830	ca. 125	ca. 100	< 10
Gippersee	16 – 28	750 – 930	540 – 700	290 – 400	75 – 150	69 – 90	6 – 11
Goaßelesee	33 – 52	590 – 1050	450 – 800	500 – 670	70 – 100	115 – 170	8 – 11

Ca/K, and Na (Table 2). More detailed information, however, is presented by the trends of the concentrations in the figures.

In general, the atmospheric pollution of the lakes is indicated by the presence of several compounds which clearly originate from anthropogenic activities. Among these, organic trace pollutants, such as polycyclic aromatic hydrocarbons, are unequivocally of anthropogenic origin and distributed in the atmosphere. Paleolimnological studies from Adirondack and Rocky Mountains lakes by Heit *et al.* (1981, 1984) as well as analyses of high altitude snow profiles of Mount Blanc (Batifol & Bouton, 1984) have revealed that in these remote environments several trace metals are of atmospheric rather than pedogenic origin. This has been proved for Pb, Cd, and As in the North American lakes, and for Cu and Zn additionally in the Mount Blanc snows. On the other hand, Al as well as Fe are thought to originate mainly from the soils and rocks of the watersheds. It may be generalized that in remote regions increases in concentrations of Pb, Cd, As, or Cu in sediments of lakes prove the impact of atmospheric pollution.

In Schwarzsee ob Sölden (Fig. 1), Zn, Cu, Pb, Cd and As show increasing concentrations towards the sediment surface demonstrating clearly the atmospheric burden. This trend cannot be seen in the sediments of the other lakes, with the exception of As in Goaßelesee and Upper Plenderlesee. The reason for the striking differences between Schwarzsee ob Sölden and the other lakes could be due to a local phenomenon. Schwarzsee is the only lake with an anoxic hypolimnion. We therefore except that bioturbation is more or less absent in this lake. Additionally, concentrations of phosphorus and chlorophyll compounds rise simultaneously which indicates that enriching components accompany the atmospheric metal burden. The strong coincidence between phosphorus and chlorophyll compounds in the upper 6 cm evidences that at least a great moiety of phosphorus was incorporated into photosynthetic active organisms. Increasing chlorophyll concentration thus seem to be a symptom of slight eutrophication rather than better preservation conditions for pigments in the uppermost sediment layers (Swain, 1985). The above mentioned anoxic conditions in Schwarzsee may account for better preservation of organic compounds in recent sediment layers. In addition, many high altitude lakes have maximum concentrations of nutrients at the sediment water interface and high light penetration. Algal biomass maxima are consequently often found near the sediment surface. High concentrations of P and pigments in the uppermost layers can be an indication of special nutrient and light conditions.

Besides eutrophication, atmospheric pollutants can also cause acidification in poorly buffered regions. Acidification of watersheds with well developed soils leads to releases of base cations (Na, K, Mg, Ca etc.) from the soils in early stages of

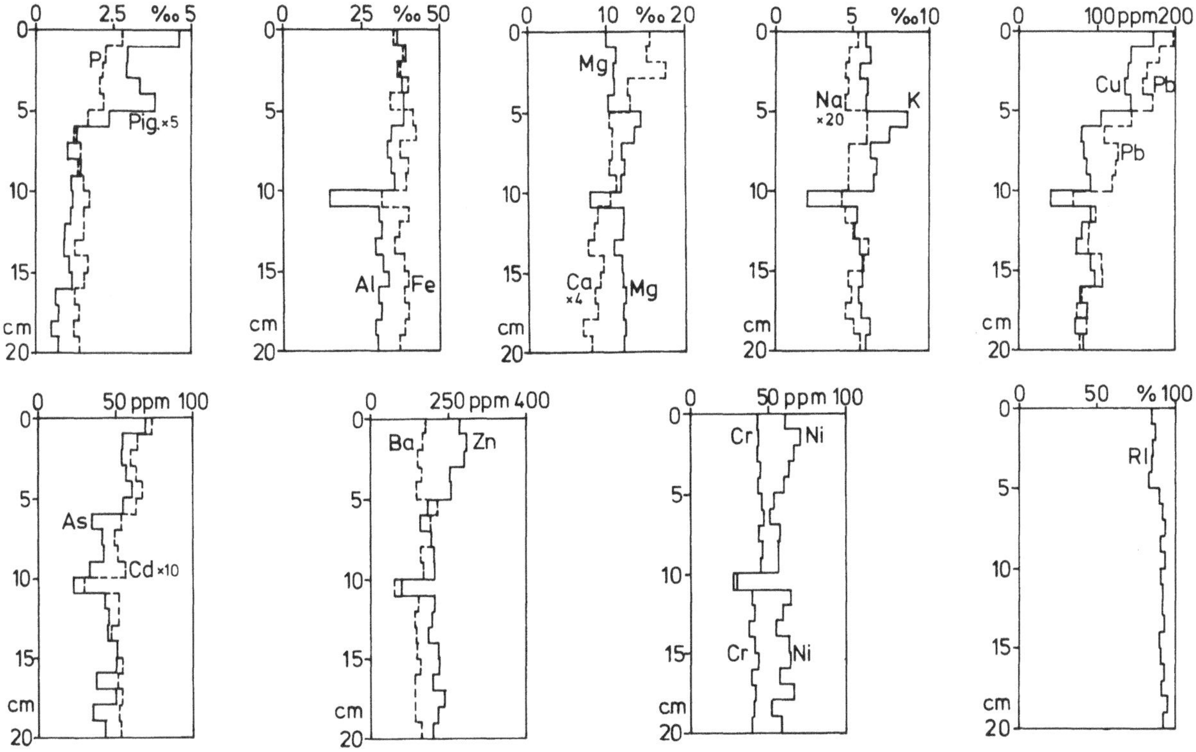

Fig. 1. Chemical sediment chronology in Schwarzsee ob Sölden.

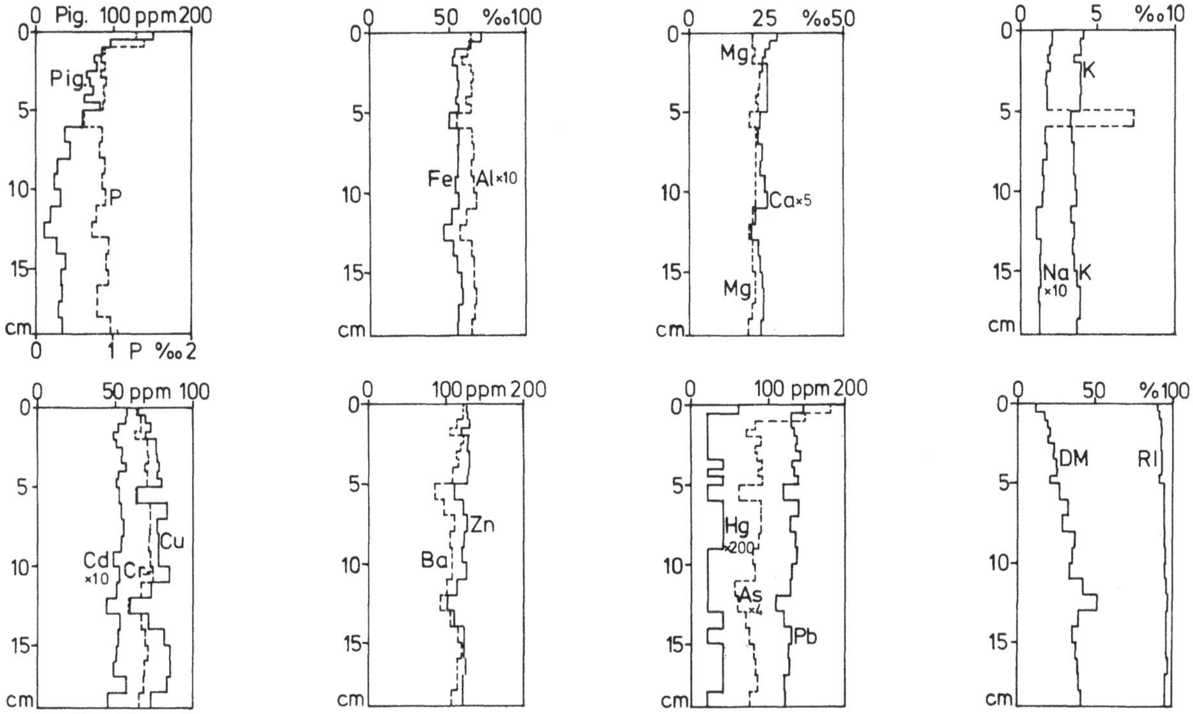

Fig. 2. Chemical sediment chronology in Upper Plenderlesee.

acidification (Galloway *et al.*, 1983). Later stages of acidification are characterized by release of cation acids, especially hydrated ions of Al and Fe (Overrein *et al.*, 1981, Dillon *et al.*, 1984). Under certain geochemical circumstances, base cations are well preserved in the sediments and even increasing acidification does not lead to release from the sediment as could be demonstrated for K in Großer Arbersee, Bavarian Forest, Fed. Rep. Germany. Consequently, changes in the contents of Al and Fe on one hand and stable base cations are worth to be considered with respect to lake acidification.

Among the metals of preponderately pedogenic origin, Fe shows more or less constant concentrations in all depths in Schwarzsee ob Sölden, whereas Al increases slightly towards the surface. Within the base cations, there is obviously no uniform trend: Mg concentrations decrease slightly, whereas Ca contents increase significantly. Na and K show roughly constant concentrations. Arzet *et al.* (in press) observed a significant shift in diatom composition towards the sediment surface. There seems to occur a drop in diatom-inferred pH from about 5.5 in 8 cm depth to about 5.1 at the surface. But the inferred pH in layers below 10 cm was not significantly above 5.1. Furthermore, taxonomical problems and uncertainties on the ecology of the rarely recorded diatoms complicate the diatom interpretation. Neither does the sediment chemistry present clear evidence. In the sediment of Schwarzsee, the sum of base cation concentrations decreases towards the top. This does not exclude eventual recent acidification of this lake for it is not known if in regions with less developed soils acidification starts really with increasing exports of base cations from the soils. Further, there is no in-

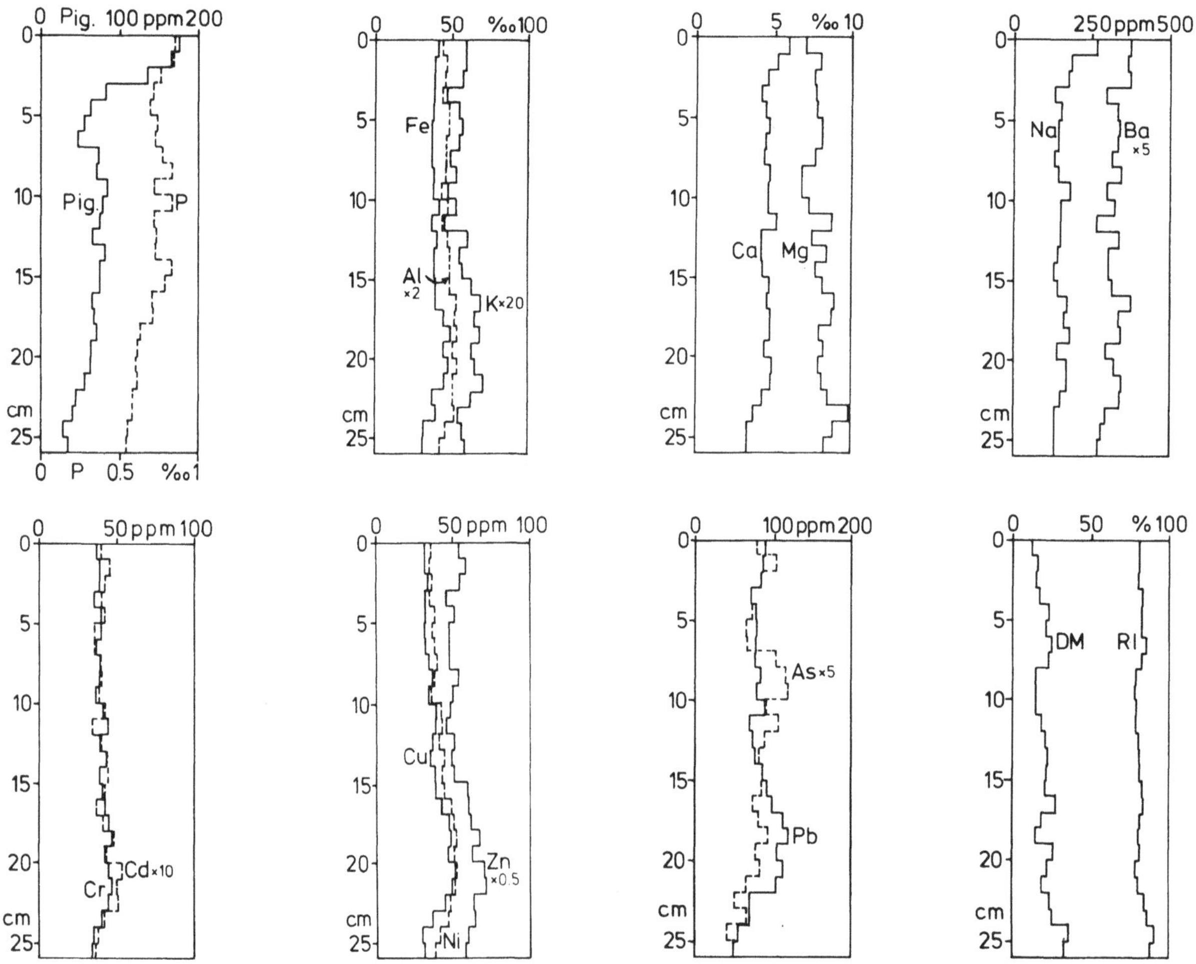

Fig. 3. Chemical sediment chronology in Gippersee.

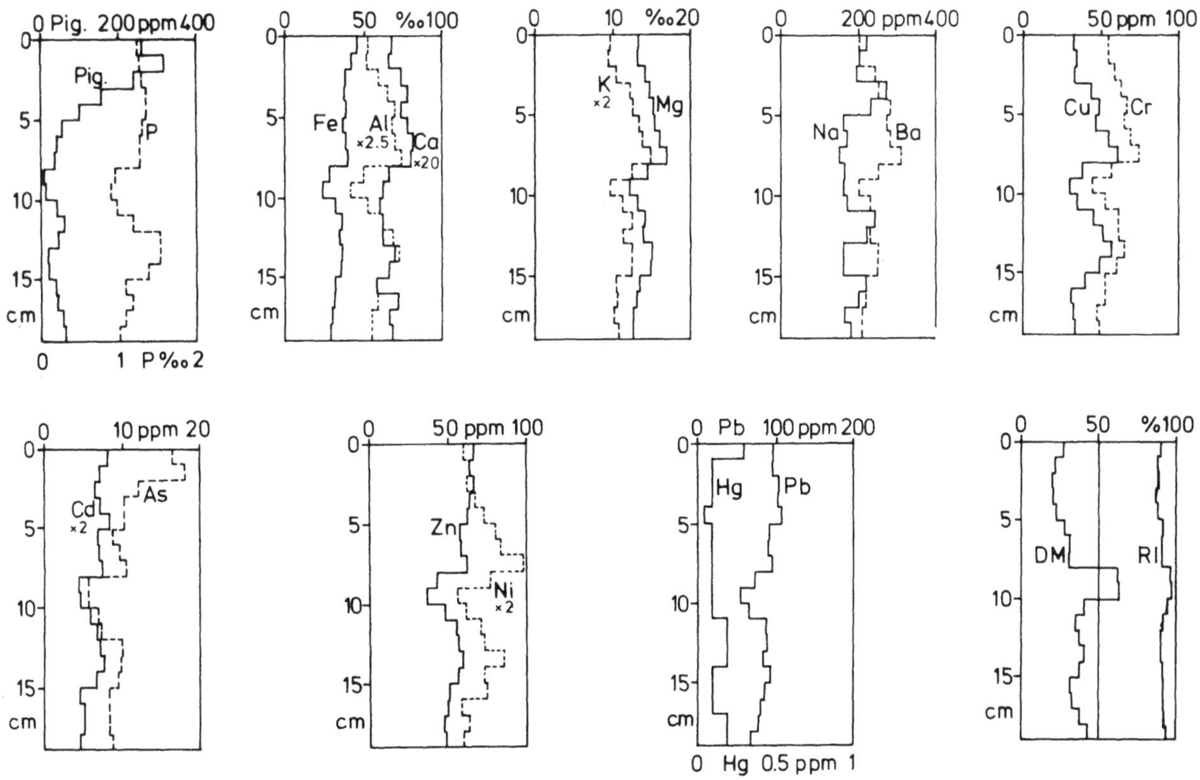

Fig. 4. Chemical sediment chronology in Goaßelesee.

formation on any retention mechanisms in Schwarzsee sediments for base cations.

As mentioned, in Schwarzsee only Al and Ca concentrations increase slightly and may serve as a potential indication of acidification. This increase, however, is obviously too weak to assess acidification.

The above general considerations on eutrophication and acidification are also valid for the other three high altitude lakes under study. Thus it is sufficient to discuss changes in their chemical stratigraphy only briefly.

In Upper Plenderlesee (Fig. 2), only the concentrations of As clearly indicates the impact of atmospheric pollutants. In this lake, too, the atmogenic loading seems to lead primarily to a slight eutrophication effect which is also evidenced by increasing numbers of the alkaliphilous diatom species, *Fragilaria brevistriata* (see Arzet *et al.*, 1986).

In Gippersee (Fig. 3), a slight eutrophication effect has most likely taken place as well (increasing phosphorus- as well as pigment-concentrations), although no clear paleoindication of atmospheric impact by increasing trace metal concentrations could be detected. The diatom flora, however, gives also proof of a slight alkalization (see Arzet *et al.*).

In the catchment of Goaßelesee, obviously there has been a historical landslide which is indicated by elevated contents of dry matter and residue on ignition around 10 cm depth as well as diminished concentrations of phosphorus, pigments and nearly all metals (Fig. 4). Recent atmospheric pollution is evidenced by As and to a lesser degree by Cd, Zn, and Pb. These metals show roughly constant concentrations in the sediment core, whereas all other metals decrease in contents towards the surface. Assuming iron is acting as a cation acid, there might have been a very slight acidification in this lake. Yet the increase in Fe-concentrations is to poor to present unequivocally evidence of acidification. This statement is congruent with a weak decrease in diatom-inferred pH (Arzet *et al.*, 1986), which, however, lies well within the confidence limits of inference.

Conclusions

All these lakes show more or less clear paleoindications of man's impacts via the atmosphere, but there is no uniform trend concerning eutrophication and/or acidification of high mountain lakes. In three lakes, this impact seems to result in slight eutrophication. Additionally, two lakes seem to show first but not yet significant signs of acidification. These statements by chemical sediment chronology are well confirmed by diatom analyses (Arzet *et al.*, in press).

Acknowledgements

The help of the laboratory staff of the Bayer. Landesamt für Wasserwirtschaft is greatly acknowledged. Thanks are due to Dr. E. Haworth for correcting the manuscript. This study was supported by the Umweltbundesamt under UFO-Plan no. Wasser 102 04 333.

References

Arzet, K., C. Steinberg, R. Psenner & N. Schulz, 1986. Diatom distribution and diatom inferred pH in the sediments of four alpine lakes. Hydrobiologia (in press).

Batifol, F. & C.F. Boutron, 1984. Atmospheric heavy metals in high altitude surface snows from Mount Blanc, French Alps. Atmosph. Envir. 18: 2507–2525.

Dillon, P.J., N.D. Yan & H.H. Harvey, 1984. Acidic deposition: effects on aquatic ecosystems. CRC Critical Rev. Envir. Contr. 13: 167–194.

Galloway, J.N., S.A. Norton & M.R. Church, 1983. Freshwater acidification from atmospheric deposition of sulfuric acid: a conceptual model. Envir. Sci. Technol. 17: 541A–545A.

German Standard Methods: Deutsche Einheitsverfahren zur Wasser-, Abwasser- und Schlammuntersuchung, 1981. Ed.: Fachgruppe Wasserchemie in der Gesellschaft Deutscher Chemiker. Verlag Chemie.

Heit, M., Y. Tan, C. Klusek & J.C. Burke, 1981. Anthropenic trace elements and polycyclic aromatic hydrocarbon levels in sediment cores from two lakes in the Adirondack acid lake region. Water Air Soil Poll. 15: 441–464.

Heit, M., C. Klusek & J. Baron, 1984. Evidence of deposition of anthropogenic pollutants in remote Rocky Mountain lakes. Water Air Soil Poll. 22: 403–416.

Overrein, L.N., H.M. Seip & A. Tollan, 1981. Acid precipitation – effects on forst and fish. Final report of the SNSF-project 1972–1980, 2nd ed., Oslo.

Steinberg, C., R. Meier, H. Emeis-Schwarz, D. Krause-Dellin & K. Arzet, 1984. Versauerung des Großen Arbersee, dokumentiert durch paläolimnologische Untersuchungen. Vom Wasser 63: 35–56.

Stumm, W. & G. Righetti, 1982. Tessiner Bergseen: Saurer Regen, saure Traufe. Neue Zürcher Zeitung, Forschung und Technik, Fernausgabe Nr. 231.

Stumm, W., J.J. Morgan & J.L. Schnoor, 1983. Saurer Regen, eine Folge der Störung hydrogeochemischer Kreisläufe. Naturwissenschaften 70: 216–223.

Swain, E.B., 1985. The effect of preservation conditions on the stratigraphic record of blue-green algal abundance. Abstract, IV. Int. Symposium Paleolimnology, Sept. 1985, Ossiach, Carinthia, Austria.

Accepted 18 March 1986.

Hydrobiologia 143: 349–354, (1986)

Acidification history and crustacean remains: some ecological obstacles

Jens P. Nilssen & Steinar Sandøy
Zoological Division, Biological Institute, University of Oslo, P.O. Box 1050, N-0316 Oslo, Norway

Keywords: acidification, Crustacea, community structure

Abstract

Crustacean remains are not yet useful as parameters for assessing the recent acidification phenomenon, due to too few studies on their ecology. The changes observed in the sediments are related not just to changes in lake chemistry, but also to predatory and competitive interactions. Moreover, the habitat pH of the organism may differ from the measured lake pH, and may as such not be representative. Therefore, species in some acidic lakes may decrease their abundance, whereas the same species in nearby acidic lakes increase. The caution mainly concerns species with no clear pH distribution. Species predominant in acidic or alkaline lakes generally follow patterns more closely related to their physiology. Study of pelagic species, including both body size and shape measurements, should be included in the analysis. If all the ecological factors are included, animal as well as algal remains would form powerful tools to reveal lake histories.

Introduction

Analyses of plant and animal remains has made possible a refinement in the evaluation of the historical development of lakes that was hitherto inaccessible by purely chemical data. The most common biological tool to assess lake ontogeny is above all remnants of plants (resting spores, diatoms); but Cladocera and Chironomidae also are important fossils (Battarbee, 1984; Whiteside & Swindoll, in press). Recently the remnants of plants and animals have been used to reconstruct lake pH over geological time and have thus extended the use of these paleoindicators (Battarbee, 1984; Krause-Delling & Steinberg, 1984). There are, however, still many obstacles in the use of cladoceran remains in tracing the acidification history of lakes.

The community structure that at a specific time characterizes a lake is in a dynamic balance with the external milieu. If the external factors vary over time, e.g. through altered climatic conditions, increased input of nutrients, or fish immigration, a new equilibrium will be established and a new community will inhabit the lake. In the sediment we read the result of a complicated interplay between the organisms and their milieu, but we get no information on the living dynamics. To increase our information on this we shall have to carefully study present populations of the paleoindicators; only then shall we have some possibility of undertaking a realistic reconstruction of the history of lakes.

The aims of this paper are to comment upon some obstacles that presently restrict the use of Cladocera in paleolimnological analysis, and to suggest what should be done to overcome these. The obstacles are chiefly connected with sampling methods, habitat ecology, biogeography, relationship to biotic factors and community structure of the organisms.

Comments on methods

Methods have been carefully reviewed by Frey (1979). Nearly all investigations involve the use of a net through which an untreated or treated portion of the sediment has to pass. Such a method is appropriate for large-size remnants of Cladocera like head shield, body shell, and the postabdomen of large-sized species, but can be inappropriate for species that primarily leave claws in the sediments, like some pelagic species (e.g. *Holopedium gibberum*), and members of the family Macrothricidae. Paterson (1985, pers. comm.) claims that the use even of a 30 μm net would lead to loss of very important remnants of the above species and groups. We suggest avoiding the use of coarse nets and perhaps using a net as small as 10–15 μm. One should also not overmanipulate the sample to the extent that animal remains are being lost or displaced. Moreover, one should investigate remnants of the whole cladoceran community, not concentrate solely on the family Chydoridae as has been common up to now. In Fennoscandian lakes, for example, the pelagic cladoceran community most frequently includes the genera: *Daphnia, Bosmina, Ceriodaphnia, Diaphanosoma, Holopedium, Leptodora* and *Bythotrephes*. All these genera are recoverable from the sediments (Paterson, 1984; Cotten, 1985), and their use will greatly enhance the knowledge of lake histories, including biotic factors (Cotton, 1985). This requires more time being spent on individual samples, but the result will be increased information from the separate strata. It will also be important to measure the size of specific species and record the variation of phenotypes, especially within *Bosmina* (cf. Kerfoot, 1974, 1981; Nilssen, 1978, 1984; Hofmann, 1984). The remains of Macrothricidae are important to recover, since within this group we have species characteristic of acidic waters, like *Streblocerus serricaudatus* and *Acantholeberis curvirostris* (Sandøy & Nilssen, 1985), partly due to a physiological ion exchange ability in such water (Potts & Fryer, 1979). Remnants of copepods have also been recorded from sediments (Deevey, 1964, own observations), and could also be of interest in lake histories.

Habitat ecology and abiotic factors

The habitat ecology of Crustacea as paleoindicators has been reviewed many times and we will therefore comment upon some few points of interest for the present paper. Cladocera are found in freshwater lakes under nearly all chemical conditions. The most frequent chemical parameters that have been measured include conductivity, specific ions, pH and alkalinity (Whiteside, 1970; Crisman, 1980; Sandøy & Nilssen, 1985). Many studies show highest cladoceran densities at low levels of total dissolved solids (TDS), as in Canadian lakes (Chengalath, 1982). It should be borne in mind that the present most important paleoindicators, the Chydoridae, live in the littoral, therefore littoral habitat heterogeneity, including periphytic growth of filamentous algae, will probably be a better predictor of Cladocera distribution than TDS. It is also intuitively difficult to understand why high TDS in itself should lead to lowever diversity of the littoral fauna, unless predation for some other factors limit macrophyte growth more severely in such waters and Chydoridae decrease because of this.

Chydoridae also show a specific relationship to pH, in that diversity tends to increase as pH goes below 6.0 (Whiteside & Swindoll, in press). This correlation with pH probably has an indirect cause, perhaps involving decreasing fish predation below pH 6.0 (see Nilssen *et al.*, 1984). Coloured lakes have low pH, and fish abundance is usually low in such lakes in Fennoscandia. Uimonen-Simola (1985) also found that the sedimention rate of Chydoridae increased as pH went from 6.0 to 4.7, co-occurring with an increase in the ratio of pelagic to littoral chydorid remains. This could also be due to low fish predation (see Nilssen, 1978). Nilssen *et al.* (1984) in experiments on many freshwater animals using ^{22}Na and TH_2O, found that major pelagic taxa, like fish and daphnids were physiologically vulnerable to acidic environments. Copepods, water mites and *Chaoborus* showed good sodium balance, primarily because their internal fluids were more effectively insulated from the ambient water. However, life history interactions involving predations might more directly affect the distribution of these animals in many natural environments.

It may be of importance to study habitat characteristics in greater detail, especially the organic content of lake water and the seasonal distribution of pH in the whole water column. In acidified Norwegian lakes, hypolimnetic pH frequently differed by one pH unit from conditions in the upper waters (Hindar & Nilssen, 1984). Since many species used as paleoindicators also inhabit the hypolimnion, these species may be subject to a higher pH than actually measured at the lake surface. These Norwegian lakes often are humic and coloured. It is known that organic material partly chelates vulnerable metals so that their potential effect on water organisms is lowered. Hence the biota of acidic clear water lakes may be more strongly affected than the biota of acidic humic lakes (Nilssen, 1976; Raddum *et al.,* 1980).

Attempts have been made to place many aquatic species within the so-called saprobic system that is thought to indicate specific water qualities (Sládeček, 1973). This is of very questionable use with many cladocerans since their distribution in perhaps the majority of cases is related to such parameters as predation, which has no place in the saprobic system.

Biogeography

Comparisons between Nearctic and Palearctic core records have treated many taxa as if they were the same species. For the family Chydoridae this is specifically questionable, since there is reason to believe that species have a much more restricted distribution than earlier thought (Frey, 1982, 1985; Changalath, 1985). Species given the same name in the two biogeographical regions may thus turn out to be different species, or may even belong to different species groups (Frey, 1982, 1985). The so-called cosmopolitanism of species is now being strongly questioned (Frey, 1982, 1985). Moreover, even if the species were the same in the two areas, they could still exhibit different ecologies. *Alonella nana,* is in Fennoscandia characteristic of acidic waters (Sandøy & Nilssen, 1985), and increase towards the sediment surface in recently acidified lakes (Uimonen-Simola, 1985). This species decreases towards the surface in recently acidified lakes in North America (Paterson, 1984). Similarly, *Alona intermedia* is more commonly found in less acidic lakes in Fennoscandia (Sandøy & Nilssen, 1985) and disappears towards the sediment surface in some recently acidified lakes (Uimonen-Simola, 1985). The species is common towards the surface in acidified American lakes (Paterson, 1984).

With increasing altitude, more of the so-called pioneer species of Chydoridae are found. They exhibit a wide geographical distribution and are found in a majority of environments (cf. Goulden, 1964).

Single species analysis

Interpretation of single species distribution in a sediment core should be made with care, especially within the group Chydoridae (see below under Community structure). The stratigraphy of species which have been shown to exhibit a physiological response to acidified lakes (cf. Potts & Fryer, 1979; Nilssen *et al.,* 1984) can help in the interpretation of other data. Thus the distribution of the acid vulnerable *Daphnia longispina* in sediments from a number of lakes in a restricted area in S Norway showed the same pattern of recent disappearance, irrespective of whether the catchment was influenced by forestry, pasture or not; this supports the hypothesis of a regional acidification. Morphological changes within species (cf. Kerfoot, 1974, 1981; Hofmann, 1984) is also helpful in tracing lake ontogeny. Changes in morphology of *Bosmina* remains have been used to detect changes in major lake habitat (water transparency, water level changes), the relative importance of invertebrate and vertebrate predation, and perhaps trophic status, although it is argued that succession of recent *Bosmina* may be more tied to predatory interaction than to trophic state (Nilssen, 1978). Some Chydoridae, notably *Alona guttata* and *A. rustica,* frequently have 'bubbles' in their body shell. It seems that these forms become more common in acidic lakes or towards the top of the sediment in recently acidified lakes (Sandøy & Nilssen, unpubl. data).

To summarize, single species analysis should be used with care or abandoned altogether within the Chydoridae, but can be helpful with pelagic species to reveal lake ontogeny. A careful analysis of changes in body size and shape should also be carried out.

Biotic factors

Biotic interactions within a lake generally are much more time consuming to assess than are abiotic factors. As shown earlier, Cladocera have physiological limits, and the physiogical boundaries set by these limits establish the abiotic habitat within which predation and competition operate. Detailed reviews of biotic factors have been done earlier and are outside the frame of this paper (Kerfoot, 1974, 1981; Nilssen, 1978; Kitchell & Kitchell, 1980;

Whiteside & Swindoll, in press). Here we will mainly concentrate on biotic factors associated with acidification. Nilssen *et al.* (1984) investigated some reasons for species replacements in acidified lakes and concluded that two contrasting invertebrate groups take over the role of fish. The first group includes relatively unselective invertebrate predators (e.g. *Chaoborus*), and prey abundance here is restricted to periods when the predators do not possess predatory stages or temporarily leave the habitat as non-aquatic adults. With such predators, large erratic fluctuations in prey abundance will occur. In the presence of stabilizing fish predators, these invertebrate predators were usually themselves predator controlled, and prey numbers were more stable. The second group of invertebrate predators includes less voracious types (e.g. *Heterocope saliens,* pelagic Corixidae) which exert a smaller influence on the community. Here the coexistence of predator and prey is common.

A delicate balance of invertebrate and vertebrate predators is found in many lakes (cf. Kerfoot, 1975), a state which implies more difficulties in the interpretation of findings from a core (cf. Nilssen, 1978). For cladocerans, many biotic factors influence seasonal and long-term changes in population structure (Whiteside *et al.,* 1978; Lynch, 1980; Brambilla, 1982). Most of these biotic influences are impossible to reconstruct from the sediments.

Goulden (1971) tried to investigate the reasons for seasonal changes in benthic Chydoridae, and found that the summer decreases were due to predation by chironomid larvae. Goulden (1971) mentioned many other benthic and littoral predators. Straškrába (1963, 1965, 1967) has shown that the removal of fish from the littoral leads to a replacement of smaller cladoceran species (*Chydorus, Pleuroxus* spp.) by larger ones (*Eurycercus, Sida, Simocephalus*) analogous to that observed in plankton communities. Competition must also be considered: for example Kerfoot (1974, 1984) reported species and morph deplacement in *Bosmina* over geological time and Nilssen (1984) found evidence that disappearance of the acid vulnerable *Daphnia* within the last 20–40 years followed an increase in the potential competitor *Bosmina.*

To summarize, when analyzing a sediment core we must be aware that the majority of the finer biotic adjustments of species to predators (colour, eye pigmentation, egg number and shape) are not being detected in our samples (see Lynch (1980) for a comprehensive review on Cladocera evolutionary ecology). We should therefore avoid oversimplified generalization without a thorough knowledge of the ecology of the species being investigated.

Influences on community structure

It is difficult to isolate factors responsible for community structure in present-day lake communities. Careful analysis of all available evidence is needed to interpret the paleoassemblages. Whiteside & Swindoll (in press) states: 'Paleolimnologists must consider time/space dimensions. It is impossible to ascertain the spatial relationship of fossil assemblages. It is also ill advised for an investigator to invoke arguments that require detailed knowledge of spatial patterns in paleolimnological assemblages'. Although these statements are appropriate criticism of the frequent non-ecological approach to treating samples, they sound too pessimistic (see Nilssen, 1978). It is urgent that paleolimnologists include physiological and ecological data in their work; a core is not reflective of a static relationship, but expresses the final result of

Table 1. Species and community characteristics that could be used in studying recent development of acidified lakes. A: relationship to physical factors (pH), B: potential vulnerability to fish predation, C: to invertebrate predation, D: fate in acidified lakes. N.V. = not vulnerable, FISH = fish predation, INV = invertebrate predation. In Macrothricidae the two species: *Streblocerus serricaudatus* and *Acantholeberis curvirostris* are included.

Taxa	A: physical	B: affecting size	C: affecting shape	D: Reaction
Bosminidae	N.V.	FISH	INV	B & C (increase)
Daphnidae	V.	FISH	INV	A, B & C (decrease)
Holopedium gibberum	N.V.	FISH	?	B (increase)
Diaphanosoma brachyurum	N.V.	FISH	?	B (increase)
Macrotricidae	N.V.	?	?	? (increase)
Chydoridae	N.V.	FISH (large species)	?	B (?)

a community being acted upon by abiotic and biotic factors.

Most communities possess a great amount of plasticity, which is responsible for the frequent lack of agreement with respect to the fates of individual species in sedimentary records. Thus, a species in some acidic lakes may increase in abundance, whereas the same species in neighbouring lakes decreases (cf. Vimonen-Simola, 1985). Such 'contradictions' usually involve so-called indifferent species, species recorded over a wide pH-range. Most species assemblages (plankton or littoral) contain a great deal of plasticity to absorb variation in single species abundances, both seasonally and long-term.

To summarize, Table 1 indicates the species and community characteristics that should be used in studying the recent development of acidified lakes.

References

Battarbee, R. W., 1984. Diatom analysis and the acidification of lakes. Phil. Trans. r. Soc. Lond B 305: 451–477.

Brambilla, D. J., 1982. Seasonal variation of egg size and number in a *Daphnia pulex* population. Hydrobiologia 97: 233–248.

Chengalath, R., 1982. A faunistic and ecological survey of the littoral Cladocera of Canada. Can. J. Zool. 60: 2668–2682.

Chengalath, R., 1985, in press. The distribution of chydorid Cladocera in Canada. Dev. Hydrobiol.

Crisman, T. L., 1980. Chydorid cladoceran assemblages from subtropical Florida. pp. 657–668, In W. C. Kerfoot (ed.), Evolution and ecology of zooplankton communities. University Press of New England.

Cotten, C. A., 1985, in press. Remains of non-chydorid cladocerans from surficial sediments in eastern Finnish lakes. Dev. Hydrobiol.

Deevey, E. S. jr., 1964. Preliminary account of fossilization in Rogers Lake. Verh. int. Ver. Limnol. 15: 981–992.

Frey, D. G., 1979. Cladoceran analysis. In Paleolimnological changes in the temperate zone in the last 15000 years. IGCP project 158B, Lake and mire environments, Vol. II pp. 227–257. Lund.

Frey, D. G., 1982. Questions concerning cosmopolitanism in Cladocera. Arch. Hydrobiol. 93: 484–502.

Frey, D. G., 1985, in press. The taxonomy and biogeography of Cladocera. Dev. Hydrobiol.

Goulden, C. E., 1964. The history of the cladoceran fauna of Esthwaite Water (England) and its limnological significance. Arch. Hydrobiol. 60: 1–52.

Goulden, C. E., 1971. Environmental control of the abundance and distribution of the chydorid Cladocera. Limnol. Oceanogr. 16: 320–331.

Hindar, A. & J. P. Nilssen, 1984 (eds). Årsrapport, Gjerstad 1982/84. Rep. Norw. Liming Project 1980–84, Rep. 21: 1–124.

Hofmann, W., 1984. Postglacial morphological variation in *Bosmina longispina* Leydig (Crustacea, Cladocera) from the Grosser Plöner See (north Germany) and its taxonomical implications. Z. zool. Syst. Evol. forsch. 22: 294–301.

Keen, R., 1976. Population dynamics of the chydorid Cladocera of a southern Michigan marl lake. Hydrobiologia 48: 269–276.

Kerfoot, W. C., 1974. Egg-size cycle of a cladoceran. Ecology 56: 1259–1270.

Kerfoot, W. C., 1975. The divergence of adjacent populations. Ecology 56: 1298–1313.

Kerfoot, W. C., 1981. Long-term replacement cycles in cladoceran communities: A history of predation. Ecology 62: 216–233.

Kitchell, J. A. & J. F. Kitchell, 1980. Size-selective predation, light transmission, and oxygen stratification: Evidence from the recent sediments of manipulated lakes. Limnol. Oceanogr. 25: 389–402.

Krause-Dellin, D. & C. Steinberg, 1984. Evidence of lake acidification by a novel biological pH-meter. Envir. Tech. Letts 5: 403–406.

Lynch, M., 1980. The evolution of cladoceran life histories. Quart. Rev. Biol. 55: 23–42.

Nilssen, J. P., 1976. Community analysis and altitudinal distribution of limnetic Entomostraca from different areas in southern Norway. Pol. Arch. Hydrobiol. 23: 105–122.

Nilssen, J. P., 1978. Selective vertebrate and invertebrate predation – some paleolimnological implications. Pol. Arch. Hydrobiol. 25: 307–320.

Nilssen, J. P., 1984. An ecological jig-saw puzzle: reconstructing aquatic biogeography and pH in an acidified region. Res. Rep. Freshwat. Inst. Drottningholm 61: 139–147.

Nilssen, J. P., T. Østdahl & W. T. W. Potts, 1984. Species replacements in acidified lakes: physiology, predation or competition. Res. Rep. Freshwat. Inst. Drottningholm 61: 148–153.

Potts, W. T. W. & G. Fryer, 1979. The effects of pH and salt content on sodium balance in *Daphnia magna* and *Acantholeberis curvirostris*. J. comp. Physiol. 129: 289–294.

Paterson, M. J., 1984. Paleolimnological analysis of recently acidified lakes in the Adirondack mountains. Unpubl. Ph.D. thesis, Univ. Tenessee, Bloomington.

Raddum, G. G., A. Hobæk, E. R. Lømsland & T. Johnsen, 1980. Phytoplankton and zooplankton in acidified lakes in south Norway. In D. Drabløs & A. Tollan, Ecological impact of acid precipitation, pp. 332–333, SNSF-Project, Ås-NLH.

Sandøy, S. & J. P. Nilssen, 1985, in press. A geographical survey of littoral Cladocera in Norway and their use in paleolimnology. Dev. Hydrobiol.

Sládeček, V., 1973. Water quality measurements from the biological point of view. Ergeb. Limnol. 3: 1–273.

Straškrába, M., 1963. The share of the littoral region in the productivity of two fishponds in southern Bohemia. Rožpr. Česk. Akad. Véd Řada. Mat. Přiř. Véd 73: 3–61.

Straškrába, M., 1965. The effect of fish on the number of invertebrates in ponds and streams. Mitt. int. Ver. Limnol. 13: 106–127.

Straškrába, M., 1967. Quantitative study on the littoral zooplankton of the Poltruba backwater with an attempt to disclose the effect on fish. Rožpr. Česk. Akad. Véd Řada. Mat. Přiř. Véd 77: 7–34.

Uimonen-Simola, P., 1985, in press. Cladocera in small Finnish lakes. Dev. Hydrobiol.
Williams, J. B., 1982. Temporal and spatial patterns of abundance of the Chydoridae (Cladocera) in Lake Itasca, Minnesota. Ecology 63: 345–353.
Whiteside, M. C., 1970. Danish chydorid Cladocera: modern ecology and core studies. Ecol. Monogr. 40: 79–118.
Whiteside, M. C., J. B. Williams & C. P. White, 1978. Seasonal abundance and pattern of chydorid Cladocera in mud and vegetative habitats. Ecology 59: 1177–1188.
Whiteside, M. C. & M. R. Swindoll, in press. Guidelines and limitations to cladoceran paleoecological interpretations. Paleogeography, paleoclim. paleoecol.

Accepted 18 March 1986.

Hydrobiologia 143: 355–360, (1986)

Recent and long-term acidification of Upper Wallface Pond (N.Y.) as indicated by mallomonadacean microfossils

Catherine E. Christie & John P. Smol
Department of Biology, Queen's University, Kingston, Ontario, Canada, K7L 3N6

Keywords: lake acidification, pH, Adirondack Mountains, Mallomonadaceae, Chrysophyceae, acidic precipitation, Holocene

Abstract

The postglacial distribution of mallomonadacean scales is described from Upper Wallface Pond, a small, acidic lake in the High Peaks Region of the Adirondack Mountains. Similar to studies in other lake regions, Mallomonadaceae were present at only very low numbers during the pond's early postglacial history, but then populations developed rapidly about 11000 B.P. The generalist *Mallomonas crassisquama* dominated the scale assemblage in all but the most recent sediments. Changes in species composition and abundance that occurred prior to this century can be related to events occurring in the pond's catchment, such as successions in terrestrial vegetation. The most striking changes in mallomonadacaean distributions occurred over the last 30 years, with the dominance of taxa tolerant of very acid (pH < 5.0) waters. The most likely cause for this recent decline in lakewater pH is acidic precipitation.

Introduction

The lakes and ponds of the Adirondack mountain region of New York are currently under intensive study because many receive (Barrie & Halse, 1984) and apparently have been affected by acidic precipitation (Pfeiffer & Festa, 1980). Since little historical data are available on past changes in lakewater pH, paleolimnological techniques have been developed to reconstruct and interpret past lake conditions. Analysis of stratigraphic profiles of diatom populations has been a popular approach for inferring past pH (Battarbee, 1984), and these techniques have been succesfully applied to Adirondack lakes (e.g. Charles, 1984, 1985, and unpublished data). However, euplanktonic diatoms appear to be excluded in Adirondack waters of pH < 5.5 (Charles, 1985), as is the case in many lake regions (Battarbee, 1984). In these acidic lakes, limnological reconstructions using diatoms are based solely on littoral and benthic taxa.

Mallomonadacean chrysophytes, on the other hand, are euplanktonic, and are often found in large numbers in acidic lakes. Their siliceous, species-specific scales, which are preserved in lake sediments, can be used to document past populations (e.g. Munch, 1980; Smol, 1980, 1986; Smol *et al.*, 1984a; Tolonen *et al.*, 1986). Moreover, many taxa are characteristic of restrictive pH regimes (Smol *et al.*, 1984b; Smol, 1986). Fossil scales may therefore represent the only algal microfossils that reflect changes in past phytoplankton populations in these acidic lakes.

The recent sediments of several Adirondack lakes have been studied to document limnological changes that could be related to changes in the lakes' watersheds or airsheds (Charles, 1984; Smol *et al.*, 1984a; Smol, 1986). Declines in inferred lakewater pH over the last few decades have been related to the atmospheric deposition of strong, inorganic acids. However, relatively little is known about long-term (several thousands of years) changes in lakewater characteristics. Recently, Whitehead *et al.* (1986) reconstructed long-term changes in lakewater chemistry (primarily pH) for three lakes located on an elevational gradient in the

High Peaks Region of the Adirondack Mountains. Diatoms were used to infer past limnological changes, whilst changes in the lakes' drainage basins were reconstructed using pollen and plant macrofossil distributions. This study indicated that the three lakes were initially slightly alkaline, then acidified during the late-glacial and the early Holocene. Inferred pH values appeared to stabilize about 8–9000 years ago for the two higher elevation lakes (Lake Arnold and Upper Wallface Pond) and about 6000 years ago for the lowest elevation lake (Heart Lake). Whitehead and his co-workers argued that these changes were the result of a complex of interdependent variables related to climatic and vegetational shifts, as well as to associated changes in soil biogeochemistry. Subsequent analyses of diatoms in the recent sediments of one of these lakes (Upper Wallface Pond) indicated that this lake underwent additional acidification in recent years (Charles *et al.*, 1986).

In this paper we describe the postglacial distribution of Mallomonadaceae in Upper Wallface Pond. Throughout most of this lake's history (ca. the last 10000 years) euplanktonic diatoms were only present at trace levels. This paper represents the first study that has used chrysophycean microfossils to infer long-term patterns of pH change in an acid sensitive lake. We have also conducted a detailed investigation of the recent sediments to document limnological changes occurring over the last century.

Study area

Upper Wallface Pond (44°08′47′′N, 74°03′15′′W) is a small lake (surface area = 5.5 ha, maximum depth = 9.0 m, mean depth = 3.8 m) at an elevation of 947.9 m asl in the High Peaks Region of the Adirondack Mountains. Its 58 ha drainage basin is underlain by anorthosite bedrock, covered by thin tills, which supports a sub-alpine, spruce-fir-paper birch forest. The forests in the catchment have never been logged and the presence of large red spruce indicates that it is unlikely that major forest fires have had any direct effect on the lake in recent history. Because there are no roads in the area, the lake is largely inaccessible to the general public and cultural impact in the catchment has been minimal. Recent limnological surveys (Charles, 1982, 1985) indicate that the lake has negative alkalinity (−2 eq/L) and is acidic (pH < 4.7–5.0). Additional details concerning the study area are presented in Charles (1985) and Whitehead *et al.* (1986). The pond has apparently been fishless since the early 1970's, and fish survey records indicate poor growth of brook trout from as early as 1963 (Paterson, 1985).

Materials and methods

Mallomonadacean scales are siliceous; they can therefore be studied using the same preparative techniques developed for fossil diatoms. A detailed description of coring and laboratory procedures is outlined in Charles (1982, 1984) and Whitehead *et al.* (1986). Additional techniques used specifically for identifying and enumerating mallomonadacean microfossils are discussed in Smol (1986). The recent sediments were dated by S. A. Norton and G. Blake of the University of Maine at Orono using Pb-210 chronology (CRS model).

Results and discussion

Late-glacial and Holocene changes

The long-term changes in mallomonadacean distributions are discussed within the context of the pollen zones delineated by Whitehead *et al.* (1986) and summarized in Table 1. Interpretations of the associated limnological changes, which are discussed by Whitehead and his co-workers, are not re-iterated here.

A total of 22 mallomonadacean taxa were recorded in the Upper Wallface sediments. The distributions of only those species that accounted for more than 2% of the total scale sum at any stratigraphic level are presented here (Fig. 1).

Mallomonadacean scales were not recorded in sediments deposited prior to about 10000 B.P., coinciding with the tundra (HP1) and spruce-dominating (HP2) pollen zones. Similar distribution patterns have been recorded in the late-glacial and early Holecene sediments of several Canadian lakes (Smol, 1979, 1982; Smol & Boucherle, 1985). Whitehead *et al.* (1986) estimated that the pH during zones HP1 and HP2 was slightly alkaline to cir-

Table 1. Vegetational succession in the Upper Wallface Pond region (from Whitehead *et al.*, 1986).

Pollen zone	Time period (years B.P.)	Vegetation trends
HP 1	prior to 11000	– treeless, high percentage of herbs
HP 2	11000 – 9700	– introduction of spruce, then decline at 10500 B.P., alder dominates
HP 3	9700 – 7000	– fir-birch forests develop, hemlock introduced at about 8000 B.P.
HP 4	7000 – 4800	– fir-birch forests persist, hemlock present
HP 5	4800 – 2000	– hemlock declines at 4800 B.P., shift to hardwood (beech, maple, birch) dominance
HP 6	2000 to the present	– establishment of existing spruce-fir-birch forests

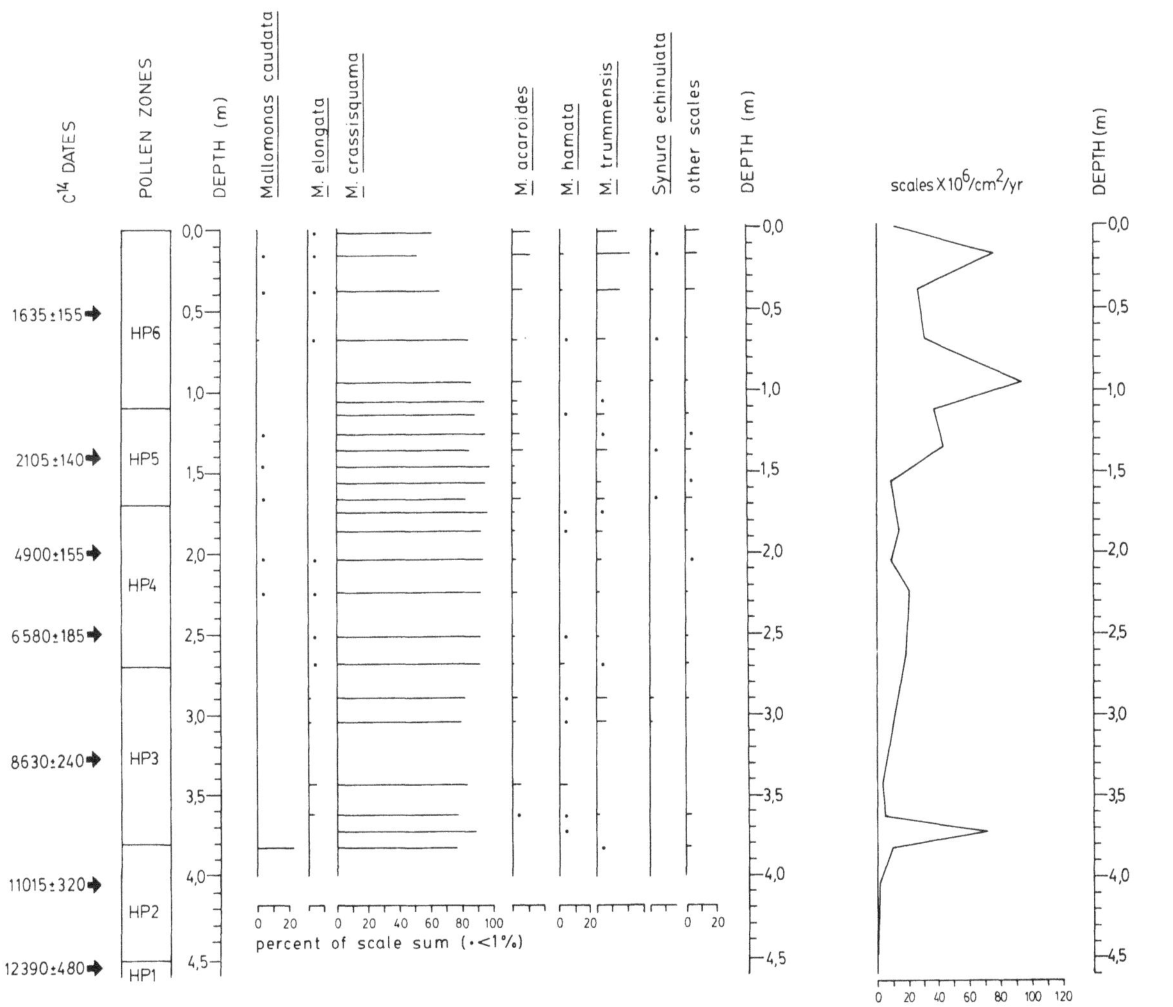

Fig. 1. The postglacial distribution of mallomonadacean scales from Upper Wallface Pond. Pollen zones and radiocarbon dates are from Whitehead *et al.* (1986).

cumneutral. This in itself would not preclude mallomonadacean growth because many taxa thrive in alkaline waters (cf. Smol & Boucherle, 1985). The reasons why Mallomonadaceae maintain such low populations during a lake's early development are not known; however, it is noteworthy that, although Chrysophyceae often dominate the phytoplankton in the High Arctic, no Mallomonadaceae have yet been identified from an ongoing survey of ponds and lakes on Ellesmere Island (N.W.T., Canada) and northwestern Greenland (Smol, unpublished data), and no fossil scales were recorded in the 9000 year history of a lake on Ellesmere Island (Smol, 1983). Similar environmental conditions may have occurred during the early postglacial history of Upper Wallface Pond.

Mallomonadacean populations increase strikingly near the transition to pollen zone HP3 (Fig. 1), when the vegetation shifted to a mixed deciduous-coniferous forest. The increase in scale accumulation rates may partly be related to the decreased competition from euplanktonic diatoms, which were present earlier in the lake's history but had declined precipitously at this time. However, other variables must be involved because a similar increase in scales was recorded in Canadian lakes where euplanktonic diatoms increased about this time (Smol, 1979, 1982; Smol & Boucherle, 1985).

Whitehead *et al.* (1986) estimated that the pH was still circumneutral near the transition to, and the early part of, zone HP3. This interpretation is corroborated by the presence of fossil scales such as *Mallomonas caudata* Iwanoff *em.* Krieger, a circumneutral to alkaliphilous taxon (Smol *et al.*, 1984b). *Mallomonas elongata* Reverdin, another alkaliphilous species (*ibid.*), is also more common during the early stages of zone HP3 than at any other time in the lake's history.

Whitehead *et al.* (1986) postulated that, following the initial decline in alkalinity, the inferred pH of Upper Wallface Pond has remained relatively stable at about 5.5, although some small fluctuations may have occurred. The Mallomonadaceae reinforce these interpretations. *Mallomonas caudata* declines to trace levels, as does *M. elongata.* For the remainder of the profile, the scale assemblage is overwhelmingly dominated by the generalist *M. crassisquama* (Asmund) Fott, the most common *Mallomonas* species in the world (Cronberg & Kristiansen, 1980; Nicholls, 1982). Not surprisingly, it was also the most common taxon recorded in a survey of 38 Adirondack lakes, although it was rarely common in waters with a pH <5.0 (Smol *et al.,* 1984b). The other species recorded are only sparsely represented, however *M. acaroides* Perty *em.* Iwanoff and *M. trummensis* Cronberg tend to increase towards the surface of the core. The former taxon appears to be an acidophil in the Adirondacks (Smol *et al.*, 1984b), whereas the latter species has been known to reach high concentrations in circumneutral waters (Cronberg, 1975). Whitehead *et al.* (*ibid.*) speculated that lakewater pH may have declined slightly over the last 2000 years; the Mallomonadaceae provide limited support to this hypothesis.

In contrast to the species composition, the accumulation rate of scales fluctuates widely during the Holocene, with a marked increase beginning in zone HP5 and continuing into zone HP6. This approximately corresponds to the 'hemlock decline', a stratigraphic period when hemlock was reduced to low levels, perhaps due to a pathogen (Davis, 1981). Increased nutrient input into the lake would have been a likely result (Whitehead & Crisman, 1978; Boucherle *et al.*, 1986), with subsequent increases in mallomonadaceaen populations (Fig. 1). A similar pattern was recorded at this time in Found Lake (Smol, 1979), a small Canadian lake from a similar geologic setting as Upper Wallface Pond; but an opposite trend was found in Little Round Lake, a calcareous, meromictic lake (Smol & Boucherle, 1985). Further paleoecological studies will be required to interpret more accurately these stratigraphic changes.

Recent changes

In contrast to most of the Holocene, marked changes in mallomonadacean species composition occur over the last few decades (Fig. 2). These shifts are consistent with an interpretation of recent lake acidification to the present pH of 4.7 to 5.0.

The generalist *Mallomonas crassisquama* dominates assemblages below the 8 cm level (circa A.D. 1910). Above this point acidophilic taxa such as *Synura echinulata* Korsh. and *M. acaroides* increase in relative frequency. Concurrent changes in the diatom flora, such as an increase in *Fragilaria acidobiontica* Charles, also indicate a decline in lakewater pH (Charles *et al.*, 1986). The trend to

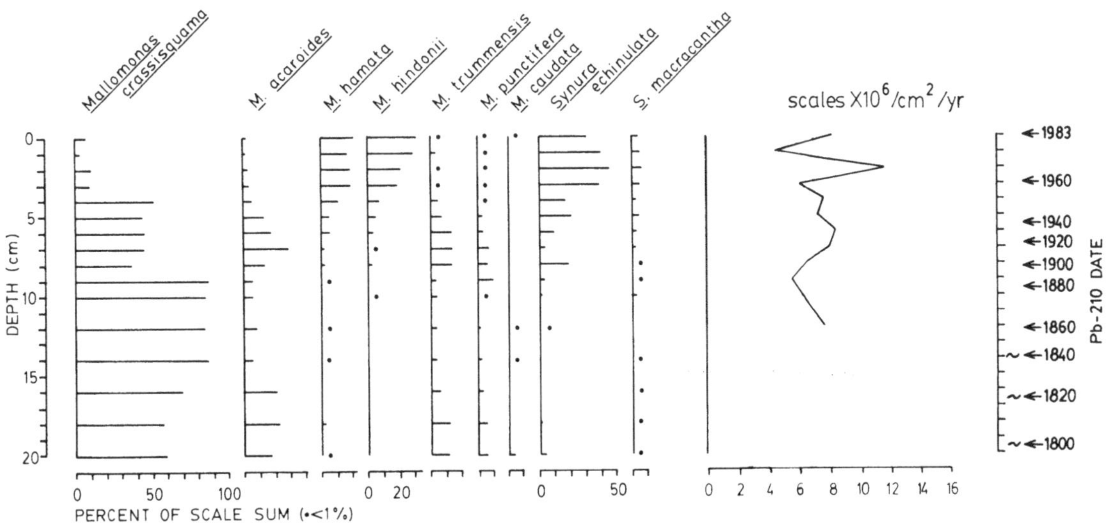

Fig. 2. Mallomonadacean scale distributions from the recent sediments of Upper Wallface Pond. The Pb-210 dates were provided by S. Norton and G. Blake. Dates prior to 1850 are increasingly approximate, and therefore accumulation rates are not calculated for these sediments.

more acidic conditions becomes more apparent near the 5 cm level (circa 1945) where acidobiontic Mallomonadaceae, such as *M. hamata* Asmund, *M. hindonii* Nicholls, and *S. macracantha* (Petersen et Hansen) Asmund, increase in relative frequency, with compensitory declines in circumneutral taxa such as *M. trummensis* and *M. punctifera* Korsh., and the generalist *M. crassisquama.* Almost identical species shifts have been recorded in the recent sediments of other Adirondack lakes that were suspected of having recently acidified (Smol *et al.*, 1984a; Smol, 1986). Increased abundance of acidobiontic diatoms coincide with these changes, and Charles *et al.* (1986) estimate that the pH of Upper Wallface Pond declined at this time to its present low levels. Stratigraphic changes in fossil cladoceran assemblages can also be related to recent pH changes (Paterson, 1985). Because of the absence of any cultural activity in the lake's catchment, the atmospheric deposition of strong acids is the most likely cause for this pH shift.

The accumulation rates of scales remain relatively constant at about 6×10^6 scales/cm^2/year throughout most of the pond's recent history (Fig. 2). In the uppermost 2 cm there appear to be oscillations in the accumulation rates, but these may simply reflect anomalies in our estimates of the sedimentation rates, which may occur in a lake's uppermost sediments. In any event, no sustainable trends are evident. In Big Moose Lake there appeared to be a modest increase in Mallomonadaceae coincident with acidification, and this was attributed to reduced competition from euplanktonic diatoms (Smol, 1986). However, in Upper Wallface Pond, euplanktonic diatoms had been excluded about 10000 years earlier, and therefore Mallomonadaceae would not have had any advantage from reduced diatom competition in recent times.

Summary and conclusions

The mallomonadacean data from the Upper Wallface Pond stratigraphy tend to reinforce the interpretations made with diatoms. A shift to more acidic conditions occurred about 10000 B.P., after which time the mallomonadacean species composition was remarkably stable until the most recent sediments. However, in contrast to a steady accumulation rate of diatom frustules (Whitehead *et al.*, 1986), estimates of past mallomonadacean standing crop fluctuated widely during the Holocene. Causative factors are still speculative, but shifts in the lake's trophic status related to terrestrial changes in its catchment may have been influential. Marked increases in taxa characteristic of very acidic waters have occurred over the last seventy years, coinciding with proposed periods of anthropogenically related acidic deposition. These species changes were similar to those recorded in the sediments of other Adirondack lakes that are presently acidic.

Acknowledgements

This paper is contribution No. 9 to the project 'Paleoecological Investigation of Recent Lake Acidification (PIRLA)' funded by the Electric Power Research Institute (RP-2174-10). We thank Drs. D. F. Charles and D. R. Whitehead (Indiana University) for providing the sediment samples used in this study, and Dr. S. A. Norton and Ms. G. Blake (University of Maine at Orono) for letting us use their Pb-210 data. Drs. Charles and Whitehead provided useful comments on the manuscript.

References

Barrie, L. A. & J. M. Halse, 1984. The spatial distribution of precipitation acidity and major ion wet deposition in North America during 1980. Tellus 36B: 333–355.

Battarbee, R. W., 1984. Diatom analysis and the acidification of lakes. Phil. Trans. r. Soc., Lond. 305: 451–477.

Charles, D. F., 1982. Studies of Adirondack mountain (New York) lakes: Limnological characteristics and sediments diatom – water chemistry relationships. Ph.D. thesis, Indiana University, Bloomington, 428 pp.

Charles, D. F., 1984. Recent pH history of Big Moose Lake (Adirondack Mountains, New York, U.S.A.) inferred from sediment diatom assemblages. Verh. int. Ver. Limnol. 22: 559–566.

Charles, D. F., 1985. Relationships between surface sediment diatom assemblages and lakewater characteristics in Adirondack lakes. Ecology 66: 994–1011.

Charles, D. F., D. R. Whitehead, D. Anderson, R. Bienert, K. Camburn, R. Cook, T. Crisman, R. B. Davis, B. Fry, R. A. Hites, J. S. Kakl, J. Kingston, R. Kreis, Jr., M. J. Mitchell, S. A. Norton, L. Roll, J. P. Smol, P. R. Sweets, A. Uutala, J. White, M. Whiting & R. Wise, 1986. The PIRLA Project (Paleoecological Investigation of Recent Lake Acidification): Preliminary results for the Adirondacks, New England, N. Great Lakes States, and N. Florida. Wat. Air Soil Pollut. (under review).

Cronberg, G., 1975. *Mallomonas trummenensis* nov. sp. (Chrysophyceae) studied by means of scanning and transmission microscopy. Bot. Notiser. 128: 69–72.

Cronberg, G. & J. Kristiansen, 1980. Synuraceae and other Chrysophyceae from central Smaland, Sweden. Bot. Notiser 133: 595–618.

Davis, M. B., 1981. Outbreaks of forest pathogens in quaternary history. Proc. IV Internat. Conf. Palynology, 3: 216–227.

Munch, C. S., 1980. Fossil diatoms and scales of Chrysophyceae in the recent history of Hall Lake, Washington. Freshwat. Biol. 10: 61–66.

Nicholls, K., 1982. Mallomonas species (Chrysophyceae) from Ontario, Canada including descriptions of two species. Nova Hedwigia 34: 80–124.

Paterson, M. J., 1985. Paleolimnological reconstruction of cladoceran response to presumed acidification of four lakes in the Adirondack mountains (New York). M.S. thesis, Indiana University, Bloomington, 202 pp.

Pfeiffer, M. H. & P. J. Festa, 1980. Acidity status of lakes in the Adirondack region of New York in relation to fish resources. N.Y. State Department of Environmental Conservation, Albany, N.Y. 73 pp.

Smol, J. P., 1979. Paleolimnology of selected Precambrian shield lakes. M.Sc. thesis, Brock University, St. Catherines, 225 pp.

Smol, J. P., 1980. Fossil synuracean (Chrysophyceae) scales in lake sediments: A new group of paleoindicators. Can. J. Bot. 58: 458–465.

Smol, J. P., 1982. Postglacial changes in fossil algal assemblages from three Canadian lakes. Ph.D. thesis, Queen's University, Kingston, 151 pp.

Smol, J. P., 1983. Paleophycology of a high arctic lake near Cape Herschel, Ellesmere Island. Can. J. Bot. 61: 2195–2204.

Smol, J. P. 1986. Crysophycean microfossils as indicators of lakewater pH. In J. P. Smol *et al.* (eds.), Diatoms and Lake Acidity. Dr. W. Junk Publ., Dordrecht, 275–287.

Smol, J. P. & M. M. Boucherele, 1985. Postglacial changes in algal and cladoceran assemblages in Little Round Lake, Ontario. Arch. Hydrobiol. 103: 25–49.

Smol, J. P., D. F. Charles & D. R. Whitehead, 1984a. Mallomonadacean microfossils provide evidence of recent lake acidification. Nature 307: 628–630.

Smol, J. P., D. F. Charles & D. R. Whitehead, 1984b. Mallomonadacean (Chrysophyceae) assemblages and their relationships with limnological characteristics in 38 Adirondack (N.Y.) lakes. Can. J. Bot. 62: 911–992.

Tolonen, K., Liukkonen, R. Harjula & A. Patila, 1986. Acidification of small lakes in Finland documented by sedimentary diatom and chrysophycean remains. In J. P. Smol *et al.* (eds.), Diatoms and Lake Acidity. Dr. W. Junk Publ., Dordrecht, 169–199.

Whitehead, D. R. & Crisman, T., 1978. Paleolimnological studies of small New England (U.S.A.) ponds. Part 1. Late glacial and postglacial trophic oscillations. Pol. Arch. Hydrobiol. 25: 471–481.

Whitehead, D. R., D. F. Charles, S. E. Reed, S. T. Jackson & M. C. Sheehan, 1986. Late-glacial and Holocene acidity changes in Adirondack (N.Y.) lakes. In J. P. Smol *et al.* (eds.), Diatoms and Lake Acidity, Dr. W. Junk Publ., Dordrecht, 251–274.

Accepted 18 March 1986.

Hydrobiologia 143: 361–369, (1986)

Recent sedimentation in a freshwater reservoir in SW-Finland: The diatomological and chemical evidence of sediments on the development of the water body.

Matti Räsänen
University of Turku, Department of Quaternary Geology, SF-20500 Turku, Finland

Keywords: freshwater reservoir, isolation, recent sedimentation, redeposition, diatom ecology, paleolimnology

Abstract

A freshwater reservoir was dammed from the Baltic Sea in 1965. The recent sediments and the processes of accumulation in the reservoir have been investigated. The physical, chemical and diatom analyses of the sediments were carried out at 5 sampling points in order to dinstinquish the freshwater sediments from the underlying brackish sediments as well as to date the freshwater sediments. The pattern of sedimentation in the reservoir is described on the basis of these results and conclusions are drawn regarding the development of the water body. In interpretating the diatomological and the chemical results from the recent sediments certain common features are highlighted.

Introduction

The recent sediments and their sedimentation in a freshwater reservoir near the town of Uusikaupunki, on the southwest coast of Finland have been studied. The reservoir was dammed in 1965 through the construction of a series of earth-banks which link several islands and thus engulf the two former open seabays.

The limnological evolution of the reservoir is well known. There exist chemical and physical data which document the freshening and concurrent acidification of the water body and its later neutralization. On the basis of this information it is worthwhile investigating the changes in the sedimented diatom communities in order to provide information as to how the diatom flora has reacted to the changed environmental parameters. We also derived information about the sedimentation process and how the diatom thanatocoenosis in the bottom deposits were actually formed. The dating of the sediments was carried out through the correlation of changes observed in the sediment diatom communities with known alteration in the water quality.

The substantial alteration in the limnological character of the reservoir has its causes in the watershed area. It is partly regulated by the annual weather conditions and partly by the human actions in the area. Chemical and physical analyses of the sediments were carried out so that these results together with the diatomological studies could provide new information concerning the real causative factors of the past development in the reservoir.

Study area

The reservoir (Fig. 1) is 37 km^2 in area, 165×10^6 m^3 in volume and has a mean depth of only 4 m. The watershed area of the reservoir is about 500 km^2; 86% of this area is drained by the river Sirppujoki which releases its current on the eastern shore of Velhovesi. The residence time of the reservoir is 16 months.

Marine litorina-clays cover about 25% of the drainage area of the river Sirppujoki, the clays are acid because they contain sulphides and 10.8% of the catchment area is classified as having acid sulphate soils (Palko *et al.,* 1985). Because there are so

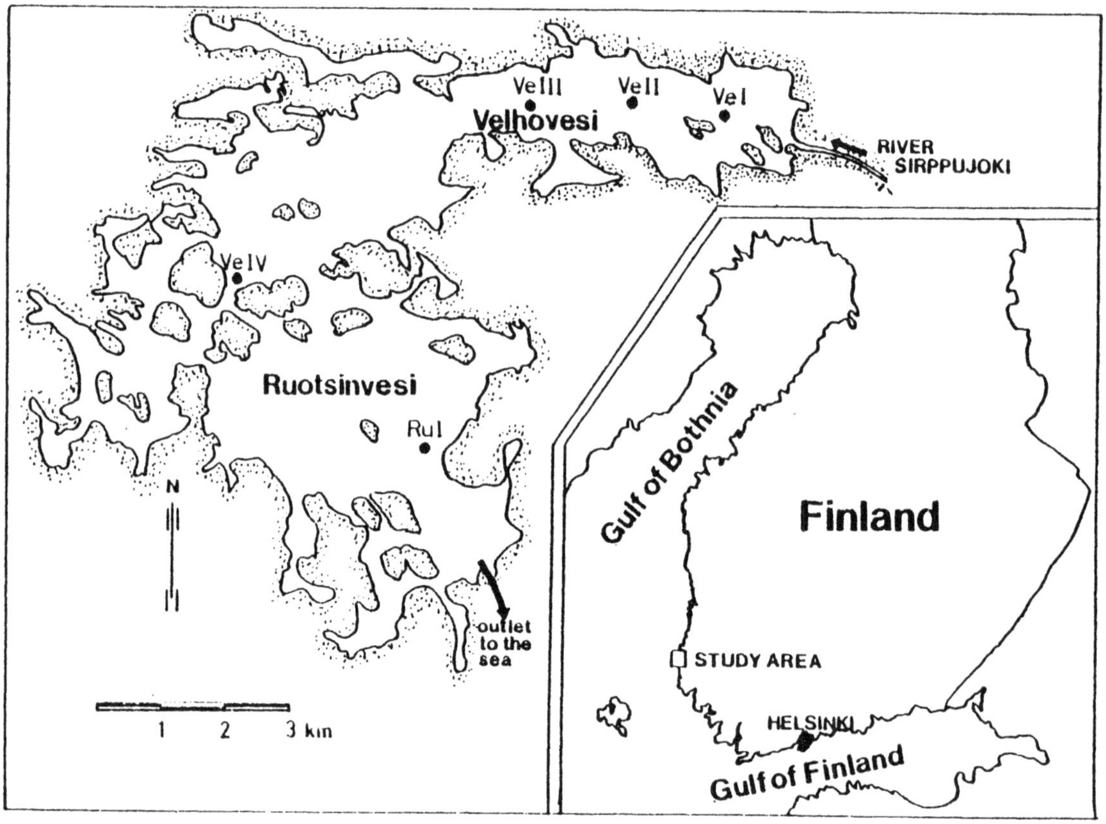

Fig. 1: Study area and the location of sediment sampling points.

many acidic finegrained sediments in the catchment area, river discharge is periodically very acidic.

This acidic discharge from the catchment area has determined the biological development in the freshwater reservoir during the past twenty years. The quality of the water in the reservoir depends on the weather conditions in the area, particularly during the summer (Palko *et al.,* 1985). If the weather is very rainy during the summer, less sulphide is oxidized to sulphate and the discharged water is not so acidic. When there is a long period without rain during the summer, the ground-water level sinks and the litorina-clays are oxidized very deeply. Afterwards in the autumn or following spring surface runoff leaches these well oxidized sediments. Thus the runoff waters become acidic and the pH in the reservoir decreases.

The brackish sea water began to be replaced by fresh water after the damming in 1965. As a result of the freshening, the buffer capacity of the reservoir water decreased and the pH-value of the water consequently decreased to about 5 because of the acidic water discharge.

The water quality subsequently remained almost constant for nearly 10 years. There were extremely high concentrations of ions in the water with low pH-value, especially manganese (0.6–1.6 mg/l), sulphate (100–150 mg/l) and chloride (80–40 mg/l). However, the concentrations of iron and nutrients were low. Productivity in the plankton was insignificant and the water was clear. In the beginning of 1980s, the pH-value of the water increased to neutral values because of wet summers and the concentrations of ions that earlier occurred abundantly concurrently decreased to normal values. The productivity in plankton has now increased.

Methods

The final investigations were carried out from five cores (Fig. 1) sampled by the freezing method (Saarnisto, 1975). The samples Velhovesi I and Velhovesi III were taken in November 1984, the rest in February 1983.

The content of water and dry matter of the sediments was ascertained by drying the sediment at 105 °C overnight. The content of organic matter was determined as loss on ignition at 550 °C. The chemical analyses was carried out according to Bengtsson (1979) and the total concentration of elements was determined through a Perkin Elmer atomic absorption spectrofotometer.

The absolute diatom analysis was carried out according to Battarbee (1982). The identification of species were ascertained following the publications of Cleve-Euler (1951–1955), Foged (1974, 1977, 1980), Hustedt (1927–1930a, 1930b), Mölder & Tynni (1967–1973) and Tynni (1975–1980). The observations were made with a Leitz Orthoplan microscope using 1200-fold magnification and phase contrast. 200–400 diatom frustules were identified from each depth in sediment.

Results

In the cores at Velhovesi a transformation from gyttja clay to clay gyttja takes place at sampling points Ve I, II, III and IV at 45, 25, 18 and 22 cm depth respectively. At these levels the dry weight decreases and the content of organic matter increases and irregular but mostly horizontal laminations appear in the clay gyttja. The lithostrathigraphic alterations are followed by an increase in the Mn, Ca, Zn and Cu concentrations and a concurrent decrease in Mg, K and Na concentrations. In the cores of sampling points Ve IV and RU I, the lithological change are slight, as are the alterations in physical and chemical parameters. The sediments are homogenous at these two sampling points with the marks of turbulent sedimentation.

The purpose of the diatom analysis was firstly to differentiate between the freshwater deposits in the reservoir from the underlying brackish post litorina sediments. 73 of the most abundant taxa are shown in Fig. 2 as percentages of the total sum of the frustules, at the sampling point Ve I.

The halophobous species occur in very small amounts from 45 cm depth upwards in the sediment. Indifferent species appear like halophobous but less regularly, mostly increasing in amount from 45 cm level upwards. Halophilous species are most abundant below the depth of 45 cm as do mesohalobous species. As a whole, these changes indicate the alteration in the diatom flora, which is a decrease in salinity, caused by the damming of the water body in 1965.

The change of the diatom communities in the transition zone from the brackish sediment to a freshwater deposit at Ve I is similar to all the other sampling points. Only the relative abundance of the species varies. The abundance of the halophobous species diminishes in the cores from the mouth of the river Sirppujoki to Ruotsinvesi; at the same time as the abundance of mesohalobous and halophilous species increases. This phenomenon is apparent in both the marine and freshwater sediments and is a result of the greater marine influence at Ruotsinvesi than at Velhovesi.

The isolation horizons at the sampling points Ve I, II and III are at the same levels as in which gyttja clay changed to clay gyttja/gyttja, thus dating the lithostrathigraphical changes to 1965. At the sampling points Ve IV and Ru I similar lithostrathigraphical changes are observed but they are nearer the sediment surface, although the diatoms show that the isolation occurred at a deeper level.

The mean annual rates of sedimentation during the freshwater period at the sampling points are as follows: Velhovesi I = 2.25 cm, Velhovesi II = 1.39 cm, Velhovesi III = 0.90 cm, Velhovesi IV = 1.22 cm, Ruotsinvesi I = 1.39 cm. The rates of sedimentation decrease westwards as the distance from the mouth of river Sirppujoki increases. The sedimentation is higher again at the sampling points Ve IV and Ru I as they are located in depressions.

The concentrations of diatoms in the dry matter are highest in the surface sediment at all the sampling points except Ve I (Fig. 3). The concentrations are highest in the delta area and they decrease towards Ruotsinvesi. In all the cores, concentrations in freshwater sediment are higher than in the brackish sediment.

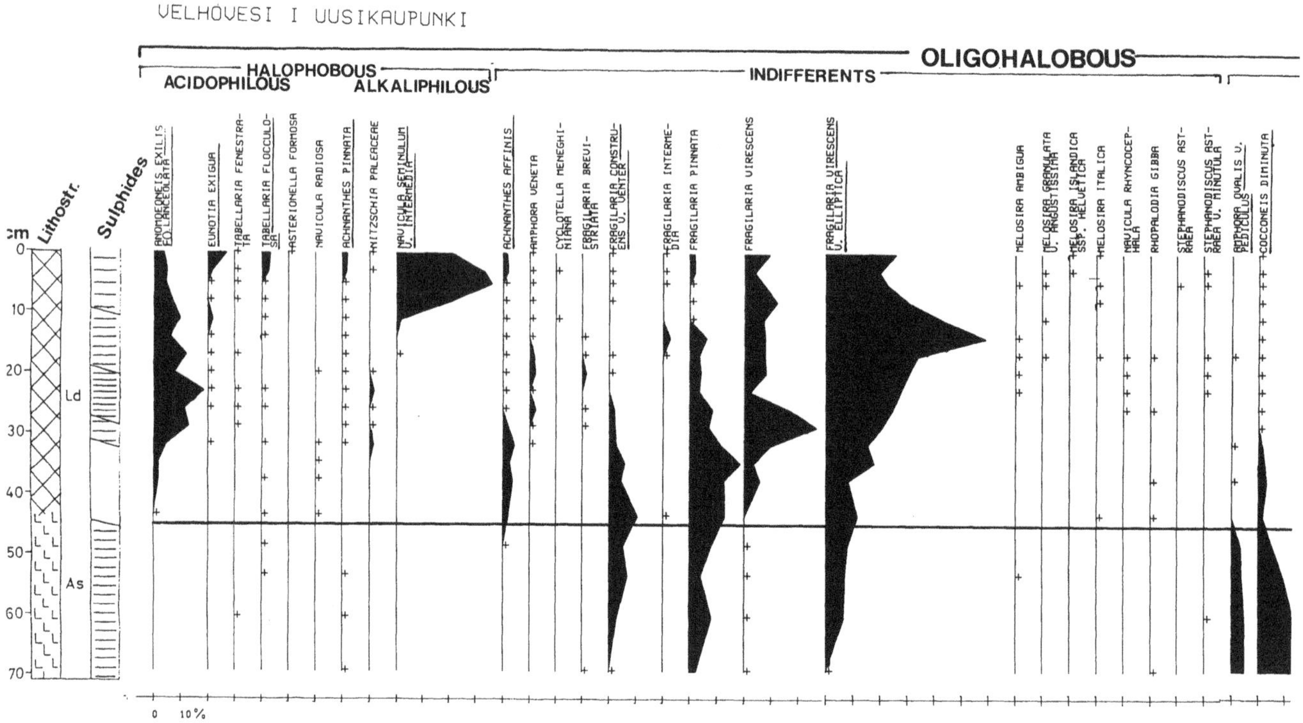

Fig. 2: Diatom diagram from sampling point Velhovesi I. As = gyttja clay, Ld = clay gyttja. Horizontal line indicates isolation horizon.

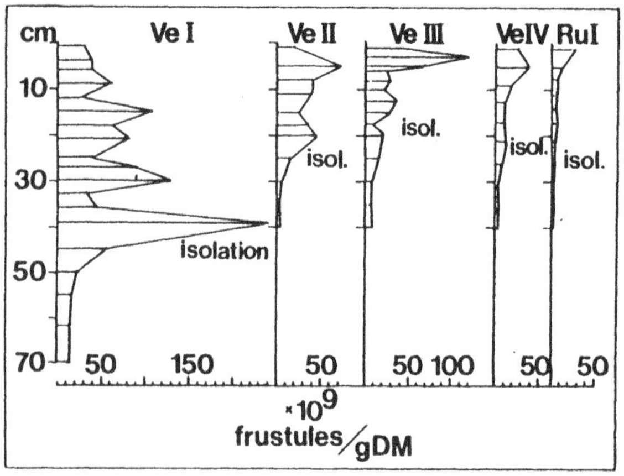

Fig. 3: Absolute diatom concentrations in dry matter of the sediments.

Discussion

The areal and temporal alteration of the diatom flora in the freshwater sediments of the five sampling points is depicted in Fig. 4, where the data of 33 of the most common species in every core are analysed by detrended correspondence analysis.

The ordination shows clearly the gradient from the mouth of the river Sirppujoki to Ruotsinvesi. The separate sampling points are located in the gradient according to their distance from the mouth of the river Sirppujoki. The gradient is distinct when scanning the isolation communities and is well shown in the further temporal development, until the communities in uppermost sediments are very much alike.

The ordination shows that in the area of the freshwater reservoir the different kind of diatom thanatocoenoses in the sediments describe the same kind of past development. The most important factor which accounts for this phenomenon is the distance from the mouth of the river Sirppujoki.

The occurrence of three dominant species is described in the freshwater sediments at the different sampling points in Fig. 5. Their abundancies are compared to the measured water conductivity and Mn-concentration in the reservoir water. The

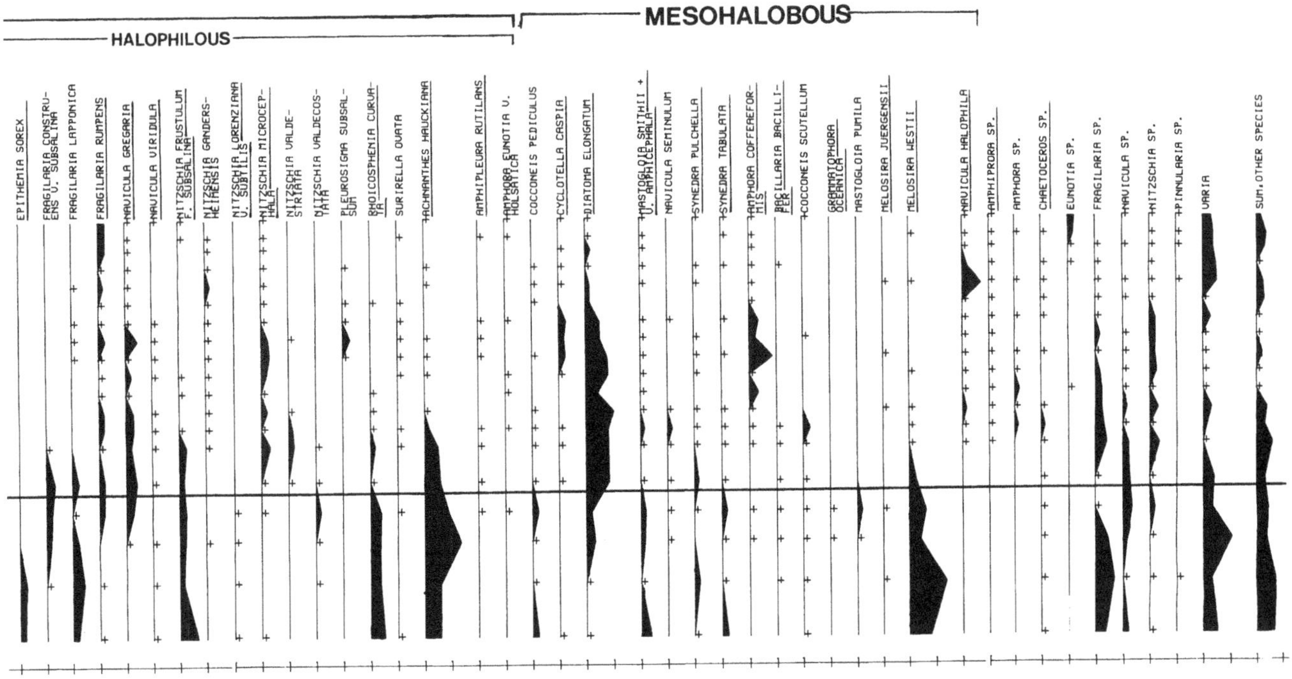

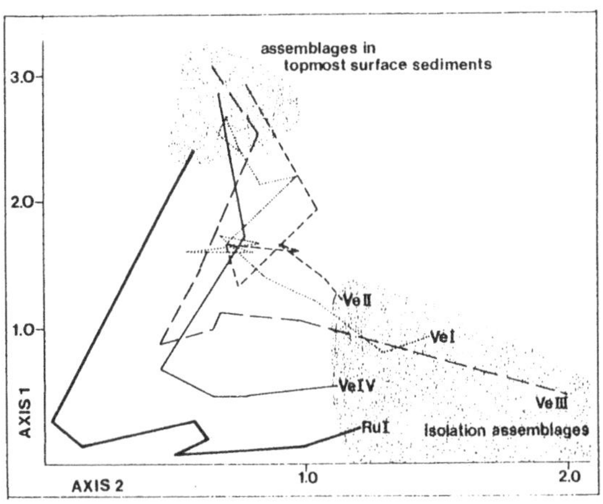

Fig. 4: An ordination constructed by detrended correspondence analysis to describe the areal and temporal alteration of the diatom communities at separate cores in the area of freshwater reservoir.

appearance of halophobous acidophilous *Anomoeoneis exilis fo. lanceolata* A. Mayer in the area of Velhovesi and mesohalophobous *Cyclotella caspia* at Ruotsinvesi in the diatom flora indicate the freshening of the water immediately after isolation. Halophobous alkaliphililous *Navicula seminulum v. intermedia* Hust becomes dominant in the flora when the reservoir water loses its acidity and become neutralized in 1981. Its occurrence is identical at each core. These horizons date the sediment: otherwise the sedimentation rate in Fig. 5 is supposed to be linear.

The difference between the water chemistry parameters depicted in Fig. 5 is very slight between Velhovesi and Ruotsinvesi, but there is a clear difference in the dominant species of the area. The reason for this floral gradient can partly be attributed to the morphological differences between shallower Velhovesi area and deeper Ruotsinvesi area. Deeper waters favor the planktonic *C. caspia* in Ruotsinvesi. The same effect is evident in the occurrence of the epiphytic *Fragilaria* species which are more abundant in shallower Velhovesi area. The occurrence of a greater number of life habitats for diatoms in the Velhovesi area is also the reason for the higher diatom concentrations in the sediments (Fig. 3). Most of the autochthonous diatoms sedimented are planktonic at the sampling points Ve IV and Ru I.

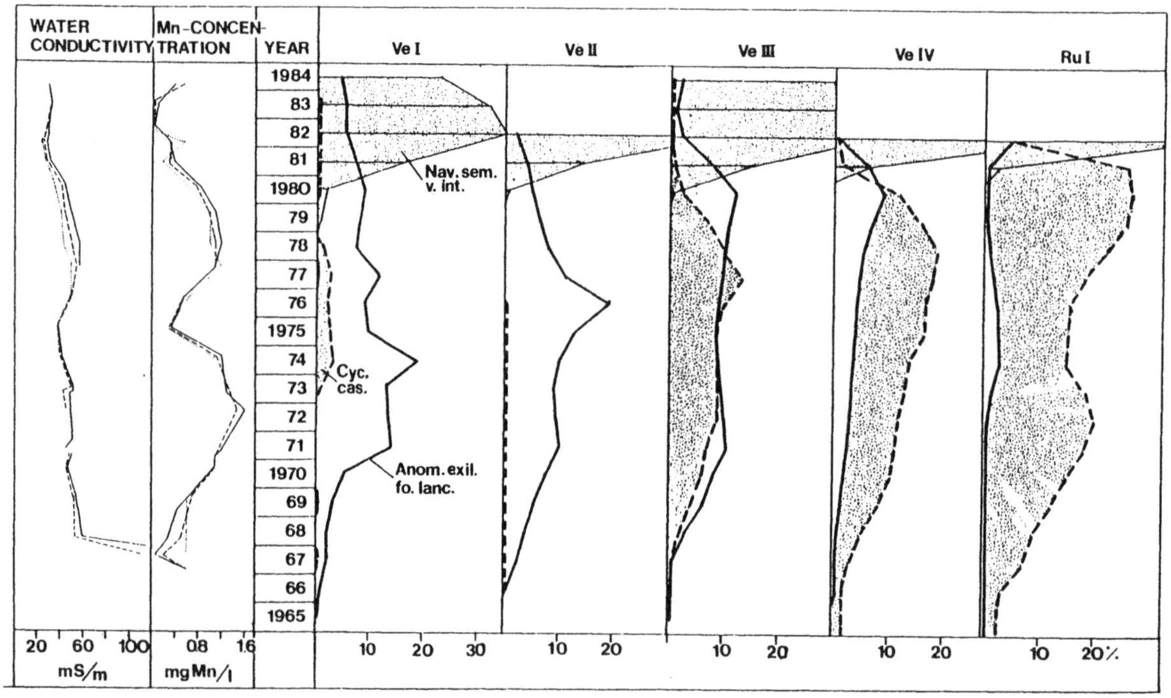

Fig. 5: The occurrence of three dominant diatom species in the freshwater sediments of separate cores, compared with the measured changes of water conductivity and Mn-concentration, on a linear time scale. The water chemistry data is described from three separate sampling points, two located at Velhovesi area and one at Ruotsinvesi. Values are measured during autumn fall.

Fig. 5 shows the water quality alteration documented in the occurrence of these three species in the sediments extremely well. On this basis the freshwater sediment is exactly dated at every sampling point and the differences in the physical and chemical nature of the sediments deposited concurrently can be compared.

The sedimentation in the reservoir can be described from the variation of dry weight and the content of organic matter in the sediments. These parameters are focused against a linear time scale in Fig. 6.

Much allochthonous mineral and organic matter is sedimented in the are of Velhovesi (Hinneri, 1974) from the discharge of the river Sirppujoki. The structure of the sediment is loose and the dry weight is low. The mean rates 2.25 – 0.90 cm/yr (Velhovesi I – III) of sedimentation are high and laminations therefore formed in the sediments even in a waterdepth of 4 meters.

The sedimentation is quite different at the sampling points Ve IV and Ru I. The cores were both sampled from the bottom of depressions 15 and 24 m deep respectively and in the field it was already observed that some structures exist in the sediments which indicate mixing. Only the top 8 cm at the sampling point Ve IV is laminated. It is obvious that the sediments in the depressions are a result of continuous turbulent mixing. As the biological production in the reservoir has been very low the content of organic matter in the sediments is minor. A larger part of deposits at these places is redeposited clays from the upper sediment layers. That is the reason for smooth changes in the dry matter and the organic matter content after isolation.

The effect of the redeposited sediments is particularly well shown in the diatom results of these two cores. The occurrence of several mesohalobous species continues above the level of isolation. They still occur in abundance, redeposited in the freshwater sediment.

The effect of the redeposition is also shown in Fig. 3, where the concentration of diatoms per dry

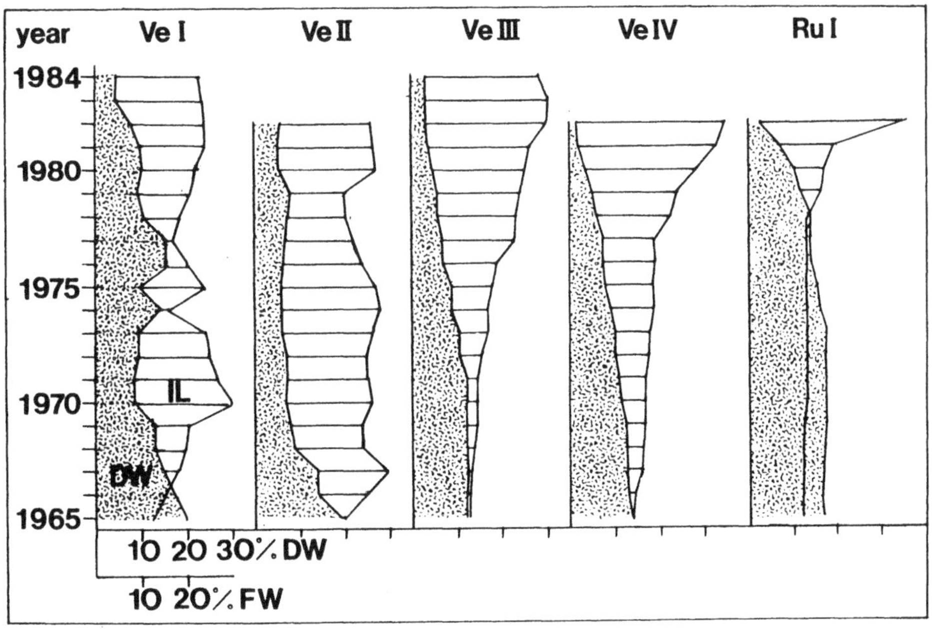

Fig. 6: Variation of dry matter and organic matter at dated sediment levels in each core.

matter increases only minimally after isolation at the sampling points Ve IV and Ru I. The minor increase is caused by the small primary production in plankton, in which *C. caspia* has had a dominant role among the diatoms. The diatom frustules are otherwise mostly redeposited.

The dominant species *C. caspia* and *A. exilis f. lanceolata* have been continuously sedimented from the upper watermass to the bottom at Ve IV and Ru I. The level at which the sedimentation in freshwater conditions has occurred has thus been fingerprinted.

Mn, Ca, Zn, and Cu increase in the freshwater sediment particularly in the area of Velhovesi. Cu, Ca and Zn have good correlations with organic matter, 0.866***, 0.685***, and 0.455*** respectively ($n = 52$). Mn in the freshwater sediment has no significant correlation with organic matter.

The enrichment of Cu, Ca and Zn can be explained by the higher content of organic mater in the freshwater sediment (mean 21.4%) than in the brackish sediment (mean 11.8%). The concentrations of Cu and Ca are not especially high when compared to the values determined from freshwater sediments (Cato, 1977). The concentrations of Zn (mean 454 mg/kg) are however high and comparable to concentrations in sediments from many polluted areas (Cato, 1977, Håkanson, 1977, Kansanen, 1985).

The anomalous high concentrations of Mn (mean 1127 mg/kg) and Zn in the freshwater sediment are a result of the strong leaching effect of the acid water discharge from the litorina-sediments in the catchment area. Zn and Mn are particularly water soluble in pH-values under 5 (Alasaarela, 1983) and the water discharge of the river Sirppujoki has often these values. These elements have been very effectively leached, then sedimented and concentrated into the freshwater sediment. Mn seems to be deposited in the form of hydroxides and not with organic matter as is probably the case with Zn.

Conclusions

The sediment of the area of Velhovesi provide the best information about the development of the reservoir. They are undisturbed with high sedimentation rates, lie near the river outlet and consist of materials, brought by the river the water discharge of which determines the water quality in the reservoir. The variation in the nature of discharge waters during the last twenty years is recorded in these sediments.

Certain analysed parameters from the sampling points Ve I, II and III and the measured water chemistry data are summarized in Fig. 7. The variations in the amount of acidophilous *A. exilis f. lanceolata* and mesohalobous *C. caspia* are very closely related to the changes in the water conductivity and the Mn-concentration that are strongly negatively correlated to the pH-history. The species most frequently occur when the water is very acid and Mn-concentrations are high during the years 1970–1980. At the sampling point Ve I the frequency of *A. exilis f. lanceolata* concurs very closely with the variation in water Mn-concentration. *N. seminulum v. intermedia* indicates clearly the neutralization of the waterbody and the decrease in Mn-concentrations in 1981.

The annual influxes of Mn and Zn are shown in Fig. 7. These elements are abundant in the water discharge of the river Sirppujoki (Isotalo 1971) and their variation is closely related to the pH-value of the waters. The annual accumulation of Mn and Zn decrease from the peak influxes in the beginning of the reservoir history in 1960s. This is due to dry summers and the fact that large scale dredging was carried along the riversides of Sirppujoki in the early 1960s and that several lakes were drained in 1966.

Wide areas of unleached acid sulphate soil was exposed as a result of these activities. During the following dry summers great quantities of Mn and Zn from these areas were leached into the reservoir. Leaching decreased subsequently to about uniform level and the influxes of Mn an Zn to the sediments decreased. It is noticeable that these influxes do not increase in later years from 1976–1979, although the measured pH-values have been at the same level or even lower than in the beginning of the decade. The nature of the deposited material has changed from the beginning of 1970s to the end of the decade and influxes of Mn and Zn have decreased. The trend in influxes is very analogous to the trend of measured Mn-concentrations in the water and is strong proof of the decreasing input of Mn into the reservoir caused from decreased amount of leachable compounds in the catchment area.

The development of the water quality in the freshwater reservoir has thus two processes which effect its evolution: firstly, the annual weather conditions which effect the water quality very quickly and secondly, the decrease of the amount of leachable compounds in the acid sulphate soils in the catchment area.

Substantial areal differences in sedimentation have been shown in this study. In order to interpret the history of a water body, it is necessary to have

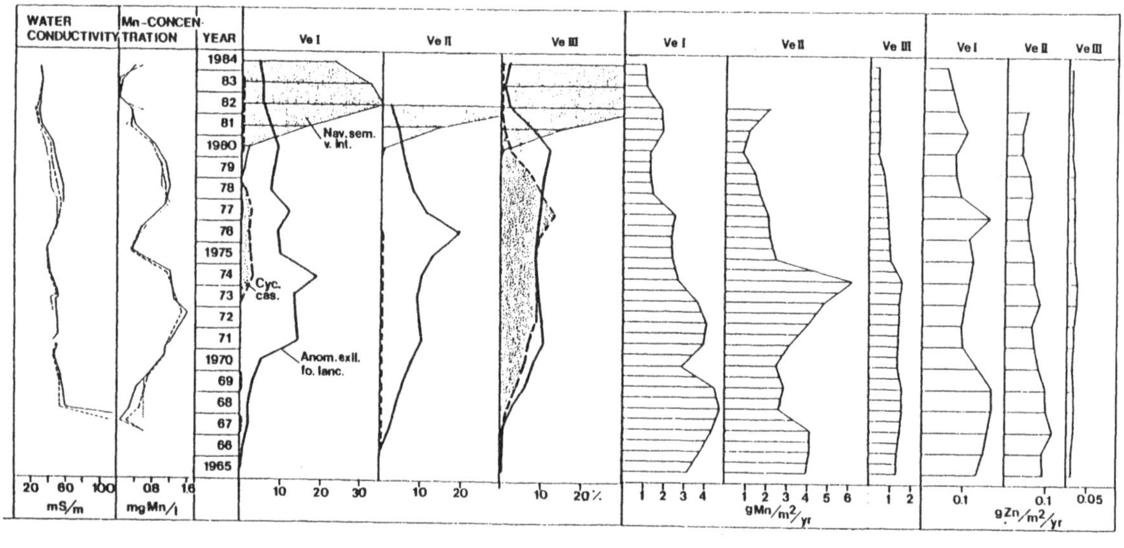

Fig. 7: Measured water chemistry data from freshwater reservoir focused against the three concurrently sedimented diatom species and compared with the annual influxes of manganese and zinc, in three cores at Velhovesi area.

several exactly dated sediment cores in the water basin.

In the freshwater reservoir there were two gradients which controlled the forming of the diatom assemblages in sediments. Firstly, the morphological differences between Velhovesi and Ruotsinvesi were apparent from the greater amount of littoral species in the sediments at Velhovesi. Secondly, their differences in water chemistry (e.g. 10 to 20 mg/l in Cl-concentration). Lastly, it was important to distinguish the redeposited diatoms from the autochthonous flora.

The chemical results depended completely on the relation between allochthonous, autochthonous and the s.c. 'autochthonous' redeposited material. In this case, the variation of allochthonous and redeposited material was of great importance because of the minor productivity of organic matter in the reservoir. Only the sediments, in which the allochthonous component was large enough, could be used in the interpretation of water-body development. This component of the sediment contained information from the catchment area.

The role of allochthonous material has been stressed in this case because the production of autochthonous material was very low. In this respect, the Uusikaupunki freshwater reservoir differs greatly from the lakes with small catchment areas where production of autochthonous organic matter is more dominant. The reservoir has features which are analogous to the sedimentation in the marine estuaries or the estuaries in large lakes.

Acknowledgements

I wish to express my thanks to Dr. R. Tynni for help in determination of certain diatom species and to K. A. Msindai, M. Sci., and Mr. S. Malmberg who kindly revised the English manuscript.

References

Alasaarela, E., 1980. Acidity problems caused on flood control works of the river Kyrönjoki. Publications of the Water Research Institute National Board of Waters, Finland. 49: 3–16.

Batterbee, R. W., 1982. The use of electronically counted microspheres in absolute diatom analysis. Limnol. Oceanogr. 27: 184–188.

Bengtsson, L., 1979. Chemical analysis. In B. E. Berglund (ed.), Paleohydrological changes in the temperate zone in the last 15 000 years, vol. 2: 113–132. Lund 340 pp.

Cato, I., 1977. Recent Sedimentological and Geochemical Conditions and Pollution Problems in Two Marine Areas in South-Western Sweden. Striae vol. 6, Uppsala. 158 pp.

Cleve-Euler, A., 1951-1955. Die diatomeen von Sweden und Finnland. (1–5). Kunsl. Svenska Vetenskapsakademins Handlingar 4. Ser., 2: 1 (1951); 3: 3 (1952); 4: 1 (1953); 5: 4 (1955).

Foged, N., 1974. Freshwater diatoms in Iceland. Bibliotheca Phycologica 15. J. Cramer, Vaduz, 118 pp.

Foged, N., 1977. Freshwater diatoms in Ireland. Bibliotheca Phycologica 34. J. Cramer, Vaduz, 221 pp.

Foged, N., 1980. Diatoms in Öland, Sweden. Bibliotheca Phycologica 49. J. Cramer, Vaduz, 192 pp.

Hinneri, S., 1974. Enrichment of elements especially heavy metals in recent sediments of the freshwater reservoir of Uusikaupunki, SW coastland of Finland. Ann. Univ. Turku (A II) 56: 1–30.

Hustedt, F., 1927-1930a. Die Kiselalgen Deutschland, Österreichs und der Sweitz. I. Teil. Rabenhorst: Kryptogamenflora, 7. Leipzig.

Hustedt, F., 1930b. Bacillariophyta (Diatomeae). Bascher: Süsswasserflora, 10. Jena, 466 pp.

Håkanson, L., 1977. Sediments as Indicators of Contamination – Investigations in the Four Largest Swedish Lakes. SNW PM 839 NLU Rapport 92: 1–159.

Isotalo, I., 1971. Mangaanin kulkeutumisesta Uudenkaupungin makeavesialtaaseen. Vesitalous 1: 7–8.

Kansanen, P. H., 1985. Assessment of pollution history from recent sediments in Lake Vanajavesi, southern Finland. I. Selection of representative profiles, their dating and chemostratigraphy. Ann. Zool. Fennici 22: 13–55.

Mölder, K. & R. Tynni, 1967-1973. Über Finnlands rezente und subfossile Diatomeen, (1–8). Compt. Rend. geol. Finlande 39: 199–217. Bull. Geol. Soc. Finland, 40: 151–170, 41: 235–251, 42: 129–144, 43: 203–220, 44: 141–159. 45: 159–171.

Palko, J., Räsänen, M. & Alasaarela, 1985. Happamien sulfaattimaiden esiintyminen ja vaikutus veden laatuun Sirppujoen vesistölalueella. National Board of Waters, Finland. Reports. 260.

Saarnisto, M., 1975. Pehmeiden järvisedimenttien näytteenottoon soveltuva jäähdytysmenetelmä. Geologi 27: 37–39.

Tynni, R., 1975-1980. Über Finnlands rezente und subfossile Diatomeen 8–11. Geol. Surv. Finland, Bull. 274, 284, 296, 312.

Accepted 18 March 1986.

Hydrobiologia 143: 371–377, (1986)

Recent sedimentological conditions in the delta of the river Kyrönjoki, Western Finland

Heikkilä, Raimo
University of Helsinki, Department of Geography, Hallituskatu 11, SF-00100 Helsinki, Finland

Keywords: river delta, recent sedimentation, sedimentological conditions, acidity, Fe/Mn ratio

Abstract

The estuary of the River Kyrönjoki was echo sounded and, on the basis of these results the delta was delimited and areas of erosion and accumulation were separated. Samples of the sediment surface (0–2 cm) from 65 sites in the estuary were analysed for organic content, C, N, P, Ca, Fe, Mn, Pb, Cu, Zn, Cd and Hg. On the basis of the chemical analyses the delta was divided into two parts according to the sedimentary conditions: (I) In the inner part, where acid fresh water of *c.* pH 5.5 from the river is predominant, the conditions were reducing. The Fe/Mn ratio of the sediment was high. The heavy metal content was usually low. The Hg content in the organic compounds of the sediment was, due to the acidity, higher than in the outer part. (II) In the outer part of the delta, where neutral brackish water of pH *c.* 7.0 dominates, the sedimentary conditions were mostly oxidizing. The Fe/Mn ratio was very low. The heavy metal and organic content was mostly high in the accumulation areas. This is because the fine-grained allochthonous material, which has a high heavy metal and organic content, flocculates and precipitates when it reaches the brackish water of the Gulf of Bothnia, and the salinity and the pH are higher.

1. Introduction

The sediments and sedimentation of river deltas have not been studied very extensively in Finland during recent decades. Rosberg (1895) studied the morphology of the deltas of the rivers flowing into the Gulf of Bothnia, and described the delta of the River Kyrönjoki as modified, with wide alluvial plains. Säntti (1954) also studied the then recent development of the delta of the River Kokemäenjoki. The sediments of the Baltic Sea and Finnish lakes have, moreover, recently been studied from many different points of view (e.g. Tulkki, 1977; Hallberg, 1979; Niemistö & Voipio, 1981; Simola, 1983; Tolonen & Meriläinen, 1983; Kansanen & Jaakkola, 1985). The delta of the Kyrönjoki was again intensively studied in the 1980s (Meriläinen, 1984a, 1984b, 1985a, 1985b; Heikkilä, 1985), and the aim of the current study was to find out the regional differences in the recent sedimentary conditions in this estuary of the Kyrönjoki.

2. Study area

The estuary of the river Kyrönjoki is located some 30 km north of Vaasa, a city on the west coast of Finland. The present delta is purely submarine, without any flood plains, due to artificial embankments in the downstream area and the estuary, which prevent floods (cf. Rosberg, 1895).

The drainage area of the river Kyrönjoki is 4920 km^2, of which 47% is made up of forests, 26% of peatlands, 24% of arable fields, 2% of built-up areas and 1% of lakes. In the river valleys the dominating soils are clay and silt. In the watershed areas peat and glacial till cover wide areas. Glaciofluvial sand and gravel exist only in limited areas near the watersheds. Most of the fields and built-up areas lie along the riversides. The isostatic land upheaval in the delta is 8 $mm \cdot a^{-1}$, and in the upstream area 7 $mm \cdot a^{-1}$ (Kääriäinen, 1975).

The discharge of the river Kyrönjoki fluctuates over a wide range (NQ:MQ:HQ = 1:43:528 m^3/s).

Spring, summer and autumn floods are frequent. The river channel has been cleared many times since the 1600s to prevent floods and to make the use of the riversides for agriculture possible. Since 1963, four reservoirs and four power stations (with daily regulation) have been built.

The water of the river Kyrönjoki contains a lot of humic substances due to drainage from extensive peatland areas, mostly drained for forestry or peat harvesting (Kenttämies, 1981). Furthermore, the water is often very acid, especially during flood periods, due to the cultivation of the sulphide-bearing Litorina clays in the downstream area along the riversides (Eronen, 1974; Alasaarela, 1982).

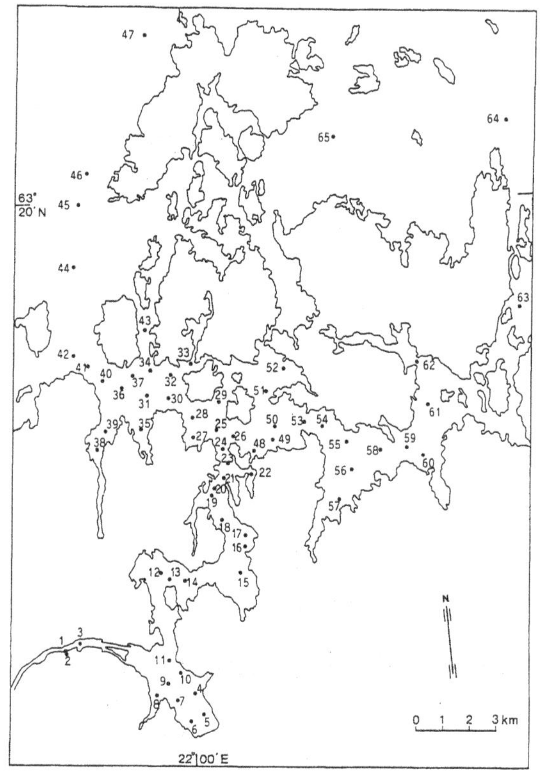

Fig. 1. Study area and sampling sites.

3. Material and methods

The material of this study consists of samples of the sediment surface, the uppermost 2 cm, taken with a Kajak sampler from 65 sites in the estuary of the river Kyrönjoki (Fig. 1), and 100 km of echo sounding lines.

The sediment samples were dried to constant weight at +50 °C, and then ignited at +550 °C for 3 hours. The organic content was determined as loss-on-ignition. The ignited samples were extracted using 37% HCl. Total Fe, Mn, Pb, Cu, Zn, Cd, Hg and Ca were analysed by AAS. Total P was determined spectrophotometrically using the molybdenum blue method. Total C and N were determined with an HP 185B CHN analyzer.

The estuary was echo sounded with an Atlas Monograph 58 echo sounder. Plain soft bottoms were determined to be areas of accumulation, and hard bottoms areas of erosion.

Two samples (1 and 2, Fig. 1) were dated according to the annual growth rate of *Carex acuta* L. in a flooded meadow. One sample (22, Fig. 1) was dated using annual laminae (See Simola, 1983), and one (38, Fig. 1) according to the depth of a severe change in sediment chemistry due to damming in 1957.

4. Results

4.1. Delimitation of the delta

The delta of the river Kyrönjoki was delimited according to the echo soundings, and areas of erosion and accumulation were separated (Fig. 2). Plain soft bottoms were determined to be areas of accumulation, and hard bottoms areas of erosion. Transportation areas could not be separated from the accumulation areas. In the shallow areas, where wave erosion is weak due to the sheltering islands, only the middle of the river channel is an erosion area due to the fast flow of water, while the bays and shallowest beach areas are areas of accumulation. In the deeper NW parts of the delta, which are open to winds from the NW and W, the shallow areas near the shores are eroded by waves, and the deepest points (6–30 m) are usually areas of accumulation (Fig. 2, Brydsten, 1985). The outer limit of the delta was determined as the outer limit of foreset layers in the deep areas. Bottomset layers exist even further out. During high flood periods the water from the river Kyrönjoki flows up to 30 km out from the outer limit of the delta.

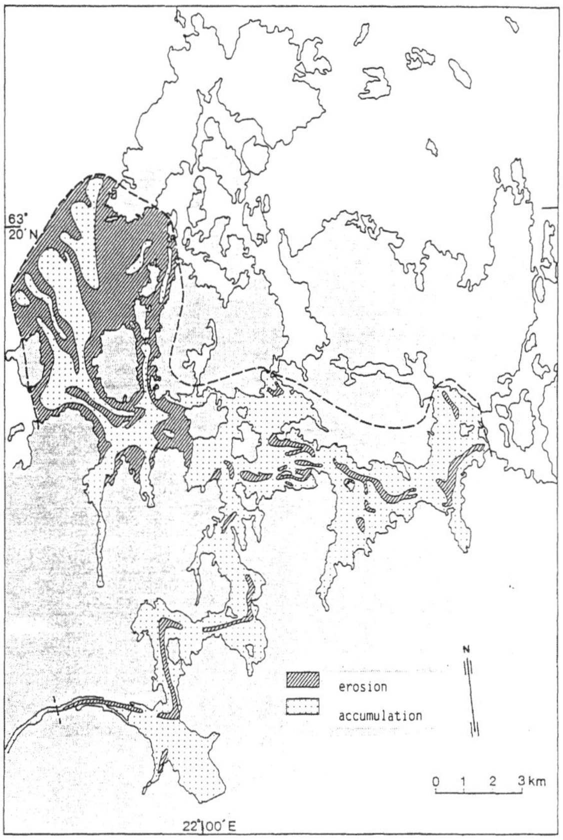

Fig. 2. Delimitation of the delta of the river Kyrönjoki (dashed line), and its subdivision into erosion and accumulation areas.

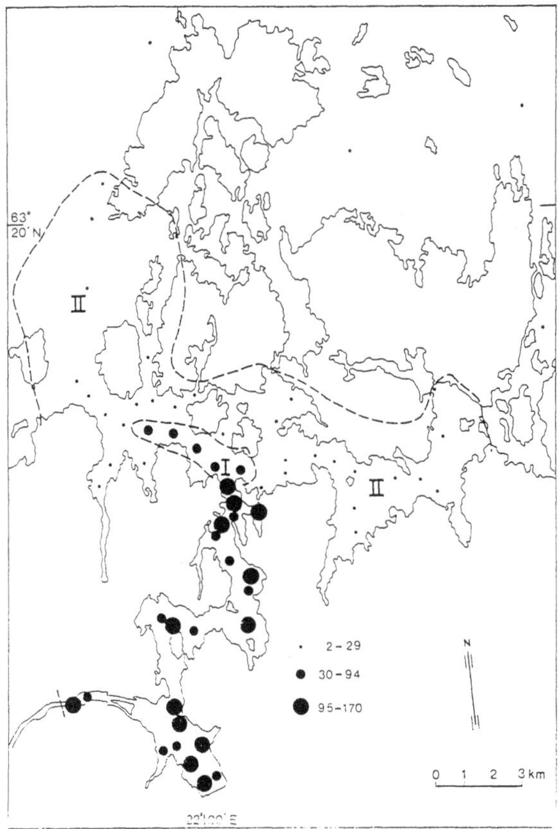

Fig. 3. Fe/Mn ratio in the sediment surface (0–2 cm) in the delta of the river Kyrönjoki, and the subdivision of the delta. I = the inner part, and II = the outer part of the delta.

Table 1. Some characteristics of the sediment surface (0–2 cm) in the delta of the river Kyrönjoki. I = the inner part, and II = the outer part of the delta (see Fig. 3).

	I				II			
	Min.	Max.	Mean	SD	Min.	Max.	Mean	SD
org. %	2.8	29.3	13.1	5.7	8.4	29.4	16.6	5.4
C %	2.1	9.1	5.7	2.0	3.6	10.0	5.9	1.9
N %	0.1	0.6	0.36	0.14	0.2	0.7	0.47	0.15
P mg/g	3.1	11.4	7.9	1.98	6.5	10.3	8.5	1.13
Ca ppm	120	1890	1160	430	610	1990	1620	320
Fe %	2.1	5.2	3.3	0.7	3.0	6.3	4.3	0.8
Mn %	0.02	0.07	0.04	0.01	0.14	1.70	0.55	0.30
Fe/Mn	44	170	98	34	2	29	10	5
Pb ppm	8	53	25	10	14	41	30	8
Cu ppm	8	33	23	6	20	42	29	5
Zn ppm	58	231	122	32	177	409	303	68
Cu/Zn	0.12	0.33	0.20	0.04	0.07	0.14	0.10	0.02
Cd ppm	1.2	2.6	1.8	0.4	1.2	3.6	2.2	0.6
Hg ppm	0.02	0.18	0.08	0.04	0.01	0.12	0.07	0.03
Hg ppm/org.	0.20	1.68	0.68	0.31	0.07	0.85	0.45	0.18

Table 2. Significant correlations between organic content and analysed elements in the sediment surface (0–2 cm) in the inner part of the delta of the river Kyrönjoki (see Fig. 3).

	org.	C	N	P	Ca	Fe	Mn	Pb	Cu	Zn	Cd
C	0.938***										
N	0.632**	0.596**									
P	0.686***	0.686***	0.616**								
Ca	-	-	-	0.371*							
Fe	-	-	-	0.487**	-						
Mn	-0.404*	-	-	-	-	0.395*					
Pb	0.412*	-	-	-	-	-	-				
Cu	0.480**	0.719***	0.457*	0.578***	-	-	-	-			
Zn	0.511**	0.604**	-	0.396*	-	0.367*	-	0.367*	0.672***		
Cd	0.563**	0.625**	0.453*	0.417*	-	-	-	0.591**	0.588**	0.659***	
Hg	0.602***	0.603**	-	-	-	-	-	-	0.669***	0.383*	-

Table 3. Significant correlations between organic content and analysed elements in the sediment surface (0–2 cm) in the outer part of the delta of the river Kyrönjoki (see Fig. 3).

	org.	C	N	P	Ca	Fe	Mn	Pb	Cu	Zn	Cd
C	0.970***										
N	0.904***	0.881***									
P	-	-	-								
Ca	-	-	-	-							
Fe	0.359*	0.570**	0.607**	-	0.465**						
Mn	0.542***	-	0.529**	-	-	-					
Pb	0.326*	-	-	-	-	-	-				
Cu	0.874***	0.859***	0.783***	-	-	-	0.434**	0.462**			
Zn	0.785***	0.721***	0.821***	-	-	0.483**	0.570***	-	0.770***		
Cd	-	-	-	-	-	-	0.543**	-	-	0.496**	
Hg	0.559**	0.564**	-	-	-	0.423*	-	-	0.402*	0.371*	-

4.2. Subdivision of the delta

The delta has been divided into two parts according to the sediment chemistry, especially the redox conditions (Fig. 3).

(I) In the inner part, fresh water from the river Kyrönjoki dominates. The water is quite acid (pH usually about 5.5, except in late summer when it may increase to 6.5 (Meriläinen, 1984b)). The sedimentary conditions are reducing, as indicated by high Fe/Mn ratios (44–170) and high Cu/Zn ratios (0.14–0.33) (e.g. Simola, 1983; Hallberg, 1974; Vuorinen, 1978; see Table 1). The heavy metal content is low, and except for Hg it does not significantly differ from the natural background. The Hg content is at its highest in the delta in the very acid area of Vassor Bay (sites 4–7).

(II) In the outer part of the delta, brackish water of neutral pH usually dominates (Meriläinen, 1985b). Sedimentary conditions are mainly oxidizing, as indicated by low Fe/Mn ratios (2–29), and low Cu/Zn ratios (0.07–0.14). The heavy metal content is clearly higher than the natural background. Moreover, the mean organic content of the sediment is higher than in the inner part of the delta (Table 1).

4.3. Chemical composition

Most of the analysed elements correlate significantly with each other (Tables 2 and 3). Carbon and nitrogen are strongly bound with the organic matter, and due to the acidity of the water of the river Kyrönjoki, and lack of carbonate rocks in the drainage area, all carbon can be regarded as organic. The C/N ratio is fairly high all over the delta; in almost all cases it is over 10, which shows that most of the sediment is allochthonous (Hansen, 1961).

Most of the elements analysed correlate significantly with the organic content, and they can be regarded as more or less bound with the organic matter of the sediment. Of the heavy metals, Cu and Zn have the highest correlation coefficients when compared with the organic content. In the outer part of the delta P, Ca and Pb, and in the inner part Ca, Pb, Fe and Mn do not correlate significantly with the organic content, and they are evidently not bound with the organic matter (Tables 2 and 3).

The Cu and Zn content of the sediment surface have a regional pattern in the delta, very much like that of the organic content (Fig. 4). Cu and Zn content is high in the most acid sites in the inner part of the delta, and in the area dominated by brackish water. The Cu content is everywhere fairly low, not more than 2–3 times higher than the natural background in the delta, and it does not significantly differ from the natural background given e.g. by Förstner (1976:98), Niemistö & Voipio (1981:90) and Håkanson *et al.* (1984:132).

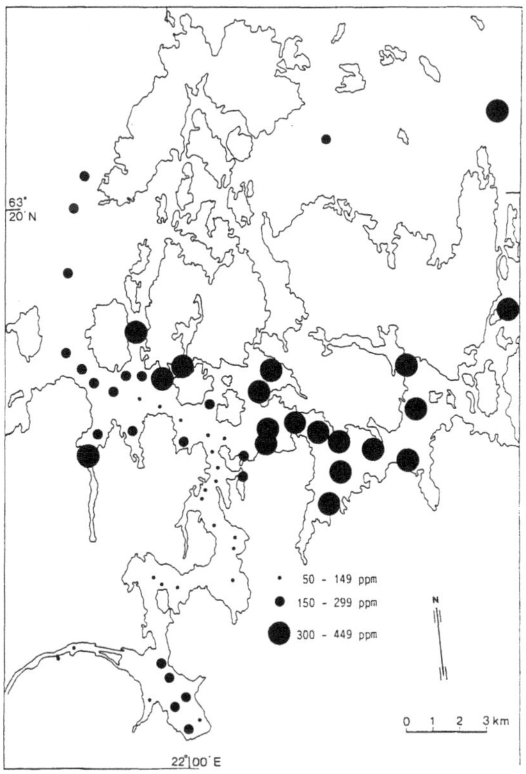

Fig. 4. Zn content in the sediment surface (0–2 cm) in the delta of the river Kyrönjoki.

The Zn content of the sediment surface in the area dominated by brackish water is relatively high, 6–8 times higher than the natural background. The Zn content is still a lot lower than in areas polluted by waste waters from chemical or metallurgical plants (e.g. Förstner, 1976; Koivo & Oravainen, 1982; Kansanen & Jaakkola, 1985). In the areas dominated by fresh water from the river, the Zn content is usually fairly low, and it does not significantly differ from the natural background given by Tolonen & Meriläinen (1983) and Håkanson *et al.* (1984).

5. Discussion

The regional differences in the sediments of the delta of the river Kyrönjoki are mainly caused by two factors: differences in acidity and salinity.

When fresh water from the river mixes with brackish water in the delta, the salinity is increased outwards from the river mouth (Meriläinen, 1985b). When the salinity increases, even though only to 3‰, the suspended fine-grained material, clay and particulate organic matter, flocculates (e.g. Calles, 1983). Then the particle size grows and the suspended matter precipitates rapidly. Due to this precipitation most of the fine-grained particulate material is deposited in the delta, and not further out in the Gulf of Bothnia. The fine-grained matter precipitated mostly in the outer part of the delta has a higher heavy metal and organic content than the coarser matter precipitating in the inner part of the delta (e.g. Keulder, 1982).

On the other hand, the fresh water from the river is very acid. Its pH is usually about 5.5, and it sinks below 5.0 during flood periods, which may last 2–3 weeks. In limited areas (sites 6 and 38, Fig. 1), where water is pumped from fields onto acid Litorina clay, the pH of the water is between 3 and 4 (Alasaarela, 1982; Meriläinen, 1984b). Many chemical constituents are soluble in the river water due to the acidity. When the river water mixes with the neutral brackish water, the pH rises, and some of the constituents dissolved in the acid water are precipitated.

Due to the acidity, the sedimentary conditions are highly reducing in the inner part of the delta and this is expressed by the high Fe/Mn and Cu/Zn ratios. The oxygen saturation of the river water is continuously quite high (Meriläinen, 1984b), and so the reducing conditions in the sediment surface are not caused by lack of oxygen. The zoobenthos is almost lacking in the inner part of the delta (Meriläinen, 1984a), and bioturbation of the sediment is negligible.

In the outer part, where fresh water from the river mixes with brackish water, the brackish water usually dominates (Meriläinen, 1985b), and the pH is clearly higher. The sedimentary conditions are mainly oxidizing, which is indicated by low Fe/Mn and Cu/Zn ratios. There is a rich zoobenthos (Meriläinen, 1984a) and bioturbation of the sediment is considerable.

Only four of the 65 samples could be dated, but it is evident, that the samples do not represent equal periods, because the sedimentation rates seem to be different in different parts of the delta. Continuous erosion and re-deposition due to the severe changes in the discharges makes it difficult to determine the sedimentation rates in the inner part of the delta. In the outer part problems are caused by strong bioturbation of the sediment. In the dated cases the annual sedimentation rates vary from 1500 to 9000 $gm^{-2}a^{-1}$ (Heikkilä, 1985).

The heavy metal content of the sediment surface is clearly higher in the outer part of the delta than in the inner part. This is due to the differences in sedimentary conditions, and also to increased intensity of human activities in the drainage basin, especially forestry drainage of peatlands and agriculture (Heikkilä, 1985).

6. Acknowledgements

I am very grateful to the following persons: Professors Matti Seppälä and Kimmo Tolonen, Pertti Sevola and Jarmo J. Meriläinen, for their invaluable advice during the work; Dr. Elizabeth Haworth for critical reading of the manuscript, and valuable comments; Professors Toive Aartolahti and Rauno Ruuhijärvi for providing excellent working facilities; Hanna Heikkilä, Kati Heikkonen, Teija Peltoniemi, Tiina Tonteri and Esa Tulisalo for laboratory analyses; Kirsti Lehto and Petri Rummukainen for technical assistance; Tony Meadows for language revision; and many others. The National Board of Waters and Ministry of the Environment financed the study.

References

Alasaarela, E., 1982. Acidity problems caused by flood control works of the river Kyrönjoki. Publ. Water Res. Inst. 49: 3–16.

Brydsten, L., 1985. Om stränder – processer, material och former (English summary: On shores – processes, material and forms). Gerum 1: 1–193.

Calles, B., 1983. Settling processes in a saline environment. Geogr. Ann. 65A: 159–166.

Eronen, M., 1974. The history of the Litorina Sea and associated holocene events. Comm. Physico-Math. 44: 79–195.

Förstner, U., 1976. Metal concentrations in freshwater sediments – natural background and cultural effects. In H. L. Golterman (ed.), Interactions between sediments and fresh water. Dr. W. Junk Publishers. The Hague.: 94–103.

Håkanson, L., I. Kulinski & H. Kvarnäs, 1984. Vattendynamik och bottendynamik i kustzonen (Abstract: Water dynamics and bottom dynamics in Swedish coastal waters). Statens naturvårdsverk PM 1905. 228 pp.

Hallberg, R. O., 1974. Paleoredox conditions in the Eastern Gotland basin during the recent centuries. Finnish mar. Res. 238: 3–16.

Hallberg, R. O., 1979. Heavy metals in the sediments of the Gulf of Bothnia. Ambio 8: 265–269.

Hansen, K., 1961. Lake types and lake sediments. Verh. int. Ver. Limnol. 14: 285–290.

Heikkilä, R., 1985 in press. Recent sedimentation in the delta of the river Kyrönjoki, western Finland. Publ. Water Res. Inst. 66.

Kansanen, P. & T. Jaakkola, 1985. Assessment of pollution history from recent sediments in Lake Vanajavesi, southern Finland. I. Selection of representative profiles, their dating and chemostratigraphy. Ann. Zool. Fenn. 22: 13–55.

Kenttämies, K., 1981. The effects on water quality of forest drainage and fertilization in peatlands. Publ. Water Res. Inst. 43: 24–31.

Keulder, P. C., 1982. Particle size distribution and chemical parameters of the sediments of a shallow turbid impoundment. Hydrobiologia 91: 341–353.

Koivo, L. & R. Oravainen, 1982. Zinc in water and sediment of two Finnish lakes. Hydrobiologia 91: 155–160.

Kääriäinen, E., 1975. Land uplift in Finland on the basis of sea level recordings. Rep. Finnish Geod. Inst. 75(5): 1–14.

Meriläinen, J. J., 1984a. Zonation of the macrozoobenthos in the Kyrönjoki estuary in the Bothnian Bay, Finland. Ann. Zool. Fenn. 21: 89–104.

Meriläinen, J. J., 1984b. Macrophyte vegetation at the River Kyrönjoki estuary in 1982. Publ. Water Res. Inst. 57: 61–76.

Meriläinen, J. J., 1985a, in press. Loads of nutrients, organic matter and suspended solids discharged by the River Kyrönjoki into the Bothnian Bay. Publ. Water Res. Inst. 66.

Meriläinen, J. J., 1985b. Spreading of river waters and behaviour of different fractions of particulate matter and dissolved organic matter in the non-tidal Kyrönjoki estuary, Bothnian Bay. Aqua Fennica 15: 53–64.

Niemistö, L. & A. Voipio, 1981. Notes on the sediment studies in the Finnish pollution research in the Baltic Sea. Rapp. P.-v. Reun. Cons. int. Explor. Mer. 181: 87–92.

Rosberg, J. E., 1895. Bottenvikens finska deltan. Akademisk afhandling. Helsingfors, Finland. 255 pp.

Simola, H., 1983. Limnological effects of peatland drainage and fertilization as reflected in the varved sediment of a deep lake. Hydrobiologia 106: 43–57.

Säntti, A. A., 1954. Die rezente Entwicklung des Kokemäenjoki-deltas. Publ. Inst. Geogr. Univ. Turkuensis 29. 61 pp.

Tolonen, K. & J. Meriläinen, 1983. Sedimentary chemistry of a small polluted lake, Gallträsk, S. Finland. Hydrobiologia 103: 309–318.

Tulkki, P., 1977. The bottom of the Bothnian Bay. Finnish Mar. Res. 241: 5–89.

Vuorinen, J., 1978. The influence of prior land use on the sediments of a small lake. Polskie Arch. Hydrobiol. 25: 443–451.

Accepted 18 March 1986.

Hydrobiologia 143: 379–385, (1986)

Concentration and annual accumulation values of heavy metals in lake sediments: Their significance in studies of the history of heavy metal pollution

Ingemar Renberg
Department of Ecological Botany, Umeå university, S-901 87 Umeå, Sweden

Keywords: palaeolimnology, sediment, heavy metal pollution

Abstract

The first part of the paper discusses the significance of using either concentration or accumulation values for expressing the results of investigations of lake sediment cores aimed at studying the history of heavy metal pollution. Neither the values for heavy metal concentration in the lake sediment, whether expressed per gram dry sediment or per gram soluble (organic) sediment, nor the values for their total annual accumulation per unit area of the lake bottom, can, on their own, provide an accurate picture of past pollution conditions, but when considered in combination they render a fairly reliable and detailed interpretation. The second part of the paper deals with Pb, Cd and Hg analyses of cores of varved sediments from several lakes in N. Sweden. Pb and Cd pollution increased during the second half of the 19th century. In most lakes, Hg pollution seems first to have started during the 20th century. Marked increases in both concentration and accumulation of heavy metals took place during the 20th century. For recent decades, the estimated accumulation rates of heavy metals from anthropogenic sources are: Pb 0.5–1.5 μg cm^{-2} yr^{-1}, Cd 15–30 ng cm^{-2} yr^{-1} and Hg 1–2 ng cm^{-2} yr^{-1}. Higher values were recorded in lakes affected by local emissions.

Introduction

Environmental pollution by heavy metals from mining activities, from industrial and domestic effluents, and from the combustion of fossil fuels, has been much discussed during recent decades. In attempts to estimate the pre-industrial, natural level of heavy metals and to study the historical development of environmental pollution, several different types of naturally-occurring biological and geological records have been analysed (reviews in MARC Report 31). Lake sediments have frequently been used (see e.g. Alderton in MARC Report 35: 1–95). In most cases, the sediment cores have been divided up into consecutive, often half or centimetre-thick samples, which have then been analysed and the results of the heavy metal analyses expressed simply in terms of the dry weight of each sediment sample, e.g. as μg metal per gram dry sediment. This may be a rather imprecise way of expressing the results for the purpose of ascertaining the history of pollution. A more reliable way is often to express the results on the basis of the content of the organic or acid-soluble fraction of the sediment sample. An alternative approach to the problem is to estimate the net annual accumulation of the heavy metal on the lake bottom, e.g. as the annual accumulation per cm^2. This method, however, presupposes use of an adequate core-sampling and sub-sampling technique and the possibility of establishing a detailed dating for the sediment core. The present paper discusses: (i) the advantages and disadvantages of using either concentration or accumulation values for expressing the results of investigations of lake sediments aimed at studying the historical course of environmental pollution, and (ii) the history of lake pollution by lead, cadmium and mercury in N. Sweden. Analyses of cores from varved lake sediments, form the basis for these discussions.

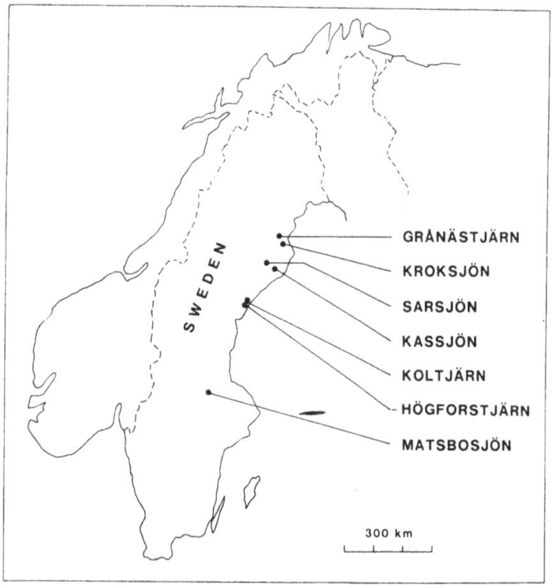

Fig. 1. Map showing the location of the investigated lakes.

Material and methods

In a varved lake sediment (annually laminated sediment) the material deposited on the lake bottom during the course of a single year is preserved intact, no subsequent mixing with the older, underlying, sediment occurring. Each year's deposit is clearly distinguishable and the samples are easily dated by making varve-counts. The type of varved sediment dealt with here has been described in detail in a previous paper (Renberg, 1981a).

Cores of varved sediments, spanning recent centuries, were taken with a crust-freeze sampler, a method which yields undisturbed cores of the unconsolidated surface sediments by freezing the material *in situ* (Renberg, 1981b). The cores were sampled and analysed in 1979 (except Koltjärn which was sampled in 1983). Pre-historic sediment samples from two lakes, taken with a Russian peat corer, have also been analysed.

Consecutive samples, each comprising a known number of years and representing a certain area of the lake bottom, were cut from the frozen cores. Each core extended from the sediment-water interface to 20–30 cm. The sampling and subsampling techniques used have been described in detail elsewhere (Renberg, 1981b, 1985).

Each sediment sample was dried, weighed, and then boiled in a mixture of conc. HNO_3 and $HClO_4$, or HNO_3 and a little 4% $KMnO_4$-solution. The latter method, used for Hg, broadly follows that described by Håkansson (1975). After wet oxidation each sample was centrifuged and the supernatant used for the heavy metal analyses. Mercury was determined by flameless spectroscopy and lead and cadmium by AAS (graphite furnace). The residue, mainly composed of mineral grains and diatom valves, was carefully washed, dried and weighed. This fraction is termed the 'insoluble' fraction of the sediment. The 'soluble' fraction of each sample is the value given by the difference between the initial dry weight and the dry weight of the insoluble fraction. The soluble fraction consists of organic material and various acid-soluble precipitates such as iron hydroxides.

Results and discussion

As an introduction to the discussion about the reliability of concentration and accumulation values when attempting to ascertain the heavy metal pollution history of an area, the results of analyses of the lead and soluble and insoluble sediment in a core from Lake Grånästjärn are presented in detail. Subsequently, the course of lead, cadmium and mercury pollution of the northern Swedish environment, as indicated by the results from varved sediments from several lakes, is discussed.

Lake Grånästjärn

The annual accumulation rates of both the soluble and insoluble fractions of the lake sediment in any lake vary from year to year, as well as fluctuating over longer periods of time. This is particularly true in recent times, when there has been a great increase in human influence on the lakes and their surroundings, but it is also a natural phenomenon. This has been clearly revealed by studies of varved lake sediments (Segerström *et al.*, 1984; Simola *et al.*, 1984). Ditching, ploughing, road construction, and changes in forestry practice etc, have all resulted, from time to time, in increases in the influx of allochthonous organic and inorganic material to the lakes. The inflow of both agricultural and sewage effluents has led to increases in lake productivity. These factors have led to changes in the basic composition of the sediments deposited during the past century which have, in turn, affected the concentrations and accumulation rates of heavy metals.

In Lake Grånästjärn, the accumulation of mineral material reached its maximum during the 20th century, with two pronounced peaks, one covering a period of years in the early 1900's and a second around 1940 (Fig. 2b). Both these peaks were due to an artificial lowering of the lake waterlevel and to ditching operations. The lead content per gram dry weight of the sediment samples from both these periods are the lowest recorded throughout the entire core (Fig. 2c). The reason for this is not that lead pollution of the lake was unusually low during these two periods, but that greater quantities of mineral grains were deposited on the lake bottom, producing a diluted lead content when expressed per gram dry sediment. It is obvious that, since the mineral matter accumulation has varied markedly with time, any results of heavy metal analyses expressed in this manner will yield a false impression of the history of lead pollution in the area.

A second and more reliable picture of the history of lead pollution is obtained when the lead analysis results are expressed in terms of the soluble fraction of the sediment samples (Fig. 2d). But this picture is only valid if no major fluctuations in the annual influx of allochthonous material have taken place, that the lead concentration in this allochthonous material has always remained constant and, furthermore, that lake productivity has also remained constant. In the case of Lake Grånästjärn, as in most lakes, none of these requirements is fulfilled. For example, the peak in the accumulation of soluble sediment during the period 1930–38 was caused both by a lowering of the lake water-level and by drainage of an adjacent peatland. The latter activity, in particular, must have led to an increased influx into the lake of old organic material, probably with a lower lead content than the contemporaneous material being deposited, and can be expected to have produced a spurious decrease in the lead content when the values are expressed per gram soluble sediment. The same effect could also have been produced by a slight increase in the annual accumulation rate of the autochthonous organic material caused by increased lake productivity. This is because the major part of the lead influx into the lake is derived from atmospheric fall-out, which either accumulates in particulate form, becomes bound to particles suspended in the lake water which is then deposited on the lake bottom, or becomes bound to the allochthonous organic matter. Thus any increase in the amount of autochthonous organic matter or non-polluted allochthonous organic matter in the sediment will result in the dilution of the atmospheric flux of a heavy metal as recorded in the sediment column.

The third way of expressing the results of the lead analyses (Fig. 2e), in terms of its annual accumula-

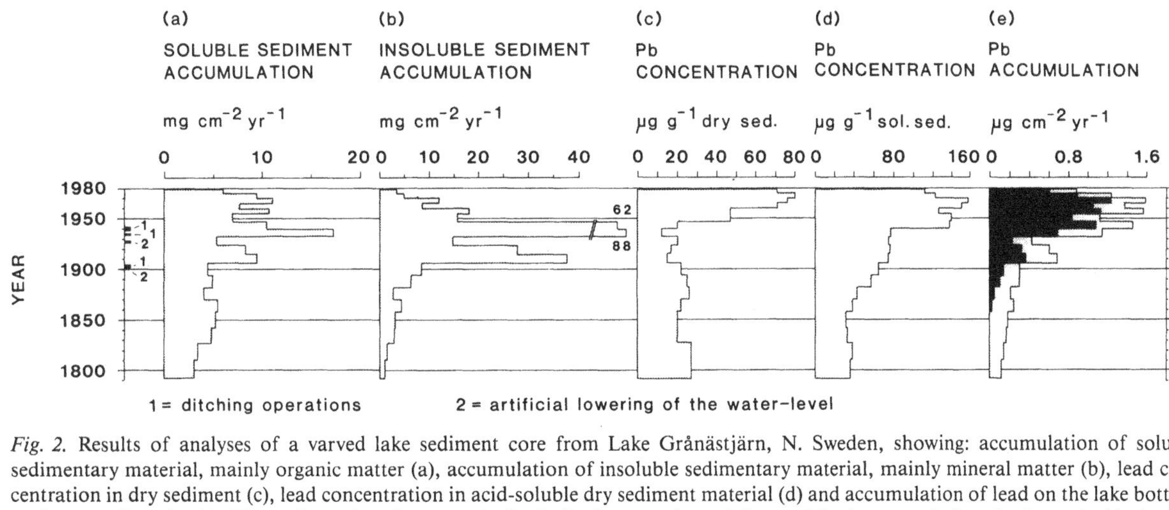

Fig. 2. Results of analyses of a varved lake sediment core from Lake Grånästjärn, N. Sweden, showing: accumulation of soluble sedimentary material, mainly organic matter (a), accumulation of insoluble sedimentary material, mainly mineral matter (b), lead concentration in dry sediment (c), lead concentration in acid-soluble dry sediment material (d) and accumulation of lead on the lake bottom at the sampling site (e). The estimated anthropogenically derived proportion of the total lead accumulation, is shown in black.

tion per unit area of the lake bottom, is independent of any variation in the accumulation rate of the autochthonous organic material, providing that the lead influx to the lake is in the form of direct deposition of particulate lead from the atmosphere, or as lead bound to allochthonous material. However, the accumulation rate of lead will be affected by any variation in the influx of allochthonous material. Any increased accumulation of organic material, such as the above-mentioned peaks which were recorded during the periods 1900–1920 and 1930–38, involves a certain extra accumulation of lead, as shown in Fig. 2e, even when the allochthonous material itself has a very low lead concentration. Values for this naturally-derived lead accumulation cannot be calculated accurately, since the lead concentration in the allochthonous material is unknown. However, for each analysed sample of the core from L. Grånästjärn, a rough value for the natural lead accumulation per cm^2 and year can be obtained by multiplying the accumulation rate of soluble sediment by what is presumed to be its normal background (pre-industrial) lead concentration (here a mean value from analyses of pre-1850 sediment has been used). A reasonable value for the anthropogenically derived proportion of the annual lead accumulation can then be calculated (Fig. 2e).

To sum up, neither the lead concentration values, expressed per gram dry sediment or per gram dry soluble sediment, nor the total annual lead accumulation values per unit area of the lake bottom, are able, on their own, to provide an accurate picture of the history of lead pollution. But when considered together they enable a fairly reliable and detailed interpretation of the lead analysis results to be made.

Without considering any changes in the lead content of the sediment with time due to mobility and diagenetic processes, the conclusions about lead pollution in the Grånästjärn area that can be drawn from these results are: (i) The lead deposi-

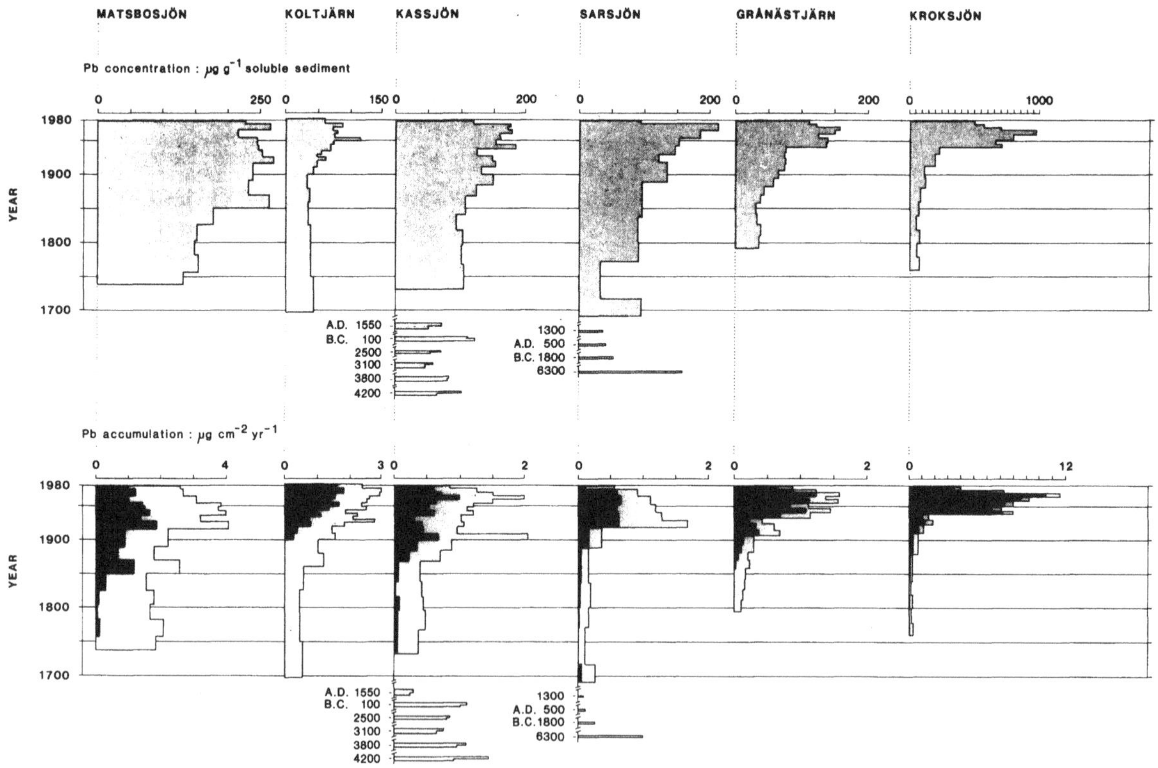

Fig. 3. Results of lead analyses of varved sediments from six Swedish lakes showing: lead concentration in acid-soluble dry sediment (above) and lead accumulation on lake bottom (below). Anthropogenic proportion of lead, shown in black.

tion from anthropogenic sources increased during the later half of the 19th century. A marked increase in the volume of lead in the environment took place between some time in the 1930's and about 1970, whereafter there are signs of a reduction. (ii) Compared to pre-industrial times, the lead concentration in the soluble sediment fraction had doubled by 1900, and quadrupled by 1970. The lead accumulation increased from 0.2 μg cm^{-2} yr^{-1} to a maximum value of 1.6 μg cm^{-2} yr^{-1}, of which about 1.2 μg cm^{-2} yr^{-1} is supposed to have been derived from anthropogenic sources. The estimated accumulation values during the late 1970's have declined to below 1 μg cm^{-2} yr^{-1}.

Other lakes

Results of analyses of lead, cadmium and mercury in the samples of varved sediments taken from L. Grånästjärn and six other lakes are shown in Figs. 3, 4 and 5.

Both the lead concentration and accumulation values clearly indicate that an increase in lead pollution began in N. Sweden as early as the second half of the 19th century (Fig. 3). Pb curves rised during the 20th century, but with a slight decline being visible during the past few years in most lakes. In all the lakes, the concentration of lead in the soluble sediment fraction has at least doubled between the beginning of the 19th century and the middle of the 20th. During the same period, in most lakes the annual lead accumulation rate per unit area of the lake bottom has increased 2–5 fold, or even more. This was due to increases in the deposition of both organic-bound lead from natural sources and of anthropogenically derived lead. The anthropogenic lead accumulation values for the past few decades lie between 0.5 and 1.5 μg cm^{-2} yr^{-1}, except in L. Kroksjön, where values about ten times higher were recorded.

The volume of cadmium deposited in the lakes also increased during the 19th century and has con-

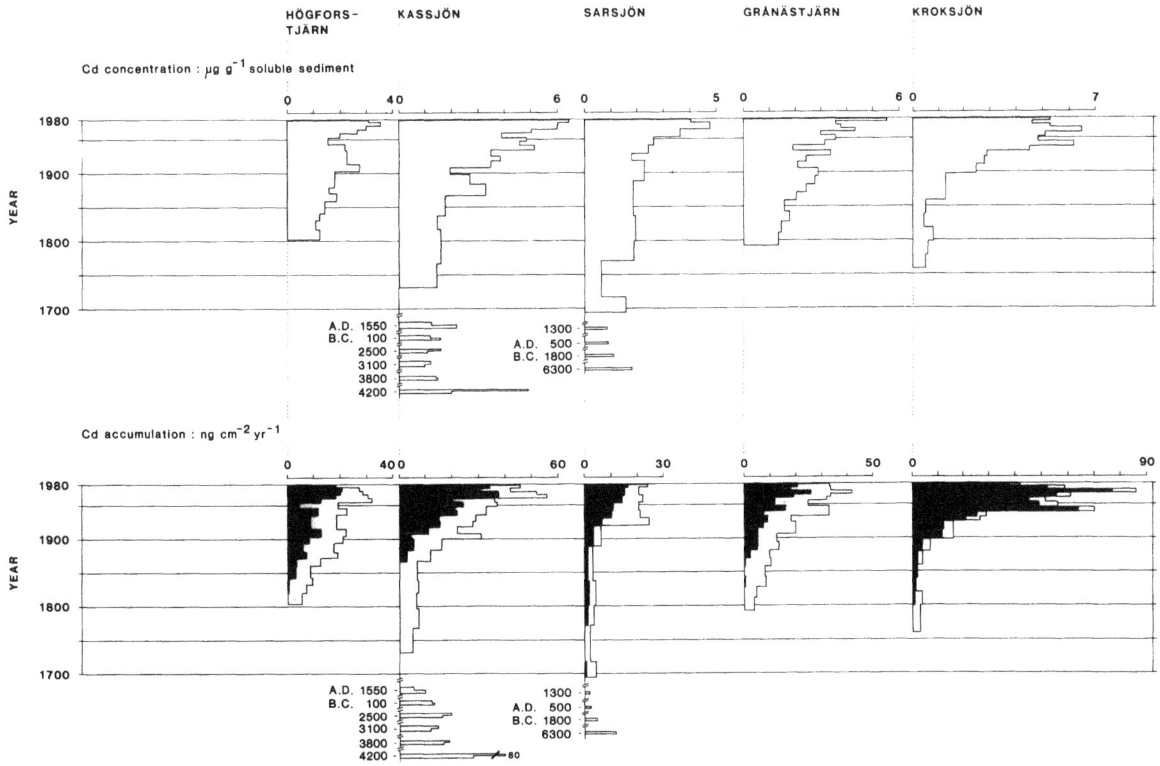

Fig. 4. Results of cadmium analyses of varved sediments from five Swedish lakes showing: cadmium concentration in acid-soluble dry sediment (above) and cadmium accumulation on lake bottom (below). Anthropogenic proportion of cadmium, shown in black.

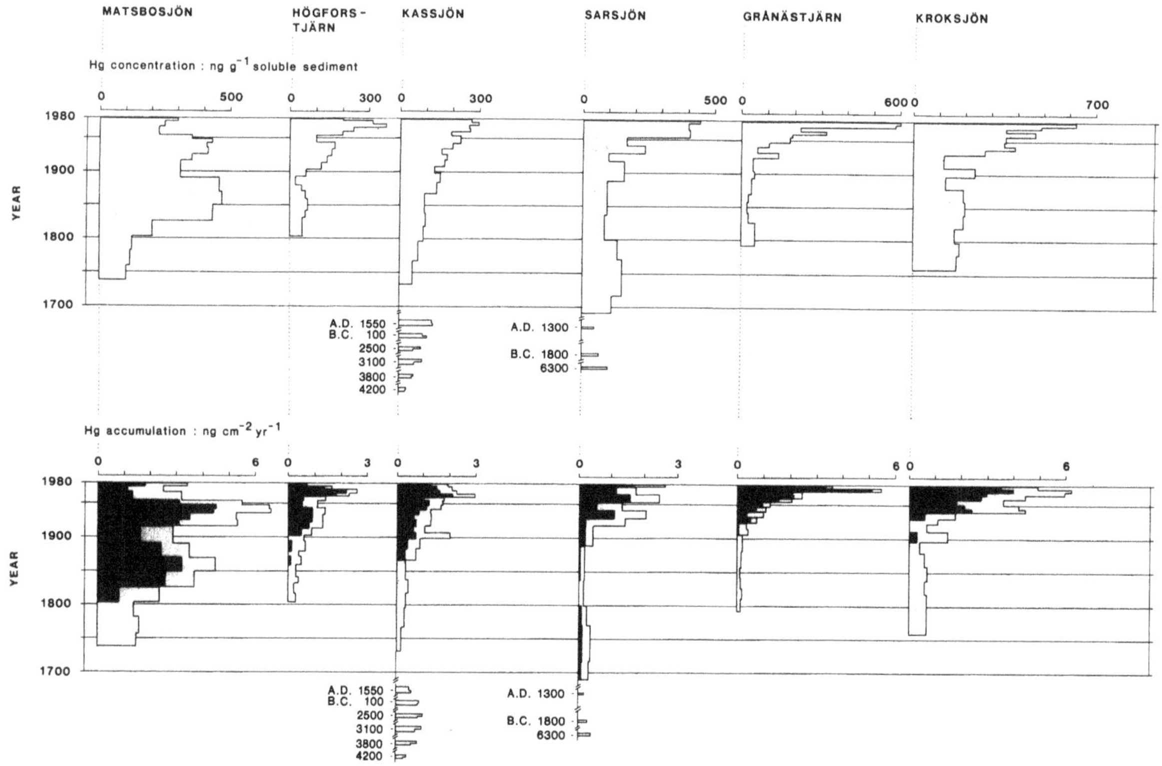

Fig. 5. Results of mercury analyses of varved sediments from five Swedish lakes showing: mercury concentration in acid-soluble dry sediment (above) and mercury accumulation on lake bottom (below). Anthropogenic proportion of mercury, shown in black.

tinued to increase up to the sampling year (1979), except in L. Kroksjön (Fig. 4). The recent values for cadmium per gram soluble sediment are about three times higher than the values for the 18th and early 19th centuries (in L. Kroksjön ten times higher). The increase in the annual accumulation rate of cadmium during the same period of time has been, on average about fivefold, and by an even greater factor in L. Kroksjön. The anthropogenic cadmium accumulation rates during the past few decades have been 15 to 30 ng cm^{-2} yr^{-1} in four of the lakes, and slightly higher in L. Kroksjön.

Mercury pollution seems to have started later than the pollution by lead and cadmium (Fig. 5). No significant increase in either the mercury concentration in the soluble sediment fraction, or in the accumulation rate of anthropogenically derived mercury occurs until some time during the present century, although L. Matsbosjön forms an exception, showing high mercury values during the 19th century. Recent values for mercury concentrations in the soluble sediment are 4–10 times higher than those recorded in 18th and 19th centuries. The annual accumulation rates of mercury have increased 2–10 fold. The anthropogenic mercury accumulation during the past few decades has varied from 1 to 2 ng cm^{-2} yr^{-1} in four of the lakes and from 3 to 5 ng in L. Grånästjärn and L. Kroksjön.

Accumulation of heavy metals in lake sediments during the 19th century probably increased as a consequence of an increase in the fall-out of airborne pollutants, due to the progressive industrialisation of Sweden and Europe. Increased deposition, during the 19th century, of carbonaceous particles characteristic of fossil fuel combustion, has also been recorded in the sediments of L. Grånästjärn and L. Koltjärn (Renberg & Wik, 1985). The marked increase in the accumulation of heavy metals in the lakes during the 20th century is consistent with the known development of industry and with the increased combustion of fossil fuels, although domestic and agricultural sources within the lake catchment areas also may have contributed to these increases.

The high 19th century value for mercury, and perhaps also that for lead, in L. Matsbosjön may indicate an early onset there of heavy metal pollution. This is a very likely explanation, when it is considered that this lake is situated in an old mining district.

The lakes Kroksjön and Grånästjärn are also almost certainly affected by local emissions of heavy metals, both lakes being situated not very far from a large-scale smelting works, which utilize complex sulphide ores rich in lead, cadmium and mercury. The initial establishment of this industry, during the 1930's, is detectable in the sediments, particularly in L. Kroksjön, which lies in one of the prevailing wind directions from the works. The reduction of lead and cadmium emissions into the atmosphere during the 1970's is also clearly recorded in the sedimentary record.

Acknowledgements

I thank the National Swedish Environmental Protection Board for funding this work and Gunnar Pohl for carrying out chemical analyses.

References

Håkansson, L., 1975. Kvicksilver i Vänern – nuläge och prognos (Engl. summary: Mercury in Lake Vänern – present status and prognosis). Statens Naturvårdsverk PM 563: 1–123.

MARC Report 31, 1985. Historical monitoring. Monitoring and Assessment Research Centre, University of London.

Renberg, I., 1981a. Formation, structure and visual appearance of iron-rich, varved lake sediments. Verh. int. Ver. Limnol. 21: 94–101.

Renberg, I., 1981b. Improved methods for sampling, photographing and varve-counting of varved lake sediments. Boreas 10: 255–258.

Renberg, I., 1985. Influences of acidification on the sediment chemistry of Lake Gårdsjön, SW Sweden. Ecol. Bull. 37: 246–250.

Renberg, I. & M. Wik, 1985. Carbonaceous particles in lake sediments – pollutants from fossil fuel combustion. Ambio 14: 161–163.

Segerström, U., I. Renberg & J.-E. Wallin, 1984. Annual sediment accumulation and land use history; investigations of varved lake sediments. Verh. int. Ver. Limnol. 22: 1396–1403.

Simola, H., P. Huttunen & J. Meriläinen, 1984. Varve-dated eutrophication history of a small lake. Verh. int. Ver. Limnol. 22: 1404–1408.

Accepted 18 March 1986.

Hydrobiologia 143: 387–394, (1986)

Sedimentary records of carbonaceous particles from fossil fuel combustion

Maria Wik,[1] Ingemar Renberg[1] & Judi Darley[2]
[1]*Department of Ecological Botany, Umeå University, S-901 87 Umeå, Sweden*
[2]*Palaeoecology Research Unit, Department of Geography, University College London, 26 Bedford Way London WC1H OAP, England*

Keywords: palaeolimnology, sediment, carbonaceous particles, oil and coal combustion.

Abstract

Carbonaceous particles produced by fossil fuel combustion can be found in considerable amounts in recent lake sediments. As these particles contain elemental carbon they are resistant to chemical decomposition and therefore both well preserved in sediments and possible to quantify. Sediment samples can be oxidized with H_2O_2 and digested with HF without the particles being destroyed. The pioneers in studying carbonaceous particles in lake sediments in relation to fossil fuel combustion were J. J. Griffin and E. D. Goldberg. They measured elemental carbon concentrations in Lake Michigan sediments, mainly by infrared assay. On the basis of these analyses, size distribution measurements and also morphological studies of single particles they concluded that the carbonaceous particle record reflected the onset of industrial activity and the increased intensities of fossil fuel combustion during the twentieth century. Similar results have been obtained from another lake in the USA by B. K. Kothari and M. Wahlen. We have counted spherical carbonaceous particles (larger than 5–10 μm), which are characteristic for oil and coal burning, in several lake sediments. In Swedish lakes, the annual accumulation of coarse carbonaceous spheres in varved sediments and the concentration stratigraphy in non-varved sediments, follow the same main pattern as statistical data for the Swedish coal and oil combustion over the last two centuries. Coarse carbonaceous spheres in two sediment profiles from Scottish lakes have also been counted. As for the USA and Sweden the sedimentary record was found to reflect the history of fossil fuel combustion.

Introduction

Poor historical data for lake acidification have increased the interest in lake sediment records as a means of studying the pH history of lakes and assessing the relative importance of different causes of acidification. Diatoms have thus became important indicators of past pH conditions (cf. Battarbee, 1984), and metals, sulphur, polycyclic aromatic hydrocarbons (cf. Alderton, 1985) and magnetic parameters (Oldfield *et al.*, 1983) have been used as indicators of the deposition history of airborne pollutants. Much of these pollutants is derived from fossil fuel combustion, which is an important source of acid precipitation and usually thought to be the main cause of surface water acidification. In addition carbonaceous particles derived from fossil fuel combustion are found in recent sediments. These particles are mainly composed of elemental carbon and are therefore chemically very stable. Since they are particulate they can be better stratigraphic indicators than some trace metals and sulphur, which, under certain conditions, are liable to chemical remobilization (Kahl & Norton, 1983, Holdren *et al.*, 1984).

Carbonaceous particle formation and transport

Several kinds of carbonaceous particles are formed during combustion processes. Submicrometer sized spherical particles are mainly generated via pyrolysis of fuel molecules that condense and then grow through surface growth and coagulation and often aggregate to form chainlike structures (Haynes & Wagner, 1981). Coarser particles are generated from fuel particles or drops that are incompletely burnt (Peterson, 1972, McCrone & Delly, 1973). Coarse irregular particles are generated by wood, coal and oil combustion. Oil and coal combustion, however, also produces particles with a characteristic spheroidal shape. As much as 90% of the particles formed during oil combustion are such spheres (Shen *et al.,* 1977, Bacci et al., 1983). Their morphology, structure and surface characteristics can indicate an origin from an oil, coal or wood source (Griffin & Goldberg, 1981).

The quantity of carbonaceous particles emitted from a combustion source depends on several factors, such as type of fuel, burner construction and its degree of burning effiency, and the kind of emission control devices used (Goldstein & Siegmund, 1976, Ericson *et al.,* 1983). The atmospheric lifetime of carbonaceous particles is also controlled by several factors, including aerodynamic diameter and meteorological conditions. Coarse particles are removed by sedimentation while precipitation scavenging is the dominant removal mechanism for smaller particles (Ogren & Charlson, 1983). Carbonaceous particles can be transported over large distances and have for example been reported from the Arctic (Rosen *et al.,* 1981). When carbonaceous particles are deposited in sediments they are preserved and can be detected.

Carbonaceous particles in sediments

Charcoal particles have been counted simultaneously with pollen in several palaeoecological studies (Tolonen, 1986), but the first workers to make detailed analysis of carbonaceous particles in lake sediments and associate the results with anthropogenic airborne emissions were Griffin, Goldberg and colleagues. They developed a spectrometric method for quantitative determination of elemental carbon applicable for sediments (Smith *et al.,* 1975) and have published a series of papers dealing with results from the analysis of a sediment core from Lake Michigan. They compared carbonaceous particles extracted from the sediments with carbonaceous particles formed during combustion of oil, coal and wood (Griffin & Goldberg, 1979), and made a quantitative analysis and morphological study of coarse particles extracted from the sediments (Griffin & Goldberg, 1981). Subsequently they sought relationships between these results and metal and mineral contents of the sediments (Goldberg *et al.,* 1981) and used Stokes' settling techniques to separate the carbonaceous particles into different size classes (Griffin & Goldberg, 1983).

Fig. 1. shows results from the analysis of coarse carbonaceous particles in the core from Lake Michigan (Griffin & Goldberg, 1981). The carbonaceous particles were isolated from the sediment by oxidation of organic matter and dissolution of soluble salts with H_2O_2, KOH and HCl and break down of siliceous minerals with HF (Smith *et al.,* 1975). They found that before 1900 the elemental carbon concentrations were lower than those in more recently deposited strata and were relatively constant. The majority of particles originated from biomass burning. 1900 to 1930 was a transition period, when carbon concentration was doubled and some of the particles had morphologies indicating a coal source. Another doubling of the elemental carbon concentration took place between 1930 and 1960. During this period the majority of the carbonaceous particles were porous spheres with textures indicating coal combustion. About 1950 the first particles originating from oil combustion occurred but after 1960 the elemental carbon content declined, probably due to improved emission controls (Griffin & Goldberg, 1981).

Kothari and Wahlen (1984) have studied the concentration and morphology of carbonaceous particles (>38 μm) in seven sediment samples from Green Lake (N.Y., USA) to determine the flux and distribution of particles from oil, coal and wood combustion. The sediment samples were treated with HCl, HNO_3 and HF and the carbonaceous particle contents were determined by weighing. They found a general decrease of carbonaceous particle concentrations with core depth and concluded that the trend was similar to the one found for energy usage in the USA. Sediment samples deposited between 1590–1823 mostly contained carbonaceous particles from wood burning. In samples deposited between 1845–1861, particles from both fossil fuel and wood combustion were found in roughly similar numbers. Carbonaceous

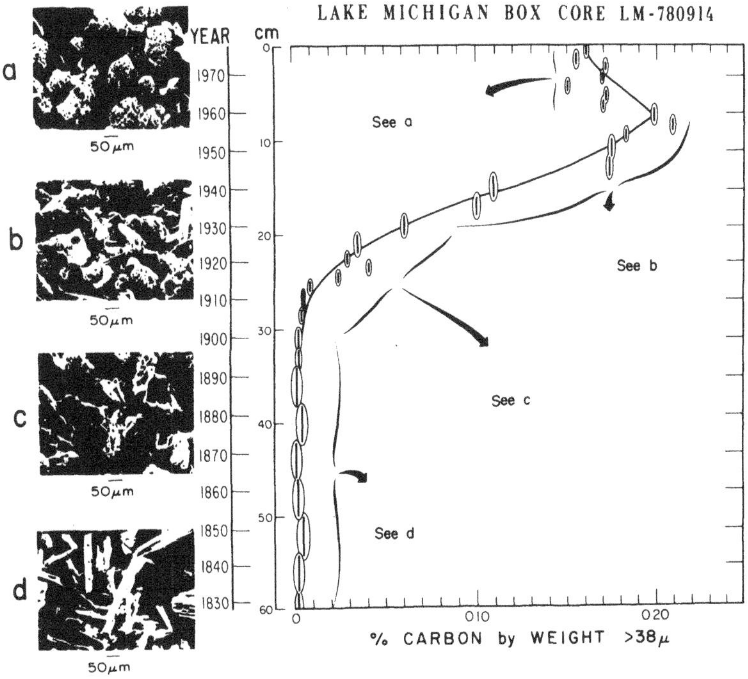

Fig. 1. Charcoal particles in a core collected from southeastern Lake Michigan. The SEM-photos show the particle morphologies common to four periods: (a) is representative of the particles in the sediments of the post 1960 period; (b) 1930–1960; (c) 1900–1930, and (d) pre-1900 period. (From Griffin, J. J. & E. D. Goldberg, 1981).

particles from oil and coal combustion were dominant in the three top sediment samples deposited between 1905–1979.

We have studied carbonaceous particles in sediment cores from Swedish and Scottish lakes and have focused on coarse spherical particles that are unique for fossil fuel combustion according to Griffin & Goldberg (1981), Fig. 2. In our previous papers we have used the term 'soot spheres'. However, this name can be interpreted wrongly, as 'soot' most often refers to submicrometer sized particles. Therefore, we have used the term 'coarse carbonaceous spheres' in this paper. The particles have been counted using a stereomicroscope (×50). Particles with a diameter larger than 5 to 10 μm are detectable. The preparation technique is rather similar to the method used in Lake Michigan (Smith *et al.*, 1975), using 32% H_2O_2 to oxidize organic matter, but only in exceptional cases has HF been used to digest silicates (Renberg & Wik, 1985b).

In Sweden lakes with both varved and non-varved sediments have been analysed (Renberg & Wik, 1985a), Fig. 3. The results from Koltjärn and Grånästjärn, two lakes with varved sediments, show several interesting features (Fig. 4). Although both lakes are situated in areas that were industrialized rather late, the accumulation of coarse carbonaceous spheres had already begun during the 18th century. The accumulation rate increased during the middle of the 19th century at least in the sediment of Grånästjärn and perhaps also in Koltjärn. A firm conclusion about the latter lake would require further analysis since the samples from this lake were more difficult to count due to a high mineral grain content. It is likely that the 19th century increase in the carbonaceous sphere accumulation rate reflects the increased use of coal following the Industrial Revolution in Sweden and Europe. A sudden increase in the accumulation of carbonaceous spheres took place during the 1930's. Since this increase is represented by only one sample in Grånästjärn we paid no regard to this peak (Renberg & Wik, 1985a), but interpreted the steep increase in Koltjärn as being the result of emissions from some local sources (e.g. the pulp industry that expanded in the coastal provinces at that

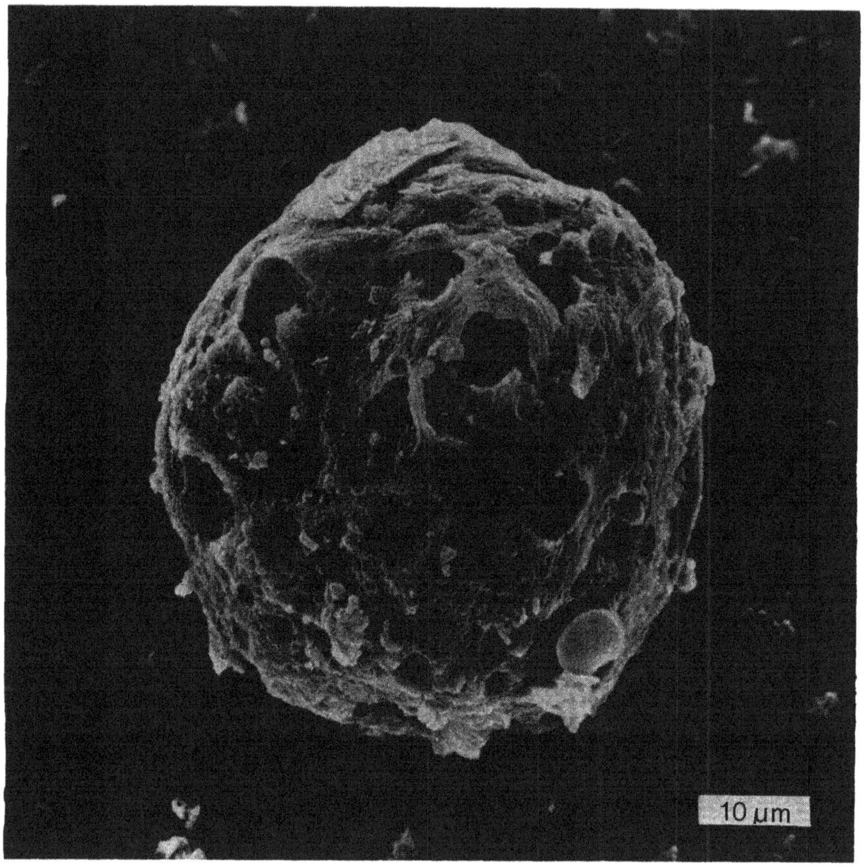

Fig. 2. Coarse carbonaceous sphere prepared from a lake sediment. (From Renberg, I. & M. Wik, 1985a).

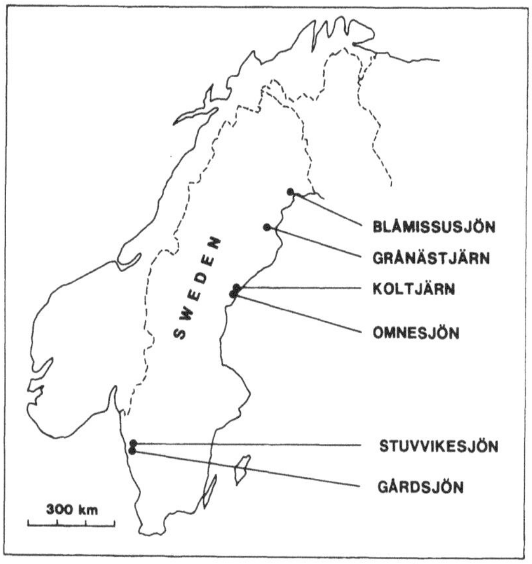

Fig. 3. Map showing the location of the investigated lakes in Sweden.

time). A possibility that cannot be ruled out, however, is that what we see here is a more general prewar peak caused by increased combustion of oil and also coal. In both cores a marked increase in the accumulation of carbonaceous spheres was recorded after World War II and a maximum was reached at about 1970 before the oil crisis.

The similarity between the accumulation pattern of coarse carbonaceous spheres in these varved sediments and the historical data for Swedish fossil fuel combustion (Fig. 4) indicates that the sediment records quite accurately reflect the deposition history of coarse particulate pollutants from fossil fuel combustion (Renberg & Wik, 1985a). It is, however, important to remember that combustion techniques and emission controls have been improved, stacks built taller etc. particularly during the last few decades, all changing the amounts and distribution of the emissions.

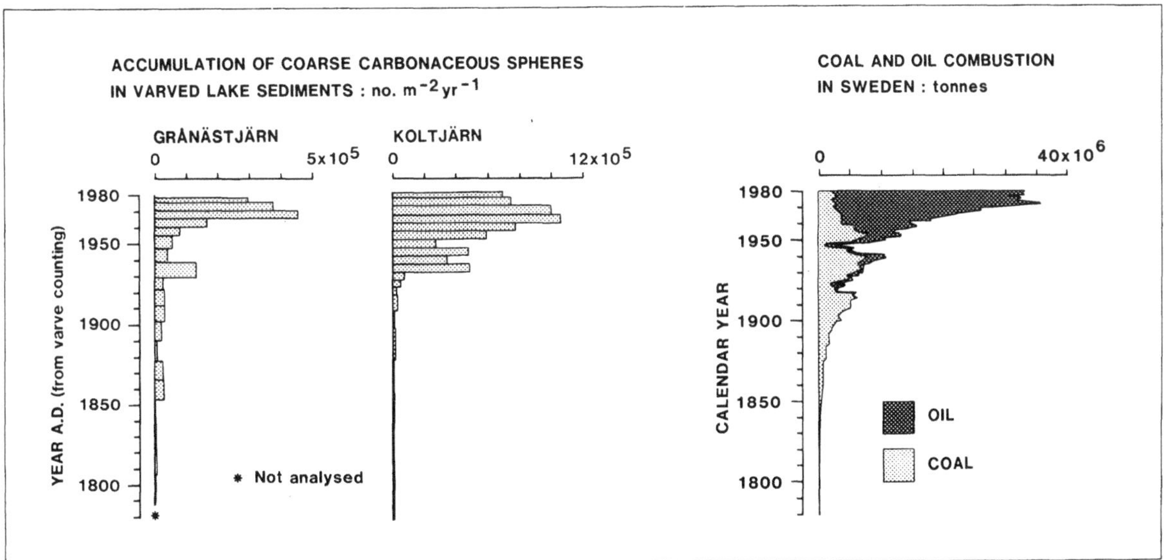

Fig. 4. Diagrams showing the accumulation of coarse carbonaceous spheres in two Swedish lakes with varved sediments and, to the right, the oil and coal combustion in Sweden, based on data for import, production and export of coal and mineral oils. (Slightly modified from Renberg, I. & M. Wik, 1985a).

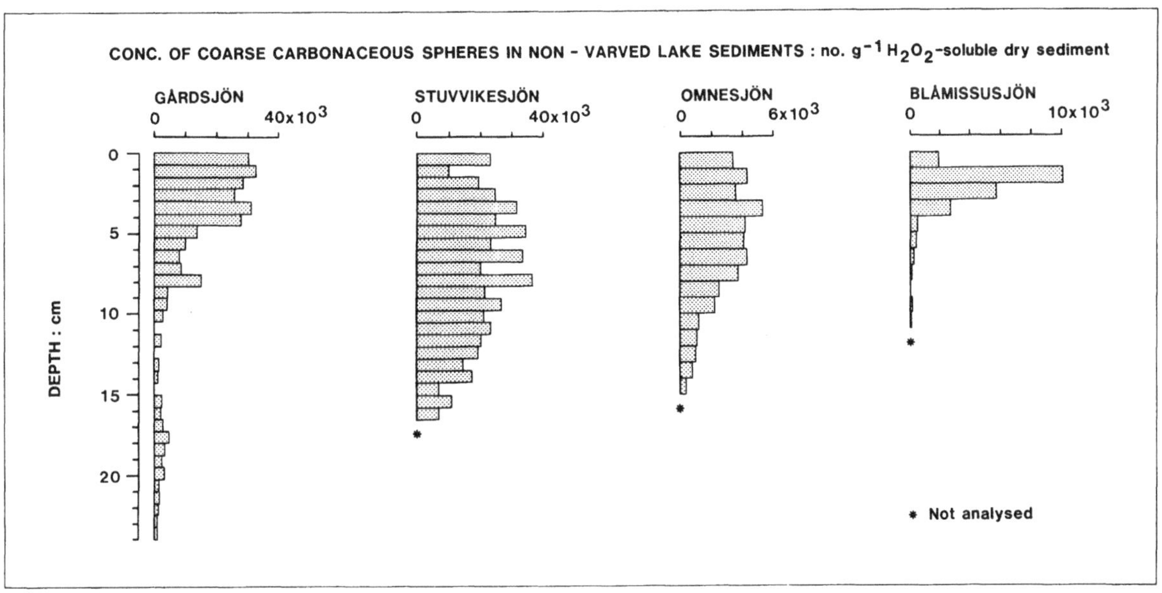

Fig. 5. Diagrams showing the concentration of coarse carbonaceous spheres in non-varved sediments from four lakes in Sweden. Varying sediment accumulation rates explain the differing extensions of the diagrams along the y-axis (depth). (Slightly modified from Renberg, I. & M. Wik, 1985a).

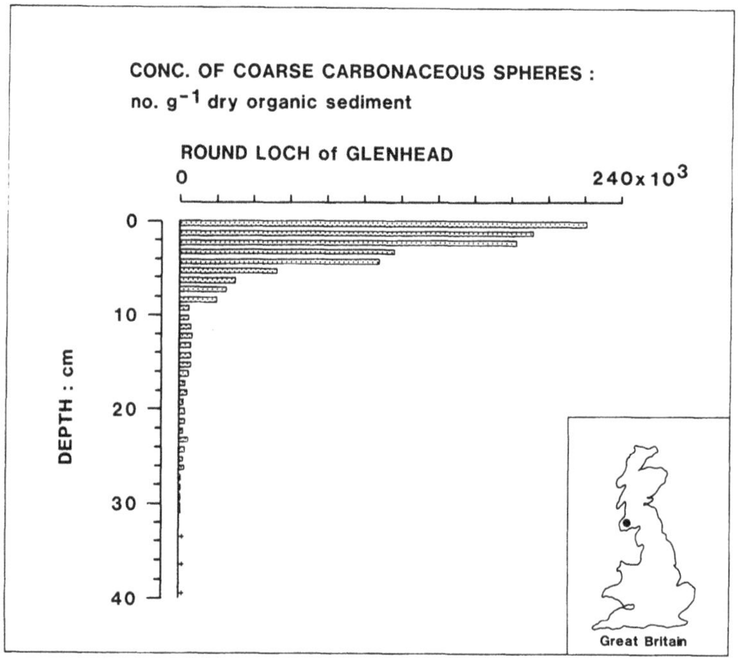

Fig. 6. Diagram showing the concentration of coarse carbonaceous spheres in a Scottish lake sediment. For lake location, see inserted map. (Redrawn from Darley, J. 1985).

The records of coarse carbonaceous spheres in non-varved cores from four lakes, situated in both northern and southern Sweden, are shown in Fig. 5. The main features of these diagrams are the same as those obtained from varved cores. However, the pattern is smoothed due to bioturbation and varying sediment accumulation rates in the lakes (Renberg & Wik, 1985a).

The investigations in Great Britain have also revealed high concentrations of coarse carbonaceous particles in lakes in Galloway, Scotland (Battarbee *et al.*, in prep.). Although some particles occur in the pre-1930 sediments the main increase in concentration is confined to the uppermost 10 cm (from the 1930's onwards) (Fig. 6), a trend that closely matches the increase in coal and oil combustion from power stations (Darley, 1985) and matches the trend of rapidly increasing acidity in the Galloway lochs from 1940–1970 (Flower *et al.*, in press).

All these results from USA, Sweden and Great Britain indicate that the carbonaceous particles in the sediments can be used as stratigraphic indicators for deposition of particulate pollutants from fossil fuel combustion.

Correlations between carbonaceous particles and other pollutants

Besides carbon, carbonaceous particles from fossil fuel combustion contain other elements and compounds. The concentrations of various trace elements in carbonaceous particles vary, as do their concentrations in different fossil fuels. In stack soot samples from oil fired boilers Al, Ca, Cr, Cu, Fe, K, Mg, Mn, Na, Ni, P, Pb, Si, S, Ti, V and Zn have been detected by energy dispersive x-ray analysis (Cheng *et al.*, 1976, Shen *et al.*, 1977, Bacci *et al.*, 1983). Ca, Fe, S, Si and V are usually major elements. The elements are either incorporated into the particle matrix or sorbed onto the surfaces of the particles. Coarse carbonaceous particles are very porous and their large surface area result in a very high sorbing capacity (Del Monte *et al.*, 1984, Griest & Tomkins, 1984, Schure *et al.*, 1985).

Goldberg *et al.* (1981) found that inputs of metals (Sn, Cr, Ni, Pb, Cu, Co, Cd, Zn and Fe) to a L. Michigan sediment core covaried with the fluxes of carbonaceous particles whose morphologies and surface characteristics were indicative of different combustion processes (oil, coal and wood burning). Of course, that does not necessarily mean that the carbonaceous particles and all metals are closely associated or even have the same source. Some of the metals may only indirectly be associated with energy-producing processes (Goldberg *et al.*, 1981).

During coal combustion fly ash particles with magnetic propterties are produced (Oldfield *et al.*, 1984) and have been reported in a variety of sedimentary environments (Doyle *et al.*, 1976, Goldberg *et al.*, 1981, Oldfield *et al.*, 1983). Correlation between the magnetic stratigraphy of ombrotrophic peats and the history of regional industrialization have been observed in investigations from different parts of the world (Oldfield *et al.*, 1984). Iron is a major trace element in coarse carbonaceous spheres and it is not unusual to find carbonaceous spheres with magnetic properties in sediments. This suggests that there might be a connection between carbonaceous particles and magnetic parameters of sediments.

Polycyclic aromatic hydrocarbons (PAH) are produced by combustion. Analyses by Griest *et al.* (1984) of stack ash from a coal fired power plant indicate that PAH are strongly sorbed by fly ash and carbonaceous particles appear to be responsible for this effect. Significant correlation between PAH and elemental carbon have also been observed in aerosol samples (Lindskog & Brorström, 1981). Several authors have documented increased concentrations of PAH in sediments from the industrial time period and attribute that to the combustion of fossil fuels (cf. Alderton, 1985).

In investigations of particulate aerosols it has been noticed that high concentrations of elemental carbon are often accompanied by high sulphate concentrations (Brosset & Nyberg, 1971). Ogren & Charlson (1984) have demonstrated a significant correlation between wet deposition fluxes of excess sulphate and elemental carbon (particles $< 5\ \mu m$), in rain water samples collected in Sweden. It is likely that carabonaceous particle records in lake sediments to some extent reflect the excess sulphate deposition and, thus, also the history of acid precipitation. Investigations in L. Gårdsjön, SW Sweden have shown a stratigraphic correlation between the coarse carbonaceous sphere and diatom records of the sediment (Renberg, 1984). The pH curve, inferred from the sub-fossil diatoms, started to decrease concomitant with a marked increase in the accumulation of coarse carbonaceous spheres.

Acknowledgements

This work was supported by the National Swedish Environmental Protection Board and the Swedish Natural Science Research Council.

References

Alderton, D. H. M., 1985. Sediments. In Historical monitoring, MARC Report 31. Monitoring and Assessment Research Centre, University of London: 1–95.

Bacci, P., M. Del Monte, A. Longhetto, A. Piano, F. Prodi, P. Redaelli, C. Sabbioni & A. Ventura, 1983. Characterization of the particulate emission by a large oil fuel fired power plant. J. Aerosol Sci. 14: 557–572.

Battarbee, R. W., 1984. Diatom analysis and the acidification of lakes. Phil. Trans. r. Soc., Lond. B 305: 451–477.

Battarbee, R. W., et al. (in prep.)

Brosset, C. & A. Nyberg, 1971. Investigations of soot and particle-borne sulfur in Sweden. In H. M. Englund & W. T. Beery (eds), Proceedings of the Second International Clean Air Congress. Academic Press, New York: 481–489.

Cheng, R. J., V. A. Mohnen, T. T. Shen, M. Current & J. B. Hudson, 1976. Characterization of particulates from power plants. J. Air Pollut. Cont. Ass. 26: 787–790.

Darley, J., 1985. Particulate soot in Galloway lake sediments: its application as an indicator of environmental change and as a technique for dating recent sediments. Department of Geography, University College London, B 301 Project: 1–46.

Del Monte, M., C. Sabbioni, A. Ventura & G. Zappia, 1984. Crystal growth from carbonaceous particles. Sci. Total Envir. 36: 247–254.

Doyle, L. J., T. L. Hopkins & P. R. Betzer, 1976. Black magnetic spherule fallout in the eastern Gulf of Mexico. Science 194:1157–1159.

Ericson, S.-O., M. Skaar, J. Tollin & T. Åbyhammar, 1983. Rening av rökgaser från koleldade anläggningar. Projekt Kol-Hälsa-Miljö, Teknisk Rapport 57: 1–371.

Flower, R. J., R. W. Battarbee & P. G. Appelby, in press. J. Ecol.

Goldberg, E. D., V. F. Hodge, J. J. Griffin & M. Koide, 1981. Impact of fossil fuel combustion on the sediments of Lake Michigan. Envir. Sci. Technol. 15: 466–471.

Goldstein, H. L. & C. W. Siegmund, 1976. Influence of heavy fuel oil composition and boiler combustion conditions on particulate emissions. Envir. Sci. Technol. 10: 1109–1114.

Griest, W. H. & B. A. Tomkins, 1984. Carbonaceous particles in coal combustion stack ash and their interaction with polycyclic aromatic hydrocarbons. Sci. Total Envir. 36: 209–214.

Griffin, J. J. & E. D. Goldberg, 1979. Morphologies and origin of elemental carbon in the environment. Science 206: 563–565.

Griffin, J. J. & E. D. Goldberg, 1981. Sphericity as a characteristic of solids from fossil fuel burning in a Lake Michigan sediment Geochim. Cosmochim. Acta 45: 763–769.

Griffin, J. J. & E. D. Goldberg, 1983. Impact of fossil fuel combustion on sediments of Lake Michigan: a reprise. Envir. Sci. Technol. 17: 244–245.

Haynes, B. S. & H. G. Wagner, 1981. Soot formation. Prog. Energy Combust. Sci. 7: 229–273.

Holdren, G. R. Jr., T. M. Brunelle, G. Matisoff & M. Wahlen,

1984. Timing the increase in atmospheric sulphur deposition in the Adirondack Mountains. Nature 311: 245–248.

Kahl, J. S. & S. A. Norton, 1983. Metal input and mobilisation in two acid-stressed lake watersheds in Maine. Land and Water Resources Center, University of Maine at Orono, Completion Report Project A-053-Me: 1–70.

Kothari, B. K. & M. Wahlen, 1984. Concentration and surface morphology of charcoal particles in sediments of Green Lake, N.Y: implications regarding the use of energy in the past. Northeastern Environ. Sci. 3: 24–29.

Lindskog, A. & E. Brorström, 1981. Determination of polycyclic aromatic hydrocarbons in airborne particulate matter in Göteborg. Nordic PAH Project, Report 11: 1–15.

McCrone, W. G. & J. G. Delly, 1973. The Particle Atlas, II. Ann Arbor Science, Ann Arbor, 267 pp.

Ogren, J. A. & R. J. Charlson, 1983. Elemental carbon in the atmosphere: cycle and lifetime. Tellus 35B: 241–254.

Ogren, J. A. & R. J. Charlson, 1984. Wet deposition of elemental carbon and sulfate in Sweden. Tellus 36B: 262–271.

Oldfield, F., C. Barnosky, E. B. Leopold & J. P. Smith, 1983. Mineral magnetic studies of lake sediments: A brief review. Hydrobiologia 103: 37–44.

Oldfield, F., J. Bloemendal, L. Barker, A. Hunt, J. M. Jones, M. H. D. Jones, R. Maxted, N. Richardson, J. Sahota & K. Tolonen, 1984. Magnetic measurements of atmospheric particulates and ombrotrophic peat: a review. In S. A. Norton (ed.), Proceedings of a Workshop on Paleolimnological Studies of the History and Effects of Acid Precipitation. University of Maine at Orono: 409–441.

Peterson, K. F., 1972. Stoftbildning vid oljeeldning. Dr Thesis, The Royal Institute of Technology: 1–229.

Renberg, I., 1984. Förbränning och försurning går hand i hand. Forskning och Framsteg 4/84: 36–40.

Renberg, I. & M. Wik, 1985a. Carbonaceous particles in lake sediments-pollutants from fossil fuel combustion. Ambio 14: 161–163.

Renberg, I. & M. Wik, 1985b. Soot particle counting in recent lake sediments; an indirect dating method. Ecol. Bull. 37: 53–57.

Rosen, H., T. Novakov & B. A. Bodhaine, 1981. Soot in the Artic. Atmos. Environ. 15: 1371–1374.

Schure, M. R., P. A. Soltys, D. F. S. Natusch & T. Mauney, 1985. Surface area and porosity of coal fly ash. Envir. Sci. Technol. 19: 82–86.

Shen, T., R. J. Cheng, V. A. Mohnen, M. Current & J. B. Hudson, 1977. Characterization of differences between oil-fired and coal-fired power plant emissions. International Clean Air Congress, 4. Tokyo: 386–391.

Smith, D. M., J. J. Griffin & E. D. Goldberg, 1975. Spectrometric method for the quantitative determination of elemental carbon. Analyt. Chem. 47: 233–238.

Tolonen, K., 1986. Charred particle analysis. In B. E. Berglund (ed.), Handbook of Holocene Palaeoecology and Palaeohydrology. John Wiley & Sons Ltd., Chichester: 485–496.

Accepted 18 March 1986.

Hydrobiologia 143: 395–400, (1986)

Oligochaete cocoon remains as evidence of past lake pollution

Carla Bonacina[1], Giuliano Bonomi[1] & Carlo Monti[2]
[1]*C. N. R.-Istituto Italiano di Idrobiologia, Largo Vittorio Tonolli 50–52, I–28048 Pallanza, Novara, Italy*
[2]*Università di Bologna, Istituto di Zoologia, Via S. Giacomo 9, I–40126 Bologna, Italy*

Keywords: paleolimnology, oligochaete cocoons, industrial pollution

Abstract

Cores taken in L. Orta, a large, deep ammonia and copper polluted lake in North Italy, revealed the presence of sub-fossil cocoons of a tubificid (*Spirosperma ferox*) and of a lumbriculid (*Stylodrilus lemani*?) in sediment layers corresponding to a period prior to the beginning of heavy industrial pollution (1926). The findings point out the possible use of sub-fossil cocoons of some aquatic Oligochaeta to indicate past lake conditions.

Introduction

Although the identification of aquatic Oligochaeta has undergone noticeable progress in recent years, nevertheless the possibility of identifying to species all their life-stages is still very difficult. Indeed the taxonomical keys make almost exclusive use of characters of the mature individuals and the species description very rarely mentions those of the cocoons. The latter are, on the other hand, of great importance for the study of field and laboratory populations (Adreani *et al.* 1981; Adreani & Bonomi, 1980). They are laid during the seasonal reproduction and their egg content depends on temperature, population density and food supply (Timm, 1974; Adreani *et al.*, 1984; Monti, in press).

Special clitellar cells secrete the cocoon wall (Grove & Cowley, 1927; Suzutani-Shiota, 1980; Morris, 1983), which appears to be made up of a kind of scleroprotein and to be resistant to various enzymes and acids (Jaana *et al.*, 1980). The cocoon plug is probably the product of a special type of clitellar glandular cell (Suzutani-Shiota, 1980); a different mechanism, however, is suggested by Kaster (1980), who believes the plug to be a mucous secretion of some pharingeal cells. Most cocoons seem to undergo a rapid decomposition in the sediments after the hatching of the new born worms. Indeed empty cocoons of the commoner species are only rarely found in grab samples during benthos investigations. This, together with the relatively low number of cocoons in the small core sections that are normally used in paleolimnological investigations, has led to the fact that aquatic oligochaete cocoons are not normally recognized as paleorecords (see, e.g. Frey, 1964).

However, the whole subject is still to be investigated as it is clear to aquatic oligochaete students that at least some species, e.g. species of the genus *Spirosperma* (syn. *Peloscolex*, see Holmquist, 1978, 1979) and related genera (e.g. some species of the brackish-marine genus *Tubificoides*) lay hard-wall cocoons that resist decomposition over a long period of time. To our knowledge, however, their paleolimnological records are still missing.

Outline history of Lake Orta pollution

Lake Orta is a deep, warm-monomictic lake in the Italian southalpine district; its main pollution source is the effluent from a rayon factory (Fig. 1), which was established at the southern end of the

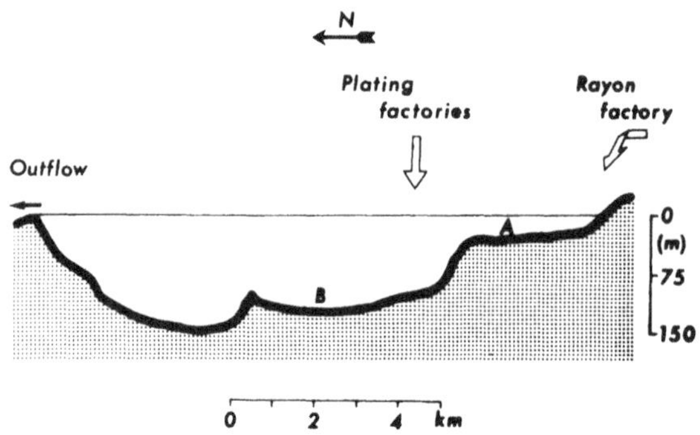

Fig. 1. Longitudinal profile of Lake Orta; A and B are the coring stations.

lake in 1926. Under the effect of enormous loading of copper and ammonium salts, the bulk of phyto- and zooplankton and profundal benthos disappeared in a few years (Monti, 1930). The huge ammonium loading and the in-lake oxidation of ammonia nitrogen caused a gradual nitrate and ammonia accumulation (up to mean values of ca 5.5 mg $N-NO_3$ l^{-1}; Fig. 2), a consequently striking lowering of pH (down to 3.8), and an increased oxygen hypolimnetic deficit (Bonacina & Bonomi, 1984).

Since the fifties an additional pollution source has been represented by many small bathroom accessory factories, which discharged heavy metals, anionic detergents and cyanides into the lake. In recent years the loading of these last pollutants has decreased, due to the adoption of many treatment plants; however the key moment in the recovery of Lake Orta was the construction of a very efficient plant at the rayon factory during 1980–1981, which

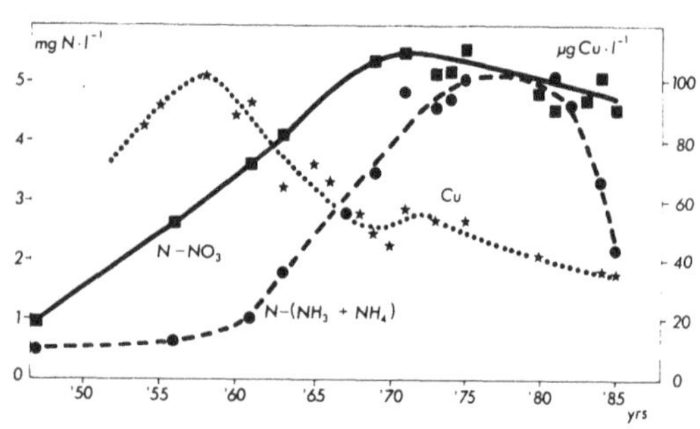

Fig. 2. Long-term trend of total copper, ammonia and nitrate nitrogen concentrations in Lake Orta.

lowered the ammonia concentration in the lake to its present levels of about 2 mg $N-NH_4$ l^{-1}.

Materials and methods

During a population dynamics investigation on *Tubifex tubifex* at station A (Fig. 1), the sieved sediment obtained by means of a tall automatic Ekman grab, immediately revealed the presence of ovoidal, whitish objects, that under the dissection microscope turned out to be empty cocoons, most probably produced by the tubificid *Spirosperma ferox*. The same cocoons were present in Jenkin cores collected both at station A and B (Fig. 1), indicating that the cocoons are presumably present all over the lake floor. A more detailed core analysis was therefore planned. Six replicate core samples were taken in the central part of the lake (station B; Fig. 1) by means of a weighted Jenkin mud sampler and an additional longer core was obtained by means of a light weight piston corer (Murray, 1976).

The six replicate Jenkin cores were cut into 2 cm sections immediately after coring; the piston core was deep-frozen in the lab and subsequently cut into 2 cm sections to a depth of 30 cm and 5 cm sections from 30 to 60 cm. Five out of the six Jenkin cores were used for cocoon analysis as well as the longer core; the sixth Jenkin core served for the count of Cladocera remains. All the biological material was counted at the microscope after sieving through the same sieve (200 μm) used for our benthos studies. The copper content of sediments at station B was recalculated from previous data by Provini & Gaggino (1985).

Results and discussion

Three types of oligochaete cocoons were found in the collected cores: living *Tubifex tubifex* cocoons, remains of cocoons of the tubificid *Spirosperma ferox* and of a lumbriculid (*Stylodrilus lemani*?) (Fig. 3).

The living cocoons of *Tubifex tubifex* are obviously those laid by the new population of this tubificid that, for some years has settled in the profundal of the lake (Bonacina & Bonomi, 1984 and in press; Bonacina *et al.*, in press a and b). They

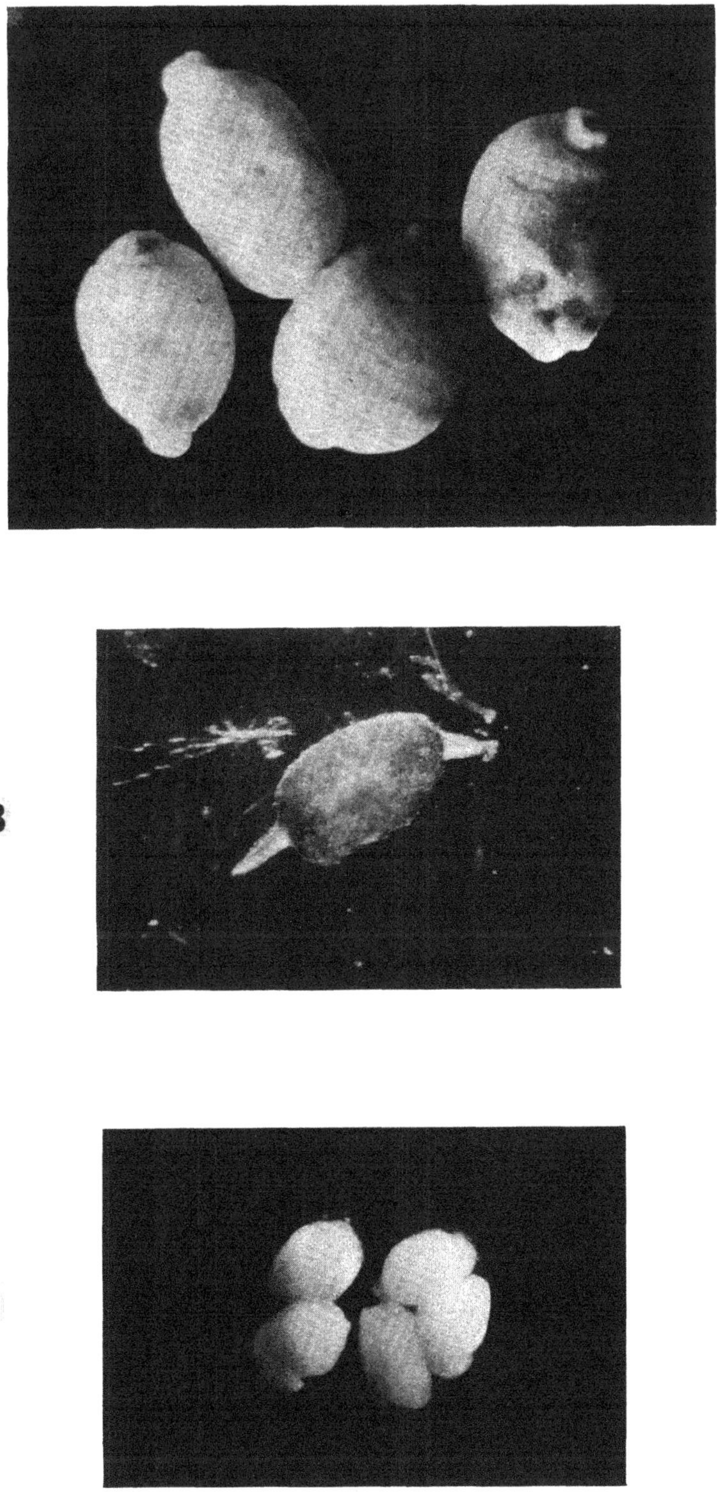

A

B

C

Fig. 3. Oligochaete cocoons from Lake Orta sediments. A: *Spirosperma ferox*; B: *Stylodrilus lemani* (?); C: living cocoons of *Tubifex tubifex*.

were found down to 12 cm at station B and down to 17 cm at station A, where their abundance, as well as that of the free living worms, was greater than in the central basin.

The sub-fossil remains of cocoons of the tubificid *Spirosperma ferox* (Fig. 3A) were indentified by comparison with the cocoons produced by laboratory cultured cohorts of *Spirosperma ferox*, which were used during a previous investigation on this species (Adreani *et al.*, 1984). They resemble the lab produced cocoons in shape, dimensions, and the typical considerable thickness of the walls.

A third type of cocoons was that which is shown in Fig. 3B; we were able to identify these cocoons as produced by a representative of the family Lumbriculidae on the basis of our laboratory cultures of *Stylodrilus lemani* (syn. *Bythonomus lemani*), but up to now we have not attempted any generic or specific identification. The estimation of the abundance of cocoon remains was attempted only for *Spirosperma ferox*, as those of *Tubifex tubifex* are of no concern in this paper and the lumbriculid cocoons were too rare. However, we have indications that the latter may be more abundant in the south basin.

As shown in Fig. 4 the core section at 12–14 cm

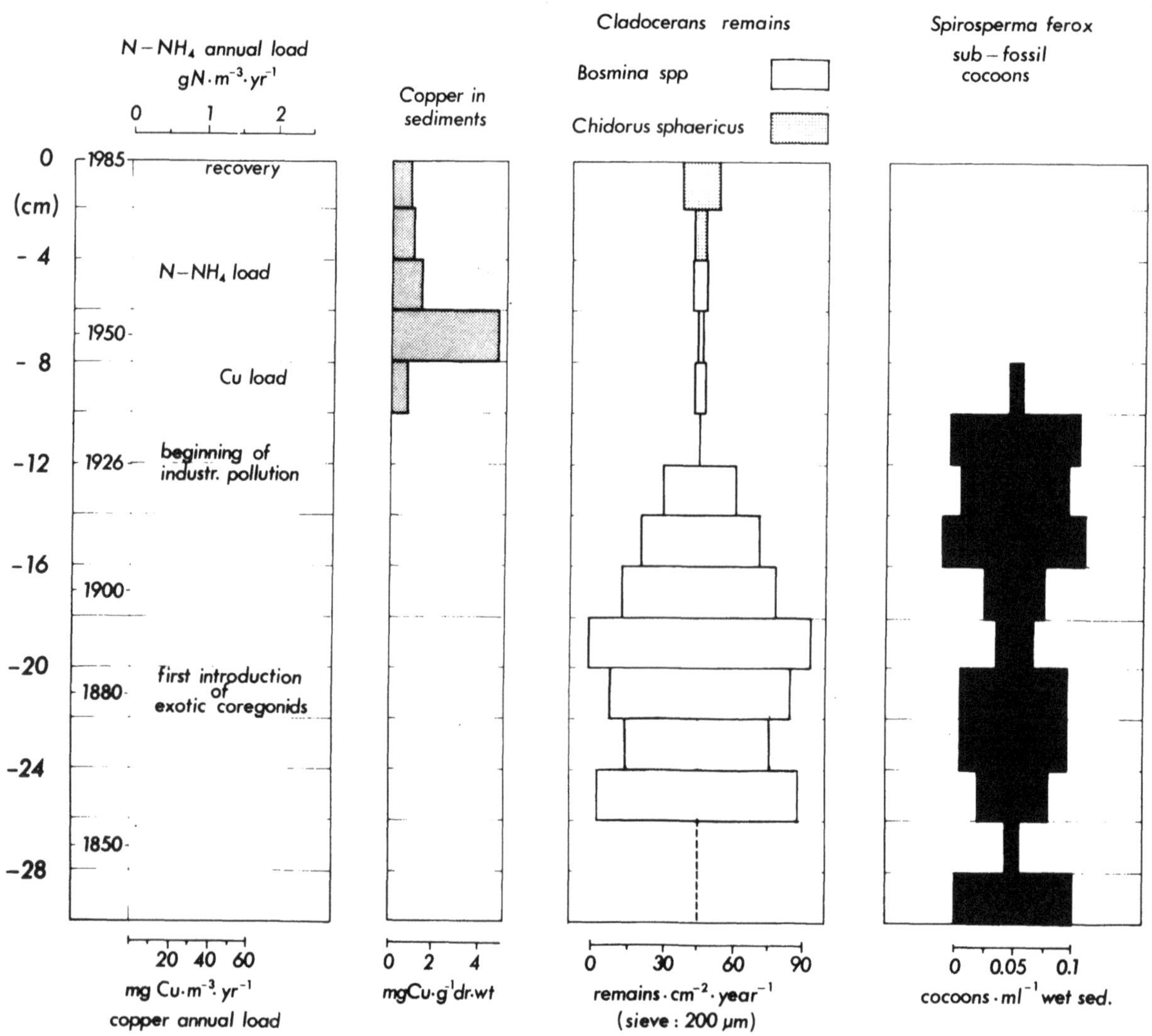

Fig. 4. Lake Orta: ammonia and copper annual load; copper content of sediments at station B; estimated deposition rate of cladoceran remains, and stratigraphy of *Spirosperma ferox* sub-fossil cocoons.

below the sediment surface is the last one with abundant cladoceran remains; we consider therefore that it corresponds to a period of time during which the industrial pollution began; this implies a net accumulation rate af about 2 mm year^{-1}, which is in good agreement with previous estimates (Corbella *et al.*, 1985; Adams *et al.*, 1978) but not with the ^{137}Cs estimations (3.2 mm year^{-1}) for the same lake station by Provini & Gaggino (1985). With an apparent delay of some years the *S. ferox* cocoons remains disappear from the sediment, testifying to the fact that its population was drastically affected. The cocoon abundance is rather low (max value: 0.12 cocoons ml^{-1} wet sediment) and, as attested to by the inspection of the longer core, which we took at the same station B, they are again missing in the deeper section, between −35 and −70 cm. The sediments also contain, in the upper four centimeters (Fig. 4), remains of the chydorid *Chydorus sphaericus*, which in recent years has developed a truly pelagic population (Bonacina & Bonomi, 1984).

Our results not only confirm the drastic changes in the biota of Lake Orta as a consequence of the beginning of the industrial pollution, but also supply interesting paleolimnological records of aquatic Oligochaeta. The discovered cocoons reveal the coexistence in the profundal zone of *S. ferox* and *S. lemani* (?) which may be considered indicative of oligotrophic or oligo-mesotrophic conditions (Milbrink, 1973, 1980, 1983; Adreani *et al.*, 1984). This is in accordance with the results of previous investigations, which concluded that Lake Orta was oligotrophic (Adams *et al.*, 1978). The 'association' *S. ferox*, *S. lemani* and *Psammoryctides barbatus* was described as dominant in the profundal of Lake Maggiore in 1953–1954 (Corbella *et al.*, 1956) when the lake was still poorly productive but, as the lake trophic status evolved to mesotrophic (Bonomi *et al.*, 1970, 1979) there was a shift to a new association where *S. lemani* and *S. ferox* were occasionally found and the dominant species in the profundal community was *Tubifex tubifex* and *Potamothrix hammoniensis* (Bonomi, 1967; Bonomi *et al.*, 1979).

The presence of *S. ferox* cocoons down to about −35 cm, i.e. approximately 150 years B.P., could be used, if we admit that the population was also present in the profundal during a period prior to 1.5 centuries ago, as an indication of the decomposition rate of such cocoons. However, although these findings need further confirmation through additional corings, we think that it is likely that this species was not present in preceeding periods. Indeed, the cocoons do not show signs of progressive deterioration with depth; furthermore Niessen (pers. commun.) found cocoon remains, that we identified as those of *S. ferox*, in cores taken in Lake Lugano, a glacial lake in the sudalpine lake district. He detected these cocoons down to about 7 m from the present sediments, i.e. as far back as about 10000 years B.P., so demonstrating that they can resist decomposition for an extremely long time.

We are not willing to speculate further on the possible reasons why this population may have become established about 150 years ago in L. Orta. Indeed, in spite of the recent progress in this subject, the setting of a lake typology based on Oligochaeta and the understanding of the factors that govern their succession and distribution are still uncertain. Students of Oligochaeta should continue to culture a larger number of species and, together with taxonomists, produce detailed descriptions of cocoon morphology, so allowing the paleolimnologists to make use of these interesting biological records

Acknowledgements

We thank Dr Frank Niessen, Geologische Institut ETH Zürich, for allowing us to mention his unpublished results and the Bemberg Company for the permission to publish data on the copper and ammonium loadings; we are deeply indebted to Miss Sandra Spence who revised the English text.

References

Adams, M. S., P. Guilizzoni & S. Adams, 1978. Sedimentary pigments and recent primary productivity in northern Italian lakes. Mem. Ist. Ital. Idrobiol. 36: 267–285.

Adreani, L., C. Bonacina & G. Bonomi, 1981. Production and population dynamics in profundal lacustrine Oligochaeta. Verh. int. Ver. Limnol. 21: 967–974.

Adreani, L., C. Bonacina, G. Bonomi & C. Monti, 1984. Cohort cultures of *Psammoryctides barbatus* (Grube) and *Spirosperma ferox* Eisen: a tool for a better understanding of demographic strategies in tubificids. Hydrobiologia 115: 113–119.

Adreani, L. & G. Bonomi, 1980. Elementi per una dinamica di popolazione dei tubificidi. Atti 3 Congresso Associazione Italiana Oceanografia e Limnologia: 355–366.

Bonacina, C., 1970. Il Lago d'Orta: ulteriore evoluzione della situazione chimica e della struttura della biocenosi planctonica. Mem. Ist. Ital. Idrobiol. 26: 141–204.

Bonacina, C. & G. Bonomi, 1974. La conoscenza dell'origine ed evoluzione dell'inquinamento del Lago d'Orta come base per la formulazione di strumenti previsionali indispensabili ad una politica di interventi. Atti 1 Convengo Internazionale sull'Ambiente e sulla Crisi dell'Energia, Torino, 8–12 Maggio 1974: vol 4, 25 pp + 13 figs.

Bonacina, C & G. Bonomi, 1984. I grandiosi effetti ambientali determinati dalle prime fasi del disinquinamento del Lago d'Orta. Documenta Ist. Ital. Idrobiol 2: 1–24.

Bonacina, C. & G. Bonomi, in press. Advances in the recovery of Lake Orta after three years of functioning of the new treatment plants. Proc. Int. Congr. 'Lakes Pollution and Recovery' Rome, 15th–18th April, 1985.

Bonacina, C., G. Bonomi & C. Monti, in press a. Population dynamics of *Tubifex tubifex*, first settler in the profundal macrobenthos of a copper and ammonia polluted recovering lake (Lake Orta, N. Italy). Proc. Third International Symposium on Aquatic Oligochaeta, Hamburg, 31. 9–5.10 1985.

Bonacina, C., G. Bonomi & C. Monti, in press b. Progress in cohort cultures of tubificids. Proc. Third International Symposium on Aquatic Oligochaeta, Hamburg, 31.9–5.10 1985.

Bonacina, C., G. Bonomi & D. Ruggiu, 1973. Reduction of the industrial pollution of Lake Orta (N. Italy): an attempt to evaluate its consequences. Mem. Ist. Ital. Idrobiol. 30: 149–168.

Bonomi, G., 1967. L'evoluzione recente del Lago Maggiore rivelata dalle cospicue modificazione del macrobenton profondo. Mem. Ist. Ital. Idrobiol. 21: 196–212.

Bonomi, G., A. Calderoni and R. Mosello, 1979. Some remarks on the recent evolution of the deep Italian subalpine lakes. Symp. Biol. Hung. 19: 87–111.

Bonomi, G., M. Gerletti, E. Indri & L. Tonolli, 1970. Report on Lake Maggiore. In: O.C.S.E. Symposium on Large Lakes and Inpoundments. Uppsala 13–16 May 1968: 299–341.

Corbella, C., V. & L. Tonolli, 1958. I sedimenti del Lago d'Orta testimoni di una disastrosa polluzione cupro-ammoniacale. Mem. Ist. Ital. Idrobiol. 10: 9–50.

Frey, D. G., 1964. Remains of animals in Quaternary lake and bog sediments and their interpretation. Arch. Hydrobiol. Beih. Ergebn. Limnol. 2: 1–114.

Grove, A. J. & L. F. Cowley, 1927. The relation of the glandular elements of the clitellum of the brandling worm (Eisenia foetida, Sav.) to the secretion of the cocoon. Q. J. Micro. Sci. 71: 31–46.

Holmquist, C., 1978. Revision of the genus Peloscolex (Oligochaeta, Tubificidae). 1. Morphological and anatomical scrutiny; with discussion on the generic level. Zool. Scr. 7: 187–208.

Holmquist, C., 1979. Revision of the genus Peloscolex (Oligochaeta, Tubificidae). 2. Scrutiny of the species. Zool. Scr. 8: 37–60.

Jaana, H., T. Shimizu, C. Suzutani-Shiota & R. Yasumura, 1980. Fine structure and chemical properties of the cocoon wall of the Tubifex. Zool. Magazine 89:130–137.

Kaster, J., 1978. The life history of *Tubifex tubifex* Müller (Oligochaeta, Tubificidae). Ph.D. Thesis, University of Colorado, IXX + 176 pp.

Milbrink, G., 1973. On the use of indicator communities of Tubificidae and some Lumbriculidae in the assessment of water pollution in swedish lakes. Zoon 1: 125–139.

Milbrink, G., 1980. Oligochaeta communities in pollution biology: the european situation with special reference to lakes in Scandinavia. In R. Brinkhurst and D. Cook (eds), Aquatic Oligochaete Biology: 433–456.

Milbrink, G., 1983. An improved environmental index based on the relative abundance of oligochaete species. Hydrobiologia 102: 89–97.

Monti, C., in press. Population regulation of *Psammoryctides barbatus* Grube (Oligochaeta: Tubificidae). Mem. Ist. Ital. Idrobiol. 42 (submitted).

Monti, R., 1930. La graduale estinzione della vita nel limnobio del Lago d'Orta. Rend. Ist. lomb. Sci. Lett. 63: 3–22.

Morris, G. M., 1983. The cocoon-producing cells of *Eisenia foetida* (Annelida, Oligochaeta): a hystochemical and ultrastructural study. J. Morphol. 167: 41–50.

Murray, D. A., 1976. A light weight corer for sampling subaqueous deposits. Limnol. Oceanogr. 21: 341–344.

Provini, A. & G. F. Gaggino, 1985. Sediment as a record of copper pollution in Lake Orta. Verh. int. Ver. Limnol. 22(4): 2390–2393.

Suzutani-Shiota, C., 1980. Ultrastructural study on cocoon formation in the freshwater Oligochaeta Tubifex hattai. J. Morphol. 164: 25–38.

Timm, T., 1974. On the life cycle of the aquatic Oligochaeta in aquaria (in russian). Inst. Zoology and Botany of the Academy of Sciences of the SSR of Estonia. Hydrobiological Investigations 6: 97–118.

Accepted 18 March 1986.

Hydrobiologia 143: 401–405, (1986)

Industrial tailings in Lake Traunsee (Salzkammergut, Austria)

Jens Müller[1], Jürgen Schneider[2] & Michael Sturm[3]
[1]*Lehrstuhl für Geologie, Technische Universität München, Lichtenbergstraße 4, D-8046 Garching, FRG*
[2]*Institut für Geologie und Dynamik der Lithosphäre, Goldschmidtstraße 3, D-3400 Göttingen, FRG*
[3]*EAWAG-ETH Zürich, Überlandstraße 133, CH-8600 Dübendorf, Switzerland*

Keywords: lake sediments, turbidites, tailings, geochemistry, environmental hazards

Abstract

Tailings from salt work and a soda work have been pumped into the southern part of Traunsee (bay of Ebensee) for more than 50 years. Solid wastes have accumulated to form a pile >40 m high with a total volume of $>3\times10^6$ m^3. The sandy silty mud consists of calcite and other phases of $CaCO_3$, brucite, CaO resp. $Ca(OH)_2$ and gypsum. The chemical environment of the mud is characterized by extreme alkaline pore water (pH>10 up to 12.5; total amount of pore water: 3×10^6 t) and highly negative Eh-values. These values result from the pumped tailings and from postdepositional chemical processes such as dissolution resp. reaction of CaO, brucite and gypsum and precipitation of $CaCO_3$.

The high water content of the mud and the steep slopes (>15°) are responsible for frequent debris flows near the waste inlet, and occasional turbidity currents flow down to the distal basin floor. One particular current may transport up to 4×10^5 m^3 of alkaline mud as far as 6 km from the inlet into the profundal basin plain, intercalating with natural lake sediments. Sedimentation rates in the northern part of Traunsee within the last decades are isotopically measured as 0.4–0.5 cm/a. In the central basin the sedimentation rates within the last 50 years are in the range of 2–3 cm/a. Up to 75% of the profundal sediments consist of industrial tailings and turbidites. Thus, within a substantial portion of the profundal zone (about 15%) the benthic fauna and the chemical environment of the sediments are influenced by the alkaline tailings.

Introduction

Salt mining has a long tradition in the Salzkammergut area reaching back into prehistoric time. Mining and refinement techniques have changed in the course of times leading to the foundation of salt works in the year 1607 in Ebensee at the southernmost end of Traunsee. Availability of primary resources such as salt and carbonate initiated the establishment of soda works, which started production in 1885.

Solid and dissolved wastes caused by the production of salt and of soda were first dumped into the river Traun or deposited on the southern shores of Traunsee. The disposal system was changed, however, when reduced fish catches and the build-up of a chemocline in Traunsee were observed. The present system, which has been in use for more than 50 years, consists of a sewer pipe where the combined wastes of the soda and the salt works are pumped into the lake after being diluted with lake water (Fig. 1). The dumping site is situated in the bay of Ebensee adjacent to the delta of the river Traun.

Presently solid wastes when leaving the works consist primarily of gypsum ($CaSO_4 \times 2 H_2O$), brucite ($Mg(OH)_2$), calcite ($CaCO_3$) and of CaO resp. $Ca(OH)_2$ (Fig. 1). Disposal rates are regulated by government authorities. They allow a total of 140000 t of solid waste per year. This amount greatly exceeds the annual suspended and bedload from the river Traun of ca. 80000 t (Müller, Sossau & Zeh, 1983). The Traun is the major natural source of allochthonous sediment for the Traunsee. Actual output of wastes ranged generally between 60000

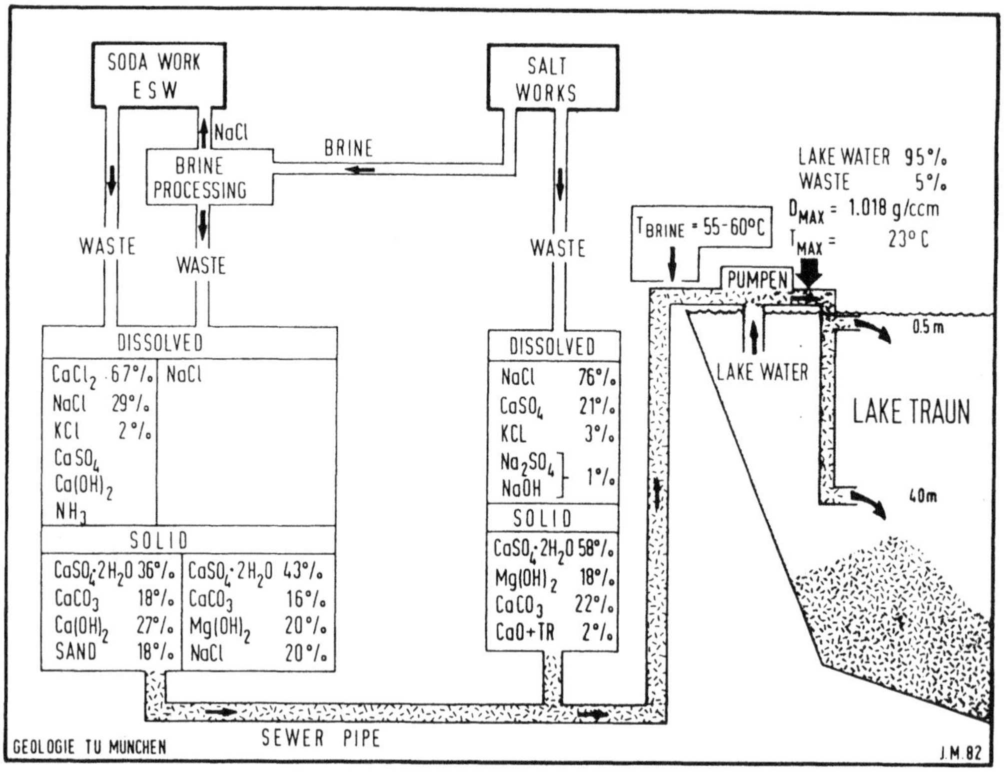

Fig. 1. Schematic diagram showing the composition and the mode of dumping of the dissolved and solid tailings deriving from soda works and salt works in Ebensee. Data from the two factories (Sept. 1983). TR – clay residue. (Modified after Müller & Schneider, 1984).

and 70000 t/a according to statements by the respective industries, but reached more than 90000 t/a during periods of NH_3 production in the past. Disposal of dissolved wastes and its effects on the hydrography and chemistry of Traunsee are not treated within this report. They were discussed in detail by Pechlaner & Sossau (1982).

Accumulation and distribution of tailings in Lake Traun

When entering the lake after dilution with lake water some of the components of the tailings dissolve on their way to the lake bottom. Thus, mineralogical and chemical analyses of core material taken in the bay of Ebensee showed pronounced differences in composition when compared with the material at source.

Tailings as accumulating on the lake floor consist primarily of calcite (60–80%) and other phases of $CaCO_3$ (aragonite, vaterite) while gypsum and brucite are minor components which can only be detected within the uppermost sediments (<0.5 m). Reaction resp. dissolution continues under burial as is indicated by interstitial water analyses and increased pH, ranging from 10 to 12.5. Alkaline pore waters trapped within the tailings are due to the dissolution of $Mg(OH)_2$ and of CaO resp. $Ca(OH)_2$ which constitute the crucial components of the industrial waste.

The abundance of calcite within the tailings is partly the result of dissolution of other compounds. However, substantial amounts of calcite are precipitated in the lake water below the inlets and on the lake floor. Precipitation is favoured by the alkaline pH-values leading to shifts in the carbonate system and to the dominance of CO_3^- as the sole carbonate ion species. Ca^{++} is present in the lake water; additional supplies come from the dissolution of gypsum and of $Ca(OH)_2$. Sampling of suspended solids in the bay of Ebensee thus showed the presence of abundant 'autochthonous' calcite crystals. Calculations based on comparisons of actual output of calcite with the amount present in the bay of Ebensee indicate that about half of the calcite is of autochthonous origin and thus due to the alkalinity of the tailings.

Within the past 50 years 3.1×10^6 m^3 of alkaline tailings have accumulated in the bay of Ebensee forming a >40 m high pile (Fig. 2). Water content averages 60% which gives a total of $>3 \times 10^6$ t of

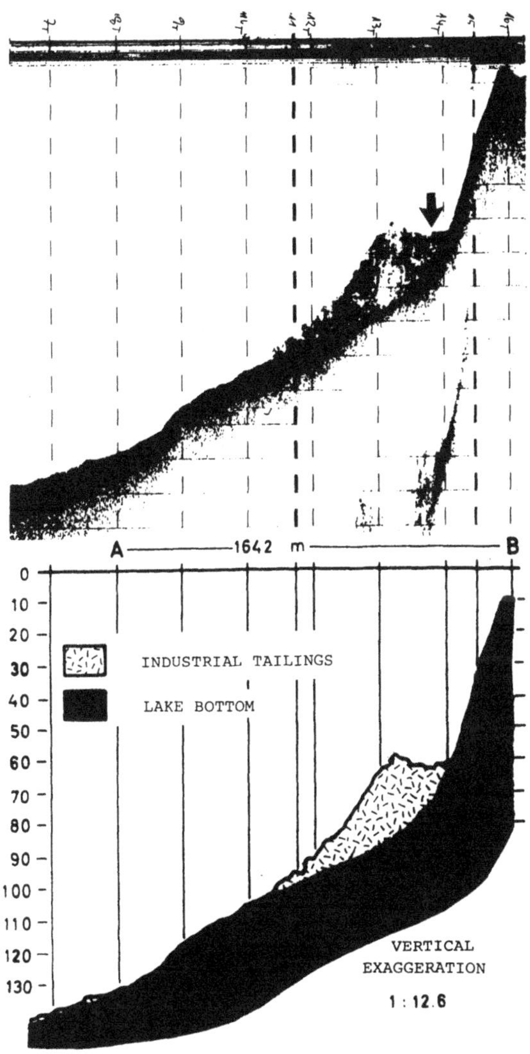

Fig. 2. Accumulation of tailings in the bay of Ebensee. Upper section: original 3.5 KHz record, profile runs NE (point A) to SW (point B). Arrow points to fault caused by slumping. Lower section: schematic graph illustrating natural lake bottom topography and superimposed industrial tailings. Depth scale in meters. (Modified after Müller & Schneider, 1984).

alkaline pore water trapped within the tailings.

Surface morphology, internal structures and distribution patterns document the instability of the deposits. The slope angle of the tailings ranged from >15° at the top to >5° at the foot of the accumulation. This is sufficient for the initiation of creeping, slumping and turbidity flows downslope into distal portions of the lake basin.

Triggering of these mass movements can be accomplished in several ways: oversteepening of slopes, earthquakes, sudden release of pore waters due to overburden or compaction, erosion of tailings by turbidity currents from the Traun delta, undercurrents during floods of the river Traun or slumping in the area of the adjacent delta foreset can cause the release of turbidity currents (Fig. 3).

Cores from the central basin of Traunsee showed numerous turbidites (up to 25 cm thickness) consisting of tailings material intercalated with normal sediments and turbidites derived from the river Traun.

These turbidites have been deposited up to 6 km away from the inlet in the bay of Ebensee and

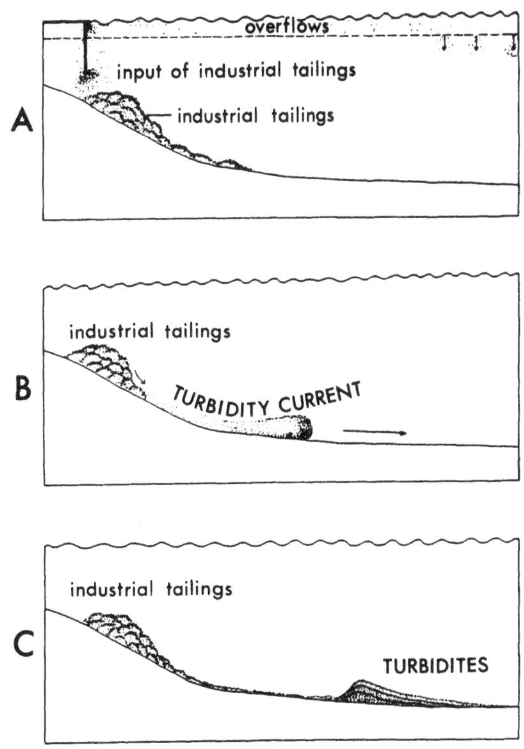

Fig. 3. Generalized diagram of turbidity current transport of tailings. (Modified after Müller & Schneider, 1984).

spread over large portions of the central basin. Presently 15% of the lake floor are influenced by the tailings (Fig. 4). In one case, a single turbidite transported 4×10^5 m^3 into the central basin. Locally, these turbidites contribute 75% of sediments deposited in this area within the past 50 years, increasing the sedimentation rate by a factor of >5.

Hazards

Sediment transport by turbidity currents is a natural phenomenon. However, these specific tailings turbidites represent a special problem due to their alkaline pore water. pH-measurements in cores containing tailings turbidites showed values ranging from 8 to 10, while those of normal sediments are usually <8 (Fig. 5).

Several surveys and investigations (i.e. Hamann, 1954; Löffler, 1983) have shown that the benthic fauna is influenced by the tailings. Ostracods for example are missing in those areas where tailings have accumulated. The consequences for the ecosystem have yet to be studied.

Total slumping of the tailings would cover the profundal area in a thickness of >1 m. Even when covered later on by natural sediments, it can be expected that benthic life will be disturbed for substantial time periods by upward diffusion of alkaline pore waters.

The impact of tailings on the natural sedimentation processes and on the natural sedimentary environment remains to be investigated in the future. The observed increases in electrolyte content of the lake water (Pechlaner & Sossau, 1982) can lead to enhanced flocculation of allochthonous sediment particles causing a rapid settling to the lake floor. Rapid rates of accumulation would therefore influence rates and modes of organic decay especially in the southern part of Traunsee. Another aspect concerns the dissolution of silicates (quartz, opal, feldspars and clay minerals) by alkaline pore waters leading to a mobilization of otherwise fixed or adsorbed heavy metals. The largest part of the tailings is accumulating on the unstable foreset slopes of the Traun delta with slope angles ranging from 20°–30°. Possibilities have thus to be considered where slumping of tailings initiates mass slides in the delta area. Under these circumstances one has to expect damage in the shore areas similar to those

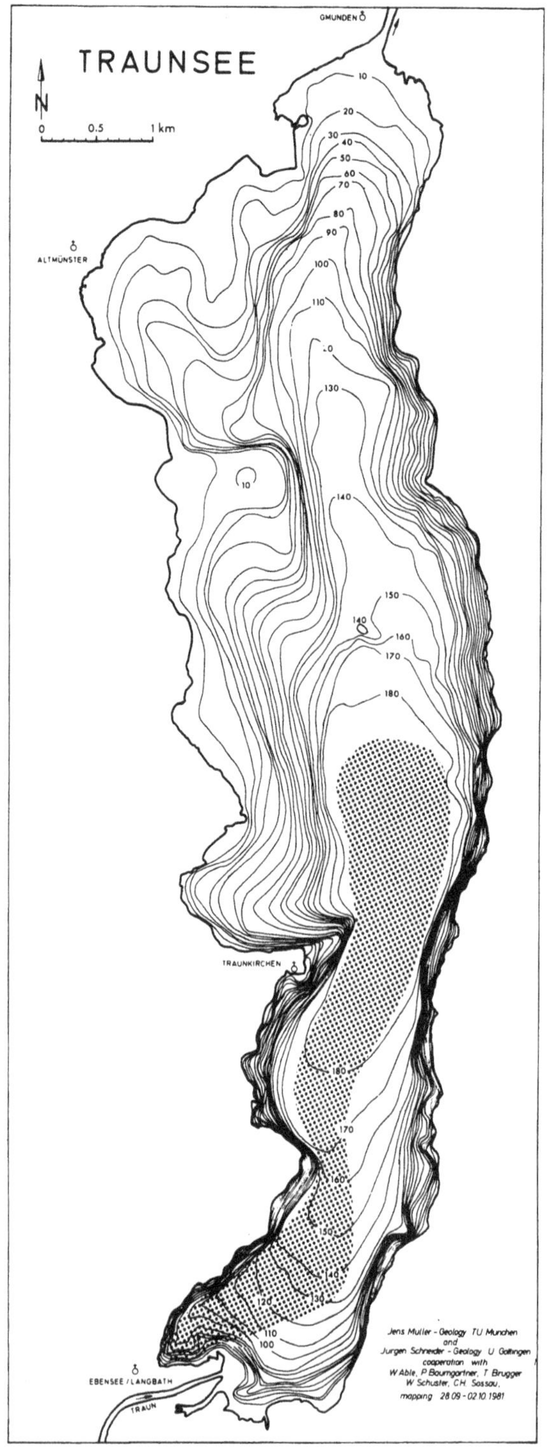

Fig. 4. Bathymetric map of Traunsee showing the distribution of tailings (shaded area) on the lake bottom. (Modified after Müller & Schneider, 1984).

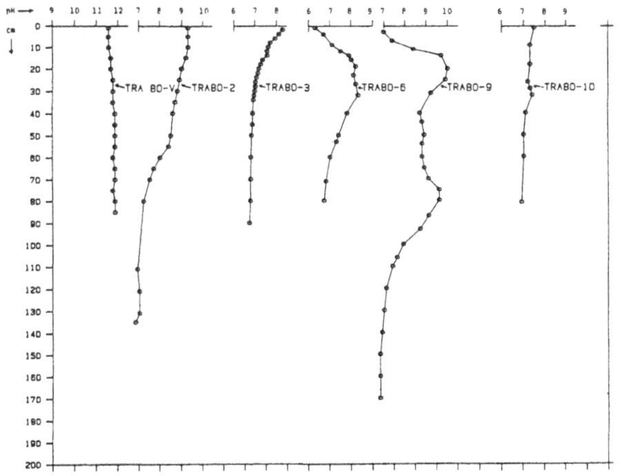

Fig. 5. pH-profiles versus core depth of different cores from Traunsee. pH-values >8 document industrial tailings turbidities (after Sturm & Müller, 1984).

described by Kelts (1978) in Swiss lakes or by Gennesseaux, Mauffret & Pautot (1980) on the Mediterranean coast. In the latter case artificial sediments accumulated on delta deposits slumped into the sea causing a turbidity flow with speeds exceeding 40 km/h destroying submarine cables at a distance of 100 km from its starting point. At the same time waves in excess of several meters flooded the coast.

Acknowledgements

The investigations were supported by a grant of the 'Amt der Oberösterreichischen Landesregierung' (AZ: Bau 2-12106/18-1980/Wei/St.).

References

Gennesseaux, M., A. Mauffret & G. Pautot, 1980. Les glissements sous-marins de la pente continentale niçoise et la rupture de cables en mer Ligure (Mediterranée occidentale). C. r. Acad. Sci. Paris 290: 959–962.

Hamann, H., 1954. Beiträge zur Biologie und Ermittlungen zu den Fischereiverhältnissen des Traunsees (1952–53). Biol. Stat. Fischereiwesen, Linz, 246 pp.

Kelts, K., 1978. Geological and sedimentary evolution of Lakes Zürich and Zug, Switzerland. Dissertation ETH Zürich, 242 pp.

Löffler, H., 1983. Changes of the benthic fauna of the profundal zone of Traunsee (Austria) due to salt mining activities. Hydrobiologia 103: 135–139.

Müller, J. & J. Schneider, 1984. Die Industrieschlammablagerungen in der Bucht von Ebensee und im Profundal des Traunsees (Oberösterreich). Limnologische Untersuchung Traunsee-Traun Amt, O. Ö. Landesregierung Bericht Nr. 12a: 100 pp.

Müller, J., C. Sossau & U. Zeh, 1983. Die Schwebstoffe der Traun und des Traunsees. Limnologische Untersuchung Traunsee-Traun, Amt O. Ö. Landesregierung Bericht Nr. 11: 50 pp.

Pechlaner, R. & C. Sossau, 1982. Die Ergebnisse der fünfjährigen Studie 'Limnologische Untersuchung Traunsee-Traun', Amt O. Ö. Landesregierung Bericht Nr. 13: 234 pp.

Sturm, M. & J. Müller, 1984. Die Untersuchung langer Sedimentprofile und die Verbreitung von Turbiditen im Traunsee. Limnologische Untersuchung Traunsee-Traun, Amt O. Ö. Landesregierung Bericht Nr. 12: 97–131.

Accepted 18 March 1986.

Hydrobiologia 143: 407–410, (1986)

Palynological investigations in sediments of ancient lake Duvensee, Schleswig-Holstein (North Germany)

Fritz-Rudolf Averdieck
Institut für Ur- und Frühgeschichte, Universität Kiel, Olshausenstraße 40, D-2300 Kiel 1, FRG

Keywords: palaeolimnology, pollen analyses, lake development, prehistoric lakeside settlements

Abstract

The Duvensee originated before the Alleroed in the Late Glacial, and had its largest areal coverage during the Preboreal. After the lake retreat, which began in early Boreal times, the marginal shore areas and nearshore islands were repeatedly inhabited by man in the early Mesolithicum. Archaeological excavations of human settlements and pollen analyses of sediment cores show evidence of lake level fluctuations in the ensuing period. The results disclose that shallow water sediments such as lake marls, algal muds and coarse detrital gyttjas predominate in the sequence. In keeping with the shallow water conditions, strong lateral facies changes were observed in the cores. The early Holocene deposits have almost the same thickness irrespective of their position in the shallow or deep parts of the lake basin. The subaerial exposure of the nearshore and island areas sometimes resulted in fern and reed peats. The last remnants of the lake, which was drained in 1850, lay in the marginal areas over shallow water sediments.

Introduction

One of the most famous discoveries of early mesolithic settlements in northwest Europe was made at Duvensee, a lake formed behind a terminal moraine after the melting of the trapped glacial ice from the last (Weichselian) glaciation. The area therefore became a type locality for civilisation during the 'Duvensee Stage' of the Mesolithicum (Schwantes, 1925). All that remains today of the former lake is a bog. Our investigations indicate that during the Late Glacial and early Holocene, the lake covered and area of about 4.25 km^2. Its bathymetry was very uneven with numerous deeps and shoals. In the late Preboreal, the water level sank. Mostly the nearshore shallow water areas and some shoals were quickly filled in and raised by reed and sedge peat, and became islands.

According to the results of excavations carried out since 1924 (Schwantes *et al.,* 1925; Schwantes, 1928; Schwabedissen, 1949; Bokelmann, 1971; 1980; 1981; Bokelmann *et al.,* 1981) some elevated areas in the northwestern part of the lake (Fig. 1) have repeatedly been inhabited by man during mesolithic stages. The first settlement was discovered during the course of geological mapping by K. Gripp in 1923 (Schwantes *et al.,* 1925). It consisted of floors of bark from pine and birch laid on the wet peat and which had served as camp sites. The number of discoveries of such settlements grew in the following years to a total of over 15. The use of the camp sites was probably seasonal and of short duration. Some sites, however, were repeatedly settled, as evidenced by five bark floors superimposed on one another in settlement no. 5. The former inhabitants of the Duvensee settlements gathered hazel nuts. For these there were accommodating grills. Finds of bones and antlers in the excavations were seldom.

With higher water levels, the buoyant peat on which the settlements were found floated and became islands. Younger gyttja together with downwashed artefacts and charcoal were shifted beneath them (Bokelmann, 1970; Schütrumpf, 1981).

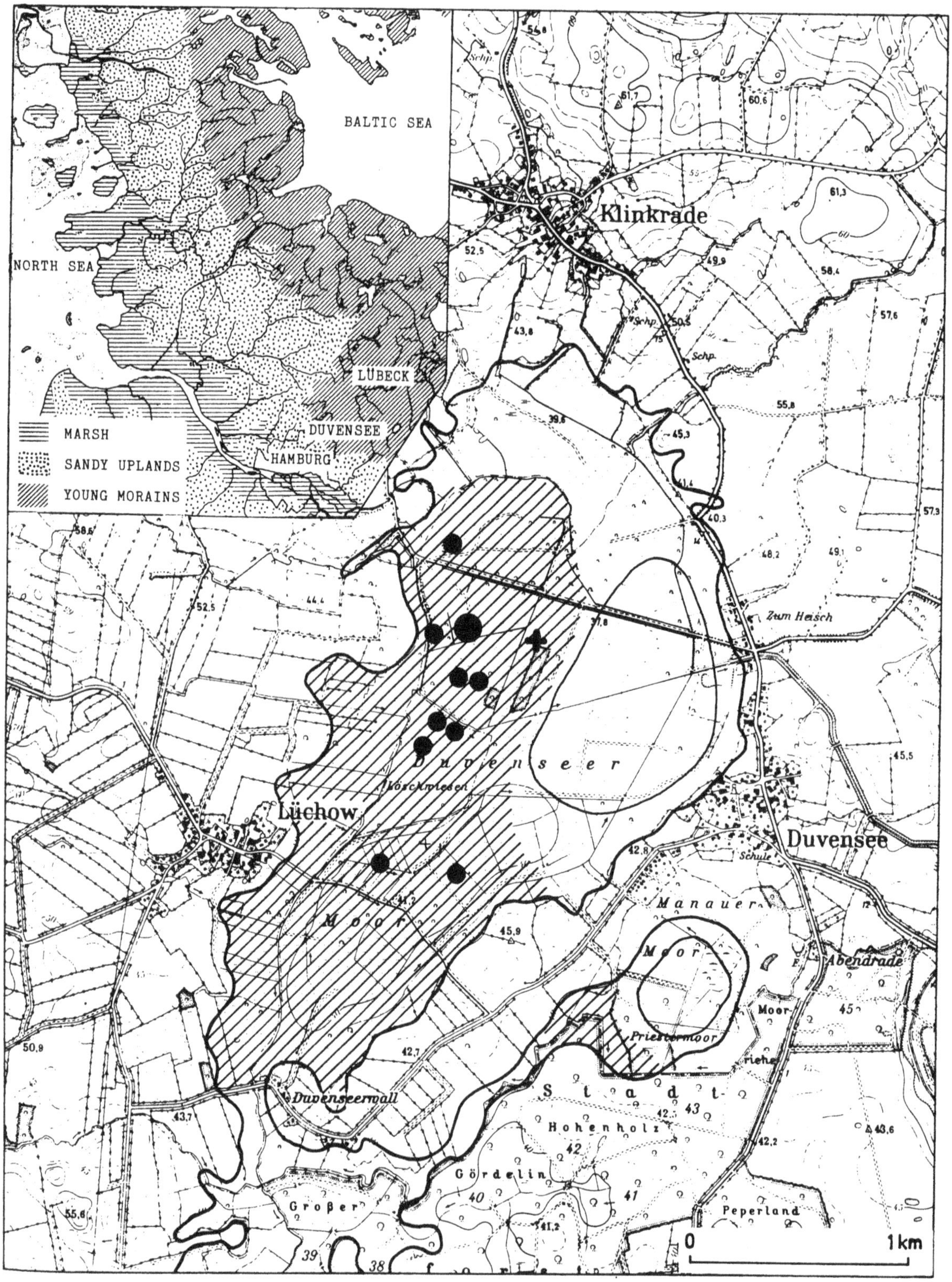

Fig. 1. Duvensee area. The Preboreal lake (thick line) and therein the two residual lakes existing till 1774. Dots signify mesolithic settlements, the largest encompasses the settlements nos. 1–5. The location of the standard profile cored in 1984 is shown with a cross. Raised bogs are hatched. Inset: geological map of Schleswig-Holstein.

Pollen analysis, which was already applied to the Duvensee sediments in 1927 (Kolumbe in Schwantes, 1928), showed that the settlements were spread over a time-span from the end of the Preboreal to the younger Boreal (Averdieck, 1981; Schütrumpf, 1981). The dynamics of the pollen curves of the Preboreal and the Boreal, especially that of *Corylus*, yielded finer differentiations than the radiocarbon method. Radiocarbon dates tend to favour the beginning of the settlements at about 7000 B.C.

Materials and methods

Besides petrographical and pollen-analytical investigations of the archaeological excavations, more than 50 locations which were areally distributed over the moor (Fig. 1) were also cored with the Puerckhauer sampler. In one of the deepest spots only 300 m distant from the settlements, a standard profile was cored with a Livingstone core sampler. The cored material has been scrutinised for pollen and ostracods. Sediment properties and lithological characteristics have also been studied.

Results and discussion

The sediment succession near the shorelines generally exhibits a gradual shallowing of the deposition environment. In the lake basin proper, the sequence is more complicated due to the fluctuating water levels during the Preboreal to Boreal. The sediment layers and climatic periods of the standard profile are shown in Fig. 2.

The majority of the deposits are typical shallow water sediments, such as calcareous gyttja, algal gyttja and coarse detrital gyttja. However, the intercalations of fine detrital gyttja within the above sediments indicate higher water levels in the Boreal and the older Atlanticum. Sedimentation in the western part of the lake from the Boreal to the beginning of the Subatlanticum led to progressively shallower waters and resulted in a raised bog (Fig. 1).

In the east and south, two remnant lakes persisted up to the last century. These lakes are not located over the maximal depths of the Pleistocene basal lake floor, but on the contrary, in the marginal positions of the original lake. These lakes were authentically recorded since the middle of the 16th century. Although the recent development of the lakes has been described by Funck (1963), no notes were given on their water depths. In the larger northern Duvensee which then covered an area of about 0.75 km², a fisherman was drowned in 1567, so that its depth must have been considerable. In 1774, this lake was artificially lowered to an area of

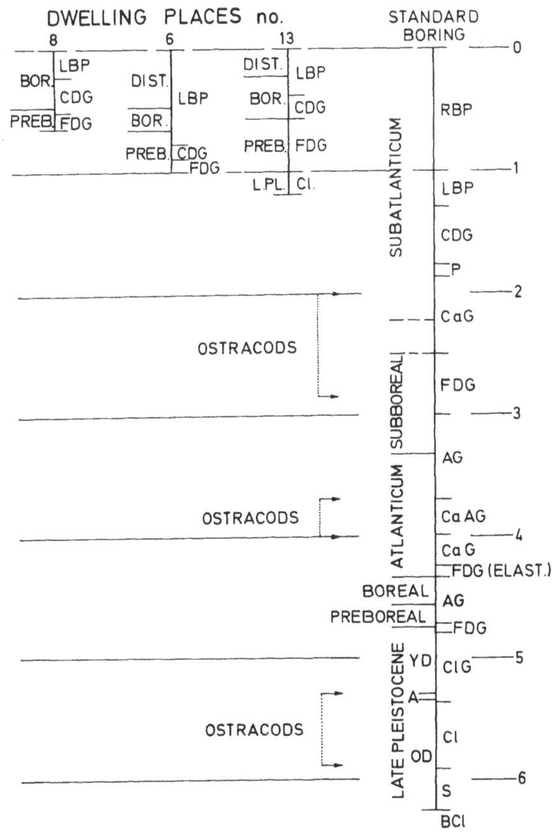

Fig. 2. Sediment and peat sequences and their chronology in some excavated settlements and in the standard profile.

Symbols:	A G	algal gyttja
	B Cl	boulder clay
	Ca G	calcareous gyttja
	Ca A G	calcareous algal gyttja
	C D G	coarse detrital gyttja
	DIST.	disturbed
	F D G	fine detrital gyttja
	L B P	low bog peat
	P	peat
	R B P	raised bog peat
	S	sand/silt

about 0.6 km^2, which means that it was reduced by about 20% for the gain and amelioration of landed properties. In 1850, both lakes were laid dry by deepening the ditch draining the Duvensee by nearly 3 m, another indication of the deepness of these lakes. This drainage was carried out especially for peat mining, after which the whole area was put to agricultural use.

A detailed description of the sediment succession and the sedimentary facies relationships between the different parts of the Duvensee area is still lacking. The sediment cores have shown that in spite of similar depositional conditions, strong lateral variations in sediment facies are present. The complicated alternations observed in the standard profile attest to multiple changes in lake level. The Duvensee is apparently very sensitive to such fluctuations and this is reflected in its sediment record. The early Holocene deposits provide a special problem of interpretation. In spite of their higher position on the mesolithic islands, their thicknesses are hardly less than in the standard profile of the deep basin and are sometimes greater. Because subsequent erosion cannot normally be made responsible in a small enclosed lake at reasonable depth, it must be assumed that during the depositional period of these sediments, the water depth in the middle of the lake was not deeper than above the shoals near the shore. On the other hand, a remote possibility exists that the basin floor was high and artificially formed by remnant dead ice, which prevented deposition of thick sedimentary sequences. Beneath the Preboreal gyttja in the basin, Late Glacial sediments more than 1 m thick were cored.

Our objective is to correlate through analysis and other scientific investigations, the sediment sequence or lake levels. For example, the ostracod studies carried out by Günther (1986) are very promising for the establishment of ancient temperature changes and the extent of the lake.

Acknowledgements

I am grateful to Miss M. Neve, Institut für Ur- and Frühgeschichte der Universität, Kiel, for pollen counts, to Dr. K. Bokelmann, Schleswig-Holsteinisches Landesmuseum für Vor- und Frühgeschichte in Schleswig, to Dr. J. Günther, Bad Krozingen, to Dr. W. Hofman, Limnologisches Institut der Max-Planck Gesellschaft in Plön, to Prof. H. Willkomm, Institut für Reine und Angewandte Kernphysik der Universität, Kiel, and to Dr. Kyaw-Winn, Geologisch-Paläontologisches Institut der Universität, Kiel, for discussions, criticsm and support.

References

Averdieck, F.-R., 1981. Ein palynologischer Beitrag zur Grabung Duvensee, Wohnplatz 6, 1975. Kölner Jahrbuch für Vor- und Frühgeschichte 15: 189–190.

Bokelmann, K., 1970. Nachweis einer borealzeitlichen schwimmenden Insel im Duvensee-Moor. Eiszeitalter und Gegenwart 21: 188–189.

Bokelmann, K., 1971. Duvensee, ein Wohnplatz des Mesolithikums in Schleswig-Holstein, und die Duvenseegruppe. Offa 28: 5–26.

Bokelmann, K., 1980. Duvensee, Wohnplatz 6. Neue Befunde zur mesolithischen Sammelwirtschaft im 7. vorchristlichen Jahrtausend. Die Heimat 87: 320–333.

Bokelmann, K., 1981. Eine neue borealzeitliche Fundstelle in Schleswig-Holstein. Kölner Jahrbuch für Vor- und Frühgeschichte 15: 181–188.

Bokkelmann, K., F.-R. Averdieck & H. Willkomm, 1981. Duvensee, Wohnplatz 8. Neue Aspekte zur Sammelwirtschaft im frühen Mesolithikum. Offa 38: 21–40.

Funck, H., 1963. Die Entwässerung des Duvensees. Schriftenreihe Heimatbund und Geschichtsverein Herzogtum Lauenburg 10, 135 pp.

Günther, J., 1986. Ostracod fauna of Duvensee, an ancient lake in Northern Germany. Hydrobiologia, this volume.

Schütrumpf, R., 1981. Der pollenanalytische Nachweis einer 'schwimmenden Insel' und die Anzahl der Siedlungsphasen am mittelsteinzeitlichen Fundplatz Duvensee, Kr. Herzogtum Lauenburg/Schleswig-Holstein. Kölner Jahrbuch für Vor- und Frühgeschichte 15: 161–180.

Schwabedissen, H., 1949. Die Bedeutung der Moorarchäologie für die Urgeschichtsforschung, Offa 8: 46–74.

Schwantes, G., 1928. Nordisches Paläolithikum und Mesolithikum. Mitteilungen aus dem Museum für Völkerkunde Hamburg 13: 159–252.

Schwantes, G., K. Gripp & M. Beyle, 1925. Der frühmesolithische Wohnplatz von Duvensee. Prähistorische Zeitschrift 16: 173–177.

Accepted 18 March 1986.

Hydrobiologia 143: 411–416, (1986)

Ostracod fauna of Duvensee, an ancient lake in Northern Germany

Jutta Günther
Schönbergstraße 20, D-7812 Bad Krozingen, FRG

Keywords: paleolimnology, Ostracoda, lake development

Abstract

About 4000 ostracod valves have been analysed from a sediment core of a cultivated bog NW of Hamburg. The deepest deposits are sands free of subfossils, the uppermost layers consist of peat. In the sediment in between, there are three layers containing undestroyed valves. The ostracod assemblies of the older claygyttja (Late-glacial) and the two younger, fine detritus deposits (Post-glacial) are strikingly different.

Cytherissa lacustris, Candona neglecta, Ilyocypris bradyi, Herpetocyrpis reptans, and two *Limnocythere* species were found in the Late-glacial layers. Changes in abundance of these species indicate alterations in climate, lake ground, water inflow and waterlevels. Postglacial layers are rich in *Metacypris cordata* valves associated with numerous *Candona* species, *L. inopinata, Darwinula stevensi,* and *Cyclocypris laevis.* These species are characteristic of a small lake with a large littoral zone. A similar development in ostracod assemblages is found in middle Europe by Absolon (1973), although the main *Candona* species is not *C. neglecta* but *C. candida.* Ostracod analysis cannot demonstrate an influence of anthropogenic impacts on ostracod successions.

Introduction

Not much is known of subfossil ostracods of lake deposits in Schleswig-Holstein. In central Europe the 'Candida' fauna of late glacial sediments are superseded by the 'Cordata' fauna (Absolon, 1973). Is a similar succession to be found in Northern Germany? Duvensee sediments are ideal to answer the question whether the succession is caused by climatic alterations or by anthropogenic influences, because settlement signs are found in Pre-Boreal and Boreal (Averdieck, 1983).

Material and methods

Coring, using a Livingstone piston sampler, and pollen analysis were carried out by Averdieck (1983). The sediment core is 6,90 m long. The lowest metre is composed of marl and coarse sands, and is overlaid by a deposit of limy clay and silt (up to 5.90 m). It shows increasing amounts of detritus. Above 4.80 m there are organic sediments, containing biogenic lime, laminated in most places. Above 1.30 m there are bog deposits, above 1.00 m cultivated high moor (see Fig. 1).

Ostracods were collected and preserved in Franke cells from the sieved fraction > 200 μm of 10–31 g fresh sediment. Papers from Absolon (1973, 1978) Diebel & Pietrzeniuk (1977), Hiller (1972), Klie (1938), Ohmert (1979), and Petkovski (1977) were used for identification of the species. They also include information about the ecology of ostracods.

Results

In late-glacial sediments ostracod valves are better preserved than in postglacial ones, where undestroyed valves could only be found in a few samples.

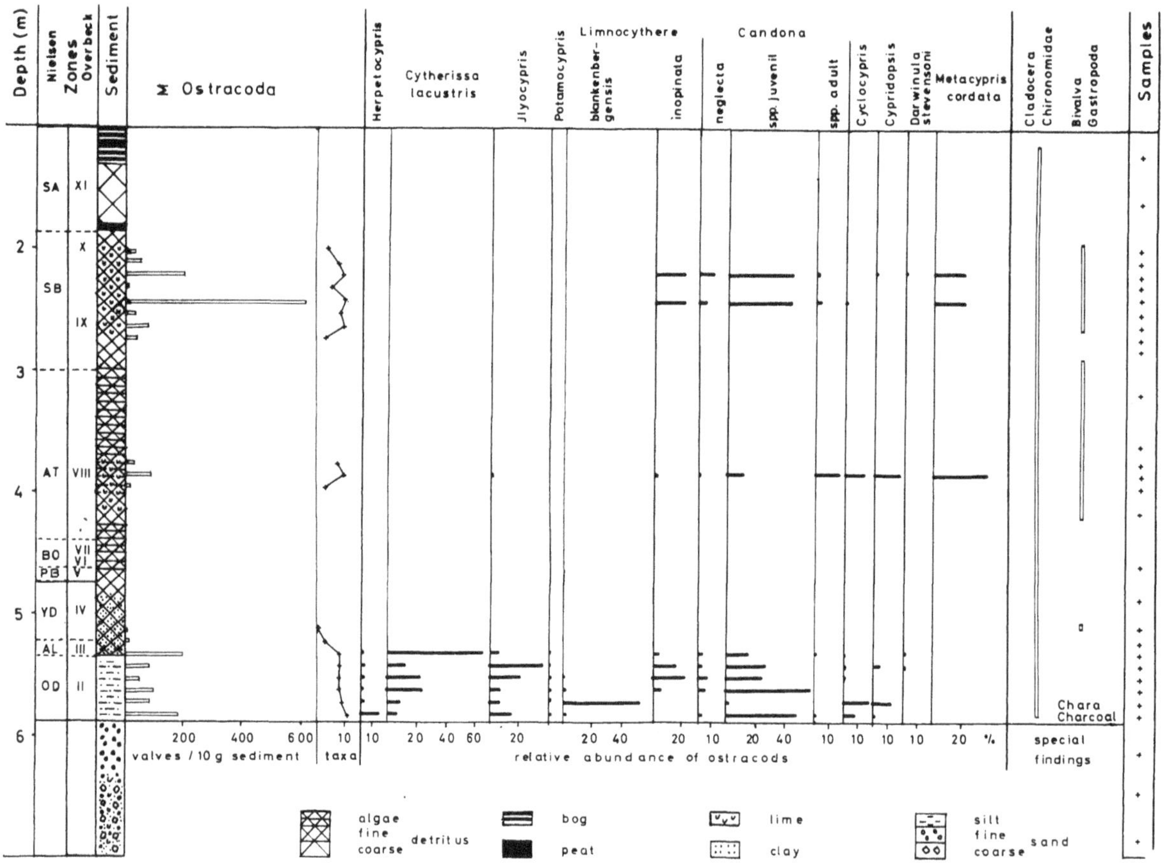

Fig. 1. Distribution of ostracods. Occurrence of cladocers, chironomids, molluscs, *Chara* residues, and charcoal. *Candona* spp. includes *C. candida, C. protzi, C. marchica,* and *C. compressa. Ilyocypris* includes *I.* sp. and *I. bradyi. Cyclocypris* includes *C. ovum* and *C. laevis.*

As Fig. 1 shows, only 19 of the 30 samples contained any ostracods, and 93% of the total 3629 valves were found in only nine samples. Six belong to late-glacial layers – Older Dryas and Alleröd – (zone II and III), one to the Atlanticum (zone VIII), and two to Sub-Boreal (zone IX and X). For more details about ostracod material and systematics see Table 1 and Fig. 3.

The lowest layers containing subfossils are characterized by high frequencies of *Candona neglecta* Sars, *Ilyocypris sp.* Brady & Norman, *Herpetocypris reptans* (Baird), and *Cytherissa lacustris* (Sars). This species assembley – together with *Potamocypris cf. villosa* Jurine – is found (see Fig. 1) during the whole late-glacial. These frequencies decrease temporarily from 88% to 18% favouring *Limnocythere 'blankenbergensis'* Diebel (1968) (see Fig. 3), *Cyclocypris ovum* Jurine, *Cypridopsis cf. vidua* Müller. Afterwards the earlier species community reasserts itself, although *H. reptans* (Baird) and *Ilyocypris sp.* sensu Brady & Norman valves decrease in number and *Limnocythere inopinata* (Baird) (see Fig. 3, G–I) and *Darwinula stevensoni* Brady & Robertson are found in the sediment for the first time *C. neglecta* Sars, *Ilyocypris bradyi* Sars, *C. lacustris* Sars, (see Fig. 3) and *L. inopinata* Baird still constitute more than 95% of the ostracod valves. In the Alleröd *C. lacustris* (Sars) dominates with a relative abundance of almost 70%. The greater number of complete carapaces are remarkable. *C. lacustris* (Sars) disappears in Younger Dryas (zone IV).

Table 1. Ostracoda, found in Duvensee sediments.

Taxa	Number of valves	Preservation + well, – bad	Occurrence	Dimensions of valves mm
Candona candida (Müller, 1776)	22♀, 400 juv.	–	late-glacial postglacial	LV, ♀, l=1.02, h=0.58
Candona neglecta Sars, 1887	147♀, 41♂, 400 juv.	– +	late-glacial postglacial	LV, ♀, l=1.15, h=0.58 LV, ♂, l=1.20, h=0.67
Candona compressa (Koch, 1837)	12 ad., 8 juv.	–	Atlanticum	LV, l=0.92, h=0.56
Candona marchica Hartwig, 1899	32♀, 7♂ 70 juv.	– +	postglacial	LV, ♀, l=0.98, h=0.57 RV, ♂, l=1.06, h=0.66
Candona protzi Hartwig, 1898	63♀, 12♂, 70 juv.	–	postglacial	LV, ♀, l=0.98, h=0.46 RV, ♂, l=0.96, h=0.50
Cyclocypris laevis (Müller, 1776)	47	– +	postglacial	RV, l=0.53, h=0.35
Cyclocypris ovum (Jurine, 1820)	75	– +	late-glacial	LV, l=0.49, h=0.34
Cypria cf. opthalmica (Jurine, 1820)	3	–	Y. Dryas Sub-Boreal	
Cypridopsis cf. vidua (Müller, 1776)	8 ad., 86 juv.	– +	late-glacial postglacial	RV, l=0.67, h=0.39
Cytherissa lacustris (Sars, 1863) Fig. 3, A–C.	277 ad., 260 juv.	+	late-glacial	LV, l=0.88, h=0.53
Darwinula stevensoni (Brady & Robertson, 1870)	23 ad., juv.	–	Alleröd postglacial	LV, l=0.72, h=0.24
Herpetocypris reptans (Baird, 1835)	23 ad., 69 juv.	–	late-glacial	LV, l=1.80, h=0.83
Ilyocypris bradyi (Sars, 1890)	80 ad., 162 juv.	– +	late-glacial	LV, l=0.90, h=0.47 RV, l=0.89, h=0.45
Ilyocypris sp. sensu Brady & Norman (1889)	11 ad., 23 juv.	– +	late-glacial	LV, l=0.95, h=0.53 RV, l=0.91, h=0.50
Limnocythere 'blankenbergensis' Diebel, 1968 (put to *L. (L.)staplini* (Gutentag & Benson, 1962) non *L. staplini* Bhatia, 1968 – by Kazmina (1975)), Fig. 3, D–F.	153 –	+	Older Dryas	LV, l=0.48, h=0.25
Limnocythere inopinata (Baird, 1843), Fig. 3, G–M.	396 ad., 46 juv.	+ –	late-glacial: postglacial:	LV, l=0.58, h=0.33 LV, l=0.62, h=0.34
Metacypris cordata Brady & Robertson, 1870, Fig. 3, N+O.	461♀, ♂, juv.	– +	postglacial	LV, ♀, l=0.53, h=0.31 LV, ♂, l=0.43, h=0.28
Potamocypris cf. villosa (Jurine, 1820)	19 juv. mostly	–	late-glacial	LV, l=0.68, h=0.40 RV, l=0.64, h=0.44

LV = left valve, RV = right valve, l = length, h = height Y. Dryas = Younger Dryas.

There are obvious differences in the Atlanticum. Here *Metacypris cordata* Brady and Robertson (see Fig. 3) predominates, associated with *Cyclocypris laevis* (Müller) and *Cypridopsis cf. vidua* (Müller). The *Candona* species *C. neglecta* Sars, *C. candida* (Müller), *C. protzi* Hartwig, *C. marchica* Hartwig, and *C. compressa* (Koch) are also found. In the Sub-Boreal the most abundant ostracod is *M. cordata* Brady & Robertson, accompanied by high abundance of *L. inopinata* (Baird) (see Fig. 3, K–M) and an increase of *C. neglecta* Sars (see Fig. 1).

Discussion

Looking at the most abundant ostracods (Fig. 2), what can they tell about the ontogeny of the ancient waters? *H. reptans* is most abundant at the beginning of lake development. It loves digging in the mud in shallow places near the shore. The oecology of *L. 'blankenbergensis'* the next extremely highly abundant species, is not known. It is found in the samples with high content of HCl-soluble ash (determined by Hofmann, Plön, in litt.). If this *Limnocythere* belongs to the *Limnocythere stapli-*

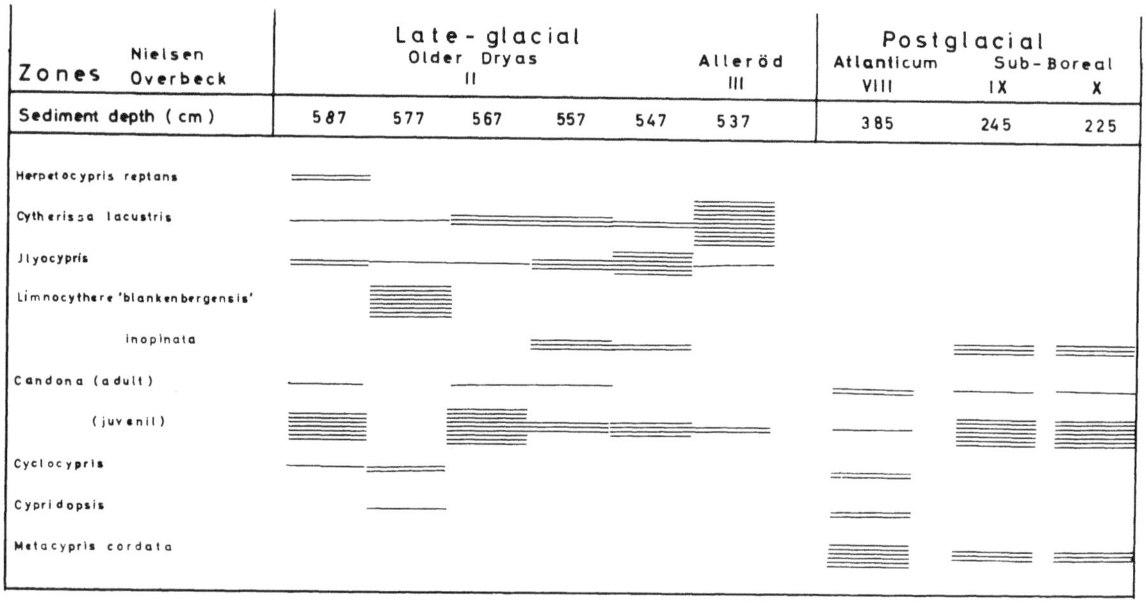

Fig. 2. Distribution of the predominant ostracods. Each line denotes a relative abundance >6%.

ni group, as Kazmina (1975) assumes, then there are some hints with regard to its oecology (Delorme, 1969; Forrester, 1983), but this still has to be confirmed. Together with *L. 'blankenbergensis'*, *Cyclocypris* and *Cypridopsis* occur in high abundancies also. They are known as good swimmers, which love littoral regions, rich in plants. Findings of *Chara* (Averdieck, pers. comm.) in the sample (see Fig. 1) supports this. *I. bradyi* has its predominance next. It likes springs or the areas near them. With the beginning of Alleröd high abundancies of *C. lacustris* are obvious. It tolerates large sizes and amounts of detritus and prefers cool, deep water. This ostracod succession gives the idea of a lake with an increasing water level: shallow water, poor in plants – littoral with *Chara* meadows – water inflow from springs – profundal.

In the postglacial, the ostracod assemblies show species representative of small water bodies, and lakes rich in plants. Differences in abundance of species are due more to the state of preservation than to alterations of ostracod associations. The ancient lake always has been rich in carbonates, as valves, shells, and amorphic calcareous layers show. Decomposition of shells seems to correlate with the productivity of the lake. As most ostracods live on the benthos, the oxygen content must have been sufficient. There are signs, however, of a reduced oxygen supply in postglacial. *C. ovum* predominant in late-glacial, is succeeded by *C. laevis*, needing less oxygen (Scharf, 1980).

The development of Duvensee is similar to that found in middle European lacustrine sediments by Absolon (1973). There a 'Candida' fauna in late-glacial is followed by a 'Cordata' fauna. The most abundant *Candona* in the late-glacial of Duvensee is not *C. candida* but *C. neglecta*. The postglacial findings are in accordance with Absolon's, where *M. cordata* exists in high abundance. In North German sediments (Poolsee, Alt Fresenburg) we also find a 'Neglecta' fauna but a 'Cordata' fauna is not always obvious.

Are the disturbances in the ostracod succession caused by anthropogeneous influences? Charcoal was also found in the oldest microfossiliferous layers. If this is a sign of the first period of clearing by fire, then the lake has been influenced since the beginning of its development by anthropogenic impact. During development in the late-glacial, no further evidence is given except the development of *Cytherissa lacustris* with the high population and its obviously fast fossilisation. At the same time when *C.*

Fig. 3. Cytherissa lacustris (Sars, 1863). A: dorsal, B: RV, C: LV. *Limnocythere 'blankenbergensis'* Diebel 1968. D: RV, E: LV, F: dorsal. *Limnocythere inopinata* (Baird 1843). Late-glacial G: dorsal, H: RV, I: LV. Post-glacial K: dorsal, L: RV, M: LV. *Metacypris cordata* Brady & Robertson, 1870. N: dorsal, female, O: dorsal, male. Scale: – 0.5 mm. Abbreviations as for Table 1.

lacustris disappears, the sediment gets less mineralic and organic matter is increasing as measured by loss on ignition, (Hoffman, Plön, in litt.). Here molluscs were found in the sediment for the first time (see Fig. 1). Löffler (1963) assumed, that *C. lacustris* is an indicator species for changes in the lake sediment due to cultural influences. More likely the disappearance of *C. lacustris* is caused here by climatic alterations. In the warmer post-glacial climate and the decreasing waterbody *C. lacustris* had no chance of survival.

Acknowledgments

I am indebted to Mrs. E. Ganß and Dr. D. Keyser for REM photomicrographs and Mr. H. Petersen, Zoologisches Institut und Museum der Universität Hamburg, Dokumentation, for nomenclature indication and supply of literature, Mr. B. Chandler, Bad Krozingen, for reviewing the manuscript, and Dr. W. Ohmert, Geologisches Landesamt Freiburg, for helpful discussions.

References

Absolon, A., 1973. Ostrakoden aus einigen Profilen spät- und postglazialer Karbonatablagerungen in Mitteleuropa. Mitt. Bayr. Staatssamml. Paläont. hist. Geol. 13: 47–94.

Absolon, A., 1978. Die Gattung *Candona* im Quartär von Europa. Rozpravy Ceskoslovenske Akademie Ved., Rada Matematickych A Prirodnich Ved, 88: 3–73.

Averdieck, F.-R., 1986. Palynogical investigations in sediments of ancient lake Duvensee, Schleswig-Holstein. Hydrobiologia, this volume.

Delorme, L. D., 1969. Ostracodes as Quarternary paleoecological indicators. Can. J. Earth Sci. 6: 1471–1476.

Diebel, K., 1968. Neue Limnocythere-Arten (Ostracoda) aus dem deutschen Pleistozän. Monatsber. deutsch. Akad. Wissensch. 10: 519–538.

Diebel, K. & E. Pietrzeniuk, 1977. Ostrakoden aus dem Travertin von Taubach bei Weimar. Quartärpaläontologie 2: 119–137.

Forrester, M., 1983. Relationship of two lacustrine ostracod species to solute composition and salinity: Implications for paleohydrochemistry. Geology 11: 435–438.

Hiller, D., 1972. Untersuchungen zur Biologie und zur Ökologie limnischer Ostrakoden aus der Umgebung von Hamburg. Arch. Hydrobiol. Suppl. 40: 400–497.

Kazmina, T. A., 1975. Stratigrafia i Ostrakody Pliocena i Rannego Pleistocena Juga Zapadno-Sibirskoj Ravniny. Trudy Inst. Geol. Geofis Sib. Otd. 264: 1–108. (In Russian).

Klie, W., 1938. Ostracoda, Muschelkrebse. In: Dahl, Die Tierwelt Deutschlands. Jena 34 (3): 1–230.

Löffler, H., 1969. Recent and subfossil distribution of *Cytherissa lacustris* (Ostracoda) in Lake Constance. Mitt. int. Ver. Limnol. 17: 240–251.

Ohmert, W., 1979. Die Ostrakoden der Kernbohrung Eurach 1 (Riß-Eem). Geologica Bavarica 80: 127–158.

Petkovski, T. K., 1977. Ostrakodenfauna des Mindelsees (S.W. Deutschland). Acta Musei Macedonici Scientiarum Naturalium, Tom. 15, 3 (128): 49–94.

Scharf, B. W., 1980. Zur rezenten Muschelkrebsfauna der Eifelmaare (Crustacea: Ostracoda). Mitt. Pollichia 68: 185–204.

Accepted 18 March 1986.

Hydrobiologia 143: 417–424, (1986)

Effects of peatbog ditching in lakes: Problems in paleolimnological interpretation

A. Liehu[1], O. Sandman[2] & H. Simola[3]
[1] *Technical Research Center of Finland, Reactor Laboratory, Otakaari 3A, SF-02150, Espoo, Finland*
[2] *National Board of Waters, Mikkeli Water District Office, P.O.Box 77, SF-50101, Mikkeli, Finland*
[3] *Univ. Joensuu, Karelian Institute, Sect. Ecology, P.O.Box 111, SF-80101, Joensuu, Finland*

Keywords: paleolimnology, peatbog ditching, Pb-210, Hg, diatoms

Abstract

Recent sediment stratigraphy in two lakes with extensive artificial drainage of surrounding peatbogs was studied by ^{210}Pb and ^{137}Cs dating and chemical and diatom analyses. In contrast to many other lakes with similar ditching history, the visual appearance of the recent sediments in both of these lakes is homogeneous. In the larger lake, Kyyvesi, the sediment accumulation rate seems to have increased, but no major changes in the chemical properties or diatom stratigraphy are to be seen. In the organic *dy* sediment of Suuri-Rostuvi the ignition residue increases and there is a mass occurrence of *Asterionella ralfsii*, an acidophilic plankton diatom, clearly due to ditching and fertilization of the catchment bogs in 1970. Mercury content is increased in both lakes in the sediment layer corresponding to the ditchings – more abruptly in Suuri-Rostuvi, up to 2.4 $\mu g\ g^{-1}$ DW. In Kyyvesi a first Hg maximum was dated to the 1930's – possibly related to agricultural use of Hg as fungicide (seed dressing) in the large partly cultivated drainage area.

The limnologic effects and stratigraphic manifestation of extensive peatbog ditching in geomorphically, geologically and hydrologically different drainage basins are discussed.

Introduction

Due to cool and humid climate and generally level relief, peatbogs (peat thickness > 0.5 m) cover about 30% of total land area in Finland. Of the more than 10×10^6 hectares, some 2/3rds have been artificially drained largely during the past two decades and mainly in order to improve forestry; about 10% of the peatbog area has been converted into arable fields. Most of the virgin peatlands are situated in the north, so in many parts of southern Finland practically all peatland has been ditched.

The potential limnological effects of extensive peatbog ditching are many: Silting of lake basins, increase in organic loading, increase of nutrients (including leaching of artificial fertilizers added for forest improvement) and mobilization of heavy metals, particularly Hg, are all matters of great concern (Heikurainen *et al.*, 1978; Karsisto & Ravela, 1971; Kenttämies, 1981; Simola, 1983; Simola & Lodenius, 1982).

As it is, the main part of the nearly total peatland drainage in South Finland had been done before extensive research was started to evaluate its environmental effects. We have analyzed paleolimnologically a number of ditching effluent recipients, and present and discuss here two cases that proved to be difficult to interpret.

The studied profiles were dated by ^{210}Pb and ^{137}Cs. Both these methods have potential fallacies in the present type of lakes, where the hydrology of the basin has been altered within the datable period.

Study sites, material and methods

Both our study lakes are naturally dys-oligotrophic softwater lakes. The smaller one,

Table 1. Basin and drainage area characteristics of the study lakes and chemical data of their winter epilimnion conditions.

Lake	Suuri-Rostuvi	Kyyvesi	
Elevation a.s. l	186	108	
Lake area km^2	0.48	133	
Max depth m	12	34	
Volume 10^6 m^3	1.9	720	
Theor. residence time d	220	850	
Drainage area km^2	12	715	
Subbasins		Lietlahti	Koiraselkä
Drainage area km^2		420	96
Lakes %	12	27	7
Cultivated fields %	0	9	8
Ditched peatlands %	50	9	22
Winter epilimnion			
pH	5.0	6.5	6.0
Water colour mg Pt l^{-1}	200	50	150
P $\mu g\ l^{-1}$	90	8	25

Suuri-Rostuvi is a genuine forest lake with no human settlement nearby, while the larger Kyyvesi is moderately influenced by farming on its big drainage area. Both lakes suffer from hypolimnetic oxygen deficiency during stratification periods due to the allochthonous organic loading. The lakes and their drainage areas are portrayed in Table 1.

Suuri-Rostuvi is a headwater lake; the first sedimentation basin for most of its catchment. The land belongs to the National Board of Forestry that since 1967 has conducted in this district a programme to study the economics of maximized use of forest technology (Sevola, 1984). All of the peatbogs, 530 ha, in the drainage area of Suuri-Rostuvi were ditched in 1970, fertilized the next winter, and re-ditched in 1981. Part of the forests around the lake have been logged earlier this century. Suuri-Rostuvi drains into the lake Polvijärvi, the recent paleolimnology of which has been studied earlier by Simola (1982, 1983; Simola & Lodenius, 1982). Rönkkö & Simola (1986) discuss the differences between the recent diatom stratigraphies in these two lakes.

Kyyvesi is the first larger lake in the Mäntyharju watercourse, which is a chain of lakes typical of the Finnish lake district. It is a complex lake formed of several subbasins. The present results are mainly from the Koiraselkä basin, which is more sheltered and has a lesser throughflow than the other sampling site, Lietlahti. Water depth at the Koiraselkä sampling site is 29 m; in Lietlahti it is 34 m.

The water quality of Kyyvesi has been monitored regularly by the National Board of Waters since 1960. Slight increase of the water colour has been observed in those subbasins, including Koiraselkä, that receive direct runoff from ditched peatlands. The history of peatland ditchings in the drainage area of Kyyvesi is shown in Fig. 1. Before 1960, the ditchings were mainly done by manpower on mod-

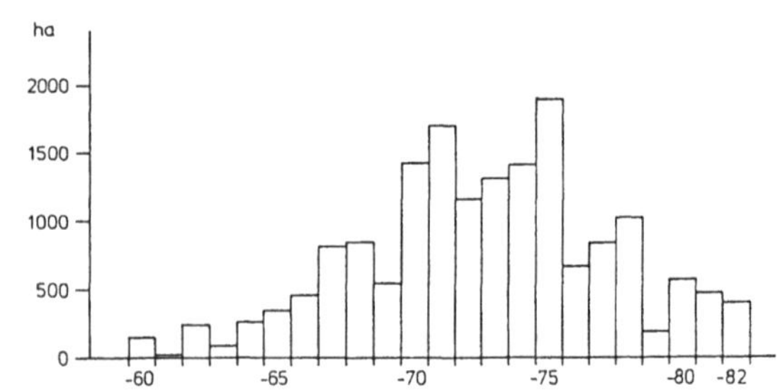

Fig. 1. Ditching of peatlands on the drainage area of Kyyvesi since 1960.

est scale, the pattern of motorized ditchings shown here is more or less representative for whole Finland: steady increase from year to year until the early 1970's, followed by decline because no more virgin peatlands exist.

Sediments of both the lakes have been sampled on several occasions, Kyyvesi with a mini-Mackereth type piston sampler and Suuri-Rostuvi with a Kajak gravity corer and a freeze-corer. For chemical analyses and for dating, the fresh cores were extruded mainly in the field and sliced into 1.0–3.0 cm subsamples.

Isotopic dating of the cores was done at the Reactor Laboratory of the Technical Research Center of Finland. ^{210}Pb was determined via its alpha-active grand-daughter ^{210}Po (Häsänen, 1977). Age calculation for the profiles was carried out by the C.R.S. model (Appleby & Oldfield, 1978). The dating cores of Kyyvesi were too short to reach the background level of supported ^{210}Pb activity, which was estimated at 2.0 dpm g^{-1} on the basis of other studies in the same area. ^{137}Cs activity of the sediment was determined directly from dried samples using Ge(Li) detector and a multichannel analyzer.

The chemical analyses for the samples of Kyyvesi were done by the Mikkeli Water District Laboratory and the Central Laboratory of National Bd of Waters in Helsinki, those of Suuri-Rostuvi by the Water Laboratory of Karelian Institute in Joensuu. Water content, loss-on-ignition, N, P and Zn were analyzed by standard methods (National Bd of Waters, 1981), Hg of Kyyvesi was analyzed by cold vapour extraction with a Perkin Elmer Hg-analyzer and that of Suuri-Rostuvi by a JIC 511 gold film mercury analyzer.

The diatom analysis of Kyyvesi was done of 1 cm slices of fresh sediment, that of Suuri-Rostuvi (Rönkkö & Simola, 1986) of cubical 1 cm pieces sawn off from two overlapping frozen cores. The organic matter was destroyed by H_2O_2, and the slides were mounted in Hyrax. At least 300 valves were counted at each analyzed level.

Results

Suuri-Rostuvi

Both ^{210}Pb and ^{137}Cs indicate an average sediment increment of some 4 mm yr^{-1} during the past twenty years in Suuri-Rostuvi (Fig. 2). A comparison of diatom stratigraphies in two overlapping cores showed that the sediment may be easily mixed (either in situ or during sampling; Rönkkö & Simola, 1986). Thus, the isotope dates are to be taken only as approximations for events observed in other cores.

The sediment of Suuri-Rostuvi is loose and homogeneous *dy*, the water content in the somewhat consolidated sediment below 12 cm is 94% (Fig. 2). The ignition residue of dry sediment is 35% at 20 cm, rises gradually, along with increase in plankton diatoms (Fig. 3) to 55% at 4–6 cm and declines to the surface. The carbon:nitrogen -ratio, a measure of allochthonous organic inputs is at minimum at 5–6 cm (11.9), but even here clearly

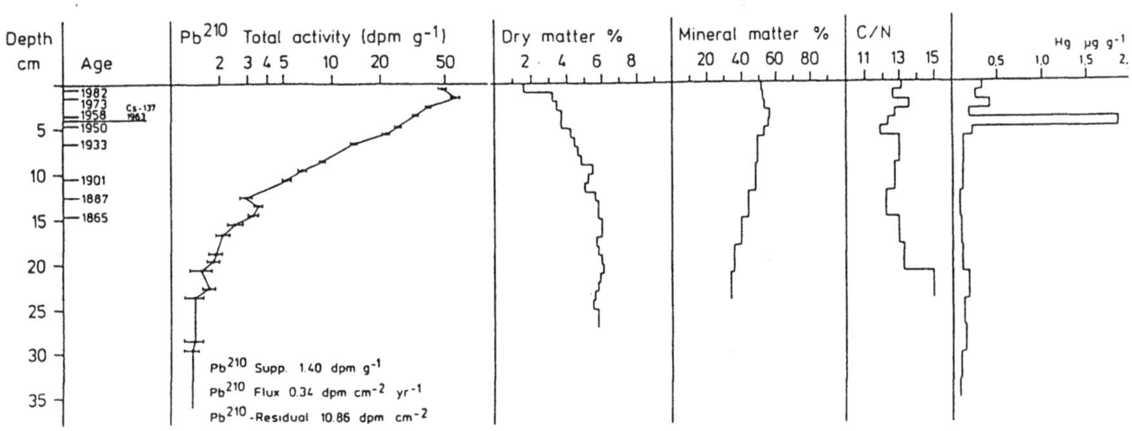

Fig. 2. ^{210}Pb, ^{137}Cs-dates, dry matter, loss-on-ignition, C/N -ratio and Hg content of the sediment of Suuri-Rostuvi.

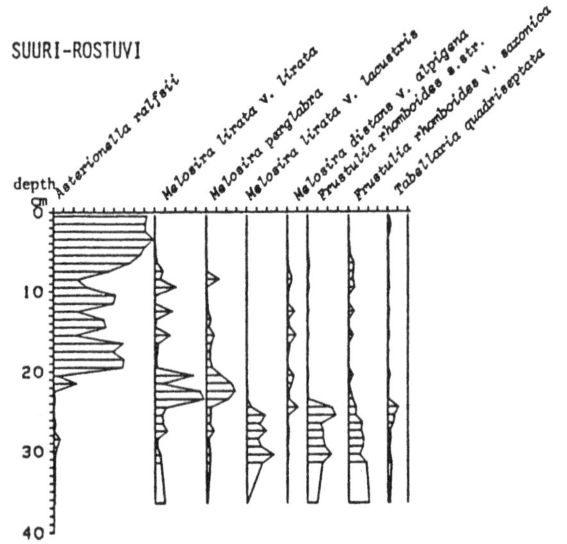

Fig. 3. Percentage diatom diagram of Suuri-Rostuvi; selected taxa only. One tic on the x-axis equals 10%.

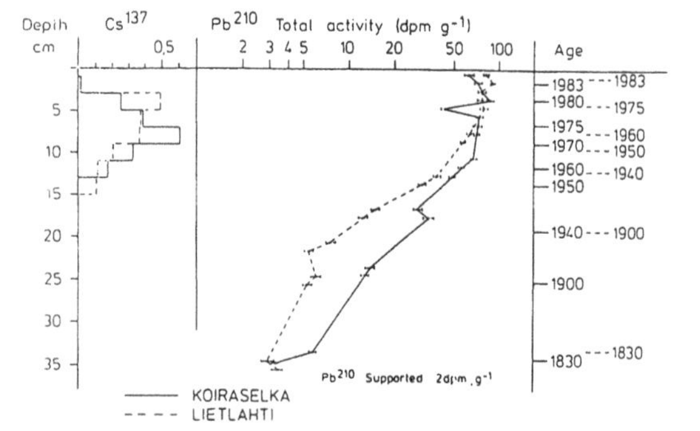

Fig. 4. ^{210}Pb, ^{137}Cs profiles and ^{210}Pb dating of sediments in two basins of Kyyvesi.

above the value 10 that would distinguish *gyttja* from *dy* (Hansen, 1961).

Mercury concentration is rather steadily 0.1 – 0.2 μg g^{-1} (dry weight) in the profile up until an abrupt peak of tenfold concentration at 4 – 5 cm level, above which it remains higher than the background (Fig. 2).

A first change of the diatoms occurs below 30 cm, when meroplanktonic *Melosira* species gain dominance over a predominantly littoral flora. *Asterionella ralfsii*, a plankton diatom typical for acidic, humic lakes increases from 23 cm upwards and reaches an overwhelming dominance in the top 7 cm (99% in 3 – 4 cm). According to the rough isotope dates, the planktonic increase has started around the turn of the century, and the maximal abundance of *A. ralfsii* seems to coincide with the ditching and fertilization of the drainage area bogs.

Kyyvesi

The ^{210}Pb chronology of the two basins of Kyyvesi, both indicate increase of sedimentation rate for the past three decades. Furthermore, in Koiraselkä there appears a recent episode (4 – 5 cm interval) when the ^{210}Pb concentration has been diluted by increased sedimentation. The recent sediments of both basins have unusually high total residual unsupported ^{210}Pb activities (Appleby & Oldfield, 1983; Oldfield & Appleby, 1984; Authors, in prep). The total unsupported ^{210}Pb content is 69 dpm cm^{-2} in Lietlahti and 42 dpm cm^{-2} in Koiraselkä (1.1 and 0.7 Bq cm^{-2}, respectively). The corresponding flux values are 2.5 dpm cm^{-2} yr^{-1} in Lietlahti and 1.4 dpm cm^{-2} yr^{-1} in Koiraselkä (Fig. 4). The residual and flux values thus differ markedly from each other in the two subbasins of Lake Kyyvesi.

The peak ^{137}Cs concentration (1960 – 65) in Koiraselkä is in the interval 7 – 9 cm and in Lietlahti in 3 – 5 cm; in both cases the shape of the distribution suggests possible upward movement of the isotope due to mixing or diffusion. However, the ^{137}Cs maxima are in rough accordance with the ^{210}Pb chronology (Fig. 4).

The loss-on-ignition of the profundal sediment of Koiraselkä varies in the range 40 – 50% while that of Leitlahti is only 20 – 30%. There is an ignition residue maximum in 6 – 8 cm in Koiraselkä and a similar one in 5 – 6 cm in Lietlahti. The datings suggest that these could reflect inorganic erosion of the early ditching period. The generally high phosphorus concentrations seem to follow the changes in the organic content, yet there is a definite increase of P from 5 cm towards the surface (Fig. 5). The conspicuous P minimum in 31 – 33 cm is possibly related with lowering of the lake level by nearly 2 m, done in 1865 – 70 (Blomqvist, 1911). The ^{210}Pb date for the 31 – 33 cm interval is 1840 ± 40.

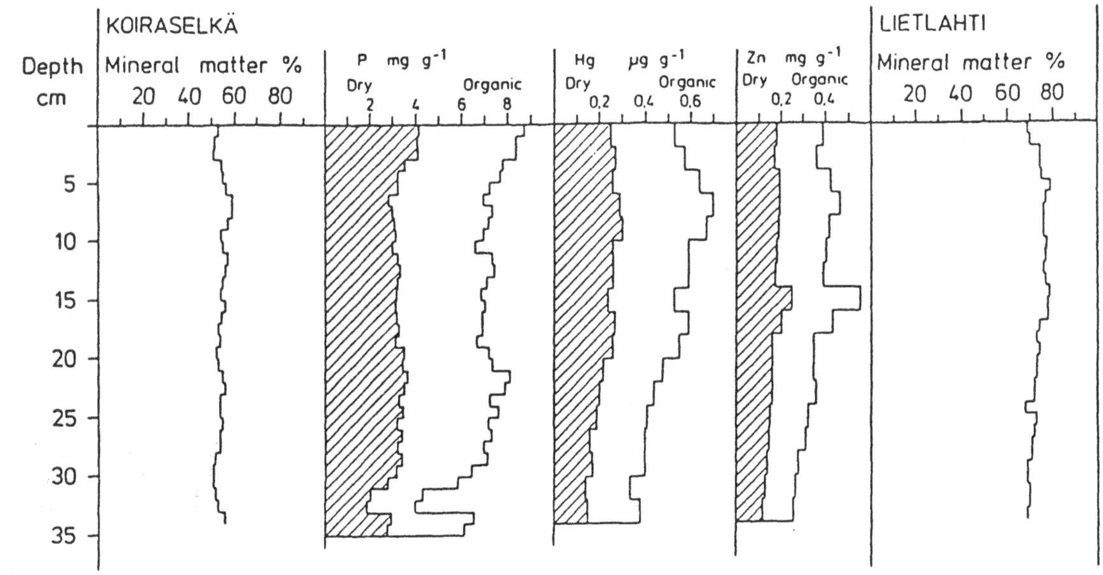

Fig. 5. Ignition residue profiles in the two sampling sites in Kyyvesi, and chemical stratigraphy of Koiraselkä. The concentrations of P, Hg and Zn have been calculated in both dry and organic weight of the sediment.

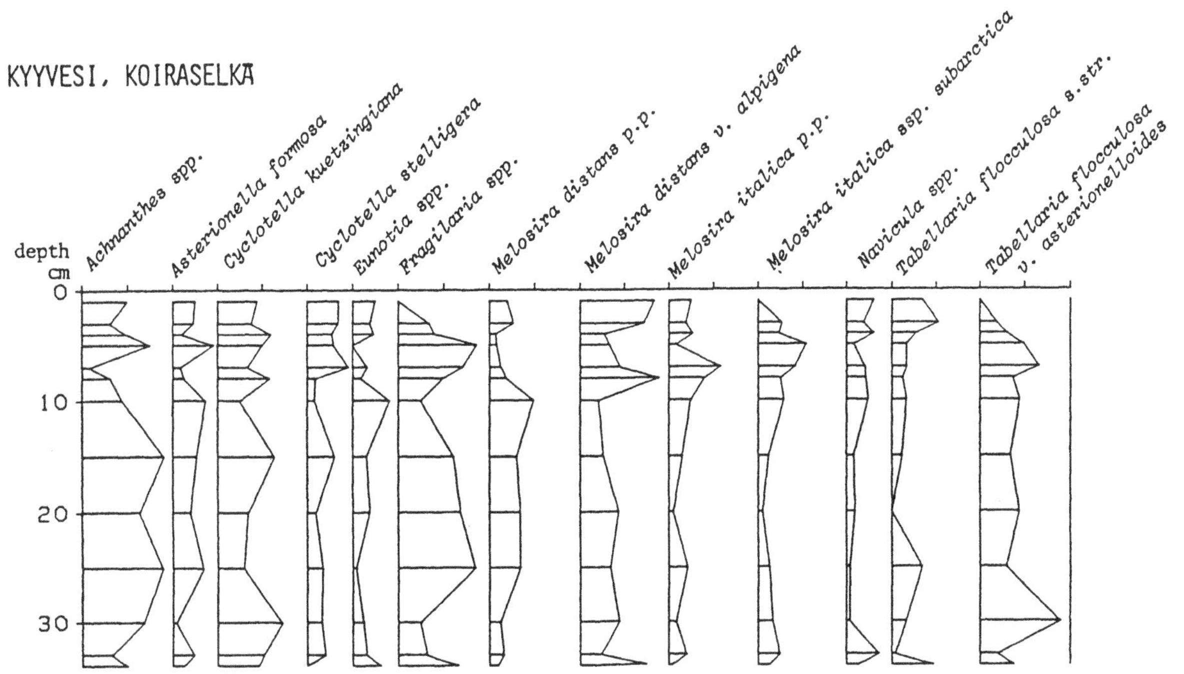

Fig. 6. Summary diatom stratigraphy in the sediment of Koiraselkä basin in Lake Kyyvesi. One tic on x-axis equals 10%. *Melosira distans p.p.* refers to the collective *M. distans* -group excluding var. *alpigena*.

In Koiraselkä, there are two maxima of both Zn and Hg: the first in the middle of our 34 cm-profile (Hg at 16–20 cm and Zn at 14–16 cm), and a second at 4–10 cm, which is more prominent for Hg than for Zn (Fig. 5).

The diatom stratigraphy of Koiraselkä is shown in Fig. 6. The flora is rather diverse throughout the profile, and quite typical for a large oligo-mesotrophic lake. About half of the frustules belong to planktonic forms and the rest are mainly epiphytes. There are prominent stratigraphic changes in the species proportions, but apparently no ecologically significant trends in the floral history.

Discussion

Preliminary results of a comparison of sedimentary ^{210}Pb data in a number of Finnish lakes indicate possible excess of unsupported ^{210}Pb in lakes that have extensively ditched large catchments (Authors, in prep). If correct, this finding would cast doubt upon C.R.S. -dating of profiles that contain a major change in the hydrological conditions due to recent bog ditchings. One factor of potential significance is the hydrological flushing rate of the lake: Kyyvesi is a large lake in relation to its drainage area, so the water residence time is long (Table 1), and the sedimentation of catchment-derived unsupported ^{210}Pb obviously effective, while in the smaller Suuri-Rostuvi, a bigger proportion of the unsupported ^{210}Pb may be flushed through the lake downstream (cf. Eakins *et al.*, 1984). As to the ^{137}Cs dating, several recent papers have shown that it may give erroneous results due to the mobility of the isotope (e.g. Longmore *et al.*, 1983; Davis *et al.*, 1984). In the present cases, the two independent isotope datings seem, however, to give consistent results, that also give a reasonable backbone for the paleolimnologic interpretations. Yet, we have felt it better to avoid too strict reference to the isotope dates. Consequently, our interpretations remain somewhat tentative.

The surface maximum of phosphorus in Kyyvesi starts from 5 cm depth upwards, so it could be partly a fossil feature, truly indicating increase in P sedimentation and not only due to the well-known enrichment of P to the oxic sediment surface (Mortimer, 1941). Similar but stronger P increase was observed in the top 5 cm in Pyhäselkä, a large lake with a 2700 km^2 drainage area (Simola *et al.*, 1985). The P of Pyhäselkä was analyzed by stepwise fractionation (Engstrom & Wright, 1984), and the increase was chiefly found in the weak-alkali soluble fraction, which led to the conclusion that the surficial P is mainly bound to humic material.

Zinc, a widely used anti-corrosive appears to be a general indicator of the use of industrial products even in rural areas (Håkanson, 1981; Evans *et al.*, 1983). Its stratigraphy in Kyyvesi is due to general and weak pollution, without industrial point-sources. The minimum in the bimodal stratigraphy, which is similar in some other lakes (Sandman *et al.*, in prep.), possibly signals the general depression of agriculture and industry during the 1940 as consequence of war.

The stratigraphy of mercury in Kyyvesi is quite similar to that of zinc, while in Suuri-Rostuvi the single maximum of Hg is much more prominent and more recent than the gradual increase in Kyyvesi. Hg has been much used in agriculture as fungicide – the annual consumption of seed-dressing Hg in Finland was at maximum over 4000 kg in 1963, and slight pollution of crop fields has been observed (Rautapää *et al.*, 1979). Hg is also an industrial atmospheric pollutant (Aston *et al.*, 1973). In Finland, elevated Hg concentrations have frequently been encountered in new man-made lakes and lakes with extensive ditchings or water-level regulation (Lodenius & Seppänen, 1982; Lodenius *et al.*, 1983; Simola & Lodenius, 1982; Verta & Rekolainen, 1985). However, Hg increase has also been observed in some rock and esker lakes with very small catchments (Tolonen & Jaakkola, 1983; Verta & Rekolainen, 1984). It appears that the soil Hg that is easily mobilized by any hydrological change could be largely of recent atmospheric origin. The maximum Hg concentration in Suuri-Rostuvi, 1.9 μg g^{-1}, is three times that in the downstream lake Polvijärvi, where it seems to be of the very same origin (Simola & Lodenius, 1982).

The diatom stratigraphy of Suuri-Rostuvi includes a drastic shift in species proportions but, as noted by Rönkkö & Simola (1986), no actual changes in the species composition. The changes show a stepwise increase in the planktonic diatoms that dilutes the percentage share of the littoral forms. The recent superdominance of *Asterionella ralfsii* is quite probably a consequence of the well-documented ditching and fertilization of the surrounding peatbogs. The former limnological changes, to which the diatom community has reacted could also be due to forest management and related events – e.g. logging of the shore forest or lake level regulation to facilitate log floating downstream; measures that would enhance wind-induced turbulence of the water.

Comparison of the two cases treated here shows that a lake ecosystem may react by different ways to extensive artificial drainage on its catchment. The present lakes appeared difficult, because there is no visual stratigraphic change indicating ditch erosion, which is typical for the sediments of many other lakes (Sandman *et al.*, in prep.). The appearance of a clear marker horizon is dependent on the basin size – distance from the ditch erosion inflow to the sampling depression – and on the type of the ditch system – whether or not the lower reaches of the ditches have been excavated down to the mineral soil.

A general problem in paleolimnological analysis that aims at assessing the effects of peatbog ditching is to distinguish the consequences of other events of erosion-inducing land use. The paleolimnological signal of modern agriculture is largely similar, but it often starts earlier and appears more persistent than the maybe stronger but more instantaneous peatland ditching effects.

Acknowledgements

We thank Jukka Rönkkö for the diatom analysis of Suuri-Rostuvi and the technical personnel of our Institutes for skillful assistance at various parts of the work. The study has been partly aided by a grant of The Academy of Finland, Bd of Environmental Science to HS (grant 06/084).

References

Appleby, P. & F. Oldfield, 1978. The calculation of lead-210 dates assuming a constant rate of supply of unsupported ^{210}Pb to the sediment. Catena 5: 1–8.

Appleby, P. & F. Oldfield, 1983. The assessment of ^{210}Pb data from sites with varying sediment accumulation rates. Hydrobiologia 103: 29–35.

Aston, S. R., D. Bruty, R. Chester & R. Padgham, 1973. Mercury in lake sediments: a possible indicator of technological growth. Nature 241: 450–451.

Blomqvist, E., 1911. Kymijoki ja sen vesistö. Lisiä Suomen hydrologiaan 2. Suomen Tie- ja Vesirakennusten Ylihallituksen julkaisu. Tilgmann, Helsinki, 235 pp.

Davis, R., C. Hess, S. Norton, D. Hanson, K. Hoagland & D. Anderson, 1984. ^{137}Cs and ^{210}Pb dating of sediments from softwater lakes in New England (U.S.A.) and Scandinavia, a failure of ^{137}Cs dating. Chem. Geol. 44: 151–185.

Eakins, J., R. Cambray, K. Chambers & A. Lally, 1984. The transfer of natural and artificial radionuclides to Brotherswater from its catchment. In E. Haworth & J. W. G. Lund (eds.), Lake sediments and environmental history. Leicester Univ. Press, Leicester: 125–144.

Engstrom, D. & H. Wright, 1984. Chemical stratigraphy of lake sediments as a record of environmental change. In E. Haworth & J. W. G. Lund (eds.), Lake sediments and environmental history. Leicester Univ. Press, Leicester: 11–67.

Evans, H., P. Smith & P. Dillon, 1983. Anthropogenic zinc and cadmium burdens in sediments of selected southern Ontario lakes. Can. J. Fish. aquat. Sci. 40: 570–579.

Hansen, K., 1961. Lake types and lake sediments. Verh. int. Ver. Limnol. 14: 285–290.

Heikurainen, L., K. Kenttämies & J. Laine, 1978. The environmental effects of forest drainage. Suo 29: 49–58.

Håkanson, L., 1981. Sjösedimenten i recipientkontrollen – principer, processer och praktiska exempel (Summary: Lake sediments in aquatic pollution control programs; principles, processes and practical examples). Statens Naturvårdsverk, Rapport 1398: 1–242.

Karsisto, K. & H. Ravela, 1971. Washing away of phosphorus and potassium from areas drained for forestry and topdressed at different times of year. Suo 22: 39–46.

Kenttämies, K., 1981. The effects of water quality of forest drainage and fertilization in a peatland area in central Finland. Publ. Water Res. Inst., Nat. Bd of Waters, Finland 43: 24–31.

Lodenius, M. & A. Seppänen, 1982. Hair mercury contents and fish eating habits of people living near a Finnish man-made lake. Chemosphere 11: 755–759.

Lodenius, M., A. Seppänen & M. Herranen, 1983. Accumulation of mercury in fish and man from reservoirs in northern Finland. Wat. Air Soil Pollut. 19: 237–246.

Longmore, M., B. O'Leary & C. Rose, 1983. Caesium-137 profiles in the sediments of a partial-meromictic lake on Great Sandy Island (Fraser Island), Queensland, Australia. Hydrobiologia 103: 21–27.

Mortimer, C. H., 1941. The exchange of dissolved substances between mud and water in lakes. J. Ecol. 29: 280–329.

National Bd of Waters, 1981. Vesihallinnon analyysimenetelmät. Vesihallituksen Tiedotuksia (Raports of Nat. Bd of Waters, Finland) 213: 1–136.

Oldfield, F. & P. Appleby, 1984. Empirical testing of ^{210}Pb dating models for lake sediments. In E. Haworth & J. W. G. Lund (eds.), Lake sediments and environmental history. Leicester Univ. Press, Leicester: 93–124.

Rautapää, J., H. Pyysalo, P. Ravio & H. Blomqvist, 1979. Mercury in cultivated soil in Finland. Ann. agric. Fenniae 18: 87–91.

Rönkkö, J. & H. Simola, 1986. Geological control upon the floral manifestation of eutrophication of two headwater lakes. Univ. Joensuu, Publ. Karelian Inst. (in print).

Sevola, Y., 1983. Metsähallinnon Nurmeksen hoitoalueen voimaperäinen puunkasvatus: seurantajärjestelmä ja tuloksia (Summary: Intensive timber growing in a state forest district: monitoring system and results). Folia Forestalia 574: 1–83.
Simola, H., 1982. En skoggssjös eutrofieringshistoria tolkad ur dess årsvarviga sediment. Univ. Lund, Dept. Quaternary geology, Report 22: 107–112.
Simola, H., 1983. Limnological effects of peatland drainage and fertilization as reflected in the varved sediment of a deep lake. Hydrobiologia 106: 43–57.
Simola, H. & M. Lodenius, 1982. Recent increase in mercury sedimentation in a forest lake attributable to peatland drainage. Bull. Env. Contam. Toxicology 29: 298–305.
Simola, H., E. Grönlund & J. Rönkkö, 1985. Pyhäselän sedimentaatioalueet ja sedimentin pintaosien stratigrafia (Summary: Sediment distribution and stratigraphy of recent sediment in the basin of Pyhäselkä, East Finland). Univ. Joensuu, Publ. Karelian Inst. (in print).
Verta, M. & S. Rekolainen, 1985. Ilmaperäisen elohopean, metsäojituksien ja tekojärvien rakentamisen vaikutus kalojen elohopeapitoisuuteen (Summary: Effects of airborne Hg, forest ditching and lake construction on mercury content of fish). Nat. Bd of Waters, Finland. Working papers 320: 1–42.

Accepted 18 March 1986.

Hydrobiologia 143: 425–431, (1986)

Paleolimnological reflections of fiber-plant retting in the sediment of a small clear-water lake

Elisabeth Grönlund, Heikki Simola & Pertti Huttunen
University Joensuu, Karelian Institute and Department of Biology, SF-80101 Joensuu, Finland

Keywords: paleolimnology, eutrophication, varved sediment, pollen, *Drepanocladus*

Abstract

Likolampi is a small groundwater kettle-hole lake in Ilomantsi, Eastern Finland. At 11 m water depth the uppermost 51 cm of sediment contains about 390 varves. A thin layer of chrysophyte statospores is the main structural unit in these; in some varves it is followed by a layer of cf. *Drepanocladus fluitans* moss spores. Below the laminated sequence, two layers of moss detritus (110–87 cm, 81–51 cm) alternate with greyish brown fine detritus. Pollen analysis reveals that the start of the deposition of varved sediment coincides with the beginning of intensive fiber plant retting with large quantities of *Cannabis/Humulus* -type pollen and even regular occurrence of the insect-pollinated *Linum usitatissimum.* The cf. *Drepanocladus* spores are almost absent during the retting period (c. 1590-1900 by the varves), but very common before and after it. *Drepanocladus fluitans* is capable of assimilating dissolved CO_2 in water, so it thrives at low pH and is typical for oligotrophic clear-water lakes. We conclude that the stratigraphy of *Drepanocladus* fragments and spores is a true negative indicator of cultural influence in Likolampi.

Introduction

The present study is one part in a long-term research project conducted at Karelian Institute, Section of ecology and Department of Biology of the University of Joensuu, the objective of which is to work out by paleolimnological methods the history of agriculture in East Finland (Vuorinen 1978, Simola *et al.* 1984, 1985, Huttunen & Simola in press).

In these studies we have made use of the fact that annually laminated sediments are rather common in small Finnish lakes (O'Sullivan 1983). Cultural eutrophication, even very mild, often initiates the formation of sediment laminations in deep enough lakes, due to increased sedimentation and hypolimnetic periods of anoxia.

The name of our study site, Likolampi, means 'retting pond' and refers to the traditional method of macerating fiber plants (hemp, flax) by retting them in water (Soininen, 1974). Until the beginning if this century, fiber plant retting was a common practice throughout the country, and the eutrophication of small water bodies through retting was early recognized by Finnish limnologists (Hagman, 1916). There are some 300 lakes named Likolampi in Finland (The Onomastic Division of the Finnish Research Centre for Domestic Languages).

Study site, material and methods

Lake Likolampi is situated in Ilomantsi, the easternmost district of North Karelia and Finland (62° 44′N, 31° 00′E). It is a small kettle-hole seepage lake in the esker Putkelansärkkä. Its surface area is 1.04 hectares and maximum depth 11 m. The Putkelansärkkä esker runs in a N-S direction in a level terrain, where extensive oligotrophic peatlands dominate. In the proximity of Likolampi, the esker is bordered on both sides by water; the Koitajoki river in the east and Lake Mekrijärvi in the west, both of which have their water-level at about 144 m a.s.l. Likolampi is separated from Mekrijärvi by an esker ridge only 40 m broad, but its water-level lies nevertheless some 3 m above that of Mekrijärvi. North of Likolampi the esker rises to the height of 185 m a.s.l.; it seems likely that the topography causes an actual flow of groundwater that flushes Likolampi through seepage (even though it lacks a surface outlet), and also that hydrologic changes may cause water-level fluctuations within it.

A late-glacial or early post-glacial loess layer covers the soils in this area (Rainio, 1978). Therefore the esker itself has been fertile enough to be exploited by slash-and-burn agriculture (this is indicated by some place-names). Nowadays, there is only one small farm with some four hectares of permanent fields in the vicinity of Likolampi. Two larger villages, Mekrijärvi in the north and Putkela in the south, are both situated 3–4 km away from it. There is a gravel-pit on the eastern side of Likolampi, probably opened when a new road causeway was constructed on the western side of the esker in the late 1970's.

The sediment of Likolampi has been sampled by freeze-coring (Huttunen & Meriläinen, 1978) and by the Russian peat sampler down to 200 cm. Aiming primarily to clarify the phases of cultural influence in the lake, we have made a close-interval analysis of pollen and other microfossils according to standard methods (Faegri & Iversen, 1964). At least 500 arboreal pollen were counted at each analyzed level. In addition, in the laminated upper part of the core, some tape peel preparations (Simola, 1977) were made in order to investigate the structure of the laminae.

The present limnology of Likolampi has been studied in August 1983 by a field course in ecology (Univ. Joensuu, Mekrijärvi Biol. Station).

Results

The limnologic study of Likolampi on 8.–9. August 1983 (Ecology field course at Mekrijärvi Biol. Station) revealed characteristics of an oligotrophic, clearwater lake at an advanced stage of summer stratification. The epilimnion was isothermal at +20.5 C down to 3 m, metalimnion spanned the interval 4–7 m, and hypolimnetic anoxia was observed at 10 m in +4.9 C.

The studied sediment sequence 0–200 cm can be divided into three zones:

I. 200–110 cm: Greyish brown fine detritus mixed with a variable amount of coarse plant fragments, and mainly with a vaguely discernible horizontal bedding. The sequence 190–200 cm contains more silt than the sediment above, and at 195 cm there is a 5 mm thick layer of vivianite.

II. 110–51 cm: Brownish and mainly organic coarse detritus predominantly composed of moss fragments: the interval 87–81 cm is a layer of fine detritus.

III. 51–0 cm: Mainly organic, regularly laminated fine detritus; algal gyttja, with varying shades of yellowish, greenish and reddish brown colour. About 390 laminations are discernible, their thickness varies from 0.5 to 3 mm (Fig. 1A). Their micro-stratigraphic structure

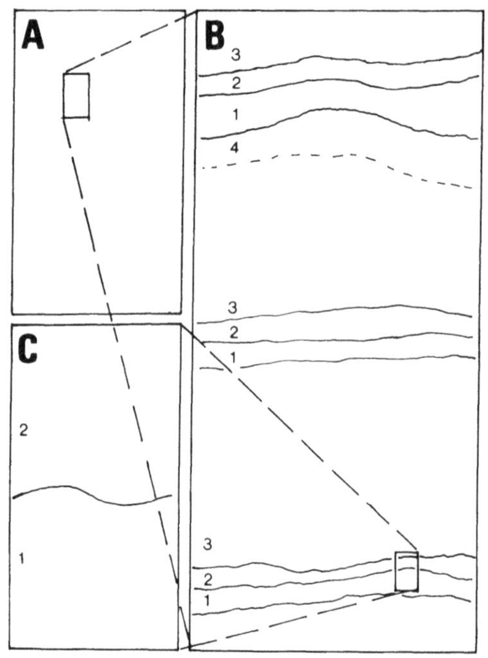

Fig. 1. Scheme and explanation (above) of the stratigraphic details in, and the relationships between the photographs A–C (opposing page). A: Photograph of the uppermost sediment in a frozen core of Likolampi. B: Compound microphotograph of a tape peel slide, showing two of the recent sediment varves. The vertical extent of the sequence is 10 mm. 1: chrysophyte statospores, 2: cf. *Drepanocladus* spores. 3: organic detritus, 4: cladoceran remains, C: A dark field microphotograph of the closely packed layers of chrysophyte statospores (1) and cf. *Drepanocladus* spores (2). Vertical extent about 0.8 mm.

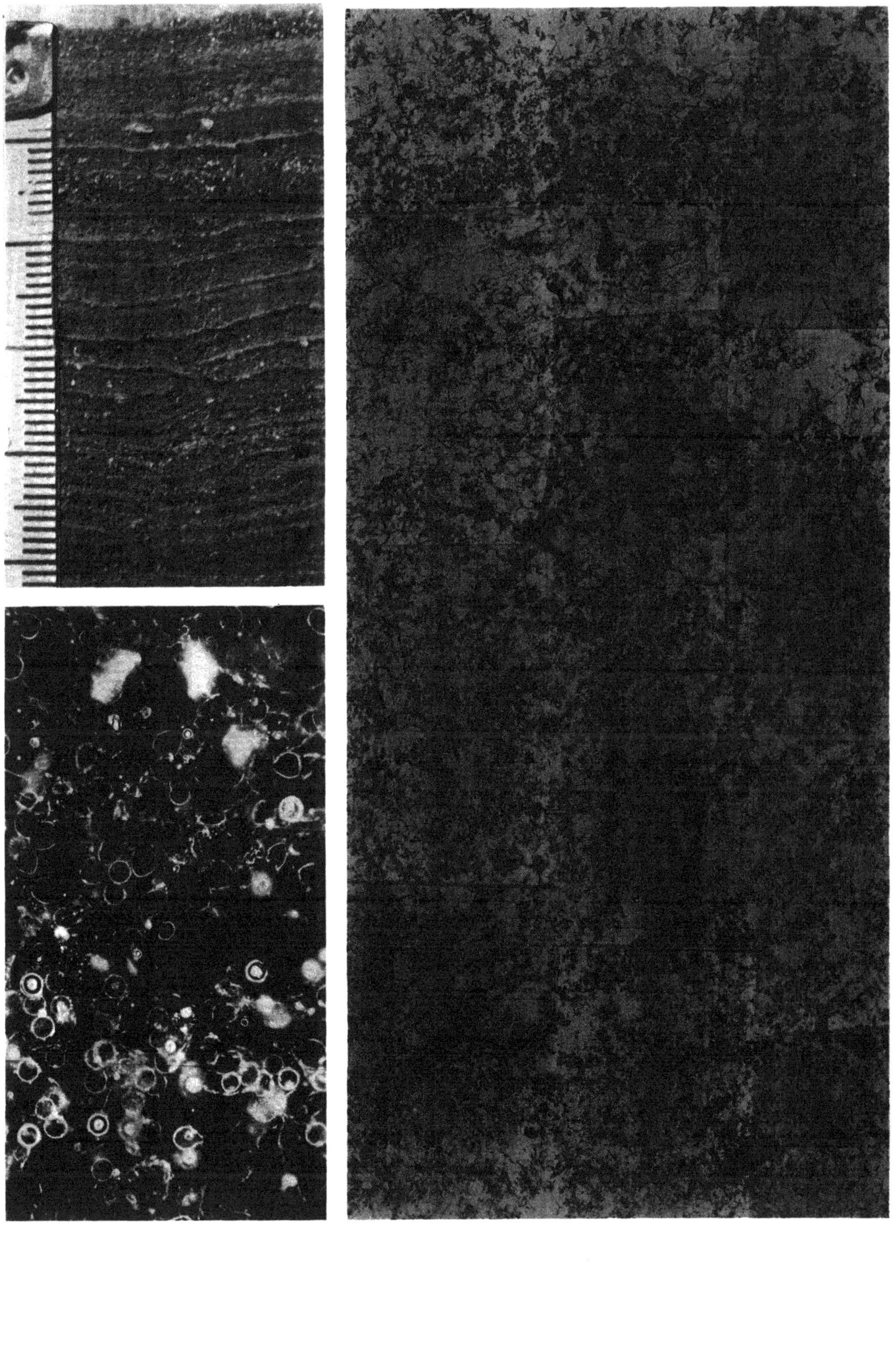

appears regular enough to justify their interpretation as annually formed varves (but see discussion). Each lamina begins with a thin layer of chrysophyte statospores – this is the feature that on visual inspection marks the borders of the laminae. In the upper part of the laminated sequence the whitish chrysophyte band is immediately followed by a dense layer of brown spores that we tentatively have identified as spores of the water moss *Drepanocladus fluitans* (L.) Warnst (Fig. 1C). Between the spore layers the sediment is made of amorphous organic matrix that contains plenty of very well preserved plant and animal microfossils. For instance, many of the cladoceran remains are not disintegrated parts but intact animals. In fact, some of the varves seem to contain a layer of entire *Bosmina* carcasses as a structural unit just below the chrysophyte layer (Fig. 1B).

Pollen analysis of the core reveals the typical features of forest history and human interference in Eastern Finland (Fig. 2). The spreading of spruce (*Picea abies* Karst.), which by numerous 14-C dates has been established at 5000–5200 BP in this area (Aartolahti 1966, Tolonen & Ruuhijärvi, 1976), is seen at the base of the profile at 182 cm. The main emphasis in our work was on the history of agriculture. The first indication of retting, a pollen grain of the *Cannabis/Humulus*-type, was encountered at 74 cm depth. From this level upwards, the general signs of slash-and-burning and pasturing are also evident: decline of spruce and increase of *Juniperus* and non-arboreals. The first cereal pollen occur at 55 cm, or slightly before the laminated sediment starts.

A drastic shift in the pollen flora occurs at 50 cm, coinciding with the change of sediment from moss detritus into varved algal gyttja. Cultural pollen increase, most notably the proportion of the *Cannabis-Humulus*-type pollen rises to nearly 30% of the total count.

In the middle parts of the laminated sequence, some flax (*Linum usitatissimum L.*) pollen is present. In contrast to hemp, flax is insect-pollinated, which accounts for the rarity of its pollen.

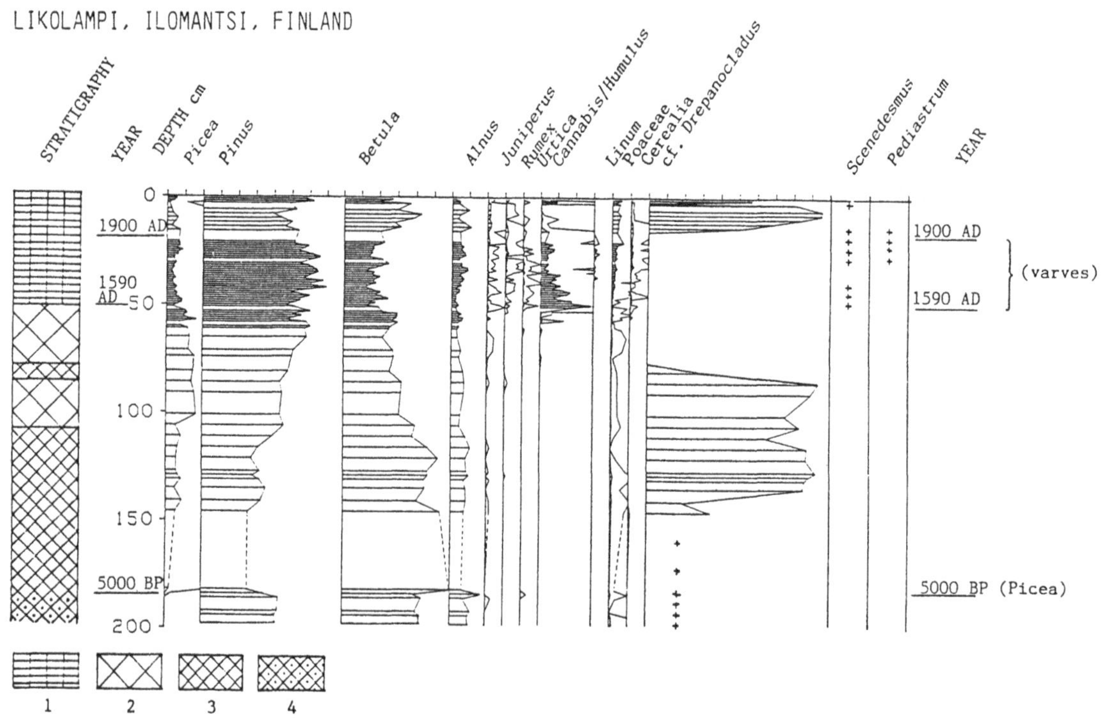

Fig. 2: Stratigraphy of pollen and other microfossils in the sediment of Likolampi. Symbols of stratigraphy – 1: varved algal gyttja, 2: coarse moss detritus, 3: organic fine detritus, in part horizontally bedded, 4: silty fine detritus.

The retting no doubt led to the development of euthrophic conditions in the lake. One indication of the change is the increase of the sedimentation rate: from the invasion of spruce to the onset of retting we have some 160 cm of sediment representing 4500 yr, i.e. 0.3 mm yr^{-1}, while in the retting period the average sedimentation rate has been roughly three times higher. Sediment accumulation rate is of course not a linear measure of productivity: the organic loading of the fiber plant mass has both directly contributed to sedimentation and led to periodic oxygen deficiency that decreases the mineralization of the lake's own organic production. A sign of ineffective decomposition is the good preservation of algal cells (e.g. *Scenedesmus, Pediastrum*) in the laminated sequence.

Along with pollen we counted the above-mentioned spores of the cf. *Drepanocladus fluitans,* which at some levels outnumbers arboreal pollen by ratio of 20:1. This conspicuous spore is brown, smooth, about 30 μm in diameter and prolate in shape, with one pointed end and the blunt end often ripped open (Fig. 1C). The spores don't stain much with saphranine. The size and shape are consistent with herbarium material that we have examined, and agree with the description of Boros & Jarai-Komlodi (1975) of the spore of *D. fluitans*, but they lack completely the fine pilate surface structure that is typical of the fresh spores.

The cf. *Drepanocladus* spore vanishes with the first indication of fiber-plant retting at 71 cm, and returns at 14 cm, when the retting ceases. In the top 1–2 cm interval there seems to be a short return of the retting-indicators and a coinciding decline in cf. *Drepanocladus.* By varve-counting this would have happened in the late 1970's which does not seem possible. We interprete this sequence as a period of redeposition of older pollen material from the littoral zone, and very tentatively (in absence of water-level data) relate it to the opening of the gravel-pit near the lake, which could have lowered ground-water level in the area (cf. Fig. 1A, in which the respective varves contain somewhat coarser material; one of these varves also includes a layer of Cladocera (B)).

Discussion

Sediment laminations and varve chronology

Diatoms are quite scarce in the sediment of Likolampi. In this respect it differs from many other varved small lakes in Finland (Simola, 1977, Vuorinen, 1978, Tolonen, 1978b, Simola, 1983). The layer of chrysophyte statospores that marks the junction of the varves is, however, a common feature for many varves. According to Battarbee *et al.* (1979) the statospores are deposited during the spring circulation (when water currents transport material from the littoral).

The cf. *Drepanocladus fluitans* spores, when present, form a layer immediately above the statospores. If the chrysophyte statospores indeed mark the spring, the dead Cladocera below some chrysophyte layers would then stand from late winter, or indicate anoxia under the ice (Fig. 1B).

Some of the varves appear horizontally discontinuous. In the frozen samples one can see breakage of a few successive varves, where a gas bubble has escaped, leaving a small crater that has subsequently been filled by new sediment (Fig. 1A). Of some varves, or a pile of them, a larger section is missing – as if a raft of sediment had floated away, possibly suspended by methane bubbles that are formed by sediment diagenesis. The specific density of the highly organic sediment is very close to that of water, so the possible water-level fluctuations with consequent changes in hydrostatic pressure, could indeed cause reworking of the sediment. This casts a doubt on our varve-chronology: we cannot be sure that all the varves are present in a core, so a given layer of sediment is as old or older than the varve-counting indicates. If indeed the sediment tends to raft away, its later redeposition would at times increase the thickness of the new varves. As a result, this type of sediment could actually have 'genuine' varves, that are thicker than the lake's productivity would allow, and fewer than the years' cycle would have dictated.

Stratigraphic signal of the cf. Drepanocladus *spore*

In Likolampi, the spore of cf. *Drepanocladus fluitans,* appears as negative indicator of eutrophication: it occurs in very large quantities in the pre-cultural moss-detritus, but disappears almost completely when the first indications of human interference occur. Upwards in the core, it reappears when the retting seems to cease. *D. fluitans* is an oligotrophic water-moss that covers large bottom areas in many clear-water lakes. It assimilates CO_2, which allows it to thrive at low pH. Presently, *D. fluitans* does grow in Likolampi but only in small numbers, and we have not managed to collect it in the sporophytic stage. According to Finnish herbarium material (Mus. Bot., Univ. Helsinki) it produces spores regularly. The spores are shed in May-June, which would agree with the stratigraphic position of the spores just above the chrysophyte layer.

Interpretation of agricultural history

The small size and sheltered kettle-hole location of Likolampi make it a site where the very local, rather than areal pollen rain is represented (e.g. Huttunen, 1980). The lake is situated a few kilometers south of the village of Mekrijärvi, in which there were some 30 farms already in the early 1500's (Könönen & Kirkinen, 1972). Thus, the dramatic change of Likolampi (by varve-counts not later than 1590) seems to have started clearly later than the historically known establishment of agriculture in the area.

The very high proportion of *Cannabis*-type pollen (nearly 30% of the total pollen count) in the early varves deserves a comment, since in SW Finland its proportion in much more intensively cultivated areas in similar lakes rarely exceeds 10% (Huttunen & Tolonen, 1977, Huttunen, 1980, Tolonen, 1978a, 1981). The difference seems to be due to differences in the hemp-cultivation practice between western and eastern Finland. Hemp is dioecious, and the male plant is considerably smaller than the female (Kaukonen, 1946). According to Soininen (1974), in Western Finland the male plants were picked out of the field together with weeds, because only the long fibers of the female plant were wanted. In the east, no distinction was made between long and short, but all plants were utilized, and therefore much of the pollen were introduced directly into the retting-ponds.

Acknowledgements

We thank the North Karelian Cultural Fond for financial support, and the staff of Karelian Institute, Section of Ecology, for skillful assistance at various phases of the work.

References

Aartolahti, T., 1966. Über die Einwanderung und die Verhäufigung der Fichte in Finnland. Ann. bot. fenn. 3: 368–379

Battarbee, R., G. Cronberg & S. Lowry, 1980. Observations of the occurrence of scales and bristles of *Mallomonas spp.* (Chrysophyceae) in the micro-laminated sediments of a small lake in Finnish North Karelia. Hydrobiologia 71: 225–232.

Boros, A. & M. Jarai-Komlodi, 1975. An atlas of recent European moss spores. Akademiai Kiado, Budapest, 466 pp.

Faegri, K. & J. Iversen, 1964. Textbook of pollen analysis (2nd ed.). Copenhagen, 233 pp.

Hagman, N., 1916. Vähäsen pellavanliotuksen vaikutuksesta kalojen ravintoon järvissämme (Effects of flax retting of fish nutrition in our lakes). Suomen kalastuslehti 23: 17–22.

Huttunen, P., 1980. Early land use, especially the slash-and-burn cultivation in the commune of Lamni, southern Finland, interpreted mainly using pollen and charcoal analyses. Acta bot. fenn. 113: 1–45.

Huttunen, P. & J. Meriläinen, 1978. New freezing device providing large, unmixed sediment samples from lakes. Ann. bot. fenn. 15: 128–130.

Huttunen, P. & H. Simola, 1986. Heinäveden asutushistoria luonnonarkistojen valossa (Paleolimnological study of agricultural history in Heinävesi). Heinäveden pitäjän historia (History of Heinävesi commune). In press.

Huttunen, P. & K. Tolonen, 1977. Human influence in the history of Lake Lovojärvi, S. Finland. Finskt Museum 1975: 68–105.

Kaukonen, T–I., 1946. Pellavan ja hampun viljely ja muokkaus Suomessa (German summary: Anbau und Bearbeitung von Flachs und Hanf in Finnland). Kansatieteellinen arkisto 7: 1–315.

Könönen, T. & H. Kirkinen, 1969. Pohjois-Karjalan historia 1. (History of North Karelia 1). Pohjois-Karjalan historiallinen yhdistys, Joensuu, 188 pp.

O'Sullivan, P., 1983. Annually-laminated lake sediments and the study of Quaternary environmental changes – a review. Quat. Sci. Rev. 1: 245–313.

Rainio, H., 1978. Eoliskt finmo i Nord-Karelen i östra Finland. 8 Nordiske geologiske Vintermode, Copenhagen January 1978, Abstracts B: 55.

Simola, H., 1977. Diatom succession in the formation of annually laminated sediment in Lovojärvi, a small eutrophicated lake. Ann. bot. fenn. 14: 143–148.

Simola, H., 1983. Limnological effects of peatland drainage and fertilization as reflected in the varved sediment of a deep lake. Hydrobiologia 106: 43–57.

Simola, H., E. Grönlund, P. Huttunen & P. Uimonen-Simola, 1985. Pollen analytical evidence for Iron Age origin of cupstones in the Kerimäki area. Iskos 5: 527–531.

Simola, H., P. Huttunen & J. Meriläinen, 1984. Varve-dated eutrophication history of a small lake. Verh. int. Ver. Limnol. 22: 1404–1408.

Soininen, A., 1974. Vanha maataloutemme (Summary: Old traditional agriculture in Finland in the 18th and 19th centuries). Historiallisia tutkimuksia 96: 1–459.
Tolonen, K. & R. Ruuhijärvi, 1976. Standard pollen diagrams from the Salpausselkä region of Southern Finland. Ann. bot. fenn. 13: 155–196.
Tolonen, M., 1978a. Paleoecology of annually laminated sediments in Lake Ahvenainen, S. Finland I. Pollen and charcoal analyses and their relation to human impact. Ann. bot. fenn. 15: 177–208.
Tolonen, M., 1978b. Paleoecology of annually laminated sediments in Lake Ahvenainen, S. Finland. II. Comparison of dating methods. Ann. bot. fenn. 15: 208–222.
Tolonen, M., 1981. An absolute and relative pollen analytic study on prehistoric agriculture in South Finland. Ann. bot. fenn. 18: 213–220.
Vuorinen, J., 1978. The influence of prior land-use on the sediments of a small lake. Pol. Arch. Hydrobiol. 25: 443–451.

Accepted 18 March 1986.

www.ingramcontent.com/pod-product-compliance
Lightning Source LLC
Chambersburg PA
CBHW080934240926
51849CB00003BA/458
9789061936244